Methods in Molecular Biology

Series Editor
John M. Walker
School of Life and Medical Sciences
University of Hertfordshire
Hatfield, Hertfordshire, UK

For further volumes:
http://www.springer.com/series/7651

For over 35 years, biological scientists have come to rely on the research protocols and methodologies in the critically acclaimed Methods in Molecular Biology series. The series was the first to introduce the step-by-step protocols approach that has become the standard in all biomedical protocol publishing. Each protocol is provided in readily-reproducible step-by step fashion, opening with an introductory overview, a list of the materials and reagents needed to complete the experiment, and followed by a detailed procedure that is supported with a helpful notes section offering tips and tricks of the trade as well as troubleshooting advice. These hallmark features were introduced by series editor Dr. John Walker and constitute the key ingredient in each and every volume of the Methods in Molecular Biology series. Tested and trusted, comprehensive and reliable, all protocols from the series are indexed in PubMed.

Manuscript Instructions https://drive.google.com/file/d/1K4Az5vpAFVzYadJ3jKx2-QSo7Dqa_SNA/view?usp=sharing

Cryopreservation and Freeze-Drying Protocols

Fifth Edition

Edited by

Willem F. Wolkers

Unit for Reproductive Medicine - Clinic for Horses, University of Veterinary Medicine Hannover, Hannover, Germany; Biostabilization Laboratory - Lower Saxony Center for Biomedical Engineering, Implant Research and Development, University of Veterinary Medicine Hannover, Hannover, Germany

Harriëtte Oldenhof

Unit for Reproductive Medicine - Clinic for Horses, University of Veterinary Medicine Hannover, Hannover, Germany; Biostabilization Laboratory - Lower Saxony Center for Biomedical Engineering, Implant Research and Development, University of Veterinary Medicine Hannover, Hannover, Germany

Humana Press

Editors
Willem F. Wolkers
Unit for Reproductive Medicine - Clinic for Horses
University of Veterinary Medicine Hannover
Hannover, Germany

Biostabilization Laboratory - Lower Saxony Center for Biomedical Engineering, Implant Research and Development
University of Veterinary Medicine Hannover
Hannover, Germany

Harriëtte Oldenhof
Unit for Reproductive Medicine - Clinic for Horses
University of Veterinary Medicine Hannover
Hannover, Germany

Biostabilization Laboratory - Lower Saxony Center for Biomedical Engineering, Implant Research and Development
University of Veterinary Medicine Hannover
Hannover, Germany

ISSN 1064-3745 ISSN 1940-6029 (electronic)
Methods in Molecular Biology
ISBN 978-1-0716-5415-6 ISBN 978-1-0716-5416-3 (eBook)
https://doi.org/10.1007/978-1-0716-5416-3

This Humana imprint is published by the registered company Springer Science+Business Media, LLC, part of Springer Nature.
The registered company address is: 1 New York Plaza, New York, NY 10004, U.S.A.

Preface

Cryopreservation, vitrification, and freeze-drying are key technologies used to ensure off-the-shelf availability of cells and tissues used in research, medicine, and agriculture. Whereas cryopreserved and vitrified samples must be maintained at ultra-low temperatures, freeze-dried specimens can be stored at room temperature, both having their own advantages and requirements for storage and transportation.

Although many types of cells and tissues can now be cryopreserved using relatively simple and standardized protocols, cryopreservation is not a routine procedure. Notably there is no universal freezing protocol that works for all specimens. Drying is generally more damaging than freezing and is particularly challenging for cells. Further development of cryopreservation and dry preservation protocols for cells and tissues requires insights from biologists, chemists, physicists, medical scientists, as well as engineers, making this a highly interdisciplinary field.

In this edition of *Cryopreservation and Freeze-Drying Protocols*, fundamental principles underlying freezing and drying of biospecimens are introduced; we highlight emerging technologies and present a variety of practical protocols. The book is divided into four sections. Part I describes fundamental aspects of cryopreservation and freeze-drying, as well as the use of mathematical modeling, to rationally design protocols and solve cryobiological problems. In Part II, emerging technologies are presented for cooling, warming, and drying, as well as platforms for cryoprinting and the use of microfluidics. Part III highlights the importance of standard operating procedures in biorepositories and biobanks, quality assessment of stored biologics, and cold-chain logistics. The main core of this book, Part IV, presents cryopreservation, vitrification, and freeze-drying protocols for a wide variety of samples and different applications, i.e., for agricultural purposes, model organisms and biodiversity conservation, reproductive and regenerative medicine, and biologics for foods and pharmaceutics.

This book serves as a practical guide that can be used without the need to consult additional reference sources. Alongside protocols that rely on specialized equipment, it also presents practical, cost-effective alternatives. It is intended for researchers and technical staff in academia and industry with a background in life sciences, medicine, or engineering, who are interested in studying the fundamentals underlying freezing and drying or in establishing practical methods for the safe storage of biological specimens.

Hannover, Germany — *Willem F. Wolkers*
Hannover, Germany — *Harriëtte Oldenhof*

Contents

Contributors

JASON P. ACKER • *Innovation, Canadian Blood Services, Edmonton, AB, Canada; Department of Laboratory Medicine and Pathology, University of Alberta, Edmonton, AB, Canada*

CANSU AGCA • *College of Veterinary Medicine, University of Missouri, Columbia, MO, USA; Department of Pathobiology and Integrative Biomedical Sciences, University of Missouri, Columbia, MO, USA*

YUKSEL AGCA • *College of Veterinary Medicine, University of Missouri, Columbia, MO, USA; Department of Pathobiology and Integrative Biomedical Sciences, University of Missouri, Columbia, MO, USA*

DANIEL ANGEL-VELEZ • *Department of Internal Medicine, Reproduction and Population Medicine, Faculty of Veterinary Medicine, Ghent University, Merelbeke, Belgium; Research Group in Animal Sciences, Universidad CES, Medellín, Colombia*

CATARINA ANJOS • *Center for Marine Sciences (CCMAR), University of Algarve, Faro, Portugal*

FIORA ARTUSIO • *Department of Applied Science and Technology, Politecnico di Torino, Torino, Italy*

FELIZITAS BAJERSKI • *Leibniz Institute DSMZ-German Collection of Microorganisms and Cell Cultures, Braunschweig, Germany*

DANIEL BALLESTEROS • *University of Valencia, Department of Botany and Geology, Faculty of Pharmacy and Food Sciences, Burjassot, Spain; Botanical Garden of the University of Valencia, Valencia, Spain; Royal Botanic Gardens Kew, Wakehurst, Ardingly, West Sussex, UK*

ANNA BARHEINE • *Leibniz Institute DSMZ-German Collection of Microorganisms and Cell Cultures, Braunschweig, Germany*

ANTONELLO A. BARRESI • *Department of Applied Science and Technology, Politecnico di Torino, Torino, Italy*

LISA BELKE • *Leibniz Institute DSMZ-German Collection of Microorganisms and Cell Cultures, Braunschweig, Germany*

JAMES D. BENSON • *Department of Biology, University of Saskatchewan, Saskatoon, SK, Canada*

KELVIN G. M. BROCKBANK • *Preservation Technologies, LifeNet Health, North Charleston, SC, USA; Department of Bioengineering, Clemson University, Clemson, SC, USA; Department of Comparative Medicine, Medical University of South Carolina, Charleston, SC, USA*

ELSA CABRITA • *Center for Marine Sciences (CCMAR), University of Algarve, Faro, Portugal*

SARA CAMPOS • *Centro de Investigación Marina CIM-Universidade de Vigo, Ecocost Research Group - Cryolab, Vigo, Spain*

Yue Cheng • *Laboratory for Cryo-Biomedical Engineering, Department of Electronic Engineering and Information Science, School of Information Science and Technology, University of Science and Technology of China, Hefei, China*
Andrzej Ciereszko • *Team of Gamete Biology, InLife Institute of Animal Reproduction and Food Research in Olsztyn, Polish Academy of Sciences, Olsztyn, Poland*
Anthony N. Consiglio • *Department of Mechanical Engineering, University of California, Berkeley, CA, USA; BioChoric Inc., Bozeman, MT, USA*
Rachael M. Davies • *Royal Botanic Gardens Kew, Wakehurst, Ardingly, West Sussex, UK*
Mariola A. Dietrich • *Team of Gamete Biology, InLife Institute of Animal Reproduction and Food Research in Olsztyn, Polish Academy of Sciences, Olsztyn, Poland*
Rebekka Einenkel-Ulrich • *Department of Gynecological Endocrinology and Reproductive Medicine, University Hospital Bonn, Bonn, Germany*
Janet A. W. Elliott • *Department of Chemical and Materials Engineering, University of Alberta, Edmonton, AB, Canada; Department of Laboratory Medicine and Pathology, University of Alberta, Edmonton, AB, Canada*
Ali Eroglu • *Department of Neuroscience and Regenerative Medicine, Medical College of Georgia, Augusta University, Augusta, GA, USA; Department of Obstetrics and Gynecology, Medical College of Georgia, Augusta University, Augusta, GA, USA*
Binnur Eroglu • *Department of Neuroscience and Regenerative Medicine, Medical College of Georgia, Augusta University, Augusta, GA, USA*
Nasim Eskandari • *Department of Laboratory Medicine and Pathology, University of Alberta, Edmonton, AB, Canada*
Natalia Fanega-Sleziak • *Royal Botanic Gardens Kew, Wakehurst, Ardingly, West Sussex, UK; The Alliance of Bioversity International and CIAT, Leuven, Belgium*
Wolfgang Friess • *Ludwig-Maximilians-Universität, Munich, Germany*
Barry J. Fuller • *Division of Surgery & Interventional Science University College London Medical School, London, UK*
Dayong Gao • *Department of Mechanical Engineering, University of Washington, Seattle, WA, USA*
Morgan B. Giers • *School of Chemical, Biological, and Environmental Engineering, Oregon State University, Corvallis, OR, USA*
Theresa Goettling • *Ludwig-Maximilians-Universität, Munich, Germany*
Andrea Gomez-Zavaglia • *Center for Research and Development in Food Science and Technology, La Plata, Argentina*
Marion Grübe • *Genebank Department, Leibniz Institute of Plant Genetics and Crop Plant Research (IPK), Seeland, Germany*
Shaohang Hao • *Department of Mechanical Engineering, University of Washington, Seattle, WA, USA*
Maite Harguindeguy • *Coriolis Pharma Research GmbH, Martinsried, Germany*
Andrea Hawe • *Coriolis Pharma Research GmbH, Martinsried, Germany*
Pablo Heres • *Centro Oceanográfico de Gijón, Instituto Español de Oceanografía (IEO-CSIC), Gijón, Asturias, Spain*
Adam Z. Higgins • *School of Chemical, Biological, and Environmental Engineering, Oregon State University, Corvallis, OR, USA*
Daiyu Ito • *Faculty of Life and Environmental Sciences, University of Yamanashi, Yamanashi, Japan*
Eric R. James • *Vaccine Stabilization and Logistics, Sanaria Inc., Rockville, MD, USA*

YE JIN • *Department of Mechanical Engineering, University of Washington, Seattle, WA, USA*
LACEY JOHNSON • *Research and Development, Australian Red Cross Lifeblood, Sydney, NSW, Australia; School of Life Sciences, University of Technology Sydney, Sydney, NSW, Australia; School of Science, RMIT University, Melbourne, VIC, Australia*
NADR M. JOMHA • *Department of Surgery, University of Alberta, Edmonton, AB, Canada*
VICTORIA KEROS • *Department of Reproductive Medicine, Division of Gynecology and Reproduction, Karolinska University Hospital, Stockholm, Sweden; Division of Urology, Department of Clinical Science, Intervention and Technology (CLINTEC), Karolinska Institutet, Stockholm, Sweden*
OLENA KISHCHENKO • *Leibniz Institute of Plant Genetics and Crop Plant Research (IPK), Seeland, Germany; Institute of Cell Biology and Genetic Engineering, The National Academy of Sciences of Ukraine, Kyiv, Ukraine*
PAUL KROPP • *Ludwig-Maximilians-Universität, Munich, Germany*
ALBA LAGO • *Centro de Investigación Marina CIM-Universidade de Vigo, Ecocost Research Group - Cryolab, Vigo, Spain*
LEO LOU • *Department of Mechanical Engineering and Department of Bioengineering, University of California Berkeley, Berkeley, CA, USA; Department of Bioengineering, University of California Berkeley, Berkeley, CA, USA*
PABLO LOZA-ÁLVAREZ • *Institut de Ciències Fotòniques, The Barcelona Institute of Science and Technology, Castelldefels, Barcelona, Spain*
RUIDONG MA • *Department of Mechanical Engineering, University of Washington, Seattle, WA, USA*
XIAO MA • *Department of Mechanical Engineering, University of Washington, Seattle, WA, USA*
ALAN LENON MAIDA • *Department of Mechanical Engineering, University of California Berkeley, Berkeley, CA, USA*
DENESE C. MARKS • *Research and Development, Australian Red Cross Lifeblood, Sydney, NSW, Australia; Sydney Medical School, University of Sydney, Camperdown, NSW, Australia*
LEAH A. MARQUEZ-CURTIS • *Department of Chemical and Materials Engineering, University of Alberta, Edmonton, AB, Canada; Department of Laboratory Medicine and Pathology, University of Alberta, Edmonton, AB, Canada*
MONICA MARRO • *Institut de Ciències Fotòniques, The Barcelona Institute of Science and Technology, Castelldefels, Barcelona, Spain*
MARIA MARSAL • *Institute of Nanoscience and Materials of Aragon (INMA), Faculty of Science, CSIC-University of Zaragoza, Zaragoza, Spain*
IRIS MARTÍNEZ-RODERO • *Assisted Human Reproduction Unit, Clinic Institute of Gynecology, Obstetrics and Neonatology (ICGON), Hospital Clínic, Barcelona, Spain*
PAUL MATEJTSCHUK • *Formulation Science, Medicines and Healthcare products Regulatory Agency, Potters Bar, UK*
TIM MENZEN • *Coriolis Pharma Research GmbH, Martinsried, Germany*
TERESA MOGAS • *Department of Animal Medicine and Surgery, Universitat Autònoma de Barcelona, Cerdanyola del Vallès, Spain*
OLGA MYKHAILOVA • *Innovation, Canadian Blood Services, Edmonton, AB, Canada*
JAYANTHI NADARAJAN • *New Zealand Institute for Bioeconomy Science Limited, Food Industry Science Centre, Fitzherbert Science Centre, Palmerston North, New Zealand*

Manuela Nagel • *Genebank Department, Leibniz Institute of Plant Genetics and Crop Plant Research (IPK), Seeland, Germany; Leibniz Institute of Plant Genetics and Crop Plant Research (IPK), Seeland, Germany*
Joanna Nynca • *Team of Reproduction and Development in Fish, InLife Institute of Animal Reproduction and Food Research in Olsztyn, Polish Academy of Sciences, Olsztyn, Poland*
Carly Olafson • *Innovation, Canadian Blood Services, Edmonton, AB, Canada*
Harriëtte Oldenhof • *Unit for Reproductive Medicine - Clinic for Horses, University of Veterinary Medicine Hannover, Hannover, Germany; Biostabilization Laboratory - Lower Saxony Center for Biomedical Engineering, Implant Research and Development, University of Veterinary Medicine Hannover, Hannover, Germany*
Robyn J. Osborne • *Department of Surgery, Duke University Medical Center, Durham, NC, USA*
Katarzyna Palińska-Żarska • *Department of Ichthyology, Hydrobiology, and Aquatic Ecology, National Inland Fisheries Research Institute, Olsztyn, Poland*
Estefanía Paredes • *Centro de Investigación Marina CIM-Universidade de Vigo, Ecocost Research Group - Cryolab, Vigo, Spain*
Anton Peterson • *Leibniz Institute of Plant Genetics and Crop Plant Research (IPK), Seeland, Germany; Institute of Cell Biology and Genetic Engineering, The National Academy of Sciences of Ukraine, Kyiv, Ukraine*
Roberto Pisano • *Department of Applied Science and Technology, Politecnico di Torino, Torino, Italy*
Matthew J. Powell-Palm • *J. Mike Walker '66 Department of Mechanical Engineering, Texas A&M University, College Station, TX, USA; Department of Materials Science & Engineering, Texas A&M University, College Station, TX, USA; Department of Biomedical Engineering, Texas A&M University, College Station, TX, USA*
Silke Pradella • *Leibniz Institute DSMZ-German Collection of Microorganisms and Cell Cultures, Braunschweig, Germany*
Shen Ren • *Department of Mechanical Engineering, University of Washington, Seattle, WA, USA; Department of Mechanical Engineering, Seattle University, Seattle, WA, USA*
Stephanie Revoredo • *Department of Neuroscience and Regenerative Medicine, Medical College of Georgia, Augusta University, Augusta, GA, USA*
Nelson Romano • *Center for Research and Development in Food Science and Technology, La Plata, Argentina*
Boris Rubinsky • *Department of Mechanical Engineering, University of California Berkeley, Berkeley, CA, USA; Department of Mechanical Engineering and Department of Bioengineering, University of California Berkeley, Berkeley, CA, USA*
Nicole Sänger • *Department of Gynecological Endocrinology and Reproductive Medicine, University Hospital Bonn, Bonn, Germany*
Andreas Schallmoser • *Department of Gynecological Endocrinology and Reproductive Medicine, University Hospital Bonn, Bonn, Germany*
Ingo Schubert • *Leibniz Institute of Plant Genetics and Crop Plant Research (IPK), Seeland, Germany*
Elisa Serrano-Revuelta • *Department of Internal Medicine, Reproduction and Population Medicine, Faculty of Veterinary Medicine, Ghent University, Merelbeke, Belgium*

SERSHEN • *University of the Western Cape, Department of Biodiversity and Conservation Biology, Bellville, South Africa*

MOHAMMADHAMED SHAHSAVARI • *Department of Surgery, University of Alberta, Edmonton, AB, Canada*

GUANGZHU SHANG • *Laboratory for Cryo-Biomedical Engineering, Department of Electronic Engineering and Information Science, School of Information Science and Technology, University of Science and Technology of China, Hefei, China*

NADIA SHARDT • *Department of Chemical Engineering, Norwegian University of Science and Technology (NTNU), Trondheim, Norway*

ZHIQUAN SHU • *Department of Mechanical Engineering, University of Washington, Seattle, WA, USA; School of Engineering and Technology, University of Washington Tacoma, Tacoma, WA, USA*

HARALD SIEME • *Unit for Reproductive Medicine - Clinic for Horses, University of Veterinary Medicine Hannover, Hannover, Germany*

KATRIEN SMITS • *Department of Internal Medicine, Reproduction and Population Medicine, Faculty of Veterinary Medicine, Ghent University, Merelbeke, Belgium*

ANN VAN SOOM • *Department of Internal Medicine, Reproduction and Population Medicine, Faculty of Veterinary Medicine, Ghent University, Merelbeke, Belgium*

RACHEL M. THOMPSON • *School of Chemical, Biological, and Environmental Engineering, Oregon State University, Corvallis, OR, USA*

RILEY E. THOMPSON-BRANDHAGEN • *Department of Clinical Sciences, Colorado State University, Fort Collins, CO, USA; Animal Reproduction and Biotechnology Laboratory, Colorado State University, Fort Collins, CO, USA*

CONGHUI TIAN • *Laboratory for Cryo-Biomedical Engineering, Department of Electronic Engineering and Information Science, School of Information Science and Technology, University of Science and Technology of China, Hefei, China*

ANNA TÖNIßEN • *Unit for Reproductive Medicine - Clinic for Horses, University of Veterinary Medicine Hannover, Hannover, Germany*

TRACEY R. TURNER • *Innovation, Canadian Blood Services, Edmonton, AB, Canada*

SAYAKA WAKAYAMA • *Advanced Biotechnology Center, University of Yamanashi, Yamanashi, Japan*

TERUHIKO WAKAYAMA • *Advanced Biotechnology Center, University of Yamanashi, Yamanashi, Japan; Faculty of Life and Environmental Sciences, University of Yamanashi, Yamanashi, Japan*

SYLWIA WAŁDOWSKA • *Team of Reproduction and Development in Fish, InLife Institute of Animal Reproduction and Food Research in Olsztyn, Polish Academy of Sciences, Olsztyn, Poland*

SHANGPING WANG • *Department of Bioengineering, Clemson University, Clemson, SC, USA*

ZIYUAN WANG • *Department of Mechanical Engineering, University of Washington, Seattle, WA, USA*

LINNEA WARBURTON • *Department of Mechanical Engineering and Department of Bioengineering, University of California Berkeley, Berkeley, CA, USA; Department of Mechanical Engineering, University of California Berkeley, Berkeley, CA, USA*

KEVIN R. WARD • *Biopharma Group, Winchester, UK*

WILLEM F. WOLKERS • *Unit for Reproductive Medicine - Clinic for Horses, University of Veterinary Medicine Hannover, Hannover, Germany; Biostabilization Laboratory - Lower Saxony Center for Biomedical Engineering, Implant Research and Development, University of Veterinary Medicine Hannover, Hannover, Germany*

KEZHOU WU • *Department of Surgery, University of Alberta, Edmonton, AB, Canada; Sports Medicine Center, First Affiliated Hospital, Shantou University Medical College, Shantou, China*
MING HAN YU • *Department of Chemical and Materials Engineering, University of Alberta, Edmonton, AB, Canada; Department of Laboratory Medicine and Pathology, University of Alberta, Edmonton, AB, Canada*
DANIEL ŻARSKI • *Team of Reproduction and Development in Fish, InLife Institute of Animal Reproduction and Food Research in Olsztyn, Polish Academy of Sciences, Olsztyn, Poland*
GANG ZHAO • *Laboratory for Cryo-Biomedical Engineering, Department of Electronic Engineering and Information Science, School of Information Science and Technology, University of Science and Technology of China, Hefei, China*

Part I

Fundamental Aspects of Cryopreservation and Freeze-Drying

Check for updates

Chapter 1

Principles Underlying Cryopreservation and Freeze-Drying of Cells

Willem F. Wolkers and Harriëtte Oldenhof

Abstract

Cryopreservation, vitrification, and freeze-drying can be used to preserve biomolecules, cells, or tissues for extended periods until they are needed. This chapter explores the underlying principles how to preserve biological systems in the frozen state at cryogenic temperatures or in the dried state at room temperature. Freezing and drying are different stress vectors that require different types of protective agents. Freezing injury results from solution effects and physical damage if intracellular ice is formed, while drying removes non-freezable water bound to biomolecules resulting in conformational and phase changes. This chapter examines the mechanisms by which cryoprotective and lyoprotective agents work, including reducing ice crystal growth, preferential exclusion, water replacement, and formation of a highly viscous glassy state, i.e., vitrification. It is discussed how protective agents can be introduced into cells or tissues. We also highlight how insights in cell-specific membrane phase and permeability properties can help design more effective (ice-free) cryopreservation and freeze-drying protocols. In addition, a brief overview of emerging trends in cryopreservation and freeze-drying technologies is presented.

Keywords Anhydrobiosis, Cryopreservation, Cryoprotectants, Lyoprotectants, Intracellular freezing, Preferential exclusion, Solution effects, Water replacement, Membrane phase behavior, Membrane permeability, Vitrification

1 Nature's Way to Preserve Life

The physical properties of water are dependent on pressure, temperature, and the presence of solutes (see Fig. 1). Living systems rely on water for proper function. The removal of water from tissues, cells, and biomolecules through drying or freezing leads to significant stress factors, including osmotic stress and conformational changes that can disrupt cellular and biomolecular structure and function (see Fig. 2D). Additionally, ice formation during freezing can cause mechanical stress. For cells or biological systems that typically cannot survive extreme conditions, one potential solution is the development of "bioinspired" protective formulations and strategies that preserve life in frozen or dried states.

Willem F. Wolkers and Harriëtte Oldenhof (eds.), *Cryopreservation and Freeze-Drying Protocols*, Methods in Molecular Biology, vol. 3060, https://doi.org/10.1007/978-1-0716-5416-3_1,

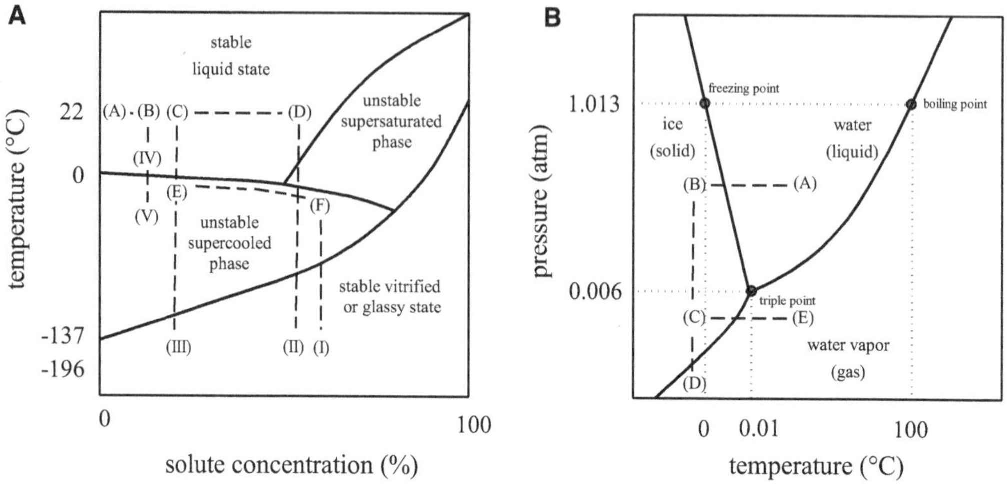

Fig. 1 Cryopreservation pathways superimposed on a solute-water state diagram of a hypothetical cryoprotectant (**A**) and the pressure temperature diagram of water to illustrate the pathways involved in freeze-drying (**B**). Axes are not drawn to scale. The solute-water state diagram shows the physical state of both solute and solvent as a function of temperature and solute concentration. The equilibrium freezing curves, the glass transition curves, and solute precipitation lines are indicated. For each preservation method (I: slow freezing, II: conventional vitrification, III: low-CPA vitrification, IV: conventional hypothermia, and V: supercooling), the thermodynamic paths are indicated. It is assumed that there is no solute precipitation during freeze concentration (path E–F) and during plunging into liquid nitrogen (path F–I). Panel **B** shows hypothetical freeze-drying paths superimposed on a water phase diagram. Freeze-drying involves first converting specimens from a liquid phase to solid (ice) by freezing to below their eutectic temperature (path A–B), then lowering pressure around specimens below the triple point by applying vacuum (path B–C), and finally converting the solid (ice) to gas (water vapor) by either further lowering the pressure below the sublimation line (path C–D) or controlled heating of specimens under vacuum (path C–E). The triple point refers to the intersection on the phase diagram, where three different phases can coexist. The schematic drawings are inspired by previous illustrations [139, 140]

Nature provides remarkable examples of life forms that can enter a reversible state of suspended animation. Anhydrobiosis, the ability of certain organisms to withstand the harmful effects of dehydration and survive in a state of near-complete dryness, is one such phenomenon [1]. This phenomenon is found across all major taxa. Although widely spread in spores, pollen, and seeds, anhydrobiosis of whole organisms is relatively rare [2, 3]. Examples include tardigrades (water bears) and resurrection plants such as *Craterostigma plantagineum* (see Fig. 2B). These organisms can be dehydrated to water contents of less than 10% of their dry weight and, upon rehydration, can resume normal metabolic functions [4, 5]. In the anhydrobiotic state, organisms stay dormant in a metabolically quiescent state for prolonged periods and most chemical reactions come to a halt [6]. However, in dry conditions, free radicals are still able to diffuse and cause chemical changes in biomolecules, leading to damage that limits the lifespan and stability of these systems.

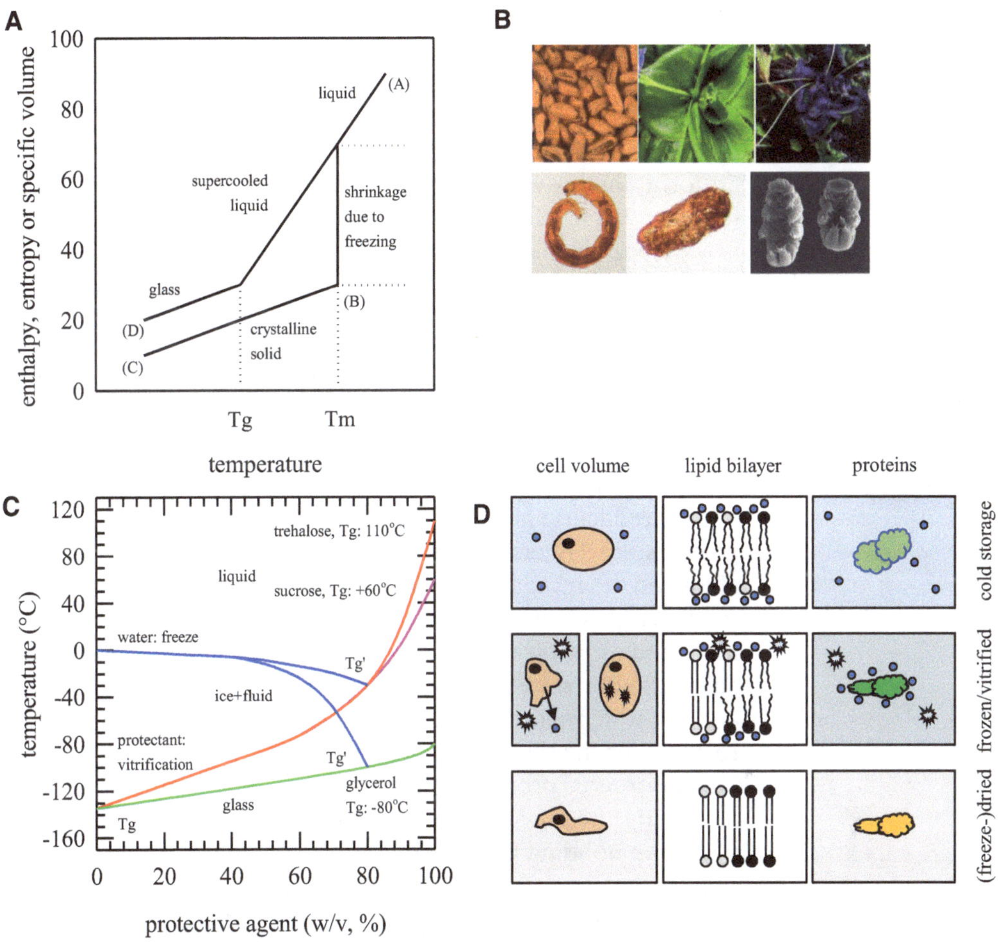

Fig. 2 First and second order phase transitions (**A**), examples of preservation in the dried state in nature (**B**), state diagram of various protective solutes in water (**C**), and impact of freezing and drying on cells and biomolecules (**D**). Panel **A** shows that a first order phase transition (liquid-crystalline solid: path A–B–C) coincides with an abrupt change in various thermodynamic parameters, including enthalpy, entropy and the specific volume, whereas a second order transition (liquid-supercooled liquid-glass: path A–D) coincides with a gradual change. Formation of a vitrified or glassy state does not involve latent heat and a change in phase state. In panel **B**, micrographs are shown of tissues and organisms that survive desiccation by vitrification including pollen and seeds, as well as anhydrobiotic organisms like the resurrection plant *Craterostigma plantagineum*, the sleeping chironomid or *Polypedilum vanderplanki*, and so-called water bears or tardigrades. Panel **C** shows solute-water state diagrams of the protectants glycerol (green), sucrose (pink), and trehalose (red) (i.e., water freezing and solute vitrification but not boiling and equilibrium saturation solubility). The plot shows the physical state of both solute and solvent (water, blue) as a function of temperature (y-axis) and solute concentration (x-axis). The equilibrium freezing (or melting) curves and the glass transition curves are indicated. The equilibrium freezing curve is also known as the freezing point depression curve. Panel **D** shows how freezing or drying affect cell volume, membranes, and proteins. Freezing and/or drying results in a decrease in cell volume, causes membranes to undergo fluid-to-gel phase changes, and proteins to undergo conformational changes. The Craterostigma pictures are taken from [64]. Seed and pollen pictures were kindly provided by Dani Ballesteros, Royal Botanic Gardens, Kew. Tardigrade pictures were made available by the Goldstein lab, UNC Chapel Hill. Takashi Okuda is acknowledged for providing *P. vanderplanki* pictures

A common feature among many anhydrobiotic organisms is the synthesis of large quantities of non-reducing disaccharides, particularly sucrose and trehalose [2]. Sucrose is predominantly found in resurrection plants, pollen, and seeds [5, 7, 8], while trehalose is common in tardigrades, brine shrimp, and nematodes [9]. These sugars form a highly viscous, glassy state when water content is reduced (see Fig. 2A, C). In this glassy state, molecular movement is greatly slowed, and chemical reactions nearly stop. It has also been suggested that sugars protect anhydrobiotic cells by substituting for the water that normally surrounds biomolecules [2]. The dried desiccation tolerant life stage of anhydrobiotic organisms have an increased glass transition temperature and reduced glass forming fragility relative to its dried desiccation intolerant life stage [10].

Sugar synthesis is often one of the first visible biochemical changes during the acquisition of desiccation tolerance. However, a general reduction in cellular metabolism and the synthesis of specific proteins, scavengers, and antioxidant are also critical for survival. In seeds and other organisms, such as tardigrades and *Polypedilum vanderplanki*, late embryogenesis abundant (LEA) proteins have been identified as lyoprotective agents [11–13]. Organisms entering an anhydrobiotic state need adequate time to prepare for desiccation, and the water content of the organism must be gradually reduced to allow for the synthesis of LEA proteins [14, 15]. These intrinsically disordered proteins help protect cells in several ways, including increasing the glass transition temperature and strengthening the hydrogen bonding network of the cytoplasmic glassy matrix, where cellular structures are embedded and preserved [16–19]. In addition, LEA proteins form membraneless condensate compartments in the cellular environment enabling anhydrobiosis by contributing to the downregulation of metabolic and developmental processes, and by selectively sequestering desiccation-sensitive molecules in the desiccating cell [20, 21].

In contrast, the complex biochemical adaptations that occur in anhydrobiotic cells during drying do not occur in ordinary mammalian cells, as they lack the necessary adaptive pathways. One of the main challenges in preserving cells and tissues that do not naturally survive freezing or drying is introducing protective agents that can enable them to withstand these extreme conditions.

2 Methods for Preservation of Cells and Tissues

Cells and tissues can be preserved through various methods; including hypothermic storage, cryopreservation, and freeze-drying (see Table 1). Hypothermic storage, typically at 4°C in a refrigerator, is generally only suitable for short-term preservation. Several cold storage solutions have been developed for organs, such as Euro Collins

Table 1

Different preservation methods for cells and tissues including hypothermic storage, cryopreservation, vitrification, and freeze-drying. Differences in applications, processing procedures, typical protective agents, and storage conditions are indicated

	Hypothermic storage	Cryopreservation	Vitrification	Freeze-drying
Process	Isothermal	Slow cooling	Rapid cooling	- Slow cooling - Sublimation - Desorption
Protective agents (i.e., examples)	Hypo-protective solutions	Dimethyl sulfoxide, glycerol (low concentration)	Dimethyl sulfoxide, propylene glycol, formamide (high concentration)	Sucrose, trehalose, hydroxyethyl starch
Storagetemperature/state	~4°C In aqueous solution	Below −150°C in frozen/glassy state Tg<−100°C	Below −150°C in glassy state Tg< −100°C	Room temperature in glassy state Tg>22°C
Storageduration	Short term	Long term		

solution, University of Wisconsin solution, Histidine-Tryptophan-Ketoglutarate solution, and Celsior [22]. These solutions are designed to mimic the intracellular environment and reduce metabolic activity, while also providing osmotic stability and scavenging agents for oxygen radicals. However, many cells experience chilling injury and are not well-suited for hypothermic storage.

Cryopreservation allows cells to be stored for longer periods at ultra-low temperatures (below −150°C), typically in liquid nitrogen or mechanical freezers. To minimize ice formation, cryoprotective agents (CPAs) like dimethyl sulfoxide (DMSO) or glycerol are added. Successful cryopreservation depends not only on the type and concentration of the CPA but also on the cooling rate. Standard cryopreservation typically uses slower cooling rates (0.1–50°C/min) and low CPA concentrations (2.5–10% v/v, up to ~1.5 M). These conditions enable cells to gradually lose intracellular water during freezing, preventing ice formation inside the cells. Ice-free cryopreservation or vitrification, on the other hand, uses rapid cooling rates (>100°C/min) and high CPA concentrations (30–50% v/v, up to ~9 M) to avoid ice formation entirely.

One of the challenges with slow freezing cryopreservation is determining the optimal cooling rate for each cell type to maximize cryosurvival [23], whereas for vitrification, the key challenge is ensuring uniform mass and heat transfer while minimizing CPA toxicity [24]. Slow freezing is routinely used for preserving cell culture lines and sperm, although cryosurvival rates can vary widely between different cell types and among individuals. Vitrification is commonly used for embryos and small mammalian and plant tissues. Currently, vitrification is the most promising method for preserving whole organs, though several major obstacles remain to be solved before this can be applied in human medicine [25, 26].

Freeze-drying, or lyophilization, is widely used in the pharmaceutical industry to preserve substances such as proteins and liposomes, as it allows samples to be stored and shipped in a stable, dried form at room temperature. However, freeze-drying is not commonly used to preserve cells or tissues because the process is more damaging than freezing primarily due to the combination of stresses caused by ice formation, sublimation, and dehydration, leading to membrane damage and protein denaturation. While freezing alone causes cryoinjury from solution effects and ice crystals, the added steps of primary (sublimation of ice) and secondary (desorption of residual water) drying in freeze-drying further reduce viability, especially without adding proper protective agents. Unlike freezing, freeze-drying removes the hydration shell surrounding biomolecules [27]. To protect cells during freeze-drying, lyoprotectants such as sucrose and trehalose are often used. The challenge lies in loading cells with these protectants, as cellular membranes are typically impermeable to them.

While freeze-dried cells may not retain full functionality, they can still be useful for certain applications. For example, freeze-dried platelets are used in topical wound healing [28] and are a promising alternative to fresh platelets to stop bleeding in thrombocytopenic patients [29]. Freeze-dried sperm can be used for fertilization via intracytoplasmic sperm injection [30]. It is unlikely that cells in tissues can survive freeze-drying, but freeze-dried acellular tissues can serve as biological scaffolds that can be used to replace diseased tissue. These scaffolds function as a starter extracellular matrix supporting the growth of new, functional tissue when seeded with a patient's own autologous cells [31].

3 Freezing and Drying Injury

3.1 Freezing Injury

Cryopreservation of cells or tissues involves using ultra-low temperatures to halt biological processes, chemical reactions, and osmotically driven mass transport processes (e.g., cellular dehydration). However, the processing pathways cells have to traverse to reach safe storage temperatures and to return back to physiological temperatures can be very damaging. First, cooling results in the conversion of liquid water to crystalline ice causing many different types of damage to a biological sample. Early theories of freezing injury proposed that ice crystals physically puncture cells and intracellular structures, destroying them through direct mechanical force [32]. Additionally, the formation of intracellular ice can damage the structure and organization of plasma and organelle membranes, leading to cell death [33, 34].

As ice forms extracellularly, the high salt concentration in the freeze-concentrated (non-frozen) solution causes intracellular water to move into the extracellular space, leading to dehydration of the cells. This process not only restores osmotic balance but also concentrates intracellular solutes and biomolecules, and brings organelles into closer proximity. This phenomenon, referred to as solution effects injury, occurs when cells are trapped in a freeze-concentrated solution of residual non-freezable water and salts [35]. While cells can tolerate a certain degree of dehydration due to freezing, exceeding specific tolerance limits results in irreversible damage and loss of viability [33, 34]. Likewise, additional damage may occur upon thawing when cells are exposed to ice recrystallization events if the rewarming rate is not fast enough. In addition, cells swell upon thawing due to osmotic imbalances [36, 37].

Biomolecules, including proteins, lipids, DNA, and RNA, may undergo conformational and phase changes during freezing.

Membrane phase transitions, in particular, and alterations in membrane domain organization during cooling and freezing, can lead to dysfunction when the system is reconstituted [38, 39]. Macromolecular assemblies, such as liposomes, can also suffer from fusion and leakage during freezing and thawing [40].

3.2 Drying Injury

Freezing and drying impose distinct stress factors on biological materials [27]. Unlike freezing, drying leads to the removal of non-freezable water surrounding biomolecules. This loss of water is often detrimental to biological samples because the structure and function of biomolecules depend on the presence of liquid water interacting with biomolecular surfaces via hydrogen bonding. While DNA and structural proteins like collagen exhibit relative resistance to drying and can often withstand drying without significant harm, other proteins may experience irreversible structural changes during drying, which can impair their function upon rehydration [41, 42]. Liposomal systems, on the other hand, cannot be dried without protective measures. During drying, lipid bilayers tend to fuse and undergo lyotropic membrane phase transitions, leading to leakage of aqueous liposomal contents [43–45].

The extent of damage associated with drying of biological materials increases with sample size and complexity. In cellular environments, biomolecules are especially vulnerable to oxidative damage, as reactive oxygen species (ROS) accumulate during drying and subsequent storage. At low water content, antioxidant defense mechanisms are compromised, increasing the likelihood of ROS reacting with biomolecules. This oxidative damage is evident as lipid peroxidation, phospholipid de-esterification [46–48], DNA breakage, and the formation of carbonyl derivatives in proteins [49]. Moreover, proteins may undergo Amadori and Maillard (browning) reactions with reducing sugars, particularly at low water levels [50, 51]. Proteins can also be degraded by proteases from lysosomes that lose membrane integrity during dehydration.

4 Mode of Action of Protectants

4.1 Cryoprotective Agents

CPAs are commonly categorized based on their ability to permeate cell membranes. Membrane-permeable CPAs are small, polar molecules that can easily cross cellular membranes through passive diffusion. Membrane-impermeable CPAs are divided into two groups: small molecules, such as sugars, which remain osmotically active, and large (bio)polymers, which have little impact on the osmolality of the surrounding medium [52, 53]. Membrane-impermeable CPAs are often used as bulking agents in combination with membrane-permeable CPAs for intracellular protection.

One of the primary mechanisms by which CPAs function is by reducing the amount of ice formed during freezing, simply by increasing the solute concentration. This effect is observed with both membrane-permeable and membrane-impermeable CPAs. Early studies demonstrated that glycerol mitigates the rise in salt concentration during freezing, thereby reducing damage caused by solution effects [35, 54]. Later, it was proposed that cryoprotective action is linked to the ability of specific agents to bind to biomolecules, thereby preventing phase transitions and conformational changes. The preferential exclusion theory explains the stabilizing effects of protective agents on proteins from a thermodynamic perspective [55]. In a solution containing water, proteins, and a CPA like DMSO, proteins tend to interact more with water than with DMSO, resulting in the exclusion of DMSO from the protein surface. This creates a thermodynamically unfavorable situation, stabilizing the native protein structure by increasing the energy barrier for unfolding and denaturation (see Fig. 3A, B). Although this theory was originally developed for dilute protein solutions at normal temperatures, it is also thought to apply to the stabilization of proteins during cooling and freezing. Similarly, membranes can be stabilized by molecules that are preferentially excluded from the membrane surface [56].

CPAs also protect cells by modulating the rate of dehydration during freezing, lowering the activation energy required for water permeation across membranes [57, 58]. This results in a more gradual dehydration process, reducing the likelihood of intracellular ice formation. Additionally, CPAs help facilitate the formation of a vitrified state, which provides further protection to cells during freezing. At low CPA concentrations and slow cooling rates, a maximally freeze-concentrated solution forms during freezing, resulting in a glassy state below a characteristic temperature, Tg' (see Fig. 2C). With higher CPA concentrations and faster cooling rates, the entire solution can enter a vitrified state. While Tg' remains constant regardless of solute concentration, the glass transition temperature (Tg) is dependent on water content. Many non-permeating CPAs function by increasing Tg, ensuring that samples reach the glassy state at higher subzero temperatures. The Tg' and Tg of sugars, proteins (e.g., albumin), synthetic polymers (e.g., polyvinylpyrrolidone [PVP], Ficoll), and polysaccharides (e.g., hydroxyethyl starch) are relatively high compared to glycerol [59] (see Fig. 2C). In addition, compounds with ice-blocking properties can be used in freezing formulations to prevent ice recrystallization during rewarming and storage under suboptimal conditions [60].

4.2 Lyoprotective Agents

Lyoprotective agents protect biological materials during both freezing and drying, and can be considered a subset of CPAs.

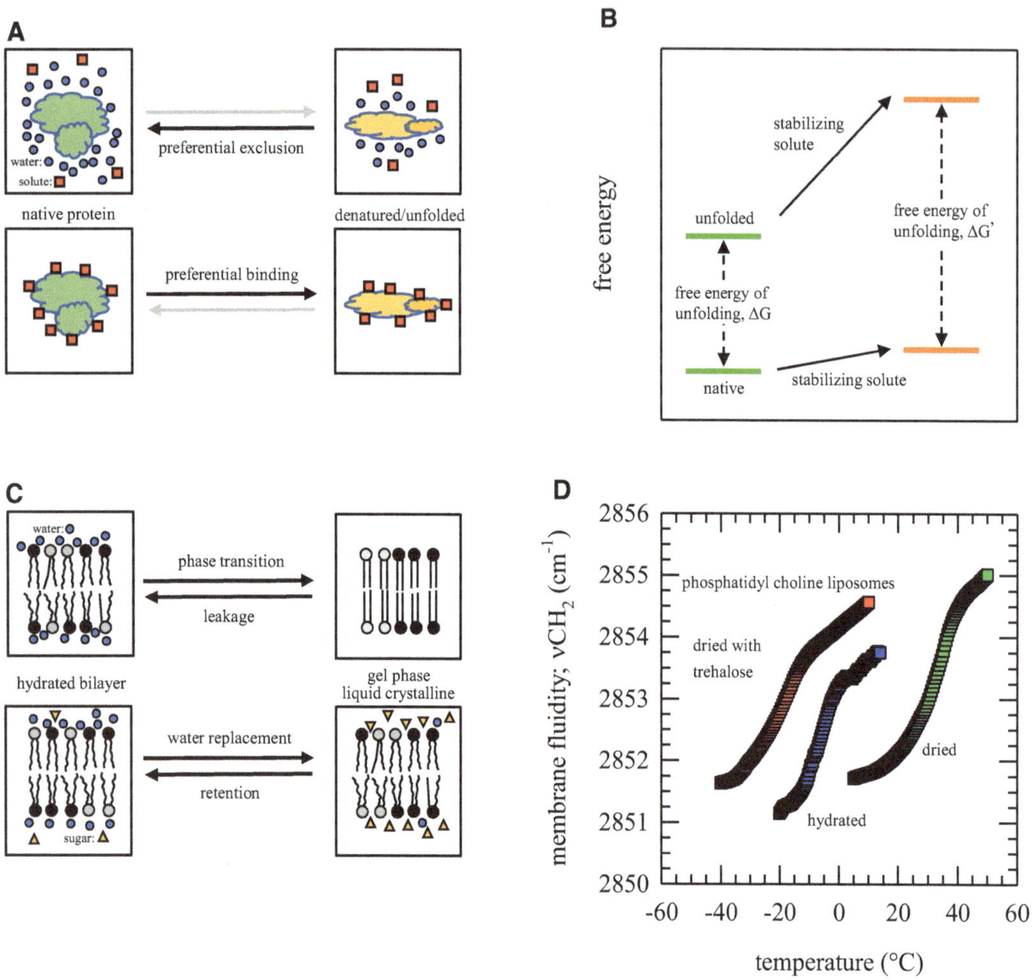

Fig. 3 Models describing modes of action of protective agents, including the preferential exclusion theory (**A, B**) and water replacement hypothesis (**C, D**). (**A**) The preferential exclusion theory postulates that solutes that are preferentially excluded from the surface of proteins stabilize the native versus the unfolded protein state. (**B**) Exclusion of solutes from the protein surface increases the energy barrier for protein denaturation. (**C**) The water replacement hypothesis postulates that protective agents such as sugars replace water normally surrounding phospholipid head groups. If water is removed from phospholipid head groups in the absence of protectants, a liquid crystalline-to-gel membrane phase transition will take place with associated leakage, whereas if membranes are dried in the presence of sugars, membranes remain in the liquid crystalline state. (**D**) Membrane phase transitions can be studied using FTIR by monitoring the CH_2-stretching vibration band position as a function of the temperature. Gel phase formation, e.g., by lowering the temperature or by drying, is visible as a shift of νCH_2 to a lower wavenumber. The phase transition temperature (Tm) of phosphatidylcholine lipids increases upon drying. When lipids are dried in the presence of sucrose, the phase transition temperature is actually lower than the Tm of hydrated lipids. Drawings and data that are presented here are adapted from [2, 55, 141]

Non-reducing disaccharides, in particular, have been widely used as lyoprotective agents in the freeze-drying of proteins and liposomal systems. Two main theories explain how sugars protect liposomes during drying. The water replacement hypothesis (see Fig. 3C, D) postulates that sugars replace the water surrounding the phospholipid head groups during drying, replacing the hydrogen bonds between water and the head groups. This prevents the lipid acyl chains come in closer proximity of each other, which prevents membranes from undergoing a liquid crystalline to gel phase transition and associated membrane leakage [2]. Alternatively, the water entrapment hypothesis proposes that sugars protect lipid membranes by trapping the water molecules around the lipid head groups [61]. These theories are not mutually exclusive, and both may contribute at different stages of a drying process. For instance, during the early stages of drying, water entrapment might be the dominant mechanism, while water replacement could occur after prolonged drying or when heat is applied to the system [62]. Additionally, it has been proposed that disaccharides reduce the hydration forces between lipid bilayers, either by adsorbing to the membrane-water interface or by excluding from it [63].

Unlike typical CPAs such as glycerol or DMSO, sugars can form a glassy state at room temperature, which ensures long-term dry storage (see Fig. 2C). A glass is a highly viscous, amorphous state that forms below a specific temperature known as Tg. In this state, biomolecules (or cellular structures) are embedded in a glassy matrix, which slows molecular mobility and reduces the likelihood of damaging reactions [64–66]. Among disaccharides, trehalose has a Tg nearly 60°C higher than sucrose, even though they have almost the same molecular weight [9]. Macromolecules like albumin and hydroxyethyl starch (HES) can be incorporated into freeze-drying formulations to increase the Tg and thereby improve storage stability [67–69]. Water acts as a plasticizer, reducing the Tg of freeze-dried samples. Both the Tg and molecular mobility in the glassy state, and thus, the storage stability of the sample, are highly dependent on the residual moisture content after freeze-drying [70].

5 Water and Solute Transport Across Cellular Membranes and Loading Cells and Tissues with Protective Agents

5.1 Water Transport Across Cellular Membranes

Membrane permeability to water (i.e., hydraulic permeability) or Lp, is dependent on cell specific properties including the membrane lipid composition. The rate at which water can pass or permeate the membrane determines the cell volume response when exposed to molar concentrations of CPAs. Furthermore, Lp at subzero temperatures determines the rate and extent of cellular dehydration during freezing. Freezing-induced dehydration actually results in removal of water from the phospholipid headgroups and hence induces a sharp membrane fluid-to-gel-phase transition [57, 71] (see Fig. 4A, B). Increased lipid packing upon ice formation, causes water molecules to encounter a more hydrophobic environment when crossing membranes, which affects Lp and the Arrhenius behavior of Lp (i.e., the activation energy of water transport, E_{Lp}) (see Fig. 4C). Addition of CPAs does not prevent the occurrence of freezing-induced membrane phase transitions but modulates membrane phase and permeability

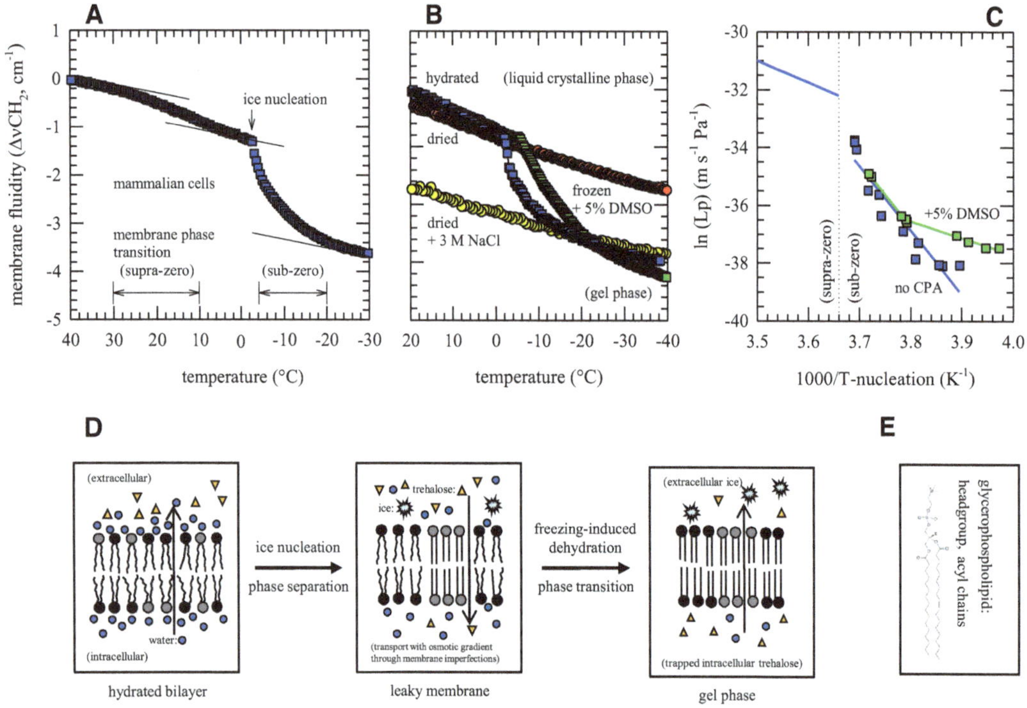

Fig. 4 Membrane phase behavior of mammalian cells (i.e., fibroblasts) at supra and subzero temperatures (**A**), and effects of freezing in the presence versus absence of CPAs (i.e., 5% DMSO), dehydration and exposure to severe hypertonic (i.e., 3 M NaCl) conditions (**B**). The effect of temperature on membrane permeability towards water (Lp) is shown in an Arrhenius plot (**C**), and a schematic presentation is given on membrane phase state changes during cooling/freezing and concomitant water and solute transport across the membrane (**D**). Panel **E** shows the structure of a typical glycerophospholipid. (**A**) FTIR studies, in which νCH_2 is plotted versus the sample temperature, revealed that cellular membranes typically exhibit a broad noncooperative phase transition at suprazero temperatures. At subzero temperatures, ice nucleation induces cellular dehydration coinciding with a sharp fluid-to-gel membrane phase transition. (**B**) The freezing-induced phase transition is not prevented by the addition of CPAs. It is especially exposure to osmotic stress and not the drying per se that results in such a phase shift. (**C**) Membrane hydraulic permeability, Lp, is affected by the temperature as well as the presence of ice and CPAs. The presence of DMSO decreases the activation energy for water transport and increases the rate of water transport, allowing cellular dehydration to continue at low subzero temperatures. Panel **D** presents a schematic presentation of the membrane phase state at physiological temperature, as well as after ice nucleation and freezing-induced dehydration. Fluid- and gel-phase lipid domains are indicated, as well as ice formation and fluxes (arrows) of water (blue circles) and impermeable solutes (orange triangles). The data presented here are adapted from previous studies by our group [39]

properties. CPAs counteract the effects of ice on water membrane permeability by increasing Lp at a given temperature and decreasing E_{Lp}. The increased Lp in the presence of CPAs thus facilitates cellular dehydration to continue down to low subzero temperatures, which in turn decreases the incidence of intracellular freezing

At suprazero temperatures, Lp can be derived by analyzing cell volume responses in anisotonic media, acquired microscopically or via electronic particle sizing measurements [72, 73]. Assessment of cell volume responses at subzero temperatures requires a special cryomicroscope [33] but also other methods can be used to determine Lp under freezing conditions including differential scanning calorimetry (DSC) [74], and Fourier transform infrared spectroscopy (FTIR) [57, 58, 71]. Subzero Lp measurements can be used to extrapolate Lpg, i.e., Lp at a reference temperature of 0°C. Lpg and the activation energy for water transport, E_{Lp}, can be used to model cell volume responses during freezing and predict optimal cooling rates for cryopreservation [23, 37, 74].

5.2 Loading Cells with CPAs

In cryopreservation, the first step typically involves loading cells with CPAs. Membrane-permeable CPAs, such as ethylene glycol, DMSO, and propylene glycol, readily cross cellular membranes. However, transferring cells into solutions containing molar concentrations of CPAs must be performed with care, as this creates an osmotic gradient that induces mass transport and cell volume changes. Because water traverses the cell membrane more rapidly than permeating CPAs (see Fig. 5B), water initially moves out of the cell, causing cell shrinkage. Subsequently, CPAs and water both enter the cell until reaching osmotic equilibrium between the intracellular and extracellular environments. This process results in a characteristic biphasic cell volume response during exposure to permeating CPAs (see Fig. 6B, D). The response is dependent on the temperature (see Fig. 7A).

The biphasic volume response can be described using transport equations that account for the membrane permeability to water (Lp) and to solutes (Ps). Several modeling approaches are available to fit experimental cell volume data, including the one-parameter (solute permeability) model [75], the two-parameter model in which water and solute transport are treated independently [76], and the three-parameter model, which incorporates solute-solvent interactions during membrane transport [77]. Once cell-specific and CPA-specific Lp and Ps values are determined, these models can be used to predict cell volume changes during CPA loading and unloading, as well as during freezing and rewarming. In this way, these models support the design of optimized cryopreservation protocols taking the osmotic tolerance limits of the cell into account [78]. It should be noted, however, that Lp and Ps values obtained under specific experimental conditions may not be universally applicable to all stages of the

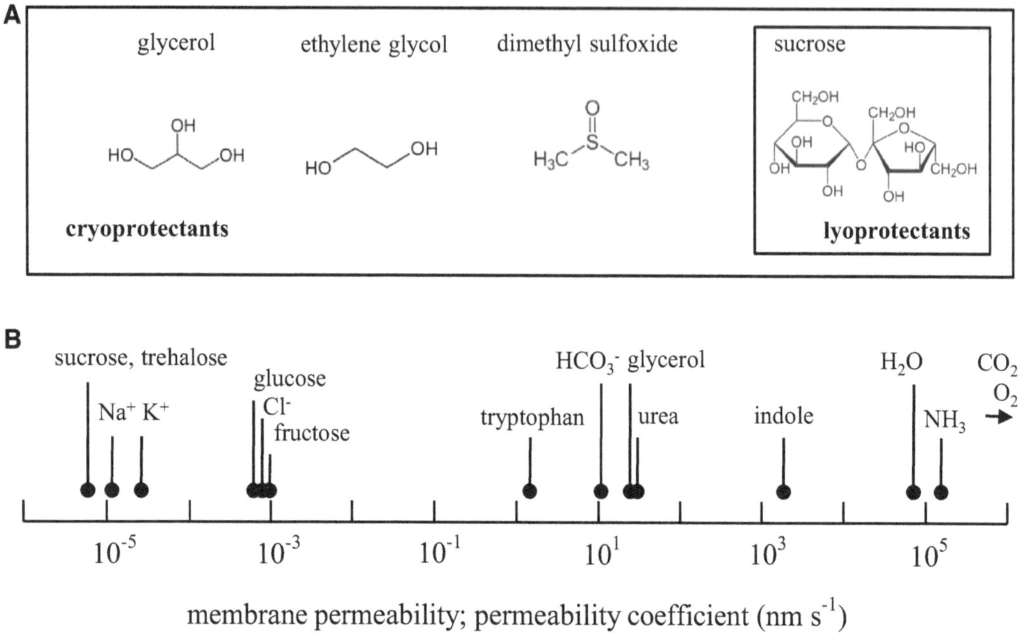

Fig. 5 Chemical structures of various cryoprotective and lyoprotective agents (**A**), and membrane permeability towards various molecules including water, glycerol, and sucrose (**B**). Membrane permeability coefficients are taken from [142]

cryopreservation process [39]. For example, the rate of water influx in a hypotonic environment often differs from the rate of water efflux under hypertonic conditions, a phenomenon known as rectification [39, 79, 80]. This difference in permeation rates is related to the lipid asymmetry of the inner and outer membrane leaflet, and is particularly relevant for predicting hypertonic cell volume responses during CPA loading and freezing versus hypotonic responses during warming, thawing, and CPA unloading (see Fig. 6B, D).

Vitrification solutions contain very high concentrations of CPAs. Due to cellular tolerance limits, cells cannot be directly exposed to full-strength vitrification solutions and must instead be equilibrated through serial exposure to increasing CPA concentrations. In the initial step, cells are typically exposed to CPA concentrations similar to those used in conventional cryopreservation, allowing intracellular CPA loading. Subsequent exposure to stronger and eventually the full-strength vitrification solution further increases intracellular CPA levels and, more importantly, dehydrates the cells prior to cooling. Dehydration in the final step raises intracellular solute concentrations and reduces the likelihood of intracellular ice formation. The duration of the final step in the full-strength vitrification solution is critical to minimize CPA-induced toxicity. During warming, CPAs are removed in a stepwise manner, often using sugars as osmotic buffers to prevent excessive cell swelling (see Figs. 6 and 7C).

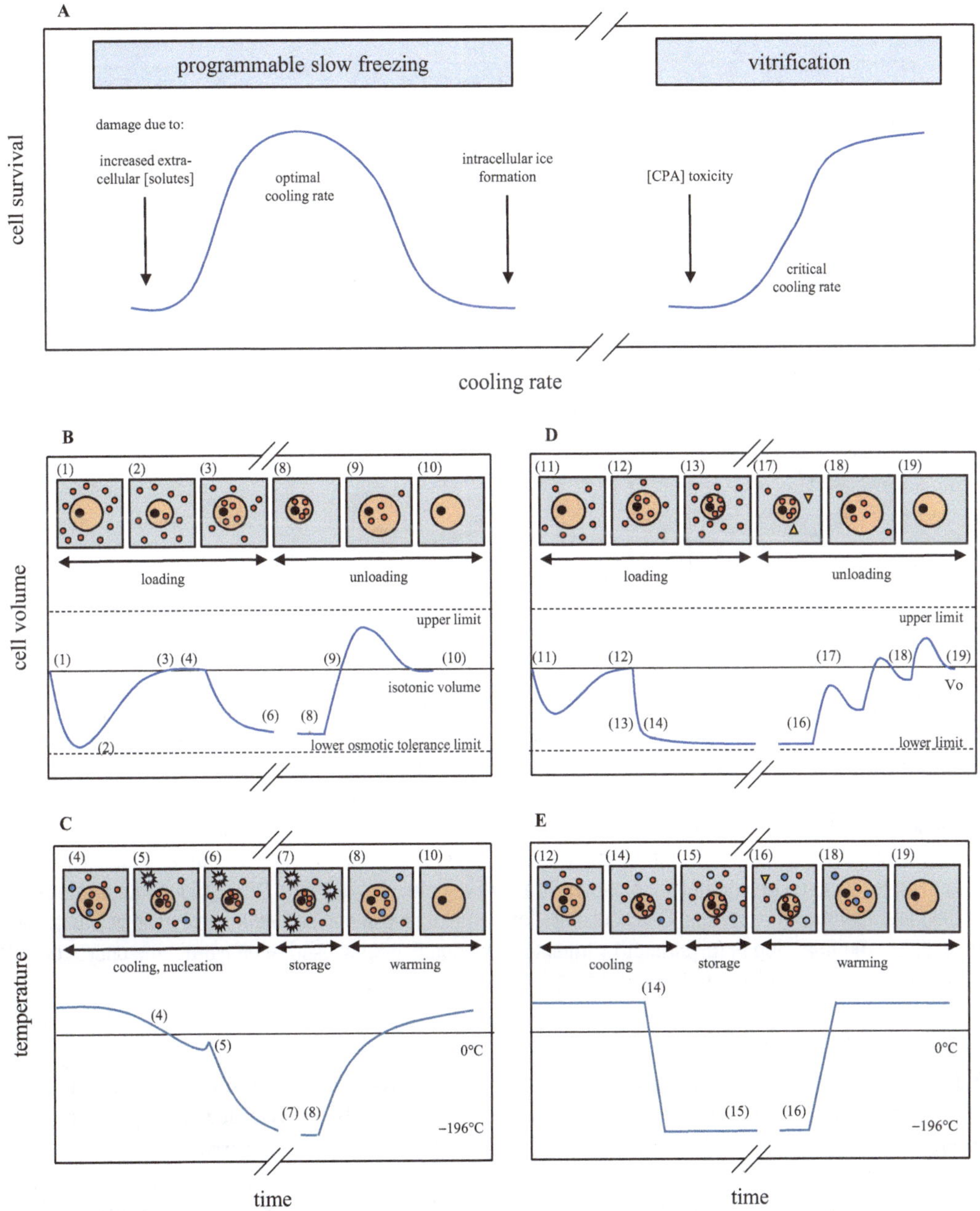

Fig. 6 Cryosurvival of cells as a function of the cooling rate (**A**), typical cell volume changes during (un)loading of cells with CPA (**B, D**) and temperature profiles (**C, E**) for typical programmable slow cooling procedures (**B, C**) and vitrification (**D, E**). (**A**) In case of programmable slow cooling cryopreservation, a typical inverted U-curve of cryosurvival is found, where at low cooling rates damage results from solute injury and damage at high cooling rates comes from intracellular ice formation. With vitrification and ultra-rapid cooling rates cryosurvival increases again, however, the use of high concentrations of cryoprotective agents reduces cryosurvival when rates are too slow. Panels **B** and **D** show cell volume responses in case of respectively slow cooling cryopreservation and vitrification. Volume excursions during typical CPA loading and warming/unloading protocols are illustrated graphically and schematically. Panels **C** and **E** show typical temperature profiles for, respectively, slow coolingcryopreservation and vitrification protocols. The drawings that are presented here are inspired by presentations from others [23, 143]

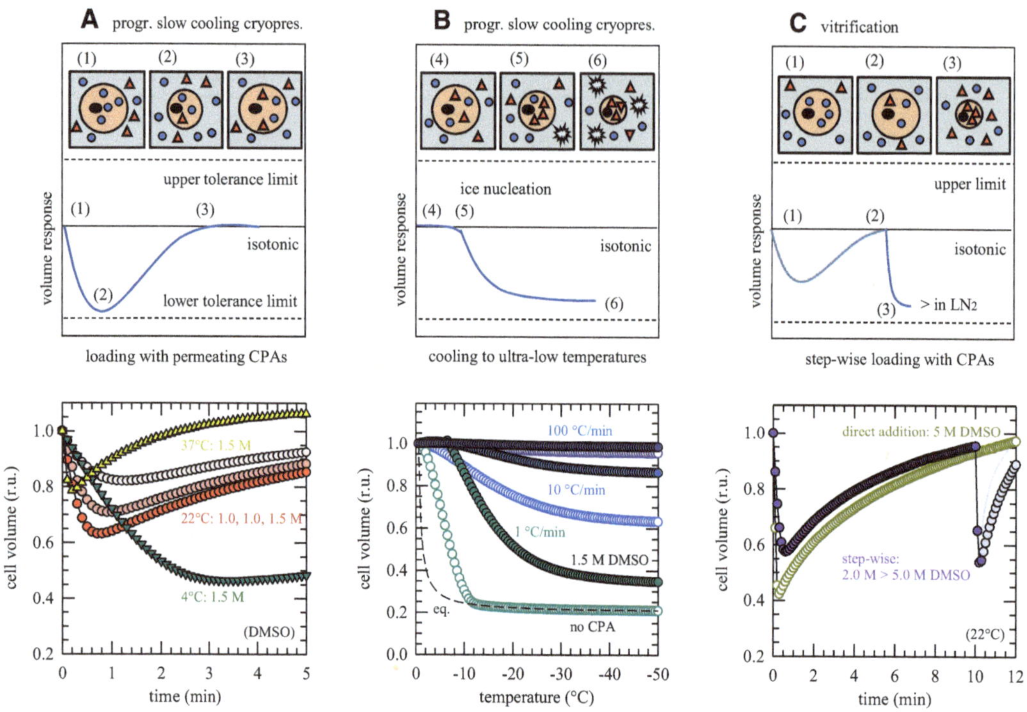

Fig. 7 Osmotic response of oocytes during loading with CPAs for programmable slow freezing procedures (**A**), freezing at different cooling rates (**B**), and during (step-wise) loading for vitrification (**C**). Cell volume changes are illustrated schematically, as obtained from mathematical simulations using models and parameters taken from the literature. Cells show a shrink swell response upon exposure to a CPA solution. The response depends on the incubation temperature and type of CPA (here DMSO is used as example). The extent of cellular dehydration during freezing is dependent on the cooling rate, and presence of CPAs. At the optimal cooling rate, cells tolerate losing up to 95% of their osmotically active volume upon freezing. Vitrification requires step-wise loading in solution with increasing CPA concentration to maintain osmotic cell volume excursions within tolerance limits. The final step in full strength vitrification solution is merely used to dehydrate the oocyte prior to rapid cooling for vitrification. Drawings and data are adapted from [37]

5.3 Loading Cells with Lyoprotective Agents

Prior to freeze-drying, lyoprotective agents such as trehalose must be introduced into mammalian cells to provide intracellular protection. A major challenge in this process is the intrinsic impermeability of cellular membranes to disaccharides (Fig. 5B). Consequently, numerous strategies have been developed to facilitate intracellular trehalose delivery and thereby improve survival during freezing and/or drying. These approaches include the use of genetically engineered pore-forming proteins [81], synthetic biopolymers that interact with cellular membranes [82], cell-penetrating peptides [83], lipophilic membrane-permeable trehalose derivatives [84], nanoparticle-mediated delivery of encapsulated trehalose [85], and thermally responsive nanocapsules for intracellular release [86]. Additional methods to promote trehalose uptake include electropermeabilization [87], sonoporation [88], osmotic shock [89], fluid-phase endocytosis

[90], and thermal cycling across membrane phase transitions at either suprazero [91] or subzero temperatures [92].

Interestingly, several studies have demonstrated that for various cell types, freezing in the presence of high extracellular trehalose concentrations can result in high post-thaw survival when optimized cooling rates are applied [68, 92–94]. In these cases, no special measures were taken to load trehalose into cells prior to freezing. Instead, trehalose appears to passively enter cells during freezing and/or thawing as a result of freezing-induced osmotic stresses combined with fluid-to-gel membrane phase transitions [92]. These phase transitions transiently disrupt membrane integrity, creating temporary defects that allow leakage of intracellular components and uptake of extracellular solutes along concentration gradients (see Fig. 4D). Such freezing-mediated intracellular loading of disaccharides has been shown to enhance the success of cell freeze-drying and to stabilize intracellular biomolecular structures, including DNA/nuclei, during dry storage [68, 92]. Moreover, unlike DMSO, trehalose is non-toxic and is, therefore, an attractive alternative protective agent, particularly in ready-to-use cell therapy applications [94].

5.4 Loading Tissues with Protective Agents

When a tissue is immersed in a solution containing cryo- or lyoprotective agents, osmotic and concentration gradient driving forces result in solute and water transport fluxes until an equilibrium state is reached. The dimensions over which mass transport takes place during protectant loading are larger compared to those during CPA loading of cells, and the outer and inner regions may have different water contents and CPA concentrations. To ensure maximum permeation and homogeneous distribution of protective molecules while minimizing the exposure time and toxicity effects, it is needed to obtain insights in diffusion kinetics of water and solutes. Various methods have been applied to study CPA permeation in tissues, including nuclear magnetic resonance [95, 96], X-ray computer tomography [97, 98], and osmometer measurements [99, 100]. FTIR and Raman spectroscopy can be used to study permeation and removal of individual CPA components in mixtures, as well as water fluxes [101–103].

6 Cryopreservation

6.1 Types of Cryoprotective Agent and CPA Toxicity

A landmark study by Polge and colleagues [104] demonstrated that the addition of glycerol allows spermatozoa to survive freezing and prolonged storage at −80°C. Glycerol remains widely used for the cryopreservation of sperm from various species [105] and human red blood cells [106, 107]. However, glycerol is not a universal cryoprotectant, and many cell types are difficult to preserve using glycerol alone.

Lovelock and Bishop [108] subsequently reported the cryoprotective properties of DMSO. Today, DMSO is the most widely used CPA for a broad range of mammalian and plant cells. Other permeating, low-molecular-weight CPAs that have been identified include ethylene glycol, propylene glycol, and dimethyl formamide [52]. A complete cryopreservation formulation typically contains physiological salts, buffering agents, nutrients, permeating CPAs, non-permeating biopolymers, antioxidants, and scavengers.

The reasons why certain CPAs are more effective for some cell types than others remain incompletely understood. One factor thought to play an important role is the relative membrane permeability of cells to water and CPAs (see Fig. 5B). The CPA concentration is also critical, as CPA toxicity generally increases with increasing concentration (see Fig. 6A). In addition, CPA-membrane interactions can destabilize cellular membranes, thereby determining the suitability of a given CPA for a specific cell type. Notably, DMSO interacts with membranes in a fundamentally different manner than glycerol: DMSO increases the membrane phase transition temperature, whereas glycerol decreases it [109, 110].

6.2 Slow Cooling Cryopreservation and Optimal Cooling Rate

Cryopreservation requires the use of CPAs; however, cooling and warming rates, as well as the ice nucleation temperature, are also critical determinants of cryopreservation outcome. The cooling rate governs the extent and kinetics of cellular dehydration during freezing. Because the rate of water transport across the cell membrane is temperature dependent, the cooling rate indirectly affects the probability of (lethal) intracellular ice formation [33]. The ice nucleation temperature determines the degree of supercooling and, consequently, the extent of cellular dehydration and likelihood of intracellular ice formation [32].

The two-factor hypothesis provides a mechanistic theory for understanding cryosurvival in slow-cooling cryopreservation protocols (see Fig. 6A). At cooling rates below the optimum, cell death is primarily attributed to the deleterious effects of prolonged exposure to high solute concentrations in the unfrozen fraction, i.e., solution effects injury. In contrast, at cooling rates higher than the optimum, survival decreases due to the increased likelihood of intracellular ice formation. It has, therefore, been postulated that each cell type exhibits an optimal cooling rate [23, 33]. Controlled-rate freezing can be used to regulate the rate and degree of cell dehydration during freezing and thereby enable cell survival. Cell volume changes during freezing can be obtained from simulations (see Fig. 7B). The extent of cellular dehydration during freezing is dependent on the cooling rate, and presence of CPAs. Notably, the optimal cooling rate at which cryosurvival is maximal has been proposed to correspond to a reduction of the osmotically active cell volume to approximately 90–95% of the original volume at −30°C [74].

In the presence of CPAs, however, optimal survival is often observed over a broad range of cooling rates rather than at a single defined rate. Optimal cooling rates depend on both the type and concentration of CPA used, and cryosurvival is governed by a complex interplay between CPA properties, CPA concentration, and cooling velocity [91, 111]. At the right combination of cooling rate and CPA concentration, many CPAs may be equally effective.

6.3 Cryopreservation by Vitrification

The damaging effects of ice formation can be avoided through vitrification, also referred to as ice-free cryopreservation, which is commonly preferred for embryo and tissues. Unlike slow-cooling approaches, vitrification does not require precise determination of an optimal cooling rate for maximal cell survival, making it inherently more suitable for specimens composed of multiple cell types. Vitrification employs high CPA concentrations in combination with rapid cooling rates, enabling samples to transition directly into an amorphous glassy state that preserves embedded structures while arresting chemical reactions [24, 112].

To mitigate CPA toxicity and remain within cellular osmotic tolerance limits, vitrification protocols typically use mixtures of permeating CPAs along with stepwise loading and unloading procedures [78, 113]. Vitrification has been successfully applied to both mammalian and plant tissues; however, its use is largely restricted to small tissues. The vitrification of larger tissues remains challenging due to limitations in achieving homogeneous CPA mass transfer and uniform heat transfer during rapid cooling and warming. Although the use of very high CPA concentrations permits lower cooling rates, this often results in reduced cell viability even before cooling, due to CPA toxicity [114].

Vitrification solutions generally consist of CPA mixtures to minimize toxicity associated with any single component and to reduce the risk of solute crystallization, which is more likely when individual CPAs are used at high concentrations. Successful vitrification requires cooling rates that are sufficiently rapid to suppress ice formation, highlighting a fundamental tradeoff between cooling rate and CPA toxicity (see Fig. 6A). While high CPA concentrations support slower cooling without ice formation, lower and less toxic concentrations require much more rapid cooling, restricting their use to smaller samples.

Achieving sufficiently rapid and uniform warming is one of the major challenges for both conventional cryopreservation and vitrification, particularly when CPA concentrations are kept low to reduce toxicity and when preserving large specimens such as organs. To address this challenge, several approaches have been developed to enable ultra-rapid and homogeneous warming, including laser warming [115], nanowarming [116], and electromagnetic heating [117].

7 Lyopreservation by Freeze-Drying

7.1 Freeze-Drying

During freeze-drying, the sample is first frozen to immobilize all components of the solution. Freezing can be performed directly on the shelves of the freeze-dryer or in an external device such as a controlled-rate freezer. During the primary drying phase, ice is removed by sublimation. This is achieved by reducing the pressure in the drying chamber to below the vapor pressure of ice at the sample temperature, thereby driving the migration of water vapor from the sample to the condenser. Throughout primary drying, the sample temperature must be maintained below the eutectic temperature, glass transition temperature, collapse temperature, and/or melting temperature to minimize structural and chemical damage [118].

Following removal of the bulk ice, a secondary drying step is employed to eliminate residual moisture that remains adsorbed to the sample matrix. This desorption process is facilitated by the controlled application of heat. After completion of secondary drying, samples can be sealed with rubber stoppers under vacuum or after back filling with an inert gas.

7.2 Formulations for Freeze-Drying of Cells

Freeze-drying of cells requires protective agents to protect them during both the freezing and drying processes. Protective formulations typically include a suitable physiological buffer solution, a protective agent such as sucrose or trehalose, and a bulking agent. Biopolymers are commonly used as bulking agents to space the cells and increase the Tg of the formulation. As mentioned earlier, specific methods can be employed to introduce protective disaccharides into the cells, and freezing itself can also facilitate the uptake of these disaccharides. A combination of albumin and trehalose, added at concentrations of 5% (w/v) and 150–300 mM, respectively, has been shown to create a stable matrix for cells protecting biomolecular structure and function [90, 119]. For example, DNA in freeze-dried sperm protected with trehalose and albumin remains stable for months, even under accelerated aging conditions [68, 69]. Similarly, sugars have been found to stabilize DNA in freeze-dried somatic cells during long-term storage [120]. Rehydration can be performed by directly adding water or by gradually rehydrating the sample under high relative humidity conditions.

7.3 Formulations for Freeze-Drying of Tissues

To freeze-dry tissues intended for use as biological scaffolds, it is essential to preserve both the tissue histoarchitecture and its biomechanical properties. In the absence of protective agents, ice crystal formation during freezing is evident as pores in the tissue after rehydration. In freeze-dried heart valves, for example, pores are still visible even if lyoprotective sugars are used at ~5% (w/v) concentrations

[121]. Pore formation can be avoided by osmotic dehydration of the tissue prior to freeze-drying. In this approach, similar to "cold" vitrification procedures, tissues are first loaded with sugars by immersion in a 5% (w/v) sugar solution, followed by exposure to a highly concentrated sugar solution (>40% w/v) to reduce the tissue water content [122].

Freeze-dried acellular tissue matrices can be used to replace diseased or malfunctioning tissues. Examples include acellular skin tissue matrices [123], decellularized heart valve conduits [121], and pericardium patches [124]. Both sucrose and trehalose can serve as protective agents in this process; however, because large quantities of sugars are required, sucrose may be preferred over trehalose. Macromolecules are generally not used for freeze-drying of tissue scaffolds, as their permeation rates are very slow [125].

8 Emerging Technologies

Emerging technologies in the fields of cryopreservation and freeze-drying are rapidly expanding the options to preserve biologics, cells, tissues, and organs. Recent advances in vitrification strategies aim to further minimize damage induced by ice recrystallization and CPA toxicity by promoting glass formation through optimized cryoprotective formulations combined with ultrarapid cooling, e.g., inkjet droplet-based cell cryopreservation [126], and, ultrarapid warming techniques including laser warming, nanowarming, Joule warming, and electromagnetic resonance warming [127]. Novel cryoprotective agents, including synthetic ice-recrystallization inhibitors and antifreeze proteins, are being developed to replace or reduce the use of toxic agents such as DMSO [128]. In addition, formulations are developed that stabilize frozen cell samples at higher subzero temperatures and during transport on dry ice [129].

Advanced microfluidic platforms are improving temperature uniformity and enable homogeneous CPA sample loading while supporting real-time process monitoring and processing larger volumes [130]. Isochoric cryopreservation represents an emerging preservation approach for biological materials and foods that seeks to eliminate ice crystal formation by maintaining a constant system volume during cooling. This method requires precise, simultaneous control of pressure, temperature, and volume to suppress ice formation throughout the freezing process [131].

Supercooled storage is an alternative strategy that can be used to maintain biological samples at subzero temperatures well below their equilibrium freezing point without ice formation. The supercooled state is metastable and susceptible to stochastic ice nucleation, with the likelihood of freezing increasing with the degree of supercooling, sample volume, and storage duration. Covering samples with an oil overlay on the air-water interface can significantly reduce surface-catalyzed ice nucleation, enabling storage of larger volumes for extended periods at lower temperatures [132].

Emerging freeze-drying technologies aim improving efficiency, product quality, and sustainability. Innovations like continuous freeze-drying systems are reducing processing times and energy consumption by enabling steady-state operations instead of batch cycles. Advances in process analytical technology, including real-time monitoring with sensors and machine learning-based control systems, are enhancing precision and consistency while minimizing product loss [133]. Microwave-assisted freeze-drying and spin freeze-drying technologies are further improving drying uniformity and preserving sensitive bioactive compounds [134, 135].

In both cryopreservation and freeze-drying applications, high-throughput screening of novel cryoprotective and lyoprotective compounds, combined with cell-specific functional and membrane permeability properties, enables machine-learning-guided protocol design [128, 136–138].

References

1. Crowe JH, Carpenter JF, Crowe LM (1998) The role of vitrification in anhydrobiosis. Annu Rev Physiol 60:73–103. https://doi.org/10.1146/annurev.physiol.60.1.73
2. Crowe JH, Hoekstra FA, Crowe LM (1992) Anhydrobiosis. Annu Rev Physiol 54:579–599. https://doi.org/10.1146/annurev.ph.54.030192.003051
3. Alpert P (2000) The discovery, scope, and puzzle of desiccation tolerance in plants. Plant Ecol 151:5–17. https://doi.org/10.1023/A:1026513800380
4. Boothby TC, Pielak GJ (2017) Intrinsically disordered proteins and desiccation tolerance: elucidating functional and mechanistic underpinnings of anhydrobiosis. BioEssays 39:1700119. https://doi.org/10.1002/bies.201700119
5. Bartels D, Schneider K, Terstappen G, Piatkowski D, Salamini F (1990) Molecular cloning of abscisic acid-modulated genes which are induced during desiccation of the resurrection plant *Craterostigma plantagineum*. Planta 181:27–34. https://doi.org/10.1007/BF00202321
6. Marks RA, Ekwealor JTB, Artur MAS, Bondi L, Boothby TC, Carmo OMS, Centeno DC, Coe KK, Dace HJW, Field S, Hutt A, Porembski S, Thalhammer A, van der Pas L, Wood AJ, Alpert P, Bartels D, Boeynaems S, Datar MN, Giese T, Seidou WI, Kirchner SM, Köhler J, Kumara UGVSS, Kyung J, Lyall R, Mishler BD, Ndongmo JBVT, Otegui MS, Reddy V, Rexroth J, Tebele SM, VanBuren R, Verdier J, Vothknecht UC, Wittenberg MF, Zokov E, Oliver MJ, Rhee SY (2025) Life on the dry side: a roadmap to understanding desiccation tolerance and accelerating translational applications. Nat Commun 16:3284. https://doi.org/10.1038/s41467-025-58656-y
7. Amuti KS, Pollard CJ (1977) Soluble carbohydrates of dry and developing seeds. Phytochemistry 16:529–532. https://doi.org/10.1016/0031-9422(77)80008-3
8. Buitink J, Walters-Vertucci C, Hoekstra FA, Leprince O (1996) Calorimetric properties of dehydrating pollen: analysis of a desiccation tolerant and an intolerant species. Plant Physiol 111:235–242. https://doi.org/10.1104/pp.111.1.235
9. Crowe JH, Crowe LM, Oliver AE, Tsvetkova N, Wolkers W, Tablin F (2001) The trehalose myth revisited: introduction to a symposium on stabilization of cells in the dry state. Cryobiology 43:89–105. https://doi.org/10.1006/cryo.2001.2353
10. Ramirez JF, Kumara UGVSS, Arulsamy N, Boothby TC (2024) Water content, transition temperature and fragility influence protection and anhydrobiotic capacity. BBA Adv 5:100115. https://doi.org/10.1016/j.bbadva.2024.100115
11. Leprince O, Pellizzaro A, Berriri S, Buitink J (2017) Late seed maturation: drying without dying. J Exp Bot 68:827–841. https://doi.org/10.1093/jxb/erw363

12. Tanaka S, Tanaka J, Miwa Y, Horikawa DD, Katayama T, Arakawa K, Toyoda A, Kubo T, Kunieda T (2015) Novel mitochondria-targeted heat-soluble proteins identified in the anhydrobiotic Tardigrade improve osmotic tolerance of human cells. PLOS One 10:e0118272. https://doi.org/10.1371/journal.pone.0118272
13. Hatanaka R, Gusev O, Cornette R, Shimura S, Kikuta S, Okada J, Okuda T, Kikawada T (2015) Diversity of the expression profiles of late embryogenesis abundant (LEA) protein encoding genes in the anhydrobiotic midge *Polypedilum vanderplanki*. Planta 242:451–459. https://doi.org/10.1007/s00425-015-2284-6
14. Hoekstra FA, Golovina EA, Tetteroo FA, Wolkers WF (2001) Induction of desiccation tolerance in plant somatic embryos: how exclusive is the protective role of sugars?. Cryobiology 43:140–150. https://doi.org/10.1006/cryo.2001.2358
15. Oldenhof H, Wolkers WF, Bowman JL, Tablin F, Crowe JH (2006) Freezing and desiccation tolerance in the moss *Physcomitrella patens*: an in situ Fourier transform infrared spectroscopic study. Biochim Biophys Acta 1760:1226–1234. https://doi.org/10.1016/j.bbagen.2006.03.025
16. Wolkers WF, Alberda M, Koornneef M, Leon-Kloosterziel KM, Hoekstra FA (1998) Properties and the glassy matrix in maturation defective mutant seed of *Arabidopsis thalania*. Plant J 16:133–143. https://doi.org/10.1046/j.1365-313x.1998.00277.x
17. Wolkers WF, Tetteroo FAA, Alberda M, Hoekstra FA (1999) Changed properties of the cytoplasmic matrix associated with desiccation tolerance of dried carrot somatic embryos. An in situ Fourier transform infrared spectroscopic study. Plant Physiol 120:153–164. https://doi.org/10.1104/pp.120.1.153
18. Wolkers WF, McCready S, Brandt WF, Lindsey GG, Hoekstra FA (2001) Isolation and characterization of a D-7 LEA protein from pollen that stabilizes glasses in vitro. Biochim Biophys Acta 1544:196–206. https://doi.org/10.1016/s0167-4838(00)00220-x
19. Olgenblum GI, Hutcheson BO, Pielak GJ, Harries D (2024) Protecting proteins from desiccation stress using molecular glasses and gels. Chem Rev 124:5668–5694. https://doi.org/10.1021/acs.chemrev.3c00752
20. Belott C, Janis B, Menze MA (2020) Liquid-liquid phase separation promotes animal desiccation tolerance. Proc Natl Acad Sci USA 117:27676–27684. https://doi.org/10.1073/pnas.2014463117
21. Elder CA, Skaggs HM, Dirk LMA, Grimm DF, Belott CJ, Wolkers WF, Oldenhof H, Uversky VN, Downie AB, Menze MA (2025) Biomolecular condensates - prerequisites for anhydrobiosis?. Protein Sci 34:e70192. https://doi.org/10.1002/pro.70192
22. Mühlbacher F, Langer F, Mittermayer C (1999) Preservation solutions for transplantation. Transpl Proc 31:2069–2070. https://doi.org/10.1016/s0041-1345(99)00265-1
23. Mazur P, Rall WF, Leibo SP (1984) Kinetics of water loss and the likelihood of intracellular freezing in mouse ova. Influence of the method of calculating the temperature dependence of water permeability. Cell Biophys 6:197–213. https://doi.org/10.1007/BF02788619
24. Fahy GM, Wowk B (2021) Principles of ice-free cryopreservation by vitrification. In: Wolkers WF, Oldenhof H (eds) Cryopreservation and freeze-drying protocols. Methods in molecular biology, vol 2180. Humana, New York, NY. https://doi.org/10.1007/978-1-0716-0783-1_2
25. Giwa S, Lewis JK, Alvarez L, Langer R, Roth AE, Church GM, Markmann JF, Sachs DH, Chandraker A, Wertheim JA, Rothblatt M, Boyden ES, Eidbo E, Lee WPA, Pomahac B, Brandacher G, Weinstock DM, Elliott G, Nelson D, Acker JP, Uygun K, Schmalz B, Weegman BP, Tocchio A, Fahy GM, Storey KB, Rubinsky B, Bischof J, Elliott JAW, Woodruff TK, Morris GJ, Demirci U, Brockbank KGM, Woods EJ, Ben RN, Baust JG, Gao D, Fuller B, Rabin Y, Kravitz DC, Taylor MJ, Toner M (2017) The promise of organ and tissue preservation to transform medicine. Nat Biotechnol 35:530–542. https://doi.org/10.1038/nbt.3889
26. Han Z, Rao JS, Gangwar L, Namsrai BE, Pasek-Allen JL, Etheridge ML, Wolf SM, Pruett TL, Bischof JC, Finger EB (2023) Vitrification and nanowarming enable long-term organ cryopreservation and life-sustaining kidney transplantation in a rat model. Nat Commun 14:3407. https://doi.org/10.1038/s41467-023-38824-8
27. Crowe JH, Carpenter JF, Crowe LM, Anchordoguy TJ (1990) Are freezing and dehydration similar stress vectors? A comparison of modes of interaction of stabilizing solutes with biomolecules. Cryobiology 27:219–231. https://doi.org/10.1016/0011-2240(90)90023-W
28. Pietramaggiori G, Kaipainen A, Ho D, Orser C, Pebley W, Rudolph A, Orgill DP (2007) Trehalose lyophilized platelets for wound healing. Wound Repair Regen 15:213–220. https://doi.org/10.1111/j.1524-475X.2007.00207.x

29. Ohanian M, Cancelas JA, Davenport R, Pullarkat V, Hervig T, Broome C, Marek K, Kelly M, Gul Z, Rugg N, Nestheide S, Kinne B, Szczepiorkowski Z, Kantarjian H, Pehta J, Biehl R, Yu A, Aung F, Antebi B, Fitzpatrick GM (2022) Freeze-dried platelets are a promising alternative in bleeding thrombocytopenic patients with hematological malignancies. Am J Hematol 97:256–266. https://doi.org/10.1002/ajh.26403
30. Wakayama T, Yanagimachi R (1998) Development of normal mice from oocytes injected with freeze-dried spermatozoa. Nat Biotechnol 7:639–641. https://doi.org/10.1038/nbt0798-639
31. Goecke T, Theodoridis K, Tudorache I, Ciubotaru A, Cebotari S, Ramm R, Höffler K, Sarikouch S, Vásquez Rivera A, Haverich A, Wolkers WF, Hilfiker A (2018) In vivo performance of freeze-dried decellularized pulmonary heart valve allo- and xenografts orthotopically implanted into juvenile sheep. Acta Biomater 68:41–52. https://doi.org/10.1016/j.actbio.2017.11.041
32. Pegg DE (2015) Principles of cryopreservation. Principles of cryopreservation. In: Wolkers WF, Oldenhof H (eds) Cryopreservation and freeze-drying protocols. Methods in molecular biology, vol 1257. Springer, New York, NY. https://doi.org/10.1007/978-1-4939-2193-5_1
33. Mazur P (1963) Kinetics of water loss from cells at subzero temperatures and the likelihood of intracellular freezing. J Gen Physiol 47:347–369. https://doi.org/10.1085/jgp.47.2.347
34. Leibo SP, McGrath JJ, Cravalho EG (1978) Microscopic observation of intracellular ice formation in unfertilized mouse ova as a function of cooling rate. Cryobiology 15:257–271. https://doi.org/10.1016/0011-2240(78)90036-6
35. Lovelock JE (1953) The haemolysis of human red blood cells by freezing and thawing. Biochim Biophys Acta 10:414–426. https://doi.org/10.1016/0006-3002(53)90273-x
36. Shuttleworth R, Higgins AZ, Eroglu A, Benson JD (2022) Comparison of dilute and nondilute osmotic equilibrium models for erythrocytes. Cryobiology 109:72–79. https://doi.org/10.1016/j.cryobiol.2022.09.001
37. Caliskan S, Liu D, Oldenhof H, Sieme H, Wolkers WF (2024) Use of membrane transport models to design cryopreservation procedures for oocytes. Anim Reprod Sci 267:107536. https://doi.org/10.1016/j.anireprosci.2024.107536
38. Steponkus PL, Uemura M, Joseph RA, Gilmour SJ, Thomashow MF (1998) Mode of action of the COR15a gene on the freezing tolerance of *Arabidopsis thaliana*. Proc Natl Acad Sci USA 95:14570–14575. https://doi.org/10.1073/pnas.95.24.14570
39. Wolkers WF, Oldenhof H, Tang F, Han J, Bigalk J, Sieme H (2019) Factors affecting the membrane permeability barrier function of cells during preservation technologies. Langmuir 35:7520–7528. https://doi.org/10.1021/acs.langmuir.8b02852
40. Anchordoguy TJ, Cecchini CA, Crowe JH, Crowe LM (1991) Insights into the cryoprotective mechanism of dimethyl sulfoxide for phospholipid bilayers. Cryobiology 28:467–473. https://doi.org/10.1016/0011-2240(91)90056-t
41. Prestrelski SJ, Tedeschi N, Arakawa T, Carpenter JF (1993) Dehydration-induced conformational transitions in proteins and their inhibition by stabilizers. Biophys J 65:661–671. https://doi.org/10.1016/S0006-3495(93)81120-2
42. Kasper JC, Winter G, Friess W (2013) Recent advances and further challenges in lyophilization. Eur J Pharm Biopharm 85:162–169. https://doi.org/10.1016/j.ejpb.2013.05.019
43. Crowe JH, Spargo BJ, Crowe LM (1987) Preservation of dry liposomes does not require retention of residual water. Proc Natl Acad Sci USA 84:1537–1540. https://doi.org/10.1073/pnas.84.6.1537
44. Crowe JH, Oliver AE, Hoekstra FA, Crowe LM (1997) Stabilization of dry membranes by mixtures of hydroxyethyl starch and glucose: the role of vitrification. Cryobiology 35:20–30. https://doi.org/10.1006/cryo.1997.2020
45. Bremer A, Wolff M, Thalhammer A, Hincha DK (2017) Folding of intrinsically disordered plant LEA proteins is driven by glycerol-induced crowding and the presence of membranes. FEBS J 284:919–936. https://doi.org/10.1111/febs.14023
46. Leprince O, Atherton NM, Deltour R, Hendry GAF (1994) The involvement of respiration in free radical processes during loss of desiccation tolerance in germinating *Zea mays* L. an electron paramagnetic resonance study. Plant Physiol 104:1333–1339. https://doi.org/10.1104/pp.104.4.1333
47. Senaratna T, McKersie BD, Borochov A (1987) Desiccation and free radical mediated changes in plant membranes. J Exp Bot 38:2005–2014. https://doi.org/10.1093/jxb/38.12.2005
48. França MB, Panek AD, Eleutherio ECA (2007) Oxidative stress and its effects during dehydration. Comp Biochem Physiol, Part A:

Mol Integr Physiol 146:621–631. https://doi.org/10.1016/j.cbpa.2006.02.030
49. Potts M (1994) Desiccation tolerance in prokaryotes. Microbiol Rev 58:755–805. https://doi.org/10.1128/mr.58.4.755-805.1994
50. Murthy UMN, Sun WQ (2000) Protein modification by Amadori and Maillard reactions during seed storage: roles of sugar hydrolysis and lipid peroxidation. J Exp Bot 51:1221–1228. https://doi.org/10.1093/jexbot/51.348.1221
51. Sun WQ, Leopold AC (1995) The Maillard reaction and oxidative stress during aging of soybean seeds. Physiol Plant 94:94–104. https://doi.org/10.1111/j.1399-3054.1995.tb00789.x
52. Elliott GD, Wang S, Fuller BJ (2017) Cryoprotectants: a review of the actions and applications of cryoprotective solutes that modulate cell recovery from ultra-low temperatures. Cryobiology 76:74–91. https://doi.org/10.1016/j.cryobiol.2017.04.004
53. Yong KW, Laouar L, Elliott JAW, Jomha NM (2020) Review of non-permeating cryoprotectants as supplements for vitrification of mammalian tissues. Cryobiology 96:1–11. https://doi.org/10.1016/j.cryobiol.2020.08.012
54. Lovelock JE (1953) The mechanism of the protective action of glycerol against haemolysis by freezing and thawing. Biochim Biophys Acta 11:28–36. https://doi.org/10.1016/0006-3002(53)90005-5
55. Arakawa T, Timasheff SN (1985) The stabilization of proteins by osmolytes. Biophys J 47:411–414. https://doi.org/10.1016/S0006-3495(85)83932-1
56. Westh P (2004) Preferential interaction of dimethyl sulfoxide and phosphatidyl choline membranes. Biochim Biophys Acta 1664:217–223. https://doi.org/10.1016/j.bbamem.2004.06.001
57. Oldenhof H, Gojowsky M, Wang S, Henke S, Yu C, Rohn K, Wolkers WF, Sieme H (2013) Osmotic stress and membrane phase changes during freezing of stallion sperm: mode of action of cryoprotective agents. Biol Reprod 88:68. https://doi.org/10.1095/biolreprod.112.104661
58. Akhoondi M, Oldenhof H, Stoll C, Sieme H, Wolkers WF (2011) Membrane hydraulic permeability changes during cooling of mammalian cells. Biochim Biophys Acta 1808:642–648. https://doi.org/10.1016/j.bbamem.2010.11.021
59. Slade L, Levine H (1991) Beyond water activity: recent advances based on an alternative approach to the assessment of food quality and safety. Crit Rev Food Sci Nutr 30:115–136. https://doi.org/10.1080/10408399109527543
60. Murray KA, Gibson MI (2022) Chemical approaches to cryopreservation. Nat Rev Chem 6:579–593. https://doi.org/10.1038/s41570-022-00407-4
61. Cordone L, Cottone G, Giuffrida S (2007) Role of residual water hydrogen bonding in sugar/water/biomolecule systems: a possible explanation for trehalose peculiarity. J Phys Condens Matter 19:205110. https://doi.org/10.1088/0953-8984/19/20/205110
62. Wolkers WF, Oldenhof H, Glasmacher B (2010) Effect of trehalose on dehydration kinetics of phospholipid vesicles, as measured in real time using ATR infrared spectroscopy. Cryobiology 61:108–114. https://doi.org/10.1016/j.cryobiol.2010.06.001
63. Wolfe J, Bryant G (1999) Freezing, drying and/or vitrification of membrane-solute-water systems. Cryobiology 39:103–129. https://doi.org/10.1006/cryo.1999.2195
64. Hoekstra FA, Golovina EA, Buitink J (2001) Mechanisms of plant desiccation tolerance. Trends Plant Sci 6:431–438. https://doi.org/10.1016/s1360-1385(01)02052-0
65. Crowe JH, Crowe LM, Wolkers WF, Oliver AE, Ma X, Auh JH, Tang M, Zhu S, Norris J, Tablin F (2005) Stabilization of dry mammalian cells: lessons from nature. Integr Comp Biol 45:810–820. https://doi.org/10.1093/icb/45.5.810
66. Buitink J, Leprince O (2004) Glass formation in plant anhydrobiotes: survival in the dry state. Cryobiology 48:215–228. https://doi.org/10.1016/j.cryobiol.2004.02.011
67. Sydykov B, Oldenhof H, Sieme H, Wolkers WF (2017) Hydrogen bonding interactions and enthalpy relaxation in sugar-protein glasses. J Pharm Sci 106:761–769. https://doi.org/10.1016/j.xphs.2016.11.003
68. Oldenhof H, Zhang M, Narten K, Bigalk J, Sydykov B, Wolkers WF, Sieme H (2017) Freezing-induced uptake of disaccharides for preservation of chromatin in freeze-dried sperm during accelerated aging. Biol Reprod 97:892–901. https://doi.org/10.1093/biolre/iox142
69. Brogna R, Fan J, Sieme H, Wolkers WF, Oldenhof H (2021) Drying and temperature induced conformational changes of nucleic acids and stallion sperm chromatin in trehalose preservation formulations. Sci Rep 11:14076. https://doi.org/10.1038/s41598-021-93569-y
70. Roos Y, Karel M (1991) Water and molecular weight effects on glass transition in amorphous carbohydrates and carbohydrate solutions.

J Food Sci 56:1676–1681. https://doi.org/10.1111/j.1365-2621.1991.tb08669.x
71. Wolkers WF, Balasubramanian SK, Ongstad EL, Zec H, Bischof JC (2007) Effects of freezing on membranes and proteins in LNCaP prostate tumor cells. Biochim Biophys Acta 1768:728–736. https://doi.org/10.1016/j.bbamem.2006.12.007
72. Agca Y, Mullen S, Liu J, Johnson-Ward J, Gould K, Chan A, Critser J (2005) Osmotic tolerance and membrane permeability characteristics of rhesus monkey (*Macaca mulatta*) spermatozoa. Cryobiology 51:1–14. https://doi.org/10.1016/j.cryobiol.2005.04.004
73. Leibo SP (1980) Water permeability and its activation energy of fertilized and unfertilized mouse ova. J Membr Biol 53:179–188. https://doi.org/10.1007/BF01868823
74. Devireddy RV, Swanlund DJ, Olin T, Vincente W, Troedsson MHT, Bischof JC, Roberts KP (2002) Cryopreservation of equine sperm: optimal cooling rates in the presence and absence of cryoprotective agents determined using differential scanning calorimetry. Biol Reprod 66:222–231. https://doi.org/10.1095/biolreprod66.1.222
75. Mazur P, Leibo SP, Miller RH (1974) Permeability of the bovine red cell to glycerol in hyperosmotic solutions at various temperatures. J Membr Biol 15:107–136. https://doi.org/10.1007/BF01870084
76. Kleinhans FW (1998) Membrane permeability modeling: Kedem-Katchalsky vs a two-parameter formalism. Cryobiology 37:271–289. https://doi.org/10.1006/cryo.1998.2135
77. Kedem O, Katchalsky A (1985) Thermodynamic analysis of the permeability of biological membranes to non-electrolytes. Biochim Biophys Acta 27:229–246. https://doi.org/10.1016/0006-3002(58)90330-5
78. Benson JD (2021) Mathematical modeling and optimization of cryopreservation in single cells. In: Wolkers WF, Oldenhof H (eds) Cryopreservation and freeze-drying protocols, Methods in molecular biology, vol 2180. Humana, New York, NY. https://doi.org/10.1007/978-1-0716-0783-1_4
79. Toupin CJ, Le Maguer M, McGann LE (1989) Permeability of human granulocytes to water: rectification of osmotic flow. Cryobiology 26:431–444. https://doi.org/10.1016/0011-2240(89)90068-0
80. Peckys DB, Kleinhans FW, Mazur P (2011) Rectification of the water permeability in COS-7 cells at 22, 10 and 0°C. PLOS One 6:e23643. https://doi.org/10.1371/journal.pone.0023643
81. Eroglu A, Russo MJ, Bieganski R, Fowler A, Cheley S, Bayley H, Toner M (2000) Intracellular trehalose improves the survival of cryopreserved mammalian cells. Nat Biotechnol 18:163–167. https://doi.org/10.1038/72608
82. Lynch AL, Chen R, Dominowski PJ, Shalaev EY, Yancey RJ, Slater NK (2010) Biopolymer mediated trehalose uptake for enhanced erythrocyte cryosurvival. Biomaterials 31:6096–6103. https://doi.org/10.1016/j.biomaterials.2010.04.020
83. Wei Y, Li C, Zhang L, Xu X (2014) Design of novel cell penetrating peptides for the delivery of trehalose into mammalian cells. Biochim Biophys Acta 1838:1911–1920. https://doi.org/10.1016/j.bbamem.2014.02.011
84. Abazari A, Meimetis LG, Budin G, Bale SS, Weissleder R, Toner M (2015) Engineered trehalose permeable to mammalian cells. PLOS One 10:e0130323. https://doi.org/10.1371/journal.pone.0130323
85. Rao W, Huang H, Wang H, Zhao S, Dumbleton J, Zhao G, He X (2015) Nanoparticle-mediated intracellular delivery enables cryopreservation of human adipose-derived stem cells using trehalose as the sole cryoprotectant. ACS Appl Mater Interfaces 7:5017–5028. https://doi.org/10.1021/acsami.5b00655
86. Zhang W, Rong J, Wang Q, He X (2009) The encapsulation and intracellular delivery of trehalose using a thermally responsive nanocapsule. Nanotechnology 20:275101. https://doi.org/10.1088/0957-4484/20/27/275101
87. Shirakashi R, Köstner CM, Müller KJ, Kürschner M, Zimmermann U, Sukhorukov VL (2002) Intracellular delivery of trehalose into mammalian cells by electropermeabilization. J Membr Biol 189:45–54. https://doi.org/10.1007/s00232-002-1003-y
88. Janis BR, Priddy MC, Otto MR, Kopechek JA, Menze MA (2021) Sonoporation enables high-throughput loading of trehalose into red blood cells. Cryobiology 98:73–79. https://doi.org/10.1016/j.cryobiol.2020.12.005
89. Puhlev I, Guo N, Brown DR, Levine F (2001) Desiccation tolerance in human cells. Cryobiology 42:207–217. https://doi.org/10.1006/cryo.2001.2324
90. Wolkers WF, Walker NJ, Tablin F, Crowe JH (2001) Human platelets loaded with trehalose survive freeze-drying. Cryobiology 42:79–87. https://doi.org/10.1006/cryo.2001.2306
91. Beattie GM, Crowe JH, Lopez AD, Cirulli V, Ricordi C, Hayek A (1997) Trehalose: a cryoprotectant that enhances recovery and preserves function of human pancreatic islets after long-term storage. Diabetes 46:519–523. https://doi.org/10.2337/diab.46.3.519
92. Zhang M, Oldenhof H, Sieme H, Wolkers WF (2016) Freezing-induced uptake

of trehalose into mammalian cells facilitates cryopreservation. Biochim Biophys Acta 1858:1400–1409. https://doi.org/10.1016/j.bbamem.2016.03.020
93. Stoll C, Holovati JL, Acker JP, Wolkers WF (2012) Synergistic effects of liposomes, trehalose and hydroxyethyl starch for cryopreservation of human erythrocytes. Biotechnol Prog 28:364–371. https://doi.org/10.1002/btpr.1519
94. Gläfke C, Akhoondi M, Oldenhof H, Sieme H, Wolkers WF (2012) Cryopreservation of platelets using trehalose: the role of membrane phase behavior during freezing. Biotechnol Prog 28:1347–1354. https://doi.org/10.1002/btpr.1600
95. Shaozhi Z, Pegg DE (2007) Analysis of the permeation of cryoprotectants in cartilage. Cryobiology 54:146–153. https://doi.org/10.1016/j.cryobiol.2006.12.001
96. Abazari A, Elliott JAW, McGann LE, Thompson RB (2012) MR spectroscopy measurement of the diffusion of dimethyl sulfoxide in articular cartilage and comparison to theoretical predictions. Osteoarthr Cartil 20:1004–1010. https://doi.org/10.1016/j.joca.2012.04.023
97. Bischof JC, Mahr B, Choi JH, Behling M, Mewes D (2007) Use of X-ray tomography to map crystalline and amorphous phases in frozen biomaterials. Ann Biomed Eng 35:292–304. https://doi.org/10.1007/s10439-006-9176-7
98. Corral A, Balcerzyk M, Parrado-Gallego Á, Fernández-Gómez I, Lamprea DR, Olmo A, Risco R (2015) Assessment of the cryoprotectant concentration inside a bulky organ for cryopreservation using X-ray computed tomography. Cryobiology 71:419–431. https://doi.org/10.1016/j.cryobiol.2015.09.007
99. Sharma R, Law GK, Rekieh K, Abazari A, Elliott JA, McGann LE, Jomha NM (2007) A novel method to measurure cryoprotectant permeation into intact articular cartilage. Cryobiology 54:196–203. https://doi.org/10.1016/j.cryobiol.2007.01.006
100. Jomha NM, Law GK, Abazari A, Rekieh K, Elliott JAW, McGann LE (2009) Permeation of several cryoprotectant agents into porcine articular cartilage. Cryobiology 58:110–114. https://doi.org/10.1016/j.cryobiol.2008.11.004
101. Vásquez-Rivera A, Sommer KK, Oldenhof H, Higgins AZ, Brockbank KGM, Hilfiker A, Wolkers WF (2018) Simultaneous monitoring of different vitrification solution components permeating into tissues. Analyst 143:420–428. https://doi.org/10.1039/c7an01576c
102. Marzi J, Biermann AC, Brauchle EM, Brockbank KGM, Stock UA, Schenke-Layland K (2019) Marker-independent in situ quantitative assessment of residual cryoprotectants in cardiac tissues. Anal Chem 91:2266–2272. https://doi.org/10.1021/acs.analchem.8b04861
103. Han J, Sydykov B, Yang H, Sieme H, Oldenhof H, Wolkers WF (2019) Spectroscopic monitoring of transport processes during loading of ovarian tissue with cryoprotective solutions. Sci Rep 9:15577. https://doi.org/10.1038/s41598-019-51903-5
104. Polge C, Smith AU, Parkes AS (1949) Revival of spermatozoa after vitrification and dehydration at low temperatures. Nature 164:666. https://doi.org/10.1038/164666a0
105. Sieme H, Oldenhof H, Wolkers WF (2015) Sperm membrane behaviour during cooling and cryopreservation. Reprod Domest Anim 50(Suppl 3):20–26. https://doi.org/10.1111/rda.12594
106. Rowe AW, Eyster E, Kellner A (1968) Liquid nitrogen preservation of red blood cells for transfusion: a low glycerol—rapid freeze procedure. Cryobiology 5:119–128. https://doi.org/10.1016/s0011-2240(68)80154-3
107. Tullis JL, Gibson JG, Sproul MT, Tinch RJ, Baudanze P (1970) Advantages of the high glycerol mechanical systems for red cell preservation: a 10-year study of stability and yield. In: Spielmann W, Seidl S (eds) Modern problems of blood preservation. Fischer, Stuttgart, pp 161–167
108. Lovelock JE, Bishop MW (1959) Prevention of freezing damage to living cells by dimethyl sulphoxide. Nature 183:1394–1395. https://doi.org/10.1038/1831394a0
109. Sydykov B, Oldenhof H, de Oliveira Barros L, Sieme H, Wolkers WF (2018) Membrane permeabilization of phosphatidylcholine liposomes induced by cryopreservation and vitrification solutions. Biochim Biophys Acta 1860:467–474. https://doi.org/10.1016/j.bbamem.2017.10.031
110. Gordeliy VI, Kiselev MA, Lesieur P, Pole AV, Teixeira J (1998) Lipid membrane structure and interactions in dimethyl sulfoxide/water mixtures. Biophys J 75:2343–2351. https://doi.org/10.1016/S0006-3495(98)77678-7
111. Oldenhof H, Bigalk J, Hettel C, Oliveira Barros L, Sydykov B, Bajcsy ÁC, Sieme H, Wolkers WF (2017) Stallion sperm cryopreservation using various permeating agents: interplay between concentration and cooling rate. Biopres Biobank 15:422–431. https://doi.org/10.1089/bio.2017.0061
112. Fahy GM, MacFarlane DR, Angell CA, Meryman HT (1984) Vitrification

as an approach to cryopreservation. Cryobiology 21:407–426. https://doi.org/10.1016/0011-2240(84)90079-8
113. Benson JD, Higgins AZ, Desai K, Eroglu A (2017) A toxicity cost function approach to optimal CPA equilibration in tissues. Cryobiology 80:144–155. https://doi.org/10.1016/j.cryobiol.2017.09.005
114. Elmoazzen HY, Poovadan A, Law GK, Elliott JA, McGann LE, Jomha NM (2007) Dimethyl sulfoxide toxicity kinetics in intact articular cartilage. Cell Tis Bank 431:125–133. https://doi.org/10.1007/s10561-006-9023-y
115. Jin B, Kleinhans FW, Mazur P (2014) Survivals of mouse oocytes approach 100% after vitrification in 3-fold diluted media and ultra-rapid warming by an IR laser pulse. Cryobiology 68:419–430. https://doi.org/10.1016/j.cryobiol.2014.03.005
116. Manuchehrabadi N, Gao Z, Zhang J, Ring HL, Shao Q, Liu F, McDermott M, Fok A, Rabin Y, Brockbank KG, Garwood M, Haynes CL, Bischof JC (2017) Improved tissue cryopreservation using inductive heating of magnetic nanoparticles. Sci Transl Med 9:pii:eaah4586. https://doi.org/10.1126/scitranslmed.aah4586
117. Luo D, Yu C, He L, Lu C, Gao D (2006) Development of a single mode electromagnetic resonant cavity for rewarming of cryopreserved biomaterials. Cryobiology 53:288–293. https://doi.org/10.1016/j.cryobiol.2006.07.001
118. Ward KR, Matejtschuk P (2021) The principles of freeze-drying and application of analytical technologies. In: Wolkers WF, Oldenhof H (eds) Cryopreservation and freeze-drying protocols, Methods in molecular biology, vol 2180. Humana, New York, NY. https://doi.org/10.1007/978-1-0716-0783-1_3
119. Ma X, Jamil K, Macrae TH, Clegg JS, Russell JM, Villeneuve TS, Euloth M, Sun Y, Crowe JH, Tablin F, Oliver AE (2005) A small stress protein acts synergistically with trehalose to confer desiccation tolerance on mammalian cells. Cryobiology 51:15–28. https://doi.org/10.1016/j.cryobiol.2005.04.007
120. Zhang M, Oldenhof H, Sydykov B, Bigalk J, Sieme H, Wolkers WF (2017) Freeze-drying of mammalian cells using trehalose: preservation of DNA integrity. Sci Rep 7:6198. https://doi.org/10.1038/s41598-017-06542-z
121. Wang S, Oldenhof H, Goecke T, Ramm R, Harder M, Haverich A, Hilfiker A, Wolkers WF (2015) Sucrose diffusion in decellularized heart valves for freeze-drying. Tissue Eng Part C Methods 21:922–931. https://doi.org/10.1089/ten.TEC.2014.0681
122. Vásquez-Rivera A, Oldenhof H, Dipresa D, Goecke T, Kouvaka A, Will F, Haverich A, Korossis S, Hilfiker A, Wolkers WF (2018) Use of sucrose to diminish pore formation in freeze-dried heart valves. Sci Rep 8:12982. https://doi.org/10.1038/s41598-018-31388-4
123. Gouk SS, Lim TM, Teoh SH, Sun WQ (2008) Alterations of human acellular tissue matrix by gamma irradiation: histology, biomechanical property, stability, in vitro cell repopulation, and remodeling. J Biomed Mater Res B Appl Biomater 84:205–217. https://doi.org/10.1002/jbm.b.30862
124. Zouhair S, Aguiari P, Iop L, Vásquez-Rivera A, Filippi A, Romanato F, Korossis S, Wolkers WF, Gerosa G (2019) Preservation strategies for decellularized pericardial scaffolds for off-the-shelf availability. Acta Biomater 84:208–221. https://doi.org/10.1016/j.actbio.2018.10.026
125. Wang S, Oldenhof H, Dai X, Haverich A, Hilfiker A, Harder M, Wolkers WF (2014) Protein stability in stored decellularized heart valve scaffolds and diffusion kinetics of protective molecules. Biochim Biophys Acta 1844:430–438. https://doi.org/10.1016/j.bbapap.2013.11.016
126. Akiyama Y (2025) Inkjet droplet-based cell cryopreservation: principles, advances, and prospects. Jpn J Appl Phys 64:050805. https://doi.org/10.35848/1347-4065/add6af
127. Pan J, Zeng Q, Peng K, Zhou Y, Shu Z (2024) Review of rewarming methods for cryopreservation. Biopreserv Biobank 22:304–311. https://doi.org/10.1089/bio.2023.0015
128. Warren MT, Biggs CI, Bissoyi A, Gibson MI, Sosso GC (2024) Data-driven discovery of potent small molecule ice recrystallisation inhibitors. Nat Commun 15:8082. https://doi.org/10.1038/s41467-024-52266-w
129. Uhlmannsiek L, Shen H, Eylers H, Martinsson G, Sieme H, Wolkers WF, Oldenhof H (2024) Preserving frozen stallion sperm on dry ice using polymers that modulate ice crystalization kinetics. Cryobiology 114:104852. https://doi.org/10.1016/j.cryobiol.2024.104852
130. Wang X, Wang E, Zhao G (2023) Advanced cryopreservation engineering strategies: the critical step to utilize stem cell products. Cell Regen 12:28. https://doi.org/10.1186/s13619-023-00173-8
131. Zarriz A, Journaux B, Powell-Palm MJ (2024) On the equilibrium limit of liquid stability in pressurized aqueous systems. Nat Commun 15:10666. https://doi.org/10.1038/s41467-024-54625-z
132. Huang H, Yarmush ML, Usta OB (2018) Long-term deep-supercooling of large-volume water and red cell suspensions via surface sealing with immiscible liquids. Nat Commun 9:3201. https://doi.org/10.1038/s41467-018-05636-0

133. Leys L, Nuytten G, Van Bockstal PJ, De Beer T (2023) Evaluation of a PAT-based in-line control system for a continuous spin freeze-drying process. Int J Pharm 641:123062. https://doi.org/10.1016/j.ijpharm.2023.123062
134. Lammens J, Goudarzi NM, Leys L, Nuytten G, Van Bockstal PJ, Vervaet C, Boone MN, De Beer T (2021) Spin freezing and its impact on pore size, tortuosity and solid state. Pharmaceutics 13:2126. https://doi.org/10.3390/pharmaceutics13122126
135. Hilmer M, Gruber S, Kis Z, Schulz M, Foerst P (2024) Design of a pilot-scale microwave freeze dryer for in situ neutron imaging. Rev Sci Instrum 95:083704. https://doi.org/10.1063/5.0213685
136. Tu F, Bhat M, Blondin P, Vincent P, Sharafi M, Benson JD (2022) Machine learning and hypothesis driven optimization of bull semen cryopreservation media. Sci Rep 12:22328. https://doi.org/10.1038/s41598-022-25104-6
137. Herve Q, Ipek N, Verwaeren J, De Beer T (2025) A deep learning approach to perform defect classification of freeze-dried product. Int J Pharm 670:125127. https://doi.org/10.1016/j.ijpharm.2024.125127
138. Murray A, Kilbride P, Gibson MI (2024) Trehalose in cryopreservation. Applications, mechanisms and intracellular delivery opportunities. RSC Med Chem 15:2980–2995. https://doi.org/10.1039/d4md00174e
139. Huang H, He X, Yarmush ML (2021) Advanced technologies for the preservation of mammalian biospecimens. Nat Biomed Eng 5:793–804. https://doi.org/10.1038/s41551-021-00784-z
140. Keskintepe L, Eroglu A (2021) Preservation of mammalian sperm by freeze-drying. In: Wolkers WF, Oldenhof H (eds) Cryopreservation and freeze-drying protocols, Methods in molecular biology, vol 2180. Humana, New York, NY. https://doi.org/10.1007/978-1-0716-0783-1_39
141. Sassi P, Caponi S, Ricci M, Morresi A, Oldenhof H, Wolkers WF, Fioretto D (2015) Infrared versus light scattering techniques to monitor the gel to liquid crystal phase transition in lipid membranes. J Raman Spectrosc 46:644–651. https://doi.org/10.1002/jrs.4702
142. Robertson RN (1983) The lively membranes. Cambridge University Press, Cambridge, NY, p 206
143. Zhao G, Zhang Z, Zhang Y, Chen Z, Niu D, Cao Y, He X (2017) A microfluidic perfusion approach for on-chip characterization of the transport properties of human oocytes. Lab Chip 17:1297–1305. https://doi.org/10.1039/c6lc01532h

Check for updates

Chapter 2

The Principles of Freeze-Drying and Application of Analytical Technologies

Kevin R. Ward and Paul Matejtschuk

Abstract

Freeze-drying is a complex process despite the relatively small number of steps involved, since the freezing, sublimation, desorption, and reconstitution processes all play a part in determining the success or otherwise of the final product qualities, and each stage can impose different stresses on a product. This is particularly the case with many fragile biological samples, which require great care in the selection of formulation additives such as protective agents and other stabilizers. Despite this, the process is widely used, not least because once any such processing stresses can be overcome, the result is typically a significantly more stable product than was the case with the starting material. Indeed, lyophilization may be considered a gentler method than conventional air-drying methods, which tend to apply heat to the product rather than starting by removing heat as is the case here. Additionally, due to the high surface area to volume ratio, freeze-dried materials tend to be drier than their conventionally dried counterparts and also rehydrate more rapidly. This chapter provides an overview of freeze-drying (lyophilization) of biological specimens with particular reference to the importance of formulation development, characterization, and cycle development factors necessary for the commercial exploitation of freeze-dried products and reviews the recent developments in analytical methods which have come to underpin modern freeze-drying practice.

Keywords Freeze-drying, Freeze drying microscopy, Lyophilization, Lyoprotectants, Process analytical technology, Sublimation, Thermal analysis

1 Introduction

1.1 General Overview

Water is essential to life, providing a universal solvent supporting biochemical activities within cells, which enables metabolism to continue and sustains all living processes. Quite simply, in the absence of water, life as we define it will cease, or at best, enter a period of dormancy in live cells, while biochemical activity will be suspended. Water determines the structure and function of biomolecules, and removal of water may lead to irreversible changes in biomolecular structure and loss of function. On the other hand, biomolecules are prone to degradation reactions when stored in water. Vaccines,

Willem F. Wolkers and Harriëtte Oldenhof (eds.), *Cryopreservation and Freeze-Drying Protocols*, Methods in Molecular Biology, vol. 3060, https://doi.org/10.1007/978-1-0716-5416-3_2,

other biological materials, and microorganisms can be stabilized by chilling or freezing. However, maintaining and transporting products in the frozen state is costly, while a freezer breakdown may result in the complete loss of valuable product [1]. Alternatively, biological specimens can be dried to increase their shelf life. Traditional drying processes typically result in marked changes in the physical and chemical properties of the product due to high solute concentration or thermal inactivation and are more appropriate for dehydrating low-cost products such as foodstuffs. Freeze-drying combines the benefits of both freezing and drying to provide a dry, biologically active, shelf-stable product that can be readily rehydrated into a solution or suspension [2, 3].

Operationally, we could define freeze-drying as a controllable method of dehydrating labile products by vacuum desiccation. Earlier accounts of freeze-drying suggested that ice was only removed by sublimation and defined this step as primary drying. The cycle was then described as being extended by secondary drying or desorption. Although these definitions are applicable to ideal systems, they incompletely define the process for typical systems that form an amorphous matrix or glass when cooled [4].

1.2 History

The principles of low temperature drying can be traced back to prehistoric times and were used by South American and Arctic peoples for preserving foodstuffs. Toward the end of the 1880s, the process was used on a laboratory scale, and the basic principles were understood at that time. Practically, the method remained a laboratory technique until the 1930s when there was a need to process heat-labile antibiotics and blood products. At this time, refrigeration and vacuum technologies had advanced sufficiently to enable the production of freeze-dryers to be developed, and since then, the process has been used industrially in both the food and pharmaceutical industries [5].

1.3 Applications of Freeze-Drying in the Pharmaceutical and Biological Sectors

Freeze-drying offers a stable and robust format with significantly reduced weight to facilitate storage and shipment; it provides stabilization for materials not readily stabilized by conventional drying technologies, which tend to use higher temperatures; the product can be sealed under vacuum or an inert gas, which can reduce oxidative denaturation; it can accommodate the need to separately dry two or more components that would be incompatible if dispensed together as a solution within a single container. In the case of the latter, freeze-drying can enable this to be practically achieved by filling and freezing one component before adding the other and completing the freezing process; an example is dual-layer collagen implants for surgical use.

A wide range of products are freeze-dried even within the biologicals and healthcare sectors, including: small molecules, where the process is used to dehydrate or concentrate reactive or heat-sensitive chemicals/pharmaceuticals; biologically active molecules including enzymes, hormones, antibiotics, vitamins, blood products, inactivated vaccines, polymerase chain reaction (PCR) components for medical diagnostics; bone and other body tissues for surgical or medical use; and living organisms for vaccine or seed culture use, which must grow and multiply to produce new progeny after drying and reconstitution. However, freeze-drying is less appropriate for:

1. Oily or non-aqueous solutions where the material has a low melting temperature and/or low vapor pressure.
2. Products that form an impervious surface skin (typically a layer of concentrated solute) or excluded liquid phase (e.g., glycerol rising to the top of the otherwise frozen structure), which can prevent vapor migration from the product.
3. Biological materials that are able to retain viability only when in the presence of additives that are incompatible with practical freeze-drying.

2 The Process of Freeze-Drying

For convenience, the freeze-drying process may be divided into a number of discrete steps that may be summarized as:

1. Freezing, which reduces thermal denaturation of the product, immobilizes solution components, and prevents foaming when the vacuum is applied. Freezing conditions may be adjusted to induce a desired ice-crystal structure within the product, which can facilitate subsequent drying.
2. Primary drying(sublimation), where conditions must be maintained in the drying chamber to sustain water migration from the product ice during drying. During primary drying, the product temperature (strictly freeze-drying interface temperature) must be maintained below the eutectic, glass transition, collapse, or melt temperature as appropriate, to minimize product damage during drying.
3. A secondary drying stage during which resident moisture absorbed or adsorbed to the apparently dry structure is removed by desorption to a level compatible with ambient storage.
4. Sealing the dried product in a vacuum or under an inert gas at the end of the process, both of which exclude the entry of reactive, destabilizing, atmospheric gases such as oxygen or carbon dioxide into the dried product and prevent the ingress of damp air into the freeze-dried product. Note that a freeze-dried product will have a vastly expanded dry surface area and

is therefore particularly sensitive to air denaturation or moisture uptake.

5. The products are then removed from the freeze-dryer, stored, and/or distributed for use prior to reconstitution for injection, application, or regrowth.

2.1 Operational Principles

A freeze-dryer (lyophilizer) will comprise a number of fundamental component parts, including a product chamber (which may house several temperature-controlled shelves), a process condenser (also known as a cold trap), and a vacuum pump (typically a rotary vane pump). Other features may vary, depending on the size and operational requirements of the system and the chosen approaches employed by equipment manufacturers, each of whom tends to have slightly different designs and engineering solutions to achieve the same goal. For example, shelf temperature control is almost invariably achieved by the use of a circulating thermal medium such as low viscosity silicone oil, which is cooled via a heat exchanger and warmed using an electrical heater, both of which are part of the refrigeration circuit. The flow pattern of the fluid, its viscosity, and the speed at which it is pumped through the shelves may vary from manufacturer to manufacturer, but the ultimate aim will be to achieve effective heat transfer into or out of the product on the shelves. On smaller dryers, a single refrigeration compressor may be employed to cool both the shelves and the process condenser, while on an industrial dryer, there may be several compressors and/or separate systems dedicated to different cooling functions within the one dryer.

There are also differences in the devices used for process monitoring: some examples of Process Analytical Technologies (PAT) are given later in this chapter, but even at the most basic level, a small research laboratory dryer may have a Pirani gauge to measure pressure (vacuum), while an industrial dryer is more likely to have a Capacitance Manometer (CM gauge) - these work in very different ways and will typically provide very different readouts even under identical conditions and when each gauge is working well and has been perfectly calibrated. Control systems work in different ways, with those on smaller dryers usually offering the operator the flexibility to change setpoints of temperature, pressure, and time at any time during the process, while at manufacturing scale, such changes are typically (and understandably) not possible, and therefore, the lyophilization cycle should be demonstrated to be sufficiently robust by the time it is employed at this scale. All of these factors must be considered as part of the scale-up process.

Freeze-drying is a complex process during which drying may proceed more or less rapidly within individual samples throughout the process batch, such that parts of the product will be frozen, whereas other areas are drying or will have dried depending on the nature of the product and stage in the cycle. The precise freezing and drying behavior will be determined by the interrelationship between the product and shelf temperature, system pressure, extent of product dryness, and variations in drying conditions throughout the cycle.

Often regarded as a gentle method of drying materials, freeze-drying is in reality a potentially damaging process where the individual process stages should be regarded as a series of interrelated stresses, each of which can damage sensitive bio-products. Damage sustained during one step in the process chain may be exacerbated at succeeding stages and even apparently trivial changes in the process. One example is that of a change in container or closure to an apparently identical one but whose freeze-drying characteristics may differ markedly from the original, such as the heat transfer characteristics of a vial or a different vapor resistance from a stopper [6]; such a change may be sufficient to transform a successful process into one that is unacceptable.

Freeze-drying. The essence of the formulation exercise should be to minimize freeze-drying damage, loss of viability, or activity. To ensure minimal losses of activity, the product may require dilution in a medium containing protective additives, specifically selected for the product or application. Although frequently described as "protectants," these additives may not be effective at all stages of the process but may protect only during particular steps in the freeze-drying cycle. At other stages, the additive may fail to protect the active component and indeed may be incompatible with the process. It is also important to appreciate that individual stages in the process can result in damage, which initially remains undetected, becoming evident only when the dried product is rehydrated [7, 8]. Particular attention must be applied to the selection and blending of the additive mixes in the formulation, and the importance of formulation will be discussed at greater length later.

Products should be formulated to ensure batch product uniformity, whereas there may be particular requirements relating to product use. In this context, vaccines freeze-dried for oral or aerosol delivery may require the inclusion of excipients that minimize damage when the dried product is exposed to moist air, since different materials can display different rates and extents of moisture uptake under identical conditions [9].

A wide range of containers can be used to freeze-dry vaccines, microorganisms, and others, including glass ampoules, rubber stoppered vials, double chambered vials, and prefilled syringes that hold both dried vaccine and diluent, bifurcated needles, and so on. Alternatively, vaccines can be dried in bulk in stainless steel or plastic trays and the resultant powder tableted, capsulated, sachet-filled, or dispensed into aerosol devices for lung or nasal delivery.

2.2 Freezing

Regarded as the first step in the process, the liquid formulation must be frozen before evacuating the chamber to induce sublimation. Freezing will:

1. Immobilize the components in the solution and prevent foaming as the vacuum is applied.
2. Reduce thermal inactivation of the dispensed formulation.
3. Induce a specific ice-crystal structure within the frozen mass, which will facilitate vapor migration from the drying cake. In short, the ice structure formed during freezing will dictate subsequent freeze-drying behavior and the ultimate morphology of the dried cake.

Ideally, freezing should minimize any adverse effects of solute concentration and result in all components being evenly distributed throughout the frozen structure. However, it may not be possible to achieve this ideal when solutions or suspensions are frozen, not least since the latter can undergo some degree of sedimentation. When addressing the freezing of aqueous solutions or suspensions, there is a need to consider both the solvent (water in the case of aqueous solutions) and solute(s) in the formulation. A comprehensive treatise on the fundamentals of freezing and annealing phenomena as they apply to freeze-drying is provided by Searles [10].

Frequently, the terms cooling and freezing are erroneously interchanged and confusion in understanding the process may occur and may be compounded by failing to distinguish between shelf or product cooling and freezing. Cooling refers to the reduction of temperature of the freeze-dryer shelves, the fluid circulating through the shelves, the vial and tray mass, the interior of the freeze-dryer, and the dispensed solution or suspension. Cooling does not assume a change in state from liquid to solid and strictly should be used to describe reducing temperature during the initial stage of freeze-drying. Freezing refers to the abrupt phase change when water freezes as ice. Except for very complex biomolecules or cold-sensitive cells, cooling in the absence of freezing (chilling) is generally not damaging to biomaterials.

When solutions or suspensions are frozen, they may cool appreciably below their thermodynamic freezing point prior to ice formation, a phenomenon defined as supercooling (sometimes referred to as undercooling or sub-cooling). The extent of supercooling depends on cooling rate, sample composition and cleanliness, dispensed fill volume, container type, method of sample cooling, and so on. Even when a simple solution is repeatedly cooled or warmed, the onset and extent of supercooling will vary from cycle to cycle. In the supercooled state, while the composition of the solution remains unchanged, the cooled liquid is thermodynamically unstable and sensitive to ice formation. As the solution is cooled to lower temperatures, the probability of ice crystallization will correspondingly increase. For optimized freeze-drying, the intention should be to induce supercooling in the suspension to encourage uniform cooling and freezing throughout the sample contents.

Freezing may be defined as the abrupt conversion of the solution or suspension into a mixture of ice and solute concentrate. It is a two-step process during which water initially nucleates, followed by the growth of the ice crystals that pervade the solute phase, resulting in a mixture of ice and solute concentrate. Under typical processing conditions, ice nucleates heterogeneously around microscopic particles within the suspension and is encouraged by reducing the temperature and agitating the supercooled suspension to increase the probability of contact between nucleating foci and water clusters. Nucleation depends on the number and physical nature of particulate impurities within the suspension or solution. Ice is a particularly effective nucleation focus and cryobiologists may deliberately seed samples with ice to induce nucleation. Other effective ice nucleators include glass shards and specifically formulated nucleation promoters. Whereas nucleation aids can be added to experimental systems, deliberate attempts to add ice inducers to pharmaceutical materials would be at variance with Good Pharmaceutical Manufacturing Practice.

In contrast to nucleation, ice growth (proliferation) is encouraged by raising the temperature, thereby decreasing the suspension viscosity. Ice nucleation and proliferation are inhibited at temperatures below the glass transition temperature (Tg'), whereas above the melting temperature (Tm), the suspension or solution will melt. The consequences and measurements of these parameters are important elements in the formulation exercise [11]. To facilitate the sublimation of water vapor from the drying mass, the ice crystals should be large, wide, and contiguous, extending from the product base toward its surface, thereby providing an optimized structure for vapor migration. Crystal structures commonly observed during freeze-drying when solutions are frozen in trays or vials include dendritic structuring, where the ice crystal branches continuously from the nucleating focus, and the spherulitic form, where sub-branching is discouraged because the solution viscosity is high, or fast rates of cooling are used.

2.2.1 Controlled Nucleation

The use of controlled (or induced) ice nucleation in freeze-drying is receiving increased attention due to the advantages it can offer in terms of increasing sublimation efficiency through control of the ice nucleation temperature, which affects product porosity and uniformity [12, 13], and may also reduce activity losses in cases where ice formation or freeze-concentration effects are known to induce protein denaturation [14–16]. Any method that allows for control of the nucleation temperature results in several degrees of freedom for the freeze-drying process. In principle, there are three parameters that can be directly controlled: the nucleation temperature, the isothermal hold time post-nucleation, and the cooling rate of the shelf post-nucleation. Methods employed in initial attempts to achieve controlled ice nucleation are numerous and are covered in a comprehensive review by Winter and Geidobler [17]; however, only three methods have been commercialized to date, namely the Pressure-Depressurization technique [12], the ice fog method, and vacuum-induced nucleation, although other emerging methods also show promise [13, 18–25].

In addition to the above benefits, controlled nucleation may also yield improvements in the uniformity of dried product appearance, a tighter range of intra- and inter- batch moisture content values, more rapid reconstitution, and more predictable scalability.

2.2.2 Shelf-Cooling Rate

The shelf-cooling rate is the simplest parameter to control, and programmed rates of cooling are standard options on research and production freeze-dryers. Because shelf temperature and product responses are not identical, defining the shelf-cooling rate will not fully define product behavior. Although we are concerned with the cooling rate achievable within each individual production container (typically a vial), this parameter is less easy to monitor compared with shelf cooling, and freeze-drying cycles generally are controlled by programmed shelf cooling rather than feedback control from the sample. Cooling rates of the product/cell suspension will vary considerably from vial and throughout the sample within the vial, and consequently, measuring the temperature of vial contents at a fixed position will give only an approximation of the sample temperature variation.

Observing the freezing pattern of a number of vials arranged on a shelf will demonstrate that while the contents of some vials will freeze slowly from the vial base, neighboring vials may remain unfrozen and supercool appreciably before freezing instantly. This random freezing pattern will reflect differences in ice structure from vial to vial and be translated into different drying geometries. In summary, freezing patterns will be related to:

1. The ice-forming potential within each container (e.g., vial or tray).
2. The relative position of the vial on the shelf causing exposure of individual containers to cold or hot spots.
3. Edge effects where samples in vials on the periphery of each shelf will be subjected to heat transmitted through the chamber walls or door.
4. The insertion of temperature probe into the container, which will induce ice crystallization.
5. The evolution of latent heat as samples freeze, which will tend to warm adjacent containers.
6. Variations in container base geometry, which may impede thermal contact between container and shelf.

The ice and solute crystal structure obtained during the freezing stage of lyophilization has a major impact on subsequent freeze-drying behavior, encouraging the sample to dry efficiently or with defects such as melt or collapse depending on freezing rate used. The preferred ice structure comprising large contiguous ice crystals is induced by cooling the sample at a slow rate of ~0.1–0.5°C/min, which may also induce the crystallization of solutes reluctant to crystallize when faster rates of cooling are used. However, a slow rate of cooling may exacerbate the development of a surface skin, which inhibits sublimation efficiency, and can also inactivate a bioproduct by prolonging exposure to the solute concentrate. Conversely, a fast rate of cooling can result in the formation of numerous, small, randomly orientated ice crystals embedded in an amorphous solute matrix, which may be difficult to freeze-dry, although snap freezing may be unavoidable for some products, such as adjuvanted vaccines or whole cells, where ice crystal growth could physically rupture the outer membrane or cell wall. Complicating the choice of freezing regimes is the fact that the optimal cooling rate cannot be sustained uniformly throughout the entire sample, particularly with increased fill depths or volumes, where temperature gradients within a single container will be more marked. Therefore, defining cooling rates often requires a compromise in sample requirements.

2.2.3 Ice Structure and Freeze Consolidation

A period of consolidation (defined as the hold or soak time) is necessary at the end of the initial cooling step to ensure that all the contents in the sample batch have frozen adequately, although excessively long hold times may only serve to increase the overall cycle time without offering any particular other advantage. Indeed, it is a fallacy to assume that the ice structure induced remains unchanged during this consolidation period, and an ice structure comprising a large number of small ice crystals, induced by rapid cooling, is thermodynamically less stable than an ice structure comprising fewer, larger crystals. The thermodynamic equilibrium can be maintained by Ostwald ripening of ice from small-to-large crystals, a process termed grain growth. Although ice structure changes take place randomly from one container to another (e.g., vial to vial), the hold period is a major factor in ice recrystallization, resulting in significant variation in crystal structure; as a consequence, sublimation efficiency from sample to sample may vary by a wider margin the longer the freezing hold period is employed.

2.2.4 Annealing

As an alternative to increasing the length of the hold time to encourage ice recrystallization, a more controlled and time-efficient method of inducing recrystallization is to anneal the frozen sample [10].

Essentially, annealing is achieved by:

1. Cooling the product to freeze the solvent (usually water) and any readily crystallizable solutes.
2. Raising the product temperature during the freezing stage to recrystallize ice from a small to a large ice crystal matrix (it should be noted that this warming phase may also crystallize solutes that are reluctant to crystallize upon initial cooling).
3. Cooling the product to the terminal hold temperature prior to chamber evacuation (although this may not be necessary if annealing has caused the critical temperature of the formulation to increase above the holding temperature).

Annealing is particularly useful to:

1. Increase ice crystal size and networking, which reduces product resistance to vapor flow (Rp) and consequently improves sublimation efficiency.
2. Crystallize solutes that are reluctant to crystallize during initial cooling.
3. Provide greater uniformity of product structure throughout the batch and between batches.
4. Integrated with rapid cooling, annealing may minimize the development of a surface skin on the product while also increasing ice crystal size, thereby facilitating sublimation.
5. Because annealing induces larger pores in the cake structure, this can aid wetting and rehydration during the reconstitution process. However, studies using controlled nucleation have demonstrated that the reduction in specific surface area can lead to secondary drying (desorption) efficiency being reduced and therefore, it may be necessary to increase the length of the secondary drying step accordingly [12, 13].

Although annealing will increase the length of the freezing stage of the cycle, and possibly also the secondary drying stage as highlighted above, overall freeze-drying cycle times may be significantly reduced because of improvements in sublimation efficiency resulting from annealing, leading to a shorter primary drying phase.

Care should be exercised when selecting temperatures and hold times for annealing, particularly when defining the upper temperature for warming. Subjecting a labile product, such as a vaccine, to temperatures above the eutectic temperature will expose the active material to a hypertonic solution as the sample partially melts, which can damage sensitive biomolecules, cells, or organisms.

2.2.5 Solute Freezing Behavior

Regardless of the precise freezing pattern, the formation of ice will concentrate the remaining solution within the container. As the proportion of ice increases within the mixture, solute concentration will correspondingly increase. In the case of an aqueous 1% (w/v) saline solution, this concentration effect will be considerable, increasing to approximately 30% (w/v) just prior to freezing, which may cause damage to biomolecules as a consequence of solute concentration exposure rather than direct damage by ice crystals. The behavior of the solute(s) within the solute concentrate depends on the nature, concentration, cooling rate, and interactions between individual solutes present in the medium and forms the basis for experimental review during a formulation development exercise [9].

Overall, four patterns of solute response are observed during freeze-drying:

1. Solute crystallizes readily, regardless of cooling rate or freezing conditions, to form a mixture of ice and solute crystals (this behavior is termed eutectic freezing).
2. Solute crystallizes, but only when the solution is subjected to a slow rate of cooling.
3. Solute crystallizes only after the solution has been annealed.
4. Solute fails to crystallize regardless of cooling rate or regime adopted, and solute remains associated with unfrozen water as a metastable amorphous mass or glass.

For a crystallizing solute, the eutectic point is the lowest temperature in a system in which a residual liquid phase and solid phase are in equilibrium. Above the eutectic point, ice and solute concentrate persist, whereas below the eutectic point, a mixture of ice and solute crystals is formed. Eutectic temperatures for aqueous solutions containing crystallizing salts are characteristic for each solute and are significantly below the freezing point of water (for example, the eutectic temperature for sodium chloride is –21.1°C). Exposing cells or proteins for prolonged periods to a eutectic solution comprising hypertonic salt concentrations can cause damage by membrane deformation or precipitation by "salting out."

The eutectic zone is the range of temperatures encompassing all the eutectic temperatures within the system. For a two-part water/solute system, the eutectic temperature is a discrete, quantifiable temperature in contrast to multi-solute systems where a eutectic zone may be observed that represents a range of temperatures where the minimum eutectic temperature is lower than that of any individual eutectic temperature in the medium.

Formulations of biologics are typically amorphous, although they may contain some salts or excipients that can crystallize under certain conditions. When exposed to temperatures above their glass transition (Tg') or collapse temperature (Tcollapse), these samples may warm during sublimation causing the amorphous mass to soften, so that the freeze-drying progresses with collapse to form a sticky, structureless residue within the vial. Less severe collapse will result in the formation of a shrunken, distorted, or split cake, although a small degree of shrinkage while maintaining the aspect ratio of the original fill volume is typical for most amorphous materials [26].

Attempts to freeze-dry cells in water or a simple salt solution typically result in poor survival. A wide range of protective media has been developed for preserving freeze-dried vaccines, including augmented growth media or sugar solutions. Carbohydrates are widely used as freeze-drying protectants either individually or in combination with other solutes. They should be chosen on the basis of experimentally determining their freeze-drying characteristics rather than on a pragmatic basis. Monosaccharides, such as glucose, provide good bio-protection during freezing and freeze-drying but exhibit a low glass transition temperature (Tg') and collapse temperatures (Tcollapse), and tend to dry with collapse when orthodox freeze-drying cycles are used. Disaccharides are effective lyo-protectants, and because they display higher glass transition temperatures than monosaccharides, typically freeze-dry successfully when conventional drying cycles are used. Reducing sugars such as lactose and glucose may induce damaging Maillard reactions, thereby compromising stability, and for this reason, non-reducing disaccharides such as sucrose or trehalose are preferred to reducing sugars [27]. The addition of salts to formulations containing sugars will markedly depress Tg' or Tcollapse [2, 28]. Morgan et al. provided a helpful review article in 2005 discussing microorganism preservation by various drying technologies highlighting factors such as growth phase and growth medium, as well as cell concentration, and bacteria type [29]. More recent publications demonstrate that the preservation of probiotics, other microbes, and cells still presents a number of challenges [30–32].

Although presenting technical difficulties such as product collapse during freeze-drying, the amorphous phase may be an essential prerequisite for stabilizing biomaterials, such as vaccines and live cells, by providing an integration of the protective additive and biomolecule, thereby minimizing damage during freezing and drying.

Collapsed cakes are not only cosmetically unacceptable but may also be poorly soluble, exhibit reduced activity, or have compromised shelf stability. Collapse may be exacerbated by the formation of a surface skin, which impedes vapor migration from the drying structure. To avoid collapse, it is necessary to maintain the sublimation interface below Tg' or Tcollapse throughout primary drying and to include excipients in the formulation that reduce the likelihood of collapse. It is therefore essential to characterize formulations during the process development program. Although collapse may cause operational difficulties during freeze-drying, the induction and maintenance of the amorphous state may be essential for protecting labile biomolecules during freezing, drying, and storage [33].

A further issue that can affect the stability of a product despite the lack of obvious loss of structure is a phenomenon known as microcollapse. This may occur when separate amorphous and crystalline phases form in the frozen structure, with each displaying independent behavior and having its own critical temperature, which will be Tg' or Tcollapse for the amorphous phase and Teu for the crystalline phase. If lyophilized between these temperatures, then the phase with the higher critical temperature may retain its physical structure while the other phase may 'microcollapse' or 'micromelt' onto it [11].

Even amorphous materials, when mixed in the same solution without any crystalline components, are not immune from phase separation, and in some cases, distinct glass transition temperatures may be observed for the multiple phases [34]. Therefore, it is always advisable to carry out some basic characterization of candidate formulations prior to lyophilization, especially for complex mixtures where the behavior may not be completely predictable even if the individual components have been well characterized.

2.2.6 Freezing in Practice

Products may be frozen in a variety of ways depending on operational requirements:

1. The liquid formulation may be frozen in a freezer or a cooling tunnel prior to transfer to the freeze-dryer for desiccation. Advantages include increased annual sample throughput because the freeze-dryer is used only for drying. Disadvantages include the greater risk of melting or contamination resulting from the need to transfer material from the freezer into the dryer.
2. The most widely used technique is to freeze the liquid formulation in the final container directly on the freeze-dryer shelf. Although this method has the disadvantage that the dryer is used for part of the cycle as a freezer, freezing and drying samples within a single machine eliminates the need to transfer samples from freezer to dryer, and therefore improves process control, as well as reducing product vulnerability; this is generally the method preferred by regulatory bodies.
3. Formulations may be frozen using controlled nucleation, where they are loaded in the liquid state and the shelf temperature is reduced to below 0°C (typically between −2°C and −8°C) prior to the mechanism of nucleation being triggered. A number of different mechanisms may be employed, as listed earlier in this chapter. Controlled nucleation assists in obtaining a more homogenous ice crystal size, which can positively impact the success and reduce the length of the freezing stage and increase the efficiency of the primary drying phase, much as with annealing as discussed earlier. Additionally, since it occurs from the surface to the base, it can also help minimize the possibility of surface skin (crust) formation.

4. In certain cases, products or samples may be processed without a defined freezing step, relying instead on the evaporative cooling effect during the chamber evacuation step to cause the liquid to freeze. While there may be an element of risk in relying on this effect, a number of companies known to the authors employ this with consistent success.
5. "LyoBeads": Here, a solution or suspension is introduced into a cryogenic liquid or onto a cold surface to form frozen droplets, which are then placed into trays or flasks for freeze-drying. Under these conditions, sublimation rates are typically very high because the thickness of the dry layer is restricted only by the bead radius, and drying proceeds in a virtually unimpeded manner from each bead. A well-established platform technology utilizing this method of freezing is the "BioBall" from Biomérieux, where defined numbers of microorganisms are counted into each droplet immediately before freezing in liquid nitrogen, using a specially designed system. Each frozen droplet is placed into an individual vial, and each tray of partially stoppered vials is then loaded into a lyo chamber for primary and secondary drying. More recently, lyobead generators have been produced commercially, such as the LyoBeadPRO™ (Biopharma Group, Winchester, UK) shown in Fig. 1, which can allow for high-throughput generation of frozen droplets that are then harvested from the cryogenic liquid and transferred to appropriate pre-cooled containers (vials, trays, custom packaging) for primary and secondary drying in a lyo chamber.

2.3 Primary and Secondary Drying

For clarity, it is usual to separate the drying cycle into primary drying (the sublimation stage) and secondary drying (mainly desorption). The first step in the drying cycle is defined as primary drying and represents the stage where ice, which constitutes between approximately 70 and 95% of the product's water content, is converted into water vapor. Sublimation is a relatively efficient process, although the precise length of primary drying will vary depending on the product formulation, solute concentration, fill depth, and the heat transfer characteristics of the container, among other factors. During primary drying, the product dries at a discrete boundary (the sublimation interface), which recedes through the product from surface to base as drying progresses (or in the case of a Lyobead, from the outside towards the center).

2.3.1 Primary Drying (Sublimation)

Under atmospheric conditions, liquid water is converted into vapor by warming, a process defined as evaporation. However, the three states of water—ice, liquid, and vapor—coexist at the triple point and illustrate that at sub-atmospheric pressures, ice can convert directly to vapor by sublimation. Ice sublimation from a frozen aqueous

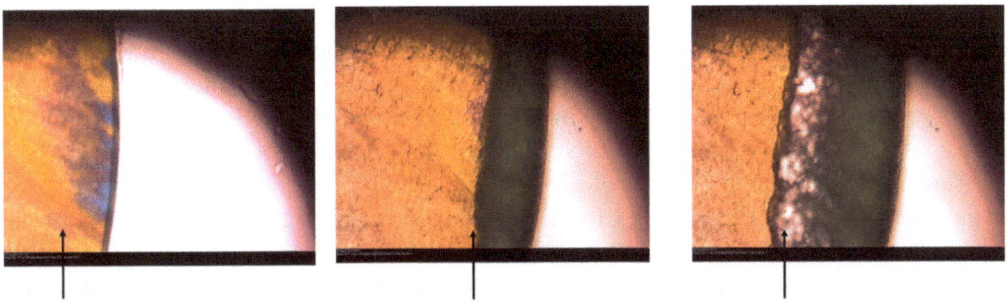

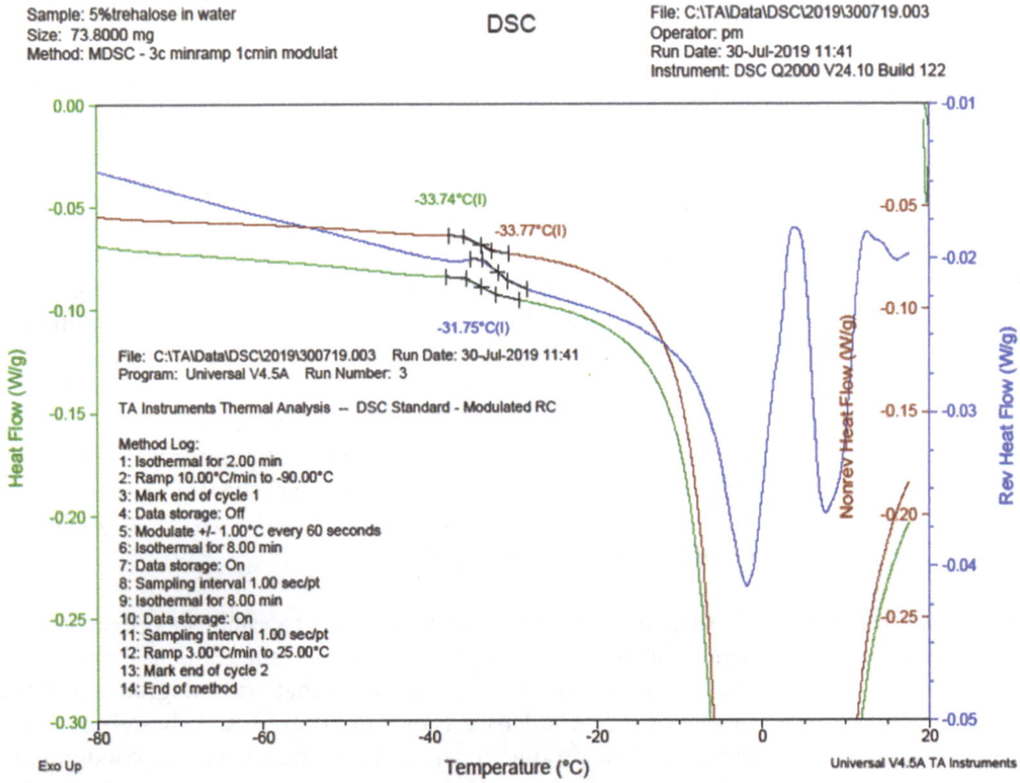

Fig. 1 The LyoBeadPRO™ (Biopharma Group, Winchester, UK) is a lyobead generator that forms droplets of a liquid formulation of a defined size range and creates frozen beads using a cryogenic liquid, which can then undergo sublimation in a lyo chamber

solution or suspension results in an open, porous, dry structure where solutes are spatially arranged as in the original solution or suspension. In contrast to evaporation, where components are concentrated as drying progresses, sublimation under vacuum minimizes concentration effects providing a dry product that is active and readily soluble.

Having frozen the liquid formulation, the next step is to dry it by subliming ice directly into water vapor. In order to initiate the sublimation process, it is essential to lower the chamber pressure so that the partial pressure of water is maintained below its corresponding vapor pressure to ensure the direct conversion of ice into water vapor.

Sublimation Rate and Chamber Pressure Conditions

Decreasing the chamber pressure will increase the rate of sublimation by reducing the gas/vapor concentration above the sample to provide minimal resistance to water molecules migrating from the product. However, reducing the system pressure beyond a certain point will be counterproductive and indeed, contrary to expectations, at very low system pressures the sublimation rate will decrease due to the reduction in the number of gas or vapor molecules in the chamber to conduct heat energy from the shelf into the product. Essentially, under high vacuum conditions, a "Thermos flask" effect is induced in the chamber, which inhibits heat transfer from the shelf. Under high-pressure (poor vacuum) conditions, heat transfer from the shelf to the product is primarily via gas/vapor conduction (convection), while in low-pressure (higher vacuum) conditions, heat transfer by gaseous conduction is reduced and energy input to the product is predominantly by radiative heat transfer, which is a relatively inefficient mechanism. As a general rule of thumb, in order to provide sufficient heat input through gaseous conduction while also allowing for some sublimation cooling and temperature gradients within each product container, the chamber will typically be maintained at between one-third and one-half of the pressure that equates to the vapor pressure of the ice in the product at its measured (or calculated) temperature, as defined by standard vapor pressure tables; however, chamber pressure may be subject to more rigorous optimization as part of a Quality by Design (QbD) approach [35].

Vapor Differential Pressure and Drying Efficiency

To sustain freeze-drying it is necessary to establish a pressure gradient from a product (higher pressure) to the condenser and finally the vacuum pump (lower pressure) so that water migrates from the product to the condenser as drying progresses. Although the temperature of the product must be higher than that of the condenser to ensure a net migration of water from the product, the system driving force represents the difference in vapor pressure (VP) rather than the difference in temperature between product and condenser, and can be calculated as the difference in VP between the two. For example, a product at −20°C has a VP = 0.78 Torr and with the condenser at −40°C (equivalent to a VP of 0.097 Torr), driving force will be 0.78–0.097 or 0.683 Torr. Little improvement in driving force is achieved by operating the condenser at −70°C. (VP = 0.002 Torr, providing a VP differential of 0.78 [product] − 0.002 [condenser] of 0.778 Torr.) The example illustrates that greater sublimation efficiency is derived by increasing product temperature rather than reducing condenser temperature, and the selection of suitable excipients that enable high processing temperatures to be used during freeze-drying without compromising product quality plays an important role in process and cycle development.

Heat and Mass Transfer

The essence of the freeze-drying process depends on maintaining a critical balance between the conversion of ice into water vapor by sublimation under vacuum and the removal of that vapor from the frozen mass. To maintain sublimation, heat energy is applied to the product to compensate for sublimation cooling. However, the heat extracted from the drying product as water vapor must be carefully balanced with the amount of energy added to the product.

Unless this equilibrium can be maintained, the product temperature will either decrease thereby reducing drying efficiency or increase, which may compromise product quality by inducing melt or collapse. This critical balance between product warming to increase drying rate and vapor extraction is defined by the heat and mass transfer equation. In the early stages of sublimation, the equilibrium is simple to maintain because the dry structure offers minimal resistance to vapor flow. However, as drying progresses and the depth of the dry layer increases, impedance to vapor flow will also increase, and the product may warm sufficiently to melt or collapse unless the process temperature is reduced. One consequence of reducing the energy input will be to reduce drying rate and prolong cycle times, but this may be unavoidable if product quality is to be preserved [36].

Cooling and Warming the Product

The shelves fitted into the freeze-dryer to support product containers may be alternatively cooled to initially freeze the product or maintain the shelf at a constant temperature throughout the drying cycle or warmed to provide energy for drying. Basically, two systems may be fitted:

1. An independent cooling coil is embedded in the shelf through which cold refrigerant is supplied (this system is termed direct expansion), and a heating element is bonded into or onto the base of the shelf. Shelf control is maintained by alternately operating either the heater or the cooler. Direct expansion systems are relatively inexpensive but fail to achieve temperature control much better than ±5°C.
2. For industrial or development activities, where shelf control to ±1°C is necessary to meet Good Manufacturing Practice (GMP) requirements, a diathermic fluid, which is invariably silicone fluid, is circulated through the shelves, and a separate refrigerator/heat exchanger maintains the diathermic fluid at a preset temperature.

The mechanism and the relative quantities of heat entering the product will depend on:

1. The nature of the product, its fill depth, consistency, and so on.
2. The dimensions and geometry of the product container and whether the container rests directly on a shelf or is supported in a tray.
3. The freeze-dryer design.
4. Chamber vacuum conditions.

Product temperature can be maintained by either raising or reducing shelf temperature or by alternating system pressure, which has the effect of improving or reducing heat transfer efficiency. Regardless of the precise system incorporated into the freeze-dryer, shelf temperature conditions may be controlled manually or programmed using a PC or microprocessor control.

The Sublimation Interface

Variously described as the drying front, freeze-drying front, and so on, macroscopically, the sublimation interface can be observed as a discrete boundary that moves through the frozen product to form an increasingly deeper layer of dried product above the frozen product. Heat is conducted from the shelf through the vial base and the frozen product layer to the sublimation front where ice is converted into water vapor. Several consequences result from this progressive recession of the sublimation front through the dry layer, which include:

1. The maintenance of the frozen zone at a low temperature because of sublimation cooling.
2. An increase in the resistance to vapor migration and a decrease in the sublimation rate as the dry layer increases in thickness.
3. Because the sublimation interface represents a zone representing maximum change of product temperature and moisture content, the interface represents the zone over which structural softening or collapse is likely to occur.
4. Water migrating from the sublimation front can reabsorb into the dried material above the sublimation interface.

Because the sublimation interface is the region where freeze-drying takes place, temperature monitoring of the interface is of paramount importance for product monitoring. However, because the sublimation front is constantly moving through the product, interface temperature cannot be effectively monitored using traditional temperature probes. Although the sublimation interface is defined as a discrete boundary, this is true only for ideal eutectic formulations, where ice crystals are large, open, and contiguous with each other. For typical amorphous formulations, such as vaccines, the sublimation front is much broader and comprises individual ice crystals imbedded in the amorphous phase. Under these conditions, although ice sublimes within the isolated crystals, the water vapor must diffuse through the amorphous phase (which is itself progressively drying) until it can migrate freely from the drying product matrix. Under these conditions, sublimation rates are much lower than those anticipated from data derived using eutectic model systems. Complicating a precise prediction of sublimation rate is the fact that fractures in the dry cake between the ice crystals can improve drying efficiency. All of these factors, including system impedances caused by the development of a surface skin on the product, have to be considered during product formulation and cycle development programs.

Notwithstanding these complications in precisely defining primary drying, sublimation is nevertheless a relatively efficient process and conditions used for primary drying include the use of shelf temperatures high enough to accelerate sublimation without comprising product quality by inducing collapse or melt, combined with high system pressures designed to optimize heat conduction from shelf into product. Removing the product when sublimation has been judged as complete will provide a vaccine that appears dry but which displays a high-moisture content that is invariably too high (e.g., 10%) to provide long-term storage stability, and the drying cycle is extended to remove additional moisture by desorption during secondary drying.

2.3.2 Secondary Drying

In contrast to primary drying, which is a dynamic process associated with high vapor flow rates, secondary drying is much less efficient with secondary drying times typically representing between 10% and 25% of the total process time but only removing a small fraction of the total water initially present in the liquid formulation. Under secondary drying conditions, the product approaches steady-state conditions where moisture is desorbed or absorbed from or into the product in response to relative humidity and shelf temperatures. Desorption is favored by increasing shelf temperature, using high-vacuum conditions in the chamber, thereby reducing the system's vapor pressure or relative humidity. Conversely, when the shelf temperature is reduced, and the vapor pressure in the system is increased by warming the condenser, dried products will reabsorb moisture and exhibit an increase in moisture content. Although product collapse during secondary drying is generally less likely than collapse during primary drying, it is possible to induce collapse in the dried matrix by exposing the product to temperatures above its glass transition temperature (Tg).

Setting a target residual moisture level for a lyophilized material can be difficult without specific prior knowledge of the material and its various components. Water will act as a plasticizer in the dried material, so based purely on the inverse relationship between residual moisture and Tg, one might conclude that the drier the material, the more shelf-stable it will be. Based on a series of enthalpic relaxation studies using DSC, Hancock et al. suggested that the optimal long-term stability is achieved when the dry state Tg is more than 40°C (and ideally 50°C) higher than the target storage temperature [37]. However, other factors may need to be taken into consideration, including whether some 'structural water' is necessary for the preservation of activity and/or chemical stability [38].

2.4 Stoppering the Product

A freeze-dried product will generally be very hygroscopic and have a very high specific surface area (typically in the order of 1–10 m^2 per gram, depending primarily on solute density and freezing conditions). Consequently, exposing the dried product to atmospheric conditions will result in reabsorption of atmospheric moisture into the product. Both water and air are damaging to a dried product, causing degradative changes resulting in poor stability; it is therefore prudent to stopper the product within the freeze-dryer prior to removal. It may also be necessary to dry stoppers prior to use or ensure appropriate low-moisture stoppers are used to prevent damage by moisture ingress from the stopper during the shelf life of the product post-lyophilization; the significance of this will be dependent on the amount and properties of the product. Stoppering under a full vacuum (low pressure) provides ideal conditions for ensuring product stability because reactive atmospheric gases are reduced to a minimum. However, injecting water into a product in a fully evacuated vial can induce foaming, which can be reduced by back-filling vials with an inert gas, such as nitrogen, before stoppering. Maintaining a certain level of vacuum e.g., 500 – 900 mbar (380–680 Torr) can help ensure a good stopper/vial seal is achieved while minimizing foaming during rehydration.

2.5 Reconstituting the Product

It is often supposed that because freeze-drying only removes water, then all products will be fully active by rehydrating only with water. This may not be the case and freeze-dried products often exhibit enhanced activity when reconstituted in an isotonic medium, such as saline, rather than water. Establishing the correct re-hydration method can be as critical as developing a suitable and robust cycle; this has been demonstrated in the freeze drying and reconstitution of red blood cells, where it was found that reconstitution with hypotonic solutions - or even water - could lead to regions of localized temporary hypotonicity during the reconstitution process, which led to erythrocyte lysis, despite the fact that the pre-lyophilized suspension had itself been significantly hypertonic [39].

3 Thermoanalytical and Microscopic Methods

Probably the most impactful change in freeze drying over the past twenty years has been the introduction of commercially available freeze-drying microscopy technology and the development of modulated differential scanning calorimetry, which has dramatically improved the ability to measure glass transition and enthalpy relaxation behavior of amorphous formulations. These technologies have transformed how freeze-drying cycles and lyo formulation development have been undertaken [11].

3.1 Freeze Drying Microscopy (FDM)

The principle of measuring the thermal properties of material under the microscope was developed in the 1960s [40] but has only become commercially available in the 2000s. Using a few microliters of sample mounted on glass slides or quartz crucibles, the programmable cycle design can identify the freezing point, and then the solidification of the freeze-dried matrix. Once a vacuum is applied, the rate of sublimation can be measured on small quantities of freeze-dried material under given conditions of vacuum and temperature, and the impact of modifying the freeze-drying cycle can be evaluated [41]. In Fig. 2A, the FDM profile of 5% trehalose is given, showing good drying at −37°C but collapse at −33°C. The freeze-drying collapse temperature can be determined for any given formulation over the course of a few experiments in typically under half a day. In Fig. 3A, the FDM images for a 5% sucrose 20 mM histidine 0.01% Tween 20 formulation are given with a collapse temperature around −35°C and in Fig. 4A, the FDM profile of a 20 mg/mL heparin (a complex polysaccharide) sample is shown, with collapse beginning around −21°C. Once robust data are obtained from such studies, these conditions can then be transferred into the freeze drying cycle for a given trial, together with knowledge of the heat transfer properties of the selected containers; this can result in a first freeze drying cycle that has a far greater chance not only of success at small scale but also in being robust, transferable, and scalable.

3.2 Modulated Differential Scanning Calorimetry (mDSC)

DSC has, for several decades, been available for measuring eutectic temperatures in predominantly ionic formulations for the freeze drying of pharma and foods. This has been well described and was a recommended approach for freeze-drying development [42]. Although this can be delivered with such simple heat flow profiles, the development of modulated DSC has allowed meaningful analysis of the weaker glass transition points of dilute amorphous formulations. As many biopharmaceuticals are formulated as amorphous materials then the determination of relevant thermo-analytical properties can only be measured using these techniques or refinements of them [43].

An illustration of the weak glass transitions in an amorphous biological formulation can be seen in Figs. 2B, 3B, and 4B. Not only can weak glass transitions be detected using the more sensitive modulated technique (note in these examples the Tg' is seen in all 3 signals - total, non-reversing- and reversing- heat flow - although the value differs by up to 2°C), but this technique can also detect the presence of crystalline materials such as inorganic salts and the crystallization of co-formulants such as mannitol, which can exist

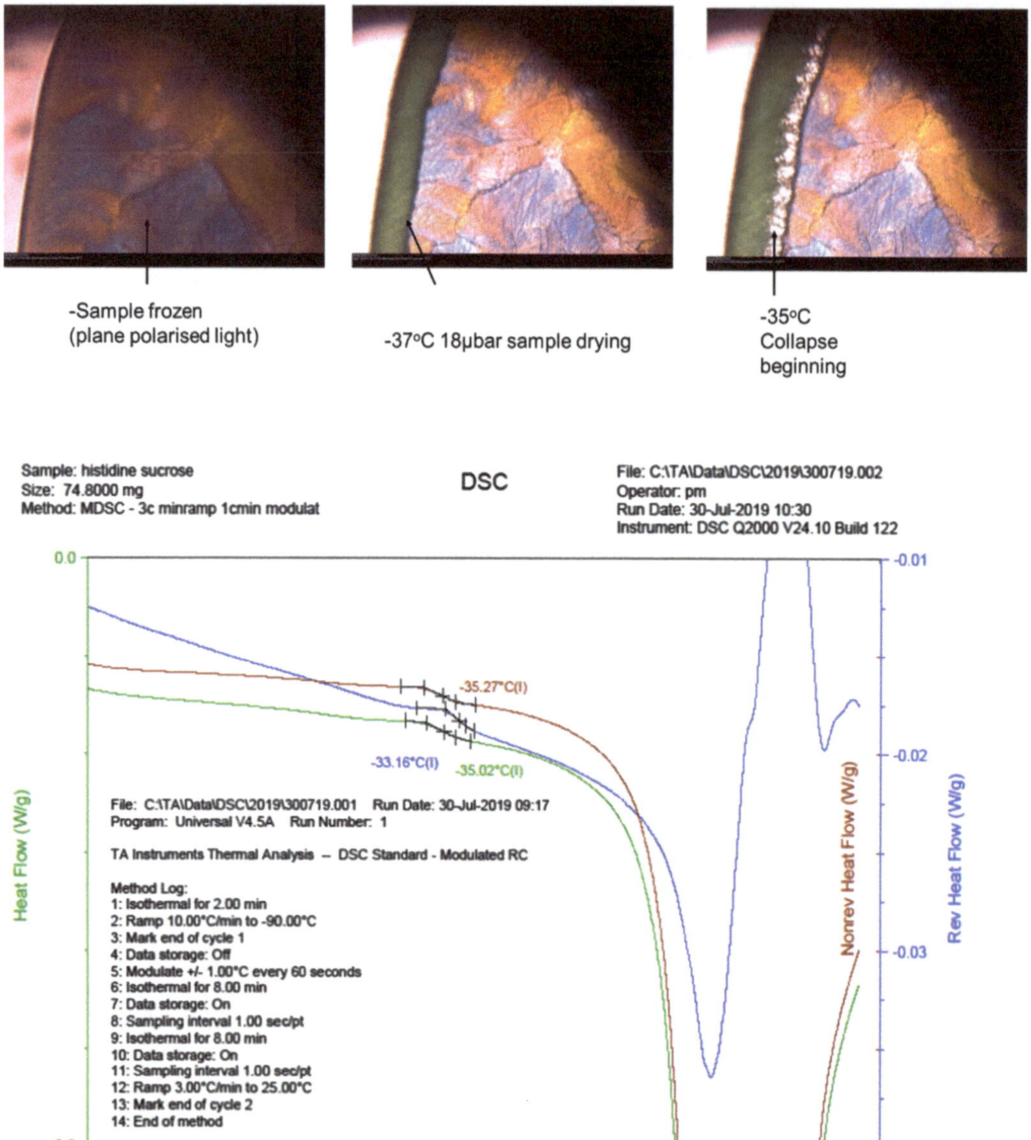

Fig. 2 Thermal analysis of 5% w/v trehalose in water showing (**A**) FDM profile from frozen (left-hand side image) to drying (middle image) to collapse at −33°C (right-hand side image), and (**B**) mDSC profile with heat flow (−33.7°C) reversing (−31.7°C) and non-reversing (−33.8°C) profiles. Sample 80 mg in high-volume steel pan, temperature ramp rate 3°C/min with 1°C/min modulation. Exothermic events shown as upward peaks

either in the amorphous form or in various crystalline states [44]. It also has the advantage of being applicable to measuring the dry state glass transition (Tg) and enthalpic relaxation of freeze-dried materials, which may be useful in predicting stability. This means that

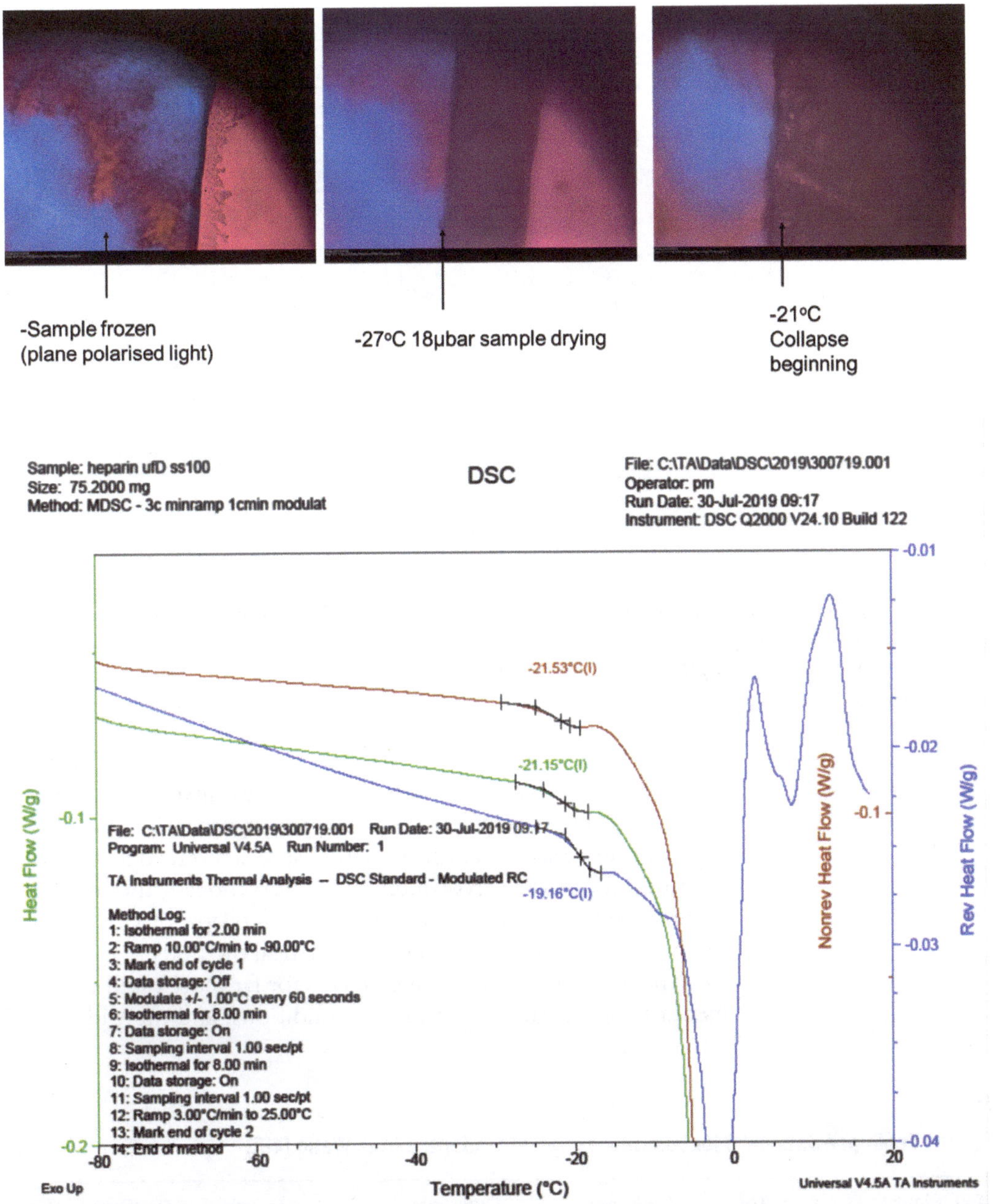

Fig. 3 Thermal analysis of 5% w/v sucrose, 20 mM histidine buffer 0.01% Tween 20 pH 6.5, showing (**A**) FDM profile from frozen (left hand side image) to drying (middle image) to collapse at −35°C (right hand side image), and (**B**) mDSC profile with total heat flow (−35.0°C), reversing (−33.2°C) and non-reversing (−35.3°C) profiles. Sample 75 mg in a high-volume steel pan, temperature ramp rate 3°C/min with 1°C/min modulation. Exothermic events shown as upward peaks

it can be used to study the impact of formulation on the stability of freeze-dried material and the impact of residual moisture content and its plasticizing effect [45]. The Tg' and Tcollapse of some typical freeze-dryingexcipients are given in Table 1.

Fig. 4 Thermal analysis of 20 mg/mL heparin in water showing (**A**) FDM profile from frozen (left-hand side image) to drying (middle image) to collapse at −21 °C (right-hand side image) and (**B**) mDSC profile with heat flow (−21.2°C) reversing (−19.2°C) and non-reversing (−21.5°C) profiles. Sample 75 mg in high-volume steel pan, temperature ramp rate 3°C/min with 1 °C/min modulation. Exothermic events shown as upward peaks

The Tcollapse and Tg' values for complex mixtures may reflect the impact of the excipients being present together: for instance, in Fig. 3, it can be seen that the Tg'/Tcollapse is several degrees lower than that for pure sucrose alone (−32°C). Indeed, the impact of thermal tempering (annealing) on the Tg'/Tcollapse of a formulation can be monitored by these thermoanalytical methods.

The use of these techniques allows for far more rapid development of appropriate freeze-drying conditions, and the widespread

Table 1
Tg' and Tcollapse data in °C for common excipients (adapted from Wang [46])

Excipient	Tg'	Tcollapse	Excipient	Tg'	Tcollapse
Sucrose	−32	−31	KH_2PO_4		−55
Sorbitol	−44	−54	Tris-HCl	−65	
Trehalose		−29	Ovalbumin	−11	−10
Glucose	−43	−41	BSA	−12	
Dextran (10 kDa)		−10	Glycine	−37	
HEPES	−63		Sodium citrate	−41	
Histidine	−33		Mannitol	−27	

availability of these two techniques in particular has revolutionized how freeze-drying development is done in the commercial sector as well as in academic research.

Powerful experimental design techniques such as Design of Experiments (DoE) can be applied to the screening of freeze-drying formulations [47] using data derived from thermal analysis or FDM, and this can be used to design a freeze-drying cycle. This type of information can be fed then into Process Analytical Technology to scale up and optimize freeze drying cycles at medium and even large scale using statistical principles and process mapping [48].

4 Scale-Up Factors and PAT

The need to operate the freeze-dryer under low-pressure conditions to convert ice directly into water vapor (sublimation) adds to the complexity and cost of dryers because the chamber holding the product must withstand the differential pressure from vacuum to atmosphere. Although a suitable vacuum pump is essential for initially evacuating the chamber and eliminating air that may leak into the dryer during operation, vacuum pumps are not capable of continuously removing water vapor subliming from the product, and a refrigerated trap (termed the process condenser) must be placed between the product and the pump to condense the moisture migrating from the drying product. In reality, it is the condenser that comprises the "pumping force" of the system. Process condensers may be incorporated into the drying chamber (referred as an internal condenser) or located in a separate chamber between the product chamber and pump (external condenser). Each geometry has advantages and disadvantages, although either design may be used. Stainless steel is typically used to fabricate research or production dryers because this metal can be cleaned by a wide range of sanitizers, including steam. For GMP manufacture, the freeze-dryer is invariably sterilized by pressurized steam, and this adds to the complexity and expense of the dryer because it must conform to the requirements to operate under these conditions (i.e., it must be a certified pressure vessel) as well as under the sub-atmospheric pressure conditions to which it will be subjected during the lyophilization process (i.e., as a vacuum vessel). Modern freeze-dryers are also fitted with internal stoppering devices for sealing vials at the end of the cycle, valves, and monitoring devices for assessing drying efficiency, and are typically computer or microprocessor-controlled so that cycles can be reproduced and evaluated for regulatory purposes. When freeze-drying a vaccine, it may be necessary to incorporate protective devices and introduce processing protocols that ensure both safe operation and prevent product cross-contamination [49].

The primary scale-up factors in freeze-drying may be defined as:

1. Equipment specification and performance limitations (e.g., refrigeration capacity, condenser trapping rate, use of different pressure gauges, chamber size, control system differences).
2. Upstream processing factors (e.g., scale of compounding and filling, use of sterile filtration).
3. Differences in personnel in R&D laboratories/pilot plants/ sterile GMP facilities.

Many of the potential effects of the above differences at various scales can be assessed in-process using Process Analytical Technologies (PAT) that are available for integration into freeze-dryers or as standalone devices that can give real-time feedback. Product temperature is conventionally monitored using a suitable temperature measuring device such as a wired thermocouple (usually Type T) or resistance thermometer (almost invariably a Pt100); however, more recently, wireless probes have become more widely used [50]. Another method of measuring the sublimation interface temperature is by the use of the non-invasive Manometric or Barometric Temperature Measurement (MTM / BTM): this is a batch method which is central to various software-based cycle development systems that calculate the "average" sublimation interface temperature of all the samples within the product chamber based on different algorithms that take into the account the size and heat transfer characteristics of the container, number of samples and the physicochemical characteristics of the formulation itself [51]. Other methods of monitoring the progress of the lyophilization process by examining "whole batch" parameters (as opposed to individual samples) include Tunable Diode Laser Absorption Spectroscopy, or TDLAS, which provides real-time data on the instantaneous mass flux, arguably a more scalable parameter than many others [52], while some methods can be applied to specific samples, such as the use of near infra-red (NIR) spectroscopy [53]. A recent comprehensive review of PAT methods in lyophilization is given by Vallan et al. [54], and a number of other techniques have been described, including Raman and infra-red spectroscopies and the use of heat flux sensors [55–59], allowing an operational design space to be identified within a freeze-drying process [60, 61].

A relatively common challenge when freeze-drying biological materials is the need to transfer a lyophilization process from basic research laboratory equipment—which may not be equipped with shelf temperature control—to more sophisticated controllable equipment that requires specific temperature, pressure, and time setpoints to be programmed. In this case, the technology transfer process effectively requires the additional step of "lyo recipe formalization," often made more challenging by virtue of the fact that the simple lyo lacks monitoring equipment. It may well be the case that secondary drying traditionally employs low-pressure setpoints merely because of operator attempts to replicate the conditions that existed in simpler equipment that lacked pressure control. It is now understood that employing pressures lower than 100ubar during secondary drying brings little or no benefit to the efficiency of the desorption process [62] and, in fact, such low pressures can lead to phenomena such as 'oil back-streaming' during secondary drying. In this stage of the cycle, for systems with an oil-filled vacuum pump, the pump pressure can effectively reach equilibrium with the condenser pressure, and with the condenser surface being typically more than 100°C colder than the pump oil, a thermal drive exists that can encourage molecular migration of oil towards the condenser [63]. This may result in a noticeable film of oil on the bottom of the condenser chamber upon completion of the cycle. Transferring a lyo process from a less controllable unit to a more controllable system might therefore present an opportunity to review the process and make improvements to the lyo recipe in terms of efficiency, repeatability, and minimizing product defects; this is also likely to make the process more easily scalable and less reliant on radiant heat input.

5 Factors Affecting Freeze-Dried Products

Freeze-dried products should be formulated to minimize storage decay and should ideally be able to tolerate storage at ambient temperatures for distribution purposes. However, it is imprudent to assume that a freeze-dried product remains immune to damage during storage simply due to the relative lack of residual water. The specific aspects and challenges associated with the stabilization of different materials by freeze-drying—including proteins, vaccines, bacteria, sperm, and bioscaffolds—are discussed in later chapters in this volume. On a more general note, there are many factors that affect the stability of almost every freeze-dried product, including temperature, moisture content, reactive gases, light, free radicals, and specific chemical reactions such as Maillard reactions.

5.1 Residual Moisture

Even when the residual moisture level in a sealed container of freeze-dried material remains constant, there may be an equilibrium between 'sorbed' (adsorbed or absorbed) water in the lyophilized cake and free water in the headspace, which may fluctuate with time and temperature [64]. For traditional therapeutics, vaccines, and diagnostics, the residual moisture content is typically expressed as a % (w/w). However, in the food industry, and for some biological materials, it is more usual to measure water activity (A_w), which relates to the freely available water in the system. Techniques such as Thermo-Gravimetric Analysis (TGA) and Karl Fischer titration (KF) are very well established and recognized by regulatory bodies for pharmaceutical products, but some non-invasive (or at least, non-destructive) methods are gaining ground.

Fourier Transform Infrared (FTIR) spectroscopy has been demonstrated to be a useful method for examining various properties of a dried material—typically a cake in a clear glass vial—including aspects of molecular structure (e.g., amide bands of proteins) and residual moisture content. Here, it is necessary to perform some initial studies to establish which of the characteristic absorbance peaks on the FTIR spectrum correspond(s) to the residual water content in the material. Additionally, the FTIR source typically does not penetrate far into the material, so calibration/validation is required in order to determine whether and how the information obtained relates to the residual moisture content in the whole sample. However, once this has been established for a particular product, it is possible to integrate an FTIR source and detector into an automated system for rapid post-lyophilization inspection. Brülls and coworkers published useful and practical information on this technique [65, 66].

An alternative to focusing on the determination of water resident in the product itself is to analyze the headspace moisture (HSM) in its sealed container, which may give a better indication of the 'free water' in the system and thus water activity (A_w). Analysis of HSM is achieved by Frequency Modulation Spectroscopy (FMS) where a laser of defined wavelength is passed through the headspace of a container (typically a clear glass vials or ampoule) and the extent to which the laser power is absorbed by the vapor in its path is measured by a detector; the FMS1400 from Lighthouse Instruments is designed specifically for this function, where it can also serve as a means of monitoring container-closure integrity (CCI) over time [67]. The United States Pharmacopeia (USP) published the guidance document USP <922> in 2021, which provides recommendations for determining water activity in pharmaceutical products. Analysis of HSM appears to be an ideal way of determining A_w for dried materials in sealed glass vials, and even for historical HSM data; if the temperature of the analysis is known, it may be possible to calculate A_w after the fact.

5.2 Thermal Properties

As highlighted above, MDSC can be used to measure the thermal properties of a freeze-dried product and also to evaluate the plasticization effect of water. However, the thermal properties of the lyophilized material (such as the dry state Tg) are not the only influence on the long-term stability of biological materials; in addition, the chemical nature of the formulation ingredients may play a role, including those components that stabilize during freezing and drying, which may not necessarily be the optimal choice for ensuring long-term stability. This has been demonstrated in a number of cases, including a study of the role of reducing sugars [19], which are known to undergo Maillard reactions with proteins in the dried state, while non-reducing sugars do not. As such, formulation will ideally take into account the dry state characteristics of the final product as well as the behavior of the formulation during the freeze-drying process itself.

5.3 Physical and Mechanical Properties

A further factor receiving significant attention is that of the physical properties of the freeze-dried product, particularly with regard to its robustness to handling and transport. It is well understood that cosmetically acceptable cakes or lyobeads are often endowed with concomitantly superior properties compared with their collapsed counterparts, such as lower residual moisture and more rapid reconstitution; however, it could be argued that this is a property that needs to extend to the entire shelf life of the product, not just immediately post-manufacture, because a product that becomes powdery during transport and handling can also lead to difficulties for the end user. As such, strides have been made to commercially develop mechanical testing devices that are suitably sensitive to quantify parameters such as the stiffness (Young's Modulus) and strength (Max Stress at Failure) of lyophilized cakes in situ, as described by Hedberg et al. [68]. One example of this is the MicroPress™ (Biopharma Group, Winchester, UK), which is designed to analyze a lyophilized cake without the need to remove it from the vial. The system can also be used to assess the mechanical properties of a lyobead by placing the sample in a custom-made stainless-steel crucible, which centers the bead and ensures that the indenter makes contact with it at the highest point (Fig. 5). Imaging techniques such as Micro-CT scanning are also available that can provide a visual representation of the interior of a freeze-dried material in three dimensions [69].

Any analytical method that can provide reliable quantification of measurable product attributes assists not only in the conventional

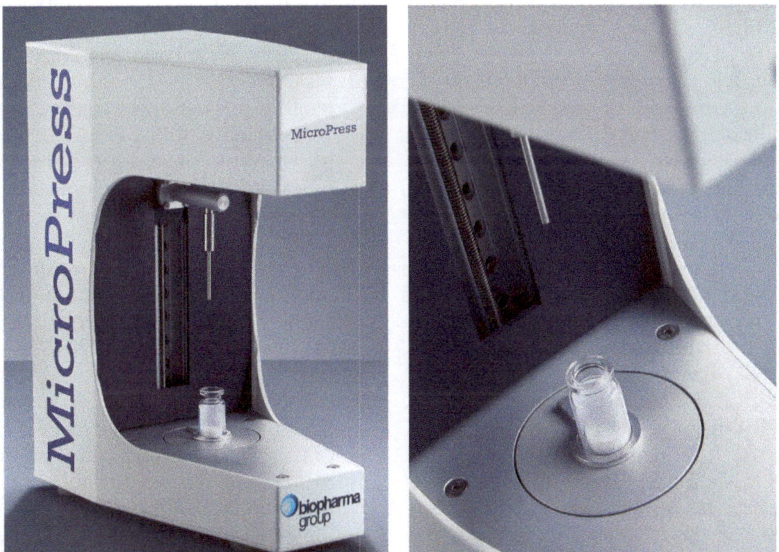

Fig. 5 The MicroPress™ (Biopharma Group, Winchester, UK) is designed to analyze various mechanical properties of lyophilized materials and other samples with low stiffness and density

approach to formulation and process development in freeze-drying, but also in the Quality by Design approach [70].

Acknowledgments

The authors would like to acknowledge colleagues at Biopharma Process Systems and MHRA Science Campus for their input and discussion on this chapter.

References

1. Fanget B, Francon A (1996) A Varicella vaccine stable at 5 degrees C. Dev Biol Stand 87:167–171
2. Adams GDJ (1995) The preservation of inocula. In: Brown MRW, Gilbert P (eds) Microbiological quality assurance: a guide towards relevance and reproducibility of inocula. CRC Press, London, UK, pp 89–119
3. Hansen LJ, Daoussi R, Vervaet C, Remon JP, De Beer TR (2015) Freeze-drying of live virus vaccines: a review. Vaccine 33:5507–5519. https://doi.org/10.1016/j.vaccine.2015.08.085
4. Bindschaedler C (1999) Lyophilization process validation. In: Rey L, May JC (eds) Freeze-drying/lyophilization of pharmaceutical and biological products. Marcel Dekker, New York, pp 373–408
5. Abla KK, Mehanna MM (2022) Freeze-drying: a flourishing strategy to fabricate stable pharmaceutical and biological products. Int J Pharm 628:122233. https://doi.org/10.1016/j.ijpharm.2022.122233
6. McAndrew TP, Hostetler D, DeGrazio FL (2019) Container and reconstitution systems for lyophilized drug products. In: Ward KR, Matejtschuk P (eds) Lyophilization of pharmaceuticals and biologicals: new technologies and approaches. Humana Press, Springer, New York, pp 193–214
7. Carpenter JF, Izutsu K-I, Randolph T (2010) Freezing- and drying- induced perturbations of protein structure and mechanisms of protein protection by stabilizing additives. In: Rey L, May JC (eds) Freeze-drying/lyophilization of pharmaceutical and biological products. 3rd ed. Informa Healthcare, New York, pp 167–197
8. Adebayo AA, Sim-Brandenburg JW, Emmel H, Olaeye DO, Niedrig M (1998) Stability of 17D Yellow fever virus vaccine using different stabilisers. Biologicals 26:309–316. https://doi.org/10.1006/biol.1998.0157
9. Matejtschuk P, Malik K, Duru C (2019) Formulation and process development for lyophilized biological reference materials. In: Ward KR, Matejtschuk P (eds) Lyophilization of pharmaceuticals and biologicals: new technologies and approaches. Humana Press, Springer, New York, pp 33–55
10. Searles JA (2010) Freezing and annealing phenomena in lyophilization. In: Rey L, May JC (eds) Freeze-drying/lyophilization of pharmaceutical and biological products. 3rd ed. Informa Healthcare, New York, pp 52–81
11. Ward KR, Matejtschuk P (2019) Characterization of formulations for freeze drying. In: Ward KR, Matejtschuk P (eds) Lyophilization of pharmaceuticals and biologicals: new technologies and approaches, Humana Press. Springer, New York, pp 1–32
12. Luoma J, Magill G, Kumar L, Yusoff Z (2019) Controlled ice nucleation using ControLyo® pressurization-depressurization method. In: Ward KR, Matejtschuk P (eds) Lyophilization of pharmaceuticals and biologicals: new technologies and approaches, Humana Press. Springer, New York, pp 57–77
13. Pisano (2019) Alternative methods of controlling nucleation in freeze drying. In: Ward KR, Matejtschuk P (eds) Lyophilization of pharmaceuticals and biologicals: new technologies and approaches, Humana Press. Springer, New York, pp 79–111
14. Bhatnagar BS, Pikal MJ, Bogner RH (2007) Study of the individual contributions of ice formation and freeze-concentration on isothermal stability of lactate dehydrogenase during freezing. J Pharm Sci 97:798–814. https://doi.org/10.1002/jps.21017
15. Awotwe-Otoo D, Agarabi C, Read EK, Lute S, Brorson KA, Khan MA, Shah RB (2013) Impact of controlled ice nucleation on process performance and quality attributes of a lyophilized monoclonal antibody. Int J Pharm 450:70–78. https://doi.org/10.1016/j.ijpharm.2013.04.041
16. Oddone I, Arsiccio A, Duru C, Malik K, Ferguson J, Pisano R, Matejtschuk P (2019) Vacuum-induced surface freezing for the

freeze-drying of the human growth hormone: how does nucleation control affect protein stability?. J Pharm Sci 109:254–263. https://doi.org/10.1016/j.xphs.2019.04.014
17. Geidobler R, Winter G (2013) Controlled ice nucleation in the field of freeze-drying: fundamentals and technology review. Eur J Pharm Biopharm 85:214–222. https://doi.org/10.1016/j.ejpb.2013.04.014
18. Schneid SC, Cohrs M, Lenger JH (2025) Scaling up controlled nucleation in freeze drying: translating vacuum-induced surface freezing from laboratory to GMP. Eur J Pharm Sci 204:106968. https://doi.org/10.1016/j.ejps.2024.106968
19. Song X, Philpott MA, Best SM, Cameron RE (2024) Controlling the architecture of freeze-dried collagen scaffolds with ultrasound-induced nucleation. Polymers 16:213. https://doi.org/10.3390/polym16020213
20. Strongrich A, Lim FJ, Kumar L, Alexeenko A (2021) Rapid depressurization based controlled ice nucleation in pharmaceutical freeze-drying: the roles of the ballast gas and the vial. J Pharm Sci 110:3639–3647. https://doi.org/10.1016/j.xphs.2021.07.011
21. Wenzel T, Gieseler M, Gieseler H (2020) Design of vacuum-induced freezing protocols for high fill volume formulations in freeze-drying: a strategic approach. J Pharm Sci 109:3035–3044. https://doi.org/10.1016/j.xphs.2020.06.025
22. Wenzel T, Gieseler M, Gieseler H (2020) Investigation of two different pressure-based controlled ice nucleation techniques in freeze-drying: the integral role of shelf temperature after nucleation in process performance and product quality. J Pharm Sci 109:2746–2756. https://doi.org/10.1016/j.xphs.2020.05.020
23. Jin J, Yurkow EJ, Adler D, Lee TC (2018) Improved freeze drying efficiency by ice nucleation proteins with ice morphology modification. Food Res Int 106:90–97. https://doi.org/10.1016/j.foodres.2017.12.028
24. Oddone I, Barresi AA, Pisano R (2017) Influence of controlled ice nucleation on the freeze-drying of pharmaceutical products: the secondary drying step. Int J Pharm 524:134–140. https://doi.org/10.1016/j.ijpharm.2017.03.077
25. Konstantinidis AK, Kuu W, Otten L, Nail SL, Sever RR (2011) Controlled nucleation in freeze-drying: effects on pore size in the dried product layer, mass transfer resistance, and primary drying rate. J Pharm Sci 100:3453–3470. https://doi.org/10.1002/jps.22561
26. Patel SM, Nail SL, Pikal MJ, Geidobler R, Winter G, Hawe A, Davagnino J, Gupta SR (2017) Lyophilized product cake appearance: what is acceptable?. J Pharm Sci 106:1706–1721. https://doi.org/10.1016/j.xphs.2017.03.014
27. Li S, Patapoff TW, Overcashier D, Hsu C, Nguyen TH, Borchardt RT (1996) Effects of reducing sugars on the chemical stability of human relaxin in the lyophilized state. J Pharm Sci 85:873–877. https://doi.org/10.1021/js950456s PMID: 8863280
28. Pikal MJ (1999) Mechanisms of protein stabilization during freeze-drying and storage: the relative importance of thermo stabilization and glassy state relaxation dynamics. In: Rey L, May JC (eds) Freeze-drying/lyophilization of pharmaceutical and biological products. Marcel Dekker, New York, pp 161–198
29. Morgan CA, Herman N, White PA, Vesey G (2006) Preservation of microrganisms by drying; a review. J Microbiol Methods 66:183–193. https://doi.org/10.1016/j.mimet.2006.02.017
30. Kiepś J, Dembczyński R (2022) Current trends in the production of probiotic formulations. Foods 11:2330. https://doi.org/10.3390/foods11152330
31. Rockinger U, Funk M, Winter G (2021) Current approaches of preservation of cells during (freeze-) drying. J Pharm Sci 110:2873–2893. https://doi.org/10.1016/j.xphs.2021.04.018
32. Prakash O, Nimonkar Y, Shouche YS (2013) Practice and prospects of microbial preservation. FEMS Microbiol Lett 339:1–9. https://doi.org/10.1111/1574-6968.12034
33. Rey LR (1999) Glimpses into the realm of freeze-drying: classic issues and new ventures. In: Rey L, May JC (eds) Freeze-drying/lyophilization of pharmaceutical and biological products. Marcel Dekker, New York, pp 1–30
34. Padilla AM, Pikal MJ (2010) Phase separation of freeze-dried amorphous solids: the occurrence and detection of multiple amorphous phases in pharmaceutical systems. In: Rey L, May JC (eds) Freeze drying/lyophilization of pharmaceuticals and biological products. 3rd ed. Informa Healthcare, New York, pp 82–111
35. Mockus LN, Paul TW, Pease NA, Harper NJ, Basu PK, Oslos EA, Sacha GA, Kuu WY, Hardwick LM, Karty JJ, Pikal MJ, Hee E, Khan MA, Nail SL (2011) Quality by design in formulation and process development for a freeze-dried, small molecule parenteral product: a case study. Pharm Dev Technol 16:549–576. https://doi.org/10.3109/10837450.2011.611138
36. Patel SM, Pikal MJ (2011) Emerging freeze-drying process development and scale-up issues. AAPS Pharm Sci Tech 12:372–378. https://doi.org/10.1208/s12249-011-9599-9
37. Hancock BC, Shamblin SL, Zografi G (1995) Molecular mobility of amorphous pharmaceutical solids below their glass transition

temperatures. Pharm Res 12:799–806. https://doi.org/10.1023/A:1016292416526
38. Ohtake S, Feng S, Shalaev E (2018) Effect of water on the chemical stability of amorphous pharmaceuticals: 2. Deamidation of peptides and proteins. J Pharm Sci 107:42–56. https://doi.org/10.1016/j.xphs.2017.09.003 ISSN 0022-3549
39. Ward K, Cowen A, Peacock T (2012) Freeze drying method. Patent WO/2012/098358
40. MacKenzie AP (1964) Apparatus for microscopic observations during freeze drying. Biodynamica 9:213–222
41. Ward K, Matejtschuk P (2010) The use of microscopy, thermal analysis and impedance measurements to establish critical formulation parameters for freeze-drying cycle development. In: Rey L, May JC (eds) Freeze drying/lyophilization of pharmaceuticals and biological products. 3rd ed. Informa Healthcare, New York, pp 112–135
42. Franks F, Auffret A (2007) Freeze-drying of pharmaceuticals and biopharmaceuticals: principles and practice. RSC Press, Cambridge UK
43. Kett V, McMahon D, Ward K (2004) Freeze-drying of protein pharmaceuticals - the application of thermal analysis. CryoLetters 25:389–404
44. Mehta M, Bhardwaj SP, Suryanarayanan R (2013) Controlling the physical form of mannitol in freeze dried systems. Eur J Pharm Biopharm 85:207–213. https://doi.org/10.1016/j.ejpb.2013.04.010
45. Duddu SP, Dal Monte PR (1997) Effect of glass transition temperature on the stability of lyophilized formulations containing a chimeric therapeutic monoclonal antibody. Pharm Res 14:591–595. https://doi.org/10.1023/a:1012144810067
46. Wang W (2000) Lyophilization and development of solid protein pharmaceuticals. Int J Pharm 203:1–60. https://doi.org/10.1016/s0378-5173(00)00423-3
47. Grant Y, Matejtschuk P, Dalby PA (2009) Rapid optimization of protein freeze drying formulations using ultra scale down and factorial design of experiment in microplates. Biotechnol Bioeng 104:957–964. https://doi.org/10.1002/bit.22448
48. Bourlès E, deLannoy F, Scutellà B, Fonseca F, Trelea IC, Passot S (2019) Scale-up of freeze-drying cycles, the use of process analytical technology (PAT), and statistical analysis. In: Ward KR, Matejtschuk P (eds) Lyophilization of pharmaceuticals and biologicals: new technologies and approaches. Humana Press, Springer, New York, pp 215–240
49. Cherry C (2019) Containment options for the freeze-drying of biological entities and potent materials. In: Ward KR, Matejtschuk P (eds) Lyophilization of pharmaceuticals and biologicals: new technologies and approaches. Humana Press, Springer, New York, pp 143–155
50. Schneid S, Gieseler H (2008) Evaluation of a new wireless temperature remote interrogation system (TEMPRIS) to measure product temperature during freeze drying. AAPS Pharm Sci Tech 9:729–739. https://doi.org/10.1208/s12249-008-9099-8
51. Tang C, Nail SL, Pikal MJ (2005) Freeze-drying process design by manometric temperature measurement: design of a smart freeze-dryer. Pharm Res 22:685–700. https://doi.org/10.1007/s11095-005-2501-2
52. Kessler WJ, Gong E (2019) Tunable diode laser absorption spectroscopy in lyophilization. In: Ward KR, Matejtschuk P (eds) Lyophilization of pharmaceuticals and biologicals: new technologies and approaches. Humana Press, Springer, New York, 113–141
53. Sylvester B, Porfire A, Van Bockstal P-J, Porav S, Achim M, De Beer T, Tomuță J (2018) Formulation of long-circulating liposomes and in-line monitoring of the freeze-drying process using a NIR spectroscopy tool. J Pharm Sci 107:139–148. https://doi.org/10.1016/j.xphs.2017.05.024
54. Vallan A, Fissore D, Pisano R, Barresi AA (2021) On the use of temperature measurements as a process analytical technology (PAT) for the monitoring of a pharmaceutical freeze-drying process. Pharmaceutics 15:861. https://doi.org/10.3390/pharmaceutics15030861
55. Carfagna M, Rosa M, Hawe A, Frieß W (2022) Design of freeze-drying cycles: the determination of heat transfer coefficient by using heat flux sensor and MicroFD. Int J Pharm 621:121763. https://doi.org/10.1016/j.ijpharm.2022.121763
56. Leys L, Nuytten G, Lammens J, Van Bockstal PJ, Corver J, Vervaet C, De Beer T (2021) A NIR-Based study of desorption kinetics during continuous spin freeze-drying. Pharmaceutics 13:2168. https://doi.org/10.3390/pharmaceutics13122168
57. Affleck RP, Khamar D, Lowerre KM, Adler N, Cullen S, Yang M, McCoy TR (2021) Near infrared and frequency modulated spectroscopy as non-invasive methods for moisture assessment of freeze-dried biologics. J Pharm Sci 110:3395–3402. https://doi.org/10.1016/j.xphs.2021.06.016
58. Sharma A, Khamar D, Cullen S, Hayden A, Hughes H (2021) Innovative drying technologies for biopharmaceuticals. Int J Pharm 609:121115. https://doi.org/10.1016/j.ijpharm.2021.121115
59. Preskar M, Korasa K, Vrbanec T, Klement D, Vrečer F, Gašperlin M (2021) Applicability of Raman and near-infrared spectroscopy in the monitoring of freeze-drying injectable ibuprofen.

Drug Dev Ind Pharm 47:758–769. https://doi.org/10.1080/03639045.2021.1934864
60. Harguindeguy M, Fissore D (2021) Micro freeze-dryer and infrared-based PAT: novel tools for primary drying design space determination of freeze-drying processes. Pharm Res 38:707–719. https://doi.org/10.1007/s11095-021-03023-x
61. Carfagna M, Rosa M, Lucke M, Hawe A, Frieß W (2020) Heat flux sensor to create a design space for freeze-drying development. Eur J Pharm Biopharm 153:84–94. https://doi.org/10.1016/j.ejpb.2020.05.028
62. Pikal MJ, Shah S, Roy ML, Putman R (1990) The secondary drying stage of freeze drying: drying kinetics as a function of temperature and chamber pressure. Int J Pharm 60:203–207. https://doi.org/10.1016/0378-5173(90)90074-E
63. Harris NS (1989) Modern vacuum practice. McGraw-Hill International (UK) Limited, pp 115–116. ISBN 0-07-707099-2
64. Cook IA, Ward KR (2011) Headspace moisture mapping and the information that can be gained about freeze-dried materials and processes. PDA J Pharm Sci Technol 65:457–467. https://doi.org/10.5731/pdajpst.2011.00760
65. Brülls M, Folestad S, Sparén A, Rasmuson A (2003) In-situ near-infrared spectroscopy monitoring of the lyophilization process. Pharm Res 20:494–499. https://doi.org/10.1023/a:1022680810474
66. Brülls M, Folestad S, Sparén A, Rasmuson A, Salomonsson J (2007) Applying spectral peak area analysis in near-infrared spectroscopy moisture assays. J Pharm Biomed Anal 44:127–136. https://doi.org/10.1016/j.jpba.2007.02.013
67. Duncan DI, Veale JR, Cook I, Ward K (2010) Using laser-based headspace moisture analysis for rapid nondestructive moisture determination of sterile freeze-dried product. White Paper. https://lighthouseinstruments.com/knowledge-center/white-paper-moisture-determination-of-freeze-dried-product/
68. Hedberg SHM, Devi S, Duralliu A, Williams DR (2019) Mechanical behavior and structure of freeze-dried cakes. In: Ward KR, Matejtschuk P (eds) Lyophilization of pharmaceuticals and biologicals: new technologies and approaches. Humana Press, Springer, New York, pp 327–351
69. Haeuser C, Goldbach P, Huwyler J, Friess W, Allmendinger A (2018) Imaging techniques to characterize appearance of freeze-dried products. J Pharm Sci 107:2810–2822. https://doi.org/10.1016/j.xphs.2018.06.025
70. Atotwe-Otoo D, Khan M (2019) Regulatory aspects of freeze-drying. In: Ward KR, Matejtschuk P (eds) Lyophilization of pharmaceuticals and biologicals: new technologies and approaches. Humana Press, Springer, New York, pp 173–192

Chapter 3

Mass Transfer in Cryopreservation at Multiple Scales: Cells, Tissues, and Organs

James D. Benson, Anthony N. Consiglio, Matthew J. Powell-Palm, Rachel M. Thompson, Morgan B. Giers and Adam Z. Higgins

Abstract

The ability to survive cryopreservation almost universally requires the use of exogenous molecules known as cryoprotective agents (CPAs), which are typically used at concentrations exceeding 1 mol/L. In the case of slow freezing, these molecules help mitigate damage associated with formation of extracellular ice. For vitrification methods, CPAs are used at even higher concentrations to completely suppress ice formation, instead promoting the formation of an amorphous glassy state throughout the sample. In either case, effective CPA delivery is critical for survival. However, CPA exposure can itself be damaging, either because of osmotically driven volume changes or inherent CPA toxicity. Therefore, a central challenge in cryopreservation is developing CPA equilibration methods that deliver sufficient CPA for protection without excessive osmotic damage or chemical toxicity. At the heart of this problem is the transport of water and CPA. Thus, mass transfer modeling can provide a foundation for rational design CPA equilibration methods. In this chapter, we review the key physical phenomena affecting transport of water and CPA at the level of cells, tissues, and organs, summarize mass transfer modeling strategies, and highlight opportunities for using these models to design improved CPA equilibration methods.

Keywords Cell, Cell membrane transport, Cryopreservation, Cryoprotectant, Diffusion, Mass transfer, Organ, Permeability, Tissue, Vascular perfusion, Vitrification

1 Mass Transfer in Single Cells

The idea to use mass transfer modeling to understand cellular fates during cryopreservation is attributed to Peter Mazur, whose seminal paper described the relationship between cooling rate and the quantity of water in the cell as a function of temperature [1]. This model allowed the prediction of the likelihood of intracellular ice formation under various cooling protocols and intersected with his conceptual model of cell injury due to cooling. Following Mazur, Liebo and Chu's two-factor hypothesis of cooling injury [2], Mazur proposed that cells should be cooled at a rate that dehydrates them so that no more than two degrees of supercooling occurs. This mathematical

Willem F. Wolkers and Harriëtte Oldenhof (eds.), *Cryopreservation and Freeze-Drying Protocols*, Methods in Molecular Biology, vol. 3060, https://doi.org/10.1007/978-1-0716-5416-3_3,

model proved useful to explain why embryo and oocytes require slower cooling rates than somatic cells. In short, the model supported the idea that these cells and small tissues must be able to lose water at nearly the same rate that the solution loses water to ice.

In this section, we will explore the evolution of this paradigm that now encompasses more sophisticated models of cell injury and cell state that capture the peculiarities of cells exposed to the highly non-physiological conditions of cryopreservation. While the focus of this chapter is on mass transport modeling, the development of these models requires a purpose. In this case, the purpose is to understand (and possibly guide the avoidance of) cell injury during cryopreservation, so we will start with cell injury models to discover the information we may need to model.

1.1 Cell Injury Models

As cryobiologists, we want to understand what damage to cells occurs during cryopreservation. As Mazur observed, during cooling this damage is due to at least two competing factors: the mechanical/physical injury due to ice formation, and the chemical injury due to solution toxicity. In fact, while not directly addressed initially by Mazur, the same two mechanisms of injury were later associated with the introduction of cryoprotective agents(CPAs), which causes both potentially damaging cell volume fluxes and cytotoxicity (c.f. [3]).

To get started, let's define a few variables related to the cell state at a time t. First, the cell volume V typically comprises cell water volume W, the volume of intracellular permeating solutes $\bar{v}_s S$, where the first term is the molar volume of the solute with S intracellular moles, and the volume of everything that does not readily cross the cell membrane V_b. This is often shown as

$$V = W + \bar{v}_s S + V_b. \tag{1}$$

This last term is often called the osmotically inactive volume, and it is usually assumed that the intracellular moles of nonpermeating solutes (e.g., ions, proteins) remains constant over the time scales of single cell CPA equilibration and/or cooling, though as we will discuss, there is some merit to including the permeability of these solutes to possibly explain anomalous behavior.

We commonly define isotonic reference values for two of these variables: V_{iso}, W_{iso}. We then use these reference values to define a normalized volume:

$$v = \frac{V}{V_{iso}} = w + \bar{v}_s s + v_b, \tag{2}$$

where $v = \frac{V}{V_{iso}}$, $w = \frac{W}{V_{iso}}$, $s = \frac{S}{V_{iso}}$, and $v_b = \frac{V_b}{V_{iso}}$.

With the basics defined, we can proceed to modeling cryopreservation injury.

Let's start with the injury due to equilibration with CPAs. First, we can observe that cells undergo osmotic stress during CPA equilibration (see Fig. 1). This osmotic stress can contribute to cell injury in two ways: cells could be catastrophically injured (through some sort of lysis) and cells can be injured over time (perhaps through the exacerbation of concentration stress, the loss of intracellular contents due to volume regulation, etc). To account for the former, the typical injury heuristic is that cells must be maintained within lower and upper osmotic tolerance limits (v_{LOTL} and v_{HOTL}, respectively). This is modeled simply by stating that if the cell volume v is held between upper and lower bounds, i.e., $v_{\mathrm{LOTL}} \leq v \leq v_{\mathrm{HOTL}}$, there should be minimal loss of viability. This classical approach is the most common. However, there has been some literature exploration of the cumulative effects of keeping cells in anisotonic solutions. The first mention that we are aware of in the context of cryobiology is in Liu et al. [4], which was refined by Olver et al. using data from urchin oocytes [5]. The idea of this cumulative model is that it should sum the time that cells spend away from isotonic volume:

$$J_{\mathrm{osm}} = \int_0^{t_f} f(|v(t) - 1|)\, dt, \quad (3)$$

where t_f is the protocol duration, and f is some real valued function of the variable $x = |v(t) - 1|$. In the case of Liu et al., f was the

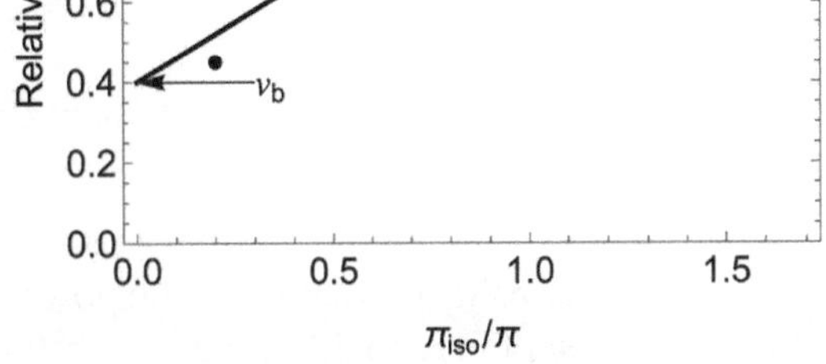

Fig. 1 Typical Boyle van't Hoff plot, showing how the equilibrium cell volume decreases in hypertonic solution and increases in hypotonic solution. Black points represent data from a cell that behaves as an ideal osmometer, and open circles represent data from a cell that may be non-ideal in the hypoosmotic region. Key parameters v_b and v_{iso} are labeled. Typically, w_{iso} is defined as v_{iso}-v_b. The regression line here has the form $v = w_{iso}\pi_{iso}\pi^{-1} + v_b$, noting that $v = V/V_{iso}$

identity function $f(x) = x$, whereas Olver et al. used a power law $f(x) = x^{\beta}$ with an exponent β that could depend on whether the cell was in a shrunken or swollen state. We note that these examples are phenomenological functions that align with heuristics that require considerable experimental data to refine. There is considerable room for improvement in the definition of f as a function of cell type, concentration, temperature, etc.

Next, even in 1971 Meryman [3] acknowledged that glycerol over time was toxic, and now CPA toxicity is considered one of the key obstacles limiting cryopreservation of cells, tissues and organs [6]. It has long been known that CPA toxicity was a function of exposure time and concentration (c.f., Fahy [7]). This leads to the natural definition of a generic "toxicity cost function":

$$J_{\mathrm{tox}} = \int_0^{t_f} g\left(s(t)/w(t)\right)\, dt, \tag{4}$$

where, now, g is some real valued function of the variable $x = s(t)/w(t)$, the intracellular CPA concentration. The simplest implementation of this model is where the toxicity accumulates as a function of time alone, and $g(x) = 1$. This model was implicit in a number of modeling and optimization papers before appearing explicitly in this form in Karlsson et al. [8] and Benson et al. [9]. We note that there is utility in defining the function this way, as it casts the problem in the context of the well-established Optimal Control Theory. Later Benson et al. [10] proposed a model that recognized the concentration dependence, setting $g(x) = x^{\alpha}$. Still later, Davidson et al. [11] include temperature dependence, setting $g(x) = \exp\left[C_0 - \frac{E_{\mathrm{tox}}}{RT}\right] x^{\alpha}$, where the exponential term represented an Arrhenius equation accounting of the toxicity of CPAs. Finally, Olver et al. [5] have used an extensive urchin oocyte dataset to explore this and other similar models, noting that while the Davidson model performed well, a model that uses extracellular instead of intracellular solute concentration performed the best:

$$J_{\mathrm{tox}} = \exp\left[C_0 - \frac{E_{\mathrm{tox}}}{RT}\right] m_e^{\exp\left[\alpha_0 - \frac{E_{\alpha}}{RT}\right]} t_{\mathrm{f}}, \tag{5}$$

where the exponential terms are the Arrhenius models for the proportionality constant and the power law constant, and we note that J_{tox}, m_e, and t_f could be the toxicity for a single step of a multistep protocol, where m_e and t_f would vary per step. As noted above, these are all phenomenological models that align with the heuristic that cell injury depends on concentration, time, and temperature, and identifying a more universal model will require considerably more experiments.

Next, many cells are sensitive to chill injury, which we define as injury due to exposure to subphysiologic temperatures in the absence of ice. Chill injury is most commonly identified in oocytes and embryo, and following the theme from above, was first explicitly modeled by Olver et al. [5], who define two candidate models:

$$J_{\text{temp}} = C_{\text{temp}} \int_0^{t_f} T_{\text{phys}} - T(t)\ dt, \tag{6}$$

and

$$J_{\text{temp}} = C_{\text{temp}} \max T_{\text{phys}} - T(t), \tag{7}$$

where T_{phys} is the lowest temperature at which cells survive without injury. In the case of urchin oocytes, the latter (time independent) model had better fits.

Finally, to complete the picture, we must combine these models. Again, this was implicitly done in many earlier modeling papers in cryobiology, but Olver et al. [5] were the first to formally define the total damage due to CPA equilibration as being the sum of Toxicity, Osmotic, and Temperature injury:

$$J_{\text{tot}} = J_{\text{tox}} + J_{\text{osm}} + J_{\text{temp}}. \tag{8}$$

This complete model, then can be used to rationally capture the damage due to CPA equilibration.

Again, we note that the models defined here and in the literature are simply phenomenological: they capture the heuristics of injury and align well with specific datasets. The utility of these models is in their ability to improve understanding of injury and perhaps guide experimental designs for minimally damaging protocols. This approach was used to great effect, for example, in human sperm cryopreservation [12], which used the simple osmotic tolerance limits and an implicit minimal time protocol. It was also used to show how to equilibrate plated endothelial cells with high concentrations of glycerol [11]. These models, however, phenomenological they are, still depend on the variables of state, that include cell volume and concentration. We will address this in the next section.

1.2 Cell State Models

To understand the modes of injury during cryopreservation, we must start by understanding the physical and chemical state of the cell. To do so we note that cells have semipermeable membranes that allow for the movement of water, the slower movement of some solutes, and the controlled movement of others. The permeability of this barrier then governs the flow of water by osmosis, which causes water to move from low- to high-osmolality compartments. During cryopreservation, these osmotic gradients are caused by the introduction and subsequent removal of cryoprotectants and the increase of the solution osmolality due to the formation of extracellular ice.

Therefore, to understand the cell state—in this case, the quantities of intracellular water and the concentration of solutes as a function of time and temperature during cryopreservation, we must understand water and solute transport in the presence of these changing extracellular environments.

While modeling of membrane mass transport of water and solutes has been in existence in the literature for at least a century [13], cryobiology presents a number of challenges that require special attention. For example, transport modeling in the literature is most often done in near-physiological concentrations (e.g., much less than 1 mol/L for mammalian cells), whereas in cryopreservation, it is not uncommon for CPA concentrations to be well above this threshold before cooling. Upon cooling and in the presence of ice, the concentration of the unfrozen liquid increases to >10 mol/L. Handling this from a modeling point of view requires careful consideration of the solution osmolality and the temperature and concentration dependence of the permeability to water and solutes. In the membrane transport literature, much attention is paid to the contribution of solute transport via selective ion channels or active transport through pumps. While important in near-physiological conditions, the sheer magnitude of the changes during cryopreservation means that these phenomena are relatively minor compared to the movement of water and CPA across the cell membrane. Therefore, while all models are wrong [14], cryobiological modelers have embraced the phenomenological nature of their models noting that they, for the most part, capture the bulk behavior of the cells well enough to estimate whether that cell should be injured due to excessive volume excursions, or to overlong CPA exposures.

As the exposure to high concentrations of permeable solutes is a necessity for cell survival during cryopreservation, and as the process of cooling in the presence of ice (sometimes called slow cooling) requires the movement of water and possibly some solutes to sufficiently dehydrate the cells, it is apparent that to understand cryobiology, one must understand the movement of water and solutes. This movement defines the time dependent quantities $w(t)$ and $s(t)$ that are at the heart of the damage models that we defined above.

1.2.1 Equilibrium Models

The most basic result of the semipermeable membrane is that cells function as osmometers: cells shrink in high concentrations and swell in low concentrations. The relationship between concentration and the ability of cells to shrink and swell is critical for modeling the cell water and solute volumes.

This response is captured by the Boyle van't Hoff relation and its more complicated analogues. While there are other ways to derive the relationship [15, 16], the Boyle van't Hoff relation was first presented as an analogue of the ideal gas law, $PV = nRT$, noting that if V is the volume of water in the cell (here denoted as W), P is the osmotic pressure (usually denoted with Π), and n is the moles of nonpermeating solute (which is assumed constant), we arrive at

$$\Pi W = \text{constant}. \tag{9}$$

This can also be expressed in terms of osmolality π by noting that osmolality is proportional to osmotic pressure. It is convenient to define this in terms of values in isotonic conditions, denoted with subscripts iso as above, yielding $\pi W = \pi_{\text{iso}} W_{\text{iso}}$, which is usually rearranged to be

$$W = \frac{W_{\text{iso}}\pi_{\text{iso}}}{\pi}. \tag{10}$$

Finally, this model gets inserted in the total volume model (Eq. 1) to yield

$$V = \frac{W_{\text{iso}}\pi_{\text{iso}}}{\pi} + \bar{v}_s S + V_b. \tag{11}$$

To inform this model a Boyle van't Hoff plot is usually made, where equilibrium total cell volume is plotted against inverse osmolality, resulting in data that allow estimation of both the osmotically inactive volume and the isosmotic cell volume. A representative Boyle van't Hoff plot is shown in Fig. 1.

The equilibrium volume behavior of cells does not always fall along this linear regression. As described in a comprehensive meta-analysis, Olver et al. showed that many cells systematically deviate from the Boyle van't Hoff model, especially in the hypotonic region (to the right of 1.0 in Fig. 1) [17]. There are a number of physiological explanations for this behavior that include the leakage of osmolytes during swelling and the resistance to swelling via cytoskeletal or extracellular turgor. In the former, the "n is constant" assumption of the Boyle van't Hoff model is broken, and in the latter the "$P = \Pi$" assumption is broken. Dealing with these special cases and deciding the cause of the deviation is not always straightforward, so the reader is encouraged to refer to other in-depth manuscripts [17, 18]. From a practical modeling point of view, the mechanism is less important than the prediction: at the end of the day, the damage models introduced above rely on the prediction of the intracellular water volume and moles of solute to arrive at the intracellular concentration, s/w. If the model for w captures the data (e.g., using a piecewise linear or quadratic function) then it can be used to inform the damage models, at the risk of neglecting the additional information that mechanistic models provide.

Finally, it is worth noting that a great many cells equilibrium volume response is as defined above, where a cell will reach and then stay at an equilibrium volume for a given nonpermeating solute extracellular osmolality. However, there are a number of cells that undergo volume regulation. This likely evolved as a protective mechanism and comes in two flavors: Regulatory Volume Increase (RVI) and Regulatory Volume Decrease (RVD), applicable in the hyperosmotic and hypoosmotic regions of the Boyle van't Hoff plot above. This behavior confounds simple estimates of the Boyle van't Hoff parameters and requires careful treatment [17, 19–22]. However, again, from a practical point of view, cells exposed to the multiosmolal concentration gradients and the (relatively) short time periods experienced during CPA equilibration are not likely able to actively transport osmolytes in sufficient quantities to significantly affect the intracellular CPA concentrations relevant to the damage models above. This is certainly an area of experimentation and modeling in cryobiology that should be more carefully explored.

1.2.2 Osmolality and Chemical Potential

The movement of molecules in liquids is caused by the gradient of constituent molecule chemical potentials. In the case of water, the chemical potential is $\mu_w = \mu_w^0 - RT\pi(\boldsymbol{m})$, where μ_w^0 is the chemical potential of pure water at standard temperature and pressure, R is the gas constant, and T is the absolute temperature, and, as defined above, π is the solution osmolality, which is a real valued function of the vector $\boldsymbol{m}$ of molalities (e.g., $\boldsymbol{m} = (m_1, m_2, \ldots, m_n)^T$ for solutes 1… n, respectively). In the case of solutes, the chemical potential of the ith solute is $\mu_i(\boldsymbol{m}) = RT(g(\boldsymbol{m}) + \Psi_i^*)$, where Ψ_i^* is a function of temperature and pressure and g is a real valued function of $\boldsymbol{m}$.

There are large bodies of literature discussing the functional forms of π and g, but in the field of cryobiology, the most common implementation is the application of the osmotic virial equation, which uses polynomial expansions with specific mixture rules in the case of multiple solutes [23]. In particular for osmolality, the osmotic virial equation is

$$\begin{aligned}\pi(\boldsymbol{m}) = \sum_{i=1}^{n} m_i + \sum_{i,j=1}^{n} \frac{B_i + B_j}{2} m_i m_j \\ + \sum_{i,j,k=1}^{n} (C_i C_j C_k)^{1/3} m_i m_j m_k,\end{aligned} \tag{12}$$

where B_i and C_i are constant "osmotic virial coefficients." The key advantage of this formulation is that these constants may be determined independently of the mixture itself, and as such allows for their determination from a simple binary (water plus solute) phase diagram, which maps the amount of freezing point depression, T_{FPD}, as a function of solute concentration, via the relationship $T_{FPD} = 1.86\pi$. Elliott et al. argue that for salts, the quadratic and cubic coefficients are effectively zero, that for most small molecule CPAs, the cubic coefficients are effectively zero, and that there is merit in including the cubic terms for large molecules such as hemoglobin [23].

The chemical potential of the *i*th solute on the other hand takes on the form

$$\mu_i(\boldsymbol{m}) = RT\left(\Psi_i^* + \ln m_i + \sum_{j=1}^{n} (B_i + B_j)\, m_j\right), \tag{13}$$

where Ψ_i^* and B_i are defined above.

In practice most modeling in single cell cryobiology simplifies the above expressions, but the nature of this simplification is best presented in the context of trans-membrane transport, discussed below.

1.2.3 Membrane Transport Models

As mentioned above, transport of water and solutes across cell membranes is driven by the gradient of the chemical potential. This spatial gradient reduces to a flux that is proportionate to the transmembrane difference of chemical potential:

$$\frac{dn_i}{dt} = P_{n_i} A \left(\mu_{n_i}^{\mathrm{e}} - \mu_{n_i}^{\mathrm{i}}\right), \tag{14}$$

where n_i is the moles of the ith species, P_{n_i} is the membrane permeability to that species, A is the cell surface area, and superscripts *e* and *i* indicate extracellular and intracellular quantities, respectively. We note that in the case of water, the equation is most often written in terms of the volume of intracellular water (rather than moles) using osmolality as the driving force (rather than chemical potential):

$$\frac{dW}{dt} = \bar{v}_w P_w A \left(\mu_w^{\mathrm{e}} - \mu_w^{\mathrm{i}}\right) = -L_p ART\left(\pi^{\mathrm{e}} - \pi^{\mathrm{i}}\right), \tag{15}$$

where L_p is the hydraulic conductivity of the cell membrane, and we note that it is the norm to include the negative sign in front of the term instead of reversing the extracellular and intracellular osmolalities.

If the parameters from the chemical potential equations above are known, then these forms can be plugged in directly. Often, however, the chemical potential functions are simplified. In the case of osmolality, the most common simplification is to assume the virial coefficients are zero, which reduces the Eq. 15 to

$$\frac{dW}{dt} = -L_p ART \sum_{j=1}^{n} \left(m_j^{\mathrm{e}} - m_j^{\mathrm{i}}\right). \tag{16}$$

In the case of membrane transport of the jth solute, we have

$$\frac{dS_j}{dt} = P_{s_j} A\,(\mu_j^{\mathrm{e}} - \mu_j^{\mathrm{i}}). \tag{17}$$

The most common simplification of this difference in chemical potentials is to replace it with $(m_j^{\mathrm{e}} - m_j^{\mathrm{i}})$. Please see [15] for a thorough discussion of the relative error associated with this common simplification.

Next, we note that the system of differential equations that describe the volume flux of water w and molar flux of solutes s is not closed. To do so, we note that $m_j^{\mathrm{i}} = S_j / W\rho_w$, where ρ_w is the density of water, and we define $m_{\mathrm{N}}^{\mathrm{i}} = N / W\rho_w$ to be the molality of intracellular nonpermeating solutes. Then, we get the system

$$\frac{dW}{dt} = -L_p ART\left(m_{\mathrm{N}}^{\mathrm{e}} + \sum_{j=1}^{n} m_j^{\mathrm{e}} - \frac{N + \sum_{j=1}^{n} S_j}{W\rho_w}\right),$$

$$\frac{dS_1}{dt} = P_{s_1} A\left(m_1^{\mathrm{e}} - \frac{S_1}{W\rho_w}\right),$$

$$\frac{dS_2}{dt} = P_{s_2} A\left(m_2^{\mathrm{e}} - \frac{S_2}{W\rho_w}\right),$$

...

$$\frac{dS_n}{dt} = P_{s_n} A\left(m_n^{\mathrm{e}} - \frac{S_n}{W\rho_w}\right). \tag{18}$$

This system is then coupled with the equations relating the cell water volume and moles of solute to total cell volume, which allow for the fitting of volume vs time experimental data to determine the parameters L_{p} and P_{s_j} and for the implementation of the damage models presented at the beginning of this section.

It is worth noting that the parameters L_{p} and P_{s_j} are concentration and temperature dependent, with concentration dependence roughly following Michaelis–Menten kinetics [24, 25] and temperature dependence following the Arrhenius model (described above for the accumulation of cytotoxic injury). For complete treatments of these, see [15].

Finally, solving this system of differential equations is usually straightforward using modern computational packages, but it is worth noting that the system becomes stiff when the extracellular concentrations are high enough to drive W to near zero. In these cases, stiff solvers may be used, or one of the methods introduced by Benson may be used to provide faster and more robust solutions [26].

For the commonly encountered condition where cells are exposed to a single permeating solute, Eq. 18 is usually reduced to

$$\frac{dW}{dt} = -L_pART\left(m_N^e + m_s^e + \frac{N+S}{W\rho_w}\right),$$

$$\frac{dS}{dt} = P_sA\left(m_s^e - \frac{S}{W\rho_w}\right). \tag{19}$$

This results in the standard "shrink-swell" curve associated with the relative difference in permeability to water and CPA, where the cell initially loses water, and then gradually regains this water as the CPA enters the cell (Fig. 2).

2 Mass Transfer in Tissues

Effective cryopreservation of tissue requires uniform penetration of CPAs throughout the sample—from the surface to the core—while managing exposure conditions to prevent osmotic stress, mechanical injury, and toxicity associated with prolonged contact with high CPA concentrations. In the following section, we explore the key factors influencing the design of CPA equilibration strategies for tissues. We

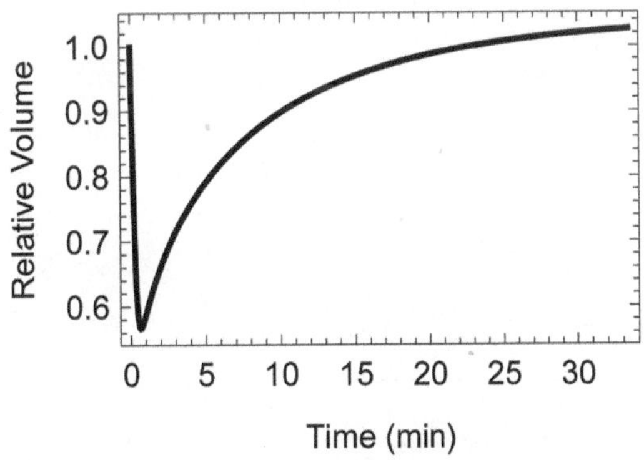

Fig. 2 Plot of relative rhesus macaque oocyte volume as a function of time after exposure to 1 mol/L ethylene glycol. Parameters were obtained from [8] and used with Eq. 19. The CPA enters the cell from the beginning of the exposure, but the initial shrink occurs until the osmolality of the intracellular and extracellular compartments are equal. At this minimum, the CPA will continue to enter the cell, increasing its intracellular osmolality and causing water to enter through osmosis

examine the typical timescales involved, review common modeling approaches for mass transport in tissues, and discuss how these models can be integrated with damage models to mathematically optimize CPA equilibration protocols.

2.1 Phenomena Relevant to CPA Equilibration in Tissues

Exposure of tissue to CPA triggers several processes, including transport within the interstitial space, movement of CPA and water across cell membranes accompanied by changes in cell volume, and shifts in the overall size of the tissue. The relative effects of these phenomena are highly tissue dependent, as the extracellular matrix composition and cell density can vary widely for different tissues (e.g., the volume occupied by cells can range from 1%–2% for intervertebral discs to ~70% for pancreatic islets). Understanding these processes is essential for developing effective strategies to achieve uniform and nondamaging CPA equilibration.

2.1.1 Transport in the Interstitial Space

CPA transport resulting from a change in external concentration is often analyzed using Fick's laws, which express the rate of diffusion in terms of the diffusion coefficient *D*. Figure 3 (left) summarizes diffusion coefficients from the literature for the four most common CPAs—DMSO, ethylene glycol, propylene glycol, and glycerol—in free solution (solid lines) and in various types of tissue (symbols). Overall, the diffusion coefficients are similar for all four CPAs. The free solution diffusion coefficient is highest at low CPA concentration and decreases as the CPA concentration increases. The apparent diffusion coefficient in tissue is expected to be lower than the free solution diffusion coefficient because the cells and extracellular matrix create obstacles in the diffusion path. Apparent diffusion coefficients in tissues are highly scattered, and in some cases, they exceed the free solution diffusion coefficient. This could be due to experimental errors or to deviations between the actual transport

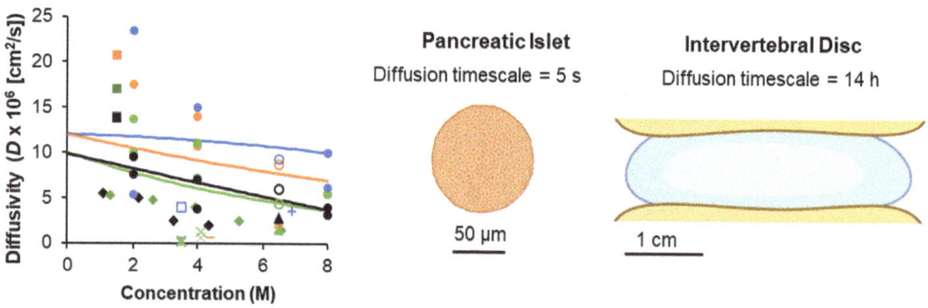

Fig. 3 CPA diffusion in tissue. Left: Diffusion coefficient of DMSO (blue), ethylene glycol (orange), propylene glycol (green), and glycerol (black) in free solution (lines [28–30]) and in various types of tissue: closed squares [31]; open squares [32]; closed circles [33]; open circles [34]; triangles [35]; diamonds [36]; × [37]; – [38]; + [39]; * [40]. Right: Timescale for diffusion in a pancreatic islet and intervertebral disc estimated using a diffusion coefficient $D = 5 \times 10^{-6}$ cm²/s. Due to the much larger size of the intervertebral disc, diffusion is expected to take approximately 10,000 times longer, underscoring the difficulty of CPA loading in large tissues

processes occurring in the tissue and the assumptions of Fick's laws. Taken together, CPA diffusion in tissues can be characterized by a diffusion coefficient of approximately $D = 5\times10^{-6}\,\text{cm}^2/\text{s}$.

Figure 3 (right) illustrates how the time required for CPA diffusion varies for tissues of different sizes. The diffusion timescale t_d can be estimated from the diffusion coefficient and the diffusion distance L: $t_d = L^2/D$. This can be thought of as the time it takes for a meaningful amount of solute with diffusion coefficient D to diffuse a distance L. Human pancreatic islets have a diameter of ~100 μm, which corresponds to a diffusion distance of 50 μm. This results in a diffusion timescale of 5 s. The intervertebral disc is the largest avascular structure in the body. It has an approximately cylindrical shape, with a diameter of 4 cm and a thickness of 1 cm [27]. The relevant diffusion distance is 0.5 cm, resulting in a timescale of 14 h, about 10,000 times longer than the time required for diffusion in a pancreatic islet. For a structure as large as the intervertebral disc, it is especially challenging (if not impossible) to load CPA by diffusion from the exterior without excessive toxicity to the cells toward the periphery due to prolonged CPA exposure.

2.1.2 Transport Across Cell Membranes

Changing the CPA concentration at the tissue surface causes transport in the interstitial space, which alters the chemical composition outside the cells within the tissue. These changes in extracellular composition drive the movement of water and CPA across the cell membrane, leading to changes in cell volume (see Sect. 1.2.3). Figure 4 illustrates the effects of CPA diffusion in the interstitial space on cell volume in a hypothetical tissue. At the surface of the tissue, the concentration changes abruptly, but the concentration in the center changes much more slowly. As a result, a cell located

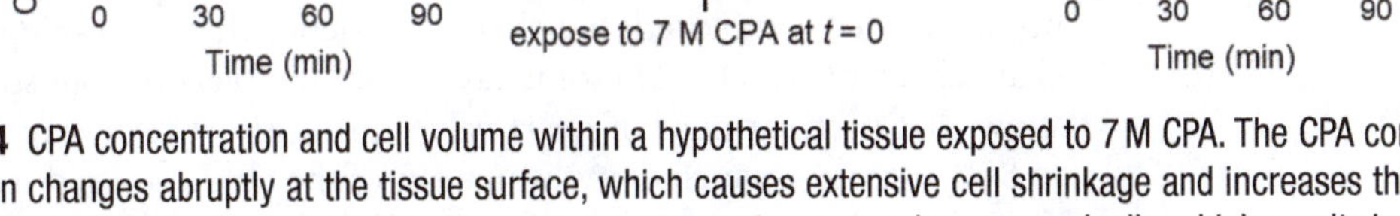

Fig. 4 CPA concentration and cell volume within a hypothetical tissue exposed to 7 M CPA. The CPA concentration changes abruptly at the tissue surface, which causes extensive cell shrinkage and increases the risk of osmotic damage. In contrast, the CPA concentration at the center changes gradually, which results in more modest cell volume changes. CPA concentrations were predicted by numerically solving the diffusion equation (see Eq. 20 below) for a 0.5 cm thick tissue slab with $D = 5\times10^{-6}\,\text{cm}^2/\text{s}$. Relative osmotically active cell volumes were predicted using the two-parameter membrane transport model (System 19) with $L_pA/W_{iso} = 2\times10^{-8}\,\text{Pa}^{-1}\,\text{s}^{-1}$ and $P_sA/W_{iso} = 0.01\,\text{s}^{-1}$

near the tissue surface is expected to experience more extreme cell volume changes, and thus, is more susceptible to osmotic damage.

2.1.3 Tissue Size Changes

Like isolated cells, tissues change overall size when exposed to CPA. Increasing the CPA concentration at the tissue surface drives movement of both water and CPA through the interstitial space but in opposite directions. CPA moves down its concentration gradient into the tissue, whereas water moves out of the tissue towards the relatively high osmolarity at the tissue surface. Because water molecules are relatively small, they are less restricted in their movement through the interstitial space, which typically leads to faster water loss than CPA entry. This causes the tissue to shrink in the initial period after CPA exposure. Tissue shrinkage after CPA exposure has been observed for various tissue types, including articular cartilage [41], cornea [42], heart valve [32], ovarian tissue [33, 34], and pancreatic islets [43]. The extent of shrinkage varies, but it can be as high as 40% [33, 43].

The initial period of tissue shrinkage is followed by reswelling, resulting in a shrink-swell response like that observed for isolated cells. In cells, shrinkage from water loss continues until the osmotic gradient is dissipated and the cell reaches a minimum volume. At this point, further CPA influx causes the cell to reswell. The driving force for this CPA influx arises in part due to trapped intracellular solutes that become concentrated upon cell shrinkage, allowing osmotic equilibrium to be established at a relatively low intracellular CPA concentration. This maintains a concentration difference that drives CPA influx. A similar phenomenon occurs in the extracellular component of tissues because they contain "fixed charges." These arise from polyanionic components in the extracellular matrix—primarily glycosaminoglycans—which attract and bind positively charged counterions. When a tissue shrinks, these fixed charges become more concentrated, much like the concentration of intracellular solutes during cell shrinkage. This maintains a driving force for CPA influx that results in tissue reswelling. The rate of reswelling depends on the fixed charge density, percent volume occupied by cells, and mechanical properties of the tissue, as well as its size. Pancreatic islets reswell over approximately 10 min [43], but ovarian tissue can take several hours to reswell to its original volume [34].

The impact of tissue size changes on CPA equilibration has not been widely explored, but there is evidence that leveraging size changes can be beneficial in some cases. For example, previous studies show that pretreatment of heart valves using short term exposure to 80% sucrose—which draws water out of the tissue and reduces its size—improves preservation of extracellular matrix structure after freeze drying and rehydration [44, 45]. We recently demonstrated the potential for improving CPA equilibration by manipulating tissue size through application of a mechanical force [46]. We used mechanical compression followed by free swelling to significantly accelerate CPA uptake in the intervertebral disc, dramatically improving cell viability after cryopreservation compared to samples loaded with CPA by diffusion alone. This swelling-mediated CPA loading process is illustrated in Fig. 5.

2.2 Models of Mass Transfer in Tissues

The most common approach for modeling mass transfer in a tissue exposed to CPA at its surface is Fick's second law of diffusion:

$$\frac{\partial c_i}{\partial t} = D_i \nabla^2 c_i \quad , \tag{20}$$

where c_i and D_i are the CPA concentration and diffusivity, respectively. This equation assumes diffusive transport through a tissue of constant size, and it neglects the effects of cell membrane transport on the local concentration in the interstitial space. It also uses the concentration gradient as the driving force for diffusion, which assumes an ideal and dilute solution. Despite these limitations, Fick's law is widely used to predict CPA transport in tissues [33–35, 37, 38, 47, 48] because of its relative simplicity and the availability of analytical solutions for various scenarios [49].

To address the limitations of Fick's law, multiple groups have developed alternative models of CPA transport in tissues. Several groups have developed mathematical models that account for the effects of cell membrane transport on interstitial diffusion [31, 50–52]. While these models offer more realistic representations of CPA transport, they still assume constant tissue volume and ideal, dilute solutions. Xu and Cui used the Maxwell-Stefan equations to predict diffusion in nondilute, nonideal mixtures [53]. However, their model does not account for the effects of cell membrane transport and assumes a constant tissue volume. Models for pancreatic islets that account for size changes have been reported by multiple

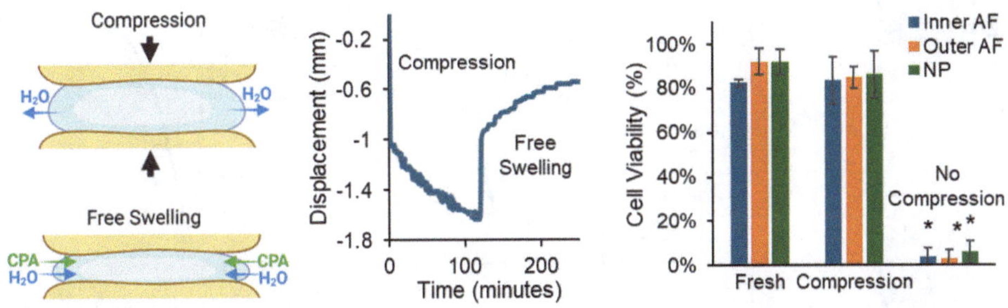

Fig. 5 Accelerating CPA loading in the intervertebral disc using mechanical compression and free swelling. Left: Compression causes the disc to contract, which expels water. After removing the compressive force, the disc expands, drawing in water and CPA. Middle: Disc height changes during compression and free swelling for a bovine intervertebral disc. Right: Compression-mediated CPA loading dramatically improved cell viability after cryopreservation of bovine intervertebral discs compared to loading using diffusion alone [46]

groups [54, 55], but these models do not consider tissue mechanical properties, and they assume ideal or dilute solutions. Abazari et al. introduced a biomechanical transport model for articular cartilage, which accounts for interstitial transport, tissue mechanical properties, and nondilute solutions [41]. However, the model does not account for cell membrane transport in the continuity equation, which may not be a reasonable assumption for cell dense tissues.

We recently developed a general model of CPA transport in tissues that addresses the major limitations of Fick's law [56]. In particular, the model accounts for nondilute solutions, interstitial transport, cell membrane transport, tissue mechanical properties, and tissue size changes. Figure 6 illustrates the modeling approach, which predicts movement of water, CPA, and salt through the interstitial space, movement of water and CPA across cell membranes, and tissue size changes due to movement of solids. The model enables prediction of the time-varying CPA concentration within the tissue, as well as changes in tissue size, yielding predictions consistent with experimental data for diverse tissue types, including pancreatic islets and articular cartilage.

2.3 Optimization of CPA Equilibration in Tissues

To optimize CPA equilibration, mass transfer predictions must be combined with models of cell damage. In general, damage is divided into two main categories: osmotic (or mechanical) damage due to size changes and chemical toxicity. For vitrification methods, the goal is to achieve a vitrifiable CPA concentration throughout the tissue without incurring excessive osmotic/mechanical damage or toxicity.

Several previous studies have used mass transfer modeling to optimize CPA equilibration for tissue vitrification. In two recent studies, Fick's law was used to predict mass transfer in articular cartilage to facilitate design of CPA equilibration methods [57, 58].

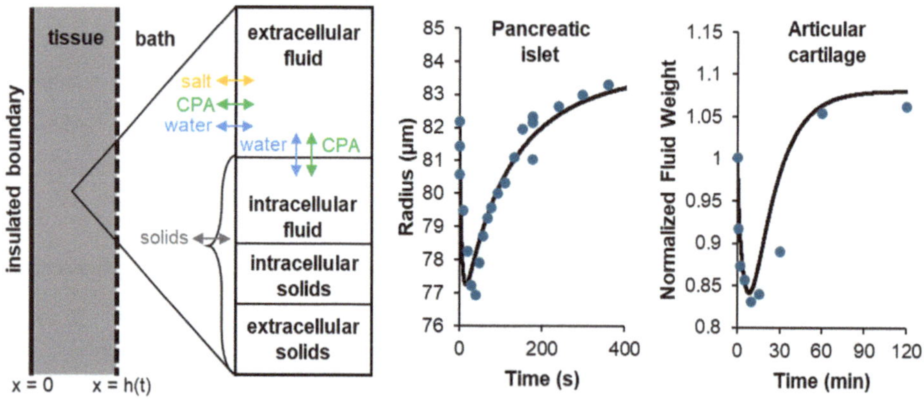

Fig. 6 General model for CPA transport in tissues that accounts for movement of water, CPA, and salt through the interstitial space, movement of water and CPA across cell membranes, and tissue size changes associated with these processes. The model enables prediction of size changes after CPA exposure in cell-dense tissues like pancreatic islets, as well as low cell density tissues like articular cartilage [56]

The predicted CPA concentration was combined with a model of vitrifiability and conceptual reasoning about osmotic damage and toxicity to optimize a multistep CPA equilibration method, and this method was demonstrated to yield high chondrocyte viability after vitrification [58]. Benson et al. used Fick's law with a toxicity cost function to identify minimally toxic CPA equilibration methods for three types of tissue [37]. To minimize toxicity and avoid osmotic damage, the time-varying CPA concentration within the tissue was used to calculate the CPA toxicity cost function, and cell volume predictions for cells on the tissue surface were used to evaluate osmotic damage. The optimal method avoids excessive cell volume changes and achieves a specified minimum CPA concentration throughout the tissue with minimal toxicity.

To our knowledge, a biomechanical model of CPA transport that can predict tissue volume changes—such as the models presented by Abazara et al. [41] and Warner et al. [56]—has not been used for optimization of CPA equilibration methods. This represents an exciting opportunity for future studies for the following reasons: (1) volume changes during CPA exposure are widely observed across numerous tissue types [32–34, 41–43]; (2) these tissue volume changes are qualitatively similar to those observed for isolated cells; (3) toxicity-optimized methods for isolated cells leverage cell size changes to reduce toxicity [10, 11, 59], and it may be possible to adapt these methods to tissues; (4) recent evidence shows that inducing tissue shrinkage by osmotically removing water can improve outcomes after heart valve freeze drying [44, 45]; and (5) we recently reported the potential for accelerating CPA loading by manipulating tissue size using mechanical compression and free swelling [46].

3 Mass Transfer in Organs

In the case of organs, CPA is typically delivered by perfusion through the vasculature. The goal is to achieve a high enough CPA concentration for vitrification without damaging the organ mechanically (via pressure, shear stress, or osmotic effects) and without excessive chemical toxicity. In this section, we give an overview of phenomena relevant to CPA equilibration in organs, examine mathematical models of organ perfusion, and explore opportunities for optimization of CPA delivery by vascular perfusion.

3.1 Phenomena Relevant to CPA Equilibration in Organs

The delivery of CPA into an organ via vascular perfusion depends on several factors, including perfusion pressure, the composition of the perfusate, vascular resistance, and the permeability of the vascular endothelium. CPA perfusion not only affects the CPA concentration in the organ, it also triggers changes in organ size and vascular resistance. To design CPA equilibration procedures for organs, it is helpful to understand these phenomena.

3.1.1 Relationship Between Pressure, Flow, and Vascular Resistance

Fluid flow through the vasculature is often expressed in terms of the perfusion resistance R_p:

$$Q = \frac{\Delta P}{R_p}, \tag{21}$$

where Q is the flow rate and ΔP is the pressure drop between the artery and vein. The perfusion resistance depends on the geometry of the vasculature (e.g., the number of capillaries open for perfusion), as well as the viscosity of the perfusate. Viscosity can vary widely for different CPAs, their concentrations, and the perfusion temperature.

The vascular structural resistance, R_v, is a measure of resistance to flow that is dependent only on the geometry of the vasculature. It is obtained by modifying Eq. 21 to account for the effects of viscosity:

$$Q = \frac{\Delta P}{\mu R_v}. \tag{22}$$

The vascular resistance of human organs can be estimated from typical values of arterial pressure, venous pressure, blood perfusion, and blood viscosity. The resulting vascular resistance values range from about 0.001 mmHg/(mL/min cP) for the lung to about 0.07 mmHg/(mL/min cP) for the heart. Vascular resistance also depends on the size of the organ. A kidney with a mass of 160 g has a vascular resistance of about 0.05 mmHg/(mL/min cP), but a kidney with 2× lower mass is expected to have 2× higher vascular resistance [60].

Organ cryopreservation involves removal of the organ from the donor and possible hypothermic storage before ex vivo perfusion to load CPA. Ischemia during this process can cause vascular resistance to increase. For example, 24 h cold ischemia has been observed to increase vascular resistance in pig kidneys by more than 4× [61]. Ischemia has been linked to swelling of endothelial cells and other cells surrounding the capillaries, which may increase vascular resistance by obstructing the capillary lumen [62]. CPA perfusion also alters vascular resistance, typically causing it to decrease during CPA loading [63] and increase during CPA removal [64]. This has been attributed to volume changes in cells and/or interstitial space surrounding the capillaries, leading to changes in the internal diameter of the capillaries [65, 66].

3.1.2 Transport Across the Capillary Endothelium

Transport across the capillary endothelium occurs either through gaps between adjacent endothelial cells or by passage through the endothelial cell membranes. The permeability of capillaries varies by organ—for example, the brain has low permeability due to the blood–brain barrier, while the lungs have more permeable capillaries with larger intercellular gaps. CPA perfusion creates gradients that drive movement of both water and CPA between the capillaries and the surrounding tissue. This can cause volume shifts that affect overall organ size and vascular resistance.

Another factor during ex vivo perfusion is water balance in the capillaries. Under normal physiological conditions, blood proteins create an osmotic driving force that counteracts the relatively high hydrostatic pressure in the capillaries. These proteins act as oncotic agents that help prevent excessive water leakage into surrounding tissues. During ex vivo perfusion, oncotic agents are often added to replicate this effect. Larger molecules that block ice or promote the formation of a glass—which are commonly used in vitrification solutions—can serve the dual role of also acting as oncotic agents [67].

3.1.3 Organ Size Changes

Organ size changes are commonly observed during perfusion with solutions that are osmotically imbalanced relative to the surrounding tissue, including CPA solutions and other hypertonic formulations. Size changes have been reported in a wide variety of organs, such as the kidney [66, 68], heart [69, 70], lung [71], brain [72], hindlimb [73, 74], and uterus [75]. In our recent work, we observed a mass change of approximately 5% in pig kidneys perfused with a 10% CPA solution [66]. Interestingly, this effect was absent in kidneys that experienced prolonged ischemia prior to perfusion, suggesting that ischemic damage to the endothelium may prevent the size change response. In contrast, much larger size changes—up to 50%—have been documented in the brain [72]. Notably, these size changes can be suppressed by pretreatment with surfactants that disrupt the blood–brain barrier [76]. Collectively, these findings support the idea that organ size changes during CPA perfusion result from movement of water and CPA across an intact capillary endothelium.

3.2 Models of Mass Transfer in Organs

Figure 7 illustrates three general strategies for modeling mass transfer during perfusion of an organ with CPA, in order of increasing complexity.

Let us first consider a simple model representing the limiting case of instantaneous equilibration of the perfusate with the surrounding tissue (Fig. 7A). This assumes a single fluid compartment in the organ with spatially uniform composition. It represents the fastest possible change in CPA concentration within the organ. For this limiting case, a material balance results in:

$$\frac{d(CV)}{dt} = QC_a - QC, \tag{23}$$

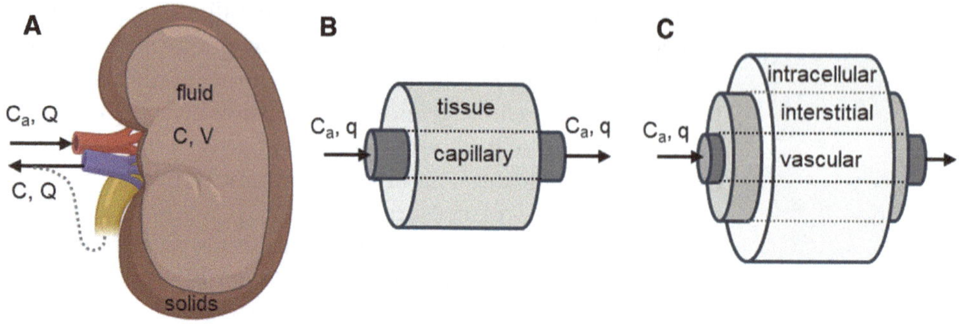

Fig. 7 Strategies for modeling mass transfer during perfusion of an organ with CPA. (**A**) Simple modeling approach that assumes a single well-mixed fluid compartment within the organ. This represents the fastest possible CPA delivery, assuming negligible resistance to mass transfer between the vasculature and the surrounding tissue. (**B**) Krogh cylinder approach, which divides the organ into multiple identical units comprising a capillary surrounded by a cylinder of tissue, and models transport between these two fluid compartments [63, 65, 77]. (**C**) More complex modeling approach that considers three fluid compartments—vascular, interstitial, and intracellular—and models transport between these compartments while also considering mechanical effects associated with organ size changes [78, 79]

where C is the CPA concentration in the organ, V is the osmotically active volume of the organ, Q is the perfusion flow rate (assumed identical at the arterial and venous ends), and C_a is the CPA concentration in the perfusate at the arterial end. If V is assumed constant and removed from the derivative, Eq. 23 can be recast as

$$\frac{dC}{dt} = \frac{1}{\tau}(C_a - C), \tag{24}$$

wherein we define the ratio of the osmotically active volume of the organ and the perfusion flow rate as a characteristic flow constant $\tau = \frac{V}{Q}$. This is the key parameter that dictates how the CPA concentration in the organ changes with time.

Current CPA loading protocols typically involve an initial ramp in CPA concentration, followed by a step change to the final vitrification concentration at the end of CPA loading. This final step is arguably the most challenging due to the competing needs of minimizing CPA toxicity and delivering sufficient CPA to prevent ice formation. For a step change in perfusate concentration, Eq. 24 can be solved, resulting in:

$$\frac{C - C_0}{C_a - C_0} = 1 - \exp\left(-\frac{t}{\tau}\right) = 1 - \exp\left(-\frac{\Delta P}{\mu R_v V} t\right), \tag{25}$$

where C_0 is the CPA concentration in the organ just prior to the step change, and we have substituted Eq. 22 for Q assuming ΔP, μ, and R_v are constant. This equation allows estimation of the extent of

CPA loading as a function of time for the limiting case illustrated in Fig. 7A based on three key variables: ΔP, which can by adjusted by altering arterial pressure; μ, which can be modified by changing CPA composition or temperature; and R_v, which can potentially be altered by vasodilators or osmotic agents that affect blood vessel lumen diameter.

Figure 8 demonstrates the importance of considering perfusate viscosity during CPA loading. Viscosity increases with CPA concentration, as illustrated in Fig. 8A for four commonly used CPAs. This rise is substantial and contributes to significantly reduced CPA delivery, especially as concentrations approach those required for vitrification. Viscosity is also strongly dependent on temperature. As shown in Fig. 8B, the viscosity of M22 increases significantly at lower temperatures. This highlights the importance of balancing the benefits of colder temperatures for reducing toxicity with the drawbacks of slower perfusion due to increased viscosity. Figure 8C presents the best-case scenario predictions for CPA transport into a human kidney at 0°C and −20°C, assuming instantaneous equilibration between the perfusate and tissue. Even under these idealized conditions, it is expected to take over 20 min to reach the full M22 concentration when perfusing at −20°C. This analysis is consistent with the challenges observed by Fahy et al. during perfusion of rabbit kidneys with M22 at −22°C [80]. In contrast, CPA delivery is predicted to be much faster at 0°C, but this comes at the cost of higher toxicity.

The simple single-compartment, well-mixed model in Fig. 7A can also be used to investigate complete perfusion loading protocols.

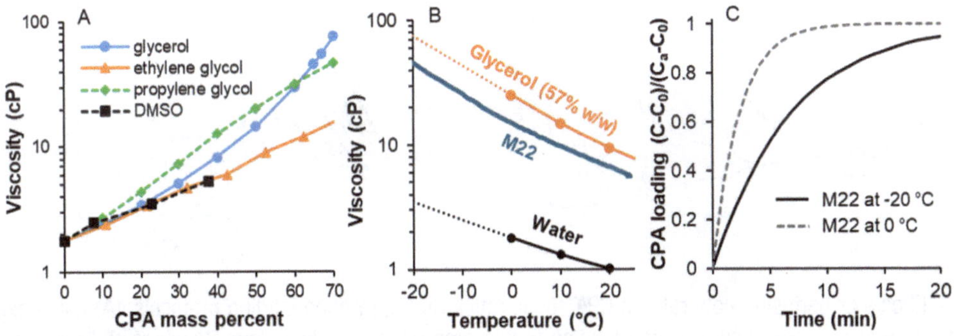

Fig. 8 Effect of viscosity on CPA delivery into an organ. (**A**) Viscosity rises exponentially with increasing CPA concentration, which is expected to slow CPA delivery, particularly as concentrations approach the >50% w/w required for vitrification. Viscosity data is at 0°C [81, 82]. (**B**) CPA solutions become substantially more viscous at lower temperatures, revealing a key tradeoff: while low temperatures can reduce toxicity, they also hinder CPA perfusion. M22 viscosity was measured using a ViscoLab 4000 viscometer from Cambridge Viscosity. Glycerol and water viscosity is from [81]. Dotted lines are extrapolations. (**C**) CPA delivery into a human kidney predicted using Eq. 25 with $\Delta P = 80$ mmHg, $V = 120$ mL, and $R_v = 0.1$ mmHg/(mL/min cP) and μ equal to the M22 viscosity at the given temperature. It takes >20 min to achieve the full M22 concentration in the kidney when perfusion is carried out at −20°C, even in the best-case scenario of instantaneous equilibration between the perfusate and the surrounding tissue

Figure 9A shows a recently reported protocol for perfusing rabbit kidneys with M22 [83]. The flow rate and arterial CPA concentration were measured throughout the perfusion process, which allows the concentration in the kidney to be estimated by numerically integrating Eq. 24. The resulting predictions show that CPA concentration in the organ lags slightly behind the arterial concentration, especially during the two step-changes in concentration near the end of CPA loading. The trends in the predicted lag, expressed as $C_a - C$, are consistent with the measured difference between the arterial and venous CPA concentrations, highlighting the potential of this simple model to facilitate interpretation of organ perfusion data.

While most organ perfusion procedures use constant perfusion pressure, it is also illustrative to consider the alternative approach of using constant perfusion flow rate. Let us consider the loading profile shown in black in Fig. 9B. Here, the loading concentration entering the artery is ramped linearly over an initial period t_1, and then held at the target full strength concentration $C_{\max}$ until an endpoint t_2.

For the first period $0 < t \leq t_1$, we may thus define our arterial concentration as $C_a = C_{\max}\frac{t}{t_1}$, and for the second period $t_1 < t \leq t_2$, $C_a = C_{\max}$. Substituting these definitions of C_a into Eq. 24 and integrating, we find:

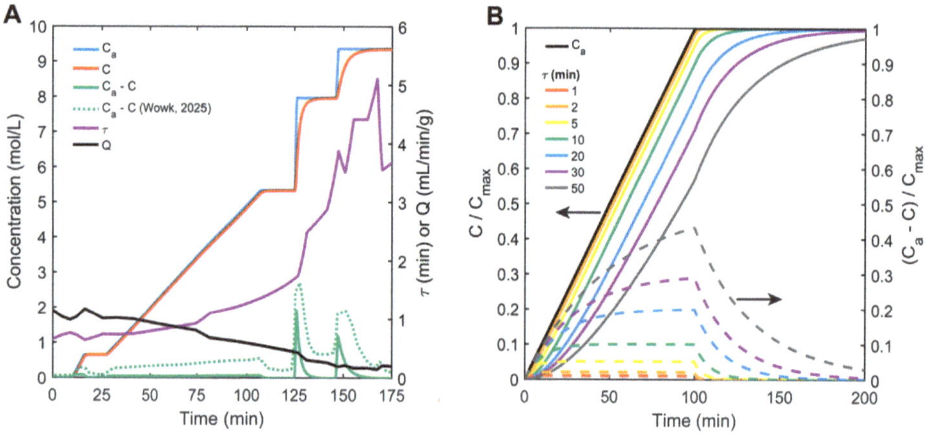

Fig. 9 Effects of perfusion flow rate on CPA equilibration during ramped loading protocols. (**A**) Using experimental flow rate (black solid line) and loading profile data (blue solid line) from Wowk et al. [83] as model inputs, the lag between the arterial load concentration and the kidney concentration can be estimated (green solid line). Comparing this estimation to the experimentally observed differences between arterial and venous concentrations (dashed green line, measured via refractometry), we see good agreement between prediction and experiment. The ~0.5 mol/L discrepancy observed may be a product of the inherent uncertainty in venous refractometry measurements, which cannot account for the nonuniform dilution of different CPAs in the effluent during perfusion. (**B**) Constant flow-rate analyses can provide further insights into CPA delivery limitations that emerge at excessive values of the characteristic flow constant τ (i.e., deficient flow rates). The solid black line shows the concentration being delivered to the organ via the artery. Solid colored lines give the concentration in the organ (left axis) for different values of τ, while dashed lines give the difference between the loading concentration and the organ concentration (right axis)

$$\frac{C(t)}{C_{\max}} = \frac{t}{t_1} - \frac{\tau}{t_1}\left(1 - e^{-\frac{t}{\tau}}\right) \quad (\text{for} 0 < t \leq t_1) \tag{26}$$

and

$$\frac{C(t)}{C_{\max}} = 1 - \frac{\tau}{t_1}\left(1 - e^{-\frac{t_1}{\tau}}\right)e^{-\frac{t-t_1}{\tau}} \quad (\text{for } t_1 < t \leq t_2) \tag{27}$$

Using these equations, we investigate the effect of characteristic flow constant τ on CPA loading time in Fig. 9B for t_1 = 100 min and t_2 = 200 min, which provides several salient insights. First, even in this conservative case of instantaneous equilibration throughout the tissue (i.e., the case in which permeation out of the vasculature and across the cell membrane are *not* rate-limiting factors), perfusion flow rates that replace the osmotically active volume less than once every ~10 min (i.e., with $\tau > \sim 10$) begin to deviate substantially from the input loading protocol, extending the required loading time, and thus the duration of potential cytotoxic CPA exposure and ischemia. Trivially, this implies that flow rates should be selected to be as fast possible, down to a flow constant approaching ~ $\tau = 1$. Practically however, these flow rates are typically dictated by the observed vascular resistance of the organ, implying that controlled-pressure perfusion of a given solution at an optimal *pressure* cannot be assumed optimal from a CPA delivery perspective, unless the flow rate also provides an acceptable value of τ.

For example, in the rabbit kidney perfusion effort shown in Fig. 9A, an average τ of ~2 was achieved at an average perfusion pressure of ~40 mm Hg [83], providing both desirable pressure *and* CPA delivery that is plausibly not flow-limited. However, in recent efforts to load rat vascular composite allografts with a glycerol-based solution for supercooling, an average τ of ~20 was achieved at an average perfusion pressure of ~35 mmHg [84], providing the desired pressure, but suboptimal, flow-limited CPA delivery.

This pressure-flow relationship can be a complex function of both CPA and carrier medium composition, and thus, the minimization of vascular resistance should be a paramount solution design priority for optimal loading. Oncotic support, impermeant concentration, and CPA permeability all provide powerful solution design levers by which to manipulate vascular resistance. We note again, however, that analyses like those shown in Fig. 9B provide insight only into how CPA delivery may be limited by *flow rate*—a limitation that will in practice prove *additive* to limitations imposed by transport between the blood vessels and the surrounding tissue.

Figure 7B illustrates the Krogh cylinder approach for modeling CPA perfusion, which was first presented in the cryobiological context by Rubinsky and Cravalho [77], and later implemented by others to help interpret and design organ perfusion methods [63, 65]. The general approach is to consider an organ as a composite of multiple identical units containing a blood vessel surrounded by a cylinder of tissue. These units are known as Krogh cylinders. This approach has been shown to predict changes in vascular resistance consistent with experimental data.

However, Krogh cylinder model predictions should be interpreted with caution because of the limitations imposed by the model assumptions. The model assumes that the composition in the fluid flowing through the capillary is identical to the composition of the fluid entering at the arterial end. This assumption may not be accurate for slow perfusion flow rates, such as for high viscosity fluids (see Fig. 8C). The assumption also precludes the use of the Krogh cylinder model to inform decisions about perfusion conditions that affect flow rate (e.g., pressure, temperature, viscosity).

Another limitation of the Krogh cylinder model is that it assumes the volume of the Krogh cylinder is constant. This is equivalent to assuming the total organ volume is constant. This assumption makes it possible to relate volume shifts from fluxes of water and CPA to changes in capillary diameter and vascular resistance (e.g., if the tissue volume goes up, the capillary volume must go down by the same amount to keep the total volume constant). As discussed above, organ volume changes are widely observed during CPA perfusion. This is expected to confound interpretation of Krogh cylinder model predictions. For example, a 10% change in organ volume can have a dramatic effect on prediction of capillary diameter, because the capillaries only occupy about 10% of the volume of the Krogh cylinder [65, 77].

The simplifying assumptions of the Krogh cylinder model may limit the accuracy of its predictions in some contexts. For example, Han et al. [65] recently reported that just 26% of the Krogh cylinder for rat kidneys is osmotically active, implying that 74% of the kidney is solids. However, the solids fraction estimated by drying of kidneys is less than 30% [85, 86], about 2.5× lower than the estimate from the Krogh cylinder model. This indicates that the Krogh cylinder model may not give an accurate physical representation of the response of an organ during CPA perfusion.

Alternative modeling approaches have been presented that capture more of the complexity of the organ response to CPA perfusion, including organ size changes [78, 79]. These models consider three fluid compartments—the vascular fluid, the interstitial fluid, and the intracellular fluid—and account for exchange of water and CPA between these compartments (Fig. 7C). The models also include biomechanical descriptions of organ size changes resulting from transport of water and CPA. While these models are promising, they have only been used in a limited way to explore their ability to predict observed trends. Further study of these models could inform the development of more effective CPA perfusion methods. It may also be fruitful in future studies to adapt complex transport models from the organ physiology literature [69] for predicting the response of organs to the non-physiological conditions encountered during CPA perfusion.

3.3 Optimization of CPA Equilibration in Organs

Optimization of organ perfusion is complex and involves decisions about perfusion pressure, perfusate composition, and temperature, while considering the effects of these factors on vascular resistance and CPA delivery rate, as well as chemical toxicity and/or mechanical damage from size changes or shear stresses. For the most part, previous studies have optimized perfusion methods using a combination of trial and error and conceptual reasoning. Recently, Han et al. [65] used the Krogh cylinder model to identify a kidney perfusion method predicted to achieve a vitrifiable CPA concentration within the organ without excessive osmotic damage and minimal toxicity. This method was shorter than the conventional method and was shown to yield comparable outcomes. These advances demonstrate the value of incorporating modeling into perfusion protocol design, and with continued development, such approaches could greatly accelerate progress toward more effective organ cryopreservation methods.

Acknowledgments

This work was supported by a grant from MTF Biologics. M22 viscosity data was provided by Soheil Kavian at Texas A&M University.

References

1. Mazur P (1963) Kinetics of water loss from cells at subzero temperatures and the likelihood of intracellular freezing. J Gen Physiol 47:347–369. https://doi.org/10.1085/jgp.47.2.347
2. Mazur P, Leibo SP, Chu EH (1972) A two-factor hypothesis of freezing injury. Evidence from Chinese hamster tissue-culture cells. Exp Cell Res 71:345–355. https://doi.org/10.1016/0014-4827(72)90303-5
3. Meryman HT (1971) Cryoprotective agents. Cryobiology 8:173–183. https://doi.org/10.1016/0011-2240(71)90024-1
4. Liu J, Mullen S, Meng Q, Critser J, Dinnyes A (2009) Determination of oocyte membrane permeability coefficients and their application to cryopreservation in a rabbit model. Cryobiology 59:127–134. https://doi.org/10.1016/j.cryobiol.2009.06.002
5. Olver DJ, Heres P, Paredes E, Benson JD (2023) Rational synthesis of total damage during cryoprotectant equilibration: modelling and experimental validation of osmomechanical, temperature, and cytotoxic damage in sea urchin (*Paracentrotus lividus*) oocytes. PeerJ 11:e15539. https://doi.org/10.7717/peerj.15539
6. Giwa S, Lewis JK, Alvarez L, Langer R, Roth AE, Church GM, Markmann JF, Sachs DH, Chandraker A, Wertheim JA, Rothblatt M, Boyden ES, Eidbo E, Lee WPA, Pomahac B, Brandacher G, Weinstock DM, Elliott G, Nelson D, Acker JP, Uygun K, Schmalz B, Weegman BP, Tocchio A, Fahy GM, Storey KB, Rubinsky B, Bischof J, Elliott JAW, Woodruff TK, Morris GJ, Demirci U, Brockbank KGM, Woods EJ, Ben RN, Baust JG, Gao DY, Fuller B, Rabin Y, Kravitz DC, Taylor MJ, Toner M (2017) The promise of organ and tissue preservation to transform medicine. Nat Biotechnol 35:530–542. https://doi.org/10.1038/nbt.3889
7. Fahy GM, MacFarlane DR, Angell CA, Meryman HT (1984) Vitrification

as an approach to cryopreservation. Cryobiology 21:407–426. https://doi.org/10.1016/0011-2240(84)90079-8
8. Karlsson JOM, Younis AI, Chan AWS, Gould KG, Eroglu A (2009) Permeability of the rhesus monkey oocyte membrane to water and common cryoprotectants. Mol Reprod Dev 76:321–333. https://doi.org/10.1002/Mrd.20956
9. Benson JD, Chicone CC, Critser JK (2011) A general model for the dynamics of cell volume, global stability, and optimal control. J Math Biol 63:339–359. https://doi.org/10.1007/s00285-010-0374-4
10. Benson JD, Kearsley AJ, Higgins AZ (2012) Mathematical optimization of procedures for cryoprotectant equilibration using a toxicity cost function. Cryobiology 64:144–151. https://doi.org/10.1016/j.cryobiol.2012.01.001
11. Davidson AF, Glasscock C, McClanahan DR, Benson JD, Higgins AZ (2015) Toxicity minimized cryoprotectant addition and removal procedures for adherent endothelial cells. PLOS One 10:e0142828. https://doi.org/10.1371/journal.pone.0142828
12. Gilmore JA, Liu J, Woods EJ, Peter AT, Critser JK (2000) Cryoprotective agent and temperature effects on human sperm membrane permeabilities: convergence of theoretical and empirical approaches for optimal cryopreservation methods. Hum Reprod 15:335–343. https://doi.org/10.1093/humrep/15.2.335
13. Jacobs MH (1927) The exchange of material between the erythrocyte and its surroundings. Harvey Lect 22:146–164
14. Box GEP, Draper NR (1987) Empirical model-building and response surfaces. John Wiley & Sons
15. Benson JD (2021) Mathematical modeling and optimization of cryopreservation in single cells. Methods Mol Biol 2180:129–172. https://doi.org/10.1007/978-1-0716-0783-1_4
16. Prickett RC, Elliott JA, Hakda S, McGann LE (2008) A non-ideal replacement for the Boyle van't Hoff equation. Cryobiology 57:130–136. https://doi.org/10.1016/j.cryobiol.2008.07.002.
17. Olver DJ, Benson JD (2023) Meta-analysis of the Boyle van't Hoff relation: turgor and leak models explain non-ideal volume equilibrium. Cryobiology 113:104581. https://doi.org/10.1016/j.cryobiol.2023.104581
18. Olver DJ, Azam I, Benson JD (2024) HepG2 cells undergo regulatory volume decrease by mechanically induced efflux of water and solutes. Biomech Model Mechanobiol 23:1781–1799. https://doi.org/10.1007/s10237-024-01868-w
19. Casula E, Traversari G, Fadda S, Klymenko OV, Kontoravdi C, Cincotti A (2019) Modelling the osmotic behaviour of human mesenchymal stem cells. Biochem Eng J 151:107296. https://doi.org/10.1016/j.bej.2019.107296
20. Hoffmann EK, Schettino T, Marshall WS (2007) The role of volume-sensitive ion transport systems in regulation of epithelial transport. Comp Biochem Physiol A Mol Integr Physiol 148:29–43. https://doi.org/10.1016/j.cbpa.2006.11.023
21. Jiang HY, Sun SX (2013) Cellular pressure and volume regulation and implications for cell mechanics. Biophys J 105:609–619. https://doi.org/10.1016/j.bpj.2013.06.021
22. Yellin F, Li YZ, eenivasan VKA, Farrell B, Johny MB, Yue D, Sun SX (2018) Electromechanics and volume dynamics in nonexcitable tissue cells. Biophys J 114:2231–2242. https://doi.org/10.1016/j.bpj.2018.03.033
23. Prickett RC, Elliott JA, McGann LE (2010) Application of the osmotic virial equation in cryobiology. Cryobiology 60:30–42. https://doi.org/10.1016/j.cryobiol.2009.07.011
24. Garcia-Martinez T, Mogas T, Mullen SF, Martinez-Rodero I, Gulieva RE, Higgins AZ (2021) Effect of cryoprotectant concentration on bovine oocyte permeability and comparison of two membrane permeability modelling approaches. Sci Rep 11:15387. https://doi.org/10.1038/s41598-021-94884-0
25. Lahmann JM, Benson JD, Higgins AZ (2018) Concentration dependence of the cell membrane permeability to cryoprotectant and water and implications for design of methods for post-thaw washing of human erythrocytes. Cryobiology 80:1–11. https://doi.org/10.1016/j.cryobiol.2017.12.003
26. Benson JD (2020) Robust numerical solution for the two parameter solute solvent transport model. Cryobiology 97:238–241. https://doi.org/10.1016/j.cryobiol.2020.06.015
27. Urban JP, Roberts S (2003) Degeneration of the intervertebral disc. Arthritis Res Ther 5:120–130. https://doi.org/10.1186/ar629
28. Tsay CS, Mchugh AJ (1992) A technique for rapid measurement of diffusion-coefficients. Ind Eng Chem Res 31:449–452. https://doi.org/10.1021/ie00001a063
29. Ternstrom G, Sjostrand A, Aly G, Jernqvist A (1996) Mutual diffusion coefficients of water plus ethylene glycol and water plus glycerol mixtures. J Chem Eng Data 41:876–879. https://doi.org/10.1021/je9501705
30. Wang MH, Soriano AN, Caparanga AR, Li MH (2010) Binary mutual diffusion coefficient of aqueous solutions of propylene glycol and dipropylene glycol. J Taiwan Inst Chem

E 41:279–285. https://doi.org/10.1016/j.jtice.2009.09.001

31. Devireddy RV (2005) Predicted permeability parameters of human ovarian tissue cells to various cryoprotectants and water. Mol Reprod Dev 70:333–343. https://doi.org/10.1002/Mrd.20209
32. Vasquez-Rivera A, Sommer KK, Oldenhof H, Higgins AZ, Brockbank KGM, Hilfiker A, Wolkers WF (2018) Simultaneous monitoring of different vitrification solution components permeating into tissues. Analyst 143:420–428. https://doi.org/10.1039/c7an01576c
33. Lotz J, Icli S, Liu D, Caliskan S, Sieme H, Wolkers WF, Oldenhof H (2021) Transport processes in equine oocytes and ovarian tissue during loading with cryoprotective solutions. Biochim Biophys Acta Gen Subj 1865:129797. https://doi.org/10.1016/j.bbagen.2020.129797
34. Han J, Sydykov B, Yang H, Sieme H, Oldenhof H, Wolkers WF (2019) Spectroscopic monitoring of transport processes during loading of ovarian tissue with cryoprotective solutions. Sci Rep 9:15577. https://doi.org/10.1038/s41598-019-51903-5
35. Jomha NM, Law GK, Abazari A, Rekieh K, Elliott JAW, McGann LE (2009) Permeation of several cryoprotectant agents into porcine articular cartilage. Cryobiology 58:110–114. https://doi.org/10.1016/j.cryobiol.2008.11.004
36. Hubel A, Bidault N, Hammer B (2002) Transport characteristics of glycerol and propylene glycol in an engineered dermal replacement. ASME Conference Proceedings IMECE2002-32557, pp 121–122. https://doi.org/10.1115/imece2002-32557
37. Benson JD, Higgins AZ, Desai K, Eroglu A (2018) A toxicity cost function approach to optimal CPA equilibration in tissues. Cryobiology 80:144–155. https://doi.org/10.1016/j.cryobiol.2017.09.005
38. Han X, Ma L, Benson J, Brown A, Critser JK (2009) Measurement of the apparent diffusivity of ethylene glycol in mouse ovaries through rapid MRI and theoretical investigation of cryoprotectant perfusion procedures. Cryobiology 58:298–302. https://doi.org/10.1016/j.cryobiol.2009.02.005
39. Mukherjee IN, Li Y, Song YC, Long RC, Sambanis A (2008) Cryoprotectant transport through articular cartilage for long-term storage: experimental and modeling studies. Osteoarthritis Cartilage 16:1379–1386. https://doi.org/10.1016/j.joca.2008.03.027
40. Higgins AZ, Lavarti R, Eroglu B, Ahmadkhani N, Benson JD, Eroglu A (2023) Permeation of individual cryoprotectants and their different combinations into mouse liver tissue. Cryobiology 111:26–29. https://doi.org/10.1016/j.cryobiol.2023.03.004
41. Abazari A, Elliott JA, Law GK, McGann LE, Jomha NM (2009) A biomechanical triphasic approach to the transport of nondilute solutions in articular cartilage. Biophys J 97:3054–3064. https://doi.org/10.1016/j.bpj.2009.08.058
42. Tapia Lishner SE, Marquez-Curtis LA, Elliott JAW (2023) Permeation kinetics of dimethyl sulfoxide in porcine corneoscleral discs. Cryobiology 113:104566. https://doi.org/10.1016/j.cryobiol.2023.104566
43. Woods EJ, Liu J, Zieger MA, Lakey JR, Critser JK (1999) Water and cryoprotectant permeability characteristics of isolated human and canine pancreatic islets. Cell Transplant 8:549–559. https://doi.org/10.1177/096368979900800510
44. Wang S, Oldenhof H, Goecke T, Ramm R, Harder M, Haverich A, Hilfiker A, Wolkers WF (2015) Sucrose diffusion in decellularized heart valves for freeze-drying. Tissue Eng Part C Methods 21:922–931. https://doi.org/10.1089/ten.TEC.2014.0681
45. Vasquez-Rivera A, Oldenhof H, Dipresa D, Goecke T, Kouvaka A, Will F, Haverich A, Korossis S, Hilfiker A, Wolkers WF (2018) Use of sucrose to diminish pore formation in freeze-dried heart valves. Sci Rep 8:12982. https://doi.org/10.1038/s41598-018-31388-4
46. Shalash W, Forcier R, Higgins AZ, Giers MB (2024) Cryopreserving the intact intervertebral disc without compromising viability. JOR Spine 7:e1351. https://doi.org/10.1002/jsp2.1351
47. Shardt N, Al-Abbasi KK, Yu H, Jomha NM, McGann LE, Elliott JAW (2016) Cryoprotectant kinetic analysis of a human articular cartilage vitrification protocol. Cryobiology 73:80–92. https://doi.org/10.1016/j.cryobiol.2016.05.007
48. Jomha NM, Elliott JA, Law GK, Maghdoori B, Forbes JF, Abazari A, Adesida AB, Laouar L, Zhou X, McGann LE (2012) Vitrification of intact human articular cartilage. Biomaterials 33:6061–6068. https://doi.org/10.1016/j.biomaterials.2012.05.007
49. Crank J (1975) The mathematics of diffusion, 2nd edn. Oxford University Press, Oxford
50. He YM, Devireddy RV (2005) An inverse approach to determine solute and solvent permeability parameters in artificial tissues. Ann Biomed Eng 33:709–718. https://doi.org/10.1007/s10439-005-1511-x
51. Bhowmick S, Khamis CA, Bischof JC (1998) Response of a liver tissue slab to a hyperosmotic sucrose boundary condition: microscale cellular and vascular level effects. Ann N Y Acad Sci 858:147–162. https://doi.org/10.1111/j.1749-6632.1998.tb10149.x

52. Cui ZF, Dykhuizen RC, Nerem RM, Sembanis A (2002) Modeling of cryopreservation of engineered tissues with one-dimensional geometry. Biotechnol Prog 18:354–361. https://doi.org/10.1021/Bp0101886
53. Xu X, Cui ZF (2003) Modeling of the co-transport of cryoprotective agents in a porous medium as a model tissue. Biotechnol Prog 19:972–981. https://doi.org/10.1021/Bp025674n
54. Benson JD, Benson CT, Critser JK (2014) Mathematical model formulation and validation of water and solute transport in whole hamster pancreatic islets. Math Biosci 254:64–75. https://doi.org/10.1016/j.mbs.2014.06.003
55. De Freitas RC, Diller KR, Lachenbruch CA, Merchant FA (1998) Network thermodynamic model of coupled transport in a multicellular tissue the islet of Langerhans. Ann N Y Acad Sci 858:191–204
56. Warner RM, Shuttleworth R, Benson JD, Eroglu A, Higgins AZ (2021) General tissue mass transfer model for cryopreservation applications. Biophys J 120:4980–4991. https://doi.org/10.1016/j.bpj.2021.10.014
57. Shardt N, Chen ZR, Yuan SC, Wu KZ, Laouar L, Jomha NM, Elliott JAW (2020) Using engineering models to shorten cryoprotectant loading time for the vitrification of articular cartilage. Cryobiology 92:180–188. https://doi.org/10.1016/j.cryobiol.2020.01.008
58. Wu K, Shardt N, Laouar L, Elliott JAW, Jomha NM (2021) Vitrification of particulated articular cartilage via calculated protocols. NPJ Regen Med 6:15. https://doi.org/10.1038/s41536-021-00123-5
59. Davidson AF, Benson JD, Higgins AZ (2014) Mathematically optimized cryoprotectant equilibration procedures for cryopreservation of human oocytes. Theor Biol Med Model 11:13. https://doi.org/10.1186/1742-4682-11-13
60. Paivarinta J, Oikonen V, Raisanen-Sokolowski A, Tolvanen T, Loyttyniemi E, Iida H, Nuutila P, Metsarinne K, Koivuviita N (2019) Renal vascular resistance is increased in patients with kidney transplant. BMC Nephrol 20:437. https://doi.org/10.1186/s12882-019-1617-2
61. Unger V, Grosse-Siestrup C, Fehrenberg C, Fischer A, Meissler M, Groneberg DA (2007) Reference values and physiological characterization of a specific isolated pig kidney perfusion model. J Occup Med Toxicol 2:1. https://doi.org/10.1186/1745-6673-2-1
62. Kloner RA, King KS, Harrington MG (2018) No-reflow phenomenon in the heart and brain. Am J Physiol Heart Circ Physiol 315:H550–H562. https://doi.org/10.1152/ajpheart.00183.2018
63. Pegg DE, Rubinsky B, Diaper MP, Lee CYC (1986) Analysis of the introduction and removal of glycerol in rabbit kidneys using a Krogh cylinder model. Cryobiology 23:150–160. https://doi.org/10.1016/0011-2240(86)90006-4
64. Pegg DE, Jacobsen IA, Diaper MP, Foreman J (1987) Perfusion of rabbit kidneys with solutions containing propane-1,2-diol. Cryobiology 24:420–428. https://doi.org/10.1016/0011-2240(87)90045-9
65. Han Z, Rao JS, Ramesh S, Hergesell J, Namsrai BE, Etheridge ML, Finger EB, Bischof JC (2023) Model-guided design and optimization of CPA perfusion protocols for whole organ cryopreservation. Ann Biomed Eng 51:2216–2228. https://doi.org/10.1007/s10439-023-03255-5
66. Warner RM, Yang J, Drake A, Lee Y, Nemanic S, Scott D, Higgins AZ (2023) Osmotic response during kidney perfusion with cryoprotectant in isotonic or hypotonic vehicle solution. PeerJ 11:e16323. https://doi.org/10.7717/peerj.16323
67. Han Z, Rao JS, Gangwar L, Namsrai BE, Pasek-Allen JL, Etheridge ML, Wolf SM, Pruett TL, Bischof JC, Finger EB (2023) Vitrification and nanowarming enable long-term organ cryopreservation and life-sustaining kidney transplantation in a rat model. Nat Commun 14:3407. https://doi.org/10.1038/s41467-023-38824-8
68. Cady B, Barner HB, Rivers Jr RJ, Haynes LL, Watkins Jr E (1966) Glycerolization of the canine kidney. I. Fluid exchanges. Cryobiology 3:76–80. https://doi.org/10.1016/s0011-2240(66)80003-2
69. Kellen MR, Bassingthwaighte JB (2003) Transient transcapillary exchange of water driven by osmotic forces in the heart. Am J Physiol Heart Circ Physiol 285:H1317–H1331. https://doi.org/10.1152/ajpheart.00587.2002
70. Vargas F, Johnson JA (1967) Permeability of rabbit heart capillaries to nonelectrolytes. Am J Physiol 213:87–93. https://doi.org/10.1152/ajplegacy.1967.213.1.87
71. Wangensteen OD, Lysaker E, Savaryn P (1977) Pulmonary capillary filtration and reflection coefficients in the adult rabbit. Microvasc Res 14:81–97. https://doi.org/10.1016/0026-2862(77)90143-1
72. Fahy GM, Takahashi T, Crane AM (1984) Histological cryoprotection of rat and rabbit brains. CryoLett 5:33–46
73. Rippe B, Haraldsson B (1986) Capillary permeability in rat hindquarters as determined by estimations of capillary reflection coefficients. Acta Physiol Scand 127:289–303. https://doi.org/10.1111/j.1748-1716.1986.tb07908.x
74. Wolf MB, Watson PD (1989) Measurement of osmotic reflection coefficient for small

molecules in cat hindlimbs. Am J Physiol 256:H282–H290. https://doi.org/10.1152/ajpheart.1989.256.1.H282
75. Farrant J (1964) Pharmacological actions and toxicity of dimethyl sulphoxide and other compounds which protect smooth muscle during freezing and thawing. J Pharm Pharmacol 16:472–483. https://doi.org/10.1111/j.2042-7158.1964.tb07496.x
76. McIntyre RL, Fahy GM (2015) Aldehyde-stabilized cryopreservation. Cryobiology 71:448–458. https://doi.org/10.1016/j.cryobiol.2015.09.003
77. Rubinsky B, Cravalho EG (1982) Transient mass-transfer processes during the perfusion of a biological organ with a cryophylactic agent solution. Cryobiology 19:70–82
78. Lachenbruch CA, Diller KR (1999) A network thermodynamic model of kidney perfusion with a cryoprotective agent. J Biomech Eng 121:574–583. https://doi.org/10.1115/1.2800856
79. Levin RL (1981) Osmotic effects of introducing and removing cryoprotectants: perfused tissues and organs. In: Viano DC (ed) Annual Meeting of the American Society of Mechanical Engineers, Washington, DC. American Society of Mechanical Engineers, New York, pp 131–134
80. Fahy GM, Wowk B, Pagotan R, Chang A, Phan J, Thomson B, Phan L (2009) Physical and biological aspects of renal vitrification. Organogenesis 5:167–175
81. Flick EW (1998) Industrial solvents handbook, 5th edn. Noyes Data Corp., Westwood, NJ
82. Eto TK, Costello BJ, Wenzel SW, White RM, Rubinsky B (1993) Viscosity sensing with lamb-wave microsensor—dimethylsulfoxide solution viscosity as a function of temperature. J Biomech Eng 115:329–331. https://doi.org/10.1115/1.2895494
83. Wowk B, Ting A, Phan J, Pagotan R, Sharif A, Chow L, Fahy GM (2025) 55 MHz constant field dielectric warming of kidneys and ovaries cryopreserved by vitrification. Cryobiology 119:105257. https://doi.org/10.1016/j.cryobiol.2025.105257
84. Filz von Reiterdank I, Dinicu AT, Rosales I, Cetrulo Jr CL, Coert JH, Mink van der Molen AB, Uygun K (2024) Supercooling preservation of vascularized composite allografts through CPA optimization, thermal tracking, and stepwise loading techniques. Sci Rep 14:22339. https://doi.org/10.1038/s41598-024-73549-8
85. Reinoso RF, Telfer BA, Rowland M (1997) Tissue water content in rats measured by desiccation. J Pharmacol Toxicol Methods 38:87–92. https://doi.org/10.1016/s1056-8719(97)00053-1
86. Forbes RM, Cooper AR, Mitchell HH (1953) The composition of the adult human body as determined by chemical analysis. J Biol Chem 203:359–366. https://doi.org/10.1016/S0021-9258(19)52646-1

Part II

Emerging Technologies

Chapter 4

Thermodynamic Principles of Emerging Cryopreservation Technologies

Matthew J. Powell-Palm and Anthony N. Consiglio

Abstract

Modern cryopreservation exists at the convergence of diverse disciplines—materials science, physical chemistry, mechanical engineering, biological engineering, etc.—and emerging technologies often draw from many of these disciplines simultaneously. Thermodynamics, as one of the foundational theories underlying both physical and biological science, provides a framework through which to understand these interdisciplinary technologies, yet the full kit of requisite thermodynamic tools is not housed within any one discipline. This Chapter aims to articulate a foundational thermodynamic approach to the description, interrogation, and design of modern cryopreservation technologies, and to review the state of the art in emerging cryopreservation technologies through the lens of this approach. We focus in particular on the management of phase change across equilibrium-driven techniques (e.g., liquidus tracking, partial freezing, isochoric freezing), kinetics-driven techniques (e.g., supercooling, ice seeding), and transport-driven techniques (e.g., directional freezing, droplet approaches), and we hope to equip the reader with a self-consistent theoretical toolkit that enables meaningful comparison of these techniques from a thermodynamic perspective.

Keywords Cryopreservation technologies, Solution thermodynamics, Ice nucleation, Supercooling, Isochoric freezing, Partial Freezing, Liquidus tracking, Directional freezing

1 Introduction

Effective cryopreservation of biological matter—be that matter a cell, a tissue, an organ, or an organism—hinges upon the uniquely complicated interplay between the thermodynamics of aqueous solutions, the nucleation and growth kinetics of ice and other solid phases, and the transport of heat and mass through spatially and chemically heterogeneous media. As such, mathematical description, interrogation, and design of cryopreservation processes require a uniquely varied blend of physical theories, drawn from disciplines ranging from physical chemistry and atmospheric science to mechanical engineering and materials science. Classical geometric (or Gibbsian) thermodynamics can provide a unifying framework to connect these disparate bases of

Willem F. Wolkers and Harriëtte Oldenhof (eds.), *Cryopreservation and Freeze-Drying Protocols*, Methods in Molecular Biology, vol. 3060, https://doi.org/10.1007/978-1-0716-5416-3_4,

knowledge, but reconciliation of different schools of thermodynamic practice in the context of cryopreservation is required.

To that end, in this Chapter, we will present a thermodynamic description of the physical aspects of cryopreservation, enabling self-consistent interrogation of the physical bases of various emergent cryopreservation technologies. In the first Section, we will establish this description in additive subsections focusing on solution thermodynamics, ice nucleation and growth kinetics, and heat and mass transport. In each subsection, we will aim to equip the user firstly with high-level physical insight applicable to emergent cryopreservation technologies, and secondly with basic predictive power over the same. In the second Section, we will review various such technologies within the context of our unified physical framework, aiming to facilitate interrogation of how the underlying thermodynamic premise of each technology may resemble or differ from the others.

For consistency with our theoretical approach, we will group these emergent technologies by the theoretical aspect—equilibrium thermodynamics, kinetics, or transport—that most principally drives them. For example, liquidus tracking, partial freezing [1], and isochoric freezing [2] all rely upon attainment of a desired state of thermodynamic equilibrium, wherein ice may be either entirely absent (liquidus tracking) or allowed to an extent prescribed by the liquidus equilibria of the solution (partial freezing and isochoric freezing). Supercooling [3] and ice seeding, by comparison, rely upon attainment of a desired nucleation rate (supercooling) or ice growth rate (seeding), which, while of course influenced by the underlying phase equilibria, are ultimately kinetic factors. Likewise, various other techniques rely upon attainment of a desired temperature profile en route to the final state of preservation, be it a profile in time (as in droplet and mesh technologies that seek to amplify cooling rates) or in space (as in directional freezing). These techniques are no less subject to the relevant phase equilibria and nucleation kinetics, but they additionally leverage transport phenomena to optimize (and in many cases circumvent) aspects of those factors.

Throughout this Chapter, we will treat these many technologies from a principally physical perspective, and we will adopt to a significant degree the terminology and philosophy of Gibbsian geometric thermodynamics. For a robust introduction to this school of thought, we recommend the seminal textbook of Herbert Callen [4].

2 Thermodynamic and Physical Principles of Ice Management in Cryopreservation

Cryopreservation, at its most fundamental, seeks to manipulate the state and action of water in and around a biological specimen such that mechanical, chemical, and biological damage are avoided at those subnormothermic temperatures at which the processes of life are slowed or arrested. While the toxicological challenge of cryopreservation is concerned principally with the biochemical action of this water, the physical challenge of cryopreservation most often involves the *phase* of this water, and its tendency to change in undesirable ways upon cooling to the sub-0°C temperatures at which the slowing of life processes becomes sufficient to enable days-, weeks-, or years-long preservation. As such, our physical description of cryopreservation is first and foremost a description of the unique behaviors of liquid water and solid ice, in the presence of various chemicals, under various temperatures and pressures, in contact with various surfaces, and subject to various gradients in time, space, energy, and matter. We will group this description into three subsections, treating relevant aspects of the equilibrium thermodynamics, ice nucleation and growth kinetics, and heat and mass transport at play.

2.1 Equilibrium Thermodynamics

We begin our description of the thermodynamics of cryopreservation with the most fundamental quantity dictating the response of water and ice to these varying conditions: the chemical potential, μ. The chemical potential, or the rate at which the free energy of a substance changes with the addition of more of that substance to a system, is the thermodynamic parameter most responsible for directing the action of water. It dictates both whether water can (and in part whether it *will*) change phase from a liquid to a solid; it dictates the extent to which (and in part *the speed* with which) crystalline ice will grow upon nucleation; and it dictates whether (and in part *how quickly*) water will migrate by diffusion or osmosis. As such, description of the chemical potential is the first step in formulating our broader description of physical cryopreservation.

2.1.1 The Chemical Potential

The chemical potential of a component i is defined as:

$$\mu_i = \left(\frac{\partial G}{\partial N_i} \right)_{T,p,N_{j \neq i}} \tag{1}$$

wherein G gives the Gibbs free energy of the system, N_i is the number of mols of component i, and subscripts outside the differentials indicate variables held constant. μ_i is most typically given in units of J/mol, and it is an intensive thermodynamic variable, which means that it is scale-invariant, and it provides a driving force for equilibration among chemical species. Analogous to how differences in temperature compel the flow of heat and differences in pressure compel the flow of volume, differences in chemical potential compel the flow of molecules, be they into new phases (as in the formation of ice) or into new locations (as in osmosis or diffusion).

For pure water (or any pure substance), the chemical potential is simply equal to the Gibbs free energy per mol. For liquid water, common hexagonal ice (ice Ih), and several high-pressure ice polymorphs, this quantity has been rigorously measured across a wide range of temperatures and pressures and parameterized into powerful and accessible equations of state. Throughout this work, we will use the open-source SeaFreeze [5] package maintained by Journaux and colleagues.

Thermodynamic equilibrium in a cryobiological system consisting of different phases (or alternatively compartments) 1 and 2 requires the condition of equality of the intensive thermodynamic variables ($\mu_1 = \mu_2$, $p_1 = p_2$, $T_1 = T_2$). Practically speaking, these conditions state that, at a given pressure, at the phase transition temperature, wherein the two phases at hand are definitionally in equilibrium, their chemical potentials must be equal. We refer to the set of interrelated intensive variables that define equilibrium between two or more phases as phase equilibria, or equilibrium coordinates.

In cryobiological problems, calculation of the chemical potential is used for three principal purposes: (1) to solve for the temperature (and/or pressure) at which water and ice are in equilibrium in a given solution, i.e., to calculate the composition-dependent melting point of ice, (2) to calculate the driving force for ice nucleation (discussed in later sections), and (3) to calculate the driving force for mass transfer during CPA loading (discussed in later sections).

2.1.2 Solution Theories

As stated, in the modern era, the chemical potentials of pure water and pure ice are well known—the challenge of cryopreservation thus falls to describing how the chemical potential of water *changes* in the presence of the myriad biologically-necessary or cryoprotective molecules, i.e., in solution. Armed with this knowledge because the chemical potential of ice does not change with solution composition (given that ice rejects solutes and forms an approximately pure crystal), liquid-ice phase equilibria may be readily determined by equating the calculated chemical potential of water in solution with the known chemical potential of ice and solving for the temperature-pressure-concentration conditions that yield that equality.

To calculate this chemical potential, we utilize geometric or Gibbsian solution thermodynamics. Originally derived in the late nineteenth and early twentieth centuries by Gibbs, Raoult, Lewis, Hildebrand, and others, and today practiced in many distinct varieties, the Gibbsian school of solution thermodynamics generally treats the free energy of a solution G as a mol-weighted average of the free energy of its constituent components plus a term capturing the effects of mixing these components together, such that

$$G = \sum_i N_i \mu_i^0 + \Delta G_{\text{mix}}. \tag{2}$$

Generally, ΔG_{mix} aims to capture the effects of entropy that accompany mixing arbitrary components together (the entropy of mixing ΔS_{mix}) and the molecular interactions that may occur between the components (the enthalpy of mixing ΔH_{mix}), such that $\Delta G_{mix} = \Delta H_{mix} - T\Delta S_{mix}$. While this framework is more or less universal within Gibbsian thermodynamics, specific solution theories proceed to quantify ΔG_{mix} in a variety of different ways, requiring different degrees of experimental input and yielding different predictive and descriptive powers. We will discuss here three key theories, chosen for their particular applicability to problems of cryobiology and their extensibility to multisolute solutions.

The simplest theory of solutions is the ideal model, which assumes that the different components in solution do not interact whatsoever ($\Delta H_{mix} = 0$), and that, for entropic purposes, the molecules may be treated as physically identical. In this case, the entropy of mixing (in units of Joules per Kelvin) is given by:

$$\Delta S_{mix}^{ideal} = -R\sum_i N_i \ln x_i. \tag{3}$$

The Gibbs free energy of the solution is then

$$G = \sum_i N_i \mu_i^0 + RT\sum_i N_i \ln x_i, \tag{4}$$

and the chemical potential of component i in an ideal solution is

$$\mu_i = \mu_i^0 + RT\ln x_i. \tag{5}$$

Critically, this result implies that, within the ideal conception, the chemical potential of a given component is compositionally dependent only on *its own* concentration, and agnostic to the specific solutes accompanying it in solution. As cryobiologists surely appreciate, this agnosticism does not bear out experimentally, considering that different cryoprotectants in reality depress the melting point of water to very differing degrees at the same molar concentration.

The ideal model is thus limited in empirical accuracy but has nonetheless proven an invaluable exploratory tool for the interrogation of generalized effects of composition on various physical processes. This utility is driven both by its simplicity and its reliance only on properties of the *pure* components, as opposed to properties of the solution itself, which enables *predictive* exploration of new solutions without synthesizing the desired solution and measuring some subset of its properties. Furthermore, the ideal model is readily extensible to arbitrarily complex n-component solutions.

Recently, Alliston et al. [6] introduced a modification to the ideal model (the size-dependent ideal solution or SIS model) to incorporate solute-specific size information whilst retaining the predictive, parameter-free nature of the ideal solution theory and its easy extensibility to multicomponent mixtures. Based on statistical mechanics arguments by Flory [7] and Hildebrand [8] and kinetic observations by Powell-Palm et al. [9], Alliston et al. replace the ideal entropy of mixing with a size-dependent entropy of mixing capturing the entropic impact of the considerable size difference between water and typical organic cryoprotectants, which often possesses ~4–10 times the molar volume of water. For a solution of components i with molar volumes v_i, and volume fraction ϕ_i, the SIS mixing entropy is

$$\Delta S_{\mathrm{mix}}^{\mathrm{SIS}} = -R \sum_i N_i \ln\left(\frac{v_i N_i}{\sum_j v_j N_j}\right) = -R \sum_i N_i \ln \phi_i. \tag{6}$$

Substituting this entropy into the expression for the Gibbs free energy and differentiating, one obtains a size-dependent chemical potential. For a general multicomponent mixture, this can be written compactly as:

$$\frac{\mu_i - \mu_i^0}{RT} = \ln \phi_i + \sum_{j \neq i} (1 - r_{ij}) \phi_j, \tag{7}$$

where $r_{ij} = v_i / v_j$ is the ratio of molar volumes. Using only bulk-material molar volumes at room temperature, the incorporation of this simple size-dependence has been shown to increase the accuracy of the ideal model by ~50% or more for binary solutions of water and cryopreservation-relevant organic molecules, both in calculation of the melting point of ice in solution and in identification of solubility limits and eutectic temperatures/compositions.

While the above theories are purely *predictive*, requiring no experimental knowledge of the solution of interest, more accurate theories can be developed using experimental input measured from a few discrete concentrations. Many such descriptive theories start from the standard definitions:

$$\mu_i = \mu_i^0 + RT \ln a_i, \tag{8}$$

$$\gamma_i = \frac{a_i}{x_i}, \tag{9}$$

wherein a_i denotes the activity and γ_i the activity coefficient of component i in solution, and is a function of both composition, temperature, and pressure (though the dependence on temperature and pressure is often very weak). The activity coefficient of a component is an empirical measure of its deviation from ideality (as $\gamma^{\mathrm{ideal}} = 1$), and activity-based descriptive theories typically seek to prescribe how γ_i varies with composition.

A convenient unifying language for many nonideal models is the excess free energy approach, which considers the free energy as sum of ideal and excess contributions: $G/N = g = g^{\text{ideal}} + g^{\text{excess}}$. The activity coefficient is thus given by

$$\ln\gamma_i = \frac{1}{RT}\left(\frac{\partial G^{\text{excess}}}{\partial N_i}\right)_{T,p,N_{j\neq i}}. \tag{10}$$

The ideal Gibbs free energy is given in Eq. 4 and the ideal entropy is given in Eq. 3. The size-dependent ideal solution theory, which maintains zero excess enthalpy, has excess entropy $(Ts^E = h^E - g^E)$ and water activity coefficient of

$$s^{\text{E,SIS}} = -R\sum_i x_i \ln\left(\frac{\phi_i}{x_i}\right), \tag{11}$$

$$\ln\gamma_{\text{w}}^{\text{SIS}} = \ln\left(\frac{\phi_{\text{w}}}{x_{\text{w}}}\right) + \sum_i \left(1 - \frac{v_{\text{w}}}{v_i}\right)\phi_i. \tag{12}$$

The regular solution theory represents the simplest energetic extension of the ideal solution theory. In its conventional form, it assumes zero excess *entropy* (no size-dependence), but allows for nonzero *enthalpy* of mixing, typically represented in terms of binary interaction parameters, χ_{ij}, between each pair of species. The excess enthalpy and water activity coefficient for a regular solution can be written as

$$h^{\text{E,regular}} = \frac{1}{2}RT\sum_{i,j}\chi_{ij}x_i x_j, \tag{13}$$

$$\ln\gamma_{\text{w}}^{\text{regular}} = \sum_i \chi_{\text{w}i}x_i^2 + \sum_{j>i}(\chi_{\text{w}i} + \chi_{\text{w}j} + \chi_{ij})x_i x_j. \tag{14}$$

For binary systems, this reduces to the familiar symmetric form involving a single χ parameter. In practice, the χ_{ij} are fitted to empirical data such as freezing points or vapor pressures. In the context of cryobiology, regular solutions are rarely used in their full generality, but the underlying idea of augmenting the ideal or SIS baseline with simple energetic interaction terms underpins many common approaches.

In cryobiological systems, the chemical potential of water is of primary interest; the activities of the other solutes are needed in fewer instances. It is, therefore, common to express the state of water in solution by the parameters known as osmolality and osmotic pressure.

Osmolality is directly related to the chemical potential of water and the water activity by

$$\pi = -\frac{\ln a_w}{M_w} = \frac{\mu_w^0 - \mu_w}{M_w RT} \tag{15}$$

where M_w is the molecular weight of water. Normal physiological osmolality lies approximately in the range 280–300 mOsm/kg. The corresponding osmotic pressure is related to π via

$$\Pi = \pi \rho_w RT \tag{16}$$

with ρ_w the density of water. Normal physiological osmotic pressure (at 37°C) is approximately 7–8 atm.

A particularly useful implementation of regular solution theory pioneered by Elliott and colleagues for cryobiologically-relevant mixtures is the osmotic virial equation [10–12]. In this model, the osmolality is expanded as a polynomial in terms of solute molalities, m_i:

$$\begin{aligned}\pi = &\sum_i k_i m_i + \sum_{i,j} \frac{B_i + B_j}{2} k_i m_i k_j m_j \\ &+ \sum_{i,j,k} (C_i C_j C_k)^{1/3} k_i m_i k_k m_j k_k m_k\end{aligned} \tag{17}$$

Here, k_i is the dissociation constant (capturing ionic dissociation of electrolytes), and B_i and C_i are the first and second virial coefficients encapsulating pairwise and triplet interactions between solutes. For each solute, the parameters k_i, B_i, and C_i are obtained by fitting single-solute osmotic or freezing-point data; multisolute behavior is then predicted via combining rules analogous to those used in regular solution theory. Requiring only binary parameters to describe multicomponent solutions, the osmotic virial equation has become a workhorse solution theory for many cryobiological modeling efforts.

Beyond the models emphasized here, a wide variety of more elaborate solution theories have been developed, which could be of use to cryobiological problems. These include classical local-composition models such as NRTL [13] (non-random two-liquid; and extensions such as eNRTL) and UNIQUAC [13] (universal quasichemical); group contribution methods built upon these such as UNIFAC [13] (UNIQUAC functional-group activity coefficients) and AIOMFAC [14] (aerosol inorganic-organic mixtures functional group activity coefficients); and quantum-chemical models such as COSMO-RS [15, 16] and COSMO-SAC [17] (conductor-like screening model). These models are widely used in chemical engineering for vapor–liquid and liquid–liquid equilibria and provide a semipredictive way to handle large families of organic solutes. As the cryobiology field moves toward rational design of multicomponent CPA cocktails (including natural deep eutectic solvents and polymer-rich formulations), such models offer a path to predictive screening of candidate mixtures without the need to commit to extensive experimental testing.

2.1.3 Phase Equilibria

Armed with these solution models of the chemical potential, we are now empowered to mathematically predict or describe the melting points, solubilities, eutectics, and other cryopreservation-relevant equilibria of aqueous cryoprotectant solutions. We will describe two simple methods by which to achieve this.

First, we may simply calculate the chemical potential of water in solution at a desired concentration, using any of the theories above, compare it to the chemical potential of ice drawn from any robust equation of state [5], and identify the conditions under which the two are equal. Figure 1A shows this method for solutions of NaCl in water, using the SeaFreeze equation of state for the chemical potential of water in ice-Ih and aqueous NaCl solutions, with the corresponding liquidus curve shown in Fig. 1B. Figure 1C, D show the predicted liquidus curves for binary solutions of glycerol or sucrose computed using the ideal, size-dependent ideal, and osmotic-virial equations.

A more compact route to determining the phase coexistence exploits the Gibbs-Helmholtz relation, which for a pure substance is written as

$$\frac{\partial}{\partial T}\left(\frac{\mu}{RT}\right)_p = -\frac{h}{RT^2} \tag{18}$$

and for the difference between liquid and solid phases as

$$\frac{\partial}{\partial T}\left(\frac{\Delta\mu^0_{\text{liquid}-\text{solid}}}{RT}\right)_p = -\frac{\Delta h^0_{\text{liquid}-\text{solid}}}{RT^2}, \tag{19}$$

where $\Delta H^0_{\text{liquid}-\text{solid}}$ is the enthalpy of fusion of pure water (i.e., ΔH^0_{f}). At the melting point of a solution, the liquid is in equilibrium with (practically) pure ice, so

$$\mu_{\text{w}}^{0,\text{ice}}(T_{\text{m}}) = \mu_{\text{w,aq}}(T_{\text{m}}, a_{\text{w}}) = \mu_{\text{w}}^{0,\text{liq}}(T_{\text{m}}) - \pi RT_{\text{m}}M_{\text{w}}, \tag{20}$$

Hence,

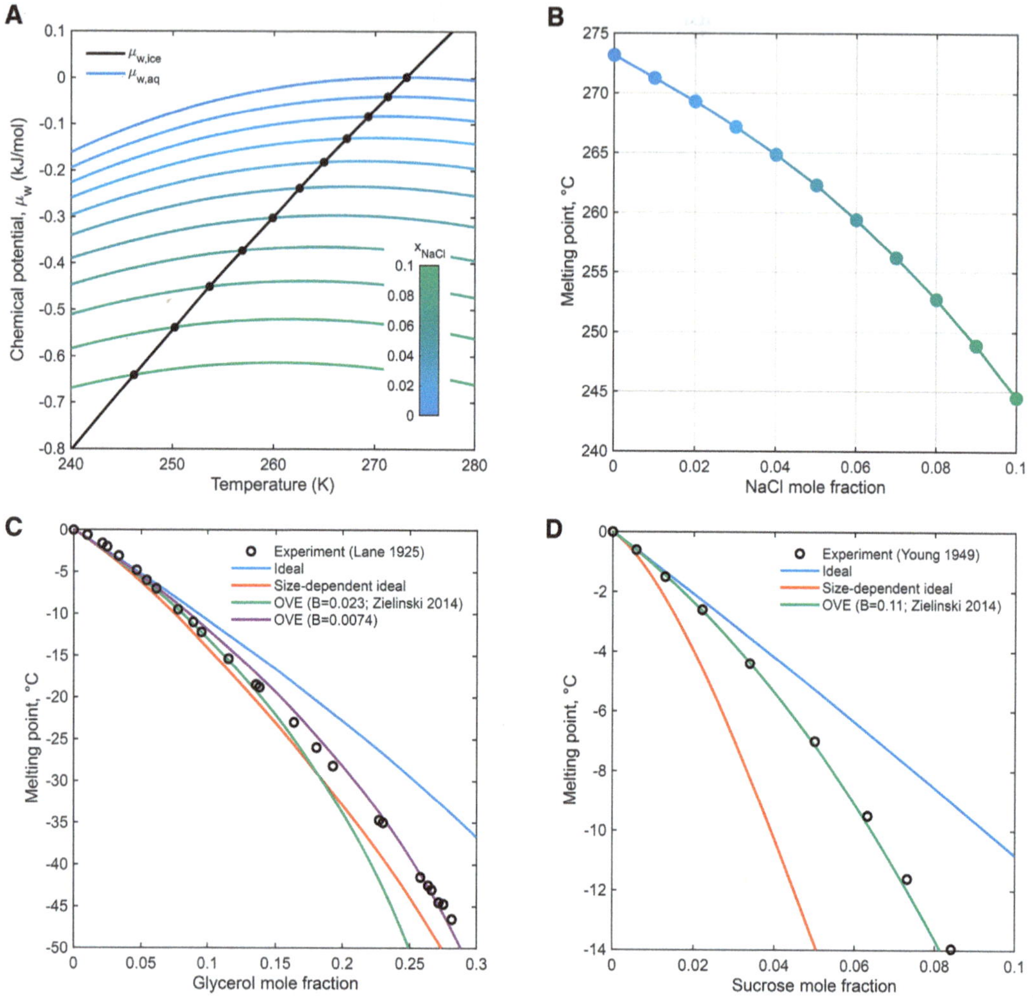

Fig. 1 Equilibrium melting points. (**A**) Chemical potential of water in solution with NaCl and in ice as a function of temperature and (**B**) corresponding liquidus curve. Intersections of the ice and water curves denote the coexistence temperature (i.e., T_m). Melting point as a function of mole fraction for a typical (**C**) permeating solute (glycerol) and (**D**) non-permeating solute (sucrose)

$$\pi = \frac{1}{M_w R} \int_{T_m}^{T_m^0} \frac{\Delta H_f^0}{T^2} dT \tag{21}$$

At the melting point, the enthalpy and entropy are related by $\Delta H_f^0 = T\Delta S_f^0$, and so, the entropy of fusion can be substituted in the place of enthalpy in these relations. This integral form of the above expression is exact for a given $\Delta H_f^0(T)$, but the temperature dependence of enthalpies may not always be known. To a first-order approximation, the enthalpy of fusion may be assumed independent of temperature ($\Delta H_f^0(T) \approx \Delta H_f^0(T_m)$) yielding the 0th-order Gibbs-Helmholtz liquidus relation:

$$\pi \approx \frac{\Delta H_f^0}{M_w R}\left(\frac{1}{T_m} - \frac{1}{T_m^0}\right). \tag{22}$$

This may also be rearranged to provide a closed-form expression for the dependence of the coexistence temperature on the osmolality (or solvent activity in the more general case):

$$T_m \approx \left(\frac{1}{T_m^0} + \frac{\pi R M_w}{\Delta H_f^0}\right)^{-1} \tag{23}$$

Due to the wide availability of tabulated enthalpy of fusion values, the 0th-order Gibbs-Helmholtz relation is particularly useful. A more accurate (though data-intensive) description involves a first-order expansion of the enthalpy of fusion via the heat capacity ($\Delta H_f(T) \approx \Delta H_f^0 + (T - T_m^0)\Delta C_p^0$), yielding

$$\pi \approx \frac{\Delta H_f^0}{M_w R}\left(\frac{1}{T_m} - \frac{1}{T_m^0}\right) + \frac{\Delta C_{p,f}^0}{M_w}\left(\frac{T_m^0}{T_m} - \ln\left(\frac{T_m^0}{T_m}\right) - 1\right) \tag{24}$$

From Eq. 22, we can also derive the cryoscopic constant for water, $K_f = -1.86$ K per osmole/kg, which is a proportionality constant relating the melting point depression (relative to pure water) to the solution osmolality:

$$K_f = \frac{R T_m^{0\,2} M_w}{\Delta H_f^0} \approx \frac{\Delta T_m}{\pi}. \tag{25}$$

With these relations, it is readily possible to compute the melting point of a solution given the osmolality (or the osmolality given the melting point). The 0th-order approximation remains accurate to within 1°C down to –28°C and both the first-order approximation and cryoscopic ratio down to roughly –52°C.

One exemplary application of the first-order Gibbs-Helmholtz relation is the computation of eutectic phase diagrams (liquidus and solidus curves inclusive) using only pure component data (ΔH_f^0 and T_m^0). Alliston et al. [6] recently reported that the size-dependent 0th-order Gibbs-Helmholtz relation provides remarkably accurate predictions of binary eutectic phase transitions in cryobiologically relevant binary solutions. Examples for glycerol and sucrose are depicted in Fig. 2.

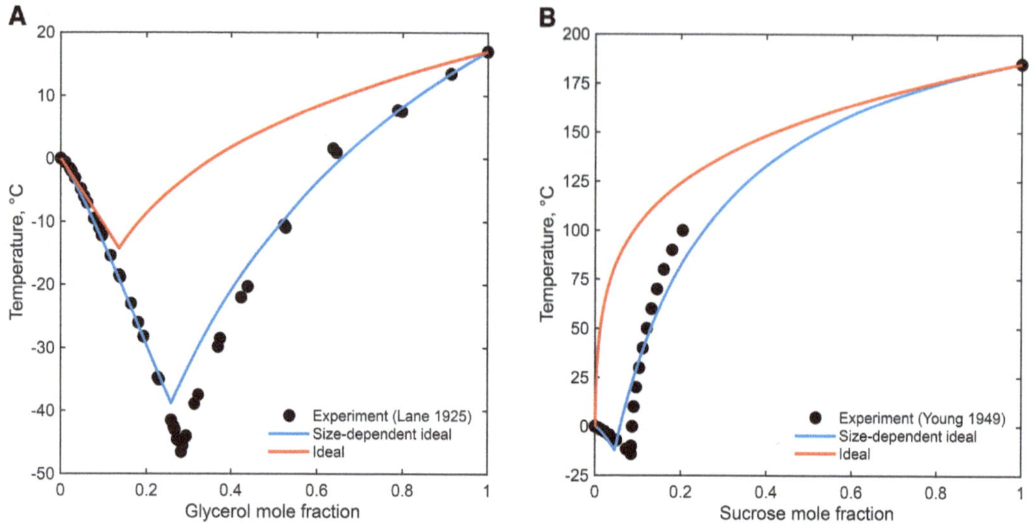

Fig. 2 Eutectic phase diagrams for binary aqueous solutions generated via size-dependent and ideal predictive models. Experimental data from the literature are shown in gray markers for (**A**) glycerol and (**B**) sucrose. Both models require only pure component properties and do not require empirical information on the solution itself

2.1.4 Isochoric Phase Equilibrium

Thus far, much of the discussion has considered systems under isobaric conditions—those in which the system is in contact with a pressure reservoir such as the atmosphere and for which volume is able to freely vary. In isochoric (constant volume or confined) systems [18, 19], the *volume* of the system is controlled instead, and the *pressure* is free to vary. This thermodynamic situation arises when a sample is cryopreserved within a rigid, hermetically sealed container absent air (or any bulk gas phase), referred to in recent literature as an isochoric chamber. Under isochoric conditions, the natural thermodynamic variables are $\{T, V, x\}$, and the according thermodynamic potential describing the system is the Helmholtz free energy. Thus, the equilibrium conditions become

$$\mu_{\mathrm{w,aq}} = \mu_{\mathrm{w,ice}} : \mu^{v}_{\mathrm{w,aq}} - v_{\mathrm{w,aq}} \left(\frac{\partial \mu^{v}_{\mathrm{w,aq}}}{\partial \overline{v}_{\mathrm{w,aq}}} \right)_{T,x} = \mu^{v}_{\mathrm{w,ice}} - \overline{v}_{\mathrm{w,ice}} \left(\frac{\partial \mu^{v}_{\mathrm{w,ice}}}{\partial \overline{v}_{\mathrm{w,ice}}} \right)_{T,x} \tag{26}$$

$$P_{\mathrm{aq}} = P_{\mathrm{ice}} : \left(\frac{\partial \mu^{v}_{\mathrm{w,aq}}}{\partial \overline{v}_{\mathrm{w,aq}}} \right)_{T,x} = \left(\frac{\partial \mu^{v}_{\mathrm{w,ice}}}{\partial \overline{v}_{\mathrm{w,ice}}} \right)_{T,x} \tag{27}$$

$$T_{aq} = T_{ice} \tag{28}$$

where $\bar{v}$ is the partial molar volume and μ_w^v is the partial molar Helmholtz free energy. These equalities are supplemented by the condition that the sum of the volumes of individual phases must equal the system volume (i.e., $v_{sys} = \left(1 - f_{ice}\right)v_{aq} + f_{ice}\, v_{ice}$, with f_{ice} as the volumetric phase fraction of ice).

Figure 3A shows a schematic convex-hull construction [20] of the Helmholtz free energy curves for several phases of water under isochoric conditions, and Fig. 3B shows the resulting isochoric T–V phase diagram for pure water. For many specific volumes, cooling at fixed v necessarily drives the system into a two-phase (ice + liquid) coexistence region, even when the corresponding isobaric path would remain single-phase. Practically, this enables preservation of biological matter at sub-0°C temperatures in the portion of the isochoric system that remains liquid, thereby protecting it from ice formation. For a thorough review of both the theory and practice of isochoric cryopreservation, we refer the reader to Consiglio et al. [19].

2.1.5 Phase Fraction and the Lever Rule

In addition to the temperature at which ice may begin to form, cryobiologists are also often interested in the *amount* of ice that will form at a given temperature. For a binary solution at constant pressure, a typical T–x phase diagram produces a two-phase liquid-ice region beneath the liquidus curve. A system with overall solute mole fraction x_0 cooled to a temperature T such that the coordinates $[T, x_0]$ are inside this two-phase region (Fig. 4A) will separate into

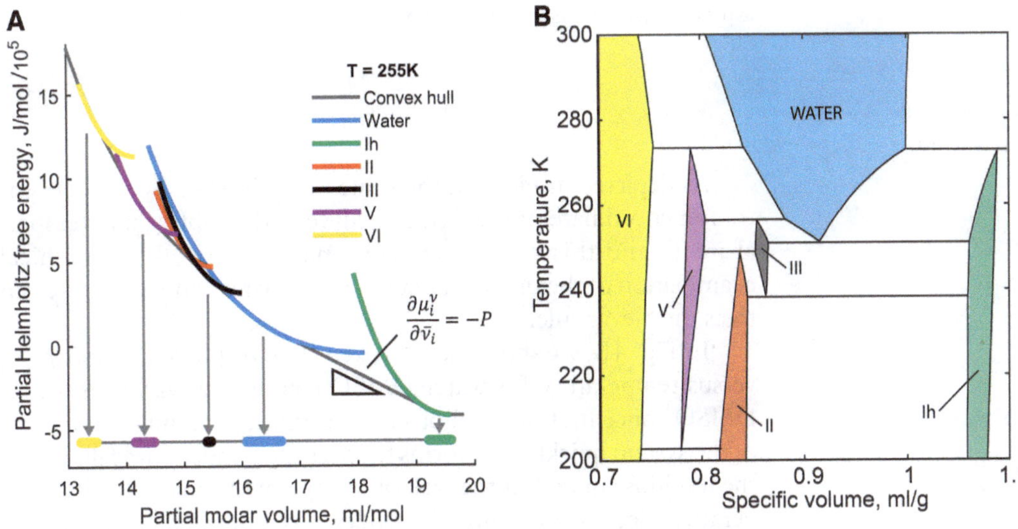

Fig. 3 Isochoric equilibrium: (**A**) Convex hull construction for determination of phase equilibria. (**B**) Isochoric T-V phase diagram for pure water. (Adapted with permission from Consiglio et al. [19])

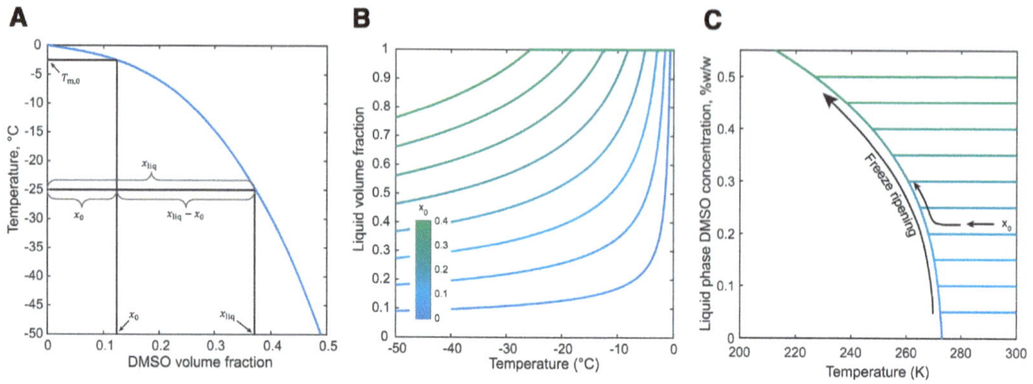

Fig. 4 Isobaric ice–liquid phase fractions and solute ripening processes. (**A**) Construction of the lever rule for a binary solution T-x phase diagram, which enables calculation of the fraction of the system that remains in the frozen or unfrozen state at a given temperature. (**B**) Unfrozen phase fraction versus temperature for various starting concentrations of DMSO in a water-DMSO solution. (**C**) According concentration of the liquid phase during progressive freezing, for the same starting concentrations indicated by the color bar in panel (**B**)

distinct phases with liquid and solid solute concentrations x_{liquid} and x_{solid}. Conservation of mass gives:

$$x_0 = f_{\text{liquid}} x_{\text{liquid}} + (1 - f_{\text{liquid}}) x_{\text{solid}} \tag{29}$$

where f_{liquid} is the phase fraction that remains liquid (in units that match the concentration, be they weight fraction, volume fraction, mol fraction, etc.). In aqueous solutions, the solute concentration in the solid ice phase is typically taken as $x_{\text{solid}} \approx 0$, because ice rejects nearly all solutes. As such, the unfrozen phase fraction f_{liquid} for aqueous solutions simplifies to

$$f_{\text{liquid}} = \frac{x_0}{x_{\text{liq}}} \tag{30}$$

As depicted in Fig. 4A, the value of x_{liq} is given by the concentration coordinate of the liquidus curve at the subfreezing temperature T, and this same expression may be derived by graphical examination of the phase diagram, using what is commonly referred to as the "lever rule."

In Fig. 4B, we show the unfrozen (liquid) phase fraction f_{liquid} versus temperature for water-DMSO solutions at various starting DMSO concentrations x_0. For small starting x_0, a significant fraction of the system quickly transforms to ice as the temperature falls below the liquidus curve, but the rate-of-change of this fraction with temperature decays continuously as deeper temperatures are reached. This behavior is explained by, Fig. 4C, which shows the corresponding liquid-phase concentration $x_{\text{liquid}}(T)$, and illustrates how solute "ripening" from rejection by the solid phase intensifies as freezing progresses. In classical slow-freezing cryopreservation protocols, this progressive concentration of the remaining liquid is deliberately exploited: as extracellular ice forms and solutes are rejected, the external solution becomes more concentrated, drawing water osmotically out of cells and thereby reducing intracellular ice formation. When carried too far, however, the same process leads to extreme extracellular solute concentrations and osmotic stresses, producing the "solution effects" injury central to Mazur's two-factor hypothesis of freezing damage [21].

In an isochoric system filled with pure water, the same logic applies, but the conserved quantity is specific volume rather than composition. For a given temperature T and total system specific volume v_{sys} lying inside the two-phase dome of the isochoric T–v diagram, the coexisting liquid and solid phases have specific volumes v_{liqiud} and v_{solid}. Conservation of volume yields,

$$v_{sys} = f_{liquid}\, v_{liquid} + (1 - f_{liquid})\, v_{solid} \tag{31}$$

So that,

$$f_{liquid} = \frac{v_{solid} - v_{sys}}{v_{solid} - v_{w}} \tag{32}$$

Figure 5 (isochoric lever rule) illustrates this construction graphically: Fig. 5A shows the isochoric lever rule on a specific-volume diagram, and Fig. 5B shows the resulting liquid fraction as a function of temperature for several choices of total specific volume. In practice, the specific volume of an isochoric system is set when the chamber is sealed—typically at 5°C–10°C, giving $v_0 \approx 1\,\text{mL/g}$. Other specific volumes can be accessed by deliberately including a vapor space (increasing v_0) or by compressing the system with a piston (decreasing v_0). If the liquid filling the isochoric chamber is not pure water, but instead an aqueous solution, a three-dimensional lever rule may be devised, which conserves both phase volume *and* phase concentration.

In summary, the tools of multiphase equilibrium thermodynamics—regardless of the specific theories used to compute chemical potential—allow us to map solution composition (and, in isochoric systems, specific volume) to melting points, eutectics, and equilibrium phase fractions. These maps in turn underpin the thermodynamic design and analysis of equilibrium-dominated cryopreservation strategies such as liquidus tracking, partial freezing, and isochoric freezing, and they set the backdrop against which kinetic and transport phenomena act in more complex protocols.

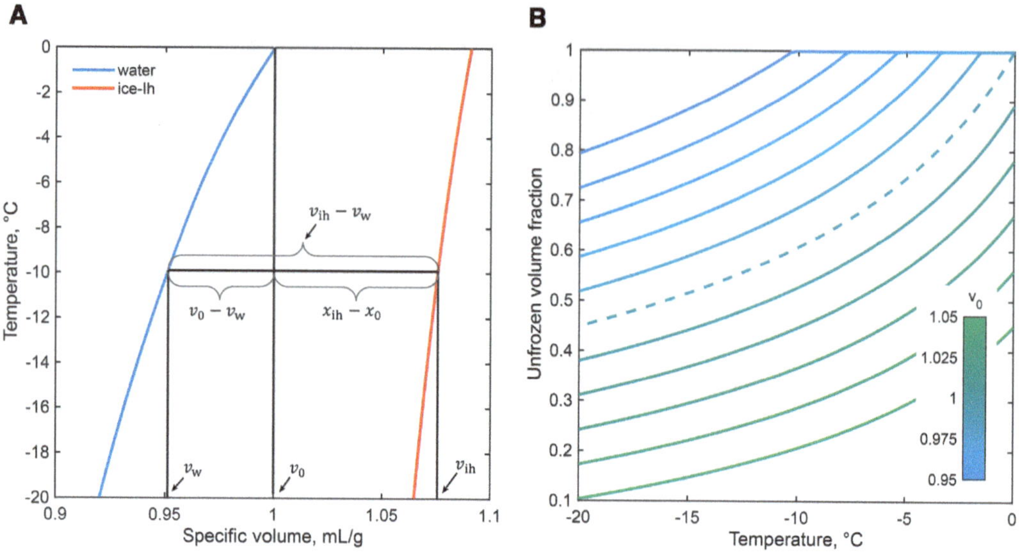

Fig. 5 Isochoric ice–liquid phase fractions for pure water. (**A**) Construction of the isochoric lever rule from the water/ice Ih temperature-specific volume phase diagram derived by Powell-Palm et al. [20]. (**B**) Fraction of isochoric system that remains liquid during isochoric freezing as a function of system specific volume

2.2 Ice Nucleation and Growth

Having established how equilibrium thermodynamics dictates the conditions under which liquid water and ice can coexist, we next consider how a system *transitions* between these phases. In practice, freezing proceeds by the stochastic formation of microscopic ice nuclei, followed by their subsequent growth. The standard framework for describing this process is classical nucleation theory (CNT), which links the statistics of ice nucleation directly to the underlying solution thermodynamics and interfacial properties. We will here review salient aspects of this theory, emphasizing both its practical applications in cryopreservation and its connection to equilibrium thermodynamics.

2.2.1 The Nucleation Barrier

The formative relations of CNT are derived by considering the change in energy in a system between a first state in which the entire system is liquid and a second state, in which a nucleus (a stable cluster of molecules) of a solid phase (here ice) has emerged (Fig. 6).

Assuming the volume of the system is much larger than that occupied by the emerging solid, the free energy of these two states may then be written as:

$$G_1 = \mu^1_{\mathrm{w,aq}} N^1_{\mathrm{w,aq}} + \mu^1_{\mathrm{s,aq}} N^1_{\mathrm{s,aq}}, \tag{33}$$

$$G_2 = \mu^2_{\mathrm{w,aq}} N^2_{\mathrm{w,aq}} + \mu^2_{\mathrm{s,aq}} N^2_{\mathrm{s,aq}} + \mu^2_{\mathrm{w,ice}} N^2_{\mathrm{w,ice}} + 4\pi R^2 \sigma_{\mathrm{iw}}. \tag{34}$$

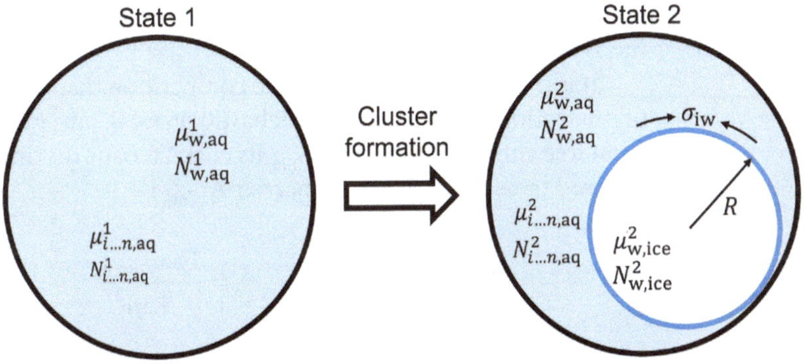

Fig. 6 Conceptual illustration of the change in free energy upon formation of a solid nucleus from a supercooled liquid

In State 1, $\mu^1_{w,aq}N^1_{w,aq}$ gives the free energy of bulk liquid water in solution, and $\mu^1_{s,aq}N^1_{s,aq}$ gives that of the solute. In State 2, the free energy of the bulk incipient ice phase is further given by $\mu^2_{w,ice}N^2_{w,ice}$.

The last term in G_2, which encodes the most critical phenomenological insight provided by CNT, gives the *surface energy* of the ice nucleus, or the energetic and entropic toll exacted upon the system in order to host two phases where previously there was one. The proportionality constant of this energy is given by the *liquid-ice interfacial free energy* σ_{iw}, which describes the energy required per unit surface area of the ice nucleus to bound it from the liquid phase.

Again, assuming a small initial ice nucleus relative to the volume of the system, we may take $\mu^1_{w,aq} \approx \mu^2_{w,aq}$ and $\mu^1_{s,aq} \approx \mu^2_{s,aq}$. It is customary (if arbitrary) also to treat the emerging ice nucleus as spherical, with volume $V = \frac{4}{3}\pi R^3 = v_{ice}N$ and surface area $SA = 4\pi R^2 = (36\pi)^{1/3}V^{2/3}$. Armed with these assumptions, we may now write the *difference* in free energy upon spontaneous formation of an ice nucleus (the difference between States 1 and 2) as:

$$\Delta G_{nuc} = G_2 - G_1 = \frac{4}{3}\pi R^3\left(\frac{\Delta\mu}{v_{ice}}\right) + 4\pi R^2\sigma_{iw}. \tag{35}$$

Critically, the two terms comprising this equation exist in competition. Given the lower free energy of solid ice relative to liquid water at any submelting temperature, the first term (often referred to as the "bulk" term) is negative, and scales with the *volume* of the growing ice phase, representing the spontaneous drive of the system to convert to the more stable phase. Meanwhile, because the segregation of an ordered crystalline phase from the disordered liquid must necessarily *reduce* the entropy of the system, the second term (often referred to as the "surface" term) must accordingly be positive, *increasing* the free energy of the system and scaling with the *surface area* of the nucleus.

It is this surface-versus-volume competition that belies the nature of nucleation as an activated stochastic process, and gives the nucleation free energy difference ΔG_{nuc} its characteristic maximum (Fig. 7), referred to as the nucleation barrier $\Delta G^{\star}_{\mathrm{nuc}}$:

$$\Delta G^{\star}_{\mathrm{nuc}} = \Delta G_{\mathrm{nuc}}|_{\frac{\partial \Delta G_{\mathrm{nuc}}}{\partial N}=0} = \frac{16\pi\sigma_{\mathrm{iw}}^3 v_{\mathrm{ice}}^2}{3\Delta\mu^2}. \tag{36}$$

The random thermal motion of water molecules in the solution causes the system to fluctuate and sample a variety of possible configurations, which may include clusters of water molecules arranged in the cubic or tetrahedral structures defining ice Ic or Ih. The cluster size corresponding to $\Delta G^{\star}_{\mathrm{nuc}}$ gives "critical" cluster size (or "critical radius" if not discretizing by number of molecules), and the fate of any spontaneously assembled cluster of molecules is prescribed by its relation to this critical size.

Any smaller ("sub-critical") cluster, once assembled by a random fluctuation, in order to then decrease its free energy and march towards equilibrium, will be compelled to dissolve, as recruitment of additional molecules would *increase* its free energy in defiance of the 2nd Law of Thermodynamics. Equally, any cluster produced by random fluctuation that is *larger* than the critical size will be compelled

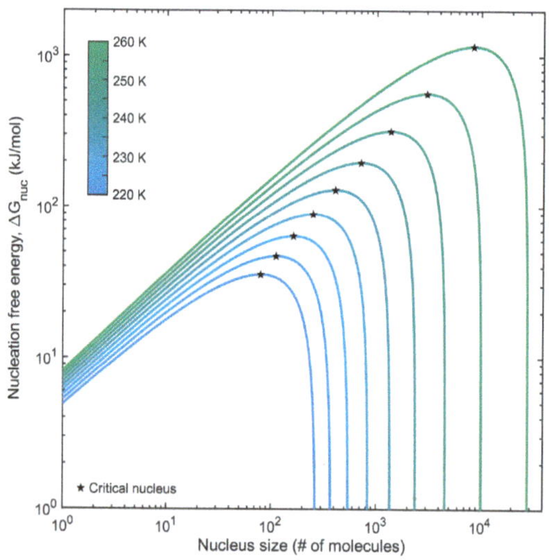

Fig. 7 Free energy change upon nucleation versus ice cluster size for pure water at various temperatures. The nucleation barrier for each temperature is marked by a black star. Consistent with expectation, the nucleation barrier grows exponentially higher as the equilibrium melting temperature is approached, at which point it reaches singularity due to the chemical potential difference in the denominator of Eq. 36 reaching zero

to *grow* continuously until equilibrium is reached, initiating bulk phase crystallization or freezing. It is this critical or supra-critical cluster that we refer to as a stable ice ***nucleus***.

2.2.2 The Nucleation Rate

The nucleation barrier $\Delta G^{*}_{\text{nuc}}$ connects the process of ice nucleation to the underlying bulk thermodynamics of the aqueous solution, via the chemical potential difference $\Delta\mu$ between solid ice and liquid water in solution, and to the surface thermodynamics of the solution via σ_{iw}. In isolation, this barrier provides only relative or trend-level insight into the ease or difficulty of ice nucleation in a given solution at a given temperature. To valorize this information, we may then incorporate this barrier as a rate constant in a probabilistic description of a stochastic activated process (i.e., one driven by random molecular fluctuation), yielding a ***nucleation rate*** J [nuclei formed per m^3/s]:

$$J = J_0 \exp\left(-\frac{\Delta G^{*}_{\text{nuc}}}{k_B T}\right) \tag{37}$$

The nucleation rate J is comprised of two general terms, the exponential driver of the process $\exp\left(-\frac{\Delta G^{*}_{\text{nuc}}}{k_B T}\right)$ and the prefactor $J_0 = \frac{Zf^*}{v_0}$. We refer to the exponential portion as the thermodynamic term, describing a simple classical Arrhenius-style kinetic process for which the characteristic constant is the nucleation barrier. This term represents the probability of a token stochastic fluctuation achieving the critical cluster size. We refer to the prefactor J_0 as the kinetic term, which in turn captures the aspects of the nucleation process that are not captured by connections to bulk thermodynamics, but instead by microscale transport processes. Here, f^* gives the attachment frequency factor, which describes the probability of a given cluster recruiting additional water molecules based on barriers to diffusion in the solution, the Zeldovich factor Z gives the likelihood that a nucleus of precisely the critical size will either dissolve or grow ("roll" to the left or right of the maximum of the nucleation free energy curve in Fig. 7), and v_0 is the molecular volume of ice.

The nucleation rate may thus be thought of as a constant competition between the kinetic and diffusive terms. At higher temperatures, where (for cryobiological applications) solution viscosity is typically relatively low (and thus diffusivity relatively high), the nucleation process is dominated by the kinetic term, and therefore, the underlying solution thermodynamics. As the temperature decreases towards the glass transition and the viscosity increases; however, the diffusive term will eventually come to dominate, preventing the recruitment of new molecules to a potentially critical nucleus, and thereby hindering the nucleation process. This transition of dominance between the two terms is embodied in temperature-time-transformation curves, which show that for any material, there exists a maximum in nucleation rate located at some temperature between that of melting and that of glass transition [22].

The rate equation J further reveals the critical role of the surface effects in the nucleation process; J is an exponential function of the water-ice interfacial free energy σ_{iw} to the third power, introducing extraordinary sensitivity to this parameter. For example, assuming typical values of J_0 for pure water [23], a 10% change in the value of σ_{iw} can change the nucleation rate by >10 orders of magnitude. This sensitivity renders accurate prediction of the precise nucleation rate difficult, and recent studies have suggested that targeting this parameter via molecular or solution design strategies may present a potent engineering lever with which to control cryobiological ice nucleation [9].

2.2.3 Growth After Nucleation

After a critical ice nucleus forms, bulk crystallization will proceed. For pure water, the growth rate of ice in this phase is often modeled using the Wilson-Frenkel relation [24]:

$$U(T) = \frac{D}{a}\left[1 - \exp\left(-\frac{|\Delta\mu|}{RT}\right)\right] \tag{38}$$

wherein D is the water diffusivity, $\Delta\mu$ is the chemical potential difference between water in the liquid phase and ice, and a is characteristic length often specified as the interfacial thickness or water molecular mean free path length. Like the nucleation rate, at small supercoolings, U is limited by the thermodynamic driving force (via $\Delta\mu$), whereas at deeper supercoolings it is limited by diffusion (via D). For cellular systems, the interplay between growth kinetics and cellular water transport rate controls whether ice remains extracellular or propagates into the intracellular space [25].

For aqueous solutions, the same basic Wilson–Frenkel concept can be applied, if we replace the pure-water driving force with the chemical-potential difference of water between the solid and the local solution at the interface. In local equilibrium models, $\mu_{\mathrm{w,ice}}(T, p)$ is set equal to $\mu_{\mathrm{w,liq}}(T, p, x_i)$ along the interface, where x_i is the interfacial solute mole fraction. As ice advances and rejects solute into the liquid, x_i departs from the bulk composition x_∞, and a solute-enriched boundary layer develops in front of the interface. The resulting local freezing-point depression reduces $\Delta\mu$ and progressively slows growth, even at fixed ambient temperature.

Mathematically, sharp-interface models treat growth in solutions by coupling the interface-kinetic relation to conservation laws for heat and solute transport. In a one-dimensional binary solution, the solute concentration $c(z, t)$ obeys a diffusion equation ahead of a moving interface $z = s(t)$, together with a Stefan condition $D\frac{\partial c}{\partial z}\Big|_{z=s^+} = V_n\,(c_l - c_s)$ that enforces solute balance at the front (here, V_n is the interface-normal velocity and c_l and c_s are the solute concentrations in the liquid and solid). When the solid phase excludes solute (typically, $c_s \approx 0$ for ice), this boundary condition ensures that growth is accompanied by a concentrated liquid layer whose thickness and composition are controlled by the competition between interface motion (V_n) and diffusion (D). In the diffusion-limited regime, V_n is set by how rapidly this boundary layer can be relaxed, whereas in the interface-limited regime V_n is controlled primarily by attachment kinetics at the ice surface.

A convenient way to summarize these regimes is to view the observed growth rate as the harmonic mean of an interface-limited velocity V_{kin} and a diffusion-limited velocity V_{diff}, i.e., $V_n^{-1} \approx V_{kin}^{-1} + V_{diff}^{-1}$. V_{kin} can be estimated from Wilson–Frenkel-type expressions that scale with the chemical-potential driving force, while V_{diff} scales with D times the ratio of the available undercooling to the liquidus slope and the characteristic diffusion length ahead of the interface. In concentrated CPA cocktails near the glass transition, D becomes so small that V_{diff} effectively vanishes and growth is arrested even though the thermodynamic driving force remains substantial. In moderately concentrated solutions at higher temperatures, by contrast, D is large enough that interface kinetics still play a significant role and growth can proceed rapidly once nucleation has occurred.

Beyond one-dimensional planar growth, solute interactions with the advancing front also determine the morphology of the ice phase. When the solute boundary layer is sufficiently strong that the local liquidus temperature ahead of the interface drops faster than the actual temperature field, a region of so-called constitutional supercooling is created. Under these conditions, small perturbations of the interface are amplified rather than damped, leading to cellular or dendritic ice structures instead of a smooth front. In cryobiological samples, this translates into complex networks of extracellular ice that can mechanically entrap cells and produce highly non-uniform solute distributions, further complicating subsequent growth and warming behavior.

After the initial growth stage, ice microstructure continues to evolve even at fixed temperature through recrystallization. In broad terms, recrystallization is a curvature-driven coarsening process in which larger crystals grow at the expense of smaller ones. Because the equilibrium melting temperature of a curved interface depends on its radius via a Gibbs-Thomson relation, small, highly curved crystals have a slightly higher chemical potential and lower melting temperature than large, gently curved ones. This difference in chemical potential drives net mass transfer from small to large crystals through melting–recrystallization cycles, leading over time to a reduction in total grain-boundary area and a progressive increase in the characteristic crystal size.

Classical Ostwald ripening theories describe this coarsening in terms of diffusion-limited transport of water through an intervening liquid or glassy matrix, predicting a characteristic length scale that grows as a power-law in time (often $t^{1/3}$ for simple systems). In CPA-rich, partially vitrified, or highly viscous solutions relevant to cryopreservation, the effective diffusivity is strongly temperature dependent, so that recrystallization is negligible at deep cryogenic temperatures but can proceed rapidly when the sample is held or passes slowly through warmer subzero ranges (e.g., between roughly −20°C and −5°C). For cells and tissues, such coarsening can be particularly damaging, because it transforms initially fine-grained, relatively benign ice networks into larger crystals that can disrupt membranes and extracellular matrices.

One of the major rationales for adding so-called ice recrystallization inhibitors (IRIs) to cryopreservation media is precisely to slow or arrest this curvature-driven coarsening. Natural anti-freeze proteins and glycoproteins, as well as synthetic polymers such as certain poly(vinyl alcohols) and related ampholytes, can adsorb selectively to particular ice crystal faces, pinning step motion and reducing the surface mobility needed for grain-boundary migration [26–29]. In continuum terms, these additives effectively reduce the kinetic coefficient associated with interface motion without necessarily changing the underlying thermodynamic driving force. The result is a dramatic suppression of recrystallization over experimentally relevant time scales, which can preserve a fine-grained ice morphology during storage and through slow warming, thereby mitigating mechanical and osmotic injury even when some ice is present.

2.2.4 Heterogeneous Nucleation

In practical cryopreservation, ice almost never nucleates homogeneously in the bulk solution. Instead, nucleation is overwhelmingly *heterogeneous*: it occurs on foreign surfaces such as container walls, extracellular matrices, cellular membranes, or deliberately added ice nucleating agents (INAs). From the perspective of classical nucleation theory, these surfaces lower the free-energy barrier for forming a critical nucleus, thereby shortening the induction time and raising the probability of nucleation at comparatively mild supercooling [30].

The simplest and most widely used description of heterogeneous nucleation is the spherical-cap model. Here, the ice embryo is assumed to form as a cap on an ideal, flat substrate rather than as a full sphere in the bulk liquid. The geometry is characterized by a contact angle θ between the ice, the liquid, and the substrate. Under this construction, the free-energy barrier for heterogeneous nucleation can be written as a simple multiplicative correction to the homogeneous barrier [30],

$$\Delta G_{het}^{*} = f(\theta)\Delta G_{hom}^{*} \quad (39)$$

where ΔG_{hom}^{*} is the barrier derived in the homogeneous case and $f(\theta)$ is the geometric factor that accounts for the reduced nucleus volume and surface area. For $\theta \to 180°$ (perfectly non-wetting substrate), $(f(\theta) \to 1)$ and the surface has no effect. For $\theta \to 0°$ (perfectly wetting substrate), $(f(\theta) \to 0)$ and the barrier may be reduced by many orders of magnitude. Because the nucleation rate depends exponentially on $\Delta G^{*}/k_{B}T$, even modest changes in θ (and thus in f) can transform nucleation from effectively impossible to nearly instantaneous under otherwise identical conditions.

To connect this barrier reduction to observable nucleation rates, we may simply replace ΔG_{hom}^{*} *with* ΔG_{het}^{*} in the CNT rate expression, yielding a heterogeneous nucleation rate [30]

$$J_{het} = J_{0}\exp\left(-\frac{f(\theta)\Delta G_{hom}^{*}}{k_{B}T}\right) \quad (40)$$

where J_0 is again the kinetic prefactor. In the immersion mode most relevant to cryobiology, nucleation is assumed to occur on active sites that are fully wetted by the solution (e.g., a nucleator particle or a patch of the container surface immersed in the liquid). The total nucleation rate is then proportional to both the site-specific rate (J_{het}) and the density (or number) of such active sites, so that both site strength (through θ or equivalent parameters) and site abundance determine the macroscopic stability of a supercooled sample.

Real cryobiological systems, however, rarely present a single, idealized contact angle, and thus are seldom adequately described by the spherical cap interpretation of heterogeneous nucleation. Instead, a given surface in contact with the liquid may host a spectrum of active sites, each characterized by its own local barrier height and kinetics. Mineral particles, polymer-coated chamber walls, proteins embedded in extracellular matrices, and bacterial ice nucleating proteins all provide distinct microenvironments for vicinal water molecules, and thus distinct effective nucleation parameters [31]. Contemporary descriptions of heterogeneous nucleation, therefore, supplement the geometric contact-angle model with active-site distributions, in which each site type is assigned a characteristic barrier (or contact angle) and an areal density. The observed nucleation statistics then arise from an extreme-value competition among these sites: nucleation almost always occurs at the *most active* site present, not at a hypothetical average site (Fig. 9C shows family of extreme value distributions describing active site variability). This framework underlies the extreme-value statistics models increasingly used to interpret cryobiological nucleation experiments and to quantify the stability of supercooled systems [32].

An alternative but complementary view, emphasized in more recent work by Barahona [31, 33, 34], focuses less on the static contact angle and more on ice templating and the dynamics of vicinal water. In this view, an active site is one that organizes nearby water into motifs structurally compatible with ice (thereby lowering the entropic cost of forming an ordered cluster) and/or modifies the local energy landscape so as to reduce the work required to assemble a critical nucleus. Crystalline substrates whose lattice spacings closely match those of basal or prism planes of ice can act as nearly perfect templates, strongly reducing the interfacial free energy and effectively lowering ΔG^{*}. At the same time, adsorption of solutes or macromolecules onto the substrate can slow local water mobility, altering the kinetic prefactor, J_0. Some authors, therefore, introduce an explicit *templating factor* that multiplies the homogeneous rate by a function of both geometric matching and local dynamical slowdown, rather than relying solely on a geometric $f(\theta)$. With the templating factor ζ varying between 0 (no templating) and 1 (full templating), a nucleation site-specific, surface area-normalized heterogeneous nucleation rate can be summarized as

$$J_{\text{het}} = \frac{Zf^{*}(\zeta)}{a_0} \exp\left(-\frac{\Delta G^{*}_{\text{het}}(\zeta)}{k_B T} \right) \tag{41}$$

in which $f^{*}(\zeta)$ is the attachment frequency factor, Z is the Zeldovich factor, a_0 is the molecular cross-sectional area of water, and $\Delta G^{*}_{\text{het}}(\zeta)$ is the surface-dependent nucleation barrier. This rate is plotted in Fig. 8 below for pure water.

From a cryobiological design standpoint, heterogeneous nucleation is both a nuisance and a tool. On one hand, unwanted active sites—including microscopic scratches or contaminants on container walls, undissolved particulates, or patches of gas-liquid interface—dramatically raise nucleation temperatures and undermine attempts to maintain supercooling or achieve vitrification. Mitigation strategies, therefore, focus on minimizing, passivating, or isolating such sites: polishing or coating surfaces, eliminating gas bubbles, overlaying solutions with immiscible liquids, or confining samples in smooth, rigid, air-free isochoric chambers (see Sect. 3.2). On the other hand, for ice seeding and partial freezing protocols, the goal is, precisely to *create* highly active heterogeneous sites in carefully chosen locations (e.g., in the extracellular space or in the perfusate), so that nucleation occurs reproducibly at the modest supercooling and in mechanically benign regions. Engineered ice nucleating agents, from mineral particles to synthetic macromolecules and bacterial ice nucleation proteins, can thus be seen as deliberate manipulations of the heterogeneous nucleation landscape, tuning both the barrier (via geometry and templating) and the site density to produce desired nucleation statistics.

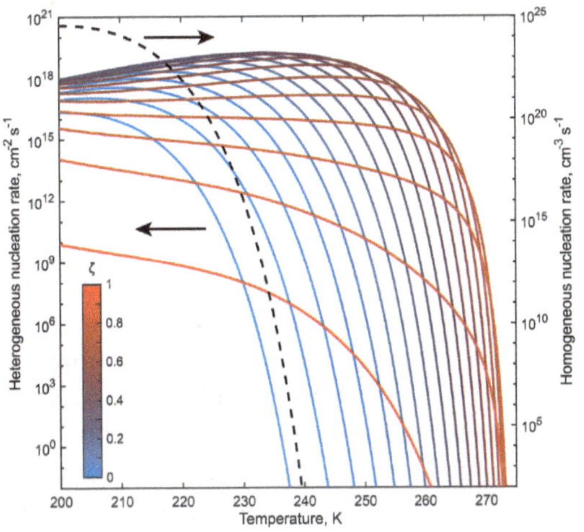

Fig. 8 Surface-scaling heterogeneous and volume-scaling homogeneous ice nucleation rates as a function of temperature. Shown for pure water and various values of Barahona's ice "templating factor" [31], which encodes the degree to which a foreign surface modifies the local free energy landscape so as to reduce the work required to assemble a critical nucleus

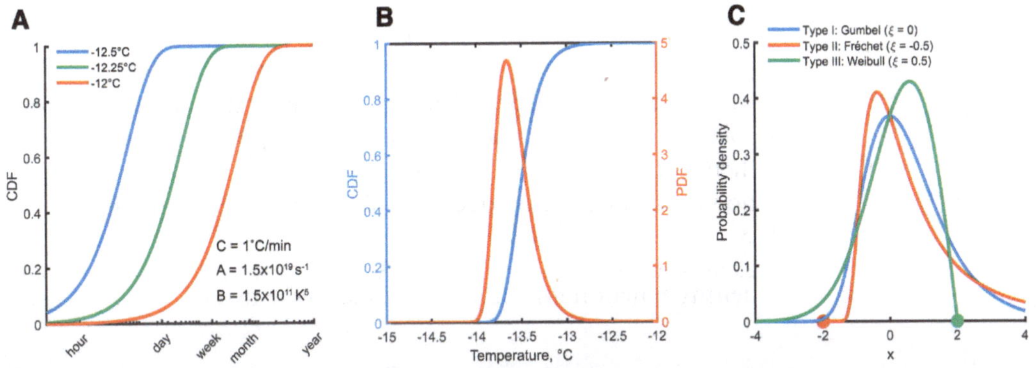

Fig. 9 Nucleation statistics (**A**) Cumulative distribution function for isothermal supercooling (i.e., freezing probability). (**B**) Survivor curve (blue/right) and probability density function for constant cooling rate. (**C**) Extreme value distributions which describe distribution of heterogeneous nucleation active sites. Adapted with permission from Consiglio et al. [32]

In summary, heterogeneous nucleation extends classical nucleation theory by incorporating the effects of real surfaces and interfaces into the free-energy barrier and kinetic prefactor. As such, whether a given cryopreservation protocol succeeds in stabilizing a supercooled or vitrified state, or in achieving a well-timed ice seeding event, may depend less on the homogeneous properties of the bulk solution and more on the nature, distribution, and management of these heterogeneous nucleation sites.

2.2.5 Statistical Aspects of Ice Nucleation

While classical nucleation theory describes nucleation as a *continuous* process in time via the nucleation rate, J, the formation of a single nucleus remains a discrete process: long, seemingly quiescent periods with no events are punctuated by individual instances of critical nucleus formation. Despite the conceptual prevalence of CNT, what the cryobiologist actually *observes* in most experiments and protocols are these discrete nucleation outcomes, e.g., "frozen" vs. "unfrozen," recorded at particular times or temperatures. Connecting these discrete and continuous views of nucleation enables marriage of laboratory outcomes to broader kinetic understanding of the kinetic phenomena at play and requires treating nucleation as a purely probabilistic process.

This may be achieved by use of Poisson statistics, which describe the probabilistic nature of memoryless stochastic events [35]. Within this framework, for a system at temperature T with an effective nucleation rate $J_{sys}(T)$, the expected number of nucleation events accumulated over a time interval $[0, t]$ is $J_{sys}(T) \cdot t$. Assuming that each possible nucleus forms independently, the probability of observing no nucleation events in that interval is

$$S(T,t) = \exp(-J_{sys}(T)t) \tag{42}$$

which is the survival probability of the metastable liquid. The probability of at least *one* nucleation event having occurred by time, t, is then $P(T,t) = 1 - S(T,t)$. Formally, this is the cumulative distribution function of a homogeneous Poisson process with rate J_{sys} (see exemplar distributions in Fig. 9A). The corresponding probability density function for the waiting time to the first event is

$$p(T,t) = \frac{\partial P}{\partial t} = J_{sys}(T)\exp(-J_{sys}(T)t) \tag{43}$$

which is an exponential distribution. The characteristic time until nucleation (the mean waiting time, often referred to as the induction time) is, therefore, the average time $\langle t \rangle$:

$$t_{nuc} = \frac{1}{J_{sys}(T)} \tag{44}$$

More generally, if one wishes to guarantee that the probability of nucleation does not exceed some small failure probability p^* during a hold at temperature, T, the maximum allowable hold time is

$$t_{p^*}(T) = -\frac{\ln(1 - p^*)}{J_{\text{sys}}(T)} \quad (45)$$

For example, $(p^* = 0.01)$ corresponds to a "1% risk" induction time; $p^* = 10^{-4}$ corresponds to the "one-in-ten-thousand" threshold often used in high-reliability engineering contexts. In most cryobiological experiments and protocols, however, the system is not held at a single temperature. Instead, the sample is cooled or warmed according to some protocol, so both T and J_{sys} vary with time. For a general time-dependent temperature history $T(t)$, the survival probability becomes

$$S(T(t)) = \exp\left(-\int_0^t J_{\text{sys}}(T(\tau))\,d\tau\right) \quad (46)$$

This is the defining form of a non-homogeneous Poisson process. A particularly important special case for cryopreservation is constant-rate cooling. If the system is cooled from the equilibrium melting point, T_{m}, at a constant rate, $R = |dT/dt|$, then time and temperature are related by $T(t) = T_{\text{m}} - Rt$, and the integral in the survival function can be re-expressed as an integral over temperature:

$$S_{\text{CCR}}(T) = \exp\left(-\frac{1}{R}\int_{T_{\text{m}}}^{T} J_{\text{sys}}(\theta)\,d\theta\right) \quad (47)$$

This is the non-homogeneous Poisson distribution in temperature, used in constant-cooling-rate experiments to extract nucleation rates from distributions of observed freezing temperatures. Differentiating with respect to temperature yields the probability density for nucleation to occur at temperature, T:

$$p_{\text{CCR}}(T) = \frac{\partial P_{\text{CCR}}}{\partial T} = \frac{J_{\text{sys}}(T)}{R}\exp\left(-\frac{1}{R}\int_{T_{\text{m}}}^{T} J_{\text{sys}}(\theta)\,d\theta\right) \quad (48)$$

An example cumulative distribution function and probability density function are shown in Fig. 9B.

The characteristic nucleation temperature, T_{nuc}, can then be defined in several equivalent ways; for example, as the mean of this distribution,

$$T_{\text{nuc}} = \int_0^{T_{\text{m}}} T p_{\text{CCR}}(T)\,dT \quad (49)$$

or, more practically, as its mode or median. In systems, where J_{sys} varies extremely rapidly with undercooling (as is typical in CNT), $p_{CCR}(T)$ becomes sharply peaked, so that the "nucleation temperature" is very well-defined even though the underlying process is stochastic.

Two additional features of the Poisson description are particularly important for cryobiology. Firstly, in the simplest homogeneous case, the system nucleation rate is proportional to the system volume, so larger samples have proportionally higher nucleation rates and correspondingly shorter induction times. This makes it much easier to maintain deep supercooling or vitrification in droplets or thin films than in organs or large tissues, even if the underlying solution thermodynamics and interfacial properties are identical. More refined models replace the single volume, V, with a sum over compartments (e.g., intracellular vs. extracellular, or separate regions of an organ), each with its own local J and effective volume.

Secondly, in heterogeneous nucleation, the system often contains a spectrum of active sites, each with its own local nucleation rate $J_i(T)$. If each site is modeled as an independent Poisson process, the system-level survival probability is product of each independent survival probability, and the system level nucleation rate is a sum of the individual active site nucleation rates. When the J_i span many orders of magnitude, nucleation almost always occurs at the single most active site present, making the observed distribution of induction times or nucleation temperatures an extreme-value statistic of the underlying site population rather than a simple reflection of an average rate. This is the basis for active-site and extreme-value models that infer the strength and abundance of heterogeneous nucleation sites from experimental nucleation statistics in supercooled cryobiological systems [32].

In practice, these statistical relations provide the bridge between measured freezing temperatures and underlying nucleation physics, and they supply the most directly useful design quantities for cryopreservation. The statistical metrics summarized here underpin many of the emerging cryopreservation technologies discussed below. They enable rational selection of cooling and warming profiles, storage temperatures, and sample geometries that keep the cumulative nucleation probability acceptably low (for supercooling and vitrification) or, conversely, guarantee prompt nucleation at desired locations and times (for ice seeding and partial freezing). We will utilize these statistics in Case Study 2 later in this work.

2.3 Heat and Mass Transport

Now having formulated a description of the thermodynamic equilibrium states that cryobiological water may occupy and the molecular-level kinetic process that facilitates transition between these states, we arrive at the final physics that connect the practitioners and protocols of cryobiology to these phenomena—heat and mass transfer.

In practice, the cryobiologist has two principal means by which to control the outcome of a cryopreservation process, in terms of the final states attained and the phase transition route followed to arrive at those states: the first is chemical design of the cryobiological solution employed; the second is choice of thermal and mass transport history (cooling rates, warming rates, CPA loading rates, etc.). Historically, cryobiology literature has addressed the latter much more systematically than the former, and (for better or worse) much of the philosophy of cryopreservation protocol design today focuses on engineering of thermal and mass flow histories. Accordingly, there are many excellent reference works addressing cryobiological heat and mass transfer, and we will refer the reader to them rather than discussing comprehensively herein. We particularly recommend the formidable recent review by Zhao and colleagues [36], a salient chapter of the recent cryopreservation-focused *Annual Review of Heat Transfer* [37], and Chaps. I-3 and II-6 of this book.

Instead, we will here highlight the oft-overlooked intersections of heat and mass transport with the equilibrium thermodynamics and phase change kinetics at play, with a particular focus on relevant transport properties affecting those frameworks.

2.3.1 Heat Transport—Key Concepts

The temperature history of a sample is one of the most important manipulable aspects of a cryopreservation protocol, underlying Mazur's famous two-factor hypothesis [21]; the process of vitrification [38]; and many other cryobiology essentials. However, the effect of this temperature history is nearly meaningless without its connection to the thermodynamics and kinetics at play, which it affects through the following broad aspects:

Temperature-Time Dependence of Ice Nucleation and Growth

As described in the preceding section, the nucleation rate and probabilistic induction time of ice formation are functions of thermal history, a fact that may be leveraged in several ways. A few examples include: (1) During the slow freezing of cells, where extracellular ice nucleation is desired, the probable induction time of nucleation may be calculated as a function of cooling profile for the extracellular and intracellular compartments separately (even distinguishing only by compartmental volume), and a cooling profile that ensures a significantly smaller induction time for the extracellular compartment may be chosen. This is the principle behind "interrupted freezing" protocols [39]. (2) During the vitrification of any biologic, the cooling profile required to reach the glass transition temperature in less than one induction time (corresponding to perfect vitrification, and the most rigorous form of the "critical cooling rate") may be estimated. (3) During supercooling, a cooling profile and storage temperature may be calculated such that the induction time for ice nucleation is much longer than the desired storage period. The cooling profiles required will be functions of both the underlying solution thermodynamics and the heterogeneous surfaces with which the system is in contact [32, 40].

Temperature-Time Dependence of Metabolism and Toxicity

The core mechanism enabling cryopreservation is the arrest of metabolism, which scales exponentially with temperature (see Fig. 15 in Case Study 2 below). The toxicity of cryoprotectants, the rate of which is driven both by the rate of metabolism and by the rate of various potential biochemical reactions and phase change processes, is likewise exponentially slowed upon cooling from normothermia. As a result, the temperature history of a given protocol is inextricably linked both to the plausible period of preservation achievable (based strictly on metabolism [41]) and to the toxicity accumulated by the sample. Thus, if the functional form of the temperature dependence of metabolism is known, metabolism arrest (i.e., preservation benefit) may thus be mathematically coupled to cryopreservation physics, as demonstrated by Consiglio et al. in the case of organ-scale supercooling [42]. Likewise, if the functional form of the temperature-dependence of toxicity rate is known, accumulated toxicity may be mathematically coupled to cryopreservation physics, as demonstrated by Benson and colleagues in the case of cellular vitrification [43–45].

Temperature-Time Dependence of Thermomechanical Stresses

Though a topic that has received comparatively minimal treatment historically, growing interest in cryopreservation of large biological samples (e.g., organs and organisms) has driven recent interest in the calculation and avoidance of thermal stresses that may accumulate due to either phase change or differential thermal contraction across a large, low-thermal-diffusivity sample during cooling. The first physics-based investigations of cryopreservation-related thermal stresses pertained to those developed during freezing [46], wherein the dynamic solidification front may drive significant local stresses. More recently, focus has turned to the considerable challenge of glass cracking during vitrification, which may manifest at even the microliter sample scale [47], and which proves especially challenging in the vitrification of macroscale tissues and organs [48, 49]. Cryopreservation thermomechanics presents a particularly rich intersection of solution thermodynamics and transport phenomena, with recent work demonstrating sharp dependences both on thermodynamic properties of the system (glass transition temperature, thermal expansion coefficient [50]), transport properties (thermal conductivity, viscosity), and precise sample geometry and localized temperature history [51].

Temperature-Dependent Thermophysical Properties

We note finally that heat transport and the thermal history of a sample are coupled to every piece of physics described heretofore by the temperature-dependence of both thermodynamic properties (i.e., first or higher-order derivatives of the free energy, such as density, entropy, chemical potential, thermal expansion coefficient, heat capacity, etc.) and transport properties (thermal conductivity, viscosity, etc.). However, while this coupling is well appreciated, there remains an acute lack of low-temperature thermophysical property data for most solutions of interest to cryobiology (and a similar lack of predictive tools to cover this data gap), presenting a meaningful challenge to empirical accurate thermophysical description of various cryopreservation processes.

Gratefully, recent works indicate a growing, field-level effort to address this critical need for low-temperature and cryobiology-relevant thermophysical data. We point the reader to several key datasets commonly used in recent cryobiological modeling: that assembled by Rabin and colleagues, relevant to thermomechanics analysis of DMSO-based vitrification media [52, 53]; by Bischof and colleagues, relevant to critical cooling and warming rates for vitrification [54]; by Elliott and colleagues, relevant to melting point depression of various cryobiological solutions [10]; by Powell-Palm and colleagues, relevant to the viscosity of many-component aqueous solutions [55]; and by Consiglio and colleagues, relevant to high-pressure phase equilibria [19, 56, 57].

2.3.2 Mass Transport—Key Concepts

Mass transport affects the practice and outcomes of cryobiology through various important processes and effects, both molecular (e.g., via the diffusivity of water in solution, which critically affects the nucleation process [58]) and biological (e.g., via the delivery of cryoprotectants across cell membranes [58]). Gratefully, a thorough treatment of cryobiological mass transport by Higgins and colleagues is included in Chap. I-3 of this book, covering delivery of cryoprotectants into cells, tissues, and organs (by both diffusion and perfusion), the fundamental role of osmotic stress in designing cryopreservation protocols, etc. For additional reading on the specific interplay between cryobiological mass transport and equilibrium solution theory, see the works of Elliott and colleagues [11, 59]. For additional reading on the specific interplay between cryobiological mass transport and ice nucleation and growth processes, see these various references [58, 60–67].

3 Emerging Techniques for Cryopreservation of Complex Biomaterials

Armed now with sufficiently diverse thermodynamic theory to analyze the physical aspects of nigh-any cryopreservation protocol, we proceed to review emergent cryopreservation approaches through the lens of this theory. Our strategy in the text to follow is to provide an accessibly comprehensive survey of new or underexplored techniques and philosophies (and especially those for which alternative review works do not yet exist); to aid the reader in identifying and understanding the relevant physical theory guiding these approaches; and, where possible, to demonstrate the power of the theoretical tools articulated in Sect. 2 to enhance both interpretation and practice of these protocols.

For consistency with our theoretical approach, we organize the emergent technologies discussed herein by the theoretical aspect—equilibrium thermodynamics, kinetics, or transport—that most principally drives their design or practice. We also include a final section dedicated to vitrification, as a prototypical technique drawing equally from all three aspects. For each section, we will review broadly the relevant techniques, then choose one on which to perform an illustrative case study in the application of relevant theory to meaningful, technique-specific analysis.

3.1 Equilibrium Techniques

While all cryopreservation protocols are beholden to relevant phase equilibria (e.g., melting points, eutectic points) and thermodynamic properties (e.g., heat capacity, thermal expansion coefficient), only a particular subset are largely decoupled from the short time-scale phenomena associated with nucleation and transport processes. For these privileged few techniques, equilibrium solution thermodynamics dominate protocol design and practice, and they thus lend themselves to relatively simple (and powerful) first-order analyses.

3.1.1 Liquidus Tracking

The technique perhaps most emblematic of the equilibrium-driven approach is *liquidus tracking*. As its name suggests, this approach aims to simultaneously introduce a cryoprotectant solution into a sample whilst reducing its temperature, such that the sample follows (in both composition and temperature) the liquidus curve of the solution (Fig. 10). By maintaining the sample at the lowest temperature at which the possibility of freezing may be systematically avoided (the liquidus temperature), this process minimizes cryoprotectant toxicity during loading to the highest degree achievable in a globally stable state and provides the fastest means of cooling an organ to sub-0°C temperatures (starting from a CPA-unloaded, supra-0°C state) without accepting *any* degree of ice formation risk.

First proposed by Farrant in 1965 [68] and powerfully realized by Elford and Walter in 1972 [69], liquidus tracking in the modern era has been nigh-exclusively implemented as a CPA-loading strategy, as opposed to a final storage method. In practice, a sample is shuttled along an equilibrium concentration-temperature path towards a target final CPA composition, from which a final cooling step (typically to beneath the glass transition temperature) may then be performed with minimal risk of freezing (even while departing then from global equilibrium). This approach was formalized by Pegg and colleagues in the preservation of cartilage [70], wherein controlled stepwise introduction of DMSO and cooling of the sample, up to a terminal

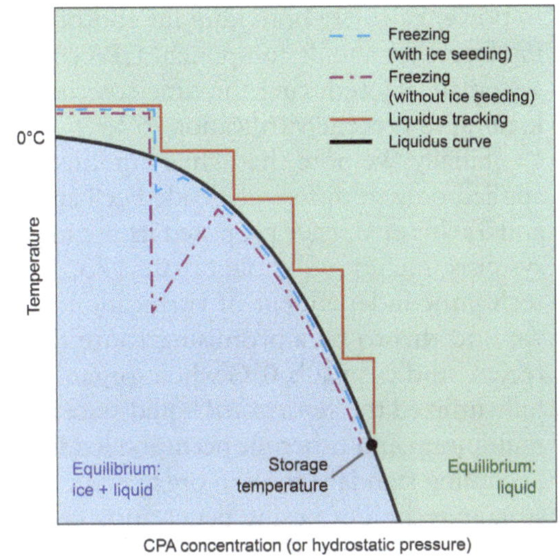

Fig. 10 Conceptual illustration of the thermodynamic paths followed by various equilibrium cryopreservation techniques. The liquidus curve, defining the melting point of ice as a function of CPA concentration or pressure, is given by the thick black line. In liquidus tracking (thin red line), a biological sample is simultaneously CPA-loaded and cooled, such that it follows the liquidus curve. This approach enables CPA-loading and sample storage at the lowest possible temperatures (and accordingly toxicities) at which the ice formation is not a risk. In partial freezing (dashed lines), a sample is loaded with an initial concentration of CPA, cooled below the liquidus, and allowed to form a two-phase liquid-ice solution. The amount of ice that will form at a given temperature is prescribed by the lever rule, and freeze-driven solute rejection will ripen the remaining liquid to the liquidus concentration at that same temperature. Seeding ice can minimize the osmotic shock accompanying this solute rejection (and mechanical damage from crystallization) upon initial nucleation from the supercooled liquid. Isochoric freezing follows an analogous thermodynamic path to partial freezing, but in either temperature-pressure space (if the liquid medium in the chamber is pure water, as is commonly used), or in 3D temperature-pressure-concentration space. We note that in partial freezing however ice forms *within* a biological sample, while in isochoric freezing ice forms *outside* of the sample

concentration of ~60% by mass and a terminal loading temperature of −70°C (approximately 10°C above the according liquidus temperature, as a margin of safety), permitted complete avoidance of ice formation during CPA equilibration, and achievement of tissue vitrification thereafter.

Pegg's approach, explored further in cartilage by his group [71], and in encapsulated 3D cell constructs by Fuller and colleagues [72], relied largely on preliminary empirical quantification of both liquidus temperatures as a function of composition, the rate or degree of CPA penetration at each step, and other protocol parameters. More recent work has advanced the technique towards its logical optimum, leveraging the imminent modelability of liquidus curves to predictively design optimal temperature-concentration protocols. In perhaps the most rigorous example of rationally-designed liquidus tracking to date, Shardt, Elliott and colleagues used solution theory (in particular multicomponent osmotic-virial equations, as described in Sect. 2.1) to power calculate both liquidus coordinates and permeation driving forces for a multicomponent CPA cocktail in articular cartilage, systematically reducing the time-toxicity profile of cryoprotectant loading to precede vitrification [59].

Finally, we note that while liquidus tracking has to date been applied only to diffusively loaded cell and tissue constructs, Wowk and Taylor et al. have proposed its use in perfusion-loaded vascular systems, including whole organs [73, 74], and as a final storage technique independent of vitrification. In light of the following, we find this to be a promising route for future exploration: (1) recent studies in sub-0°C whole-organ preservation have successfully utilized the *supercooled* liquid state for storage [3, 75–77]; (2) multipump, multithermic perfusion loading systems are increasingly becoming standard tools in organ cryopreservation; (3) metastable storage techniques using penetrating cryoprotectants only achieve ~1°C–2°C supercooling, introducing freeze-risk that may outweigh the benefit of the marginally colder temperatures enabled; and (4) a wide diversity of new cryoprotectants show minimal toxicity at concentrations (3–6 molal [78, 79]) that may enable significantly colder preservation (~−10°C to −20°C) than has yet been achieved in ice free preservation of whole organs.

3.1.2 Partial Freezing

While liquidus tracking uses thermodynamic equilibrium to ensure complete avoidance of ice in a biological sample, the same liquidus equilibria may be leveraged to *allow* freezing in the sample but systemically limit its extent. This approach, demonstrated by Ishine, Rubinsky and Lee in the 1990s [80, 81] and later named "partial freezing" by Tessier and colleagues [82–84], occupies the equilibrium corridor between liquidus tracking and classical slow-freezing, targeting temperature-composition coordinates that situate the sample beneath its liquidus temperature but above the eutectic temperature, rendering a thermodynamically-stable two-phase liquid-ice equilibrium within, with the amount of ice prescribed by the lever rule (Fig. 10).

Use of partial freezing, as opposed to liquidus tracking, is motivated by the deeper temperatures of preservation achievable per-unit cryoprotectant loaded, at the cost of bulk ice formation. However, we note that *at equilibrium*, due to solute rejection, the cryoprotectant concentrations *in the portion of the system which remains liquid* will be precisely identical to those that would enable liquidus tracking at the same temperature, raising intriguing questions as to whether the benefits of reduced load concentrations outweigh the dangers of ice formation.

This technique has been developed principally for the preservation of vascular organs, in which a cryoprotectant solution may be delivered via machine perfusion or manual flush, and this ice formation may be strategically initiated in the vascular space, where it is hypothesized that it may nucleate, grow, and reject solutes with minimal mechanical interference. This initiation has been observed to accompany risk of damage to the endothelium [80, 82], both by mechanical mechanisms related to ice growth and biochemical mechanisms related to solute ripening. However, preliminary ex vivo evaluations of mammalian livers [82, 84] and kidneys [85] nonetheless appear promising.

In order to more deeply interrogate the thermodynamic and physical implications of partial freezing, and to demonstrate a prototypical thermodynamic analysis of a two-phase liquid-ice system, we conduct a Case Study of the partial freezing technique described by Tessier et al. at the end of this section.

3.1.3 Isochoric Freezing

The vast majority of cryopreservation protocols are performed under a constant (typically atmospheric) pressure, wherein the volume of the sample may freely expand or contract in response to phase change, temperature change, composition change, or any combination thereof. In 2005, Rubinsky and colleagues proposed the opposite: to constrain the *volume* of a sample, thereby allowing its *pressure* to vary in response to these same changes [18]. This process, called isochoric (constant volume) or confined freezing, seeks to leverage hydrostatic pressure passively generated by the cooling process to depress the melting point of water, thereby protecting a biological sample from ice formation at temperatures beneath its atmospheric-pressure melting point.

In practice, isochoric freezing typically involves submerging a biological sample in an aqueous medium within a rigid, high-strength chamber; sealing the chamber absent any air within; cooling the chamber to a sub-0°C temperature; and initiating ice nucleation at a position in the chamber distant from the biologic (see Fig. 11). The growth and expansion of this ice will then, given its rigid confinement and isolation from the atmosphere, drive pressurization (on the order of 10–210 MPa) of the remaining liquid in the system (which contains the biologic), the melting point of which is in turn depressed until the ice and liquid achieve a two-phase equilibrium, as prescribed by the pressure-temperature-composition (p-T-x) liquidus curve.

As such, isochoric freezing represents the same thermodynamic principle as partial freezing: ice is deliberately nucleated and allowed to "ripen" the remaining liquid (whether in concentration, in pressure, or in both) until thermodynamic equilibrium is reached. However, two key differences distinguish these techniques. First, in isochoric freezing, the limited formation of ice within the system occurs *outside* the biological sample, as opposed to within it, thereby providing ice-free preservation at sub-0°C temperatures. Second, isochoric freezing may be practiced with or without cryoprotectants, as pressure depresses the melting point of liquid water universally. As such, cryoprotectants in isochoric freezing carry only

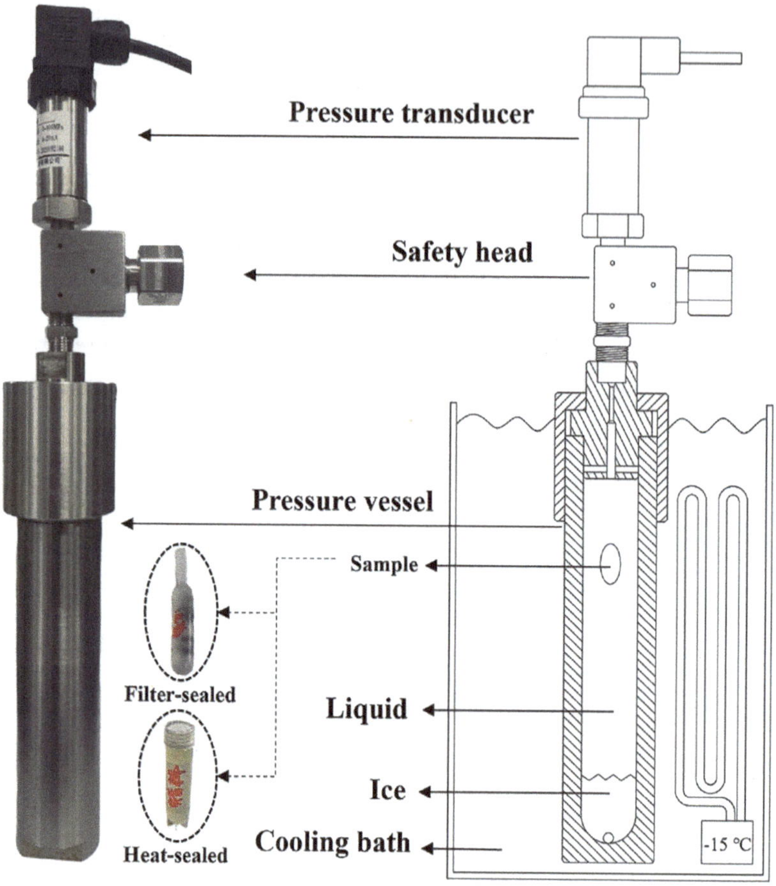

Fig. 11 Isochoric freezing chamber and schematic (adapted with permission from Lyu et al. [86])

a portion of the burden of ice protection, enabling ice-free preservation at lower cryoprotectant concentrations at a given temperature [87, 88]. However, just as the temperature advantages of partial freezing are weighed against the potential dangers of ice formation, so the advantages of isochoric freezing are weighed against the potential dangers of pressure.

For a deeper discussion of the principles of isochoric freezing, in addition to a comprehensive summary of its application to multiscale biologics ranging from mammalian hearts to fresh food products, we refer the reader to our recent review of this topic [19].

3.1.4 Case Study 1: Thermodynamic Analysis of Chemical Evolution During Partial Freezing

Partial freezing, in the limit of prompt nucleation and slow cooling rates, lends itself to straightforward analysis with the toolkit of equilibrium solution thermodynamics. In this Case Study, we will demonstrate simple applications of this toolkit to the interrogation of a partial freezing solution and protocol reported by Tessier et al.

[1], and in doing so illustrate routes towards rational design of partial freezing solutions and protocols. A general process is as follows:

1. Select a multisolute solution thermodynamics model to describe the chemical potential of the partial freezing solution of interest and utilize it to calculate the melting point.
2. Acknowledging that ice forms a pure solid phase, effectively removing water the liquid solution, calculate the partial freezing-relevant liquidus curve by iteratively decrementing the concentration of water whilst holding all other component concentrations constant, and recalculating the melting point for each decrement.
3. Use this liquidus curve and the lever rule to calculate the amount of ice that will form at a given subfreezing temperature, as a fraction of the total system mass or volume. Calculate the according concentrations of total or individual solutes by removing the calculated amount of ice from the initial liquid water content of the solution and recalculating the concentrations.
4. Use these insights to choose an appropriate storage temperature, based on desired limits of ice formation, terminal concentrations of cryoprotectants or electrolytes in the unfrozen liquid phase, etc. Alternatively, use the steps above to ***design*** a solution that provides optimal values of the aforementioned.

Using the partial freezing solution composition provided by Tessier et al., we show prototypical outputs of this process in Fig. 12. For each of the three variants investigated by the authors (which differ only in the choice of penetrating cryoprotectant, held at 12% vol/vol), we use the multisolute osmotic virial equation to calculate the chemical potential of the liquid, and its relation to calculate the melting point (see Sect. 2.1). Solution recipe and all virial coefficients are shown in Table 1 (note that we approximate the 3-O-methyl-glucose employed as glucose monohydrate, as thermophysical properties of the former are not available). Electrolytes for which virial coefficients are not available are treated as ideal. We calculated the unfrozen solution osmolalities and melting points to be $\pi_0 = 3.16$ Osmol/kg and −5.9°C for propylene glycol, be $\pi_0 = 3.07$ Osmol/kg and −5.7°C for glycerol, and $\pi_0 = 3.8$ Osmol/kg and −7.1°C for ethylene glycol.

In Fig. 12, for each choice of penetrating cryoprotectant, we show the fraction of the system that freezes, the total solute concentration in the remaining unfrozen liquid, and the individual concentrations of each component as functions of temperature at thermodynamic equilibrium. Several salient insights emerge, which we may evaluate against the observations of Tessier et al. in their histological and functional evaluations of partially frozen rodent livers.

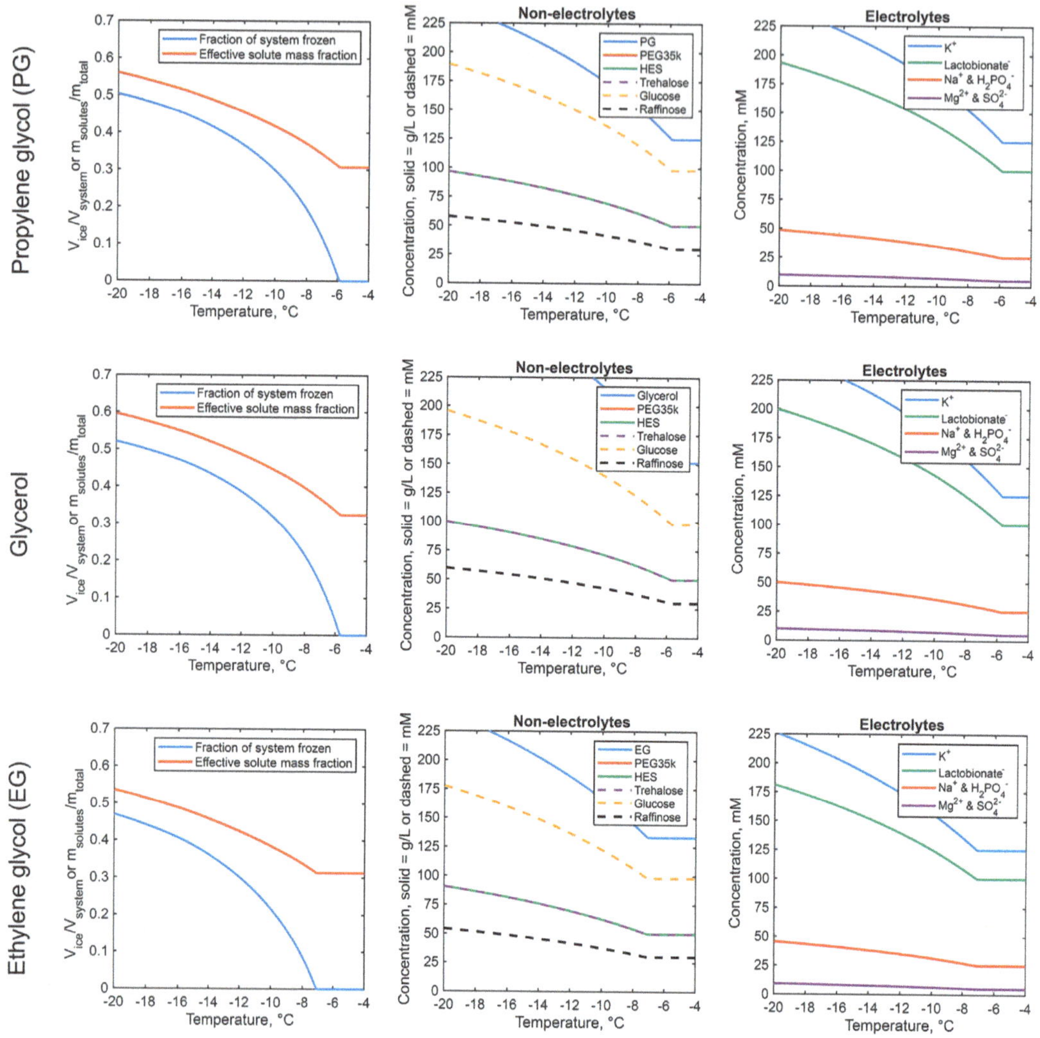

Fig. 12 Thermodynamic analysis of a partial freezing solution. We analyze the compositions reported by Tessier et al. [1], which vary only by the choice of penetrating cryoprotectant. Results for each cryoprotectant are organized by row as labeled. The first column gives the ice fraction and the effective solute mass fraction in the remaining liquid phase at thermodynamic equilibrium, as a function of temperature. The second and third columns give the freeze-driven evolution of the concentrations of each chemical component as a function of temperature

They first investigated differences between the three cryoprotectants at −10°C, observing glycerol to be broadly damaging across metrics, and the two glycols to be less so (though in different respects). Regardless of the known osmotic drawbacks of glycerol attributable to its slow penetration, these findings appear generally predictable based on Fig. 12, which shows that glycerol both allows the most ice formation and drives the largest electrolyte imbalances in the remaining liquid phase (e.g., ~178 mM potassium) at −10°C.

Table 1

Partial freezing solution and corresponding osmotic virial parameters. Solution composition from Tessier et al. [1] and virial parameters from Zelinski et al. [89] (PEG-35,000 virial coefficient from Wang et al. [90])

Component	k_{diss}	B_i	C_i	g/L	mM	mol/kg
HES	1	35.29	-1.107×10^7	50	0.25	3.38×10^{-4}
Lactobionic Acid	1	0	0	35.83	100	0.135
Potassium Phosphate monobasic	2	0	0	3.4	25	0.0338
Magnesium Sulfate (heptahydrate)	2	0	0	1.23	5	0.00674
Raffinose (pentahydrate)	1	0.08	0	17.83	30	0.0405
Potassium Hydroxide	1	0	0	5.61	100	0.135
Sodium Hydroxide, 5 N	1	0	0	1	25	0.0338
PEG-35,000	1	5200	0	50	1.4	0.00193
Trehalose (dihydrate)	1	0.12	0	18.92	50	0.0676
Glucose (monohydrate)	1	0.044	0	19.42	98	0.132
Propylene glycol	1	0.039	0	124.8	1640	2.22

Among ethylene glycol and propylene glycol, the authors observed marginally superior structural preservation (in histology) for ethylene glycol, consistent with its relatively lower ice fraction, but marginally worse functional outcomes in some cases nonetheless, which may be attributable to mild differences in biochemical toxicity or osmotic stress profile of the CPA at sub-0°C temperatures, or an insufficient normothermic perfusion timeline to observe complete recovery, which has been found in select other models to take longer than 1 h after prolonged sub-°C storage [91].

The authors proceeded to investigate propylene glycol at −10°C and −15°C, finding that storage at −15°C, despite the putative benefit in metabolism reduction (see Fig. 15D in Case Study 2), yielded more extensive hepatocellular injury, loss of liver sinusoidal endothelial cells, and ultimately poorer functional outcomes. Per Fig. 12A, this additional morphological and endothelial damage is likely a consequence of the nearly 50% increase in ice fraction (from ~0.3 to ~0.44) upon shifting from −10°C to −15°C. Intriguingly, the total intercellular space in the liver has been measured as ~30%–40% of the total liver volume [92], which may further explain the relative palatability of partial freezing at −10°C, though it is unclear whether functional degradation is driven more by mechanical aspects of ice formation, or by the rate and degree of electrolyte ripening in the remaining liquid phase (which is of course coupled to said ice). We note that in later work [84], the same authors saw somewhat improved outcomes at −15°C upon incorporating more polyethylene glycol, which may serve to retard both ice growth and the diffusion of rejected electrolytes, thereby reducing the sharpness of mechanical or osmotic shock experienced.

These simple analyses both help to interpret past experimental outcomes and suggest design approaches to for future partial freezing protocols. For example, by altering the electrolyte concentrations in the starting media, a solution could easily be designed which achieves the desired sodium/potassium balance *only at sub-0°C* equilibrium, rather than allowing the significant excursions seen here. Specific ice growth limits may also be easily targeted; different candidate cryoprotectants may easily be screened; and etc.

3.2 Kinetic Techniques

The techniques described in the preceding section are unified by their concern with achieving or maintaining stable thermodynamic equilibrium, wherefrom additional spontaneous evolution of the system (e.g., phase change, compositional change) becomes impossible. As such, from a thermophysical perspective, these techniques are driven by phenomena with long-to-infinite associated time scales, and depend only minimally on acutely transient processes. However, many cryopreservation protocols may deviate significantly from this equilibrium, passing through (or deliberately occupying) a metastable supercooled state wherefrom ice nucleation, whether desirable or to be avoided, is strongly *time* dependent, and for which temporal aspects of the protocol are key. In this sub-section, we will review several emerging techniques dominated by kinetic aspects of ice nucleation and growth, seeking to interpret each within the kinetic framework described in Sect. 2.2.

3.2.1 Ice Seeding

Kinetic techniques in general seek either to *accelerate* or *delay* the nucleation of ice upon cooling beneath the melting point. The process of *accelerating* nucleation is often referred to as "ice seeding," and is typically desired for two reasons: (1) to avoid the excessive mechanical, thermal, and osmotic damage that can occur if extracellular nucleation occurs at deeply supercooled temperatures, and (2) to avoid intracellular ice nucleation entirely, by enabling extracellular ice growth and solute rejection to gradually draw water out of (and force cryoprotectants into) the cell, thereby stabilizing it against nucleation as it reaches ever-colder temperatures.

As such, within the formalism of nucleation theory, ice seeding protocols must be designed such that the induction time of ice nucleation in the extracellular space is much shorter than that in the intracellular space. While the relative volume of a given cell compared to the extracellular space already favorably biases these induction times, this bias is typically insufficient to avoid the mechano-osmotic damage that can be associated with rapid extracellular ice growth. As such, emergent approaches to ice seeding focus on means to further shift the induction time of extracellular nucleation forward, using both passive approaches (e.g., ice nucleating agents (INAs)) or active approaches (e.g., ultrasonication). Many excellent works over the past decade have reviewed both the fundamentals of and progress in ice seeding [93–95]; as such we will here review only the most recent advances in the field.

Passive methods of ice seeding focus on increasingly effective ice nucleating agents, which leverage heterogeneous nucleation modes to induct ice formation at temperatures ever-closer to the equilibrium melting point of the solution. Historically, the *SnoMax* ice nucleating bacteria, silver iodide, and a small clutch of other nucleators were used to this end, but the past half decade has seen a rapid expansion in INA offerings. Key materials of recent interest include: (1) several varieties of naturally-occurring mineral, including various feldspars, quartz, and the "hyperactive" K-feldspar LDH1, which has been found to enable nucleation at <1°C supercooling [96]; (2) emergent classes of nature-derived water soluble organic materials, including polysaccharides from the pollen *Carpinus betulus* (or European hornbeam) [97], hydrated cholesterol [98], etc.; and (3) ice nucleating bacteria, including previously unrecognized members of the Pseudomonadaceae, Enterobacteriaceae, Xanthomonadaceae, and Bacillaceae families [99]. These molecules nucleate ice through a wide host of divergent hypothesized mechanisms, from crystalline templating/interfacial free energy reduction to specific bacterial protein interactions, significant headroom remains in our understanding of passive ice nucleation processes.

Active methods meanwhile do not involve additives to solution. One popular such method in recent years is nucleation by acoustic or ultrasonic stimulation, which is generally understood to drive cavitation effects that yield massive hyper-local fluctuations in pressure that produce ephemeral nuclei of high-pressure ice phases, which then seed hexagonal ice [100]. Ultrasonic ice seeding has shown beneficial results in the preservation of samples ranging from hepatocytes to fresh food products [101–103], though significant additional work is needed to develop a physics-first understanding of optimal sonication parameters for nucleation. As evidence, it should be noted that ultrasound is also drawing considerable recent attention as a means of supplying rapid warming for large biological samples, wherein the goal is to *avoid* ice formation en route from the vitrified state [103].

3.2.2 Supercooling

As we have established throughout this chapter, the induction of ice nucleation upon cooling beneath a solution's melting point occurs over a finite and probabilistically quantifiable time scale. If this time scale is greater than the desired preservation time for a given experiment, the solution may remain at metastable equilibrium in a supercooled liquid state, thereby enabling ice-free preservation at sub-0°C, submelting point temperatures.

Supercooling, though recently reaching new heights of popularity as means of practical stabilization have expanded, is one of the foundational concepts of cryopreservation for complex biological systems, seen both in the literature for >60 years and in nature for millennia. Supercooling as a fundamental mode of *self*-preservation in nature has been corroborated time and again across organismal complexities, from single-celled fungi (e.g., yeast, demonstrated by Franks and colleagues to stably supercool at −20°C for >16 weeks [104]) to whole mammals (e.g., the arctic ground squirrel, demonstrated to supercool at ~ −3°C for months of winter on end [105]), and has inspired the recent advent of supercooled cell [106], tissue [107], and organ preservation [3, 75–77, 108, 109].

From a physical perspective, the principal challenge of supercooling is ensuring the *stability* of the supercooled state, or the probability that nucleation will occur (i.e., the induction time will be reached) within the desired preservation period. In the modern era, select intersecting approaches to supercooling stabilization have emerged in biopreservation contexts, though many more may be imagined based on the principles described in this chapter.

Usta and colleagues discovered that the liquid-air interface contributes in a uniquely potent fashion to heterogeneous ice nucleation in supercooled solutions, and that this contribution may be reduced in large part by sealing the exposed surface of the solution with an immiscible layer of liquid (e.g., mineral oil). This principle has since been deployed extensively in the supercooling of various sensitive cell models [106, 110–114], and to a lesser degree in preservation of mammalian livers [3, 75] and kidneys [115] (wherein chemical stabilization is the dominant driver; see below).

Powell-Palm and colleagues then identified that isochoric or confined conditions (embodied by sealing the system of interest in a rigid, air-free container) further stabilized the supercooled state [40], driven by a suite of non-chemical mechanisms including the air-avoidance identified by Usta [116], isolation against vibration [117, 118] and density fluctuations (which are necessarily unconstrained in open systems with access to atmospheric pressure), and putative atomistic effects on the nucleation barrier [119]. A comparison of nucleation induction times in pure water for both these techniques (and an open, unaffected sample) are shown in Fig. 13 (adapted from Consiglio et al. [40]). Isochoric supercooling has since been deployed in the preservation of cells, engineered tissue constructs, mammalian livers, and mammalian hearts, in addition to a variety of fresh food products [19].

It is critical to note that the two mechanisms above, oil sealing and isochoric supercooling, present *purely physical, non-chemical* means of stabilizing the supercooled state—i.e., they do not require that any particular chemical composition be introduced into the supercooled biologic. However, while this non-chemical nature is no

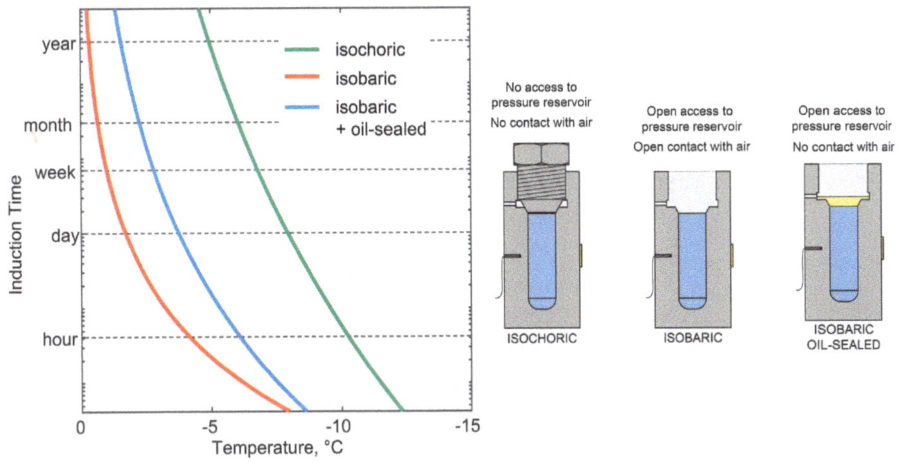

Fig. 13 Nucleation induction times as a function of temperature for isochoric (green), isobaric (red), and oil surface sealed isobaric (blue) systems (left) and corresponding schematics of systems (right). (Adapted with permission from Consiglio et al. [40])

doubt favorable from a regulatory perspective (and enables interrogation of purely temperature-dependent biopreservation phenomena, absent ice, and absent chemical confounders), nothing *precludes* these techniques from being *coupled* with chemical interventions to further stabilize supercooling, as noted previously by Taylor et al. [74].

An initial demonstration of this combination was provided by the group of Toner and Uygun, who coupled oil-sealing with penetrating cryoprotectants and high-molecular weight polymers to achieve multiday supercooled preservation of the human liver, providing a watershed moment in the demonstration of the applicability of the technique to transplant medicine. However, despite this blend of stabilizing effects, supercooling was only achieved to ~1°C beneath the melting point (−4°C storage temperature) [3]. More recently, Huang and colleagues have realized this premise to a much greater degree, obtaining supercooling to ~9°C beneath the melting point (−10°C storage) in rodent kidneys [115]. To achieve this, they combined oil-sealing with a preservation solution that includes the synthetic ice blocker 1,4-cyclohexanediol alongside multiple polymers, each at molecular weights at which they will both leave the vascular space (i.e., enter the interstitium) and plausibly provide molarity-disproportionate depression of the nucleation temperature [120, 121]. Huang and colleagues note, as we have also previously [19], that even deeper supercooling may likely be achievable by combining this chemical approach with isochoric supercooling. A survey of recent organ supercooling studies is shown in Fig. 14.

The practical utility of supercooling, especially for large, ultra-high-value biologics like transplantable human organs, is a strict function of the likelihood of ice formation at the desired storage temperature, in the precise configuration to be used. While for many years the probabilistic nature of ice nucleation had been considered an insurmountable difficulty in achievement of a reliable technique, recent work (borrowing from long-standing insights in metallurgy and phase-change thermal energy storage) has leveraged the robust mathematical describability of memoryless activated stochastic processes to powerfully constrain the nucleation probability space, enabling rational design of *safe, stable* supercooling protocols by leveraging the very stochasticity that previously appeared a liability. Using constant-cooling-rate or isothermal nucleation experiments, the theory described in Sect. 2.2.5 of this chapter may be used to calculate an induction time isopleth *for a given probability of ice* nucleation (say, 1-in-10,000), thereby enabling quantification, from an empirical, protocol- and system-specific basis, of a safe supercooling duration. In Case Study 2, we detail this thermodynamic-kinetic approach to "supercooling engineering".

3.2.3 Case Study 2: Rational Engineering Design of Supercooling Protocols

Using the statistical nucleation kinetics framework described in Sect. 2.2.5, benchtop nucleation experiments may be used to rigorously design supercooling protocols with arbitrarily high margins of safety against nucleation, or arbitrarily low probabilities of ice nucleation

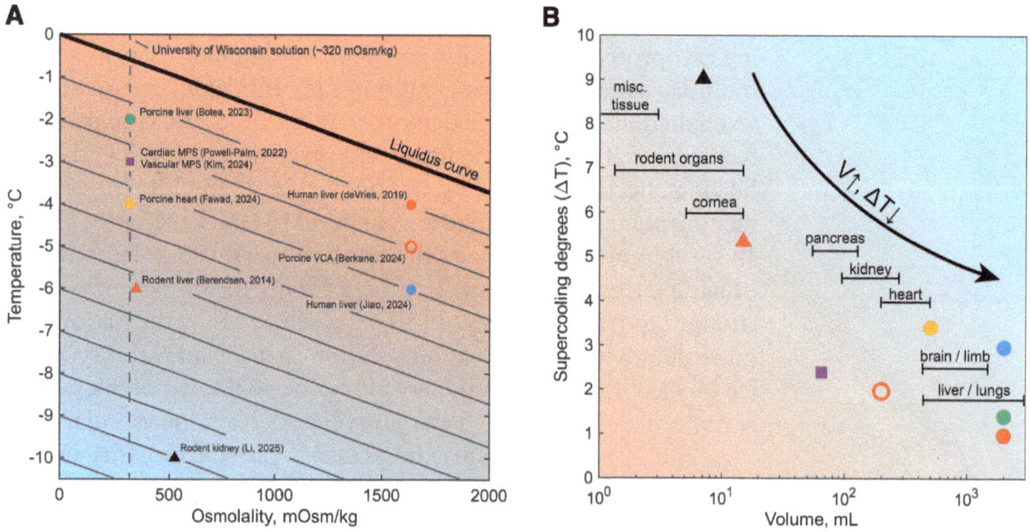

Fig. 14 Recent progress in supercooling for organ preservation. (**A**) Supercooling temperatures reported for organ and organoid models as a function of the osmolality of the solution used [3, 75–77, 91, 107–109, 115]. Thin gray lines represent supercooling isotherms, along which the distance in temperature from the melting point is constant. (**B**) Degrees of supercooling past the melting point (i.e., ΔT) for the same solutions as a function of sample volume. The volume ranges corresponding to various human organs and tissues of interest in transplantation are indicated by horizontal black bars. The exponentially inverse correlation between sample volume and supercoolability is consistent with classical nucleation theory and the extreme value interpretation of nucleation statistics [32, 122, 123]

at a desired temperature or a desired storage period. The general process is as follows:

1. For the specific container type, sample volume, and sample composition of interest, perform multiple constant cooling rate nucleation temperature tests in order to assemble a distribution of nucleation temperatures (a "survivor curve" or cumulative distribution). A minimum of ~30 replicates is recommended to robustly capture intrasample variance (i.e., using precisely the same sample repeatedly), a process that can be informed by computing confidence intervals (e.g., through bootstrapping).
   a) *Note: If these replication numbers prove untenable at full sample volume due to cooling rate or other limitations, scaled-down versions of the desired system may be used, and surface area or volume-based scaling rules may be applied to adjust the results as needed* [32].
2. Fit a non-homogeneous Poisson distribution (Eq. 47) to these data in order to extract the temperature-dependent system nucleation rate [40]. Use this rate to calculate the probability of ice nucleation at desired storage time or/and the induction time for a desired probability of nucleation as functions of temperature.
3. Repeat Steps 1 and 2 with different samples in order to capture intersample variance. Take worst case trial (most nucleation prone), average trial, or apply extreme value statistical methods in order to compute the overall freezing probability [32].
4. Identify a combination of storage temperature and storage time that provide the desired margin of safety on supercooling stability (e.g., a 1-in-10,000 chance of ice nucleation). For any given physical system/container with the same heterogeneous nucleation characteristics (i.e., same surfaces in contact with the liquid, same closure conditions, etc.) as captured in the underlying nucleation data from Step 1, extrapolate these results in sample volume, surface area, or solution composition using available physical and thermodynamic scaling rules to further explore the opportunity space.

Using data from Consiglio et al., we show sample outputs of this process in Fig. 15 (for a single sample). We use the preferred form $J = A\exp\left(-\frac{B' T_m^5}{T^3 \Delta T^2}\right)$ to capture the temperature dependence of the nucleation rate, with parameters $A = 2.01$, $B' = 1.07$, representing data for 1 L pure water contained within an isochoric container, and in direct contact only with a uniform petrolatum coating on the chamber walls (see Fig. 5 of Consiglio et al. [32]). The freezing probability is calculated as $p = 1 - e^{-Jt}$.

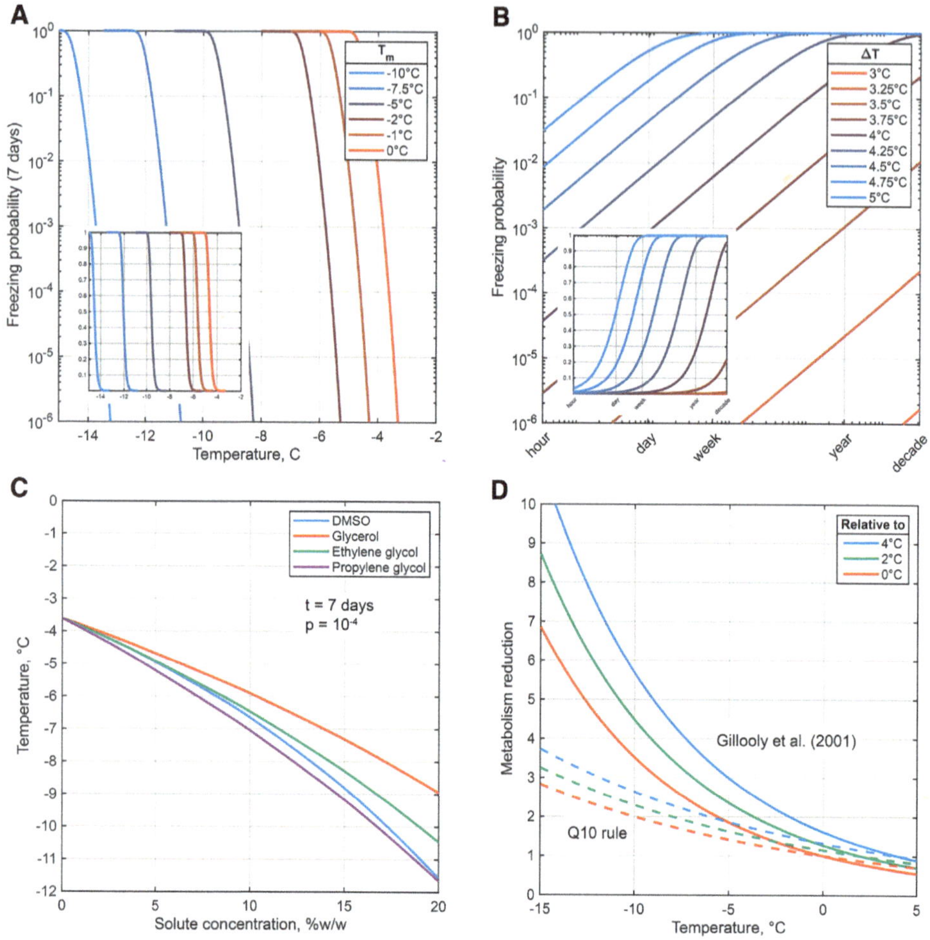

Fig. 15 Kinetics insights for the rational design of safe supercooling protocols. Panels and methodology described in detail in the text above. (**A**) Freezing probability for 7 days of supercooling at a given temperature, for solutions of varying melting temperatures. (**B**) Temperature to which a solution may be supercooled for 7 days with a 1-in-1000 chance of freezing, for aqueous binary solutions of four common cryoprotectants. (**C**) Freezing probability versus time for varying degrees of supercooling past the melting point. (**D**) Factor of metabolism reduction relative to various conventional cold storage temperatures as a function of supercooling temperature. Dashed lines represent the oft-cited (though woefully poorly supported) notion that metabolism is reduced by half for every 10°C of cooling. Solid lines represent the landmark empirical metabolism model of Gilooly et al. [41]

In Fig. 15A, we show the probability of freezing in a 7-day period as a function of solution melting point. Here, the $T_m = 0$ °C curve is calculated according to the steps above from the benchtop data described, and the $T_m < 0$ °C data utilize the assumption that, in the dilute limit, the nucleation parameters A and B are insensitive to the concentration of conventional permeating CPAs. This assumption reproduces the common observation that the ratio λ of nucleation temperature depression to melting temperature depression is equal to 1 for solutions at bulk volumes dominated by heterogeneous nucleation [40, 124]. We note however that $\lambda \sim 2$ in the same solutions when in emulsified or atmospheric droplet forms susceptible to homogeneous nucleation. Furthermore, solutions with polymeric additives may reach $\lambda \gg 2$ (whether heterogeneously or homogeneously nucleating), a phenomenon which Powell-Palm et al. have attributed to the dominant role of the configurational entropy of the liquid in prescribing the liquid–ice interfacial free energy [9].

Using the same data, in Fig. 15B, we show the probability of freezing as a function of degree of supercooling $\Delta T = T_{\text{melting}} - Tsupercooling$. In Fig. 15C we show the temperatures to which four common binary water-cryoprotectant solutions are predicted to stably supercool for 1 week with a 1-in-1000 chance of ice nucleation. Equilibrium melting temperature curves are calculated via the osmotic virial equation (see Sect. 2.1).

Finally, in Fig. 15D, we show an additional essential consideration in choosing a desired supercooling temperature - the degree of metabolic suppression putatively achieved at that temperature. Here, dashed lines show the factor by which metabolism is suppressed relative to ≥0°C classical storage temperatures according to the oft-cited "Q10 rule," which supposes that the rate of metabolism halves for every 10°C of cooling from normothermia. This "rule" is based simply on the requirement of Arrhenius scaling for the biochemical reactions of which metabolism is comprised, and not life-specific theoretical or empirical data. As a data-driven alternative, we also provide the reduction in metabolism predicted for mammals by the landmark empirical model of Gilooly et al. [41]. While their model is based on data between 0°C and 37°C, we argue that this scaling provides much better agreement with the last decade of preservation extension observed in studies on supercooling of complex biological systems.

In total, Fig. 15 conveys the rich variety of physicochemical information that may be used to rationally design safe and reliable supercooling protocols, requiring only straightforward benchtop nucleation temperature experiments and elementary statistical kinetics calculations.

3.2.4 High Pressure Freezing

High-pressure freezing (HPF) exploits the strong pressure-dependence of water's phase diagram to kinetically suppress ice formation during cooling, enabling vitrification or near-vitrification with substantially lower cryoprotectant concentrations than are required at atmospheric pressure. Raising pressure shifts the liquid–ice coexistence line and reduces the driving force for nucleation and growth at a given temperature.

In conventional HPF devices adapted from electron microscopy, small specimens (typically ≤200 μm thick) are rapidly pressurized to ~200 MPa and simultaneously cooled by a high-velocity liquid nitrogen jet or similar. Recent work shows that such platforms can be repurposed from cryo-fixation to cryopreservation [125], with 2D monolayers and 3D spheroids vitrified at ~210 MPa with only 20%–30% v/v permeating CPA plus non-penetrating polymers displaying higher post-thaw viability, metabolic activity, and retention of cell–cell junctions than comparable normal-pressure plunge-freezing protocols.

Closely related "self-pressurized rapid freezing" (SPRF) approaches achieve comparable pressures without an external pressurization system by rapidly cooling sealed capillaries or microchannels [126, 127] completely filled with aqueous media, absent air. Initial ice expansion in the confined volume raises internal pressure into the 10s to 100s MPa range, and the system follows an effectively isochoric path in which further ice growth is disfavored. HPF and SPRF thus sit at the intersection of equilibrium and kinetic considerations, where pressure first reshapes the liquidus surface, then consequently weakens the thermodynamic driving force for ice formation.

3.3 Transport-Driven Techniques

As we have argued thus far, cryopreservation, as a process dependent on the management of phase change, is driven principally by equilibrium thermodynamics and nucleation and growth. While these phenomena obviously intersect with and are dependent upon heat and mass transport in many critical ways, techniques truly *driven* by transport phenomena (i.e., moreso dependent on transport than on the underlying thermodynamics and kinetics at play) are comparatively rare. In this subsection, we will review two such techniques.

3.3.1 Directional Freezing

Directional freezing consists of advancing a biologic across a known (typically linear) temperature gradient in order to control the speed and position of the freeze front and ensure uniformity of freezing conditions. Introduced first by Rubinsky and Ikeda [128] as an analytical tool with which to investigate ice front geometry and cell–ice interactions, the technique has since been applied as a means of minimizing damage associated with ice growth during preservation of biological samples across scales.

Directional freezing devices typically utilize two constant-temperature heat sinks, one at a temperature above the melting point of the biological sample of interest and the other below, separated by a gap of known length. A conducting substrate of low thermal mass is then placed across sinks, such that the portion spanning the gap assumes a constant, approximately linear temperature gradient between the two. A biological sample in solution is then placed in intimate thermal contact with this plate, nominally fixing its temperature to its position in the direction of the gradient, and producing a freeze front perpendicular to this direction at the position corresponding to the melting point. The sample is then slowly advanced from the warm side to the cold, such that its cooling rate is a function only of its linear velocity (see schematic in Fig. 16).

By then controlling this sample velocity, directional freezing fixes the local cooling rate at the moving freeze front, converting what is ordinarily a spatially and kinetic heterogeneous freezing process into a nigh-deterministic moving-boundary problem, describable by the classical Stefan formulation [129]. Furthermore, given that latent heat and rejected solutes may both be evacuated in a prescribed direction (into the unfrozen portion), multisite nucleation and remelting related to recalescence events are suppressed, and solute concentration gradients remain approximately constant at the freeze front during freezing. In principle, directional freezing thus enables freezing of an entire biological sample at a uniform cooling rate, solute-rejection rate, etc.

In practice, this approach has been productively utilized in a variety of samples for which the optimal aspect ratio for enforcement of the linear temperature gradient by conduction (footprint dimensions » thickness) can be obtained. For adherent cellular monolayers, Braslavsky and colleagues [130] have shown that a two-step implementation incorporating an initial directional freeze to −20°C followed by conventional slow cooling to cryogenic temperatures has been shown to markedly improve post-thaw attachment and viability relative to conventional slow cooling alone. Others have shown similar benefit in the freezing of spermatozoa across multiple species and multiple cell suspensions [131–134], where uniform front propagation mitigates localized crystallization and solute pocketing that otherwise accompany uncontrolled freezing.

Directional freezing has also been investigated to a very limited degree at tissue and organ length scales, for which increasingly

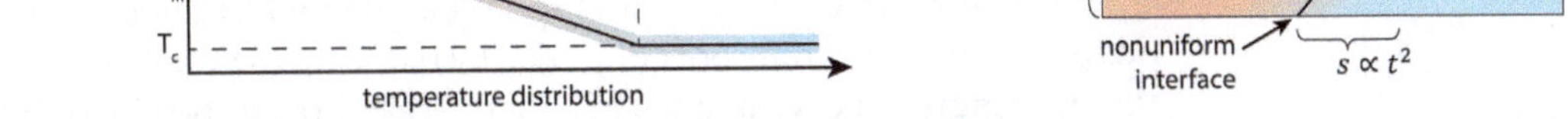

Fig. 16 Illustration of a typical directional freezing platform. A biological sample is advanced along a conductive substrate with an imposed (typically linear) temperature gradient, such that its velocity controls the rate at which the liquid–ice phase change interface moves through the sample (i.e., the freezing rate). As demonstrated by Ukpai and Rubinsky [129], while this interface is often assumed "flat", or perpendicular to the direction of the temperature gradient, it is only so for extremely thin samples, and scaling analysis suggests that uniform directional freezing of biologics with greater than ~1 cm thickness cannot be achieved using conventional single-plate conduction in a room temperature atmosphere

sophisticated stages involving multiple temperature gradients have been developed [135]. Notably, Arav and coworkers have demonstrated improvement in frozen preservation of both ovarian tissue slices [136–138] and whole sheep ovaries, some of which proceeded to produce hormonal activity upon thawing and transplantation. However, significant additional research effort is needed to establish potential utility of the technique for organ-scale biologics, and scaling analysis by Ukpai and Rubinsky suggests that samples above ~1 cm characteristic thickness cannot be directionally frozen in a uniform fashion with the standard single-conductive-substrate configuration [129].

We also note that directional freezing presents a purely physical tool for modulation of freezing behaviors, and thus may be coupled with related chemical techniques, producing for example, "directional partial freezing" or "directional isochoric freezing".

3.3.2 Spray, Droplet, and Flash Freezing

Since Birdseye's foundational discovery in the 1920s that faster cooling produces smaller ice crystals, rapid-cooling approaches have become integral to both microscale preservation of cellular products and macroscale preservation of fresh food products [139–142]. In industrial contexts, a broad class of high–heat-flux freezing technologies—including cryogenic freezers, impingement freezers [143], fluidized bed freezers, and immersion freezers [144]—leverage large convective coefficients (up to ~2000 W/m^2), thin thermal boundary layers, and varied sub-0°C refrigerants (e.g., liquid or high-velocity gaseous nitrogen, air at −40°C, ethanol mixtures, etc.) to achieve high cooling rates in mL- to L-scale food products, thereby maintaining post-thaw quality by reducing characteristic crystal size, drip loss, and textural degradation [140]. Intriguingly, while such macroscale rapid cooling platforms are omnipresent in the food preservation domain, they have not yet been widely applied to rate-dependent whole organ or large organism cryopreservation efforts (see vitrification section below), suggesting potentially low-hanging opportunities for translational engineering efforts in the near future. As noted in the excellent review by James et al. [144], this may be a consequence in part of knowledge in state-of-the-art food rapid freezing technologies being disseminated through industrial pathways, rather than through conventional scientific publishing.

At smaller scales, droplet- and spray-based rapid freezing methods pursue the same transport objective by reducing the characteristic thermal length scale. Cell-laden pL- to uL-scale droplets produced by atomizing nozzles, microfluidic droplet generators, or inkjet heads, when deposited on cryogenic substrates or into cryogenic liquids, can achieve cooling rates in excess of 10^4–10^6°C/min, sufficient to restrict crystalline ice growth to within tolerable limits even with little-to-no penetrating cryoprotectants [145–148]. For a more in-depth discussion of droplet-scale cryopreservation approaches, see Chap. II-7 of this book.

3.4 Vitrification

At long last, we reach that technique that above all others defies clean placement in one or the other of the preceding categories: vitrification. First explored for cryopreservation purposes by Luyet and Hodapp in 1938 [149], vitrification describes the process of shuttling an ice-free biological sample into and out of the glassy state, and represents the most thermophysically complex cryopreservation process yet devised. Successful practice of this technique requires an extremely delicate balance between thermodynamic, kinetic, and transport considerations, all of which grow exponentially more challenging as the biological sample grows in size, and further requires reckoning with the glass transition itself, a unique multiphysics phenomenon which, at the most fundamental levels, remains mysterious [150–152]. While a complete treatment of the theory and practice of vitrification is outside the scope of this Chapter, we will here provide some salient commentary on the current state and future prospects of vitrification. We also refer the reader to several recent reviews on **vitrification** [73, 153, 154], and to the various other Chapters of this book that address it in varying capacities (e.g., Chap. II-6, discussing the role of rewarming in vitrification; Chap. IV-30, relaying a technique for vitrification of articular cartilage).

3.4.1 Interplay Between Thermodynamics, Kinetics, and Transport

To achieve truly ice-free vitrification of a biological sample, the sample must travel through the metastable supercooled regime beneath its melting point and reach the safety of the glass transition temperature (at which dynamical arrest eliminates the possibility of phase change on human time scales) in a period less than the induction time of ice nucleation in the sample. As such, the success of vitrification in a given sample is a complex function of solution thermodynamics, which prescribe the distance in temperature that the sample must traverse under threat of ice formation (between the melting point and glass transition temperature); ice nucleation kinetics, which prescribe the induction time of nucleation within this distance, as a function of cooling rate; and transport, which prescribes the practitioner's ability to achieve a cooling rate that enables the sample to reach the glass transition before the according induction time. If the requirement of *total* ice avoidance is relaxed, ice *growth* kinetics also play a dominant role in this interplay, prescribing the degree to which any allowed nanoscale nucleation will develop into deleterious microscale or macroscale ice formation. The basic logic of this interplay is laid out in Fig. 17, and a review of thermodynamic aspects thereof is also provided by Wowk [73].

3.4.2 Approaches to Vitrification

This multiphysics interplay obviates many levers by which to design a successful vitrification process, and it is instructive, in contemplating

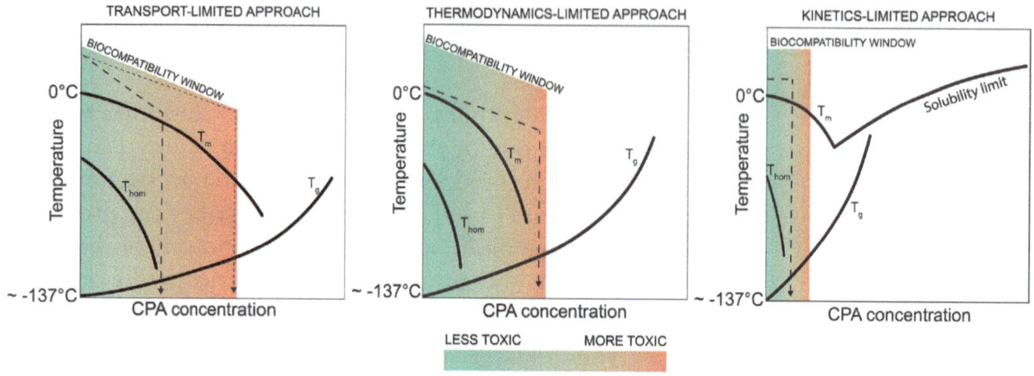

Fig. 17 In all panels, the biocompatibility window represents the range of cryoprotectant concentrations over which vitrification could be achieved with compromising the biologic through toxicity alone. The large-dashed lines represent a hypothetical temperature-concentration trajectory for a sample. Tm represents the ice-liquid melting curve; Thom represents the homogeneous ice nucleation temperature, under which rate-dependent avoidance of ice is assumed practically impossible; and T_g represents the glass transition curve. Panel (**a**) shows the conventional transport-limited approach pioneered by Fahy and colleagues, wherein the sample is shuttled through the metastable region where ice formation is possible, but where it can be avoided by cooling and warming at sufficient speeds. Panel (**b**) shows a thermodynamics-limited approach, wherein novel cryoprotectants are utilized to bring the putative intersection of the melting and glass transition curves within the biocompatibility window, thereby enabling vitrification without entering a regime where ice nucleation is possible. Panel (**c**) shows a kinetics-limited approach, where high-molecular weight polymers or other molecules specifically suppressing the nucleation of ice (as opposed to its thermodynamic stability) are used to stabilize rate-independent passage through the metastable state. We note the role of solute solubility limits only in Panel (**c**), as they often prove trivially avoidable in multi-component molecular aqueous solutions, but much more challenging when polymers or large complex molecules are incorporated

directions for the development of the field, to consider each in the limit of ideal behavior.

The transport-limited ideal is by far the best explored in recent literature, where it is well understood that in the limit of extremely fast cooling and warming, sensitivities of the vitrification process to the thermodynamic and kinetic parameters of the sample relax dramatically (Fig. 17, left). This concept has been richly demonstrated by Thorne and colleagues, who optimized a metallic sample holder and cryogen exposure protocol for bovine oocytes in order to achieve cooling rates up to 600,000°C/min and warming rates up to 200,000°C/min [155]; by Akiyama and colleagues, who developed an inkjet-based droplet extrusion platform to achieve intracellular vitrification in cells at cooling rates up to order ~1,000,000°C/min and warming rates up to order ~10,000,000°C/min [156]; by Bischof and colleagues, who utilized low-thermal mass, high-thermal conductivity meshes to achieve cooling and warming rates up to order ~100,000°C/min in coral larvae, zebrafish embryo, and other small organismal samples; and in many other works reviewed in various recent papers [157]. Transport-driven vitrification techniques have also gained recent popularity in the preservation of large-scale biologics, with particular and extensive focus on new electromagnetic and photothermal interventions during the rewarming process, as surveyed in Chap. II-6.

As a brief aside, we note that curiously little effort has been applied to maximizing *cooling rates* of the same large-scale biologics, on which, as recently pointed out by Thorne and colleagues [158], measured "critical" warming rates depend. This oversight is likely a symptom of historical definitions of successful vitrification, wherein some degree of ice formation (typically ~0.2% by volume) has been accepted. In this case, wherein potentially numerous ice nuclei are allowed to form on cooling, the *growth* of those nuclei becomes asymmetrical between cooling and warming, with the vast majority occurring during warming. This asymmetry in growth in turn gives rise to the perception that the "critical" thermal rates required for successful vitrification are much larger for warming than for cooling. However, if successful vitrification is instead defined as *complete* avoidance of crystallization, critical cooling rates and warming rates become *precisely equal*, as ice nucleation is a memoryless stochastic process (i.e., agnostic to the thermal direction in which the system is moving).

The thermodynamics-limited and kinetics-limited ideals for vitrification have been investigated much more minimally. In the thermodynamic case, we may envision a biocompatible cryoprotectant solution for which the combination of melting point depression and glass transition temperature elevation together shrink the temperature-width of the metastable zone where ice nucleation is plausible, until safe passage through this zone at achievable cooling/warming time scales becomes trivial (Fig. 17, middle). In the 1970s, this concept was powerfully demonstrated by Elford and Walter [69] in the cryopreservation and functional revival of cm-scale smooth muscle tissues using DMSO as the sole cryoprotectant and managing toxicity by liquidus tracking (we note that while Elford and Walter only stored the tissues at −80°C, the avoidance of ice for 24 h at this temperature, in addition to the concentration of DMSO used, suggest that equivalent storage beneath the glass transition would also not result in freezing). However, in recent literature, and despite the exploding recent interest in vitrification of complex, multiscale biologics, development towards (and even serious discussion of) this thermodynamic ideal has largely vanished, in favor of transport-limited strategies, pioneered by Fahy and colleagues, built around achieving "critical" cooling and warming rates needed to avoid substantial ice growth in marginally stable systems (Fig. 17, left). Notable exceptions to this absence are the review works of Taylor and colleagues [74, 159] and the mid-2000s research by Brockbank and colleagues on an ultra-concentrated (83% cryoprotectant by mass) derivative of the vitrification solutions developed by Fahy [160, 161]. We suggest, especially in light of recent progress in the discovery of new cryoprotectants that remain biocompatible at thermodynamically-relevant concentrations [162, 163], that this thermodynamics-limited ideal approach merits significant additional consideration.

Considering the past ~15 years of advances in our understanding of ice nucleation and growth processes in aqueous organic solutions, a compelling kinetics-limited ideal may also be articulated, wherein molecular or polymeric additives that interact with water and ice in increasingly *specific* ways may drive up the induction time of nucleation or inhibit post-nucleation growth, enabling ice-free cooling and warming at achievable rates in systems that retain high equilibrium melting points (Fig. 17, right). Ice recrystallization inhibitors, anti-freeze proteins, ice-modulating polymers [164, 165], and other agents affecting ice at the critical or sub-critical cluster scale provide promising routes to this end, especially in smaller-scale systems where the viscosities associated with high-molecular-weight polymers need not preclude their use at high concentrations (as it may in vascular systems loaded by perfusion). We note further that, given the dominance of heterogeneous nucleation in vitrification of bulk (≥mL-scale) systems, methods developed to suppress heterogeneous ice nucleation during *supercooled preservation* may be applied directly to vitrification, including elimination of air-liquid interfaces [40, 117], isolation from the atmospheric pressure reservoir via isochoric conditions [40], incorporation of ice-phobic surfaces, etc.

3.4.3 Toxicity, Cracking, and the Challenges in the Vitrification of Increasingly Large Systems

Each of the approaches described above has been demonstrated (to at least a nominal degree) in vitrification of *small* biological samples (generally pL- to uL-scale, and up to ~1 mL), where the intrinsically low thermal diffusivity of aqueous organic systems does not broadly prevent rapid cooling and warming; less-than-millimeter mass diffusion length scales allow effective delivery of thermodynamically-favorable cryoprotectants; and small volumes and surface areas of water within the sample minimize nucleation rates. However, in scaling to the many-mL to many-L systems relevant to organ [49, 166], whole-organism cryoconservation [167, 168], or biostasis for space travel [169], universal additional challenges emerge, including solution biocompatibility at prolonged exposure times [170], resilience against thermomechanical stress and fracture [51], and intersecting general difficulties in solution stability and deliverability.

While these problems remain far from solved, and successful vitrified cryopreservation of human-sized organs, mature organisms, etc. remains elusive, recent advances in *materials discovery* (for biocompatibility [78, 79, 162 163]; resistance to cracking [50]; profound interaction with water [16, 171, 172]; thermal rate optimization [173, 174]; viscosity optimization [55]; etc.) offer a rapidly expanding parameter space from which to formulate optimal new vitrification protocols, according to *any* of the design principles articulated above. As such, we encourage the reader to approach their own work in vitrification with an open mind, and with consideration of the rapidly evolving physical and material bases of the field.

4 Conclusion

In this chapter, we have sought to establish a self-consistent understanding of the physical principles driving all cryopreservation processes, in terms of a discrete set of accessible theories rooted ultimately in Gibbsian thermodynamics. We hope that the reader will carry from this work an increased appreciation of the fundamental similarities and differences between emerging cryopreservation technologies, and of the myriad opportunities to leverage these similarities and differences to hybridize, optimize, and otherwise improve their application of these technologies. We also hope that the extreme mismatch between the potential for thermodynamic analysis in cryopreservation and the degree of analysis that has been performed will be evident, and that this mismatch will inspire ever-increasing effort in the application of thermodynamics to the pressing, impactful problems of cryobiology.

Acknowledgments

The authors acknowledge the support of the National Science Foundation (NSF) Engineering Research Center for Advanced Technologies for Preservation of Biological Systems (ATP-Bio), NSF EEC #1941543.

References

1. Tessier SN, de Vries RJ, Pendexter CA, Cronin SEJ, Ozer S, Hafiz EOA, Raigani S, Oliveira-Costa JP, Wilks BT, Lopera Higuita M, van Gulik TM, Usta OB, Stott SL, Yeh H, Yarmush ML, Uygun K, Toner M (2022) Partial freezing of rat livers extends preservation time by 5-fold. Nat Commun 13:4008. https://doi.org/10.1038/s41467-022-31490-2.
2. Maida AL, Perez PA, Bilbao-Sainz C, Rubinsky B, Consiglio AN (2025) The thermodynamic principles of isochoric freezing pressure-aided supercooling. Cryobiology 118:105168. https://doi.org/10.1016/j.cryobiol.2024.105168
3. de Vries RJ, Tessier SN, Banik PD, Nagpal S, Cronin SEJ, Ozer S, Hafiz EOA, van Gulik TM, Yarmush ML, Markmann JF, Toner M, Yeh H, Uygun K (2019) Supercooling extends preservation time of human livers. Nat Biotechnol 37:1131–1136. https://doi.org/10.1038/s41587-019-0223-y
4. Callen HB (1985) Thermodynamics and an introduction to thermostatistics, 2nd edn. Wiley, New York
5. Journaux B, Brown JM, Pakhomova A, Collings IE, Petitgirard S, Espinoza P, Boffa Ballaran T, Vance SD, Ott J, Cova F, Garbarino G, Hanfland M (2020) Holistic approach for studying planetary hydrospheres: Gibbs representation of ices thermodynamics, elasticity, and the water phase diagram to 2300 MPa. J Geophys Res Planets 125:e2019JE006176. https://doi.org/10.1029/2019JE00617
6. Alliston SP, Dames C, Powell-Palm MJ (2025) A size-dependent ideal solution model for liquid-solid phase equilibria prediction in aqueous organic solutions. Proc Natl Acad Sci USA 122:e2415843122. https://doi.org/10.1073/pnas.2415843122

7. Flory PJ (1942) Thermodynamics of high polymer solutions. J Chem Phys 10:51–61. https://doi.org/10.1063/1.1723621
8. Hildebrand JH (1947) The entropy of solution of molecules of different size. J Chem Phys 15:225–228. https://doi.org/10.1063/1.1746484
9. Powell-Palm MJ, Smith H, Fahad MM (2024) An entropic theory of homogeneous ice nucleation in non-ionic aqueous solutions. J Chem Phys 160:101101. https://doi.org/10.1063/5.0199398
10. Elliott JAW, Prickett RC, Elmoazzen HY, Porter KR, McGann LE (2007) A multisolute osmotic virial equation for solutions of interest in biology. J Phys Chem B 111:1775–1785. https://doi.org/10.1021/jp0680342
11. Prickett RC, Elliott JAW, McGann LE (2010) Application of the osmotic virial equation in cryobiology. Cryobiology 60:30–42. https://doi.org/10.1016/j.cryobiol.2009.07.011
12. Prickett RC, Elliott JAW, McGann LE (2011) Application of the multisolute osmotic virial equation to solutions containing electrolytes. J Phys Chem B 115:14531–14543. https://doi.org/10.1021/jp206011m
13. Prausnitz JM, Lichtenthaler RN, de Azevedo EG (1999) Molecular thermodynamics of fluid-phase equilibria, 3rd edn. Prentice Hall PTR, Upper Saddle River, NJ
14. Ganbavale G, Zuend A, Marcolli C, Peter T (2015) Improved AIOMFAC model parameterisation of the temperature dependence of activity coefficients for aqueous organic mixtures. Atmos Chem Phys 15:447–493. https://doi.org/10.5194/acp-15-447-2015
15. Eckert F, Klamt A (2002) Fast solvent screening via quantum chemistry: COSMO-RS approach. AIChE J 48:369–385. https://doi.org/10.1002/aic.690480220
16. Mallya AS, Yadav P, Zakhia S, Hubel A (2025) Computational design of natural deep eutectic systems using COSMO-RS for ice control applications. ACS Sustain Chem Eng 13:14683–14692. https://doi.org/10.1021/acssuschemeng.5c01597
17. Bell IH, Mickoleit E, Hsieh CM, Lin ST, Vrabec J, Breitkopf C, Jäger A (2020) A benchmark open-source implementation of COSMO-SAC. J Chem Theory Comput 16:2635–2646. https://doi.org/10.1021/acs.jctc.9b01016
18. Rubinsky B, Perez PA, Carlson ME (2005) The thermodynamic principles of isochoric cryopreservation. Cryobiology 50:121–138. https://doi.org/10.1016/j.cryobiol.2004.12.002
19. Consiglio AN, Rubinsky B, Powell-Palm MJ (2024) A review of the physical principles of isochoric cryopreservation. Annu Rev Heat Transfer 27:93–164. https://doi.org/10.1615/AnnualRevHeatTransfer.2024054596
20. Powell-Palm MJ (2022) Calculation of a temperature-volume phase diagram of water to inform the study of isochoric freezing down to cryogenic temperatures. RSC Adv 12:20603–20609. https://doi.org/10.1039/D2RA03683E
21. Mazur P, Leibo SP, Chu EHY (1972) A two-factor hypothesis of freezing injury: evidence from Chinese hamster tissue-culture cells. Exp Cell Res 71:345–355. https://doi.org/10.1016/0014-4827(72)90303-5
22. Sutton RL (1991) Critical cooling rates to avoid ice crystallization in aqueous cryoprotectant solutions containing polymers. J Chem Soc, Faraday Trans 87:3747–3751. https://doi.org/10.1039/FT9918703747
23. Hobbs PV (1974) Ice physics. Oxford University Press, Oxford
24. Blazquez S, Sanz E (2025) On the growth rate of ices: effect of pressure and ice phase. J Chem Phys 162:194501. https://doi.org/10.1063/5.0265605
25. Toner M, Cravalho EG, Karel M (1990) Thermodynamics and kinetics of intracellular ice formation during freezing of biological cells. J Appl Phys 67:1582–1593. https://doi.org/10.1063/1.345670
26. Nada H, Furukawa Y (2012) Antifreeze proteins: computer simulation studies on the mechanism of ice growth inhibition. Polym J 44:690–698. https://doi.org/10.1038/pj.2012.13
27. Meister K, Devries AL, Bakker HJ, Drori R (2018) Antifreeze glycoproteins bind irreversibly to ice. J Am Chem Soc 140:9365–9368. https://doi.org/10.1021/jacs.8b04966
28. Biggs CI, Bailey TL, Graham Ben, Stubbs C, Fayter A, Gibson MI (2017) Polymer mimics of biomacromolecular antifreezes. Nat Commun 8:1546. https://doi.org/10.1038/s41467-017-01421-7
29. Budke C, Koop T (2006) Ice recrystallization inhibition and molecular recognition of ice faces by poly(vinyl alcohol). Chem Phys Chem 7:2601–2606. https://doi.org/10.1002/cphc.200600533
30. Pruppacher HR, Klett JD (2010) Microphysics of clouds and precipitation. Springer, Netherlands
31. Barahona D (2018) On the thermodynamic and kinetic aspects of immersion ice nucleation. Atmos Chem Phys 18:17119–17141. https://doi.org/10.5194/acp-18-17119-2018
32. Consiglio AN, Ouyang Y, Powell-Palm MJ, Rubinsky B (2023) An extreme value statistics model of heterogeneous ice nucleation

for quantifying the stability of supercooled aqueous systems. J Chem Phys 159:064511. https://doi.org/10.1063/5.0155494
33. Barahona D (2014) Analysis of the effect of water activity on ice formation using a new thermodynamic framework. Atmos Chem Phys 14:7665–7680. https://doi.org/10.5194/acp-14-7665-2014
34. Barahona D (2015) Thermodynamic derivation of the activation energy for ice nucleation. Atmos Chem Phys 15:13819–13831. https://doi.org/10.5194/acp-15-13819-2015
35. Koop T, Luo B, Biermann UM, Crutzen PJ, Peter T (1997) Freezing of HNO3/H2SO4/H2O solutions at stratospheric temperatures: nucleation statistics and experiments. J Phys Chem A 101:1117–1133. https://doi.org/10.1021/jp9626531
36. Liu Y, He L, Zhao G (2025) Multiscale heat and mass transfer in biomaterial cryopreservation. Int J Heat Mass Transfer 252:127414. https://doi.org/10.1016/j.ijheatmasstransfer.2025.127414
37. Guo Z, Zhan L, Finger EB, Hays TS, Toner M, Bischof JC (2024) Enhanced heat transfer for improved ice-free cryopreservation—interplay between cooling and rewarming. Annu Rev Heat Transfer 27:195–244. https://doi.org/10.1615/AnnualRevHeatTransfer.2024054907
38. Fahy GM, MacFarlane DR, Angell CA, Meryman HT (1984) Vitrification as an approach to cryopreservation. Cryobiology 21:407–426. https://doi.org/10.1016/0011-2240(84)90079-8
39. Ashrafi E, Elliott JAW (2025) Interrupted cooling protocols in cryopreservation: a review of fundamentals, methods and experimental outcomes for cells in suspension. Cryobiology 121:105339. https://doi.org/10.1016/j.cryobiol.2025.105339
40. Consiglio AN, Lilley D, Prasher R, Rubinsky B, Powell-Palm MJ (2022) Methods to stabilize aqueous supercooling identified by use of an isochoric nucleation detection (INDe) device. Cryobiology 106:91–101. https://doi.org/10.1016/j.cryobiol.2022.03.003
41. Gillooly JF, Brown JH, West GB, Savage VM, Charnov EL (2001) Effects of size and temperature on metabolic rate. Science 293:2248–2251. https://doi.org/10.1126/science.1061967
42. Consiglio AN, Rubinsky B, Powell-Palm MJ (2022) Relating metabolism suppression and nucleation probability during supercooled biopreservation. J Biomech Eng 144:074504. https://doi.org/10.1115/1.4054217
43. Benson JD, Higgins AZ, Desai K, Eroglu A (2018) A toxicity cost function approach to optimal CPA equilibration in tissues. Cryobiology 80:144–155. https://doi.org/10.1016/j.cryobiol.2017.09.005
44. Olver DJ, Heres P, Paredes E, Benson JD (2023) Rational synthesis of total damage during cryoprotectant equilibration: modelling and experimental validation of osmomechanical, temperature, and cytotoxic damage in sea urchin (*Paracentrotus lividus*) oocytes. PeerJ 9:e15539. https://doi.org/10.7717/peerj.15539
45. Warner RM, Brown KS, Benson JD, Eroglu A, Higgins AZ (2022) Multiple cryoprotectant toxicity model for vitrification solution optimization. Cryobiology 108:1–9. https://doi.org/10.1016/j.cryobiol.2022.09.002
46. Rubinsky B, Cravalho EG, Mikic B (1980) Thermal stresses in frozen organs. Cryobiology 17:66–73. https://doi.org/10.1016/0011-2240(80)90009-7
47. Shaw JM, Kuleshova LL, MacFarlane DR, Trounson AO (1997) Vitrification properties of solutions of ethylene glycol in saline containing PVP, Ficoll, or dextran. Cryobiology 35:219–229. https://doi.org/10.1006/cryo.1997.2043
48. Ehrlich LE, Fahy GM, Wowk BG, Malen JA, Rabin Y (2018) Thermal analyses of a human kidney and a rabbit kidney during cryopreservation by vitrification. J Biomech Eng 140:0110051–0110058. https://doi.org/10.1115/1.4037406
49. Han Z, Rao JS, Gangwar L, Namsrai BE, Pasek-Allen JL, Etheridge ML, Wolf SM, Pruett TL, Bischof JC, Finger EB (2023) Vitrification and nanowarming enable long-term organ cryopreservation and life-sustaining kidney transplantation in a rat model. Nat Commun 14:3407. https://doi.org/10.1038/s41467-023-38824-8
50. Kavian S, Sellers R, Sanchez GA, Alvarez C, Aguilar G, Powell-Palm MJ (2025) Higher glass transition temperatures reduce thermal stress cracking in aqueous solutions relevant to cryopreservation. Sci Rep 15:27903. https://doi.org/10.1038/s41598-025-13295-7
51. Rabin Y (2024) Thermomechanics modeling and visualization of physical effects to improve cryopreservation by vitrification. Annu Rev Heat Transfer 27:245–282. https://doi.org/10.1615/AnnualRevHeatTransfer.2024053892
52. Solanki PK, Rabin Y (2022) Perspective: temperature-dependent density and thermal expansion of cryoprotective agents. CryoLetters 43:1–9. https://doi.org/10.54680/fr22110110112
53. Noday DA, Steif PS, Rabin Y (2009) Viscosity of cryoprotective agents near glass transition: a new device, technique, and data on DMSO,

DP6, and VS55. Exp Mech 49:663–672. https://doi.org/10.1007/s11340-008-9191-8
54. Han Z, Bischof JC (2020) Critical cooling and warming rates as a function of CPA concentration. CryoLetters 41:185–193
55. Kavian S, Zarriz A, Powell-Palm MJ (2026) A large-scale dataset and physics-informed neural network for viscosity prediction in many-component aqueous and organic solutions. J Chem Phys 164:084503. https://doi.org/10.1063/5.0315421
56. Zarriz A, Journaux B, Powell-Palm MJ (2024) On the equilibrium limit of liquid stability in pressurized aqueous systems. Nat Commun 15:10666. https://doi.org/10.1038/s41467-024-54625-z
57. Chang B, Consiglio AN, Lilley D, Prasher R, Rubinsky B, Journaux B, Powell-Palm MJ (2022) On the pressure dependence of salty aqueous eutectics. Cell Rep Phys Sci 3:100856. https://doi.org/10.1016/j.xcrp.2022.100856
58. Karlsson JOM, Cravalho EG, Toner M (1994) A model of diffusion-limited ice growth inside biological cells during freezing. J Appl Phys 75:4442–4455. https://doi.org/10.1063/1.355959
59. Shardt N, Chen Z, Yuan SC, Wu K, Laouar L, Jomha NM, Elliott JAW (2020) Using engineering models to shorten cryoprotectant loading time for the vitrification of articular cartilage. Cryobiology 92:180–188. https://doi.org/10.1016/j.cryobiol.2020.01.008
60. Kangas J, Hogan CJ (2024) Mathematical treatments of ice formation and vitrification during cryopreservation. Annu Rev Heat Transfer 27:25–69. https://doi.org/10.1615/AnnualRevHeatTransfer.2024053903
61. Boutron P (1986) Comparison with the theory of the kinetics and extent of ice crystallization and of the glass-forming tendency in aqueous cryoprotective solutions. Cryobiology 23:88–102. https://doi.org/10.1016/0011-2240(86)90022-2
62. Murray BJ (2008) Inhibition of ice crystallisation in highly viscous aqueous organic acid droplets. Atmos Chem Phys 8:5423–5433. https://doi.org/10.5194/acp-8-5423-2008
63. Bronshteyn VL, Steponkus PL (1995) Nucleation and growth of ice crystals in concentrated solutions of ethylene glycol. Cryobiology 32:1–22. https://doi.org/10.1006/cryo.1995.1001
64. Hey JM, Macfarlane DR (1996) Crystallization of ice in aqueous solutions of glycerol and dimethyl sulfoxide. 1. A comparison of mechanisms. Cryobiology 33:205–216. https://doi.org/10.1006/cryo.1996.0021
65. Hey JM, MacFarlane DR (1998) Crystallization of ice in aqueous solutions of glycerol and dimethyl sulfoxide 2: ice crystal growth kinetics. Cryobiology 37:119–130. https://doi.org/10.1006/cryo.1998.2108
66. Sei T, Gonda T, Arima Y (2002) Growth rate and morphology of ice crystals growing in a solution of trehalose and water. J Cryst Growth 240:218–229. https://doi.org/10.1016/S0022-0248(02)00875-8
67. Vigier G, Thollet G, Vassoille R (1987) Cubic and hexagonal ice formation in water-glycerol mixture (50% w/w). J Cryst Growth 84:309–315. https://doi.org/10.1016/0022-0248(87)90146-1
68. Farrant J (1965) Mechanism of cell damage during freezing and thawing and its prevention. Nature 205:1284–1287. https://doi.org/10.1038/2051284a0
69. Elford BC, Walter CA (1972) Preservation of structure and function of smooth muscle cooled to −79°C in unfrozen aqueous media. Nature New Biol 236:58–60. https://doi.org/10.1038/newbio236058a0
70. Pegg DE, Wang L, Vaughan D (2006) Cryopreservation of articular cartilage. Part 3: the liquidus-tracking method. Cryobiology 52:360–368. https://doi.org/10.1016/j.cryobiol.2020.01.004
71. Kay AG, Hoyland JA, Rooney P, Kearney JN, Pegg DE (2015) A liquidus tracking approach to the cryopreservation of human cartilage allografts. Cryobiology 71:77–84. https://doi.org/10.1016/j.cryobiol.2015.05.005
72. Puschmann E, Selden C, Butler S, Fuller B (2017) Liquidus tracking: large scale preservation of encapsulated 3-D cell cultures using a vitrification machine. Cryobiology 76:65–73. https://doi.org/10.1016/j.cryobiol.2017.04.006
73. Wowk B (2010) Thermodynamic aspects of vitrification. Cryobiology 60:11–22. https://doi.org/10.1016/j.cryobiol.2009.05.007
74. Taylor MJ, Weegman BP, Baicu SC, Giwa SE (2019) New approaches to cryopreservation of cells, tissues, and organs. Transfus Med Hemother 46:197–215. https://doi.org/10.1159/000499453
75. Berendsen TA, Bruinsma BG, Puts CF, Saeidi N, Usta OB, Uygun BE, Izamis ML, Toner M, Yarmush ML, Uygun K (2014) Supercooling enables long-term transplantation survival following 4 days of liver preservation. Nat Med 20:790–793. https://doi.org/10.1038/nm.3588
76. Botea F, Năstase G, Herlea V, Chang TT, Şerban A, Barcu A, Rubinsky B, Popescu I (2023) An exploratory study on isochoric supercooling preservation of the pig liver.

Biochem Biophys Rep 34:101485. https://doi.org/10.1016/j.bbrep.2023.101485
77. Fawad M, Elde S, Heng E, Krishnan A, Consiglio AN, Vergel M, Alnasir D, Choi A, Simmons J, MacArthur J, Shudo Y, Ruaengsri C, Woo J, Powell-Palm MJ, Guenthart B (2025) First sub-zero cardiac allograft preservation and reanimation following 24-hours of isochoric supercooled preservation. J Heart Lung Transplant 44:S36. https://doi.org/10.1016/j.healun.2025.02.080
78. Ahmadkhani N, Benson JD, Eroglu A, Higgins AZ (2025) High throughput method for simultaneous screening of membrane permeability and toxicity for discovery of new cryoprotective agents. Sci Rep 15:1862. https://doi.org/10.1038/s41598-025-85509-x
79. Ahmadkhani N, Sugden C, Mayo AT, Higgins AZ (2025) Screening for cryoprotective agent toxicity and toxicity reduction in mixtures at subambient temperatures. Cryobiology 121:105315. https://doi.org/10.1016/j.cryobiol.2025.105315
80. Ishine N, Rubinsky B, Lee CY (1999) A histological analysis of liver injury in freezing storage. Cryobiology 39:271–277. https://doi.org/10.1006/cryo.1999.2205
81. Ishine N, Rubinsky B, Lee CY (2000) Transplantation of mammalian livers following freezing: vascular damage and functional recovery. Cryobiology 40:84–89. https://doi.org/10.1006/cryo.1999.2225
82. Tessier SN, de Vries RJ, Pendexter CA, Cronin SEJ, Ozer S, Hafiz EOA, Raigani S, Oliveira-Costa JP, Wilks BT, Lopera Higuita M, van Gulik TM, Usta OB, Stott SL, Yeh H, Yarmush ML, Uygun K, Toner M (2022) Partial freezing of rat livers extends preservation time by five-fold. Nat Commun 13:4008. https://doi.org/10.1038/s41467-022-31490-2
83. Tessier SN, Haque O, Pendexter CA, Cronin SEJ, Hafiz EOA, Weng L, Yeh H, Markmann JF, Taylor MJ, Fahy GM, Toner M, Uygun K (2022) The role of antifreeze glycoprotein (AFGP) and polyvinyl alcohol/polyglycerol (X/Z-1000) as ice modulators during partial freezing of rat livers. Front Phys 10:1033613. https://doi.org/10.3389/fphy.2022.1033613
84. Ozgur OS, Taggart M, Mojoudi M, Pendexter C, Filz von Reiterdank I, Kharga A, Yeh H, Toner M, Longchamp A, Tessier SN, Uygun K (2024) Optimized partial freezing protocol enables 10-day storage of rat livers. Sci Rep 14:25260. https://doi.org/10.1038/s41598-024-76674-6
85. Taggart M, Hassan M, Ozgur O, Lyon A, Mojoudi M, Cutrone A, Taveras C, Toner M, Tessier S, Markmann J, Yeh H, Longchamp A, Uygun K (2025) Preservation of discarded human kidneys for 10 days at subzero temperature. Am J Transplant 25:S14–S15. https://doi.org/10.1016/j.ajt.2024.12.035
86. Lyu C, Luo G, An R, Wang D, Rubinsky B (2024) Inactivation of *Bacillus cereus* spores by isochoric freezing. Food Control 161:110383. https://doi.org/10.1016/j.foodcont.2024.110383
87. Wan L, Powell-Palm MJ, Lee C, Gupta A, Weegman BP, Clemens MG, Rubinsky B (2018) Preservation of rat hearts in subfreezing temperature isochoric conditions to −8 °C and 78 MPa. Biochem Biophys Res Commun 496:852–857. https://doi.org/10.1016/j.bbrc.2018.01.140
88. Wan L, Powell-Palm MJ, Clemens MG, Rubinsky B (2019) Time-dependent effects of pressure during preservation of rat hearts in an isochoric system at subfreezing temperatures. CryoLetters 40:64–70
89. Zielinski MW, McGann LE, Nychka JA, Elliott JAW (2014) Comparison of non-ideal solution theories for multi-solute solutions in cryobiology and tabulation of required coefficients. Cryobiology 69:305–317. https://doi.org/10.1016/j.cryobiol.2014.08.005
90. Wang SC, Wang CK, Chang FM, Tsao HK (2002) Second virial coefficients of poly(ethylene glycol) in aqueous solutions at freezing point. Macromolecules 35:9551–9555. https://doi.org/10.1021/ma025663a
91. Kim Y, Goswami I, Gill E, Mahmoodi SR, Consiglio AN, Velazquez J, Nieman G, Alburo AAA, Woods B, Ellis BW, Filz von Reiterdank I, Uygun K, Uygun BE, Rubinsky B, Healy KE (2025) Vascular microphysiological system for investigating endothelial barrier function during organ preservation and reperfusion. Small 21:e2410168. https://doi.org/10.1002/smll.202410168
92. Kapanen MK, Halavaara JT, Häkkinen AM (2005) Open four-compartment model in the measurement of liver perfusion. Acad Radiol 12:1542–1550. https://doi.org/10.1016/j.acra.2005.07.001
93. Morris GJ, Acton E (2013) Controlled ice nucleation in cryopreservation—a review. Cryobiology 66:85–92. https://doi.org/10.1016/j.cryobiol.2012.11.007
94. He Z, Liu K, Wang J (2018) Bioinspired materials for controlling ice nucleation, growth, and recrystallization. Acc Chem Res 51:1082–1091. https://doi.org/10.1021/acs.accounts.7b00528
95. Zhang Z, Liu XY (2018) Control of ice nucleation: freezing and antifreeze strategies. Chem Soc Rev 47:7116–7139. https://doi.org/10.1039/c8cs00626a

96. Harrison AD, Whale TF, Carpenter MA, Holden MA, Neve L, O'Sullivan D, Vergara Temprado J, Murray BJ (2016) Not all feldspars are equal: a survey of ice nucleating properties across the feldspar group of minerals. Atmos Chem Phys 16:10927–10940. https://doi.org/10.5194/acp-16-10927-2016
97. Murray KA, Kinney NLH, Griffiths CA, Hasan M, Gibson MI, Whale TF (2022) Pollen derived macromolecules serve as a new class of ice-nucleating cryoprotectants. Sci Rep 12:12295. https://doi.org/10.1038/s41598-022-15545-4
98. Sosso GC, Whale TF, Holden MA, Pedevilla P, Murray BJ, Michaelides A (2018) Unravelling the origins of ice nucleation on organic crystals. Chem Sci 9:8077–8088. https://doi.org/10.1039/c8sc02753f
99. Failor KC, Schmale 3rd DG, Vinatzer BA, Monteil CL (2017) Ice nucleation active bacteria in precipitation are genetically diverse and nucleate ice by employing different mechanisms. ISME J 11:2740–2753. https://doi.org/10.1038/ismej.2017.124
100. Consiglio A, Ukpai G, Rubinsky B, Powell-Palm MJ (2020) Suppression of cavitation-induced nucleation in systems under isochoric confinement. Phys Rev Res 2:023350. https://doi.org/10.1103/PhysRevResearch.2.023350
101. Peng Z, Li W, Liu B, Wang J (2025) Ultrasonic ice seeding as an effective technique for Hep-G2 cell cryopreservation. CryoLetters 46:345–354. https://doi.org/10.54680/fr25510110412
102. Tian Y, Zhang P, Zhu Z, Sun DW (2020) Development of a single/dual-frequency orthogonal ultrasound-assisted rapid freezing technique and its effects on quality attributes of frozen potatoes. J Food Eng 286:110112. https://doi.org/10.1016/j.jfoodeng.2020.110112
103. Alcalá E, Encabo L, Barroso F, Puentes A, Risco I, Risco R (2023) Sound waves for solving the problem of recrystallization in cryopreservation. Sci Rep 13:7603. https://doi.org/10.1038/s41598-023-34681-z
104. Franks F, Wakabayashi T, Mathias SF (1987) Nucleation kinetics of ice in undercooled yeast cells: long-term stability against freezing. Microbiology 133:2807–2815. https://doi.org/10.1099/00221287-133-10-2807
105. Barnes BM (1989) Freeze avoidance in a mammal: body temperatures below 0°C in an arctic hibernator. Science 244:1593–1595. https://doi.org/10.1126/science.2740905
106. Usta OB, Kim Y, Ozer S, Bruinsma BG, Lee J, Demir E, Berendsen TA, Puts CF, Izamis ML, Uygun K, Uygun BE, Yarmush ML (2013) Supercooling as a viable non-freezing cell preservation method of rat hepatocytes. PLOS One 8:e69334. https://doi.org/10.1371/journal.pone.0069334
107. Powell-Palm MJ, Charwat V, Charrez B, Siemons B, Healy KE, Rubinsky B (2021) Isochoric supercooled preservation and revival of human cardiac microtissues. Commun Biol 4:1118. https://doi.org/10.1038/s42003-021-02650-9
108. Berkane Y, Filz von Reiterdank I, Tawa P, Charlès L, Goutard M, Dinicu AT, Toner M, Bertheuil N, Mink van der Molen AB, Coert JH, Lellouch AG, Randolph MA, Cetrulo Jr CL, Uygun K (2024) VCA supercooling in a swine partial hindlimb model. Sci Rep 14:12618. https://doi.org/10.1038/s41598-024-63041-8
109. Jiao X, Li Y, Chen Z, Zhang Q, He R, Huang Y, Zuo Z (2024) Targeting the PDE3B-cAMP-autophagy axis prevents liver injury in long-term supercooling liver preservation. Sci Transl Med 16:eadk0636. https://doi.org/10.1126/scitranslmed.adk0636
110. Isiksacan Z, William N, Senturk R, Boudreau L, Wooning C, Castellanos E, Isiksacan S, Yarmush ML, Acker JP, Usta OB (2024) Extended supercooled storage of red blood cells. Commun Biol 7:765. https://doi.org/10.1038/s42003-024-06463-4
111. Puts CF, Berendsen TA, Bruinsma BG, Ozer S, Luitje M, Usta OB, Yarmush ML, Uygun K (2015) Polyethylene glycol protects primary hepatocytes during supercooling preservation. Cryobiology 71:125–129. https://doi.org/10.1016/j.cryobiol.2015.04.010
112. Isiksacan Z, William N, Yarmush ML, Acker JP, Usta OB (2023) Long-term preservation of red blood cells through deep supercooling. Cryobiology 113:104654. https://doi.org/10.1016/j.cryobiol.2023.104654
113. Huang H, Rey-Bedón C, Yarmush ML, Usta OB (2020) Deep-supercooling for extended preservation of adipose-derived stem cells. Cryobiology 92:67–75. https://doi.org/10.1016/j.cryobiol.2019.11.004
114. William N, Isiksacan Z, Mykhailova O, Olafson C, Yarmush ML, Usta OB, Acker JP (2023) Comparing two extracellular additives to facilitate extended storage of red blood cells in a supercooled state. Front Physiol 14:1165330. https://doi.org/10.3389/fphys.2023.1165330
115. Li Y, Li Z, Peng L, Huang H (2025) Deep-supercooling preservation for rat kidneys. Transplantation 110:e363–e373. https://doi.org/10.1097/TP.0000000000005550
116. Huang H, Yarmush ML, Usta OB (2018) Long-term deep-supercooling of large-volume water and

red cell suspensions via surface sealing with immiscible liquids. Nat Commun 9:3201. https://doi.org/10.1038/s41467-018-05636-0
117. Powell-Palm MJ, Koh-Bell A, Rubinsky B (2020) Isochoric conditions enhance stability of metastable supercooled water. Appl Phys Lett 116:123702. https://doi.org/10.1063/1.5145334
118. Wang L, Meng H, Wang F, Liu H (2024) Ice nucleation mechanisms and the maintenance of supercooling in water under mechanical vibration. Results Phys 59:107581. https://doi.org/10.1016/j.rinp.2024.107581
119. Powell-Palm MJ, Rubinsky B, Sun W (2020) Freezing water at constant volume and under confinement. Commun Phys 3:39. https://doi.org/10.1038/s42005-020-0303-9
120. Kimizuka N, Viriyarattanasak C, Suzuki T (2008) Ice nucleation and supercooling behavior of polymer aqueous solutions. Cryobiology 56:80–87. https://doi.org/10.1016/j.cryobiol.2007.10.179
121. Zobrist B, Weers U, Koop T (2003) Ice nucleation in aqueous solutions of poly[ethylene glycol] with different molar mass. J Chem Phys 118:10254–10261. https://doi.org/10.1063/1.1571818
122. Bardsley WE, Khatep MM (1984) A general model for temperature of heterogeneous nucleation of supercooled water droplets. J Atm Sci 41:856–862. https://doi.org/10.1175/1520-0469
123. Langham EJ, Mason BJ (1958) The heterogeneous and homogeneous nucleation of supercooled water. Proc R Soc Lond A Math Phys Sci 247:493–504. https://doi.org/10.1098/rspa.1958.0207
124. Deck LT, Wittenberg L, Mazzotti M (2023) Thermodynamics explains how solution composition affects the kinetics of stochastic ice nucleation. J Phys Chem Lett 14:5993–6000. https://doi.org/10.1021/acs.jpclett.3c01371
125. Song F, Sato M, Toyama Y, Ishikawa T, Tokito F, Katsuda T, Toyooka K, Sakai Y, Nishikawa M (2026) A proof-of-concept study on high-pressure freezing for cryopreservation. PNAS Nexus 5:pgag065. https://doi.org/10.1093/pnasnexus/pgag065
126. Han HM, Huebinger J, Grabenbauer M (2012) Self-pressurized rapid freezing (SPRF) as a simple fixation method for cryo-electron microscopy of vitreous sections. J Struct Biol 178:84–87. https://doi.org/10.1016/j.jsb.2012.04.001
127. Grabenbauer M, Han H, Huebinger J (2014) Cryo-fixation by self-pressurized rapid freezing. In: Kuo (ed) Electron microscopy, methods in molecular biology. Humana Press, Totowa, NJ, pp 173–191. https://doi.org/10.1007/978-1-62703-776-1_9
128. Rubinsky B, Ikeda M (1985) A cryomicroscope using directional solidification for the controlled freezing of biological material. Cryobiology 22:55–68. https://doi.org/10.1016/0011-2240(85)90008-2
129. Ukpai G, Rubinsky B (2020) A mathematical analysis of directional solidification of aqueous sSolutions. J Heat Transfer 142:022401. https://doi.org/10.1115/1.4045312
130. Bahari L, Bein A, Yashunsky V, Braslavsky I (2018) Directional freezing for the cryopreservation of adherent mammalian cells on a substrate. PLOS One 13:e0192265. https://doi.org/10.1371/journal.pone.0192265
131. Saragusty J (2015) Directional freezing for large volume cryopreservation. In: Wolkers W, Oldenhof H (eds) Cryopreservation and freeze-drying protocols, Methods in Molecular Biology 1257. Springer, New York, NY, pp 381–397. https://doi.org/10.1007/978-1-4939-2193-5_19
132. Si W, Hildebrandt TB, Reid C, Krieg R, Ji W, Fassbender M, Hermes R (2006) The successful double cryopreservation of rabbit (*Oryctolagus cuniculus*) semen in large volume using the directional freezing technique with reduced concentration of cryoprotectant. Theriogenology 65:788–798. https://doi.org/10.1016/j.theriogenology.2005.06.010
133. Si W, Lu Y, He X, Ji S, Niu Y, Tan T, Ji W (2010) Directional freezing as an alternative method for cryopreserving rhesus macaque (*Macaca mulatta*) sperm. Theriogenology 74:1431–1438. https://doi.org/10.1016/j.theriogenology.2010.06.015
134. O'Brien JK, Robeck TR (2010) Preservation of beluga (*Delphinapterus leucas*) spermatozoa using a trehalose-based cryodiluent and directional freezing technology. Reprod Fertil Dev 22:653–663. https://doi.org/10.1071/RD09176
135. Arav A (1996) Device and methods for multigradient directional cooling and warming of biological samples. U.S. Patent No. 5,873,254
136. Revel A, Elami A, Bor A, Yavin S, Natan Y, Arav A (2004) Whole sheep ovary cryopreservation and transplantation. Fertil Steril 82:1714–1715. https://doi.org/10.1016/j.fertnstert.2004.06.046
137. Arav A, Revel A, Nathan Y, Bor A, Gacitua H, Yavin S, Gavish Z, Uri M, Elami A (2005) Oocyte recovery, embryo development and ovarian function after cryopreservation and transplantation of whole sheep ovary. Hum Reprod 20:3554–3559. https://doi.org/10.1093/humrep/dei278

138. Maffei S, Pennarossa G, Brevini TA, Arav A, Gandolfi F (2014) Beneficial effect of directional freezing on *in vitro* viability of cryopreserved sheep whole ovaries and ovarian cortical slices. Hum Reprod 29:114–124. https://doi.org/10.1093/humrep/det377
139. Birdseye C (1927) Method of preparing food products. U.S. Patent No. 1,773,079
140. Sun D (2013) Handbook of frozen food processing and packaging food oral processing: fundamentals of eating and sensory perception. Taylor & Francis
141. Jiang L, Liu D, Wang W, Lv R, Yu S, Zhou J (2025) Advancements and perspectives of novel freezing and thawing technologies effects on meat: a review. Food Res Int 204:115942. https://doi.org/10.1016/j.foodres.2025.115942
142. Loayza-Salazar S, Siche R, Vegas C, Chávez-Llerena RT, Encina-Zelada CR, Calla-Florez M, Comettant-Rabanal R (2024) Novel technologies in the freezing process and their impact on the quality of fruits and vegetables. Food Eng Rev 16:371–395. https://doi.org/10.1007/s12393-024-09371-9
143. Salvadori VO, Mascheroni RH (2002) Analysis of impingement freezers performance. J Food Eng 54:133–140. https://doi.org/10.1016/S0260-8774(01)00198-4
144. James C, Purnell G, James SJ (2015) A review of novel and innovative food freezing technologies. Food Bioproc Tech 8:1616–1634. https://doi.org/10.1007/s11947-015-1542-8
145. Akiyama Y, Shinose M, Watanabe H, Yamada S, Kanda Y (2019) Cryoprotectant-free cryopreservation of mammalian cells by superflash freezing. Proc Natl Acad Sci USA 116:7738–7743. https://doi.org/10.1073/pnas.1808645116
146. Yu Z, Garcia AS, Johnston KP, Williams RO (2004) Spray freezing into liquid nitrogen for highly stable protein nanostructured microparticles. Eur J Pharm Biopharm 58:529–537. https://doi.org/10.1016/j.ejpb.2004.04.018
147. Zhou X, Liu Z, Liang XM, Shu Z, Du P, Gao D (2013) Theoretical investigations of a novel microfluidic cooling/warming system for cell vitrification cryopreservation. Int J Heat Mass Transf 65:381–388. https://doi.org/10.1016/j.ijheatmasstransfer.2013.06.022
148. Shi M, Ling K, Yong KW, Li Y, Feng S, Zhang X, Pingguan-Murphy B, Lu TJ, Xu F (2015) High-throughput non-contact vitrification of cell-laden droplets based on cell printing. Sci Rep 5:17928. https://doi.org/10.1038/srep17928
149. Luyet BJ, Hodapp EL (1938) Revival of frog's spermatozoa vitrified in liquid air. Proc Soc Exp Biol Med 39:433–434. https://doi.org/10.3181/00379727-39-10229P
150. Berthier L, Reichman DR (2023) Modern computational studies of the glass transition. Nat Rev Phys 5:102–116. https://doi.org/10.1038/s42254-022-00548-x
151. Kivelson SA, Tarjus G (2008) In search of a theory of supercooled liquids. Nat Mater 7:831–833. https://doi.org/10.1038/nmat2304
152. Lunkenheimer P, Loidl A, Riechers B, Zaccone A, Samwer K (2023) Thermal expansion and the glass transition. Nat Phys 19:694–699. https://doi.org/10.1038/s41567-022-01920-5
153. Fahy GM, Wowk B (2015) Principles of cryopreservation by vitrification. In: Wolkers WF, Oldenhof H (eds) Cryopreservation and freeze-drying protocols. Methods in Molecular Biology 1257. Springer, New York, NY. https://doi.org/10.1007/978-1-4939-2193-5_2
154. Amini M, Benson JD (2023) Technologies for vitrification based cryopreservation. Bioengineering 10:508. https://doi.org/10.3390/bioengineering10050508
155. Abdelhady AW, Mittan-Moreau DW, Crane PL, McLeod MJ, Cheong SH, Thorne RE (2024) Ice formation and its elimination in cryopreservation of oocytes. Sci Rep 14:18809. https://doi.org/10.1038/s41598-024-69528-8
156. Takigawa T, Watanabe H, Akiyama Y (2024) Disaccharide-assisted inkjet freezing for improved cell viability. Cryobiology 116:104932. https://doi.org/10.1016/j.cryobiol.2024.104932
157. Joshi P, Taggart M, Uygun K, Sandlin RD (2024) Specimen containers and technologies for achieving vitrification through ultra-fast cooling. Annu Rev Heat Transfer 27:165–194. https://doi.org/10.1615/AnnualRevHeatTransfer.2024056141
158. Hopkins JB, Badeau R, Warkentin M, Thorne RE (2012) Effect of common cryoprotectants on critical warming rates and ice formation in aqueous solutions. Cryobiology 65:169–178. https://doi.org/10.1016/j.cryobiol.2012.05.010
159. Taylor MJ, Song YC, Brockbank KGM (2004) Vitrification in tissue preservation: new developments. In: Fuller BJ, Lane N, Benson EE (eds) Life in the frozen. State CRC Press, pp 629–668. https://doi.org/10.1201/9780203647073
160. Huber AJ, Brockbank KG, Aberle T, Schleicher M, Chen ZZ, Greene ED, Lisy M, Stock UA

(2012) Development of a simplified ice-free cryopreservation method for heart valves employing VS83, an 83% cryoprotectant formulation. Biopreserv Biobank 10:479–484. https://doi.org/10.1089/bio.2012.0006
161. Brockbank KGM, Chen Z, Greene ED, Campbell LH (2021) Vitrification of heart valve tissues. In: Wolkers WF, Oldenhof H (eds) Methods in cryopreservation and freeze-drying, Methods in Molecular Biology, Springer, pp 593–605. https://doi.org/10.1007/978-1-4939-2193-5_20
162. Jaskiewicz JJ, Callahan-Muller A, Gaby-Biegel N, Glover Z, Sandlin RD (2025) Validation of a high-throughput screening assay for the characterization of cryoprotective agent toxicity. bioRxiv. https://doi.org/10.1101/2025.05.26.654916
163. Ahmadkhani N, Sugden C, Benson JD, Eroglu A, Higgins AZ (2025) High-throughput evaluation of cryoprotective agents for mixture effects that reduce toxicity. Cryobiology 121:105316. https://doi.org/10.1016/j.cryobiol.2025.105316
164. Balcerzak AK, Capicciotti CJ, Briard JG, Ben RN (2014) Designing ice recrystallization inhibitors: from antifreeze (glyco)proteins to small molecules. RSC Adv 4:42682–42696. https://doi.org/10.1039/C4RA06893A
165. Chang T, Zhao G (2021) Ice inhibition for cryopreservation: materials, strategies, and challenges. Adv Sci 8:2002425. https://doi.org/10.1002/advs.202002425
166. Gangwar L, Han Z, Scheithauer C, Namsrai BE, Kantesaria S, Goldstein R, Etheridge ML, Finger EB, Bischof JC (2025) Physical vitrification and nanowarming at liter-scale CPA volumes: toward organ cryopreservation. Nat Commun 16:8511. https://doi.org/10.1038/s41467-025-63483-2
167. Powell-Palm MJ, Henley EM, Consiglio AN, Lager C, Chang B, Perry R, Fitzgerald K, Daly J, Rubinsky B, Hagedorn M (2023) Cryopreservation and revival of Hawaiian stony corals using isochoric vitrification. Nat Commun 14:4859. https://doi.org/10.1038/s41467-023-40500-w
168. Hagedorn M, Parenti LR, Craddock RA, Comizzoli P, Mabee P, Meinke B, Wolf SM, Bischof JC, Sandlin RD, Tessier SN, Toner M (2024) Safeguarding Earth's biodiversity by creating a lunar biorepository. BioScience 74:561–566. https://doi.org/10.1093/biosci/biae058
169. Feys R, Uygun K, Filz von Reiterdank I, Wolf SM, Isasi R (2024) Biopreservation beyond the biosphere: exploring the ethical, legal & social implications of suspended animation in space. J Law Med Ethics 52:648–665. https://doi.org/10.1017/jme.2024.148
170. Best BP (2015) Cryoprotectant toxicity: facts, issues, and questions. Rejuvenation Res 18:422–436. https://doi.org/10.1089/rej.2014.1656
171. Hornberger K, Li R, Duarte ARC, Hubel A (2021) Natural deep eutectic systems for nature-inspired cryopreservation of cells. AIChE J 67:e17085. https://doi.org/10.1002/aic.17085
172. Mallya AS, Yadav P, Zakhia S, Hubel A (2025) Low-temperature characterization of sugar alcohol-based type V deep eutectic solvents for anti-icing and cryopreservation applications. Sustain Chem Pharm 48:102256. https://doi.org/10.1016/j.scp.2025.102256
173. Alvarez C, Berrospe-Rodriguez C, Wu C, Pasek-Allen J, Khosla K, Bischof J, Mangolini L, Aguilar G (2022) Photothermal heating of titanium nitride nanomaterials for fast and uniform laser warming of cryopreserved biomaterials. Front Bioeng Biotechnol 10:957481. https://doi.org/10.3389/fbioe.2022.957481
174. Oziri OJ, Rao JS, Scheithauer C, Han Z, Kantesaria S, Tobolt D, Liang H, Etheridge ML, Yin Y, Finger EB, Bischof JC (2025) Scalable purification of iron oxide nanoparticles for organ cryopreservation and transplantation. Small 21:e04910. https://doi.org/10.1002/smll.202504910

Check for updates

Chapter 5

Emerging Manufacturing Technologies in Freeze-Drying of Pharmaceuticals

Roberto Pisano, Fiora Artusio and Antonello A. Barresi

Abstract

Freeze-drying of pharmaceutical products is an energy-intensive operation in pharmaceutical manufacturing. Emerging process-intensification and continuous-manufacturing technologies are now reshaping this traditionally static process. Among these, microwave-assisted freeze-drying, suspended-vial freeze-drying, and spin freeze-drying have demonstrated substantial reductions in cycle time, i.e., up to 80%–90%, while maintaining product stability and uniformity. These approaches enhance heat and mass transfer by replacing conduction through fixed shelves with volumetric or radiative energy delivery and thin-film geometries. Integrated Process Analytical Technology tools, coupled with mechanistic models and digital twins, enable real-time monitoring, predictive control, and potentially real-time release testing. The suspended-vial and spin freeze-drying platforms further enable continuous operation through modular design and automated vial transfer, aligning with the broader transition toward Pharmaceutical Manufacturing 4.0. Although regulatory adoption requires demonstrating sterility assurance, data integrity, and robust PAT validation, these technologies offer a credible path toward faster, safer, and more sustainable lyophilization.

Keywords Continuous manufacturing, Freeze-drying, Microwave-assisted freeze-drying, Pharmaceutical manufacturing, Spin freeze-drying, Suspended-vial freeze-drying

1 Introduction

Pharmaceutical manufacturing is undergoing a profound transformation, shifting from traditional batch-oriented technologies to continuous manufacturing (CM) systems that integrate process steps into uninterrupted, digitally controlled flows [1–5]. This transition is not a transient technological trend but a structural evolution, deeply rooted in regulatory reform, scientific progress, and economic necessity. It redefines the way quality is monitored and assured throughout the entire product lifecycle.

Willem F. Wolkers and Harriëtte Oldenhof (eds.), *Cryopreservation and Freeze-Drying Protocols*, Methods in Molecular Biology, vol. 3060, https://doi.org/10.1007/978-1-0716-5416-3_5,

1.1 From Batch to Continuous: The Evolution of Pharmaceutical Manufacturing

For decades, the pharmaceutical industry has depended on batch manufacturing technologies: self-contained unit operations executed sequentially, often in geographically distinct facilities. Each batch undergoes intermediate storage, manual transfers, and off-line quality testing, resulting in extended lead times and supply chain inefficiencies. Typical production cycles for a single drug product may span over a year, including testing, packaging, and logistics [1, 6, 7].

By contrast, continuous manufacturing integrates synthesis, formulation, and packaging into a single flow path where raw materials are continuously fed, and finished products are continuously discharged. The system operates at or near steady state, with process parameters controlled by feedback control systems and process analytical technology (PAT) tools. These technologies, ranging from near-infrared (NIR) and Raman spectroscopy to laser scattering and acoustic sensors, result in continuous monitoring of critical quality attributes (CQAs), thereby replacing traditional end-product testing with in-process verification [1–3, 8, 9].

This paradigm shift has been actively encouraged by regulatory authorities. The U.S. Food and Drug Administration (FDA) launched its *Pharmaceutical Quality for the 21st Century* initiative in 2002, followed by the *Process Analytical Technology (PAT) Guidance* (2004) and the *Emerging Technology Program* (2014). Together, these frameworks introduced the philosophy of quality by design (QbD), subsequently extended to quality by control (QbC), where real-time analytical and control systems guarantee consistent product performance [3, 5, 6]. Parallel initiatives by the European Medicines Agency (EMA) and the International Council for Harmonization (ICH) culminated in the publication of *ICH Q8–Q13*, providing harmonized guidance for the design, validation, and control of continuous pharmaceutical operations [5, 6].

1.2 Scientific and Industrial Progress

Academic and industrial collaboration has played a decisive role in advancing continuous manufacturing technologies. Major pharmaceutical companies, e.g., Novartis, Eli Lilly, Pfizer, Johnson & Johnson, and Vertex, have invested in pilot plants and commercial lines that integrate synthesis and formulation [1, 2, 7, 10, 11]. Vertex's *Orkambi* and Johnson & Johnson's *Prezista* (darunavir) were the first FDA-approved drug products produced using fully continuous lines [12–14].

The Novartis-MIT Center for Continuous Manufacturing has demonstrated the feasibility of fully integrated systems linking chemical synthesis, crystallization, filtration, drying, and tableting in a single end-to-end process [15]. These achievements show that CM can deliver consistent quality and cost advantages while maintaining regulatory compliance. The success of such demonstrations has catalyzed the formation of global research consortia, including CMAC (UK), RCPE (Austria), and SSPC (Ireland), all dedicated to developing robust models, PAT frameworks, and scalable equipment [1, 2, 15]. Continuous operation provides numerous benefits over batch processing:

i. Reduced cycle time, as periods for loading, testing, and cleaning are minimized;
ii. Improved reproducibility, since steady-state operation reduces transient variation;
iii. Enhanced safety, due to smaller reactor volumes and better transport phenomena control;
iv. Lower cost and environmental impact, through optimized use of resources and energy;
v. Simplified scale-up, as throughput is increased by numbering up identical units rather than enlarging the equipment.

1.3 Economic and Environmental Motivation

The financial and environmental implications of continuous manufacturing are substantial. Economic models comparing continuous and batch operations consistently show higher net present value and shorter payback periods for CM facilities. Rossi [4] demonstrated that under realistic capital and tax scenarios, continuous systems outperform batch counterparts both for brand-name and generic drugs, primarily because of reduced equipment size, energy efficiency, and modular expandability. From an environmental perspective, CM contributes significantly to the sustainability goals articulated by regulatory and industrial initiatives such as Green Chemistry and Sustainable Pharmaceutical Manufacturing [5]. Continuous processes typically reduce energy consumption by 30%–50%, water usage by 25%–40%, and solvent waste by up to 60% compared with equivalent batch operations [5]. Furthermore, smaller equipment footprints enable localized production facilities, named modular microfactories, that reduce transportation costs and enhance supply-chain resilience, an issue highlighted during the COVID-19 pandemic [2, 5].

1.4 Regulatory Harmonization and Quality by Control

The acceptance of continuous manufacturing by global regulators marks an unprecedented moment in pharmaceutical history. The ICH Q13 guideline explicitly recognizes that continuous processes can be applied to both drug substance and drug product manufacturing, provided a *state of control* is demonstrated and maintained [6]. This *state of control* requires a clear definition of steady-state boundaries, validated PAT tools, and comprehensive risk management plans covering the identification and removal of nonconforming material and the management of process upsets. In practice, this approach establishes real-time release testing (RTRT) as a new regulatory standard. By continuously verifying CQAs during production, RTRT eliminates the need for lengthy off-line assays and accelerates product disposition. The FDA's Emerging Technology Team and

EMA's Innovation Task Force provide [16] a pathway for early dialog between manufacturers and regulators, facilitating the validation of new analytical and control strategies, as well as incentivizing the use of process models [3, 5, 6].

1.5 Integration of Enabling Continuous Technologies

The feasibility of CM relies on the convergence of multiple enabling technologies. Flow chemistry provides the foundation for continuous drug substance production. Using plug-flow, packed-bed, and microreactor systems, chemical transformations are performed with more efficient heat and mass transfer, precise residence-time control, and the ability to sequence multiple reactions in a continuous manner without isolating intermediates [1, 3]. In parallel, continuous solids processing (e.g., wet and dry granulation, blending, and tableting) has matured through modular systems capable of self-regulated operations. Advanced sensors and digital control architectures connect these unit operations into fully integrated lines. Machine learning algorithms and digital twins simulate process behavior, predict deviations, and enable model-predictive control, thus reducing operator intervention.

1.6 The Case of Freeze-Drying

Among all pharmaceutical unit operations, freeze-drying (lyophilization) stands out as one of the most challenging to convert into continuous mode. It is essential for the stabilization of thermolabile products; however, it remains time-consuming and energy-intensive. Typical batch lyophilization cycles last from 48 to 120 h, depending on product formulation and vial size. Conventional freeze-drying is performed in large, static chambers under high vacuum. Heat is delivered primarily by conduction through the shelf-vial contact. The process comprises three sequential stages: freezing, primary (sublimation), and secondary drying (desorption).

A major limitation of batch freeze-drying is the heterogeneity of heat transfer within the vial batch. Edge vials receive more radiant energy from chamber walls, leading to uneven product temperatures and residual moisture. Moreover, stochastic ice nucleation during freezing generates variability in pore size and mass-transfer resistance, affecting drying kinetics and product morphology. These sources of nonuniformity translate into within-batch and batch-to-batch variation in cake structure, reconstitution time, and product stability.

Another challenge lies in the operational inefficiency of large batch dryers. Loading, CIP/SIP cycles, condenser defrosting, and leak testing extend downtime between runs. The cumulative effect is a process that, while robust, is slow, energy-intensive, and difficult to scale linearly. Consequently, research has increasingly focused on process intensification and continuous lyophilization technologies capable of reducing cycle time and improving process control.

The limitations of conventional batch lyophilization have driven the development of several intensified and continuous approaches, all aimed at shortening the drying cycle and improving process control while maintaining product quality and sterility. Among these, three technologies stand out as particularly promising: microwave-assisted freeze-drying (MFD), suspended-vial freeze-drying (SVFD), and spin freeze-drying (SPFD). These developments differ in mechanism and engineering configuration but share a common aim: to decouple heat transfer from static shelf conduction and introduce dynamic or volumetric energy delivery.

While each of these technologies addresses the limitations of traditional lyophilization from a different angle, they share several defining principles:

i. Enhanced heat and mass transfer through volumetric, radiative, or thin-film geometries;
ii. Process intensification reduces total drying time by up to 80%–90%;
iii. Integration of PAT and model-based control enabling real-time monitoring of product state;
iv. Potential for modular and continuous operation compatible with aseptic pharmaceutical manufacturing.

The convergence of these innovations, supported by digital twins and predictive control, is gradually transforming lyophilization from a static operation into a data-driven, continuously controlled process [7].

From a regulatory perspective, continuous freeze-drying is no longer viewed as an experimental curiosity but as a credible component of modern manufacturing strategy. The ICH Q13 guideline [6] explicitly recognizes continuous processes for both drug substance and drug product manufacture, provided that a state of control and traceability is ensured, and this is true also for the freeze-drying process. The FDA's Emerging Technology Program and the EMA's Innovation Task Force now provide pre-submission dialogs that facilitate the validation of novel lyophilization equipment.

Nevertheless, regulatory adoption demands robust evidence of process understanding, validated PAT sensors, and traceable data integrity. Continuous systems must demonstrate that they can maintain sterility across multiple load-lock interfaces and that process deviations can be detected and mitigated in real time. For this reason, it would be desirable to establish pilot installations of technologies such as SVFD and SPFD, operated within collaborative consortia that bring together academia, equipment suppliers, and pharmaceutical companies to generate the datasets required to build regulatory confidence. Industrial adoption will likely proceed incrementally. Hybrid configurations, combining semi-continuous drying modules with traditional freeze-dryers, are expected to dominate in the short term. These systems will enable integration of PAT, digital control, and accelerated heating strategies without requiring radical changes to existing cleanroom infrastructure.

The long-term vision is consistent with the broader evolution toward Pharmaceutical Manufacturing 4.0, defined by digital connectivity, modular design, and sustainability. Fully continuous lyophilization lines, integrated with continuous upstream and downstream processes, could enable decentralized production of biologics and vaccines in modular facilities distributed across multiple regions. Such a networked manufacturing architecture would strengthen global supply resilience, lower energy consumption, and allow a rapid response to public health emergencies.

2 Microwave-Assisted Freeze-Drying

2.1 Principles of Microwave Technology in Freeze-Drying

Microwave drying (MWD) is increasingly recognized as a next-generation drying technology for pharmaceuticals owing to its ability to generate heat directly within the material rather than transferring it from an external source. The term covers several related processes: atmospheric microwave drying (AMD), microwave vacuum drying (MVD), and microwave freeze-drying (MFD). These applications differ mainly in operating pressure and temperature but share the same physical principle: the direct volumetric transfer of electromagnetic energy into the sample.

Microwave energy induces dipolar rotation and ionic conduction within polar materials. When exposed to an alternating electromagnetic field, molecules possessing a permanent dipole moment oscillate rapidly, leading to molecular friction and uniform volumetric heating. Unlike conventional freeze-drying, where heat transfer is dominated by conduction from the shelf through the vial bottom through a thin gas layer, microwaves deliver energy throughout the product volume, significantly enhancing heat-transfer efficiency.

The volumetric power dissipated, P_d, depends on the dielectric properties of the material (i.e., its dielectric constant and loss factor) and on the frequency of the applied field, f, as well as on the electric-field strength, $|E|$:

$$P_d = 2\pi f \varepsilon_0 \varepsilon_r \ |E|^2$$

where ε_0 is the vacuum permittivity and ε_r the dielectric loss factor and is a function of f. The effectiveness of microwave heating, therefore, depends strongly on the dielectric behavior of the formulation, which in turn varies with composition, temperature, and water content.

Microwave heating interacts differently with ice and liquid water. The loss factor of ice is much lower than that of unfrozen water, resulting in preferential energy absorption by the residual liquid phase. As drying proceeds and the amount of free water decreases, the overall absorption declines, leading to a form of self-regulation: the process naturally supplies less energy as the sample becomes drier.

Early implementations of microwave heating in drying aimed to accelerate water removal in combination with convective or vacuum methods. In MVD, the reduced pressure lowers the boiling point of water, enabling rapid evaporation at moderate temperatures. The volumetric energy delivery of microwaves further enhances drying kinetics, leading to the formation of highly porous structures with low mass-transfer resistance and rapid rehydration, comparable to those achieved by freeze-drying [17]. However, the efficiency of microwave drying decreases when the sample contains a large amount of liquid water, making it more suitable for the final stages of dehydration.

Microwave-assisted freeze-drying extends these principles to sublimation. The product is frozen prior to irradiation, and ice sublimates under reduced pressure while microwave energy supplies the latent heat of sublimation. This concept was first proposed over 60 years ago [18] and demonstrated that sublimation rates could be dramatically increased relative to conventional freeze-drying. MFD therefore combines the low-temperature benefits of lyophilization with the process-intensification potential of volumetric heating.

Nevertheless, MFD presents several challenges. Because microwaves are electromagnetic fields, they can cause nonuniform heating and the formation of local hot spots where the field intensity is high. Overheating or corona discharge may occur if the electric-field strength exceeds the breakdown threshold of the gas phase [19, 20]. This aspect is further discussed below. Optimal operation is therefore achieved close to, but below, the onset of discharge, maximizing energy efficiency while avoiding product degradation.

Microwave systems typically operate at standard industrial frequencies, the most common being 2.45 GHz, corresponding to a wavelength of approximately 12 cm. The penetration depth increases as the frequency decreases and as water converts to ice. In frozen materials below about –5°C, the penetration depth can reach several centimetres, allowing efficient energy delivery to substantial sample volumes [21]. Lower frequencies (e.g., 915 MHz) further increase penetration but raise the risk of electrical discharge [22].

In atmospheric microwave freeze-drying (AFD), sublimation occurs at subfreezing temperatures (typically between –3°C and –10°C) under atmospheric pressure. The drying rate is limited by internal heat- and mass-transfer resistances, but microwave irradiation can reduce process time by about 50% compared to conventional freeze-drying [23]. However, most pharmaceutical applications rely on vacuum conditions to ensure low product temperatures and controlled ice sublimation.

In summary, microwaves transform the thermal-energy supply mechanism of freeze-drying. Instead of transferring heat by conduction or radiation, energy is generated directly within the product. It may be noted that both cases have been considered: energy supplied only by MW, or addition of MW heating to a conventional system (which is also named hybrid freeze-drying by some authors); here, the terms MFD or MW-assisted freeze-drying will be used indifferently. The challenge, in any case, is to achieve spatially uniform energy deposition and to control temperature to avoid degradation of thermosensitive materials such as proteins and vaccines.

2.2 Current Applications, Benefits, and Limitations of Microwave-Assisted Freeze-Drying

The application of microwaves to freeze-drying has expanded rapidly, but its use in the pharmaceutical industry remains at an early stage. The technology offers major potential advantages: significant reductions in drying time, improved energy efficiency, and the possibility of continuous or on-demand manufacturing. However, these benefits are balanced by challenges concerning uniformity, process control, and product quality.

2.2.1 Advantages and Process Efficiency

Compared with conventional freeze-drying (CFD), MFD delivers energy volumetrically, heating the entire product mass rather than only its bottom and/or peripheral surfaces. Ice sublimation therefore proceeds throughout the volume within the microwave penetration depth, leading to markedly shorter drying times. Reductions exceeding 80% have been reported while maintaining comparable product stability [24]. The rapid and direct heat input allows flexible cycle design, enabling adaptive manufacturing concepts where the process can be dynamically adjusted to batch size or facility throughput.

Microwave heating also facilitates fast process control because the power output of modern generators can be modulated almost instantaneously. This contrasts with the sluggish thermal response of conventional shelf-based systems, which regulate temperature via a silicone-oil loop. The fast response of MFD enables closed-loop temperature control, and power modulation can prevent overheating and corona effects by adjusting energy input based on real-time product temperature measurements.

2.2.2 Product Quality and Stability

The main concern regarding pharmaceutical MFD is its potential impact on biomolecules' stability. Several studies have explored protein and vaccine formulations to evaluate structural integrity and biological activity after microwave exposure. Ambros et al. [25] demonstrated that lactic acid bacteria maintained survival and membrane integrity after MFD. Wang et al. [26] compared immunoglobulin Y (IgY) dried by MFD and CFD and observed comparable activity retention when trehalose was included as a stabilizer, with a significantly shorter process time in MFD. Similar findings were reported for monoclonal antibodies (mAbs): Gitter et al. [27, 28] obtained drying-time reductions exceeding 70% while preserving structural integrity and acceptable cake appearance. Härdter et al. [29] later extended these results to six mAb formulations, confirming that microwave-dried products exhibited good long-term stability and elegant morphology.

The effect of protein concentration was also investigated. At high antibody concentrations, microwave radiation can induce aggregate formation through direct excitation of polar residues, leading to higher subvisible-particle counts [30]. Cooling during irradiation and the use of higher sugar-to-protein ratios mitigate these effects. In general, MFD seems most beneficial for formulations with low API concentration and high excipient content, such as vaccines or RNA-based products, whereas its suitability for concentrated antibody formulations remains limited.

2.2.3 Uniformity and Batch Consistency

Achieving uniform heating remains one of the major technological barriers. Because the microwave field inside a freeze-dryer forms a complex standing-wave pattern, local field maxima and minima cause uneven temperature distributions. In contrast to CFD, where edge-vials usually warm faster, in MFD, central vials can become warmer than those at the periphery. Despite these differences, several research groups have achieved moisture distributions comparable to conventional cycles by improving field uniformity. For instance, solid-state semiconductor microwave generators, which permit precise control of frequency and phase, have enhanced spatial uniformity relative to magnetron sources [24, 28].

Another aspect often overlooked is the heating of the container itself. Glass possesses a dielectric-loss coefficient two orders of magnitude greater than ice and, thus, absorbs microwaves strongly. Significant wall-to-product temperature gradients (approx. up to 40°C) have been measured [31]. This phenomenon is critical and should be considered when modeling heat transfer and designing process control strategies.

2.2.4 Corona Discharge and Safety

Microwave fields can produce plasma discharge at high electric-field intensity, particularly under low-pressure conditions. This phenomenon, also known as the *corona effect*, leads to gas ionization, local overheating, and product damage. Its onset depends on pressure, humidity, and applied power. Avoiding corona discharge requires dynamic power modulation based on the instantaneous dielectric properties of the material [22, 32]. Recent work demonstrated that adjusting microwave power in response to process feedback can prevent discharge while maintaining high drying efficiency [24].

2.2.5 Microwave Frequency Selection

Microwave frequencies approved for industrial heating include 915 MHz, 2.45 GHz, and 5.8 GHz, but the optimal frequency for freeze-drying may differ. Higher frequencies increase the number of excited modes within the cavity, improving uniformity. Abdelraheem et al. [33] applied statistical electromagnetics to select frequencies

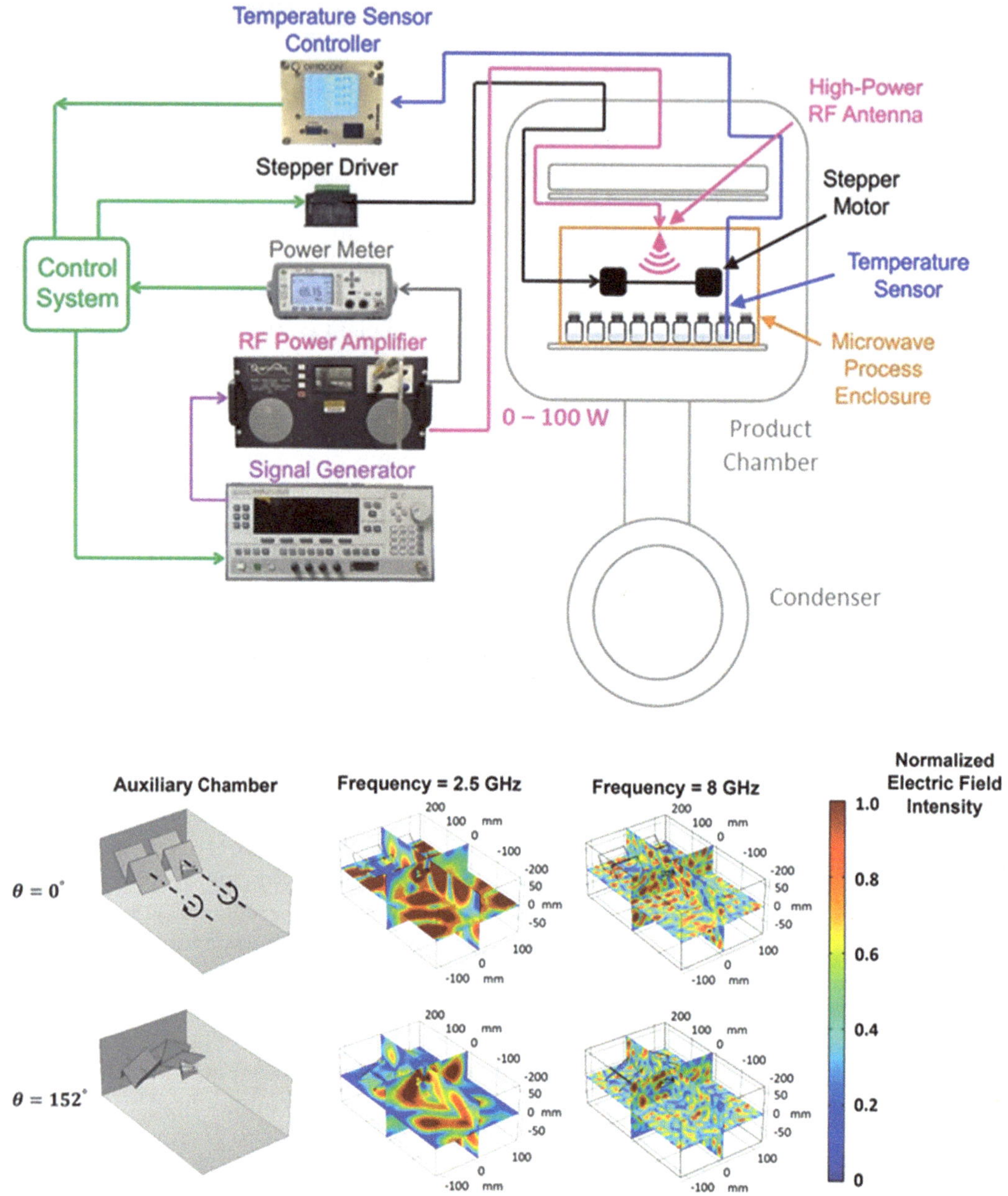

Fig. 1 Schematic of an MFD prototype and normalized electric field distributions inside the chamber as simulated using COMSOL for scatterer rotation angle $\theta = 0$ or 152° and excitation frequencies = 2.5 or 8 GHz. Figures reprinted from [31]

between 8 GHz and 18 GHz, achieving a threefold reduction in primary drying time compared with conventional freeze-drying. This approach has been realized in prototypes combining frequency stirring with closed-loop control of power input [31], as shown in Fig. 1.

2.2.6 Process Control and Monitoring

Microwave heating offers unprecedented opportunities for real-time process control. Because the electromagnetic field responds almost instantaneously to heat power adjustments, temperature can be regulated via proportional-integral (PI) feedback control loops to maintain target conditions. In a collaboration between Purdue University, Merck, and IMA Life, a radio-frequency lyophilization cycle was successfully operated under PI control, adjusting power dynamically as product temperature approached its limit value [34]. In such systems, conventional thermocouples cannot be used because they interfere with the field; instead, fiber-optic sensors provide accurate, microwave-immune temperature measurements with resolutions of 0.01°C and response times of 10 ms [31].

2.2.7 Design of Customized GMP Equipment

Historically, MFD research relied on prototype systems that lacked pharmaceutical functionality, such as in-chamber freezing and stoppering. These limitations prevented direct comparison with GMP-compliant freeze-dryers [29]. Recently, new designs have emerged. Härdter et al. [29] retrofitted a standard freeze-dryer with flat microwave antennas, enabling scalable and uniform radiation without product or source rotation. Another group [31] developed a quasi-random-field (qRF) system operating between 8 GHz and 18 GHz with mechanically stirred scatterers to randomize the field. The combination of a horn antenna and electromagnetic stirring achieved a highly uniform energy distribution, reducing drying time by half compared with conventional operation.

3 Continuous Manufacturing in Freeze-Drying of Unit-Dose Pharmaceuticals

Continuous freeze-drying of unit-dose injectable pharmaceuticals represents a major technological advancement over conventional batch lyophilization. Traditional batch freeze-drying, while well established, remains inherently limited by its long cycle times, vial-to-vial variability, and complex scale-up requirements. These limitations not only increase the risk of contamination and batch rejection but also constrain manufacturing flexibility and responsiveness to fluctuating production demands.

In contrast, continuous freeze-drying decouples and automates the main process stages (i.e., freezing, primary drying, and secondary drying) under tightly controlled and reproducible conditions. As discussed in Pisano et al. [7], by enabling the continuous transfer of vials through dedicated process zones, the approach minimizes human intervention, reduces downtime between stages, and facilitates real-time monitoring and control. As a result, continuous technologies offer the potential to dramatically improve process efficiency, product uniformity, and throughput while maintaining the stringent quality standards required for parenteral drug manufacturing and regulatory authorities.

Continuous freeze-drying redefines the process architecture. Rather than performing freezing, primary drying, and secondary drying sequentially within a single chamber, the process is spatially decomposed into dedicated, continuously operated units. Each stage proceeds under independently controlled conditions, ensuring that all vials undergo comparable thermal and pressure conditions. This modular design enables uninterrupted material flow and supports the implementation of advanced monitoring and feedback control strategies. This concept agrees with the broader evolution of pharmaceutical manufacturing toward continuous, data-driven production paradigms. Within the frameworks of QbD and PAT, regulatory agencies now emphasize building quality into the entire process rather than verifying it solely through end-point testing. Continuous freeze-drying naturally fits within this philosophy: it allows direct integration of real-time measurements, model-based control, and traceable process analytics that strengthen both process understanding and regulatory robustness.

Over the past decade, two technologies have gained particular attention as viable platforms for continuous pharmaceutical lyophilization: the suspended-vial system (SVFD) and spin freeze-drying (SPFD). These two concepts differ substantially in design and operating principles.

3.1 Suspended-Vial Freeze-Drying

3.1.1 Concept and Principles

Suspended-vial freeze-drying (SVFD), first proposed by Capozzi et al. [35], was developed to overcome fundamental drawbacks of batch processing, e.g., spatial nonuniformity of heat transfer and complex scalability, by removing the direct contact between vials and shelves. In SVFD, each vial is suspended in a controlled environment through which it moves continuously during freezing and drying. In its first design, the vials were held by their necks on a moving track and transported through sequential process modules operating at different pressures and temperatures, separated by load-lock chambers that maintain isolation between modules. This configuration eliminates the geometric and conductive constraints of shelf contact, enabling heat transfer primarily by radiation and, during freezing, forced gas convection. As a consequence, each vial experiences nearly identical thermal and pressure conditions, which ensures homogeneous sublimation and uniform product morphology.

This concept forms the backbone of recent continuous lyophilization platforms developed jointly by Politecnico di Torino and Massachusetts Institute of Technology, integrating forced-convection freezing chambers, magnetically levitated vial motion systems, and in-line monitoring of temperature and mass [36]. In contrast to spin freeze-drying, which modifies the product morphology, the suspended-vial approach preserves the conventional vial-based dosage form, facilitating regulatory acceptance and compatibility with existing fill-finish infrastructures.

3.1.2 System Architecture

The suspended-vial continuous freeze-dryer comprises four main functional sections [36]:

i. A filling and loading station;
ii. A freezing module;
iii. A primary and secondary drying tunnel;
iv. An unloading and capping unit.

Each section is connected through load-lock modules that allow pressure transitions without compromising sterility. The design enables the steady progression of vials along the process line under vacuum, thus resulting in true continuous manufacturing. A schematic of the technology is shown in Fig. 2.

Freezing module: the freezing module is subdivided into three zones:

i. a conditioning chamber, where vials are cooled by forced convection using nitrogen gas at controlled temperature and flow rate;
ii. a nucleation chamber, in which controlled nucleation is triggered, typically through vacuum-induced surface freezing (VISF) or an impulse of cold gas;
iii. a solidification chamber, where the ice front propagates until complete solidification.

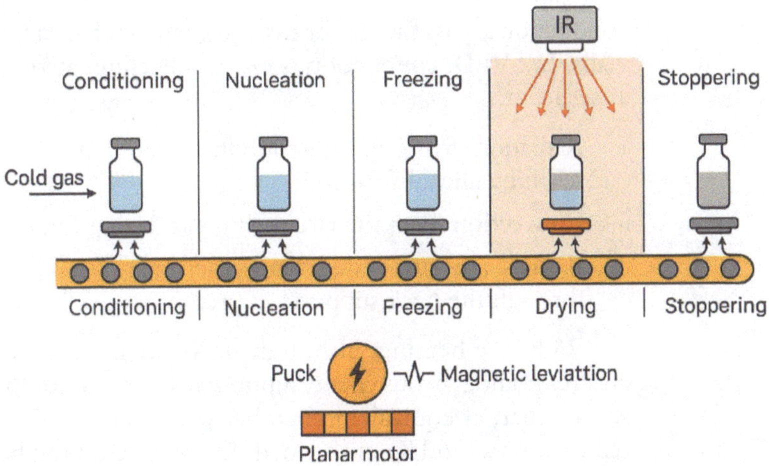

Fig. 2 Schematic of the suspended-vial continuous lyophilizer. Modules for conditioning, nucleation, freezing, drying, and stoppering, showing vial suspension on a magnetically levitated motion system and load-lock connections are shown

Forced gas convection plays a central role in maintaining uniform temperature. Computational fluid dynamics analyses and physical prototypes have shown that cross-flow nitrogen at −30 to −50°C can cool 10R vials from 25°C to −1°C in less than 20 min, with vial-to-vial temperature deviation below 0.5°C [37]. When a thermal impulse (<−30°C gas) is applied, nucleation occurs almost simultaneously in all vials within 30 s, dramatically reducing vial-to-vial heterogeneity.

Drying tunnel: after freezing, the vials are transferred through a load-lock module to a vacuum tunnel where primary and secondary drying occur. In this section, heat transfer is dominated by radiation from temperature-controlled walls. The tunnel is divided into modular segments, each thermally regulated to maintain a desired product-temperature trajectory. Computational fluid dynamics-assisted design ensures uniform pressure (<10 Pa) and minimal temperature gradients along the 2.5 m tunnel length [38]. Condensers are connected laterally to manage the continuous removal of water vapor; multiple condensers operate alternately to allow regeneration during ongoing production.

Each vial is held in a compliant wire or spring suspension system that both supports its weight and serves as a mass-sensing element [39]. Cameras mounted outside the chamber track markers on the wires, allowing real-time measurement of vial mass with a precision better than 50 mg per vial. This system enables direct measurement of sublimation rate and drying-end detection without perturbing the process.

3.1.3 Heat- and Mass-Transfer Mechanisms

The transition from shelf-based to suspended geometry changes the dominant heat-transfer mechanisms. In batch freeze-drying, heat is provided via conduction through the shelf–vial contact, gas conduction across the vial-bottom gap, and radiation from chamber walls. In SVFD, conduction is practically eliminated; heat transfer arises from:

i. Radiation from the surrounding walls and fins, uniformly reaching all vial surfaces;
ii. Convection from the circulating gas during freezing;
iii. If vials are suspended from their neck, minor conduction through the neck supports, if present (typically <2%).

Radiative heating, although slower than direct conduction, ensures homogeneous energy input. Experimental comparisons have shown that, at equivalent chamber pressures (5–10 Pa) and shelf temperatures (263 K), suspended vials receive a mean heat flux of ≈ 0.20 W per vial, independent of pressure, whereas contact-vials in batch dryers receive 0.10–0.30 W with large spatial variability [35]. The independence of heat flux from pressure allows operation at lower pressures, thereby increasing the driving force for ice sublimation and shortening cycle time.

In the case of freezing, forced convection markedly enhances cooling efficiency. Raising gas velocity from natural convection to 1 $m s^{-1}$ reduces freezing time by ≈1.5 h, while maintaining uniform solidification [37]. The uniformity of heat removal results in larger, more consistent ice crystals (mean pore diameter ≈70 versus 40 μ m for shelf freezing) and a narrower pore-size distribution, yielding lower mass-transfer resistance during drying [35]. Radiative uniformity also eliminates the edge-vial effect. Measurements show negligible variation in sublimation rate across the batch; vial-to-vial variation in drying time decreases from ≈5 h (batch) to <1 h (suspended) at comparable conditions.

3.1.4 Monitoring and Process Analytical Technologies

The SVFD platform integrates per-vial, in-line analytics. Three complementary PAT tools have been demonstrated [36]:

i. *Mass sensors*: each vial is supported by a compliant helical-spring or flexural-beam element whose deflection is tracked optically through chamber windows [39]. The resulting displacement correlates directly with vial mass, enabling continuous measurement of sublimation rate of each vial throughout the cycle. Median mass-measurement error is ≈6 mg under vacuum, sufficient to resolve drying-rate changes and detect individual drying endpoints.
ii. *Thermal imaging*: FLIR A35 infrared cameras mounted on each side of the vacuum tunnel capture vial-surface temperatures through ZnSe windows [36]. Calibration using cryogenic references allows accuracy within ±1°C down to −70°C. This provides spatially resolved temperature fields across all vials, supporting real-time control of wall temperature and chamber pressure.
iii. *Supervisory control system*: all sensors are coordinated by a MATLAB-based supervisory controller that manages tray motion, zone transitions, and feedback loops to maintain the prescribed thermal profile [36]. High-frequency data from the cameras and mass sensors feed model-based algorithms that estimate the residual-water mass and automatically trigger secondary-drying or stoppering actions.

Measuring mass and temperature of every vial also enables RTRT: when each vial's drying endpoint is independently verified, the batch concept becomes obsolete.

3.1.5 Design and Modeling of the Vacuum System

The extended drying tunnel demands precise pressure control and efficient vapor removal. Stratta et al. [38] developed a modular CFD-based design methodology in which each vacuum component (elbows, pipes, condensers) is modeled individually and concatenated analytically to predict overall pressure distribution. Validation against experimental tunnels (250 × 125 × 2500 mm) confirmed pressure uniformity below 10 Pa throughout.

Multiple condensers maintained at 77 K are attached in parallel to manage continuous sublimation loads of approximately 90 $g h^{-1}$. During regeneration, one condenser is isolated while others remain active, ensuring uninterrupted operation. The narrow geometry of the tunnel also improves radiative uniformity by increasing view factors between vials and walls.

Simulation and experimental data showed that, under representative loads, the pressure drop along the tunnel is <1 Pa, and the total vapor-removal rate exceeds 0.3 $g s^{-1}$. These values are consistent with product drying times ≤10 h for 3 mL vials.

3.1.6 Performance and Product Quality

Experimental demonstrations comparing SVFD and batch freeze-drying consistently show superior performance of the former:

i. *Cycle time*: primary drying time is reduced by a factor of 2–4; total cycle duration (including filling and unloading) decreases up to sixfold because ancillary operations occur in parallel [35].
ii. *Thermal uniformity*: maximum temperature deviation across vials <0.5 K during primary drying.
iii. *Morphology*: larger, more homogeneous pores and reduced dried-cake resistance to vapor flow.
iv. *Residual moisture*: final moisture at 2 wt% with reduced vial-to-vial standard deviation (±0.3%) [36].
v. *Stability*: protein formulations and enzymatic activity (e.g., catalase, myoglobin) remain unaltered compared with batch references [36].

These improvements arise from uniform radiative heating and from the ability to maintain lower chamber pressure without compromising energy delivery. Furthermore, controlled nucleation (VISF or impulse gas) results in reproducible ice morphology, directly translating into consistent reconstitution times and cake appearance.

3.1.7 Integration with Continuous Manufacturing

The modular design of the SVFD system allows seamless integration into end-to-end continuous pharmaceutical production lines. The levitated-tray transport system employed by Burcat et al. [36] moves vials through conditioning, freezing, drying, and stoppering zones without physical contact, eliminating particle generation and enabling aseptic operation. This motion system, combined with flexible load-lock airlocks, allows throughput scaling simply by adding or extending modules. Laboratory- and pilot-scale units share identical geometry, preserving heat-transfer characteristics across scales and, thus, eliminating the problem of process scale-up. Real-time PAT data (mass, temperature) inform a model-predictive control algorithm that adjusts wall-temperature set-points to maintain the product temperature just below its collapse temperature. This strategy optimizes power input and minimizes cycle time while guaranteeing product quality integrity.

3.1.8 Industrialization of SVFD

The SVFD approach offers several advantages over both batch and alternative continuous concepts:

i. Elimination of shelf contact removes the primary source of heat-transfer variability.
ii. Lower chamber pressures can be used, accelerating sublimation.
iii. Inline PAT allows real-time control and potential real-time release.
iv. Modular architecture facilitates scaling and maintenance.
v. Energy efficiency improves due to smaller equipment volume and continuous operation.

However, there still remain some challenges:

i. Engineering complexity of maintaining sterility across multiple load-locks.
ii. Requirement for robust vacuum seals during tray motion.
iii. Potential radiative shadowing between adjacent vials, though mitigated by the reflective chamber design.
iv. Need for extensive validation of PAT calibration for regulatory acceptance.

Ongoing research focuses on refining the mechanical design of suspension systems, improving condenser-regeneration logistics, and developing digital twins coupling electromagnetic, thermal, and mass-transfer models to support predictive control. While regulatory and industrial adoption will require further standardization and validation, the convergence of automation, PAT, and model-based control now positions the SVFD technology as a practical pathway toward continuous freeze-drying of parenteral drugs.

3.2 Spin Freeze-Drying

3.2.1 Concept and Principles

Spin freeze-drying (SPFD) aimed to overcome the inherent inefficiencies of batch freeze-drying by decoupling freezing and drying and enabling truly continuous operation [40]. Vials containing a liquid formulation are rapidly rotated around their longitudinal axis while exposed to a stream of cryogenic gas. The combination of centrifugal force and convective cooling spreads the liquid as a thin layer along the inner wall of the vial, producing a hollow cylindrical frozen product. The resulting frozen film has a thickness below 1 mm, i.e., an order of magnitude smaller than that of conventionally frozen vials. As a result, the product's resistance to vapor flow is smaller, and the drying time is shorter [41]. Typical drying times are reduced by

80%–90 % [42]. In this configuration, infrared (IR) radiation provides a noncontact energy input during drying, responding to the limited energy transfer through the shelf-vial contact [43].

3.2.2 System Architecture

A spin freeze-dryer consists of four main functional modules: (i) the filling and capping unit, (ii) the spin-freezing chamber, (iii) the drying tunnel, and (iv) the unloading and stoppering module, as shown in Fig. 3. These are interconnected by pressure-controlled load-locks that maintain isolation between atmospheric and vacuum conditions [44].

During spin freezing, vials are clamped by their necks and rotated at high speed (2000–6000 rpm) under a controlled stream of sterile nitrogen or air. The gas temperature, typically between −30°C and −80°C, is regulated through proportional–integral–derivative (PID) control. This feedback control loop adjusts both gas temperature and flow rate to follow the desired cooling trajectory [45].

After freezing, vials pass through the load-lock into the drying module, which is maintained under vacuum. In this module, each vial remains in a fixed, dedicated position where it receives heat from an infrared source. This configuration ensures that vials do not move during the drying phase. Functionally, the system operates as multiple parallel batch units, yet it delivers an overall continuous flow: vials enter and exit the module in a staggered manner according to their drying times, maintaining uninterrupted throughput. The IR heaters are individually controlled by a supervisory algorithm that modulates power based on product temperature feedback. Radiant energy is transferred through IR-transparent windows, e.g., CaF_2 or ZnSe, that maintain sterility while transmitting IR efficiently. Lastly, vials slowly rotate around their axis to guarantee uniform heat transfer [43, 46]. Once drying is completed, vials are stoppered and unloaded from the equipment.

3.2.3 Spin Freezing

The hydrodynamics of spin freezing is crucial for determining the uniformity of the frozen layer. As the vial rotates, a competition arises between gravitational, viscous, and centrifugal forces, quantified by the nondimensional gravitational parameter (λ) which can be considered as a modified version of the Galilei number [47, 48]:

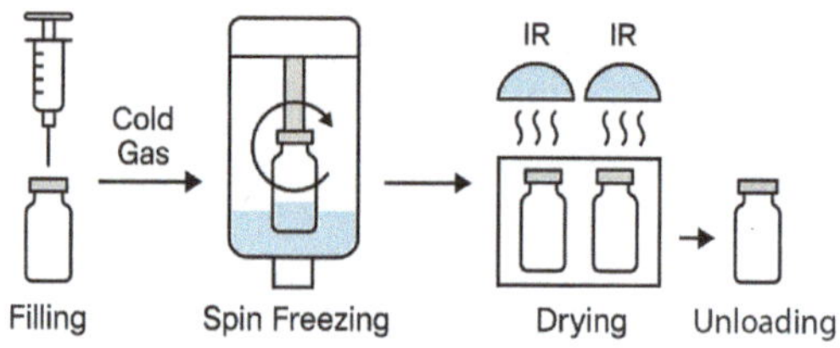

Fig. 3 Schematic of the continuous spin freeze-drying line, showing modules for filling, freezing, IR drying tunnel, and stoppering/unloading

$$\lambda = \frac{A^2 \rho g R}{\mu \omega}$$

where A is the filling fraction, ρ is the mass density of the liquid being frozen in kg m^{-3}, g is the acceleration due to gravity in m s^{-2}, R is the radius of the vial in m, μ is the dynamic viscosity in kg $m^{-1} s^{-1}$, and ω the rotation velocity in rad s^{-1}.

Depending on the value of λ, the formation of thin-film inside the rotating cylinder is classified as follows:

i. $\lambda \leq 2$: uniform thin film along the inner surface
ii. $2 \leq \lambda \leq 5$: accumulation of liquid from the rising side of the cylinder
iii. $\lambda \geq 5$: thin film is created from the stagnant liquid pool at the bottom.

Experimental work by Srinivasan and Raja [49] revealed that when $\lambda \leq 2$, a uniform annular pattern forms, yielding the most favorable structure for subsequent drying. At lower rotational speeds or smaller fill volumes, spiral or patchy ice structures emerge, resulting in less homogeneous sublimation fronts. A later study by Srinivasan et al. [50] demonstrated that vials frozen under spin motion produced thinner ice layers (~1 mm) with surface areas twice as large as static-frozen counterparts. This annular morphology significantly improves both the heat transfer, by increasing the vial-to-product heat transfer area, and reducing the overall product resistance to vapor flow, reducing overall drying time by up to 80%. The authors also asserted that the spin-freezing process reduces supercooling, meaning nucleation occurs earlier and more uniformly than with static freezing. However, this assertion is not supported by any data and remains speculative.

3.2.4 Cooling Behavior and Heat Transfer

The cooling dynamics of spin-freezing differ from those of static freezing. Srinivasan et al. [50] compared the cooling curves of water and skimmed milk samples frozen under both modes. The time required to reach −22°C decreased from 720 s in static freezing to 120 s in spin freezing, corresponding to a sixfold reduction in cooling time. The temperature evolution exhibits three distinct stages: pre-cooling, phase change, and subcooling. However, the phase transition in spin-frozen vials proceeds almost immediately once the freezing point is reached. The higher convective heat flux, promoted by forced gas flow and enhanced heat transfer area, ensures rapid solidification, resulting in smaller ice crystals. Srinivasan et al. [50] also confirmed that the product thickness in spin freezing is approximately 88% lower than in static freezing (1 vs. 8.8 mm), while the ice-vapor surface area is roughly doubled. These parameters collectively yield faster drying.

3.2.5 Desorption Kinetics and Secondary Drying

While much early work focused on the primary drying phase, recent studies have highlighted the importance of optimizing secondary drying. Using near-infrared spectroscopy, Leys et al. [51] developed an inline method for quantifying residual moisture in real time during continuous spin freeze-drying. The partial least squares calibration, validated against Karl Fischer titration, achieved an accuracy of 0.2% w/w, enabling precise monitoring of desorption.

Leys et al. [51] also showed that product temperature exerts the strongest influence on desorption, followed by cooling rate and fill volume, while chamber pressure (10–30 Pa) has a negligible effect. The rate of desorption, thus, depends primarily on diffusion within the dry matrix rather than vapor transport. These findings, corroborated by Leys and De Beer [52], demonstrate that secondary drying can be significantly accelerated by controlling both freezing and heating parameters. The combination of the thin-layer geometry, high specific surface area, and IR-driven heating allows the residual moisture content to reach below 2% in less than 2 h; this result is a remarkable improvement over batch freeze-drying cycles that often exceed 10 h.

3.2.6 Modeling and Process Understanding

Mechanistic models are now central to understand and optimize spin freeze-drying. Nuytten et al. [45] developed a physics-based model describing the energy balance during spin freezing, capable of predicting vial and product temperature profiles. The model incorporates uncertainty and sensitivity analyses, demonstrating that 96% of predictive variability originates from uncertainties in the heat-transfer coefficient and gas temperature.

In the drying phase, Van Bockstal et al. [43] proposed a mechanistic model for radiative heat transfer in spin-frozen vials. Their work introduced dynamic IR-heater control strategies that adjust power input according to drying progress, ensuring product temperature remains below the collapse temperature. The model successfully predicted the drying endpoint with less than 10% deviation from experimental results.

Leys and De Beer [52] extended this framework to secondary drying, incorporating moisture-dependent activation energy to account for nonlinear desorption behavior. The calibrated semi-mechanistic model predicted final moisture content with $<$ 0.1% error across various conditions.

Vanbillemont et al. [44] added a structural dimension through 4D micro-computed tomography, quantifying intra-vial variations in dried-layer resistance. When integrated into heat- and mass-transfer models, these data enable the prediction of spatial drying heterogeneities.

3.2.7 Spin Drying in Other Contexts

The physical principles underpinning spin freeze-drying align closely with those used in the broader field of spin-assisted desiccation.

Table 1
Key performance differences between conventional batch freeze-drying and spin freeze-drying

Parameter	Batch freeze-drying	Spin freeze-drying
Ice thickness, mm	8–10	0.8–1.0
Surface area	1×	2–3×
Primary drying time, h	10–30	1–3
Secondary drying time, h	5–10	1–2
Residual moisture, %	>1%	<0.3%
Product temperature variation, K	>3	<0.5
Energy efficiency	1×	≈5×
Morphology	Variable, nonuniform	Elegant, homogeneous

Chakraborty et al. [53] proposed a spin-drying method for preserving mammalian cells in trehalose solutions, achieving uniform submicron sugar films within 1 min of spinning. The authors showed that SPFD maintained >95% membrane integrity after drying, compared with <10% for conventional drying. This evidence reinforces the notion that rapid, uniform thin-film drying minimizes osmotic stress and local thermal gradients. Whether the objective is biological preservation or pharmaceutical drying, controlled centrifugal spreading, and fast freezing and drying yield homogeneous products with minimal internal gradients.

3.2.8 Performance Comparison and Process Advantages

As shown in Table 1, when compared directly with conventional batch lyophilization, spin freeze-drying exhibits substantial performance advantages across nearly all metrics [40, 43, 49, 50].

These gains are largely attributed to (i) thin-film geometry, (ii) uniform radiative heating, and (iii) inline PAT integration. Moreover, scaling up SPFD does not require redesigning equipment geometry; throughput can simply be increased by "numbering up" modules in parallel while preserving identical thermal conditions.

3.2.9 Challenges and Future Outlook

Several challenges must be addressed for SPFD full industrial adoption. Engineering complexities include the mechanical design of high-speed spinning systems compatible with aseptic environments and reliable sealing across load-lock transitions. Radiative heat transfer within the drying tunnel also presents optimization challenges: IR shadowing between adjacent vials and reflection from metallic surfaces may cause minor nonuniformities. Computational fluid dynamics and radiative modeling could be used to optimize the equipment design to mitigate these effects.

From a regulatory perspective, extensive validation of PAT tools and mechanistic models is necessary before RTRT can be accepted. Furthermore, the annular morphology of the spin freeze-dried product neces sitates the development of a customized reconstitution protocol to prevent, for instance, the liquid from coming into contact with the stopper during reconstitution or to ensure that the entire product shell adhering to the inner wall of the vial is fully reconstituted. Future directions include hybrid heating methods combining radiative and conductive modes to further accelerate drying, adaptive model-predictive control algorithms for autonomous operation, and integration of artificial intelligence for anomaly detection and predictive maintenance. Recent studies are also focusing on extending SPFD technology to temperature-sensitive formulations such as mRNA vaccines, viral vectors, and protein–lipid complexes [54].

4 Concluding Remarks

Continuous freeze-drying stands at the intersection of process engineering innovation and regulatory modernization. By combining new heating modes (microwave and infrared), advanced vial configuration geometries (suspended and spin), and digital control architectures, it offers a route toward faster, safer, and more sustainable lyophilization. This trajectory is irreversible, although significant challenges remain, e.g., in regulatory qualification, cost, and operator training. The transition from static batch dryers to model-driven and continuously monitored systems represents one of the most transformative developments in pharmaceutical manufacturing since the introduction of GMP itself.

5 List of Symbols

A Filling fraction

$|E|$ Magnitude of the incident electromagnetic wave, $J^{1/2}s^{-1/2}m^{-3/2}$

f Frequency of the applied field, s^{-1}

g The acceleration due to gravity, $m\,s^{-2}$

P_d Dissipated power density, $J\,s^{-1}m^{-3}$

R Vial radius, *m*

6 Greek Letters

ε_0 Free-space peremittivity, -

ε_r Dielectric loss factor of the product, -

θ Scatterer rotation angle, °

λ	Galilei number
μ	Dynamic viscosity, $kg\,m^{-1}s^{-1}$
ρ	Mass density, $kg\,m^{-3}$
ω	Rotation velocity in rad s^{-1}

7 Abbreviations

AFD	Atmospheric microwave freeze-drying
AMD	Atmospheric microwave drying
CFD	Conventional freeze-drying
CIP	Cleaning in place
CM	Continuous manufacturing
CQA	Critical quality attribute
EMA	European Medicines Agency
FDA	Food Drug Administration
ICH	International Council for Harmonization
IgY	Immunoglobulin Y
IR	InfraRed
mAb	Monoclonal antibody
MFD	Microwave freeze-drying
MVD	Microwave vacuum drying
MWD	Microwave drying
NIR	Near InfraRed
PAT	Process analytical technology
PI	Proportional integral
PID	Proportional integral derivative
QbC	Quality by control
QbD	Quality by design
RTRT	Real-time release testing
SIP	Sterilization in place
SPFD	Spin freeze-drying
SVFD	Suspended-vial freeze-drying
VISF	Vacuum-induced surface freezing

References

1. Burcham CL, Florence AJ, Johnson MD (2018) Continuous manufacturing in pharmaceutical process development and manufacturing. Ann Rev Chem Biomol Eng 9:253–281.https://doi.org/10.1146/annurev-chembioeng-060817-084355
2. Domokos A, Nagy B, Szilágyi B, Marosi G, Nagy ZK (2021) Integrated continuous pharmaceutical technologies—a review. Org Process Res Dev 25:721–739. https://doi.org/10.1021/acs.oprd.0c00504
3. Lee SL, O'Connor TF, Yang X, Cruz CN, Chatterjee S, Madurawe RD, Moore CMV, Yu LX, Woodcock J (2015) Modernizing pharmaceutical manufacturing: from batch to continuous production. J Pharm Innov 10:191–199. https://doi.org/10.1007/s12247-015-9215-8
4. Rossi CV (2022) A comparative investment analysis of batch versus continuous pharmaceutical manufacturing technologies. J Pharm Innov 17:1373–1391. https://doi.org/10.1007/s12247-021-09612-y

5. Alemie AA, Siraj EA, Yayehrad AT, Tafere C, Tessema TA, Belete A (2025) Continuous pharmaceutical manufacturing and its contemporary regulatory insights. Discover Appl Sci 7:1057. https://doi.org/10.1007/s42452-025-07712-9
6. International Council for Harmonisation (2023) ICH Q13—Continuous manufacturing of drug substances and drug products. International Council for Harmonisation, Geneva, Switzerland. Online document, available at: https://database.ich.org/sites/default/files/ICH_Q13_Step4_Guideline_2022_1116.pdf
7. Pisano R, Arsiccio A, Capozzi LC, Trout BL (2019) Achieving continuous manufacturing in lyophilisation: technologies and approaches. Eur J Pharm Biopharm 142:265–279. https://doi.org/10.1016/j.ejpb.2019.06.027
8. Artusio F, Barresi AA, Pisano R (2026) Emerging technologies in pharmaceutical freeze-drying. In: Smith GS, Fissore D (Eds.) Process analytical technology for pharmaceutical freeze-drying. John Wiley & Sons, Inc., Hoboken, New Jersey, ch. 1154
9. Richards P (2007) MIT and Novartis in new partnership aimed at transforming pharmaceutical manufacturing. MIT News Office. Retrieved 15 Feb 2025, from https://news.mit.edu/2007/novartis-0928. Accessed 2 Nov 2025
10. Continuous Manufacturing and Advanced Crystallization Center CMAC Hub. University of Strathclyde. Retrieved 15 Feb 2025, from https://www.cmac.ac.uk/. Accessed 2 Nov 2025
11. Research Center for Pharmaceutical Engineering (RCPE). Retrieved 15 Feb 2025, from https://www.rcpe.at/. Accessed 2 Nov 2025
12. Schaber SD, Gerogiorgis DI, Ramachandran R, Evans JMB, Barton PI, Trout BL (2011) Economic analysis of integrated continuous and batch pharmaceutical manufacturing: a case study. Ind Eng Chem Res 50:10083–10092. https://doi.org/10.1021/ie2006752
13. Plumb K (2005) Continuous processing in the pharmaceutical industry: changing the mind set. Chem Eng Res Des 83(A6):730–738. https://doi.org/10.1205/cherd.04359
14. Yu LX, Kopcha M (2017) The future of pharmaceutical quality and the path to continuous manufacturing. J Pharm Sci 106:738–744. https://doi.org/10.1016/j.ijpharm.2017.06.039
15. Mascia S, Heider PL, Zhang H, Lakerveld R, Benyahia B, Barton PI, Braatz RD, Cooney CL, Evans JMB, Jamison TF, Jensen KF, Myerson AS, Trout BL (2013) End-to-end continuous manufacturing of pharmaceuticals: integrated synthesis, purification, and final dosage formation. Angew Chem Int Ed 52:12359–12363. https://doi.org/10.1002/anie.201305429
16. European Medicines Agency. Supporting innovation. https://www.ema.europa.eu/en/human-regulatory-overview/research-development/supporting-innovation. Accessed 2 Nov 2025
17. Drouzas AE, Tsami E, Saravacos GD (1999) Microwave/vacuum drying of model fruit gels. J Food Eng 39:117–122. https://doi.org/10.1016/S0260-8774(98)00133-2
18. Copson DA (1958) Microwave sublimation of foods. Food Technol 12:270–272
19. Ma YH, Peltre PR (1975) Freeze dehydration by microwave energy: part I. AIChE J 21:335–344. https://doi.org/10.1002/aic.690210215
20. Pester RP, Arsem HB, Ma YH (1977) Applications of microwave heating to freeze-drying. AIChE Symp Ser 73:131–133
21. Fan K, Zhang M, Mujumdar AS (2019) Recent developments in high-efficiency freeze-drying assisted by microwave. Crit Rev Food Sci Nutr 59:1357–1366. https://doi.org/10.1080/10408398.2017.1420624
22. Li L, Zhang M, Wang W (2019) A novel low-frequency microwave-assisted pulse-spouted bed freeze-drying of Chinese yam. Food Bioprod Proc 118:217–226. https://doi.org/10.1016/j.fbp.2019.09.012
23. Eikevik TM, Alves-Filho O, Bantle M (2012) Microwave-assisted atmospheric freeze-drying of green peas: a case study. Drying Technol 30:1592–1599. https://doi.org/10.1080/07373937.2012.700671
24. Bhambhani A, Stanbro J, Roth D, Sullivan E, Jones M, Evans R, Blue J (2021) Evaluation of microwave vacuum drying as an alternative to freeze-drying of biologics and vaccines. AAPS Pharm Sci Tech 22:52. https://doi.org/10.1208/s12249-020-01912-9
25. Ambros S, Mayer R, Schumann B, Kulozik U (2018) Microwave freeze-drying of lactic acid bacteria: influence of process parameters on drying behaviour and viability. Innov Food Sci Emerg Technol 48:90–98. https://doi.org/10.1016/j.ifset.2018.05.020
26. Wang Z, Duan X, Li L, Ren G, Wu T, Chen J, Ang Y, Guo J, Zhao M (2023) Effects of freeze-drying and microwave vacuum freeze-drying on the activity of IgY: from the perspective of protein structure. Drying Technol 41:222–232. https://doi.org/10.1080/07373937.2021.2015373
27. Gitter JH, Geidobler R, Presser I, Winter G (2018) Significant drying-time reduction using microwave-assisted freeze-drying

for a monoclonal antibody. J Pharm Sci 107:2538–2543. https://doi.org/10.1016/j.xphs.2018.05.023

28. Gitter JH, Geidobler R, Presser I, Winter G (2019) Microwave-assisted freeze-drying of monoclonal antibodies: product quality aspects and storage stability. Pharmaceutics 11:674. https://doi.org/10.3390/pharmaceutics11120674
29. Härdter N, Geidobler R, Presser I, Winter G (2023) Accelerated production of biopharmaceuticals via microwave-assisted freeze-drying (MFD). Pharmaceutics 15:1342. https://doi.org/10.3390/pharmaceutics15051342
30. Härdter N, Geidobler R, Presser I, Winter G (2023) Microwave-assisted freeze-drying: impact of microwave radiation on the quality of high-concentration antibody formulations. Pharmaceutics 15:2783. https://doi.org/10.3390/pharmaceutics15122783
31. Alexeenko AA, Darwish A, Strongrich D, Kazarin P, Patil C, Tower CT, Wheeler IS, Munson E, Zhou Q, Narsimhan V, Yoon K, Nail SL, Cofer A, Stanbro J, Renawala H, Roth D, DeMarco F, Griffiths J, Peroulis D (2025) Randomized-field microwave-assisted pharmaceutical lyophilization with closed-loop control. Sci Rep 15: 10536. https://doi.org/10.1038/s41598-025-91642-4
32. Duan X, Liu LL, Ren GY, Zhu WX (2013) Microwave freeze drying of type I collagen from bovine bone. Drying Technol 31:1701–1706. https://doi.org/10.1080/07373937.2013.793704
33. Abdelraheem A, Tukra R, Kazarin P, Sinanis MD, Topp EM, Alexeenko A, Peroulis D (2022) Statistical electromagnetics for industrial pharmaceutical lyophilization. PNAS Nexus 1:pgac052. https://doi.org/10.1093/pnasnexus/pgac052
34. Kalinke I, Röder J, Unterbuchberger G, Kulozik U (2025) Microwave-assisted freeze drying: the role of power input and temperature control on energy efficiency and uniformity. J Food Eng 390:112410. https://doi.org/10.1016/j.jfoodeng.2024.112410
35. Capozzi LC, Trout BL, Pisano R (2019) From batch to continuous: freeze-drying of suspended vials for pharmaceuticals in unit-doses. Ind Eng Chem Res 58:1635–1649. https://doi.org/10.1021/acs.iecr.8b02886
36. Burcat SJ, Kadambi RP, Stratta L, Braatz RD, Pisano R, Slocum AH, Trout BL (2026) Continuous lyophilization of suspended vials with per-vial online analytics. J Pharm Sci 115:104157. https://doi.org/10.1016/j.xphs.2026.104157
37. Burcat SJ, Kadambi RP, Stratta L, Braatz RD, Pisano R, Slocum AH, Trout BL (2025) Forced gas convection for uniform freezing of lyophilization vials. J Pharm Innov 20:1–17. https://doi.org/10.1007/s12247-025-10037-0
38. Stratta L, Kadambi RP, Pasero L, Burcat SJ, Braatz RD, Trout BL, Slocum AH, Pisano R (2026) Modular design of vacuum systems for lyophilization. Ind Eng Chem Res 65(11):6066–6076. https://doi.org/10.1021/acs.iecr.5c02756
39. Burcat SJ, Kadambi RP, Stratta L, Braatz RD, Pisano R, Trout BL, Slocum AH (2025) Vacuum compatible spring wire system for mass measurement of vials during lyophilization. Precis Eng 96:729–744. https://doi.org/10.1016/j.precisioneng.2025.07.007
40. De Meyer L, Van Bockstal PJ, Corver J, Vervaet C, Remon JP, De Beer T (2015) Evaluation of spin freezing versus conventional freezing as part of a continuous pharmaceutical freeze-drying concept for unit doses. Int J Pharm 496:75–85. https://doi.org/10.1016/j.ijpharm.2015.05.025
41. Lammens J, Mortier STFC, De Meyer L, Vanbillemont B, Van Bockstal PJ, Van Herck S, Corver J, Nopens I, Vervaet C, De Beer T (2018) The relevance of shear, sedimentation and diffusion during spin freezing as potential first step of a continuous freeze-drying process for unit doses. Int J Pharm 539:1–10. https://doi.org/10.1016/j.ijpharm.2018.01.009
42. Leys L, Vanbillemont B, Van Bockstal PJ, Lammens J, Nuytten G, Corver J, Vervaet C, De Beer T (2020) A primary drying model-based comparison of conventional batch freeze-drying to continuous spin-freeze-drying for unit doses. Eur J Pharm Biopharm 157:97–107. https://doi.org/10.1016/j.ejpb.2020.09.009
43. Van Bockstal PJ, Mortier STFC, De Meyer L, Corver J, Vervaet C, Nopens I, De Beer T (2017) Mechanistic modelling of infrared mediated energy transfer during the primary drying step of a continuous freeze-drying process. Eur J Pharm Biopharm 114:11–21. https://doi.org/10.1016/j.ejpb.2017.01.001
44. Vanbillemont B, Lammens J, Goethals W, Vervaet C, Boone MN, De Beer T (2020) 4D micro-computed X-ray tomography as a tool to determine critical process and product information of spin freeze-dried unit doses. Pharmaceutics 12:430. https://doi.org/10.3390/pharmaceutics12050430
45. Nuytten G, Ríos Revatta S, Van Bockstal PJ, Kumar A, Lammens J, Leys L, Vanbillemont B, Corver J, Vervaet C, De Beer T (2021) Development and application of a mechanistic cooling and freezing model of the spin freezing step within the framework of continuous freeze-drying. Pharmaceutics 13:2076. https://doi.org/10.3390/pharmaceutics13122076

46. Lammens J, Moazami Goudarzi N, Leys L, Nuytten G, Van Bockstal PJ, Vervaet C, Boone MN, De Beer T (2021) Spin freezing and its impact on pore size, tortuosity and solid state. Pharmaceutics 13:2126. https://doi.org/10.3390/pharmaceutics13122126
47. Srinivasan G, Raja B (2020) Evaluation of immersion-contact type heat transfer for continuous pharmaceutical spin freeze-drying process. J Food Process Eng 43:e13153. https://doi.org/10.1111/jfpe.13153
48. Ashmore J, Hosol AE, Stone HA (2003) The effect of surface tension on rimming flows in a partially filled rotating cylinder. J Fluid Mech 479:65–98. https://doi.org/10.1017/S0022112002003312
49. Srinivasan G, Raja B (2019) An experimental study of drying behaviour in ice patterns formed during spin freezing and its influence on the freeze-drying process. Heat Mass Transf 55:2469–2481. https://doi.org/10.1007/s00231-019-02596-z
50. Srinivasan G, Muneeshwaran M, Wang CC, Raja B (2022) An experimental investigation on the cooling curve and drying behaviour of static and spin-frozen samples in freeze–drying process. J Therm Anal Calorim 147:11221–11230. https://doi.org/10.1007/s10973-022-11325-4
51. Leys L, Nuytten G, Lammens J, Van Bockstal PJ, Corver J, Vervaet C, De Beer T (2021) A NIR-based study of desorption kinetics during continuous spin freeze-drying. Pharmaceutics 13:2168. https://doi.org/10.3390/pharmaceutics13122168
52. Leys L, De Beer T (2024) Optimising the secondary drying phase of a continuous spin-freeze-drying process: a semi-mechanistic modelling approach. Int J Pharm 664:124597. https://doi.org/10.1016/j.ijpharm.2024.124597
53. Chakraborty N, Menze MA, Malsam J, Aksan A, Hand SC, Toner M (2011) Cryopreservation of spin-dried mammalian cells. PLoS ONE 6:e24916. https://doi.org/10.1371/journal.pone.0024916
54. Lamoot A, Lammens J, De Lombaerde E, Zhong Z, Gontsarik M, Chen Y, De Beer TRM, De Geest BG (2023) Successful batch and continuous lyophilization of mRNA LNP formulations depend on cryoprotectants and ionizable lipids. Biomater Sci 11:4327–4334. https://doi.org/10.1039/D2BM02031A

Chapter 6

Advanced Rewarming Technologies for Cryopreservation

Ziyuan Wang, Shaohang Hao, Ye Jin, Xiao Ma, Ruidong Ma, Shen Ren, Dayong Gao and Zhiquan Shu

Abstract

Cryopreservation is a strategy for long-term preservation of biological samples by storing them at low temperatures while maintaining viability and functionality over an extended period. However, widespread adoption of cryopreservation remains limited by significant challenges, particularly those associated with injuries incurred during the rewarming of large or complex biological samples. Effective rewarming requires rapid and uniform heating to prevent damage from ice recrystallization and thermal stress. However, the current gold standard warming method, water bath warming, relies on convective heat transfer through the sample surface, which is insufficient to meet these critical requirements. Consequently, extensive research has explored alternative heating modalities to overcome these limitations. This chapter reviews recent advancements in rewarming technologies, including nanowarming, Joule heating, laser heating, and dielectric heating, single-mode electromagnetic resonance (SMER) heating and capacitive heating, highlighting their unique advantages, mechanisms, and application scenarios.

Keywords Capacitive heating, Cryopreservation, Dielectric heating, Electromagnetic rewarming, Fast and uniform heating, Joule heating, Laser heating, Nanowarming, Single-mode electromagnetic resonance (SMER) rewarming

1 Importance of Fast and Uniform Rewarming for Cryopreservation

While current cryopreservation methods are mostly effective for small volume and homogenous cell suspensions, preservation of larger, structurally complex cells, tissues, and organs remains a significant challenge. This challenge stems from the fact that vitrification, an ultralow-temperature ice-free technique, is the predominant technique used to preserve complex and large biological samples. Samples require that both the cooling and warming phases surpass the respective critical cooling rate (CCR) and critical warming rate (CWR) of the specific cryoprotectant solutions (CPA) or biospecimens [1]. While modern cooling rate-controlled freezers are capable of achieving the necessary cooling rate, the rewarming process poses a greater challenge. Achieving a high enough warming rate is

Willem F. Wolkers and Harriëtte Oldenhof (eds.), *Cryopreservation and Freeze-Drying Protocols*, Methods in Molecular Biology, vol. 3060, https://doi.org/10.1007/978-1-0716-5416-3_6,

particularly problematic for larger specimens compared to cooling, as the CWR typically exceeds the CCR by an order of magnitude or more [1]. Failure to attain the CWR can result in devitrification or ice recrystallization (IRC), a lethal ice formation phenomenon for biological samples. Furthermore, conventional rewarming methods, which rely on surface conduction or convection, often result in nonuniform temperature distributions within samples, inducing steep thermal gradients, mechanical stress, and potential fracturing of biospecimens. Thus, there is a critical need for rewarming technologies capable of delivering both rapid and spatially uniform heating to avoid ice formation and thermal stress-induced fracture.

To address this challenge, several innovative heating strategies have been explored in recent years, including nanowarming, Joule heating, laser heating, dielectric heating, SMER rewarming, and capacitive heating [2]. Each offers distinct advantages and considerable promise, but each also has important limitations and requires further research to overcome them.

2 Nanowarming in Cryopreservation

2.1 Introduction of Nanowarming

To achieve rapid and uniform heating, a novel technique known as "nanowarming" has been developed. Nanowarming employs silica-coated iron oxide nanoparticles (sIONPs), which are perfusion-loaded and uniformly distributed within the cryoprotectant solution throughout the sample's vasculature [3–5]. Upon exposure to an alternating radiofrequency (RF) magnetic field, these nanoparticles generate heat internally, enabling rapid and uniform rewarming of bulk vitrified samples. Typically, nanowarming can achieve heating rates ranging from ~60°C to ~321°C per minute, while maintaining acceptable temperature gradients within the vitrified sample. These conditions are sufficient to prevent devitrification and minimize thermomechanical stress during the rewarming process [5–9].

2.2 Mechanism of Nanowarming

Nanowarming is a volumetric rewarming approach based on magnetic inductive heating, first introduced and developed by Bischof and collaborators [5, 6]. This technique involves the use of cryoprotectant solutions containing biocompatible nanoparticles, which are perfused into the organ during CPA loading. Once the CPA and nanoparticles are uniformly distributed within the vasculature, the organ is vitrified using a cooling rate-controlled freezer designed to exceed the CCR, thereby avoiding ice crystallization. For rewarming, samples are placed at the center of an RF coil. Upon exposure to an alternating magnetic field (AMF), as illustrated in Fig. 1, the nanoparticles act as localized heat generators, enabling rapid and uniform rewarming of vitrified samples [6, 10].

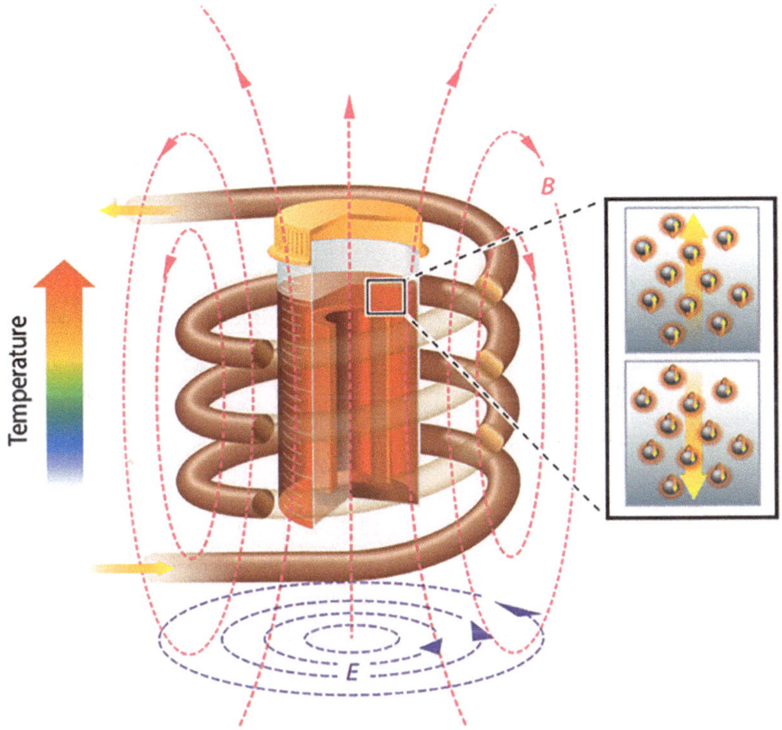

Fig. 1 Schematic illustration of the nanowarming system setup and its working mechanism [6]. (Reproduced with permission. Copyright 2017, The American Association for the Advancement of Science)

The most commonly used nanoparticles are sIONPs and mesoporous silica-coated iron oxide nanoparticles (msIONPs), which are biocompatible and specifically designed to maintain colloidal stability in CPA solutions while preventing biological aggregation [3, 4]. The silica shell protects the magnetic core from degradation and aggregation in highly viscous CPA solutions, while also allowing long-term dispersion stability and biocompatibility [4]. Studies have shown that msIONPs and sIONPs exhibit specific absorption rates (SAR) of up to approximately 392 W/g Fe in VS55, while maintaining colloidal stability for at least 6 months and demonstrating minimal cytotoxicity in human dermal fibroblast (HDF) cells [3, 4]. Recently, more colloidally stable magnetic particles have been developed, such as superparamagnetic iron oxide nanoparticles (SPIONs), which have demonstrated high heating performance, achieving heating rates of up to 321°C per minute under an AMF (42.5 kA/m and 278 kHz) in 20 mL VS55 samples [9]. Of note, the rewarming rate can be controlled by adjusting the concentration of nanoparticles in the CPA solutions, the magnetic properties of nanoparticles, and the amplitude and frequency of the AMF used to induce heat generation, as illustrated in Fig. 2. This tunability suggests that nanowarming is

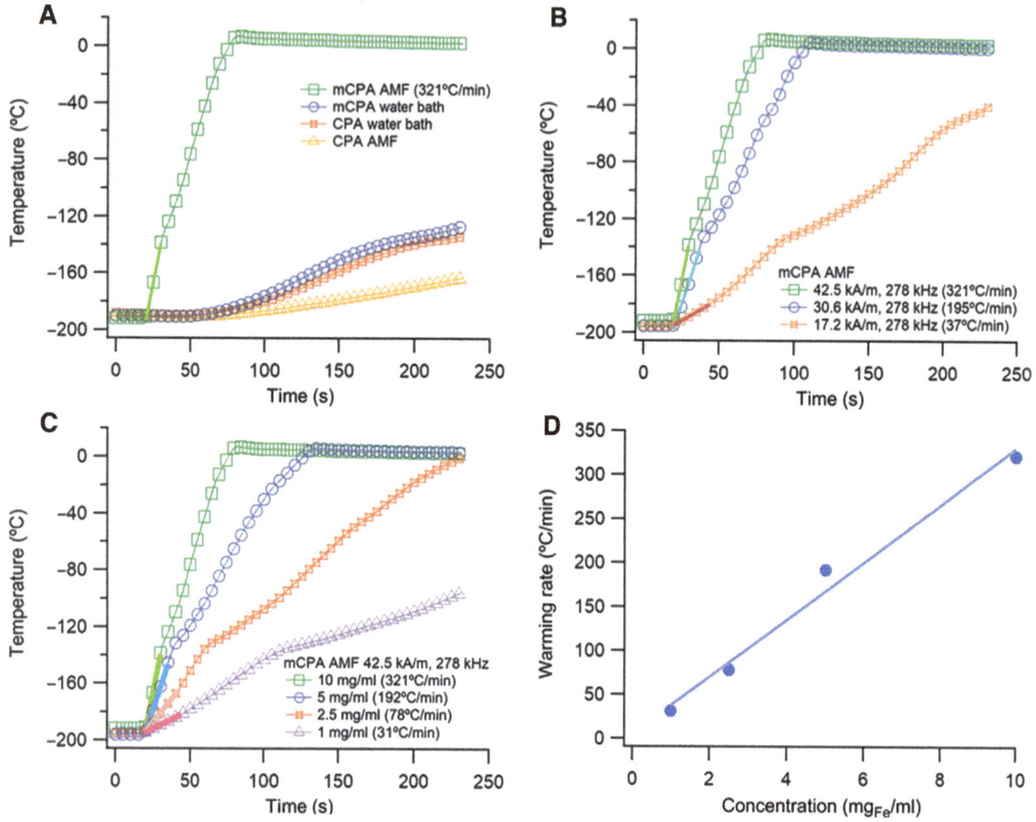

Fig. 2 Magnetic CPAs exhibiting high and tunable heating rates in nanowarming. (**A**) Comparison of warming rates for CPA and mCPA solutions (10 mg Fe/mL SPIONs) measured in a 37°C water bath and under an AMF (42.5 kA/m, 278 kHz), with mCPA nanowarming under AMF reaching 321°C/min. (**B**) Tunable heating performance of mCPA (10 mg Fe/mL) achieved through controlled nanowarming under different AMF conditions. (**C**) Temperature profiles of nanowarming in VS55 at varying SPION concentrations (**D**) Relationship between SPION concentration and nanowarming rate, showing a positive correlation [9]. (Reproduced with permission. Copyright 2021, The American Association for the Advancement of Science)

a scalable approach with potential application at the human-organ scale [6, 9].

Moreover, RF coil power supplies ranging from 1 to 15 kW are commonly selected based on the size of the RF coil and the volume of the biological sample, which typically ranges from 1 mL to 80 mL [5, 6, 8, 11]. Nanowarming employs AMF in the range of 180 and 360 kHz, which allows effective electromagnetic field penetration into biological samples with minimal attenuation.

2.3 Nanowarming Setup and Protocol

The nanowarming system consists of an RF power supply and an RF coil with multiple turns that generates the AMF. The power level and coil size are selected based on the volume of the sample to be rewarmed, as the heating rate can be tuned by adjusting the amplitude and frequency of the AMF. Figure 3 illustrates the system's

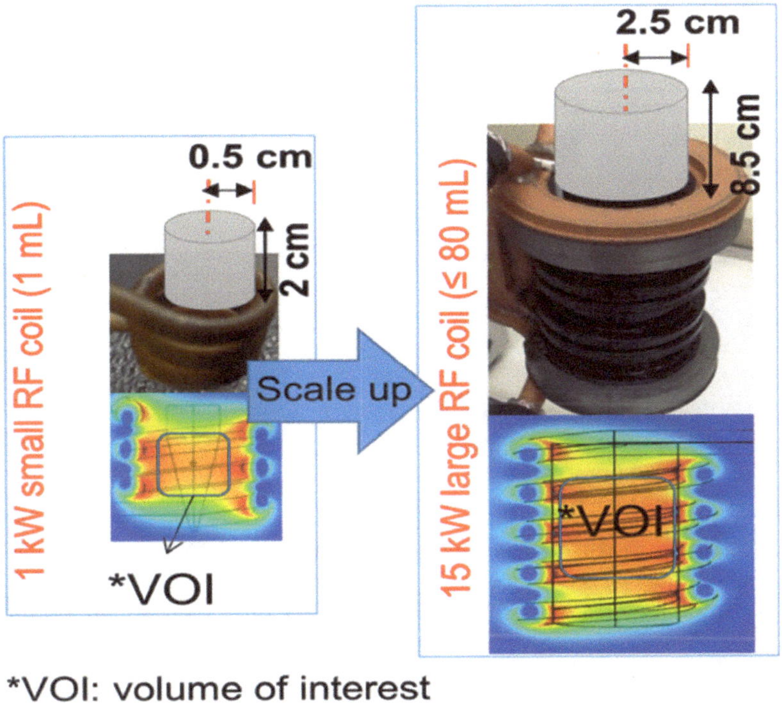

Fig. 3 Schematic of an RF heating system scaled from 1 kW to 15 kW for heating up to 80 mL with increased SAR [6]. (Reproduced with permission. Copyright 2017, The American Association for the Advancement of Science)

scalability, showing configurations ranging from a 1 kW RF coil to a 15-kW coil, suitable for rewarming sample volumes from 1 to 80 mL. The simulated temperature distribution at the center of the coil demonstrates a uniform heating region, supporting consistent thermal performance across different sample sizes.

Several critical steps are required to ensure successful nanowarming in cryopreservation, as illustrated in Fig. 4. Following organ recovery from the donor, magnetic nanoparticles suspended in CPA solutions must be delivered uniformly throughout the organ via vascular perfusion. This step is essential not only to ensure homogeneous distribution of nanoparticles for consistent volumetric heating during rewarming, but also to deliver an adequate concentration of CPA to all regions of the organ to support successful vitrification. Proper perfusion ensures that both the nanoparticles and CPAs reach the microvasculature uniformly, minimizing the risk of ice formation during cooling and ensuring effective nanowarming upon rewarming [8].

The general protocol for loading and unloading CPAs and nanoparticles is well established. Initially, CPAs are perfused into the biological samples under controlled hypothermic conditions (0–4°C) using a stepwise loading protocol (Fig. 4B). This method is designed

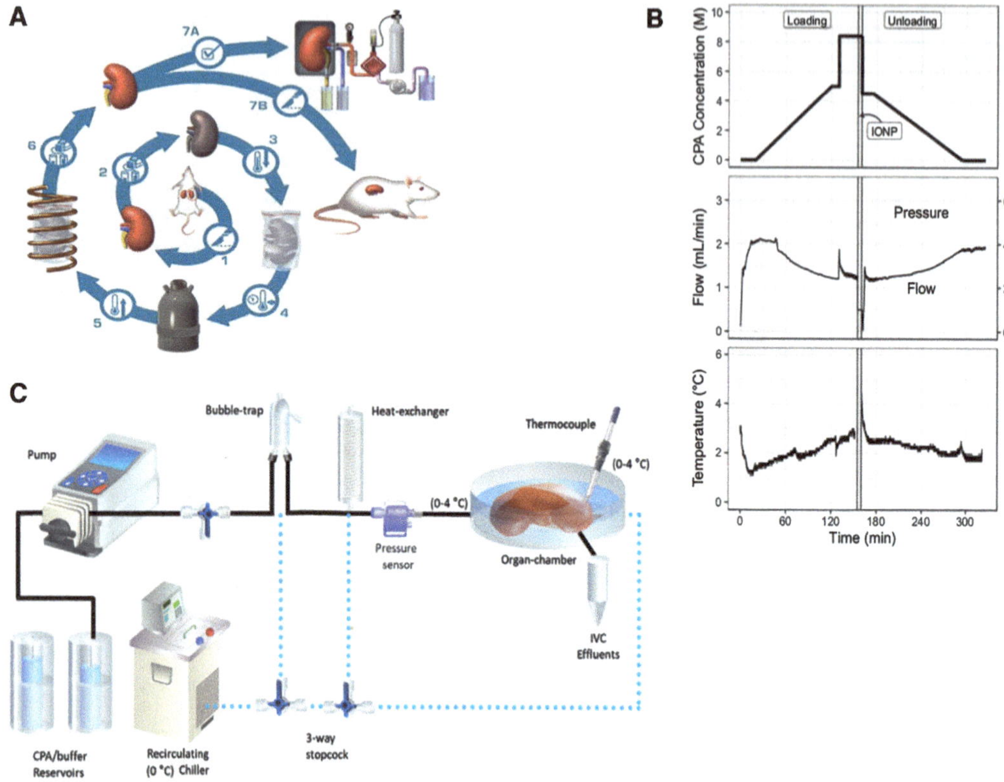

Fig. 4 (**A**) Sequential steps in the nanowarming process for rat kidney cryopreservation. (1) Organ retrieval from donor; (2) CPA/IONPs loading; (3) Rapid cooling to vitrified state; (4) Cryogenic storage at −150°C; (5) Rapid and uniform rewarming via RF field; (6) CPA and IONPs removal; (7A) normothermic machine perfusion (NMP) assessment; (7B) Transplantation [8]. (Reproduced with permission. Copyright 2023, Springer Nature). (**B**) Profiles of CPA concentration, perfusion pressure, arterial flow, and temperature during CPA/IONP loading and unloading in rat kidneys [8]. (**C**) Hypothermic perfusion system setup for rat kidney cryopreservation [7]. (Reproduced with permission. Copyright 2021, John Wiley and Sons)

to gradually increase CPA concentration, thereby minimizing osmotic stress while providing sufficient time for both CPA and nanoparticle diffusion throughout the biological tissue [7, 8].

Figure 4B presents the protocol for the pre-phase of nanowarming. Han et al. demonstrated that CPA concentration gradually increased from 0 to intermediate levels (e.g., 5 M) at a controlled rate, followed by a hold period to allow osmotic equilibration, and then further increased to full-strength CPA (e.g., 8.4 M VMP in LM-5). IONPs are introduced in the final step of full-strength CPA perfusion [8]. The perfusion system used in nanowarming (Fig. 4C) incorporates a pressure-regulated pump that enables stepwise CPA loading with precise control over both flow rate and pressure. A pressure sensor works in coordination with the pump to maintain a perfusion pressure of approximately 40 mmHg during CPA loading, which is then increased to 60 mmHg during nanoparticle perfusion to minimize barotrauma and avoid damage to internal vascular structures [7, 8]. Following CPA and nanoparticle loading, the organ is transferred to a CPA bath containing 4 mg Fe/mL IONPs for equilibration prior to vitrification [8].

Following CPA and nanoparticle loading, the biological samples are vitrified using a controlled-rate freezer at cooling rates that exceed the CCR of the specific CPA formulation. Furthermore, micro-computed tomography (micro-CT) is employed to verify both the uniform distribution of nanoparticles and the vitrified state of the samples [5, 6, 8]. Vitrified samples exhibit a characteristic radiodensity of approximately 520 Hounsfield units (HU) [8], providing a clear and quantitative measure to confirm whether vitrification has been successfully achieved.

After vitrification, the samples are stored at approximately −150°C until rewarming is required [7]. To initiate nanowarming, the vitrified sample is transferred into a RF coil (Fig. 1). Once positioned within the coil, the vascular-distributed nanoparticles are excited by an externally applied AMF, generating internal heat through magnetic inductive heating processes [6]. This volumetric heating approach enables rapid and uniform rewarming throughout the entire sample, minimizing the surface-to-core temperature gradients commonly associated with conventional convective warming. Following nanowarming, the samples are moved to post-thaw assessment using normothermic machine perfusion or, ultimately, prepared for transplantation [8].

2.4 Nanowarming Applications

Several studies to date have demonstrated the use of nanowarming for rewarming various types and sizes of biological samples. Etheridge et al. established a critical proof-of-concept by showing that RF-excited magnetic nanoparticles can uniformly and rapidly rewarm 1 mL samples of vitrified VS55 and 6 M glycerol solutions containing magnetic nanoparticles (mNPs) at rates as high as 300°C per minute [5]. Building upon prior work, Manuchehrabadi et al. demonstrated successful nanowarming of biological systems ranging from 1 mL to 50 mL, including human dermal fibroblast cells, porcine arteries, and porcine aortic heart valve leaflet tissues in VS55 containing msIONPs. They achieved uniform and rapid rewarming at rates exceeding 130°C per minute. In addition, nanowarming of 80 mL systems without arteries was performed at rates of approximately 112°C per minute [6]. In a small-scale organ study, Anirudh Sharma et al. reported the successful recovery of a vitrified rat kidney using nanowarming in 20 mL of VS55 solution containing sIONPs, achieving a uniform rewarming rate of 63.7°C per minute [7].

Chiu-Lam et al. successfully applied nanowarming to whole rat hearts using an AMF at 42.5 kA/m and 278 kHz, along with newly developed SPIONs with improved colloidal stability. A maximum heating rate of up to 321°C per minute was achieved in a 20 mL vitrified sample [9]. More recently, Gao et al. reported the application of nanowarming to a rat heart model in 10 mL of vitrified VS55 solution containing sIONPs. A uniform rewarming rate of approximately 65°C per minute was achieved, with a temperature gradient of less than 20°C across the entire vitrified heart, as shown in Fig. 5.

Sharma et al. conducted a study on a rat liver model using 25–30 mL of CPA solution composed of 40% (v/v) EG and 0.6 M sucrose in EC solution, supplemented with sIONPs. Under a 15 kW AMF coil operating at 63 kA/m and 185 kHz, a rewarming rate of approximately 61°C per minute was achieved, exceeding the CPA's CWR of ~45°C per minute. No crystallization was observed during the rewarming process [14]. Additional successful demonstrations of nanowarming have been reported in various tissues and organs, including stem cells and ovarian tissue in a sheep model [15, 16]. The combination of nanowarming with hydrogel encapsulation has also shown promise for enabling successful cryopreservation at lower CPA concentrations, as demonstrated in studies involving mouse preantral follicles [17] and stem cells [16], thereby reducing exposure to CPA-induced toxicity. Notably, Han et al. conducted a study on rat kidneys vitrified in 20 mL of VMP solution containing sIONPs, achieving a uniform and rapid rewarming rate of approximately 78°C per minute. This study demonstrated that vitrified rat kidneys could be cryogenically stored for up to 100 days and successfully rewarmed via nanowarming, resulting in full restoration of renal function following transplantation. It represents the first reproducible proof-of-concept for long-term organ banking with functional recovery [8].

2.5 Discussion

Nanowarming offers several advantages over traditional rewarming methods. By enabling rapid and uniform volumetric heating, it eliminates thermal gradients between the center and surface of the sample that can induce mechanical stress and fracture. Additionally,

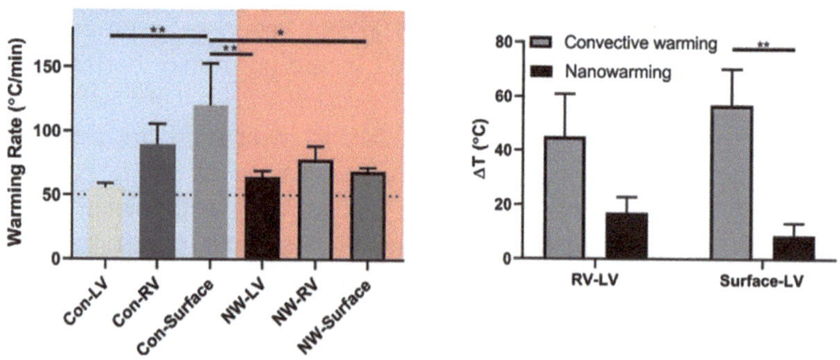

Fig. 5 Warming rates of hearts rewarmed by convection and nanowarming (left); Maximum temperature differences between fiber optic probes during convection and nanowarming rewarming (right). Con-LV convective, left ventricle, Con-RV convective, right ventricle, Con-surface convective, surface, NW-LV nanowarmed, left ventricle, NW-RV nanowarmed, right ventricle, NW-surface nanowarmed, surface [13]. (Reproduced with permission. Copyright 2021, John Wiley and Sons)

it achieves rewarming rates that exceed the CWR required for commonly used CPAs such as VS55, VMP, and M22, thereby preventing devitrification and preserving tissue integrity. RF waves at relatively low frequencies can penetrate biological tissues efficiently with minimal attenuation. When combined with the uniform distribution of nanoparticles throughout the vasculature, this allows nanowarming to be inherently scalable and largely independent of sample size. With adequate power supply and proper coil design, this approach can be extended from small rodent organs to potentially human-scale organs. Moreover, the implementation of nanowarming significantly improves post-thaw viability and functionality compared to conventional techniques and has even restored full renal function post-transplantation. Despite its advantages, several limitations are also noted. Nonuniform nanoparticle distribution throughout complex, dense tissues remains a challenge, as vascular heterogeneities can lead to either insufficient or excessive nanoparticle accumulation, compromising uniform volumetric heating. In addition, although most nanoparticles are successfully washed out during CPA unloading, residual particles remain in some samples. The long-term biocompatibility of these residues and potential immune responses remain unclear and require further investigation. Moreover, engineering challenges persist in generating sufficiently strong and homogeneous RF magnetic fields to support rapid and uniform rewarming at the human-organ scale.

3 Joule Heating in Cryopreservation

3.1 Introduction of Joule Heating

Joule heating is a resistive heating technique that directly converts electrical energy into thermal energy within conductive materials. Unlike conventional convective methods, Joule heating can be precisely engineered to match sample geometry and thermal properties through modulation of applied voltage, pulse duration, and conductor configuration [18]. This approach in cryopreservation enables rapid, spatially uniform, and controllable heating. It ensures consistent thermal homogeneity across a wide range of sample dimensions, from submillimeter cellular monolayers to multi-millimeter tissue sections, while facilitating programmable and highly reproducible warming protocols [19].

Joule heating has been employed in the rewarming of cryopreserved specimens, particularly for small-scale samples such as cells. Recent advancements—such as the development of microfabricated electric heating chips (EHCs) and optimized mesh-based systems—have significantly expanded the applicability of Joule heating to a wide range of biological materials, from subcellular constructs to multi-millimeter tissues. This section reviews the physical mechanisms, experimental implementation, and scalability of Joule heating in the context of cryopreserved biospecimens.

3.2 Mechanism and Experimental Operation of Joule Heating

Joule heating for cryopreservation involves placing cryopreserved biological samples into close contact with an electrical conductor. A precisely controlled voltage pulse is then applied to induce resistive (Joule) heating within the conductor, which in turn transfers thermal energy to the sample [20].

The experimental procedure typically begins by loading the biological specimens into CPA solutions, which may be used for either vitrification or conventional slow-freezing approaches. The specimens are then loaded onto conductive materials, such as stainless-steel mesh, foil, or microfabricated EHCs. The choice of conductive material is critical for optimal performance. Stainless steel mesh or foil is commonly used due to its moderate electrical resistivity, corrosion resistance, mechanical robustness, and acceptable biocompatibility [20]. For micron-scale specimens such as adherent cell monolayers or Drosophila embryos, a single conductive layer is generally sufficient to achieve uniform heating. In contrast, millimeter-scale samples such as tissue slices typically require a two-sided "sandwich" configuration to enable bidirectional heat transfer and minimize axial thermal gradients. Recent developments include thin-film EHCs, which integrate micron-scale resistive elements on disposable substrates. These methods improve heater–sample conformity and simplify handling [21]. Although metals such as copper and gold offer superior conductivity, their susceptibility to corrosion or high material costs has limited their use to prototype applications.

3.3 Recent Advancements of Joule Heating

3.3.1 Scalability: from Cellular Systems to Tissues

The scalability of Joule heating is more accurately defined by its adaptability to various sample types and geometries within a constrained size range, rather than by any inherent capability for unrestricted scaling to whole-organ dimensions [22]. Experimental studies have demonstrated its effective application across a broad range of sample sizes, highlighting its flexibility. Proof-of-concept demonstrations have spanned sample thicknesses from adherent cells to kidney slices. As illustrated in Fig. 6, the plunge-cooling and Joule-heating of CPA-loaded samples show how heat diffusion time scales with sample thickness, underscoring the challenge of maintaining uniform rewarming across increasing dimensions [20].

Furthermore, Joule heating systems can be configured to process multiple small samples simultaneously by scaling up the conductive mesh area and proportionally increasing the applied voltage, thus enabling the simultaneous rewarming of larger sample batches. This adaptability underscores the technique's versatility for diverse biomedical research and clinical applications within its effective operational scale.

However, significant challenges remain in extending Joule heating to bulk specimens larger than 100 mL in volume, such as intact human organs [23]. Because thermal transfer is governed by conduction from the heating element into the tissue, the time required to achieve uniform internal temperatures increases with sample size, leading to substantial thermal gradients that may compromise tissue integrity and post-warmed viability [20].

3.3.2 Achieving Ultra-Rapid and Uniform Heating

In Joule heating-based rewarming, electrical energy is dissipated within high-conductivity elements in direct contact with the frozen specimen, enabling rapid volumetric heat transfer and minimizing internal temperature gradients. Reported warming rates range from approximately 5×10^4 °C/min for submilliliter vitrified samples using

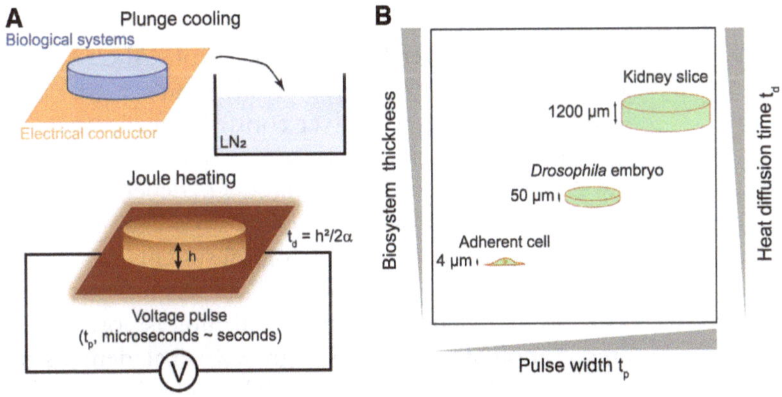

Fig. 6 (**A**) Plunge-cooling and Joule-heating of CPA-loaded samples. (**B**) Model systems show that heat diffusion time increases with sample thickness [20]. (Reproduced with permission. Copyright 2022, Nature Communications)

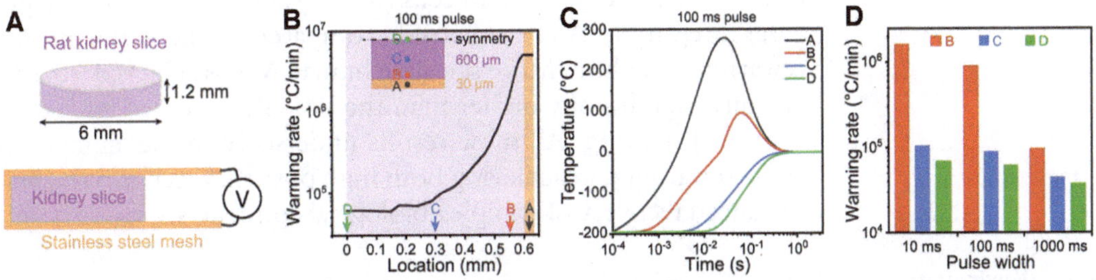

Fig. 7 Kidney slice cryopreservation using Joule heating [20]. (**A**) CPA-loaded rat kidney slice (1.2 mm) in stainless steel mesh, plunge-cooled, and Joule-rewarmed. (**B**) Warming rate distribution across mesh (A) and tissue points (B–D). (**C**) Temperature profiles at points A–D for 100 ms heating. (**D**) Warming rates at points A–D for 10, 100, and 1000 ms heating. (Reproduced with permission. Copyright 2022, Nature Communications)

foil-based systems to up to 6×10^8 °C/min for monolayer cell cultures on microheater chips [20, 21]. Figure 7 demonstrates this in a rat kidney slice model, where CPA-loaded tissue embedded in stainless steel mesh exhibited different warming rate distributions and temperature profiles depending on pulse duration, highlighting the trade-off between heating rate and thermal uniformity.

Minimizing thermal gradients is central to Joule heating system design, as large gradients induce stress and risk cracking in vitrified samples. Numerical simulations and scaling analyses by Zhan et al. showed that, for uniform rewarming, the voltage pulse width should be comparable to the characteristic heat diffusion time (t_d) of the biosystem, but not exceed roughly an order of magnitude beyond td [20]. If the pulse width is shorter than td, incomplete heat diffusion produces steep local temperature gradients and localized overheating. In contrast, excessively long pulses lower the average warming rate, potentially dropping below the critical warming threshold required to avoid devitrification and recrystallization.

Although partial ice formation can occur during rewarming, particularly at lower CPA concentrations, the high warming rates achieved by Joule heating typically limit ice growth to nonlethal levels. This ability to deliver controllable, spatially uniform, and extremely rapid heating not only ensures survival under stringent rewarming conditions but also reduces dependence on highly cytotoxic CPA concentrations, thereby broadening the applicability of vitrification across diverse biological systems.

3.3.3 Enhanced Biological Viability and Functional Recovery

The ultimate benchmark for rewarming is post-thaw viability and preservation of functional integrity. Xu et al. demonstrated this with multiple biological systems loaded into Joule heating-based EHCs, vitrified by plunge immersion in liquid nitrogen, and rewarmed either conventionally (37–40°C water bath) or via Joule heating [21].

Under Joule heating, A549 lung adenocarcinoma cells exhibited 97.2% viability, compared with 93.0% in the water-bath group. As shown in Fig. 8, Joule heating produced higher viability in A549 cells and mesenchymal stem cells (MSCs) and preserved the structural integrity of A549 spheroids compared to the conventional rewarming method. More sensitive human MSCs achieved 93.18% viability, significantly higher than the 86.83% observed with conventional thawing. All these results underscore the advantages of Joule rewarming in achieving both high post-thaw cell survival and enhanced recovery of complex biological constructs.

3.4 Discussion

While Joule heating provides fast and uniform heating, especially for complex biological systems, it still requires solving several important challenges and limitations. A primary limitation lies in achieving spatiotemporal temperature uniformity, especially in large or geometrically complex samples [23]. Although Joule heating provides high

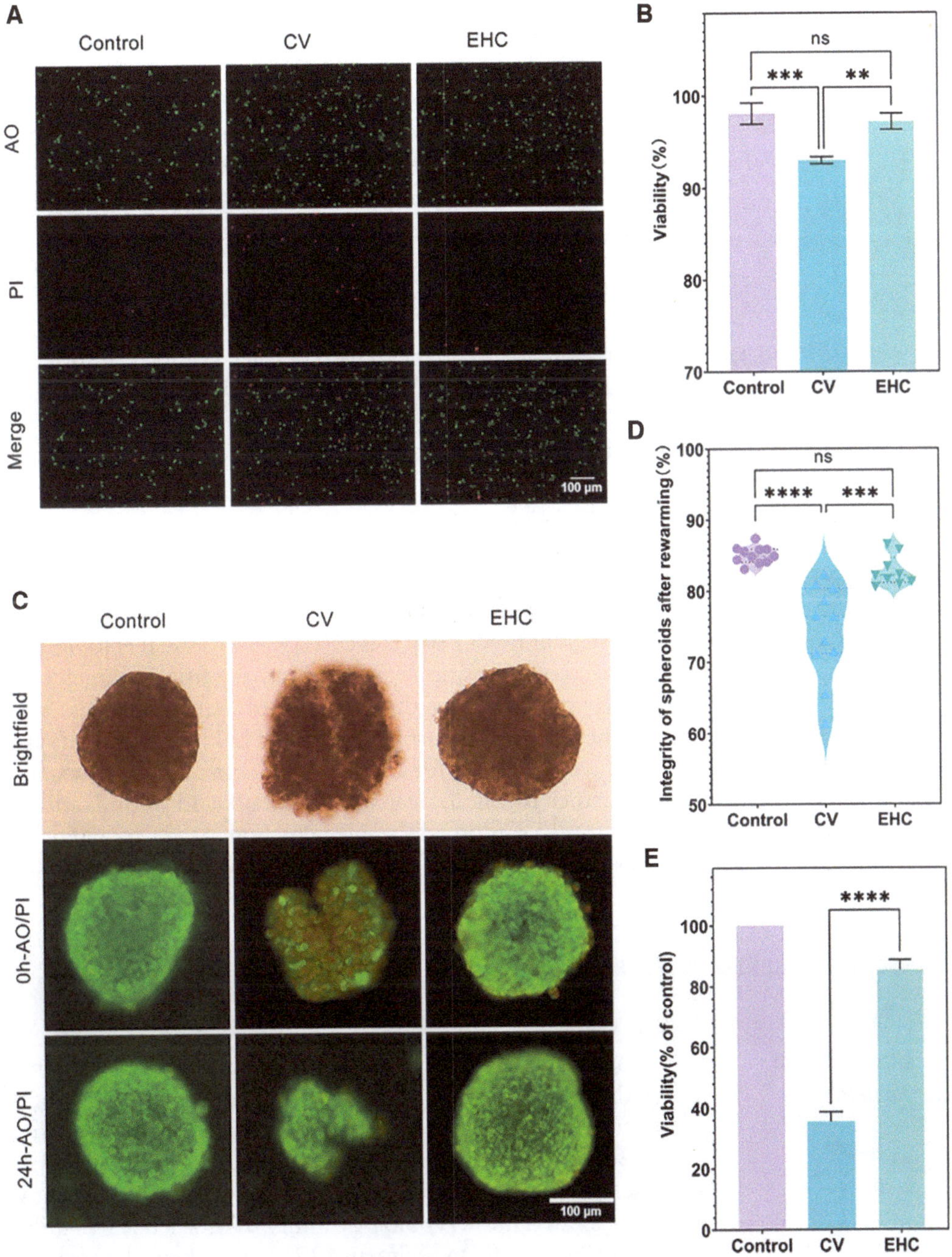

Fig. 8 Cell viability and spheroids integrity of A549s post-cryopreservation under different conditions [21]. (Reproduced with permission. Copyright 2023, Langmuir)

average heating rates, its reliance on thermal conduction from surface or embedded heaters limits performance in bulk tissues due to their inherently low thermal conductivity [24]. Rapid pulses shorter than the tissue's thermal diffusion time can create steep temperature gradients, with surface temperatures substantially exceeding core temperatures, as demonstrated in rat kidney slice models. These gradients induce differential thermal expansion, risking internal mechanical stress, cracking, and structural damage in vitrified samples [22].

The interface between electrodes and ion-rich CPA solutions presents challenges in electrode stability and biocompatibility. Issues such as material fatigue, corrosion, and surface degradation can impair heating efficiency and contaminate samples. For example, copper electrodes in microfluidic systems degrade rapidly, releasing toxic metallic ions [25].

Strong pulsed electric fields used in Joule heating can also induce undesirable nonthermal effects and interact with polar molecules in cells or CPAs [20]. Intense local heating combined with electric fields near heater surfaces can degrade CPAs, generating reactive species and altering their physicochemical properties, including conductivity and viscosity [25].

In summary, Joule heating offers an efficient, rapid, and contamination-free thermal delivery approach because heating occurs internally within the conductor rather than by direct contact with external baths or fluids [20]. It is currently most suitable for intermediate-scale cryopreservation of samples ranging from millimeter-sized constructs up to tissues on the order of 1 to 2 cm in thickness, where uniform heat transfer can still be achieved [21]. However, its application remains limited for large or geometrically complex specimens. Further technological advancements and optimization are required to enable its broader implementation in long-term biological cryopreservation.

4 Laser Heating in Cryopreservation

4.1 Introduction of Laser Heating

Laser-based technologies have been extensively applied in biomedical fields prior to their introduction in cryobiology, with notable applications including laser-assisted cryosurgery for cancer treatment [26], laser-induced hyperthermia for tumor ablation [27], and laser-assisted drug delivery [28]. However, laser heating was not incorporated into cryobiology until 2014, when it was utilized for rewarming vitrified specimens [29]. Since then, the technique has evolved through the development of advanced laser absorbers and its extension to a broader range of biological samples. Within cryobiology, laser rewarming is primarily employed to achieve ultra-rapid heating of small-scale specimens and to investigate the thermal behavior and physical properties of CPAs.

4.2 Mechanism of Laser Heating

Laser rewarming utilizes pulsed infrared (IR) lasers to irradiate cryopreserved samples, with a photothermal agent, such as India ink or nanoparticles, introduced into the solution to enhance energy absorption. These agents efficiently convert absorbed laser energy into heat through photothermal conversion, which is then rapidly transferred to the surrounding medium and biological material, enabling spatially uniform and highly localized rewarming [29, 30]. Compared to traditional convective or conductive rewarming methods, laser-based rewarming is a relatively recent innovation in cryobiology. It offers key advantages in heating precision, scalability, and ultrafast rewarming rates, particularly beneficial for small-volume vitrified specimens where avoiding devitrification is critical.

4.3 Recent Advancements of Laser Heating

4.3.1 Laser Heating for Rewarming in Cryopreservation

In 2014, laser rewarming in cryobiology was reported by Jin et al., where IR lasers were employed to achieve rapid and uniform rewarming of vitrified specimens [29, 31]. In their study, the CPA solution composed of ethylene glycol (EG), acetamide, sucrose, Ficoll, and phosphate-buffered saline (PBS) was used. To enable efficient photothermal conversion, a low concentration of carbon black (India ink) was added as a laser absorber, which was subsequently confirmed to be nontoxic to cells. As illustrated in Fig. 13A, their experimental setup involved pulsed IR laser irradiation of CPA-containing microdroplets. The technique achieved a peak rewarming rate of 1×10^7 °C/min with a 1.8 μL maximum volume, sufficient to avoid devitrification and minimize cellular damage during the warming process. By systematically varying CPA concentrations and rewarming rates, they demonstrated a dramatic improvement in post-thaw cell viability, achieving survival rates of up to 96% with laser rewarming, compared to only 28% with conventional convective rewarming under identical cooling and CPA conditions (Fig. 9B). Notably, the laser-based approach expanded the range of CPA concentrations that could be used effectively while maintaining high viability, thereby reducing dependence on highly concentrated, and potentially cytotoxic, CPAs. This study marked a significant advancement in cryobiology, demonstrating the feasibility and efficacy of using pulsed IR lasers combined with optical absorbers to achieve ultra-rapid, volumetric rewarming of small-volume samples.

Building on the promising results of early studies, research on laser rewarming in cryobiology has rapidly expanded. In 2017, Khosla et al. conducted a pioneering study in which they utilized gold nanorods (GNRs) as photothermal absorbers to rewarm vitrified millimeter-scale biological systems, including zebrafish embryos and human dermal fibroblasts, using propylene glycol (PG)-based CPAs [30, 32, 33]. The system achieved a rewarming rate of 1.4×10^7°C/min, significantly enhancing post-warming survival of zebrafish embryo compared to conventional convective rewarming. The same group later extended this approach to the cryopreservation of human umbilical cord blood stem cells and pancreatic islets, reporting cell viabilities exceeding 95% despite using 20% lower CPA concentrations and successfully preserving sample volumes up to 4 μL [34, 35].

Laser rewarming has also been applied to environmental conservation. GNP-mediated laser rewarming was employed to cryopreserve coral larvae and tissues, contributing to coral reef biodiversity preservation [36–39]. Figure 10 illustrates the schematic of the laser warming procedure used for the rewarming of cryopreserved coral larvae [39]. This approach enabled, for the first time, successful rewarming of cryopreserved coral larvae, with preserved larval sizes reaching up to 500 μm and later extended to adult coral tissues.

Following Khosla et al.'s foundational work, various research groups explored alternative nanoparticle-based absorbers to further optimize rewarming efficiency and biocompatibility. In 2018, Panhwar et al. utilized graphene oxide (GO) and molybdenum disulfide (MoS_2) nanosheets (NSs) with high photothermal conversion efficiency to inhibit ice crystal formation during rewarming, significantly improving the viability of human umbilical vein endothelial cells (HUVECs) [40]. The cell viability can be up to 92% with a volume of up to 0.15 mL.

In 2020, Yi et al. synthesized liquid metal nanoparticles composed of 75.5% gallium and 24.5% indium, achieving high photothermal conversion efficiency and low cytotoxicity. Their system yielded a post-thaw viability of 78% for human bone marrow-derived mesenchymal stem cells with a volume of 50 μL, compared to just 25% using traditional rewarming methods [41].

In 2021, Chang et al. introduced a composite photothermal agent combining tungsten diselenide (WSe_2) with polyvinylpyrrolidone (PVP) for rewarming HUVEC-laden constructs under low CPA concentrations [42]. The 2D WSe_2 nanomaterial provided efficient laser absorption, while PVP served to mitigate cryoinjury

Fig. 9 (**A**) Experimental setup and laser rewarming apparatus used in the study by Jin et al. [29]. The system integrates a LaserStar pulsed infrared laser with a custom-designed working chamber, enabling controlled transitions between cooling and laser rewarming processes for vitrified samples. (**B**) Survival rate of mouse oocytes under laser rewarming (1×10^7 °C/min) and conventional rewarming (1.2×10^5 °C/min). (Reproduced with permission. Copyright 2014, Elsevier)

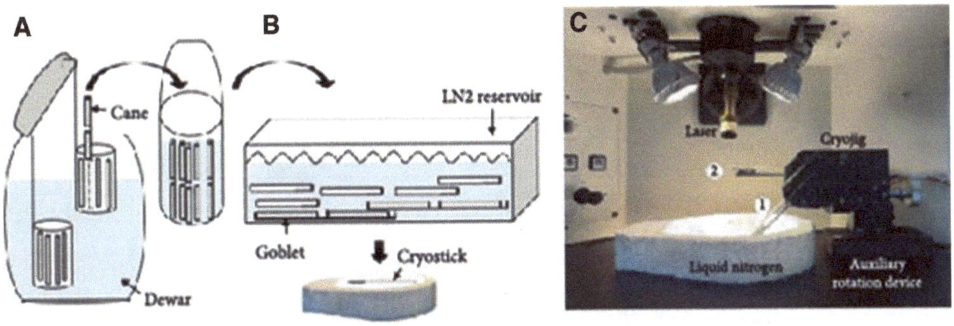

Fig. 10 Schematic of the laser warming procedure for coral larvae [39]. (**A**) Cryopreserved samples were first transferred from a cryogenic dewar into a liquid nitrogen (LN_2) reservoir. (**B**) The cryostick was then detached from the goblet and placed in a Styrofoam container filled with LN_2. (**C**) Prior to rewarming, the cryojig was pre-cooled in LN_2, and the cryostick—kept continuously submerged—was loaded into the cryojig. The jig was mounted on a mechanical arm that moved the sample from a downward resting position (1) to an upward target position (2), where it was exposed to a single laser pulse for rewarming. The tip of the jig moved at a speed of 834 mm/s during the transition. (Reproduced with permission. Copyright 2024, John Wiley & Sons)

by suppressing ice crystal formation. This synergistic formulation demonstrated improved photothermal performance and biocompatibility. Finally, it achieved a cell viability of HUVECs up to 83.4% in the laden constructs.

In 2022, Alvarez et al. reported that titanium nitride (TiN) nanoparticles achieved higher heating rates and greater thermal uniformity than gold nanoparticles during laser rewarming [43]. Using TiN, they attained 96% viability in cryopreserved human dermal fibroblasts (HDFs). They proved that TiN is a cheaper and more reliable alternative to gold nanoparticles when the volume is 1 μL. Additionally, TiN exhibited superior colloidal stability during sonication, making it a promising candidate for combined laser-sonication cryopreservation strategies.

That same year, Cao et al. introduced MXene $Ti_3C_2T_x$, a 2D nanomaterial that not only demonstrated high photothermal efficiency and strong inhibition of ice crystal formation and devitrification, but also exhibited synergistic antibacterial properties [44]. This multifunctionality positions MXene as a compelling candidate for sterile, next-generation cryopreservation platforms. It achieve 80.9% of cell viability on Stem-Cell-Laden Hydrogel Constructs with a volume of 10 μL. Table 1 shows a summary of cryopreservation and rewarming with laser heating.

4.3.2 Laser Heating for Characterization of CPA Properties

Laser heating offers an exceptionally broad range of rewarming rates, from 0.4×10^5 to 1×10^7°C/min, making it a powerful tool not only for rewarming vitrified specimens but also for investigating the thermal and physical properties of CPAs [45]. Owing to its high thermal ramp capability, laser-based methods have been employed to systematically evaluate optimal CPA formulations and rewarming

Table 1
Recent studies and results of laser heating

Research	Biosample	Sample size	Viability	Laser absorber
Jin et al. (2014) [29]	Mouse oocytes and a two-cell embryo	1.8 μL	96%	Carbon black
Khosla et al. (2017) [33]	Zebrafish embryo	4 μL	95%	Gold nanorods
Panhwar et al. (2018) [40]	Human umbilical vein endothelial cells (HUVEC)	0.15 mL	92%	Graphene oxide
Yi et al. (2020) [41]	Human bone marrow-derived mesenchymal stem cells	50 μL	78%	Metal nanoparticles composite
Chang et al. (2021) [42]	HUVEC-laden	325 μm	83.4%	Tungsten diselenide and polyvinylpyrrolidone
Alvarez et al. (2022) [43]	Human dermal fibroblasts (HDFs)	1 μL	96%	Titanium nitride (TiN) nanoparticles
Cao et al. (2022) [44]	Stem-Cell-Laden Hydrogel Constructs	10 μL	80.9%	MXene $Ti_3C_2T_x$

conditions. For instance, Khosla et al. utilized laser rewarming to identify optimal conditions for maximizing post-thaw viability in human dermal fibroblasts by varying CPA concentration and rewarming rate [30]. Their study used liquid nitrogen vitrification and laser-mediated rewarming of small-scale samples to assess viability under different CPA concentrations, cooling rates, and rewarming profiles. The resulting data not only informed laser-based protocols but also provided valuable insights applicable to other rewarming methods such as Joule heating.

In another study, laser calorimetry was used to determine the critical warming rates (CWRs) required to avoid devitrification in glycerol-trehalose and propylene glycol (PG)-trehalose CPA systems [45]. Based on these empirical measurements, a predictive model was developed to estimate CWRs for both single-component CPAs and complex CPA mixtures. The model serves as a useful reference for optimizing cryopreservation protocols, particularly in cases where direct experimental determination is difficult due to sample size constraints.

Further studies have applied laser calorimetry to characterize the photothermal properties of GNPs, demonstrating that parameters such as particle size, shape, and degree of aggregation significantly affect their optical absorption and heating efficiency [46, 47]. These findings are crucial for the rational design and selection of nanoparticles in future laser rewarming systems, ensuring effective and uniform thermal delivery tailored to specific sample types and scales.

4.4 Discussion

CPA solutions are essential in cryopreservation to facilitate vitrification and minimize cellular damage during cooling. In the context of laser rewarming, the inclusion of laser absorbers is critical for enhancing energy absorption and enabling rapid thermal transfer. The use of nanoparticle-based laser absorbers enables ultrafast and spatially uniform heating, which significantly improves cell viability and minimizes the risk of devitrification during rewarming. Moreover, higher rewarming rates permit the use of lower CPA concentrations, thereby reducing CPA-induced cytotoxicity and further improving post-thaw survival.

Over the past decade, research in laser-based cryopreservation has increasingly focused on developing advanced laser absorbers with enhanced photothermal efficiency, reduced toxicity, and multifunctional properties such as devitrification suppression. In recent years, efforts have extended to the synthesis of novel two-dimensional nanomaterials designed to serve as more effective and reliable laser absorbers.

Despite these advancements, laser rewarming remains constrained by the limited penetration depth of infrared light, which restricts its applicability to small-scale biological specimens—typically on the order of millimeters. Researchers are improving the sample size for laser rewarming by testing different CPA combinations and cell types. From the initial 0.2 μL to the highest 150 μL. However, compared to other thawing methods, it is still a very low sample volume. As a result, its current use is largely limited to cellular systems and small tissue fragments, posing significant challenges for the rewarming of large tissues or whole organs.

Nonetheless, the broad heating rate range achievable with laser rewarming makes it an invaluable tool for systematic investigation of critical cryobiological parameters. These include CPA thermophysical properties, critical warming rates, and post-thaw cell viability under varying conditions. Although laser rewarming is a relatively recent innovation, with just over a decade of development, it holds considerable promise for advancing both the performance and scientific understanding of cryopreservation.

5 Dielectric Heating in Cryopreservation

5.1 Introduction of Dielectric Heating

Emerging dielectric heating methods, well known for their volumetric heating capability, have shown great promise in overcoming the rewarming challenges of cryopreservation. The two recent promising dielectric heating methods are single-mode electromagnetic resonant (SMER) heating and capacitive heating. Both of them utilize megahertz-frequency electric fields to excite and rotate dipolar molecules such as water and certain CPAs for heating. This interaction rapidly converts EM energy into heat, enabling efficient volumetric rewarming. Dielectric heating is particularly advantageous for systems rich in dipolar components, especially given that water constitutes the majority of biological mass. Dielectric heating technology thus holds the potential to achieve rapid uniform heating for the successful rewarming of large and complex cryopreserved biospecimens [48].

5.2 Mechanism of Dielectric Heating

Volumetric heating through EM energy offers a distinct advantage over conventional methods, such as conduction and convection, by allowing direct energy deposition inside biological samples. At the core of this process is the excitation of dipolar molecules, including water and various CPAs such as dimethyl sulfoxide (DMSO) and glycerol, by high-frequency fields. The interaction of an alternating electric field with these dipolar molecules induces their rapid rotational and translational motion, which produces heat uniformly throughout the sample volume through molecular friction.

Dielectric heating process is mathematically described by the volumetric power density equation:

$$q_{DH} = \omega \cdot \varepsilon_r'' \cdot \varepsilon_0 \cdot E^2 \quad (1)$$

where ω is the angular frequency of the alternating electric field, ε_r'' is the relative dielectric loss of the heated material, ε_0 is the permittivity of vacuum, and E is the electric field strength.

5.3 Recent Advancements of Dielectric Heating for Cryopreservation

In the 1970s, Ketterer first utilized power-controlled EM heating to rewarm canine kidneys, reporting partial success in roughly 20 cases [50]. Around the same time, Guttman conducted experiments in which dog kidneys were frozen to −80°C and subsequently rewarmed using a commercial 1.35 kW microwave oven operating at 2.45 GHz; of the 16 kidneys treated, half of the transplanted animals survived for periods ranging from 2 to 14 months [51]. However, attempts by Pegg to replicate Guttman's approach using a similar 2.45 GHz microwave system and 14 dog kidneys failed to yield any biologically functional grafts after transplantation [52]. Additionally, Burdette introduced an EM heating system capable of single- or multi-frequency operation with integrated power monitoring. While this system was used to rewarm rabbit and dog kidneys, their studies did not extend to assessing functionality through transplantation [53].

Despite these early milestones, several technical challenges limited the success of these approaches. Most notably, the high microwave frequencies employed, such as 2.45 GHz, were not suitable for achieving uniform heating in large biological specimens like canine kidneys due to limited penetration depth and hot-spot formation. Moreover, the applied commercial microwave ovens, which operate as multimode cavities with uncontrolled, spatially heterogeneous EM field distributions, resulted in inefficient energy deposition, low heating rates, and significant thermal nonuniformity within the samples.

While many twentieth-century attempts at dielectric-heating rewarming were unsuccessful, the approach remains attractive for its ability to heat tissues and CPAs directly and volumetrically, obviating the need for special additives. Recently, two new approaches that implement dielectric heating via different applicators have shown promise for rewarming in cryopreservation: capacitive heating and SMER heating.

5.4 Capacitive Heating

Capacitive heating operates on the same principle as dielectric heating; however, it typically utilizes frequencies at the lower end of the MHz range. Conventionally, capacitor heating devices consist of two electrodes, a coupling coil, a power supply, and a temperature sensor. The first trial of using capacitive heating for organ preservation was conducted in 1964 by Lehr [54]. The use of concentric cylindrical electrode configuration, parallel plate electrode configuration, and multiple electrode configuration (Fig. 11) at 27 MHz and 440 W. The concentric cylindrical electrode configuration (Fig. 11A) is effective only for small samples. For larger masses, however, the warming rate cannot exceed 10°C/min without causing substantial overheating of tissue adjacent to the internally cooled central electrode, an effect reminiscent of RF ablation [55], a minimally invasive cancer treatment method. Figure 8C illustrates the six-electrode organ-heating system, which enables controlled warming of the organ in three designated regions (upper, middle, and lower). However, when the thawing rate exceeds 10°C/min, overheating of substantial tissue volumes becomes unavoidable. This limitation arises from inadequate temperature monitoring, as accurate placement is challenging and thermocouples cannot be reliably used in an electric field. Furthermore, at that stage, impedance matching was not incorporated into the system design.

Recently, Wowk continued to work on parallel capacitive heating (Fig. 12) with the improvement of impedance matching [56]. A rabbit kidney perfused with 50 mL of M22 vitrification solution was vitrified and rewarmed using a 27 MHz constant-frequency system operating at 500 W. A peak rewarming rate of approximately 150°C/min was achieved near –70°C, while the average rewarming rate between –100°C and –30°C was about 55°C/min. The system incorporated a forward/reverse power meter, parallel and series variable capacitors, two coupling coils, and a variable attenuator. The forward/reverse power meter monitored the incident-to-reflected power ratio, while impedance matching was achieved by adjusting the 0–20 dB variable attenuator to maintain a stable field intensity. In 2025, they further developed a 55 MHz constant dielectric warming field with curved electrode plates using 800 W [57]. Increasing the frequency from 27 MHz to 55 MHz shifted the peak warming efficiency from –60°C to –70°C. In addition, the curved electrode design provided improved electric field uniformity, enabling rewarming rates that peaked near 200°C/min for ~15 g rabbit kidneys in 45 mL of solution, and up to ~700°C/min for ~5 g porcine ovaries in 15 mL

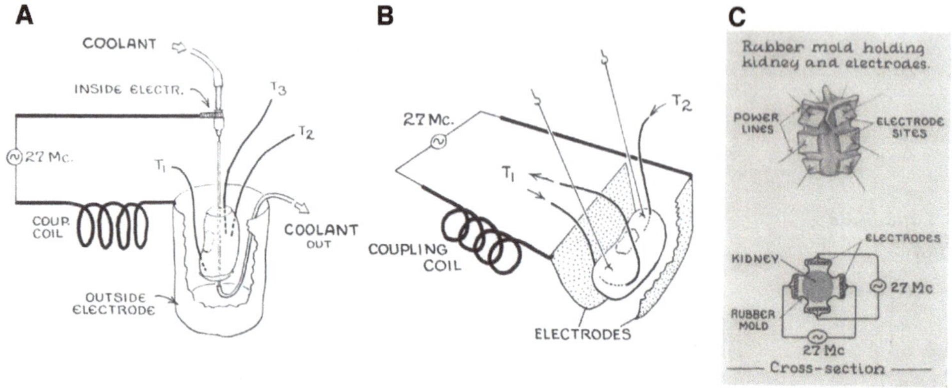

Fig. 11 Configuration of capacitive heating designed by Lehr [54]. (**A**) Concentric cylindrical electrode configuration, (**B**) parallel plate electrode configuration, (**C**) multiple electrode configuration. (Reproduced with permission. Copyright 1964, Elsevier)

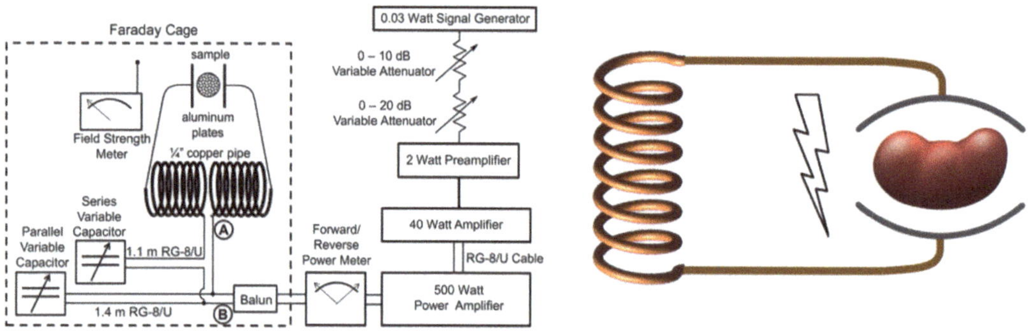

Fig. 12 Configuration of parallel capacitive design by Wowk [56, 57]. (Left) System design of the 27 MHz parallel capacitor dielectric heating. (Reproduced with permission. Copyright 2024, Elsevier). (Right) 55 MHz curved electrode dielectric heating setup. (Reproduced with permission. Copyright 2025, Elsevier)

of solution. One of the transplantation trial recipient animals has been alive for more than 17 months.

Overall, the parallel capacitor is a widely studied configuration. Both joule heating and parallel capacitive heating utilize two flat plate electrodes/conductors. The key differences between those two are frequency utilization and heating mechanism. Thus, the sample doesn't require contact with the electrode, resulting in volumetric heating with an alternative E field. The advantages of this system are high penetration depth, a high absorption rate of M22 with a designed 27 MHz frequency signal input, and impedance matching control by adjusting the variable capacitor. Later, 55 MHz was used. However, this system also holds a couple of limitations, including (1) insufficient rewarming below −100°C mentioned in Wowk's work [56], (2) an increase of energy loss when parallel electrode distance increases with increasing sample thickness, (3)

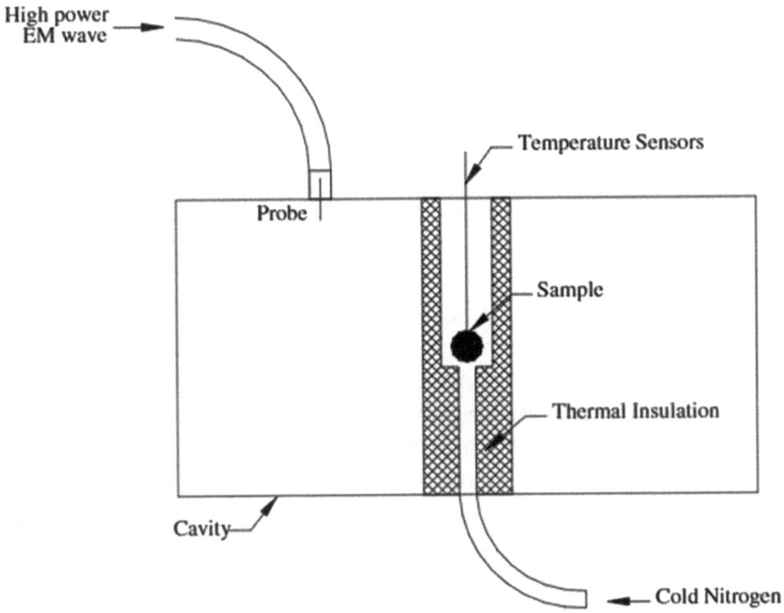

Fig. 13 Schematic diagram of the SMER cavity and cooling component/subsystem [59]. (Reproduced with permission. Copyright 2006, Elsevier)

limited power allowance to avoid voltage breakdown, (4) safety risk due to the high voltage difference between two electrodes, (5) low system quality factor Q represents low efficiency. As the sample size increases, the system demands substantially higher power, since it lacks a focusing effect.

5.5 SMER Heating

In addition to capacitive heating, the single-mode EM heating approach is another promising rewarming technology that relies on dielectric heating. Over the past 25 years, significant progress has been made in the development and refinement of SMER technology [58–62]. This approach was specifically engineered by introducing an SMER cavity coupled with a dynamic resonance tracking system, ensuring more precise and uniform energy delivery. Alongside these advancements, systematic studies on the dielectric and thermal properties of CPA solutions, combined with computational modeling, have contributed critical insights for further optimizing the SMER heating platform. More recently, these innovations have translated into successful rewarming of cryopreserved cell suspensions and tissues with larger volume, achieving both high post-thaw cell viability and functional integrity [61–63].

The following subsections (i) review SMER architectures and key applicator components, (ii) discuss optimization of CPA formulations based on dielectric properties, and (iii) examine nanoparticle additives that enhance SMER heating rate and uniformity.

5.6 SMER System

The SMER rewarming system employs a single-mode resonant cavity, offering an alternative to conventional multi-mode designs. A key benefit of using single-mode, specifically the TE101 mode, is the generation of a stable and well-defined EM field. This mode produces a strong electric field concentrated at the center of the cavity, making it ideal for uniform energy delivery to the targeted regions. To balance EM penetration depth, EM field uniformity, and compatibility with the size of large samples, a microwave frequency of 434 MHz was selected in the SMER design.

In SMER heating, the microwave signal is generated by a signal generator, which offers a broader frequency range and finer frequency tuning than conventional magnetron-based sources. However, signal generators typically output low-power signals, usually below 1 W, making them insufficient for rewarming applications on their own. To overcome this limitation, the signal is routed through a power amplifier and amplified to the required level while preserving the desired frequency.

In the original SMER design proposed by Luo et al. (Fig. 13) [59], a probe antenna mounted on the cavity wall initiates the EM field. The cavity is also equipped with a cooling mechanism, allowing samples to be pre-cooled within the same system. Briefly, a thermally insulated hollow Styrofoam cylinder is positioned at the cavity's center to hold the sample. Prior to rewarming, cold nitrogen gas can be circulated through this cylinder to reduce the sample temperature in a controlled manner.

Figure 14 shows the latest SMER design. A signal generator and power amplifier provide a high-power, tunable EM source. A spectrum analyzer monitors the cavity response and feeds data to a control module that maintains resonance. The sample is positioned at the cavity center for uniform heating.

5.7 Dielectric Properties Measurement System

The properties of heating samples in cryopreservation are dominated by the CPA solutions applied. The interaction between EM fields and CPA solution is primarily governed by their dielectric properties, which determine how efficiently EM energy is absorbed and converted into heat [64, 65]. These properties are temperature-dependent and can influence heating uniformity and efficiency. For example, as the dielectric properties of biosamples and certain CPA solutions increase with temperature, warmer regions absorb more energy, leading to uneven heating and thermal stress (the so-called thermal runaway phenomenon) that may compromise the structural integrity of cryopreserved tissues and organs [66]. Thus, understanding the dielectric behavior of biomaterials is vital for optimizing dielectric rewarming strategies, particularly in CPA-rich vitrification.

One series of measurements by the Gao group has focused on vitrification solutions with high CPA content. Dielectric loss (ε''), which reflects the ability to absorb and convert EM energy, is a critical

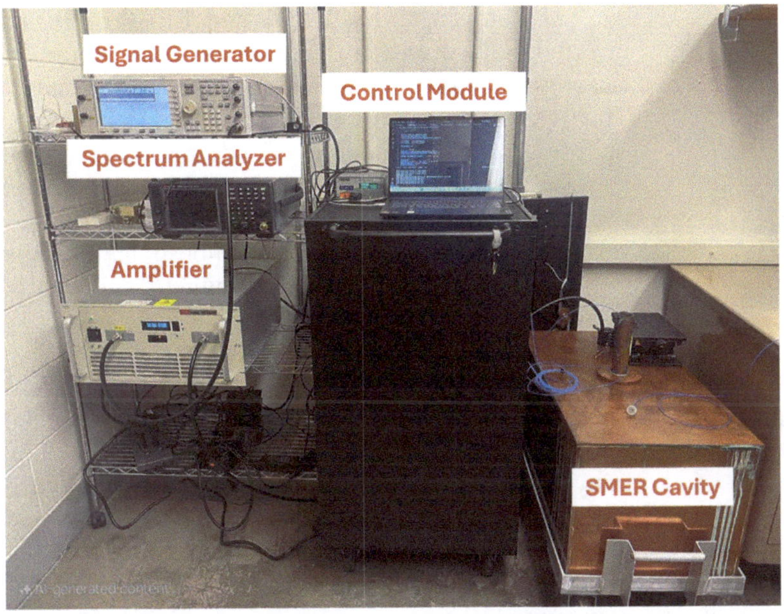

Fig. 14 Schematic of the SMER rewarming system, including a tunable high-power EM source (signal generator + power amplifier), a spectrum analyzer for cavity monitoring, a feedback control module that maintains resonance, and an SMER cavity as the applicator

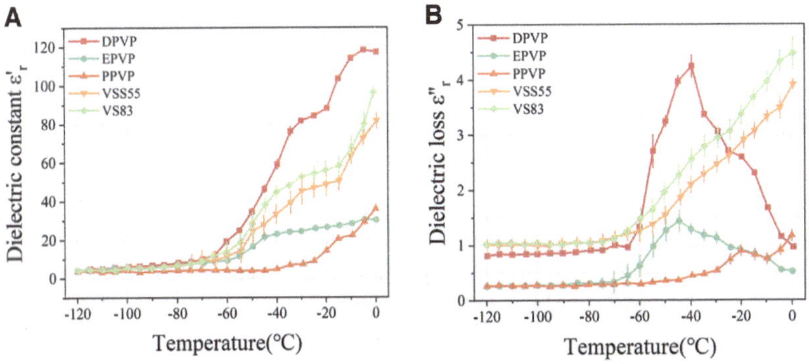

Fig. 15 Dielectric properties of vitrification solutions. (**A**) ε_r' versus the temperature for DPVP, EPVP, PVP, VS55, and VS83. The mean values ± standard deviations at each temperature point were determined based on four replicates ($n = 4$). (**B**) ε_r'' versus the temperature for DPVP, EPVP, PVP, VS55, and VS83. The mean values ± standard deviations at each temperature point were determined based on four replicates ($n = 4$) [63]

parameter for rewarming efficiency. Solutions with higher dielectric loss enable faster rewarming by absorbing more EM energy. Figure 15 presents the dielectric properties of several CPA solutions across a range of subzero temperatures, with each measurement averaged over four replicates [63]. These data were obtained by the same SMER system through perturbation theory.

The data show that DPVP and EPVP solutions exhibit an inverted "U" pattern in their dielectric loss profiles, with the loss factor increasing with temperature to a peak, then gradually declining. This specific temperature-dependent behavior helps regulate heating uniformly: as warmer regions with decreased dielectric loss become less efficient at absorbing EM energy and cooler regions absorb more, thermal uniformity across the sample is promoted. Among all tested vitrification solutions, DPVP also demonstrates the highest dielectric loss, indicating enhanced electromagnetic energy absorption and more efficient rewarming.

5.8 SMER Rewarming Enhanced by Using Magnetic Nanoparticles

Magnetic nanoparticles (MNPs) have emerged as promising agents for enhancing EM heating due to their ability to generate localized heat under oscillating magnetic fields [67]. While initially used in hyperthermia and imaging applications, MNPs are now being investigated for improving cryopreservation outcomes. Their integration into SMER systems allows dual-mode heating, combining both electric and magnetic heating mechanisms for more effective rewarming [12].

Iron oxide MNPs (10 and 15 nm) were incorporated into CPA solutions at concentrations ranging from 0.05 to 0.2 mg Fe/mL. Uniform dispersion of MNPs in the solutions is essential to ensure consistent interaction with the EM field. Transmission electron microscopy confirmed that the MNPs remained well-dispersed after freeze-thaw cycles, showing no visible aggregation.

To evaluate the impact of MNPs on SMER rewarming, fiber optic sensors were used to track temperature changes from −130°C to 0°C. The warming process was segmented into two phases: −130°C to −70°C and −70°C to 0°C, where the latter is crucial to avoid recrystallization. Without MNPs, DPVP samples rewarmed at an average rate of 105.6 ± 6.0°C/min in the critical range (−70°C to 0°C). Incorporating 10 nm MNPs at 0.1 mg Fe/mL enhanced this rate to 210.1 ± 19.7°C/min [12, 61].

The size and concentration of MNPs effects were notable: 10 nm MNPs consistently outperformed 15 nm ones. The optimal concentration was found to be 0.1 mg Fe/mL for both particle sizes. Higher concentrations led to diminished efficiency, likely due to increased inter-particle interactions or hindered EM field penetration. These findings emphasize the need to further optimize both MNP size and dosage for effective heating.

In conclusion, integration of MNPs into SMER systems offers a strategy for achieving rapid and uniform rewarming, making it a promising solution for future organ and tissue preservation applications.

5.9 Discussion

Research on dielectric heating as a solution to the rewarming challenge in cryopreservation dates to the 1970s. Two applicator classes have recently shown particular promise: (i) SMER rewarming system and (ii) capacitive heating systems. Both have demonstrated successful rewarming of biological samples in preclinical studies.

SMER employs a single-mode cavity to localize an electric-field antinode at the sample, producing high heating rates with efficient power coupling. In contrast, capacitive heating uses parallel electrodes to generate an alternating field and can be implemented at substantially lower frequencies (commonly <60 vs. ~434 MHz for SMER), which tends to enhance volumetric uniformity in large-volume or heterogeneous loads.

Despite their differences, both methods rest on the same dielectric-heating fundamentals. Optimizing CPA dielectric properties (permittivity and loss tangent) improves EM absorption and heating uniformity. In addition, each system requires a control strategy to sustain efficient power transfer, either through the resonance tracking for the SMER system or impedance matching for the capacitive heating system.

In summary, dielectric heating remains a compelling solution to the rewarming bottleneck in cryopreservation. Within this paradigm, SMER rewarming and capacitive heating represent two complementary pathways for delivering volumetric energy deposition. Both approaches show strong potential to improve rewarming rates and uniformity, thereby enhancing post-thaw outcomes.

References

1. Mazur P (1970) Cryobiology: the freezing of biological systems. Science 168:939–949. https://doi.org/10.1126/science.168.3934.939
2. Ma R, Wang Z, Shen R, Shu Z, Ming C, Gao D (2025) Rewarming strategies for cryopreservation: technological challenges and opportunities in energy conversion. Frigid Zone Med 5:91–107. https://doi.org/10.1515/fzm-2025-0010
3. Hurley KR, Ring HL, Etheridge M, Zhang J, Gao Z, Shao Q, Klein ND, Szlag VM, Chung C, Reineke TM, Garwood M, Bischof JC, Haynes CL (2016) Predictable heating and positive MRI contrast from a mesoporous silica-coated iron oxide nanoparticle. Mol Pharm 13:2172–2183. https://doi.org/10.1021/acs.molpharmaceut.5b00866
4. Gao Z, Ring HL, Sharma A, Namsrai B, Tran N, Finger EB, Garwood M, Haynes CL, Bischof JC (2020) Preparation of scalable silica-coated iron oxide nanoparticles for nanowarming. Adv Sci 7:1901624. https://doi.org/10.1002/advs.201901624
5. Etheridge ML, Xu Y, Rott L, Choi J, Glasmacher B, Bischof JC (2014) RF heating of magnetic nanoparticles improves the thawing of cryopreserved biomaterials. Technology 2:229–242. https://doi.org/10.1142/S2339547814500204
6. Manuchehrabadi N, Gao Z, Zhang J, Ring HL, Shao Q, Liu F, McDermott M, Fok A, Rabin Y, Brockbank KG, Garwood M, Haynes CL, Bischof JC (2017) Improved tissue cryopreservation using inductive heating of magnetic nanoparticles. Sci Transl Med 9:eaah4586. https://doi.org/10.1126/scitranslmed.aah4586
7. Sharma A, Rao JS, Han Z, Gangwar L, Namsrai B, Gao Z, Ring HL, Magnuson E, Etheridge M, Wowk B, Fahy GM, Garwood M, Finger EB, Bischof JC (2021) Vitrification and nanowarming of kidneys. Adv Sci 8:e2101691. https://doi.org/10.1002/advs.202101691
8. Han Z, Rao JS, Gangwar L, Namsrai BE, Pasek-Allen JL, Etheridge ML, Wolf SM, Pruett TL, Bischof JC, Finger EB (2023) Vitrification and nanowarming enable long-term organ cryopreservation and life-sustaining kidney transplantation in a rat model. Nat Commun 14:3407. https://doi.org/10.1038/s41467-023-38824-8
9. Chiu-Lam A, Staples E, Pepine CJ, Rinaldi C (2021) Perfusion, cryopreservation, and nanowarming of whole hearts using colloidally stable magnetic cryopreservation agent solutions. Sci

Adv 7:eabe3005. https://doi.org/10.1126/sciadv.abe3005

10. Bischof JC, Diller KR (2018) From nanowarming to thermoregulation: new multiscale applications of bioheat transfer. Annu Rev Biomed Eng 20:301–327. https://doi.org/10.1146/annurev-bioeng-071516-044532
11. Gangwar L, Han Z, Scheithauer C, Namsrai BE, Kantesaria S, Goldstein R, Etheridge ML, Finger EB, Bischof JC (2025) Physical vitrification and nanowarming at liter-scale CPA volumes: toward organ cryopreservation. Nat Commun 16:8511. https://doi.org/10.1038/s41467-025-63483-2
12. Pan J, Ren S, Sekar PK, Peng J, Shu Z, Zhao G, Ding W, Chen M, Gao D (2019) Investigation of electromagnetic resonance rewarming enhanced by magnetic nanoparticles for cryopreservation. Langmuir 35:7560–7570. https://doi.org/10.1021/acs.langmuir.8b03060
13. Gao Z, Namsrai B, Han Z, Joshi P, Rao JS, Ravikumar V, Sharma A, Ring HL, Idiyatullin D, Magnuson EC, Iaizzo PA, Tolkacheva EG, Garwood M, Rabin Y, Etheridge M, Finger EB, Bischof JC (2022) Vitrification and rewarming of magnetic nanoparticle-loaded rat hearts. Adv Mater Technol 7:2100873. https://doi.org/10.1002/admt.202100873
14. Sharma A, Lee CY, Namsrai BE, Han Z, Tobolt D, Rao JS, Gao Z, Etheridge ML, Garwood M, Clemens MG, Bischof JC, Finger EB (2023) Cryopreservation of whole rat livers by vitrification and nanowarming. Ann Biomed Eng 51:566–577.https://doi.org/10.1007/s10439-022-03064-2
15. Karimi S, Tabatabaei SN, Novin MG, Kazemi M, Mofarahe ZS, Ebrahimzadeh-Bideskan A (2023) Nanowarming improves survival of vitrified ovarian tissue and follicular development in a sheep model. Heliyon 9:e18828. https://doi.org/10.1016/j.heliyon.2023.e18828
16. Liu X, Zhao G, Chen Z, Panhwar F, He X (2018) Dual suppression effect of magnetic induction heating and microencapsulation on ice crystallization enables low-cryoprotectant vitrification of stem cell-alginate hydrogel constructs. ACS Appl Mater Interfaces 10:16822–16835. https://doi.org/10.1021/acsami.8b04496
17. Tian C, Shen L, Gong C, Cao Y, Shi Q, Zhao G (2022) Microencapsulation and nanowarming enables vitrification cryopreservation of mouse preantral follicles. Nat Commun 13: 7515. https://doi.org/10.1038/s41467-022-34549-2
18. Jouhara H, Żabnieńska-Góra A, Khordehgah N, Doraghi Q, Ahmad L, Norman L, Axcell B, Wrobel L, Dai S (2021) Thermoelectric generator (TEG) technologies and applications. Int J Thermofluids 9:100063. https://doi.org/10.1016/j.ijft.2021.100063
19. Pan J, Zeng Q, Peng K, Zhou Y, Shu Z (2024) Review of rewarming methods for cryopreservation. Biopreserv Biobank 22:304–311. https://doi.org/10.1089/bio.2023.0015
20. Zhan L, Han Z, Shao Q, Etheridge ML, Hays T, Bischof JC (2022) Rapid joule heating improves vitrification based cryopreservation. Nat Commun 13: 6017. https://doi.org/10.1038/s41467-022-33546-9
21. Han H, Zhan T, Cui M, Guo N, Dang H, Yang G, Shu S, He W, Xu Y (2023) Investigation of rapid rewarming chips for cryopreservation by Joule heating. Langmuir 39:11048–11062. https://doi.org/10.1021/acs.langmuir.3c01364
22. Gangwar L, Phatak SS, Etheridge M, Bischof JC (2022) A guide to successful mL to L scale vitrification and rewarming. Cryo Letters 43:316–321
23. Khaydukova IV, Ivannikova VM, Zhidkov DA, Belikov NV, Peshkova MA, Timashev PS, Tsiganov DI, Pushkarev AV (2024) Current state and challenges of tissue and organ cryopreservation in biobanking. Int J Mol Sci 25:11124. https://doi.org/10.3390/ijms252011124
24. Han Z, Sharma A, Gao Z, Carlson TW, O'Sullivan MG, Finger EB, Bischof JC (2020) Diffusion limited cryopreservation of tissue with radiofrequency heated metal forms. Adv Healthc Mater 9:e2000796. https://doi.org/10.1002/adhm.202000796
25. Yousuff CM, Tirth V, Zackria Ansar Babu Irshad M, Irshad K, Algahtani A, Islam S (2021) Numerical study of Joule heating effects on microfluidics device reliability in electrode based devices. Materials 14:5819. https://doi.org/10.3390/ma14195819
26. Martínez-Suástegui L, Duperray B, Godinez F, Guillén G, Slade A, Aguilar G (2011) Laser-assisted cryosurgery in ex vivo mice hepatic tissue: viability assays using green fluorescent protein. Ann Biomed Eng 39:636–648. https://doi.org/10.1007/s10439-010-0186-0
27. Singh R, Das K, Mishra SC (2014) Laser-induced hyperthermia of nanoshell mediated vascularized tissue - a numerical study. J Therm Biol 44:55–62. https://doi.org/10.1016/j.jtherbio.2014.07.001
28. Stumpp OF, Bedi VP, Wyatt D, Lac D, Rahman Z, Chan KF (2009) *In vivo* confocal imaging of epidermal cell migration and dermal changes post nonablative fractional resurfacing: study of the wound healing process with corroborated histopathologic evidence. J Biomed Opt 14:024018. https://doi.org/10.1117/1.3103316

29. Jin B, Kleinhans FW, Mazur P (2014) Survivals of mouse oocytes approach 100% after vitrification in 3-fold diluted media and ultra-rapid warming by an IR laser pulse. Cryobiology 68:419–430. https://doi.org/10.1016/j.cryobiol.2014.03.005
30. Khosla K, Zhan L, Bhati A, Carley-Clopton A, Hagedorn M, Bischof J (2019) Characterization of laser gold nanowarming: a platform for millimeter-scale cryopreservation. Langmuir 35:7364–7375. https://doi.org/10.1021/acs.langmuir.8b03011
31. Kleinhans FW, Mazur P (2015) Physical parameters, modeling, and methodological details in using IR laser pulses to warm frozen or vitrified cells ultra-rapidly. Cryobiology 70:195–203. https://doi.org/10.1016/j.cryobiol.2015.02.003
32. Khosla K, Wang Y, Hagedorn M, Qin Z, Bischof J (2017) Gold nanorod induced warming of embryos from the cryogenic state enhances viability. ACS Nano 11:7869–7878. https://doi.org/10.1021/acsnano.7b02216
33. Khosla K, Kangas J, Liu Y, Zhan L, Daly J, Hagedorn M, Bischof J (2020) Cryopreservation and laser nanowarming of zebrafish embryos followed by hatching and spawning. Adv Biosyst 4:e2000138. https://doi.org/10.1002/adbi.202000138
34. Zhan L, Guo SZ, Kangas J, Shao Q, Shiao M, Khosla K, Low WC, McAlpine MC, Bischof J (2021) Conduction cooling and plasmonic heating dramatically increase droplet vitrification volumes for cell cryopreservation. Adv Sci 8:2004605. https://doi.org/10.1002/advs.202004605
35. Zhan L, Rao JS, Sethia N, Slama MQ, Han Z, Tobolt D, Etheridge M, Peterson QP, Dutcher CS, Bischof JC, Finger EB (2022) Pancreatic islet cryopreservation by vitrification achieves high viability, function, recovery and clinical scalability for transplantation. Nat Med 28:798–808. https://doi.org/10.1038/s41591-022-01718-1
36. Cirino L, Wen ZH, Hsieh K, Huang CL, Leong QL, Wang LH, Chen CS, Daly J, Tsai S, Lin C (2019) First instance of settlement by cryopreserved coral larvae in symbiotic association with dinoflagellates. Sci Rep 9: 18851. https://doi.org/10.1038/s41598-019-55374-6
37. Daly J, Zuchowicz N, Nuñez Lendo CI, Khosla K, Lager C, Henley EM, Bischof J, Kleinhans FW, Lin C, Peters EC, Hagedorn M (2018) Successful cryopreservation of coral larvae using vitrification and laser warming. Sci Rep 8: 15714. https://doi.org/10.1038/s41598-018-34035-0
38. Daly J, Bouwmeester J, Perry R, Page C, Khosla K, Kangas J, Lager C, Hardy K, Bischof JC, Hagedorn M (2023) The first proof of concept demonstration of nanowarming in coral tissues. Adv Sustain Syst 7:2300303. https://doi.org/10.1002/adsu.202300303
39. Loeslakwiboon K, Hsieh WC, Huang CL, Tsai S, Lin C (2024) First cryorepository for coral larvae: safeguarding corals for future generations. Aquacult Res 2024:4887191. https://doi.org/10.1155/2024/4887191
40. Panhwar F, Chen Z, Hossain SMC, Wang M, Haider Z, Memon K, Chen P, Zhao G (2018) Near-infrared laser mediated modulation of ice crystallization by two-dimensional nanosheets enables high-survival recovery of biological cells from cryogenic temperatures. Nanoscale 10:11760–11774. https://doi.org/10.1039/c8nr01349g
41. Hou Y, Lu C, Dou M, Zhang C, Chang H, Liu J, Rao W (2020) Soft liquid metal nanoparticles achieve reduced crystal nucleation and ultrarapid rewarming for human bone marrow stromal cell and blood vessel cryopreservation. Acta Biomater 102:403–415. https://doi.org/10.1016/j.actbio.2019.11.023
42. Chang T, Moses OA, Tian C, Wang H, Song L, Zhao G (2021) Synergistic ice inhibition effect enhances rrapid freezing cryopreservation with low concentration of cryoprotectants. Adv Sci 8:2003387. https://doi.org/10.1002/advs.202003387
43. Alvarez C, Berrospe-Rodriguez C, Wu C, Pasek-Allen J, Khosla K, Bischof J, Mangolini L, Aguilar G (2022) Photothermal heating of titanium nitride nanomaterials for fast and uniform laser warming of cryopreserved biomaterials. Front Bioeng Biotechnol 10:957481. https://doi.org/10.3389/fbioe.2022.957481
44. Cao Y, Chang T, Fang C, Zhang Y, Liu H, Zhao G (2022) Inhibition effect of Ti3C2Tx MXene on ice crystals combined with laser-mediated heating facilitates high-performance cryopreservation. ACS Nano 16:8837–8850. https://doi.org/10.1021/acsnano.1c10221
45. Kangas J, Zhan L, Liu Y, Natesan H, Khosla K, Bischof J (2022) Ultra-rapid laser calorimetry for the assessment of crystallization in low-concentration cryoprotectants. J Heat Transfer 144:031207. https://doi.org/10.1115/1.4052568
46. Liu Y, Kangas J, Wang Y, Khosla K, Pasek-Allen J, Saunders A, Oldenburg S, Bischof J (2020) Photothermal conversion of gold nanoparticles for uniform pulsed laser warming of vitrified biomaterials. Nanoscale 12:12346–12356. https://doi.org/10.1039/d0nr01614d
47. Wang Y, Gao Z, Han Z, Liu Y, Yang H, Akkin T, Hogan CJ, Bischof JC (2021) Aggregation affects optical properties and photothermal heating of gold nanospheres.

Sci Rep 11: 898. https://doi.org/10.1038/s41598-020-79393-w
48. Wang Z, Shu Z, Ren S, Ma R, Kramlich J, Chen M, Gao D (2024) Development of electromagnetic warming technology for cryopreservation. Annu Rev Heat Transfer 27:319–356. https://doi.org/10.1615/AnnualRevHeatTransfer.2024055368
49. Vollmer M (2004) Physics of the microwave oven. Phys Educ 39:74–81. https://doi.org/10.1088/0031-9120/39/1/006
50. Ketterer FD, Holst HI, Lehr HB (1971) Improved viability of kidneys with microwave thawing. Cryobiology 8:395
51. Guttman FM, Lizin J, Robitaille P, Blanchard H, Turgeon-Knaack C (1977) Survival of canine kidneys after treatment with dimethyl-sulfoxide, freezing at −80°C, and thawing by microwave illumination. Cryobiology 14:559–567. https://doi.org/10.1016/0011-2240(77)90166-3
52. Pegg DE, Green CJ, Walter CA (1978) Attempted canine renal cryopreservation using dimethyl sulphoxide helium perfusion and microwave thawing. Cryobiology 15:618–626. https://doi.org/10.1016/0011-2240(78)90086-x
53. Burdette EC, Karow AM, Jeske AH (1978) Design, development, and performance of an electromagnetic illumination system for thawing cryopreserved kidneys of rabbits and dogs. Cryobiology 15:152–167. https://doi.org/10.1016/0011-2240(78)90020-2
54. Lehr HB, Berggren RB, Summers AL, Lotke PA (1964) Freezing and thawing of large organs. Cryobiology 1:194–197. https://doi.org/10.1016/0011-2240(64)90011-2
55. Zheng H, Li P, Ma R, Zhang F, Ji H, Monsky WL, Johnson E, Yang W, Ni C, Gao D, Yang X (2022) Development of a three-dimensional multi-modal perfusion-thermal electrode system for complete tumor eradication. Cancers 14:4768. https://doi.org/10.3390/cancers14194768
56. Wowk B, Phan J, Pagotan R, Galvez E, Fahy GM (2024) 27 MHz constant field dielectric warming of kidneys cryopreserved by vitrification. Cryobiology 115:104893. https://doi.org/10.1016/j.cryobiol.2024.104893
57. Wowk B, Ting A, Phan J, Pagotan R, Sharif A, Chow L, Fahy GM (2025) 55 MHz constant field dielectric warming of kidneys and ovaries cryopreserved by vitrification. Cryobiology 119:105257. https://doi.org/10.1016/j.cryobiol.2025.105257
58. Cai-Cheng L, Huai-Zhi L, Dayong G (2000) Combined electromagnetic and heat-conduction analysis of rapid rewarming of cryopreserved tissues. IEEE Trans Microwave Theory Tech 48:2185–2190. https://doi.org/10.1109/22.884213
59. Luo D, Yu C, He L, Lu C, Gao D (2006) Development of a single mode electromagnetic resonant cavity for rewarming of cryopreserved biomaterials. Cryobiology 53:288–293. https://doi.org/10.1016/j.cryobiol.2006.07.001
60. Gao D, Ren S (2020) Development of a novel electromagnetic rewarming technology for the cryopreservation of stem cells with large volume. In: Kitala D, Maurício AC (eds) Novel perspectives of stem cell manufacturing and therapies. IntechOpen, Rijeka
61. Ren S, Shu Z, Wang Z, Jin Y, Ma R, Hao S, Yang X, Gao D (2021) Successful vitreous cryopreservation of rabbit jugular vein using magnetic nanoparticles enhanced Single_Mode Electromagnetic Resonance rewarming system. Cryobiology 103:173–174. https://doi.org/10.1016/j.cryobiol.2021.11.060
62. Ren S, Shu Z, Pan J, Wang Z, Ma R, Peng J, Chen M, Gao D (2022) Single-mode electromagnetic resonance rewarming for the cryopreservation of samples with large volumes: a numerical and experimental study. Biopreserv Biobank 20:317–322. https://doi.org/10.1089/bio.2022.0107
63. Wang Z, Ren S, Shu Z, Ma R, Gao D (2022) Screening and optimization of cryoprotective agents (CPAs) for electromagnetic heating of cryopreserved biomaterials. Cryobiology 109:16. https://doi.org/10.1016/j.cryobiol.2022.11.052
64. Robinson MP, Wusteman MC, Wang L, Pegg DE (2002) Electromagnetic re-warming of cryopreserved tissues: effect of choice of cryoprotectant and sample shape on uniformity of heating. Phys Med Biol 47:2311–2325. https://doi.org/10.1088/0031-9155/47/13/309
65. Wusteman MC, Pegg DE, Robinson MP, Wang LH, Fitch P (2002) Vitrification media: toxicity, permeability, and dielectric properties. Cryobiology 44:24–37. https://doi.org/10.1016/S0011-2240(02)00002-0
66. Peng J, Ma R, Ren S, Wang Z, Shu Z, Gao D (2021) A study of thermal stress generation during the rewarming process of cryopreserved large biomaterials. Cryobiology 103:167–168. https://doi.org/10.1016/j.cryobiol.2021.11.042
67. Wang Z, Gao D, Shu Z (2024) Mechanisms, applications, and challenges of utilizing nanomaterials in cryopreservation. Adv Eng Mater 26:2400800. https://doi.org/10.1002/adem.202400800

Chapter 7

Microfluidic Platforms for Cryopreservation

Gang Zhao

Abstract

This chapter examines the application of microfluidics in cryopreservation, addressing the limitations of conventional techniques and highlighting how microfluidic approaches overcome these challenges. It delves into the fundamental principles of microfluidic cryopreservation, including microscale heat and mass transfer dynamics, precise control of cryoprotective agents (CPAs), and ultra-rapid cooling/warming. The chapter also showcases key applications, including on-chip cell membrane permeability measurement, dynamic CPA addition/removal, and integrated cryopreservation workflows on microfluidic platforms. Finally, it explores the transformative potential of microfluidics in enabling more precise, controllable, and efficient cryopreservation methods, while outlining future research directions and potential technological synergies.

Keywords Cryopreservation, Cryoprotective agents (CPAs), Cell membrane permeability, Heat and mass transfer, Microchannel, Microfluidics, Osmotic stress, Ultra-rapid cooling/warming, Vitrification

1 Introduction

1.1 Current Status and Challenges of Cryopreservation

Cryopreservation serves as a core technology across multiple domains, including biomedicine, agriculture, and biobanking, playing an indispensable role in maintaining the viability and functionality of biological specimens. In cell therapy, cryopreservation represents a pivotal technique throughout the entire workflow. Through precise cooling rate control, optimized cryoprotective agent (CPA) formulation, and sealed storage systems, it enables high cell recovery rates with functional preservation, facilitating phased manufacturing and "off-the-shelf" therapies while reducing material costs and contamination risks, thereby accelerating clinical translation of emerging therapies [1]. For assisted reproduction, cryopreservation has become the primary method for fertility preservation. By 2019, the global number of frozen embryo transfers (1,262,727 cases) significantly surpassed fresh embryo transfers (825,176 cases), with frozen embryo

Willem F. Wolkers and Harriëtte Oldenhof (eds.), *Cryopreservation and Freeze-Drying Protocols*, Methods in Molecular Biology, vol. 3060, https://doi.org/10.1007/978-1-0716-5416-3_7,

transfers demonstrating higher clinical pregnancy rates [2]. In organ transplantation, cryopreservation substantially extends donor organ preservation time, creating critical windows for cross-regional matching and surgical scheduling [3, 4]. In agriculture, cryopreservation holds strategic importance for safeguarding crop genetic diversity. The Svalbard Global Seed Vault, established by the Crop Trust in collaboration with international agricultural research centers, had archived over 1.3 million seed samples by February 2025, serving as a vital safeguard against climate change and biodiversity loss [5]. Within biobanking, standardized cryopreservation systems form the foundation for research reproducibility. Biobanks provide invaluable data resources for large-scale genomic studies, public health research, personalized medicine, and therapeutic development. The UK Biobank, for instance, stands as one of the world's largest repositories, housing 15 million biospecimens and 30 petabytes of data, while contributing to over 10,000 scientific publications—delivering critical support for related research fields [6].

However, conventional cryopreservation techniques still exhibit significant technological limitations that substantially constrain their effectiveness in modern biomedical applications. Dr. Mazur's Two-Factor Hypothesis—a foundational theory in cryobiology [7, 8]—established the critical principle that "cell cryopreservation requires precise balancing of cooling rates and CPA osmotic pressure," which remains central to optimizing cryopreservation protocols. Regarding cellular injury mechanisms, ice crystal formation represents the most destructive physical damage factor. During slow freezing, extracellular ice growth induces solution effects, causing localized solute concentration spikes that drive cellular dehydration and volumetric shrinkage [8, 9]. Conversely, excessive cooling rates prevent sufficient water efflux, leading to lethal intracellular ice formation (IIF) that directly disrupts organelles and membranes. Even when IIF isn't immediately fatal, insufficient warming rates during thawing permit small intracellular ice crystals to recrystallize into larger, damaging structures [8, 9]. Ice crystals physically perforate membranes and disrupt cytoskeletons (e.g., F-actin depolymerization), reducing mesenchymal stem cell (MSC) adhesion capacity by ~50% and impairing homing functions [10]. Osmotic stress from ice formation decreases natural killer (NK) cell IFN-γ secretion by >50% and cytotoxicity, and triggers apoptosis pathways, leading to depletion of $CD34^+$ hematopoietic stem and progenitor cells (HSPCs) [10]. Ice-damaged cells also exhibit enhanced complement activation (elevated C3b deposition on MSCs) and instant blood-mediated inflammatory reaction (IBMIR) responses, accelerating cell clearance and compromising therapeutic outcomes [10]. Chemical toxicity from CPAs constitutes another major damage source. Although dimethyl sulfoxide (Me_2SO) effectively suppresses ice formation, 10% (v/v) Me_2SO causes severe injury to human bone marrow MSCs:

inducing 10%–15% immediate post-thaw apoptosis, escalating DNA damage to 3.8-fold control levels within 48 h, and provoking cell cycle arrest with ROS elevation—ultimately impairing differentiation and migration through downregulation of adipogenic/osteogenic genes and F-actin expression [11]. Osmotic imbalance persists throughout CPA addition/removal. Rapid exposure to high-concentration CPAs triggers acute cellular dehydration from extracellular hyperosmolarity, whereas CPA removal induces dangerous swelling during hypoosmotic shock [12]. These violent volumetric fluctuations cause mechanical damage, particularly explaining the high hemolysis rates in traditional red blood cellcryopreservation (especially high-glycerol slow-freezing methods) during deglycerolization [13, 14].

At the technical operational level, conventional methods exhibit significant limitations. The manual vial-based cryopreservation approach suffers from several critical drawbacks: imprecise control over cooling rates and CPA dosing during manual handling leads to inter-vial processing variability, frequently causing osmotic shock and cellular damage [15]. Furthermore, this method has limited batch-processing capacity and lacks real-time monitoring capabilities during freezing, compromising both cell survival rates and preservation quality consistency [15]. Standardization deficiencies represent another major shortcoming, as the technique's heavy reliance on operator skill introduces substantial inter-laboratory variability. For instance, post-thaw recovery rates of vitrified oocytes may vary by 20%–30% across facilities, while implantation rates of vitrified embryo can differ by as much as 30%–40% [16, 17]. Additionally, contamination risks inherent to manual operations restrict its application in quality-critical environments. These technological bottlenecks have driven the development of more precise, controllable next-generation cryopreservation systems designed to overcome the inherent constraints of traditional methodologies.

1.2 Opportunities and Advantages of Microfluidic Technology

Microfluidic technology, leveraging its pronounced small-scale effects and precise control capabilities in heat and mass transfer, holds significant potential to provide innovative solutions to critical challenges in cryopreservation. By precisely manipulating fluid behavior at the microscale, microfluidic systems enable accurate regulation of CPA transport processes [18, 19]. At the hydrodynamic level, the stable laminar flow characteristics within microchannel (typically tens to hundreds of micrometers in width), where the Reynolds number is usually far below 100 and often less than 1.0, allow distinct parallel flow interfaces to form between different fluids [20]. Diffusion dominates mass transfer in microfluidics, facilitating the generation of precisely tunable CPA concentration gradients [18, 19, 21]. Studies demonstrate that Y-shaped microchannel-based gradient generators, by controlling CPA diffusion and laminar flow, enable gradual CPA addition and removal. This gradual adaptation minimizes osmotic shock, significantly reducing cellular damage [21, 22]. Furthermore, trapping or immobilizing cells in microperfusion chambers allows direct perfusion with programmable concentration profiles [23–25]. Controlled concentration changes can be achieved by integrating multi-inlet flow paths with on-chip mixers [26]. Experimental evidence shows that microfluidically optimized CPA addition mitigates osmotic stress responses in mouse oocytes and zygotes, yielding slower volumetric shrinkage rates and improved cellular integrity—manifested as smoother surfaces, higher sphericity, better cytoplasmic lipid retention, reduced leakage, and enhanced developmental competence (e.g., higher blastomere counts in blastocysts) [24]. Beyond cryopreservation protocols, microfluidics offers superior advantages for studying cell membrane permeability parameters compared to conventional methods. These include: precise extracellular concentration control, single-cell dynamic observation, low-cost and reusable platforms, rapid solution switching for dynamic environmental simulation, and integration with temperature control. Collectively, these features substantially enhance measurement accuracy and experimental flexibility [27].

Microfluidic technology, through its micrometer-scale channel structure and high surface-area-to-volume ratio characteristics, endows the system with extremely low thermal inertia and highly efficient heat transfer performance, enabling millisecond-level rapid temperature response (heating/cooling rates ranging from 0.1°C/s–2000°C/s) and subdegree Celsius precision temperature control (accuracy up to ±0.1°C) [28]. By integrating advanced temperature control technologies such as Joule heating, microwave radiation, or laser focusing, the system can not only achieve directional and precise heating of nanoliter droplets (regulation range 50–95°C), but also construct precise spatial temperature gradients of 6–40°C/mm [28]. The application of low thermal conductivity materials like PDMS effectively reduces thermal diffusion interference, further improving temperature control accuracy. These breakthrough thermal transfer properties bring revolutionary solutions to the field of cryopreservation: through the integration of dynamic temperature control technologies including Joule heating, Peltier cooling, and microwave/laser systems, the system has the potential to achieve ultra-rapid freezing (rates >2000°C/s) and precise rewarming (rates up to 65°C/s), thereby effectively inhibiting ice crystal formation and recrystallization damage; the uniform heat transfer characteristics enabled by microchannel structures overcome temperature gradient differences at the macroscopic scale, and when combined with real-time temperature sensing feedback systems, ensure a uniformly controlled freezing-rewarming process for samples at the nanoliter to microliter level. These innovative features successfully overcome key technical bottlenecks in traditional cryopreservation technologies, such as inaccurate temperature control and uneven heat transfer, demonstrating unique advantages in cryopreservation of single cells and precious biological samples.

Microfluidic technology demonstrates revolutionary system integration and intelligent advantages in cryopreservation applications. By integrating core functional modules, including CPA gradient addition, precise temperature regulation, programmed rewarming, viability assessment, and in-situ dynamic perfusion culture onto a single chip platform, this technology enables fully automated closed-loop control of the entire cryopreservation workflow. The system not only significantly reduces contamination risks associated with sample transfer in conventional methods, but also achieves a highly controlled cryopreservation environment through miniaturized microchannel (nanoliter-to-microliter scale) and low-thermal-conductivity materials like PDMS, combined with high-precision temperature sensors and intelligent control algorithms. Further enhanced by IoT-enabled real-time monitoring and machine-learning optimization capabilities, the platform enables high-throughput screening and the execution of personalized preservation protocols for valuable specimens, such as single cells and organoids. The "Lab-on-a-Chip" integration characteristic is driving rapid advancements in cryopreservation technology toward standardization, scalability, and intelligent automation.

1.3 Contents of This Chapter

This chapter will first elucidate the fundamental principles of microfluidic cryopreservation, with particular emphasis on the heat and mass transfer characteristics at the microscale that are critical to the cryopreservation process. Subsequently, it systematically presents how microfluidic technology can be leveraged to achieve key procedural advancements: efficient determination of cell membrane permeability parameters, dynamic and precise addition of CPAs, miniaturized sample processing, as well as chip-based precision control and highly integrated approaches for cryopreservation. Building upon these methodologies, representative application cases are provided to illustrate their practical implementation. The chapter concludes with an outlook on the application prospects of microfluidic technology in cryopreservation and explores potential technological convergence points and innovative breakthroughs. Through this comprehensive discussion, readers will not only gain a thorough understanding of the current applications of microfluidic technology in cryopreservation but also discern future developmental trajectories and the vast potential of this field.

2 Fundamentals of Microfluidic Cryopreservation

2.1 Microscale Heat and Mass Transfer

2.1.1 Unique Heat Transfer Mechanisms in Microchannel

Microchannel heat transfer mechanisms exhibit distinct characteristics that differ significantly from macroscopic systems, primarily determined by microscale effects such as high surface-to-volume ratios and laminar-dominated flows. In a microchannel with characteristic dimensions of 10–1000 μm, the Reynolds number ($Re = \rho vd/\mu$, where ρ is fluid density, v is flow velocity, d is characteristic dimension, and μ is dynamic viscosity) typically remains below 100, resulting in characteristic laminar flow conditions. This fundamental shift causes heat conduction to replace convection as the dominant heat transfer mode in microchannels [29–31]. This transition leads to significant alterations in thermal boundary layer behavior within the microchannel. The ratio of thermal boundary layer thickness (δ_T) to channel characteristic dimension (d) (δ_T/d) increases dramatically as channel dimensions decrease, causing temperature gradients to become concentrated primarily in near-wall regions [32]. Consequently, heat transfer and temperature distribution in microchannels become predominantly influenced by wall-adjacent regions, while temperature variations in central regions remain relatively minor. This unique characteristic holds crucial importance for the precise control and optimization of heat transfer processes in microchannels.

The Nusselt number ($Nu = hd/k$, where h is the convective heat transfer coefficient and k is the fluid thermal conductivity), as a key parameter characterizing microchannelheat transfer performance, exhibits distinct features under microscale conditions compared to macroscopic systems [33]. In conventional macroscopic laminar pipe flow, such as fully developed laminar heat transfer in circular pipes, the Nusselt number typically remains constant (e.g., Nu = 4.36 for constant heat flux boundary conditions and Nu = 3.66 for constant wall temperature conditions in circular pipes) and independent of Reynolds number. However, in microchannels, experimentally measured Nusselt numbers often deviate (either higher or lower) from these classical theoretical predictions and demonstrate characteristics different from macroscopic systems [33]. For instance, researchers have successfully developed microchannel heat sinks with periodically expanding-contracting cross-sections, which enhance the Nusselt number to 1.8 times that of conventional straight rectangular channels by disrupting thermal boundary layers and inducing flow separation, significantly improving heat transfer performance [34].

Another important characteristic of microchannelheat transfer is the pronounced axial conduction effect. In macroscopic systems, where characteristic dimensions are much larger than the thermal diffusion length ($L_{th} = (\alpha t)^{1/2}$, with α being thermal diffusivity and t time), axial heat conduction is typically negligible. However, under microscale conditions where channel dimensions are comparable to thermal diffusion lengths, axial conduction leads to significant temperature nonuniformity [35]. Studies show that in a 10 mm long, 100 μm wide microchannel, axial temperature gradients can reach as high as 60°C/mm. This substantial temperature gradient primarily concentrates in the channel entrance region, indicating that heat transfer in the microchannel mainly occurs during the initial stage of fluid entry.

In microfluidic cryopreservation, cell perfusion experiments (such as measuring cell membrane permeability parameters or adding/removing CPAs) in microchannels are typically conducted under isothermal conditions, allowing sufficient time for thermal equilibrium and minimizing the impact of flow heat transfer limitations. During cooling/freezing or warming/thawing processes, fluids in channels generally remain static, eliminating flow-related heat transfer issues. The advantage of microchannels lies in their ability to reshape cell suspensions into microdroplets, microcolumns, or microfilms with at least one dimension at the micrometer scale, significantly enhancing heat exchange efficiency with the external environment. Furthermore, microfluidic technology enables millisecond-scale temperature response, subdegree Celsius precision temperature control, and effective reduction of internal temperature gradients, providing more efficient and uniform thermal management solutions for cryopreservation.

2.1.2 Microscale Mass Transfer

The fundamental theory of microscale mass transfer primarily investigates substance transport behavior at micrometer or nanometer scales, characterized by the dominant role of diffusion mechanisms coupled with significant influences from surface effects, low-Reynolds-number flow, and geometric constraints [36–38]. In microscale environments such as microfluidic chips or biological cellular systems, where characteristic dimensions are typically below 100 μm, molecular diffusion becomes the principal mode of mass transport while convective effects diminish. This diffusion process can be described by Fick's first law, which states that the steady-state diffusion flux J is proportional to the concentration gradient $J = -D\nabla C$. Here, D represents the diffusion coefficient, whose magnitude depends on molecular properties, medium viscosity, and temperature, and can be estimated using the Stokes-Einstein equation: $D = k_B T / 6\pi\mu r$. For nonsteady-state diffusion processes, Fick's second law proves more applicable, expressed as:

$$\partial C/\partial t = D\nabla^2 C \tag{1}$$

This equation becomes particularly important in confined microscale systems (such as solute diffusion within microdroplets).

When fluid flow exists in microscale systems, mass transfer behavior must account for both diffusion and convection contributions. In such cases, the convection-diffusion equation is required for accurate description:

$$\partial C/\partial t + \boldsymbol{v} \cdot \nabla C = D\nabla^2 C + R \tag{2}$$

Where v represents the fluid velocity, and R denotes potential chemical reactions or source/sink terms (e.g., solute consumption by cellular metabolism).

The relative importance of diffusion versus convection is typically quantified using the Péclet number $Pe = vL/D$, where L represents the characteristic length (usually referring to the hydraulic diameter or width of microchannel in microfluidic systems). When $Pe \ll 1$, diffusion dominates, as observed in oxygen transport within static microculture droplets; whereas when $Pe \gg 1$, convective effects become more significant, such as in high-speed microfluidic flows.

Microscale mass transfer also exhibits several unique phenomena. Firstly, due to the substantial increase in surface-area-to-volume ratio, surface effects (e.g., electric double layers, hydrophobic/hydrophilic interactions) can significantly influence mass transfer; for instance, wall charges in microchannels can alter ion transport behavior through electroosmotic effects. Secondly, flows in microfluidic systems typically operate under low Reynolds number conditions ($Re \ll 1$), exhibiting laminar flow characteristics with absent turbulent mixing, thus making diffusion the primary mixing mechanism.

The dimensional effect of microfluidic chips significantly enhances diffusion-dominated mass transfer efficiency. The diffusion process, driven by Brownian motion, follows a one-dimensional diffusion model $d^2 = 2Dt$ (where d represents particle diffusion distance). Due to the quadratic relationship between distance and time, diffusion efficiency is markedly improved at the microscale. For instance, Me_2SO in dilute aqueous solution at 25°C has a diffusion coefficient of approximately $1.2–1.5 \times 10^{-9}\,m^2/s$. Based on the minimum diffusion coefficient estimation, diffusion across 1 cm in a macroscale 1 cm diameter channel would require about 41,667 s ($\approx$ 11.6 h), whereas diffusion across a proportionally equivalent distance (10 μm) in a 10 μm wide microchannel takes only $\approx$0.0417 s—representing approximately a 10^6-fold (1 million times) improvement in mass transfer rate. This highly efficient mass transfer characteristic enables microfluidic systems to rapidly establish uniform CPA concentration fields, effectively preventing cellular toxicity damage caused by localized high concentrations.

In microfluidic chips with pre-designed channel structures (such as T-junctions or Y-junctions), precise control of the inflow ratio between different concentration CPA solutions (e.g., mixing high and low concentration CPAs) enables the formation of gradually increasing concentration gradients within the channels [30, 39]. By introducing cell suspensions into this gradient environment, the samples experience a stepwise concentration increase when contacting CPA solutions, rather than direct exposure to high CPA concentrations. This approach achieves CPA addition while minimizing or even preventing cellular damage caused by sudden osmotic pressure changes [39]. Existing research has demonstrated that through physical design of microfluidic structures and hydrodynamic regulation, controlled diffusion and mixing of target molecules in microchannel flows can be realized [30]. Consequently, it becomes entirely feasible to achieve both spatial gradient distribution and temporal sequence variation of CPA concentrations within a microchannel, ultimately accomplishing progressive addition, uniform mixing, and precise elution—effectively addressing the limitations of concentration shocks and manual operations in conventional methods. This technique proves particularly suitable for low-damage cryopreservation of micro-volume samples.

2.2 Precise Delivery and Distribution Control of CPAs

2.2.1 Microfluidic Methods for Determining Cell Membrane Permeability Parameters

In cryopreservation, the precise determination of cell membrane permeability parameters to water and CPAs is critically important as it directly affects the formulation of subsequent CPA addition and removal strategies [12]. However, traditional methods (such as Colter Counter techniques) for measuring cell membrane permeability often have several limitations, such as obtaining population-averaged data [40] that cannot reflect intercellular heterogeneity. Microfluidic technology provides a powerful means to achieve this goal by creating unique microenvironments on-chip that allow precise measurements at the single-cell level [41]. The fundamental principle of such methods is to alter the solution composition surrounding cells in the microchannel of the chip to simulate CPA addition and removal, while simultaneously measuring the osmotic volume response of the cells, and then using Kedem–Katchalsky (K–K) or Two-Parameter (2-P) equations to obtain the cell membrane permeability parameters through nonlinear fitting.

Early studies in cryobiology achieved similar effects by using porous membranes to isolate cells and then changing the solution composition either on the same side as the cells (directly perfusing the cells with waste fluid flowing across the membrane) [42] or on the opposite side (indirectly changing the concentration of the solution around the cells on the other side of the membrane via transmembrane convective diffusion) [43], allowing observation of the osmotic volume response of the cells and completion of permeability parameter measurements. These devices function similarly to microfluidic devices.

The sandwich-structured microperfusion chamber can be considered one of the simplest and most practical microfluidic devices, formed by sandwiching one or two (or even multiple) silicone rubber films (with cut-out elongated channels) between two layers of transparent glass to create an enclosed microchannel. By utilizing the weak adhesion of cells to the glass substrate in an isotonic environment (without true adherent state) combined with control of the flow rate of the fluid in the channel, most cells can be kept in place without being washed away by the incoming flow [23].

PDMS microfluidic perfusion chambers are currently more widely used, with typical fabrication methods involving bonding PDMS open channels to glass substrates to form enclosed cavities. Cells can be immobilized either by coating the glass substrate in the microchannel with poly-D-lysine [44] or by directly incorporating built-in microstructures in the PDMS material to block or capture cells [12, 24, 25, 40, 41, 44, 45]. This design ensures that after introducing cell suspensions into the microchannel, cells are confined to specific areas (e.g., forming monolayers in perfusion chambers), preventing cell escape, overlap, or movement out of the focal plane that could affect imaging.

These microperfusion devices can also be conveniently integrated with temperature sensors and temperature control systems, enabling investigation of temperature-dependent changes in cell membrane permeability parameters. This provides more precise parameter references for CPA delivery during different temperature stages of cryopreservation [46, 47].

After loading cells into appropriate positions within the channel, the extracellular environment can be rapidly altered by precisely controlling the flow of different solutions (e.g., solutions with varying osmotic pressures or CPA-containing/CPA-free solutions) through the microchannel. This simulates hypertonic/hypotonic conditions or CPA addition/removal processes. For example, cells first equilibrated in an isotonic solution will undergo volume changes when rapidly exposed to a hypertonic solution due to water loss [48]. Using microscopy coupled with CCD cameras, real-time recording of cell volume changes enables calculation of key parameters, such as the water permeability coefficient, through curve fitting based on transmembrane transport models. For measuring CPA permeability parameters, the same principles apply when using CPA-containing solutions. By observing cell volume changes and CPA uptake kinetics in CPA solutions, the CPA permeability coefficient (Ps) can be determined [49].

Notably, recent studies have successfully integrated the classic electronic particle counter principle (Colter principle) with microfluidic technology. Through continuously aligned electrode pairs along microchannels on chips, researchers measured impedance changes of individual yeast cells under osmotic stress to obtain their osmotic volume changes and derived water permeability coefficients [50–53]. Such integrated microfluidic systems combine the advantages of both microfluidics and Colter principles. With ongoing technical iterations, these devices can provide, but are not limited to, cryobiological research capabilities for irregular/nonspherical cells on chips.

2.2.2 Microfluidic Strategies for Dynamic CPA Addition and Removal

The success of cryopreservation heavily depends on the CPA addition and removal processes. Conventional stepwise methods using fixed CPA concentrations often subject cells to abrupt osmotic changes, leading to osmotic shock and toxic damage [54]. Microfluidic technology addresses these challenges through continuous CPA concentration modulation, leveraging its unique advantages [12, 19, 22].

Microfluidic CPA addition strategies typically employ multi-channel architectures to create concentration gradient modules. For instance, by introducing CPA solutions of different concentrations into parallel microchannels that converge at specific junctions, continuous CPA gradients form through controlled diffusion and laminar flow characteristics [24, 25]. As cells traverse these regions, they gradually adapt to changing CPA concentrations, avoiding sudden osmotic shifts. The addition profiles can be designed with various patterns, including linear, convex, and concave configurations [55, 56]. Numerical simulations reveal distinct cellular responses to different addition profiles [56–60]. For example, concave profiles (subcategorized as low-concave and high-concave) produce varying cellular responses in terms of peak volume change (ΔV), accumulated osmotic damage (AOD), and toxic damage index (J). Optimal profile selection based on cell-specific characteristics can minimize cryoinjury.

Similarly, microfluidics enables precise CPA removal through reverse processes—gradually transitioning cells from high CPA concentrations to CPA-free solutions [21]. Post-freeze-thaw cycles typically render cell membranes more fragile and susceptible to osmotic damage. Therefore, the removal process requires particularly careful control of concentration transition rates and durations. By incorporating precisely measured membrane permeability parameters into theoretical optimization, microfluidic systems can develop scientifically-grounded CPA removal protocols that maximize post-thaw cell viability and functional integrity [21, 60]. This approach represents a significant improvement over conventional washing methods that often cause secondary osmotic injury during CPA elimination.

2.3 Ultra-Rapid Cooling and Rewarming

2.3.1 Sample Minimization Technology

In cryopreservation, the achievement of ultra-rapid cooling/warming is intrinsically linked to sample minimization technology. The fundamental principle lies in reducing sample dimensions to shorten heat transfer pathways, thereby significantly enhancing cooling/warming rates and minimizing ice crystal-induced cellular damage. From a heat transfer perspective, sample thermal conduction follows Fourier's law—larger samples require longer heat propagation times from surface to core, facilitating damaging large ice crystal formation during cooling. By contrast, reducing samples to micro/nanoscale dimensions enables faster, more uniform heat transfer.

Common sample minimization approaches in cryopreservation include droplet-based (including encapsulated), fiber-based, and thin-film techniques [61–65], all of which reduce sample geometry in at least one dimension. Droplet microfluidics typically employs specialized microchannel structures (T-junctions, flow-focusing geometries, etc.) to disperse biological samples into microdroplets. Precise control of continuous/dispersed phase flow rates and channel geometries generates highly uniform droplets at picoliter-to-nanoliter scales [66]. Alternative methods like inkjet-style printing can also produce microdroplets [65]. These miniature droplets exhibit extremely high surface-area-to-volume ratios, enabling rapid thermal exchange with cryogenic media. During cooling, samples achieve cryogenic temperatures almost instantaneously; during warming, heat penetrates quickly to avoid prolonged exposure to dangerous temperature ranges. Studies demonstrate that 40 pL droplets subjected to superfast inkjet freezing (SFF) achieve cooling rates of 3.7×10^4 °C/s, enabling CPA-free preservation of various mouse/rat cells with 86.5% recovery [65]. Microencapsulation technology packages biological samples into micrometer-scale polymer capsules, typically fabricated using microfluidic devices such as planar or coaxial flow-focusing configurations [62, 66]. These capsules combine excellent thermal conductivity with biocompatibility, protecting samples from direct environmental stress while enabling rapid thermal response. Research confirms significantly reduced IIF and improved cell viability after ultra-rapid cooling/warming of microencapsulated cells [62, 67]. Microfiber techniques offer similar benefits while enabling continuous, high-volume processing [68]. Notably, core-shell structured microcapsules/fibers provide biomimetic cryoprotection, substantially reducing requirements for permeable CPAs [62, 67, 68]. By precisely controlling microchannel depth, "thin-film" cryopreservation can be achieved by confining cell suspensions as ultra-thin layers, dramatically improving survival rates without CPAs [69].

2.3.2 Integrated Microfluidic Systems for Cryopreservation

The high integrability of microfluidic chips enables the seamless integration of sensors, heating elements, and reaction channels onto a single substrate (e.g., glass-PDMS chips), supporting the synchronous operation of multifunctional modules (e.g., fluorescence detection and temperature feedback) [70–73]. Their exceptional scalability allows flexible design of complex channel networks or heating arrays through standardized processes (e.g., photolithography, inkjet printing), while maintaining compatibility with various detection technologies (e.g., fluorescence, mass spectrometry), thereby providing modular, high-throughput solutions for chemical synthesis and bioanalysis [70–73].

Recent years have witnessed remarkable progress in on-chip temperature control and sensing, exemplified by achievements including: uniform temperature regulation across wide ranges (−3°C to 120°C,

±0.1°C precision), rapid temperature changes (heating rates up to 2000°C/s), linear temperature gradients (maximum 40°C/mm), and multi-zone independent regulation (e.g., dynamic programmable control in digital microfluidics) [46, 47]. Integration with thermoelectric coolers or cryogenic circulation systems extends these capabilities to lower temperatures, enabling complete chip-based cryopreservation systems. These systems combine ultra-rapid cooling/warming with precise programmable thermal control through deep integration of microfluidics and cryopreservation technologies, featuring multiple functional modules for comprehensive process control [46, 47, 74–77]. A typical integrated microfluidic cryopreservation system comprises three core components: the microfluidic chip, temperature control system, and detection system. The microfluidic chip, fabricated via microfabrication techniques with micron-scale channels and chambers, serves as the system centerpiece. Optimized channel layouts and structural parameters enable precise sample manipulation, including injection, positioning, and mixing [47, 78]. Material selection is critical, with thermally conductive substrates like silicon, glass, or polymers ensuring rapid heat transfer. The temperature control system, integrating miniature cooling/heating elements (Peltier devices, microheaters), enables ultra-rapid thermal modulation [46, 47, 74, 75]. Peltier devices leverage the Peltier effect to achieve rapid cooling/heating (tens of °C/s or higher) through applied current. The detection system monitors real-time sample status (temperature, cell morphology, physiological parameters) during thermal transitions using optical microscopy, Raman spectroscopy, or fluorescence detection, allowing dynamic protocol optimization.

Microfluidic systems with integrated thermal control provide unique platforms for studying nucleation and crystallization in CPA solutions. The technology generates monodisperse droplets/columns for observing phase transitions under precisely controlled temperatures. For instance, Stan et al. developed a microfluidic device producing supercooled droplets while monitoring ice nucleation temperatures at 75 Hz sampling rates, providing extensive statistical data for homogeneous/heterogeneous nucleation studies [77]. Similarly, Deck et al. employed a Microfluidic Ice Nucleation Counter (MINCZ) to investigate sucrose solution freezing, optically detecting nucleation temperature (T_m'), glass transition temperature of maximally freeze-concentrated solution (T_g'), and melting point (T_m) through droplet brightness changes [79]. These studies demonstrate microfluidics' capability for high-precision thermodynamic parameter measurement (nucleation/melting temperatures) and kinetic process quantification (crystal growth rates). The high-throughput nature facilitates rapid data acquisition, offering a comprehensive understanding of CPA solution freezing behavior—particularly valuable for optimizing complex formulations in biopharmaceuticals and the food industry, informing cryopreservation and lyophilization process design.

Integrated microfluidic cryopreservation systems also enable precise control of CPA addition/removal processes. On-chip micropumps and microvalves regulate CPA solution flow rates according to programmed protocols, achieving gradient-based CPA introduction/elimination to minimize cellular damage. Furthermore, chip-integrated cryopreservation offers high-throughput, automated processing of multiple samples simultaneously, significantly improving preservation efficiency and reliability while providing robust technical support for long-term biological sample storage and application.

3 Applications of Microfluidic Cryopreservation

Microfluidic technology, with its precise single-cell manipulation capabilities and high-accuracy regulation of temperature and concentration at cellular scales, provides a novel technical platform for cell cryopreservation research. This technology enables critical operations, including accurate measurement of cell membrane permeability parameters, gradient-based addition and removal of CPAs, and programmed cooling/warming processes, while supporting long-term on-chip cell preservation. Notably, microfluidic cryopreservation can either serve as a standalone platform completing the entire workflow from pretreatment to storage, or integrate with conventional methods to optimize key steps only—this flexible application mode fully demonstrates the technology's superior compatibility and broad applicability.

3.1 On-Chip Cell Membrane Permeability Measurement

Cell membrane permeability measurement represents a fundamental aspect of cryobiomedical research, where accuracy directly determines the optimization of cryopreservation protocols. Conventional methods, limited by macroscopic, population-based statistical analysis, struggle to capture single-cell heterogeneity and dynamic responses. The introduction of microfluidic technology—through the integration of microscale fluid control, real-time imaging, and multiphysics modeling—has advanced membrane permeability studies from qualitative description to quantitative precision, establishing a complete technical framework encompassing basic theoretical modeling, device innovation, and multi-scenario application validation.

The methods and commonly used techniques for studying cell membrane permeability based on microfluidics are shown in Fig. 1. Typically, the solution environment around cells is altered on a microfluidic chip, followed by acquisition of cellular volumetric response data, after which the 2-P equation is employed to obtain cellular permeability parameters through nonlinear fitting (Fig. 1A). It should be noted that if the solution does not contain permeable CPAs, only transmembrane water transport is involved, the 2-P equation reduces to a one-parameter (1-P) equation. Microfluidic devices commonly used for studying cellular osmotic responses can

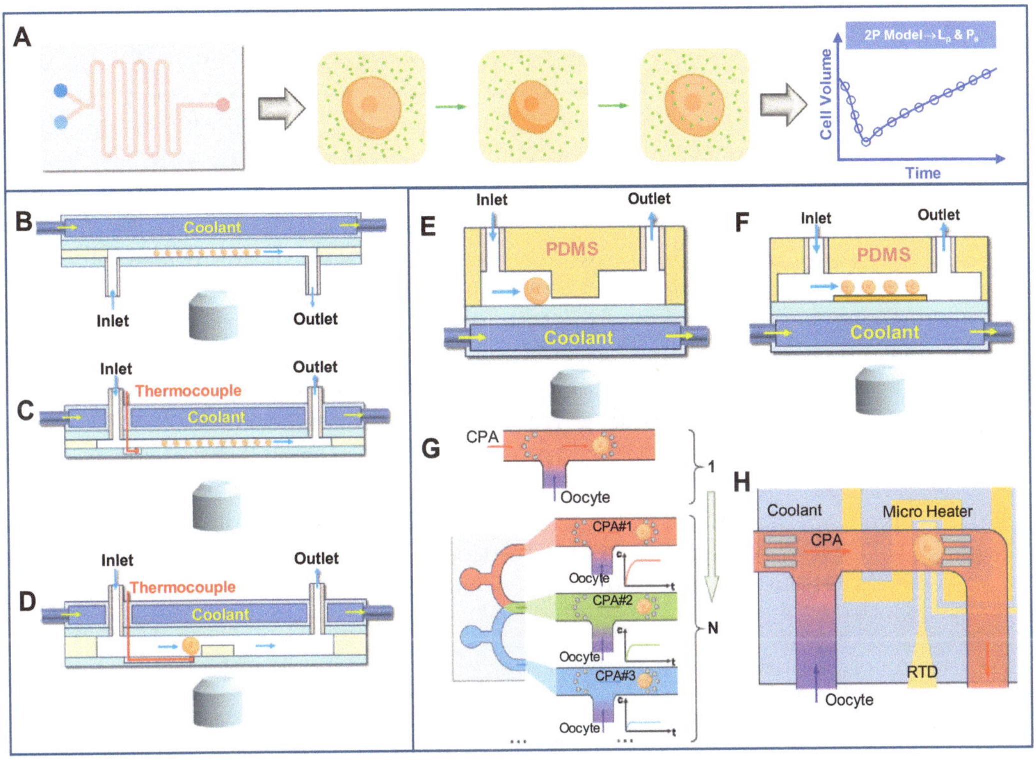

Fig. 1 Microfluidic methods for investigating cellular osmotic responses. (**A**) Fundamental workflow for measuring cell membrane permeability parameters on microfluidic chips; (**B–D**) Sandwich-structured microfluidic chips with downward-facing inlets/outlets, upward-facing inlets/outlets, and upward-facing inlets/outlets incorporating cell retention barriers, respectively; (**E–H**) PDMS-based microfluidic chips featuring integrated cell retention barriers, cell-adhesive coatings, cell capture structures, and cell capture structures with microheaters/temperature sensors, respectively. This figure was conceptually adapted from references [12, 41, 44, 45, 47, 78, 87]

be divided into two main categories: one is the sandwich-structured microperfusion chamber based on silicone membranes, usually formed by sandwiching one or multiple silicone films between two glass slides, with the silicone films featuring elongated cut-out channels (Fig. 1B–D); the other is the PDMS microchannel-based microperfusion chamber, typically consisting of a PDMS layered structure with integrated channels bonded to a glass substrate to form enclosed conduits (Fig. 1E–F). As needed, both types of perfusion chambers can incorporate circulatory cooling units for temperature control, and can also integrate microheaters and temperature sensors on the glass substrate (Fig. 1G–H).

3.1.1 Precise Characterization of Extracellular Solution Concentration Changes

Early studies of cell membrane permeability relied on the idealized step-concentration assumption, positing that extracellular solute concentrations could instantaneously reach target values. However, microfluidic perfusion experiments have revealed that actual concentration changes exhibit gradual transitions due to fluid diffusion and mixing limitations. Takamatsu et al. [23] developed a representative sandwich-structured microperfusion chamber and, for the first time, precisely measured the concentration profiles of NaCl solution at different locations within the microchannel during abrupt inflow concentration switching and linear increase processes using laser interferometry (Fig. 2A). The results demonstrated that even with abrupt inflow concentration changes, the concentration variation around cells in the microchannel did not follow a step function but rather a curvilinear gradual increase (Fig. 2A). By utilizing the experimentally determined NaCl concentration change curves in the extracellular solution, this study derived the water permeability coefficient (L_p) of cell membranes through fitting the volumetric changes during hypertonic cell dehydration (Fig. 2B). This work was among the first to highlight the critical fact that abrupt inflow concentration changes do not equate to instantaneous concentration changes in the peri-cellular microenvironment—a phenomenon dependent on channel geometry and flow parameters. This finding has profoundly influenced subsequent accurate determinations of

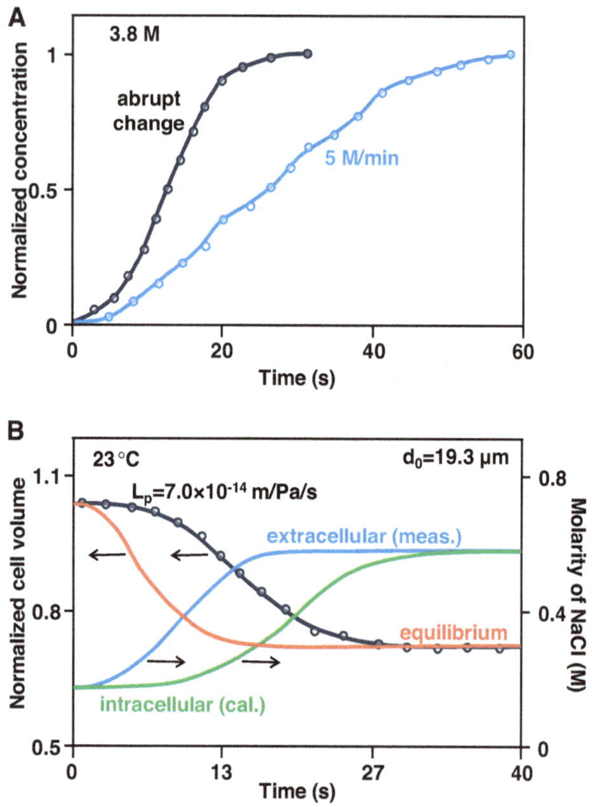

Fig. 2 High-precision determination of cell membrane water permeability coefficient. (**A**) Experimentally measured concentration change curves of extracellular solution; (**B**) Precise fitting of Lp based on the measured extracellular solution concentration change curve. (Data adapted from [23])

membrane permeability parameters and the design of novel measurement devices.

Building upon the sandwich-structured microperfusion chamber developed by Takamatsu et al. [23], Zhao [80] proposed a universal adaptive time-varying function to characterize extracellular fluid concentration changes during osmotic pressure variations for determining the hydraulic conductivity coefficient (L_p) of cell membranes (Fig. 3A). The mathematical expression of this function is:

$$C^e(t) = \begin{cases} C_0 & t < 0 \\ C_0 + \left[\left(C_f - C_0\right)/\Delta t_0\right] \cdot t & 0 \le t \le \Delta t_0 \\ C_f & t > \Delta t_0 \end{cases} \tag{3}$$

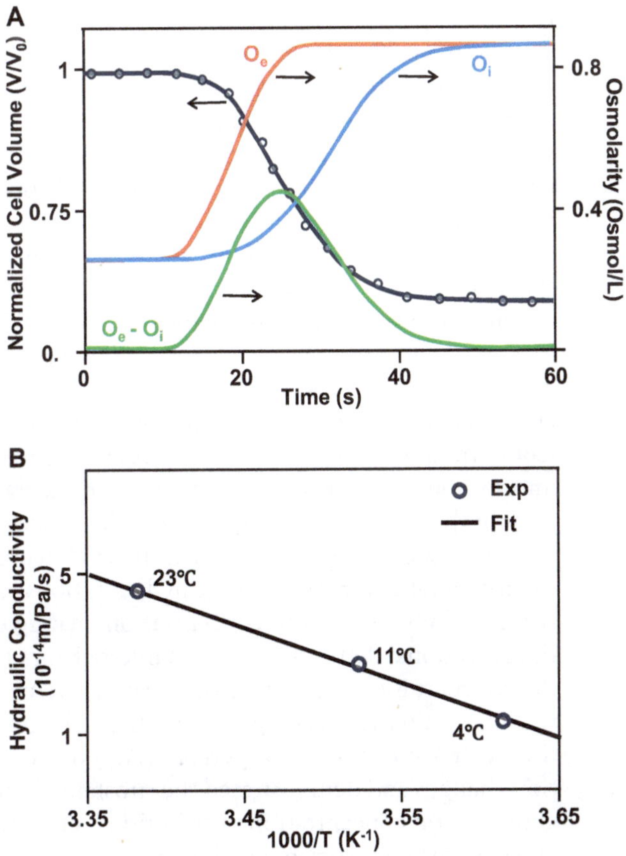

Fig. 3 Determination of water permeability coefficient of bovine carotid artery endothelial cells. (**A**) High-precision fitting of cellular volume changes using the adaptive time-varying function; (**B**) Temperature dependence of the measured water permeability coefficient Lp of bovine carotid artery endothelial cells. (Data adapted from [48])

Where $C^e(t)$ represents the extracellular solution concentration, C_0 and C_f denote the extracellular solution concentrations before and after the osmotic shift (assuming $C_f > C_0$ here), and Δt_0 indicates the duration of the osmotic shift. The concentration C can represent either nonpermeating solutes (e.g., NaCl, sucrose—for L_p measurement) or permeating CPAs (e.g., Me_2SO, glycerol—for simultaneous determination of L_p and P_s).

Experimental validation confirmed that this function significantly improves fitting accuracy, demonstrating that piecewise linear functions, particularly those with adjustable slopes, better reflect actual concentration changes compared to traditional ideal step functions. The Δt_0 value corresponding to Takamatsu's experimentally measured extracellular concentration change curve was approximately 22 s [23]. Zhao proposed constrained nonlinear fitting of three parameters (Δt_0, L_p [or L_p and P_s for CPA-containing systems], and V_b) within strictly defined upper/lower bounds [80]. The study also revealed that simultaneous fitting of the osmotically inactive volume V_b with L_p yields greater accuracy than pre-determining V_b [80]. This method applies to various osmotic experimental setups, providing a more flexible and precise solution for membrane permeability measurement. By elucidating the critical influence of concentration change kinetics on permeability parameters, the research established theoretical foundations for designing advanced microfluidic systems, driving model evolution from "static assumptions" to "dynamic measurements".

Building upon these advancements, Liu et al. [49] further refined the methodology for acquiring extracellular solution concentration profiles. By integrating finite element analysis with experimental validation, they developed a dynamic model of solution concentration changes in sandwich-structured microperfusion chambers (Fig. 4), effectively replacing the idealized step-function assumption prevalent in traditional studies [49]. The researchers achieved significant improvements in concentration prediction accuracy by directly integrating thermocouples into microfluidic devices for localized temperature monitoring, coupled with hydrodynamic simulations and experimental calibration using $CuCl_2$ solutions. The study revealed two key findings: (1) the concentration change curves exhibit distinct positional dependence within microchannel (Fig. 4A), and (2) the concentration profiles remain relatively temperature-insensitive within certain thermal ranges (Fig. 4B). This approach reconfirmed the existence of transition periods (Δt_0) during actual concentration changes and demonstrated the substantial influence of channel geometry on concentration distribution—effectively mitigating the excessive fitting errors in membrane permeability parameters caused by conventional step-function assumptions. It should be emphasized that these concentration change curves are system-specific, depending entirely on the microfluidic device's geometric characteristics and

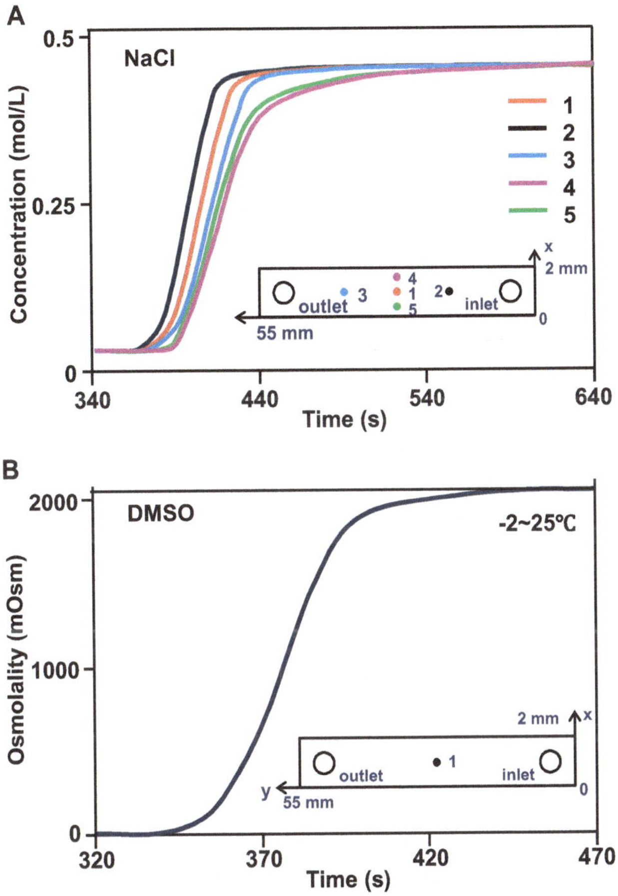

Fig. 4 Concentration change profiles of extracellular solutions in microchannel. (**A**) Position-dependent characteristics of NaCl concentration change curves; (**B**) Me_2SO concentration change curves at central locations over the temperature range of −2°C to 25°C. (Data adapted from [49])

hydrodynamic conditions. Therefore, the model cannot be simply transferred between systems, and any device redesign necessitates renewed, precise modeling.

3.1.2 Precise Description of Extracellular Solution Temperature

Early microdevices for studying cellular osmotic responses typically employed direct contact between liquid-cooling modules and microperfusion chambers, assuming the sample temperature within the microperfusion chamber equaled the coolant temperature (as shown in Fig. 1B), or alternatively placed thermocouples between the perfusion chamber and cooling module for temperature measurement. These approaches all suffered from insufficient precision in measuring the temperature of the solutions surrounding cells, potentially introducing significant errors when the set temperature was substantially below room temperature.

Liu et al. achieved real-time in situ monitoring of the cellular microenvironment temperature by designing dedicated recesses on the bottom glass slide of sandwich-structured microfluidic perfusion chambers and directly embedding thermocouples (Fig. 1C), effectively overcoming the inaccuracies inherent in traditional indirect temperature measurement through external heat exchangers [49]. This technical improvement enhanced temperature control precision to ±0.1°C, providing reliable support for temperature-dependent studies of cellular permeability within the 4°C– 25°C range.

Inspired by Fang et al.'s [46] research, Lei et al. [47] made breakthrough progress in precise temperature control at cellular scales within a microchannel. By integrating transparent ITO film microheaters with resistive temperature detectors (RTDs) on glass substrates, combined with a detachable global water bath cooling module, they achieved precise local temperature regulation across a broad range of 4°C–37°C (with fluctuations less than 0.2°C) [47] (Fig. 5). This design utilizes PID algorithms for real-time heating power adjustment, enhances temperature resolution through serpentine-arranged RTDs, reaches target temperatures within 5 s, and has verified core zone temperature uniformity (gradient less than 0.026°C/μm) through finite element simulations. This innovative design fundamentally solves the issues of temperature control lag and inhomogeneity caused by indirect measurement or single-layer structures in traditional methods, providing a temperature microenvironment with high stability and spatiotemporal resolution for cellular permeability research, significantly improving the measurement accuracy of temperature-dependent membrane permeability parameters in cryobiological experiments.

3.1.3 Applications of Sandwich-Structured Microfluidic Perfusion Chambers

The adaptive time-varying function established by Zhao [80] based on Takamatsu's experimental results for precisely describing extracellular solution concentrations [80] has been widely applied in subsequent studies of membrane permeability across various cell types at different temperatures. Zhao et al. [48] utilized sandwich-structured microfluidic perfusion chambers to investigate osmotic responses of bovine carotid artery endothelial cells (BCAECs) at different temperatures (4°C, 11°C, and 23°C) [48]. By employing image processing techniques to record dynamic changes in cell volume and performing nonlinear regression fitting with water transport equations (similar to that shown in Fig. 3A), they obtained the membrane's L_p values (0.26 ± 0.08, 0.12 ± 0.02, and 0.06 ± 0.02 μm/atm/min at 23°C, 11°C, and 4°C, respectively), activation energy E_{Lp} (47.6 kJ/mol) (Fig. 3B), and the cell's osmotically inactive volume V_b.

Wang et al. [81] employed this device to investigate the permeability characteristics of Spodoptera frugiperda 21 (Sf21) cell membranes at different temperatures, with particular focus on the dual dependence of both temperature and CPA (e.g., glycerol)

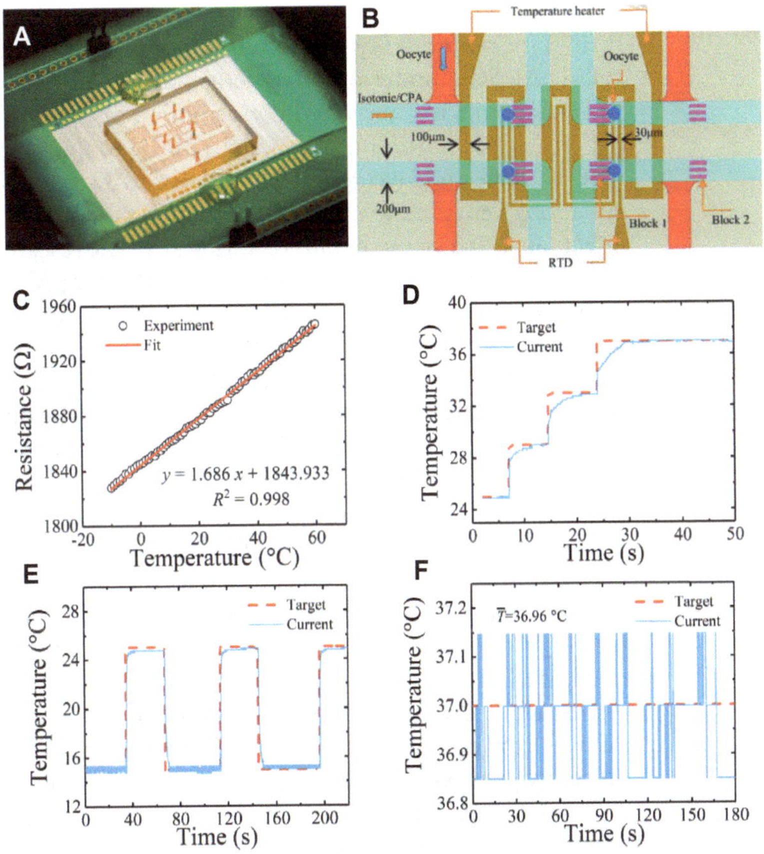

Fig. 5 PDMS microfluidic perfusion chamber integrated with MEMS temperature control module. (**A**) Physical photograph of the microfluidic chip; (**B**) Structure Design of the ITO heater and RTD; (**C**) Resistance-temperature characteristic curve of the ITO resistive temperature detector (RTD); (**D**) System response curve to stepwise temperature increases (29°C→34°C→37°C); (**E**) System response curve to rectangular pulse temperature profiles (alternating between 15°C and 25°C); (**F**) Real-time temperature fluctuation curve when the target temperature is set at 37°C. (Reproduced with permission from [47])

concentration on membrane water permeability (L_p) and glycerol permeability (P_s) [81]. Measurements of osmotically inactive volume (V_b), water permeability, and glycerol permeability in Sf21 cells were conducted at 25°C, 15°C, 5°C, and −2°C, with data fitted using both K–K and 2-P equations. The study revealed that water permeability decreased linearly with increasing glycerol concentration, with this dependence being more pronounced at higher temperatures, whereas glycerol permeability showed no significant concentration dependence. Based on experimental results, the authors proposed a modified Arrhenius relationship expressing the reference water permeability (L_{pg}) as a linear function of glycerol concentration, thereby enabling more accurate prediction of membrane permeability behavior under varying temperatures and CPA concentrations.

Later, Wang et al. [82] again utilized this apparatus to determine the membrane permeability parameters of porcine adipose-derived stem cells (pADSCs) under different temperatures and CPA conditions [82]. Experiments measured the osmotic responses of pADSCs in hypertonic phosphate-buffered saline (PBS) and three common CPAs (Me_2SO, EG, and glycerol) at 28°C, 18°C, 8°C, and −2°C. The 2-P equations were used to obtain V_b, L_p, and P_s values, with temperature effects analyzed via Arrhenius relationships. The research concluded that Me_2SO and EG were more suitable CPAs for pADSC cryopreservation, while glycerol's poor permeability at low temperatures rendered it nonideal for this application.

Liu et al. [49] further expanded the applicability of sandwich-structured microfluidic perfusion chambers by obtaining concentration profiles of extracellular solutions at different microchannel locations through finite element analysis (Fig. 4) [49]. Using this platform, they investigated membrane permeability in both Sf21 cells and porcine iliac endothelial cells (PIECs). The study measured water permeability coefficients (L_p) of 3.620×10^{-14} m/Pa/s for Sf21 cells at 25°C, and 0.824×10^{-14}m/Pa/s (L_p) with 1.879×10^{-8}m/s (P_s for Me_2SO) for PIECs at 25°C. These results reconfirmed that traditional idealized step-concentration assumptions introduce significant errors, while experimentally validated concentration distributions derived from combined experimental and finite element analysis significantly improve parameter fitting reliability. This work established an important reference for the high-precision determination of cell membrane permeability parameters.

Niu et al. [83] investigated the osmotic responses of human umbilical vein endothelial cells (HUVECs), measuring an L_p of 3.015×10^{-14}m/Pa/s and a Me_2SO P_s of 5.925×10^{-8}m/s at 25°C [83]. The study revealed that HUVECs exhibited exceptionally low permeability to Me_2SO—even at 1.5 M Me_2SO concentrations, single-step CPA removal caused unavoidable osmotic damage, while multi-step removal protocols significantly improved cell survival rates, providing optimized strategies for endothelial cell preservation [83]. Wang et al. examined the effects of iron oxide nanoparticles (Fe_3O_4) on Sf21 cell permeability, reporting a reference L_p (L_{pg}) of 0.912×10^{-14}m/Pa/s with 0.5% wt Fe_3O_4 present, where the activation energy E_{Lp} decreased to 3.364 kJ/mol and the osmotically inactive volume V_b increased with nanoparticle concentration [84]. Li et al. [85] conducted comparative studies of HepG2 cell membrane permeability during single-step versus multi-step Me_2SO addition. At 25°C, single-step addition yielded an L_p of 0.109×10^{-14}m/Pa/s and P_s of 36.127×10^{-8}m/s, while three-step addition produced values of 0.234×10^{-14}m/Pa/s (L_p) and 71.406×10^{-8}m/s (P_s) [85]. This confirmed that permeability parameters obtained from single-step protocols cannot accurately predict cellular volume changes during multi-step procedures, providing an experimental basis for optimizing HepG2 cell cryopreservation protocols.

The sandwich-structured microperfusion chambers employed in the aforementioned studies were all constructed by sandwiching silicone membranes containing straight channels between two glass layers, specifically designed for testing smaller cells typically ranging from 10 to 20 μm in diameter, as illustrated in Fig. 1B, C. Zhao and colleagues [41] enhanced this design by developing a chamber comprising two silicone membranes sandwiched between glass layers. The upper silicone membrane featured two rectangular cutouts measuring 20 mm by 2 mm, while the lower membrane contained one elongated cutout of 55 mm by 2 mm, forming a microchannel with 150 μm-deep side sections and a 50 μm-deep central region capable of trapping human oocytes approximately 120 μm in diameter, as depicted in Fig. 1D and Fig. 6 [41]. This advanced experimental system was used to measure the osmotic responses of in vitro matured human oocytes exposed to 0.56 M trehalose, 1.5 M ethylene glycol (EG), and 1.5 M propylene glycol (PG) at temperatures of 4°C, 15°C, and 25°C. Membrane transport parameters were derived through 2-P model fitting, yielding several key measurements: at 25°C, oocytes showed water permeability coefficients (L_p) of 2.45×10^{-14} m/Pa/s in trehalose solution, 1.63×10^{-14} m/Pa/s in EG, and 1.51×10^{-14} m/Pa/s in PG. The solute permeability coefficients (P_s) were measured

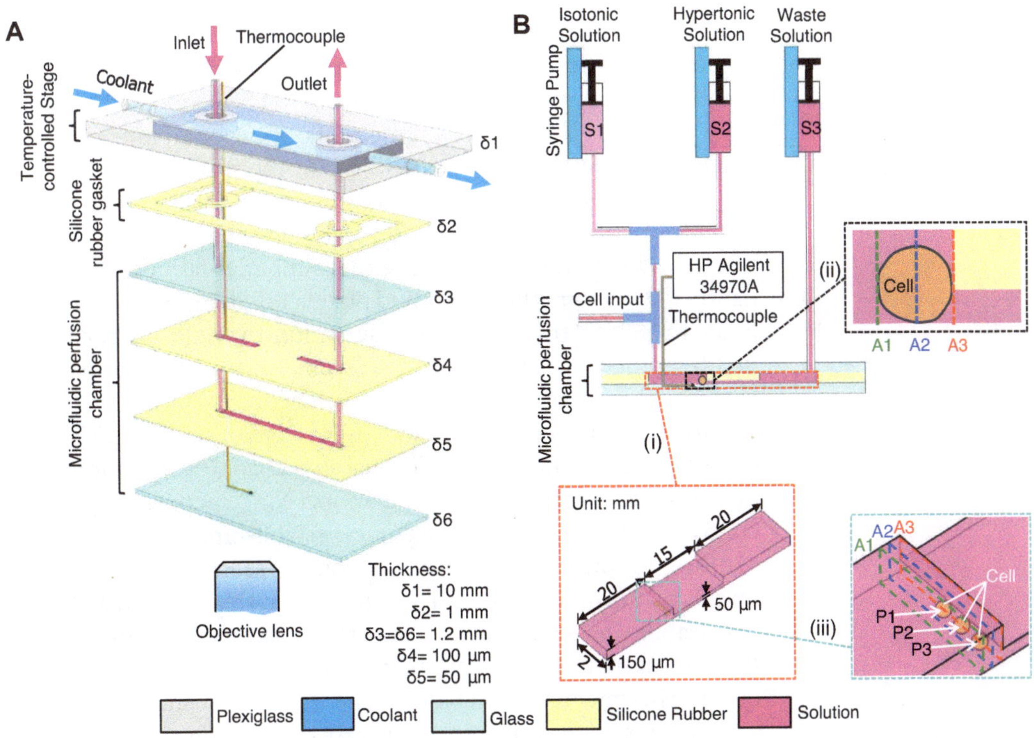

Fig. 6 Sandwich-structured microperfusion system for oocyte permeability studies. (**A**) Exploded view of the sandwich-structured microperfusion chamber; (**B**) Fluidic pathway and oocyte positioning diagram. Reproduced with permission from reference [41]

at 1.77×10^{-8}m/s and 1.26×10^{-8}m/s for PG. Arrhenius analysis revealed activation energies for water permeability (E_{Lp}) of 32.49 kJ/mol for trehalose, 43.21 kJ/mol for EG, and 53.52 kJ/mol for PG. This research achieved three significant advancements: first, it realized a genuine three-dimensional microchannel design through stacked membranes, creating channels with stepped depths; second, it marked the first application of sandwich-structured microperfusion chambers to study large cells, specifically human oocytes measuring about 120 μm in diameter; third, it successfully isolated mechanical interactions to the zona pellucida while protecting the oolemma from direct mechanical forces during testing. The innovative channel design ensured that any mechanical contact occurred only with the outer zona pellucida through specific channel structures, while the inner plasma membrane remained free from direct mechanical interaction. This approach provided more accurate measurements of membrane permeability parameters compared to previous methodologies by completely eliminating potential membrane deformation caused by mechanical forces. The study's findings were documented in reference [41], with relevant illustrations provided in Figs. 1D and 6.

The core innovation of this device lies in significantly reducing osmotic damage to oocytes by gradually increasing osmotic pressure (as opposed to traditional abrupt methods). Experimental results demonstrated that oocytes processed with this method showed no significant difference in 2PN formation rate after ICSI fertilization or 4-cell embryo development rate compared to fresh oocytes, indicating preserved physiological functionality. This technology provides a novel tool for precise measurement of oocyte membrane transport properties, offering substantial value for reproductive pharmacology, fertility preservation, and assisted reproductive technology optimization. Compared to existing devices, this system employs a silicone-glass sandwich structure that simplifies fabrication (eliminating PDMS soft lithography and plasma bonding), while achieving efficient oocyte trapping and observation through microchannel depth gradient design (150 μm → 50 μm). Integrated temperature control enhances monitoring precision. Theoretical modeling further reveals the cellular protection mechanism of gradual osmotic changes, overcoming limitations of traditional step-change assumptions. The device demonstrates superior operational convenience, temperature control reliability, and data accuracy, establishing a more efficient platform for membrane property studies of sensitive cells. Building on this platform, Chen et al. [86] discovered that oocytes with high permeability achieved 60% blastocyst formation rate post-fertilization, versus 0% for low-permeability groups, leading to a novel oocyte quality assessment method based on permeability characteristics [86].

These studies collectively advance the application of sandwich-structured microfluidic perfusion technology for measuring cellular osmotic properties. Progressing from initial device optimization for improved measurement accuracy, to exploration of permeability characteristics across cell types, and investigation of nanoparticle effects on membrane permeability, they systematically expand microfluidics applications in cryobiology and biomaterial preservation. This work provides crucial biophysical parameters and theoretical support for optimizing cellular cryopreservation protocols.

3.1.4 Applications of PDMS Microfluidic Perfusion Chambers

Although PDMS microchannels involve more complex fabrication processes and higher costs compared to sandwich-structured microperfusion chambers, their template-casting manufacturing approach enables more sophisticated three-dimensional channel architectures, demonstrating broader application potential in fields requiring precise fluid manipulation.

In 2007, Chen et al. developed a PDMS-based soft lithography microfluidic perfusion system to study osmotic behavior and membrane transport properties in rat basophilic leukemia (RBL-1) cells (Fig. 1E) [87]. This system utilized integrated step structures within a microchannel to immobilize cells while monitoring volume changes under anisotonic conditions via microscopy. Experimental measurements revealed the RBL-1 cells' osmotically inactive volume (V_b) as 41% of the initial volume, with water permeability (L_p) of 0.32 ± 0.05 μm/min/atm at 22°C and Me_2SO permeability (P_s) of $0.49 \pm 0.13 \times 10^{-3}$ cm/min. The PDMS channel's step design enabled natural cell trapping at depth transitions during flow from deep to shallow regions, facilitating osmotic response studies through perfusate modulation. Further integration with temperature-control modules (Fig. 1E) allowed investigation of murine dendritic cells (mDCs) [45]. Results showed these cells exhibited 40% osmotically inactive volume (V_b), with temperature-dependent L_p values of 0.11 ± 0.02 (12°C), 0.16 ± 0.03 (22°C), 0.95 ± 0.29 (25°C), and 1.54 ± 0.66 μm/min/atm (32°C). The Me_2SO permeability (P_s) measured $0.63 \pm 0.29 \times 10^{-3}$ cm/min at 22°C, with calculated water transport activation energy (Ea) of 2.39 kcal/mol.

Tseng et al. [44] investigated the membrane transport properties of water and Me_2SO in human megakaryocytic cell line UT-7/TPO at both above-zero and subzero temperatures using a microfluidic perfusion system (Fig. 1F) [44]. The research team developed a surface-modified microperfusion channel where cells were immobilized on glass substrates via poly-D-lysine coating. By perfusing cells with solutions of varying osmotic pressures and CPA concentrations, they recorded dynamic cell volume changes through video microscopy and analyzed data using the K–K equations. The study measured the megakaryocytes' osmotically inactive volume ($V_b = 0.15$), temperature-dependent water permeability (L_p) and Me_2SO permeability (P_s) coefficients, along with activation energies (E_a) for transmembrane transport. Results demonstrated that megakaryocytes exhibited significantly higher membrane permeability to water ($L_p = 2.62$ μ m/min/atm at 22°C) and Me_2SO ($P_s = 1.8 \times 10^{-3}$ cm/min at 22°C) compared to other mammalian cells (e.g., lymphocytes), suggesting these cells may require higher optimal cooling rates during cryopreservation to prevent excessive cellular shrinkage. Furthermore, discontinuity in Arrhenius plots between 0°C and 12°C implied potential correlations with membrane lipid phase transitions.

Lei et al. [47] achieved precise wide-range temperature control (4–37 ± 0.2°C) for oocyte permeability studies by integrating ITO thin-film heaters with temperature sensors (Fig. 1H) coupled with modular water bath cooling [47]. The platform employed a three-pillar array trapping structure capable of simultaneously immobilizing four oocytes for parallel experiments, substantially improving research efficiency. Deep learning-assisted image analysis (98.5% accuracy) quantitatively revealed significantly higher membrane permeability for 1.5 M propylene glycol (PG) ($P_s = 0.53 \pm 0.04$ μm/min) versus equivalent ethylene glycol (EG) concentrations ($P_s = 0.38 \pm 0.03$ μm/min, $p < 0.05$), providing critical evidence for CPA selection in oocyte preservation. This temperature precision methodology established fundamental technical standards for subsequent cryobiological studies of oocytes.

Chen et al. [78] developed a multi-channel microfluidic chip system that achieved synchronous capture of three oocytes and parallel perfusion with different concentration solutions through a micropillar-array design (Fig. 1G) [78]. The system employed a dual-inlet concentration gradient generator, which established linear CPA concentration gradients (0.5–1.5 M, $R^2 > 0.99$) in serpentine mixing channels by precisely controlling flow rate ratios (10 μL/min) between two input solutions. The study revealed that both water and CPA permeability coefficients (L_p and P_s) of mouse oocyte membranes increased significantly with rising concentrations ($p < 0.01$), with the 1.5 M EG group showing 42% higher L_p values compared to the 0.5 M group, demonstrating concentration-dependent permeability patterns. Utilizing an improved U-Net neural network for real-time image analysis, the system achieved 35 frames/s processing speed with 98.2% segmentation accuracy. Embryo transfer validation confirmed no significant impact on oocyte developmental potential (48.3% birth rate vs. 57.5% control, $p > 0.05$). This study established a microfluidic platform supporting simultaneous multi-concentration testing, providing crucial technical support for optimizing oocyte cryopreservation protocols.

The evolution of PDMS microfluidic perfusion technology in cellular permeability research demonstrates clear technological iteration and functional enhancement. Early studies by Chen [45] and Tseng [44] focused on methodological foundations, utilizing PDMS microchannel step structures and surface modification to achieve single-cell capture and precise permeability measurements while revealing temperature-dependent membrane transport characteristics.

Lei's [47] work represented a breakthrough in system integration, combining ITO temperature control modules with three-pillar arrays to achieve ±0.2°C temperature precision and simultaneous multi-cell observation. The more recent Chen [78] study exemplified flexible on-chip concentration control through integrated multi-channel chips incorporating gradient generation, parallel perfusion, and AI image analysis to uncover concentration-dependent permeability patterns— marking the technology's progression from fundamental parameter measurement to comprehensive physiological analysis platforms.

3.2 Microfluidic Chip-Based CPA Processing

As previously discussed, microfluidic technology enables precise spatiotemporal control of solute concentrations within a microchannel, creating time-varying concentration fields that facilitate gradual changes in peri-cellular concentrations. This ensures CPA addition/removal with minimal cellular damage.

Existing approaches can be categorized into three classes based on concentration modulation methods (Fig. 7): Class I employs on-chip premixing zones for precise concentration control, where pre-mixed solutions at target concentrations directly perfuse captured or immobilized cells in observation areas (Fig. 7A–F). Class II utilizes convective/diffusive mass transfer across porous inorganic membranes; concentration changes on one membrane side indirectly modulate solution conditions around suspended cells on the opposite side (Fig. 7G–H). Class III leverages laminar flow diffusion phenomena in microchannels to achieve gradual peri-cellular concentration changes during continuous flow (Fig. 7I–K).

3.2.1 Microfluidic Pre-Mixing and Direct Perfusion Methods

Heo et al. [25] developed a PDMS-based microfluidic platform for single-oocyte analysis (Fig. 7A–C), demonstrating that linear or complex concentration gradients for CPA addition could minimize oocyte volume changes (<10%) and osmotic damage while reducing CPA exposure time to under 15 min [25]. This foundational work highlighted microfluidics' potential for precise CPA concentration control and cellular protection. Building on this, Lai et al. [24] optimized microfluidic CPA addition by reducing cell shrinkage rates (Fig. 7D–F), proving that gradual, stable volume changes better preserved oocyte and zygote morphological integrity and developmental competence [24]. Their study reinforced the critical role of shrinkage kinetics first identified by Heo, while pioneering microfluidics' integration into practical cryopreservation workflows.

Zhou et al. [56] systematically investigated CPA addition/removal profiles, testing convex, linear, and concave concentration trajectories across nine combined protocols. Their work revealed that concave addition paired with convex removal maximized porcine oocyte survival and morula formation rates, establishing protocol synergy as a key determinant of survival. This not only validated

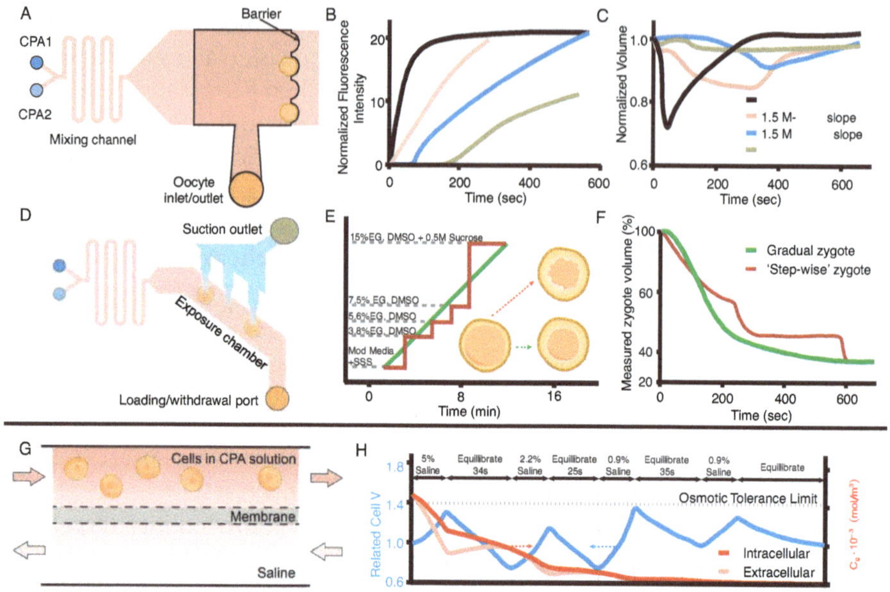

Fig. 7 Microfluidic approaches for CPA addition/removal. (**A**–**F**) Direct perfusion method; (**G**–**H**) Transmembrane transport method; (**I**–**K**) Flow-diffusion method. (Adapted from [21, 24, 25, 60]

Heo and Lai's linear approaches but also advanced complex profile optimization. The field's latest advancement came from Zhan et al. [88], with an autonomous microfluidic device incorporating fluidic transistors. Their system automated multi-step CPA processing (oocyte capture, linear ES addition, VS dehydration, and retrieval) using constant-pressure-driven fluidic timers and logic gates, eliminating manual intervention while reducing clinical translation barriers. This represented both a technical automation leap and a practical simplification of preceding methodologies.

3.2.2 Transmembrane Mass Transfer in Microfluidic Chips

Lusianti and Higgins [60] pioneered a membrane-based microfluidic device for rapid glycerol removal from cryopreserved red blood cells (Fig. 7G–H). Their design featured an AN69 dialysis membrane-separated dual channel, with mathematical modeling predicting glycerol/water transmembrane transport and cellular volume changes. Experimental results demonstrated >50% glycerol reduction in a single pass with controlled hemolysis, while theoretical analysis suggested potential complete glycerol removal within 3 min through multistage processing, offering novel clinical solutions for rapid blood product processing.

Kang et al. [89] advanced this technology with a multistep dilution-filtration device (MDFD) using layered PMMA chips and PVDF membranes. This automated system achieved 80% glycerol clearance and 90% cell viability, though with reduced recovery (~40%). Their investigation of cell density, glycerol concentration, and membrane pore size revealed tradeoffs: larger pores improved clearance but decreased recovery due to cell trapping and mechanical stress. Compared to Higgins' single-pass emphasis, Kang's design prioritized integrated multistep automation while elucidating recovery limitations inherent to membrane filtration.

These complementary studies collectively advanced microfluidic CPA removal: Lusianti [60] established foundational models and single-pass feasibility, while Kang [89] optimized operational workflows and performance parameters. Lusianti [60] prioritized speed through simplified single-stage operation, whereas Kang [89] achieved superior clearance and viability via multistep integration at the cost of reduced recovery. Together, they transformed CPA removal from theoretical modeling to practical device implementation, systematically balancing processing efficiency and cellular damage.

3.2.3 Flow-Diffusion Method in Microfluidic Chips

Song et al. [22] pioneered the application of microfluidic technology to cell cryopreservation by introducing a method that leverages laminar flow and diffusion phenomena in a microchannel to gradually load and unload CPAs, thereby minimizing osmotic shock-induced cell damage (Fig. 7I–J). Their design featured a long, narrow microchannel, and through combined theoretical modeling and experimental validation, demonstrated a significant improvement in post-thaw cell viability (average increase of 25%). The study also incorporated cell membrane transport models to quantitatively analyze the effects of different CPA addition/removal protocols, laying the groundwork for subsequent research.

Zheng et al. [21] refined this approach by systematically investigating how microfluidic-based CPA addition/removal impacts cell survival after rapid cooling (e.g., vitrification) (Fig. 8A). They designed a serpentine microchannel that optimized mixing efficiency and concentration gradient control, as confirmed by experiments and simulations. Beyond osmotic damage reduction, their work analyzed additional factors like centrifugal stress, mechanical damage, and CPA toxicity, demonstrating microfluidics' ability to mitigate these combined injuries. Importantly, Zheng's system scaled up the technology to process bulk cell populations rather than individual or small cell clusters, significantly enhancing practical applicability.

While Song's work [22] established the fundamental principle of using diffusion-dominated flow for controlled CPA exchange, Zheng's [21] improvements addressed multi-factorial damage mechanisms and scalability challenges. Both studies validated their methods through integrated modeling and experimentation, but Zheng's advancements in vitrification compatibility and throughput represented critical progress toward clinical translation.

Complementing these channel-based approaches, Farooq et al. [90] developed an innovative surface acoustic wave (SAW)-driven microfluidic platform for CPA processing [90]. This open-droplet system utilized acoustically induced convective mixing to achieve

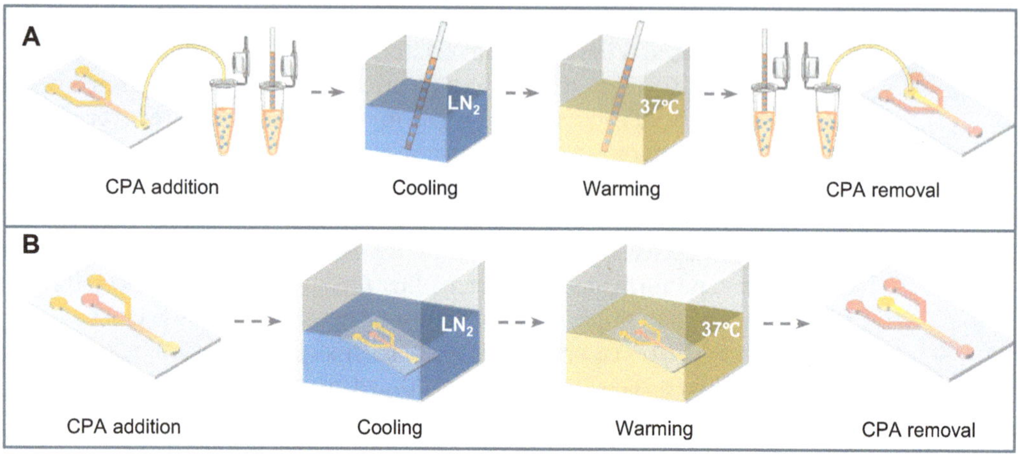

Fig. 8 On-chip cryopreservation. (**A**) On-chip CPA addition/removal for cell cryopreservation; (**B**) Integrated full-process on-chip cryopreservation

rapid (1-min) concentration adjustment of high CPA concentrations (e.g., 3 M Me_2SO), boosting cell viability by 24% compared to conventional methods. By eliminating centrifugation and enabling gradual concentration changes under physiological temperatures (37°C), the SAW technique concurrently reduced mechanical stress and osmotic shock while maintaining optimal thermal conditions.

3.3 On-Chip Cryopreservation

The application of microfluidic technology in cell freezing-thawing processes represents a transformative advancement in cryopreservation, effectively addressing critical limitations of conventional methods such as high cell damage rates and low operational standardization, through synergistic integration of microscale architecture and precision temperature control. This field has evolved along three key dimensions: from single-cell preservation to complex multicellular systems, from simple miniaturization to coordinated structural optimization and thermal regulation, and from isolated CPA handling (Fig. 8A) to fully integrated on-chip cryopreservation (Fig. 8B) workflows. These progressive innovations have not only created a powerful new technical platform for cryobiological research but have also established a standardized, high-throughput paradigm for cell preservation.

3.3.1 High-Efficiency Freeze-Thaw Driven by Minimized Feature Scale

Zou et al. [69] developed a CPA-free ultra-rapid freezing protocol specifically tailored for sperm cells. They designed PDMS microchannels with varying heights (10, 50, 100 μm) to achieve rapid freezing through extremely small sample volumes (5×10^{-3} μL) and ultrahigh surface-area-to-volume ratios. The entire freeze-thaw process was completed on-chip: sperm were injected into a PDMS microchannel and directly immersed in liquid nitrogen for CPA-free ultra-rapid freezing, followed by thawing in a 37°C incubator for 10 min before retrieval via syringe. Results demonstrated that sperm preserved in 10 μm channels exhibited comparable survival rates, motility, and DNA integrity to conventional freezing methods, while larger channels led to increased cellular damage. This study marked the first achievement of CPA-free freezing, though limited to small-volume sperm samples.

Li et al. [91] proposed a straightforward PDMS-glass chip design featuring straight channels (two dimensions: 100 and 60 μm in height). Cell suspensions were mixed with CPA containing Me_2SO, injected into the chip, and stored at −80°C. Thawing was achieved by directly placing the chip in a 37°C water bath, followed by on-chip cell evaluation using Calcein-AM/EthD-1 live/dead staining combined with microscopic imaging—eliminating the need for cell transfer while enabling viability assessment and observation of adherent growth. The study validated the survival and growth of various cell types (HUVECs, 3T3, SKBR3, and HepatG2) after preservation periods ranging from 10 days to 1 year, with viability exceeding 98%, though cell numbers significantly declined after long-term storage.

3.3.2 Integrated CPA Processing and Freeze-Thaw on Microfluidic Chips

Recent advances have demonstrated the feasibility of complete on-chip cryopreservation workflows. Li et al. [92] developed a multifunctional laser-induced graphene (LIG)/PDMS hybrid chip featuring a Y-shaped microchannel with superhydrophobic surfaces (157.8° contact angle) and photothermal properties (71.8°C in 10 s under 808 nm laser). This platform integrated CPA addition, freezing, and rapid rewarming (excluding CPA removal), achieving 92.4% survival for mouse oocytes with half-strength CPAs (7.5% EG+7.5% Me_2SO) and 88.3% without Me_2SO—surpassing conventional OPS methods (85.8%). The chip's antibacterial function enabled 100% sterilization with 30-s laser exposure.

Zhou et al. [93] pioneered full-process on-chip cryopreservation for porcine oocytes (100–150 μm) (Fig. 8B), comparing hybrid (microfluidic CPA handling + conventional carriers) and fully integrated approaches. Their Y-channel PDMS chip enabled continuous gradient CPA exchange (concave addition-convex removal), achieving 89.45% survival when combined with Cryotop carriers—outperforming stepwise methods. While PDMS-alumina ceramic chips (46 W/m·K thermal conductivity) showed better results (59.79%) than glass chips (33.40%), they didn't surpass the OPS performance.

Building on this, Xu et al. [94] achieved 86.23% survival for mouse oocytes (70 μm) using similar PDMS-alumina structures, with excellent outcomes in cellular ultrastructure, mitochondrial function, and gene expression.

These consecutive studies represent the critical evolution from partial to complete on-chip cryopreservation systems, addressing multiple technical challenges through material innovations like LIG's dual functional surfaces, improved thermal management using ceramic substrates, optimized CPA exchange protocols, and rigorous biological validation.

3.3.3 Cryopreservation with On-Chip Evaluation and Direct Application

Deutsch et al. [95] pioneered the "individual-cell-based cryo-chip" (i3C) system, featuring transparent hexagonal microwell arrays (20/100 μm diameter) constructed from polycarbonate and BK7 glass. This innovative platform enabled in situ addition and removal of CPA (10% Me_2SO) through integrated microfluidics while supporting both slow (1°C/min) and rapid (40°C/min) freezing protocols down to −80°C or liquid nitrogen temperatures. The system offered dual rewarming approaches: rapid rewarming at 30–40°C/min (~3 min) using an integrated cold stage, or conventional 37°C water bath thawing. Experimental results demonstrated comparable cell viability (91.3%) to conventional methods during slow freezing, while significantly outperforming microplates (83.6% vs. 51.4%) in rapid freezing scenarios, with exceptional cell positioning stability (>97%).

Complementing this work, Afrimzon et al. [96] conducted comprehensive functional assessments of cryopreserved cells, revealing that i3C-processed U937 cells showed no significant differences in mitochondrial membrane potential, apoptosis rates, or metabolic activity compared to unfrozen controls. Notably, the chip enabled multiparameter dynamic tracking of individual cells, including observations of mitochondrial function recovery within 1.5 h post-thaw. These studies collectively established that the i3C system, through its integration of microfluidic manipulation and rapid thermal control, could simultaneously maintain precise cell localization and functional integrity, representing a breakthrough solution for single-cell-level cryopreservation.

Advancing this technology further, Lv et al. [97] implemented a microcolumn array (5 μm spacing) design within microfluidic chips to physically trap cells during CPA exchange [97]. The chip's elliptical and circular micropillar structures allowed cell-Me_2SO mixtures to be frozen at −80°C, with post-thaw medium replacement facilitated by the pillars' cell-retention capability. Validated on breast cancer cells (MDA-MB-231 and T47D) across 15-day to 1-year preservation periods, the system demonstrated decreasing but substantial cell recovery rates over time (84% at 15 days), with the microcolumn design effectively minimizing cell loss. This work successfully combined CPA engineering with structural cell retention to streamline operational workflows, opening new possibilities for integrated experimental platforms.

3.3.4 Integrated Cryopreservation of Multicellular Constructs on Microfluidic Platforms

The field has witnessed significant progress in developing integrated solutions for preserving complex cellular architectures through innovative microfluidic designs. Kondo et al. [98] established an early PDMS-based microfluidic system integrated with culture dishes, incorporating microcavity structures for cell immobilization. Their approach enabled direct cryopreservation of multiple adherent cell types (HeLa, NIH3T3, MCF-7, and PC12) cultured on-chip using 10% Me_2SO as CPA, with subsequent storage at −80°C for 36 h. This method demonstrated superior cell viability compared to conventional dishes while maintaining critical cellular properties—PC12 cells retained differentiation capacity and HeLa cells preserved drug sensitivity profiles post-thaw.

Kim et al. [99] developed the advanced "ReadyChip" platform, which combines nanofiber membrane technology with organ-on-chip capabilities to create a true "thaw-and-use" system. Constructed from PMMA with integrated PCL/PF108 electrospun nanofibers and collagen ECM coatings, this design faithfully replicates physiological microenvironments. The optimized cryopreservation protocol employs a controlled slow-freezing rate (−1°C/min to −80°C) using a 5% Me_2SO/95% FBS CPA solution, achieving exceptional post-thaw viability exceeding 92% for HUVECs, ARPE-19, and HBEC-5i cells. Comprehensive characterization confirmed preservation of tight junction proteins (ZO-1), specific cellular markers (CD31), and critical barrier functions.

These integrated platforms represent a paradigm shift by combining cell culture, cryopreservation, and immediate application into seamless workflows. Key advantages include: maintenance of native cell-ECM interactions throughout freeze-thaw cycles, conservation of differentiation potential and pharmacological responses, standardization of experimental procedures, and significant reduction in manual cell manipulation steps. The ReadyChip system particularly stands out for its ability to maintain complex cellular architectures and functions while offering unprecedented convenience for applications ranging from drug screening to disease modeling. By addressing both preservation challenges and experimental workflow requirements, these integrated solutions are paving the way for more reliable and reproducible cell-based research and applications.

4 Summary and Outlook

Microfluidic technology has brought revolutionary breakthroughs to the field of cryopreservation. By enabling precise control of microscale heat and mass transfer, accurate addition and removal of CPAs, and minimized sample volumes with integrated chip designs, it has significantly improved cell survival rates and functional integrity during freezing. The technology demonstrates unique advantages in high-throughput single-cell permeability measurement, ultra-rapid cooling and rewarming, and the preservation of multicellular constructs. In the future, this technology will advance toward intelligent, standardized, and clinically practical applications. By integrating artificial intelligence to optimize cryopreservation protocols, developing low-toxicity biomimetic CPA materials, and innovating chip designs for thermal conductivity and biocompatibility, it will promote the establishment of a "full-process, ready-to-use" preservation system—from single cells to complex living constructs or cultures. Additionally, breakthroughs will be needed in the clinical translation of automated, closed systems, as well as the exploration of novel cryopreservation methods under chip-based physical field modulation, high/low pressure, microgravity, and other extreme conditions. Ultimately, these advancements will provide more reliable technical support for regenerative medicine, reproductive medicine, and biobanking, enabling a leap from basic research to clinical applications.

References

1. Meneghel J, Kilbride P, Morris G (2020) Cryopreservation as a key element in the successful delivery of cell-based therapies—a review. Front Med 7:592242. https://doi.org/10.3389/fmed.2020.592242
2. Adamson GD, Dyer S, Zegers-Hochschild F, Chambers G, De Mouzon J, Ishihara O, Kupka M, Baker V, Banker M, Elgindy E, Bai F, Jwa SC (2023) ICMART Preliminary World Repreliminary world report 2019. In: Paper presented at the 39th Annual Meeting of the European Society of Human Reproduction and Embryology (ESHRE), Copenhagen, Denmark, 2023, June 25–28. https://doi.org/10.1093/humrep/dead093.187
3. Cornwall W (2023) Frozen in time. Science 380:1313–1317. https://doi.org/10.1126/science.adj4220
4. Giwa S, Lewis JK, Alvarez L, Langer R, Roth AE, Church GM, Markmann JF, Sachs DH, Chandraker A, Wertheim JA, Rothblatt M, Boyden ES, Eidbo E, Lee WPA, Pomahac B, Brandacher G, Weinstock DM, Elliott G, Nelson D, Acker JP, Uygun K, Schmalz B, Weegman BP, Tocchio A, Fahy GM, Storey KB, Rubinsky B, Bischof J, Elliott JAW, Woodruff TK, Morris GJ, Demirci U, Brockbank KGM, Woods EJ, Ben RN, Baust JG, Gao D, Fuller B, Rabin Y, Kravitz DC, Taylor MJ, Toner M (2017) The promise of organ and tissue preservation to transform medicine. Nat Biotechnol 35:530–542. https://doi.org/10.1038/nbt.3889
5. Food NMoAa (2025) Svalbard Global Seed Vault. Retrieved. https://www.seedvault.no/. Accessed 6 Jun 2025
6. Biobank U (2023) UK Biobank newsletter homepage 2023/24 [Newsletter]. UK Biobank. https://www.ukbiobank.ac.uk/explore-your-participation/stay-involved/uk-biobank-newsletter-homepage-2023-24. Accessed 6 Jun 2025
7. Mazur P (1963) Kinetics of water loss from cells at subzero temperatures and likelihood of intracellular freezing. J Gen Physiol 47:347–369. https://doi.org/10.1085/jgp.47.2.347
8. Mazur P, Leibo SP, Chu EHY (1972) A two-factor hypothesis of freezing injury: evidence from Chinese hamster tissue-culture cells. Exp Cell Res 71(2):345–355. https://doi.org/10.1016/0014-4827(72)90303-5
9. Gao D, Critser JK (2000) Mechanisms of Cryoinjury in living cells. ILAR J 41:187–196. https://doi.org/10.1093/ilar.41.4.187
10. Cottle C, Porter AP, Lipat A, Turner-Lyles C, Nguyen J, Moll G, Chinnadurai R (2022) Impact of cryopreservation and freeze-thawing on therapeutic properties of mesenchymal stromal/stem cells and other common cellular therapeutics. Curr Stem Cell Rep 8:72–92. https://doi.org/10.1007/s40778-022-00212-1
11. Ding YQ, Liu S, Liu J, Jin S, Wang J (2024) Cryopreservation with DMSO affects the DNA integrity, apoptosis, cell cycle and function of human bone mesenchymal stem cells. Cryobiology 114:104847. https://doi.org/10.1016/j.cryobiol.2024.104847

12. Zhao G, Fu JP (2017) Microfluidics for cryopreservation. Biotechnol Adv 35:323–336. https://doi.org/10.1016/j.biotechadv.2017.01.006
13. Lusianti RE, Benson JD, Acker JP, Higgins AZ (2013) Rapid removal of glycerol from frozen-thawed red blood cells. Biotechnol Prog 29:609–620. https://doi.org/10.1002/btpr.1710
14. Shen LX, Guo X, Ouyang X, Huang Y, Gao D, Zhao G (2021) Fine-tuned dehydration by trehalose enables the cryopreservation of RBCs with unusually low concentrations of glycerol. J Mater Chem B 9:295–306. https://doi.org/10.1039/d0tb02426k
15. ISO (2018) ISO 20387:2018 biotechnology-biobanking-general requirements for biobanking (2018) Geneva: International Organization for Standardization. https://www.iso.org/obp/ui/en/#iso:std:iso:20387:ed-1:v1:en. Accessed 6 June 2025
16. Alpha Scientists in Reproductive Medicine and ESHRE Special Interest Group of Embryology (2011) The Istanbul consensus workshop on embryo assessment: proceedings of an expert meeting. Hum Reprod 26:1270–1283. https://doi.org/10.1093/humrep/der037
17. Schiewe MC, Anderson RE (2017) Vitrification: the pioneering past to current trends and perspectives of cryopreserving human embryos, gametes and reproductive tissue. J Biorepository Sci Appl Med 5:57–68. https://doi.org/10.2147/BSAM.S139376
18. Mata C, Longmire EK, McKenna DH, Glass KK, Hubel A (2008) Experimental study of diffusion-based extraction from a cell suspension. Microfluid Nanofluid 5:529–540. https://doi.org/10.1007/s10404-008-0265-9
19. Fleming KK, Longmire EK, Hubel A (2007) Numerical characterization of diffusion-based extraction in cell-laden flow through a microfluidic channel. J Biomech Eng 129:703–711. https://doi.org/10.1115/1.2768373
20. Whitesides GM (2006) The origins and the future of microfluidics. Nature 442:368–373. https://doi.org/10.1038/nature05058
21. Zheng Y, Zhao G, Zhang YT, Gao R (2018) On-chip loading and unloading of cryoprotectants facilitate cell cryopreservation by rapid freezing. Sens Actuators B Chem 255:647–656. https://doi.org/10.1016/j.snb.2017.08.084
22. Song YS, Moon S, Hulli L, Hasan SK, Kayaalp E, Demirci U (2009) Microfluidics for cryopreservation. Lab Chip 9:1874–1881. https://doi.org/10.1039/b823062e
23. Takamatsu H, Komori Y, Zawlodzka S, Fujii M (2004) Quantitative examination of a perfusion microscope for the study of osmotic response of cells. J Biomech Eng 126:402–409. https://doi.org/10.1115/1.1784474
24. Lai D, Ding J, Smith GW, Smith GD, Takayama S (2015) Slow and steady cell shrinkage reduces osmotic stress in bovine and murine oocyte and zygote vitrification. Hum Reprod 30:37–45. https://doi.org/10.1093/humrep/deu284
25. Heo YS, Lee HJ, Hassell BA, Irimia D, Toth TL, Elmoazzen H, Toner M (2011) Controlled loading of cryoprotectants (CPAs) to oocyte with linear and complex CPA profiles on a microfluidic platform. Lab Chip 11:3530–3537. https://doi.org/10.1039/c1lc20377k
26. Lee CY, Fu LM (2018) Recent advances and applications of micromixers. Sens Actuators B Chem 259:677–702. https://doi.org/10.1016/j.snb.2017.12.034
27. Zhao G, Huang Z, Gao D (2016) Microdevices for measurement of cell membrane biophysical properties. In: He X, Bischof JC (Eds.) Multiscale Technologies for Cryomedicine: Implementation from Nano to Macroscale. World Scientific Publishing Co. Pte. Ltd., Singapore, 221–273
28. Ko J, Lee J (2025) Advanced microfluidic systems with temperature modulation for biological applications. Biomicrofluidics 19:031301. https://doi.org/10.1063/5.0251893
29. Yager P (2025) Microfluidics tutorial: basic concepts. University of Washington. https://faculty.washington.edu/yagerp/microfluidicstutorial/basicconcepts/basicconcepts.htm
30. Fan J, Li S, Wu Z, Chen Z (2019) Diffusion and mixing in microfluidic devices. In: Santos HA, Liu D, Zhang H (eds) Microfluidics for pharmaceutical applications. William Andrew Publishing, pp 79–100
31. Kandlikar, SG, 2006. Single-phase liquid flow in minichannels and microchannels, In: Heat Transfer and Fluid Flow in Minichannels and Microchannels. Elsevier, pp. 87–136
32. Bruus H (1997) Theoretical microfluidics. Oxford University Press
33. Morini GL (2004) Single-phase convective heat transfer in microchannels: a review of experimental results. Int J Therm Sci 43:631–651. https://doi.org/10.1016/j.ijthermalsci.2004.01.003
34. Chai L, Xia G, Wang L, Zhou M, Cui Z (2013) Heat transfer enhancement in microchannel heat sinks with periodic expansion-constriction cross-sections. Int J Heat Mass Transfer 62:741–751. https://doi.org/10.1016/j.ijheatmasstransfer.2013.03.045
35. Maranzana G, Perry I, Maillet D (2004) Mini- and micro-channels: influence of axial conduction in the walls. Int J Heat Mass Transfer 47:3993–4004. https://doi.org/10.1016/j.ijheatmasstransfer.2004.04.016

36. Squires TM, Quake SR (2005) Microfluidics: fluid physics at the nanoliter scale. Rev Mod Phys 77:977–1026. https://doi.org/10.1103/revmodphys.77.977
37. Bird, RB, Stewart, WE, Lightfoot, EN, 2002. Transport phenomena, 2nd edn. Wiley, New York
38. Beebe DJ, Mensing GA, Walker GM (2002) Physics and applications of microfluidics in biology. Annu Rev Biomed Eng 4:261–286. https://doi.org/10.1146/annurev.bioeng.4.112601.125916
39. Hanna J, Hubel A, Lemke E (2012) Diffusion-based extraction of DMSO from a cell suspension in a three stream, vertical microchannel. Biotechnol Bioeng 109:2316–2324. https://doi.org/10.1002/bit.24499
40. McGrath JJ (1997) Quantitative measurement of cell membrane transport: technology and applications. Cryobiology 34:315–334. https://doi.org/10.1006/cryo.1997.2013
41. Zhao G, Zhang Z, Zhang Y, Chen Z, Niu D, Cao Y, He X (2017) A microfluidic perfusion approach for on-chip characterization of the transport properties of human oocytes. Lab Chip 17:1297–1305. https://doi.org/10.1039/c6lc01532h
42. Gao DY, Benson CT, Liu C, McGrath JJ, Critser ES, Critser JK (1996) Development of a novel microperfusion chamber for determination of cell membrane transport properties. Biophys J 71:443–450. https://doi.org/10.1016/s0006-3495(96)79246-9
43. McGrath JJ (1985) A microscope diffusion chamber for the determination of the equilibrium and non-equilibrium osmotic response of individual cells. J Microsc 139(3):249–263. https://doi.org/10.1111/j.1365-2818.1985.tb02641.x
44. Tseng HY, Sun S, Shu Z, Ding W, Reems JA, Gao D (2011) A microfluidic study of megakaryocytes membrane transport properties to water and dimethyl sulfoxide at suprazero and subzero temperatures. Biopreserv Biobank 9:355–362. https://doi.org/10.1089/bio.2011.0027
45. Chen HH, Lin EH, Heimfeld S, Gao D (2008) A microfluidic study of mouse dendritic cell membrane transport properties of water and cryoprotectants. Int J Heat Mass Transfer 51:5687–5694. https://doi.org/10.1016/j.ijheatmasstransfer.2008.04.013
46. Fang CF, Ji F, Shu Z, Gao D (2017) Determination of the temperature-dependent cell membrane permeabilities using microfluidics with integrated flow and temperature control. Lab Chip 17:951–960. https://doi.org/10.1039/c6lc01523a
47. Lei ZL, Xie D, Mbogba MK, Chen Z, Tian C, Xu L, Zhao G (2019) A microfluidic platform with cell-scale precise temperature control for simultaneous investigation of the osmotic responses of multiple oocytes. Lab Chip 19:1929–1940. https://doi.org/10.1039/c9lc00107g
48. Zhao G, Kurata K, Takamatsu H (2012) Measurement of membrane hydraulic conductivity of bovine carotid artery endothelial cells using a perfusion microscope. CryoLetters 33:232–240
49. Liu W, Zhao G, Shu ZQ, Wang T, Zhu K, Gao D (2015) High-precision approach based on microfluidic perfusion chamber for quantitative analysis of biophysical properties of cell membrane. Int J Heat Mass Transfer 86:869–879. https://doi.org/10.1016/j.ijheatmasstransfer.2015.03.038
50. Huang L, Benson JD, Almasri M (2021) Microfluidic measurement of individual cell membrane water permeability. Anal Chim Acta 1163:338441. https://doi.org/10.1016/j.aca.2021.338441
51. Huang L, Jasim I, Alkorjia O, Agca C, Oksman A, Agca Y, Goldberg DE, Benson JD, Almasri M (2023) An impedance based microfluidic sensor for evaluation of individual red blood cell solute permeability. Anal Chim Acta 1267:341226. https://doi.org/10.1016/j.aca.2023.341226
52. Wu Y, Benson JD, Critser JK, Almasri M (2010) MEMS-based Coulter counter for cell counting and sizing using multiple electrodes. J Micromech Microeng 20:085035. https://doi.org/10.1088/0960-1317/20/8/085035
53. Wu Y, Benson JD, Almasri M (2012) Micromachined Coulter counter for dynamic impedance study of time sensitive cells. Biomed Microdevices 14:739–750. https://doi.org/10.1007/s10544-012-9655-6
54. Gao D, Zhou X (2012) Prevention of lethal osmotic injury to cells during addition and removal of cryoprotective agents: theory and technology. In: Katkov II (ed) Current frontiers in cryobiology. IntechOpen, Rijeka. https://doi.org/10.5772/34473
55. Fleming Glass KK, Longmire EK, Hubel A (2008) Optimization of a microfluidic device for diffusion-based extraction of DMSO from a cell suspension. Int J Heat Mass Transfer 51:5749–5757. https://doi.org/10.1016/j.ijheatmasstransfer.2008.04.018
56. Zhou X, Du Y, Yi X, Dai J, Zhang D (2019) Optimization of cryoprotectants addition and removal protocols using microfluidic chip. Chinese J Biomed Eng 38:719–725. https://doi.org/10.3969/j.issn.0258-8021.2019.06.010

57. Yang Y, Zhou XL, Zhou NF, Shao WQ, Tao LR (2017) Numerical and experimental investigation on water-Me_2SO mixing in microfluidic device for oocyte cryopreservation. CryoLetters 38:37–42
58. Benson JD, Kearsley AJ, Higgins AZ (2012) Mathematical optimization of procedures for cryoprotectant equilibration using a toxicity cost function. Cryobiology 64:144–151. https://doi.org/10.1016/j.cryobiol.2012.01.001
59. Davidson AF, Glasscock C, McClanahan DR, Benson JD, Higgins AZ (2015) Toxicity minimized cryoprotectant addition and removal procedures for adherent endothelial cells. PLoS One 10:e0142828. https://doi.org/10.1371/journal.pone.0142828
60. Lusianti RE, Higgins AZ (2014) Continuous removal of glycerol from frozen-thawed red blood cells in a microfluidic membrane device. Biomicrofluidics 8:054124. https://doi.org/10.1063/1.4900675
61. Demirci U, Montesano G (2007) Cell encapsulating droplet vitrification. Lab Chip 7:1428. https://doi.org/10.1039/b705809h
62. Zhao G, Liu X, Zhu K, He X (2017) Hydrogel encapsulation facilitates rapid-cooling cryopreservation of stem cell-laden core-shell microcapsules as cell-biomaterial constructs. Adv Healthcare Mater 6:1700988. https://doi.org/10.1002/adhm.201700988
63. Sakai A, Engelmann F (2007) Vitrification, encapsulation-vitrification and droplet-vitrification: a review. CryoLetters 28:151–172
64. Cui M, Zhan T, Yang J, Dang H, Yang G, Han H, Liu L, Xu Y (2023) Droplet generation, vitrification, and warming for cell cryopreservation: a review. ACS Biomater Sci Eng 9:1151–1163. https://doi.org/10.1021/acsbiomaterials.2c01087
65. Akiyama Y, Shinose M, Watanabe H, Yamada S, Kanda Y (2019) Cryoprotectant-free cryopreservation of mammalian cells by superflash freezing. Proc Natl Acad Sci USA 116:7738–7743. https://doi.org/10.1073/pnas.1808645116
66. Shang L, Cheng Y, Zhao Y (2017) Emerging droplet microfluidics. Chem Rev 117:7964–8040. https://doi.org/10.1021/acs.chemrev.6b00848
67. Liu X, Zhao G, Chen Z, Panhwar F, He X (2018) Dual suppression effect of magnetic induction heating and microencapsulation on ice crystallization enables low-cryoprotectant vitrification of stem cell-alginate hydrogel constructs. ACS Appl Mater Interfaces 10:16822–16835. https://doi.org/10.1021/acsami.8b04496
68. Qin X, Chen Z, Shen L, Liu H, Ouyang X, Zhao G (2023) Core-shell microfiber encapsulation enables glycerol-free cryopreservation of RBCs with high hematocrit. Nanomicro Lett 16:3. https://doi.org/10.1007/s40820-023-01213-3
69. Zou YJ, Yin T, Chen S, Yang J, Huang W (2013) On-chip cryopreservation: a novel method for ultra-rapid cryoprotectant-free cryopreservation of small amounts of human spermatozoa. PLoS One 8:e61593. https://doi.org/10.1371/journal.pone.0061593
70. Burns MA, Johnson BN, Brahmasandra SN, Handique K, Webster JR, Krishnan M, Sammarco TS, Man PM, Jones D, Heldsinger D, Mastrangelo CH, Burke DT (1998) An integrated nanoliter DNA analysis device. Science 282:484–487. https://doi.org/10.1126/science.282.5388.484
71. Kopp MU, Mello AJd, Manz A (1998) Chemical amplification: continuous-flow PCR on a chip. Science 280:1046–1048. https://doi.org/10.1126/science.280.5366.1046
72. Hoera C, Ohla S, Shu Z, Beckert E, Nagl S, Belder D (2014) An integrated microfluidic chip enabling control and spatially resolved monitoring of temperature in micro flow reactors. Anal Bioanal Chem 407:387–396. https://doi.org/10.1007/s00216-014-8297-3
73. Miralles V, Huerre A, Malloggi F, Jullien MC (2013) A review of heating and temperature control in microfluidic systems: techniques and applications. Diagnostics 3:33–67. https://doi.org/10.3390/diagnostics3010033
74. Peng J, Fang C, Ren S, Pan J, Jia Y, Shu Z, Gao D (2019) Development of a microfluidic device with precise on-chip temperature control by integrated cooling and heating components for single cell-based analysis. Int J Heat Mass Transfer 130:660–667. https://doi.org/10.1016/j.ijheatmasstransfer.2018.10.135
75. Fang C, Lee D, Stober B, Fuller GG, Shen AQ (2015) Integrated microfluidic platform for instantaneous flow and localized temperature control. RSC Adv 5:85620–85629. https://doi.org/10.1039/c5ra19944a
76. Zhou X, Liu Z, Liang XM, Shu Z, Du P, Gao D (2013) Theoretical investigations of a novel microfluidic cooling/warming system for cell vitrification cryopreservation. Int J Heat Mass Transfer 65:381–388. https://doi.org/10.1016/j.ijheatmasstransfer.2013.06.022
77. Stan CA, Schneider GF, Shevkoplyas SS, Hashimoto M, Ibanescu M, Wiley BJ, Whitesides GM (2009) A microfluidic apparatus for the study of ice nucleation in supercooled water drops. Lab Chip 9:2293. https://doi.org/10.1039/b906198c
78. Chen ZR, Memon K, Cao Y, Zhao G (2020) A microfluidic approach for synchronous and nondestructive study of the permeability of multiple

oocytes. Microsyst Nanoeng 6:55. https://doi.org/10.1038/s41378-020-0160-4
79. Deck L-T, Shardt N, El-Bakouri I, Isenrich FN, Marcolli C, deMello AJ, Mazzotti M (2024) Monitoring aqueous sucrose solutions using droplet microfluidics: ice nucleation, growth, glass transition, and melting. Langmuir 40:6304–6316. https://doi.org/10.1021/acs.langmuir.3c03798
80. Zhao G (2010) A universal self-adaptive time-varying function for extracellular concentration during osmotic shift for curve-fitting permeability coefficients of cell membrane. CryoLetters 31:279–290
81. Wang JY, Zhu K, Zhao G, Ren J, Yue C, Gao D (2013) Dual dependence of cryobiogical properties of Sf21 cell membrane on the temperature and the concentration of the cryoprotectant. PLoS One 8:e72836. https://doi.org/10.1371/journal.pone.0072836
82. Wang JY, Zhao G, Zhang P, Wang Z, Zhang Y, Gao D, Zhou P, Cao Y (2014) Measurement of the biophysical properties of porcine adipose-derived stem cells by a microperfusion system. Cryobiology 69:442–450. https://doi.org/10.1016/j.cryobiol.2014.10.002
83. Niu D, Zhao G, Liu X, Zhou P, Cao Y (2016) Prevention of osmotic injury to human umbilical vein endothelial cells for biopreservation: a first step toward biobanking of endothelial cells for vascular tissue engineering. Tissue Eng Part C Methods 22:270–279. https://doi.org/10.1089/ten.tec.2015.0364
84. Wang JY, Zhao G, Shu Z, Zhou P, Cao Y, Gao D (2016) Effect of iron oxide nanoparticles on the permeability properties of Sf21 cells. Cryobiology 72:21–26. https://doi.org/10.1016/j.cryobiol.2015.12.002
85. Li L, Chen Z, Zhang M, Panhwar F, Gao C, Zhao G, Jin B, Ye B (2017) Cell membrane permeability coefficients determined by single-step osmotic shift are not applicable for optimization of multi-step addition of cryoprotective agents: as revealed by HepG2 cells. Cryobiology 79:82–86. https://doi.org/10.1016/j.cryobiol.2017.09.006
86. Chen ZR, Zhang Z, Guo X, Memon K, Panhwar F, Wang M, Cao Y, Zhao G (2019) Sensing cell membrane biophysical properties for detection of high quality human oocytes. ACS Sens 4:192–199. https://doi.org/10.1021/acssensors.8b01215
87. Chen HH, Purtteman JJ, Heimfeld S, Folch A, Gao D (2007) Development of a microfluidic device for determination of cell osmotic behavior and membrane transport properties. Cryobiology 55:200–209. https://doi.org/10.1016/j.cryobiol.2007.08.001
88. Zhan L, Hinnen H, Gopinathan KA, Toner M (2025) Autonomous cryoprotectant loading of the oocyte using microfluidic transistors. Device 3:100715. https://doi.org/10.1016/j.device.2025.100715
89. Kang YF, Zou LL, Qiu BS, Liang X, Sun S, Gao D, Ding W (2017) Unloading of cryoprotectants from cryoprotectant-loaded cells on a microfluidic platform. Biomed Microdevices 19:15. https://doi.org/10.1007/s10544-017-0155-6
90. Farooq U, Haider Z, Liang XM, Memon K, Hossain SMC, Zheng Y, Xu H, Qadir A, Panhwar F, Dong S, Zhao G (2019) Surface-acoustic-wave-based lab-on-chip for rapid transport of cryoprotectants across cell membrane for cryopreservation with significantly improved cell viability. Small 15:1805361. https://doi.org/10.1002/smll.201805361
91. Li L, Lv XQ, Guo H, Shi X, Liu J (2014) On-chip direct freezing and thawing of mammalian cells. RSC Adv 4:34443–34447. https://doi.org/10.1039/c4ra06972b
92. Li YF, Zhang JX, Han W, Liu B, Zhai M, Li N, Wang Z, Zhao J (2024) Multifunctional laser-induced graphene-based microfluidic chip for high-performance oocyte cryopreservation with low concentration of cryoprotectants. Adv Healthc Mater 13:2400981. https://doi.org/10.1002/adhm.202400981
93. Zhou XL, Guo YY, Yi XY et al. (2018) Experimental study of microfluidic chip for cryopreservation of oocytes. Prog Biochem Biophys 45:763–771. https://doi.org/10.16476/j.pibb.2018.0016
94. Xu SY, Lei Y, Wei ML, Fang Z, Cheng Y, Zhao G, Chen Z (2025) Transparent alumina ceramics-based microfluidic chip enables on-chip cryopreservation for mouse oocyte. Adv Funct Mater 35:2503296. https://doi.org/10.1002/adfm.202503296
95. Deutsch M, Afrimzon E, Namer Y, Shafran Y, Sobolev M, Zurgil N, Deutsch A, Howitz S, Greuner M, Thaele M, Zimmermann H, Meiser I, Ehrhart F (2010) The individual-cell-based cryo-chip for the cryopreservation, manipulation and observation of spatially identifiable cells. I: Methodology. BMC Cell Biol 11: 54. https://doi.org/10.1186/1471-2121-11-54
96. Afrimzon E, Zurgil N, Shafran Y, Ehrhart F, Namer Y, Moshkov S, Sobolev M, Deutsch A, Howitz S, Greuner M, Thaele M, Meiser I, Zimmermann H, Deutsch M (2010) The individual-cell-based cryo-chip for the cryopreservation, manipulation and observation of spatially identifiable cells. II: Functional activity of cryopreserved cells. BMC Cell Biol 11: 83. https://doi.org/10.1186/1471-2121-11-83

97. Lv XQ, Ma ZT, Guo L (2023) Freezing and thawing of cells on a microfluidic device: a simple and time-saving experimental procedure. Biosci Biotechnol Biochem 87:1478–1484. https://doi.org/10.1093/bbb/zbad123
98. Kondo E, Wada K, Hosokawa K, Maeda M (2016) Cryopreservation of adhered mammalian cells on a microfluidic device: Toward ready-to-use cell-based experimental platforms. Biotechnol Bioeng 113:237–240. https://doi.org/10.1002/bit.25704
99. Kim D, Yi S, Park BU, Lee SJ, Lee J, Kim DS, Kim HK (2025) Ready-to-use nanofiber membrane-integrated organ-on-a-chip via cryopreservation of cell-loaded microfluidic Chip. ACS Macro Lett 14:622–629. https://doi.org/10.1021/acsmacrolett.5c00129

Check for updates

Chapter 8

Temperature Controlled Cryoprinting

Leo Lou, Linnea Warburton and Boris Rubinsky

Abstract

This chapter reviews the development of temperature-controlled cryoprinting (TCC), a bioprinting approach that integrates extrusion with controlled freezing to stabilize constructs during fabrication. Conventional extrusion printing struggles to balance ink viscosity, mechanical fidelity, and biological compatibility, often necessitating secondary stabilization steps that compromise structural integrity or viability. TCC overcomes these limitations by freezing each voxel under regulated thermal conditions, maintaining structural stability throughout the printing process and eliminating the need for post-print freezing. The review traces the conceptual origins of TCC from early cryobiology research on ice physics and cell survival, through proof-of-concept demonstrations of cryogenic 3D printing, to the progressive refinement of methods including multi-layer cryolithography, freezing-modulated crosslinking, and modular cryoprinting systems. It summarizes material innovations such as alginate–agar inks and explores how freezing history influences diffusion and microstructural anisotropy. Applications are presented in tissue engineering, where TCC supports cryopreservation of cell-laden scaffolds and vascularizable tissues, and in food engineering, where it enables production of customized frozen foods with reproducible textures. The intellectual property framework that has developed alongside the research is also reviewed. Together, these studies establish TCC as an approach that unites fabrication and preservation, with implications for both biomedical and food applications.

Keywords Biomedical engineering, Cryoprinting, Food preservation, Laser-assisted bioprinting, Stereolithography bioprinting

1 3D Bioprinting Technology

The origins of three-dimensional (3D) printing trace back to the early 1980s, when American engineer Chuck Hull introduced stereolithography (SLA), the first process for layer-by-layer fabrication of solid objects from digital designs [1]. Since then, 3D printing has evolved into a versatile platform that now encompasses biological applications. When applied to biological materials, the technology is commonly referred to as bioprinting in medical contexts and biofabrication in broader settings. This emerging field holds transformative potential across multiple domains, most prominently in tissue engineering and food technology [2–6]. By enabling the precise, layer-by-layer

Willem F. Wolkers and Harriëtte Oldenhof (eds.), *Cryopreservation and Freeze-Drying Protocols*, Methods in Molecular Biology, vol. 3060, https://doi.org/10.1007/978-1-0716-5416-3_8,

assembly of complex structures composed of cells, biomaterials, and bioactive compounds, 3D printing opens new opportunities to create functional biological constructs that closely mimic the architecture and behavior of natural tissues or food products.

In tissue engineering, the integration of living cells with biocompatible scaffolds through 3D printing is advancing the development of engineered tissues and organs for regenerative medicine, drug discovery, and disease modeling [7]. Bioink formulations, consisting of living cells embedded within hydrogels or extracellular matrix analogs, are extruded to produce constructs that support cell viability, differentiation, and functional integration [8]. This strategy addresses the critical shortage of donor organs and paves the way for personalized therapies by enabling the fabrication of immunologically compatible implants from patient-derived cells. Furthermore, bioprinted tissue models are increasingly being adopted in pharmaceutical research as alternatives to animal testing, offering improved predictive accuracy for human responses [9].

Simultaneously, 3D printing is reshaping food engineering by enabling the design and fabrication of structured, customized, and nutritionally optimized food products [10–18]. Unlike conventional food-processing methods, 3D food printing offers unprecedented control over composition, texture, flavor, and appearance at the microscale. This has important implications for enhancing palatability, meeting dietary restrictions, and reducing food waste. For example, plant-based proteins, cultured cells, and other bio-based ingredients can be printed into meat analogs or novel food formats with improved sensory properties and nutritional profiles. The approach also supports advances in personalized nutrition, allowing the development of foods tailored to individual metabolic needs, allergies, or medical conditions.

Although food and biomedical applications differ in their ultimate goals, they share common challenges. Both require printable biomaterials (bioinks or food inks) with appropriate rheological, mechanical, and biological properties; robust methods to ensure sterility and scalability; and strategies to preserve functionality throughout printing and post-processing. In addition, both fields face complex regulatory, ethical, and consumer acceptance hurdles as they move closer to clinical or commercial deployment.

Recent progress in materials science, automation, artificial intelligence, and microfluidics is accelerating innovation in both domains. In tissue engineering, research is increasingly directed toward achieving vascularization, innervation, and organ-level complexity, e.g., [19]. In food technology, efforts are expanding to incorporate sustainable protein sources, enhance the sensory realism of plant-based foods, and develop on-demand food production systems for use in space missions, disaster relief, and other resource-limited settings [20].

In summary, 3D printing with biological materials represents a powerful convergence of engineering, life sciences, and digital manufacturing. By enabling precise control over structure and function, this technology is poised to revolutionize both medicine and food systems.

2 Fabrication Techniques in Bioprinting

A range of fabrication techniques has been developed for 3D bioprinting, each with distinct advantages and limitations. The dominant methods in tissue engineering include extrusion-based, inkjet, stereolithography, laser-assisted, and two-photon polymerization approaches (Table 1).

2.1 Extrusion-Based Bioprinting

Extrusion-based bioprinting deposits continuous filaments of bioink through a nozzle driven by pneumatic pressure or mechanical actuation. It is the most widely adopted technique due to its simplicity, relatively low cost, and compatibility with a broad range of biomaterials, including hydrogels and cell-laden inks [21]. Its ability to accommodate high cell densities makes it suitable for printing clinically relevant tissue constructs. However, extrusion printing faces critical challenges related to bioink rheology and printing fidelity. Bioinks with low viscosity reduce shear-induced cellular damage but often lack the mechanical strength to maintain structural integrity post-deposition, resulting in deformation or collapse. Conversely, higher viscosity inks improve shape fidelity but increase shear stresses that can compromise cell viability [22]. Achieving an optimal balance between printability, mechanical stability, and biological compatibility remains a central challenge for extrusion-based systems.

2.2 Inkjet Bioprinting

Inkjet bioprinting employs thermal or piezoelectric actuation to eject droplets of bioink onto a substrate with high spatial precision. This droplet-based approach enables fine control of cell and biomolecule placement and supports the fabrication of complex, multilayered architectures [23]. Inkjet systems offer excellent resolution and throughput but are limited by the narrow range of bioink viscosities they can handle and by risks of nozzle clogging when working with high cell densities or particulate-laden inks.

2.3 Stereolithography (SLA) Bioprinting

Stereolithography utilizes ultraviolet (UV) or visible light to selectively polymerize photosensitive bioinks, producing structures with superior resolution and surface fidelity compared to extrusion or inkjet methods. Layer-by-layer photopolymerization allows the creation of intricate, finely tuned geometries that closely replicate the microarchitecture of native tissues [24, 25]. SLA's main limitations are the requirement for photosensitive resins and concerns regarding

Table 1
Comparative overview of major bioprinting techniques

Technique	Advantages	Limitations	Typical applications
Extrusion-based bioprinting	• Simple, affordable, and widely accessible • Broad compatibility with hydrogels and cell-laden inks • Capable of printing high cell densities and large constructs	• Trade-off between viscosity, print fidelity, and cell viability • Shear stress during extrusion can damage cells • Limited resolution compared to optical methods	Tissue scaffolds, cartilage, bone, skin, and large-scale constructs
Inkjet bioprinting	• High resolution and precision • Noncontact deposition • Rapid printing of droplets with spatial control of cells and biomolecules	• Limited to low-viscosity bioinks • Risk of nozzle clogging • Less suitable for high cell densities	Patterning of cells, growth factors, drug screening models and thin tissue layers
Stereolithography (SLA) bioprinting	• Superior resolution and surface fidelity • Ability to fabricate intricate, finely tuned geometries • Rapid, layer-by-layer photopolymerization	• Requires photosensitive resins and photoinitiators • Potential cytotoxicity issues • Limited bioink diversity	Microstructured scaffolds, bone tissue, organ-on-chip systems
Laser-assisted bioprinting (LAB)	• Nozzle-free, reducing clogging issues • High cell viability due to low mechanical stress • Precise positioning of cells and biomaterials	• Expensive and technically complex • Limited throughput • Requires donor substrate preparation	High-resolution tissue constructs, skin models, vascular networks
Two-photon polymerization (TPP)	• Submicron resolution with precise micro/nanoscale control • Ideal for replicating intricate native tissue architectures • Enables fabrication of microvascular and neural networks	• Low throughput, slow printing speed • Requires specialized photopolymerizable materials • Limited scalability	Neural scaffolds, microvascular networks and high-resolution tissue models

the cytotoxicity of photoinitiators, which restricts the range of biocompatible materials available.

2.4 Laser-Assisted Bioprinting (LAB)

Laser-assisted bioprinting directs pulsed laser energy onto a donor substrate coated with bioink. Localized absorption generates a high-pressure bubble that propels discrete droplets of material onto a receiving substrate. This nozzle-free technique allows precise deposition of cells and biomaterials with minimal mechanical stress, resulting in high cell viability and accurate spatial patterning [26, 27]. LAB is particularly useful for applications requiring high-resolution cell positioning, although the setup is relatively complex and costly.

2.5 Two-Photon Polymerization (TPP)

Two-photon polymerization is an advanced bioprinting technique that employs ultrafast laser pulses to initiate polymerization through two-photon absorption within a photosensitive resin. TPP achieves submicron resolution, enabling precise fabrication of micro- and nanoscale features within tissue scaffolds [28]. Its capacity to reproduce intricate architectures makes it particularly suitable for engineering delicate structures such as microvascular networks, neural scaffolds, and other tissues where high spatial accuracy is critical. However, the method is limited by low throughput and the need for specialized photopolymerizable materials.

2.6 Summary

Among the available bioprinting techniques, extrusion-based methods remain the most widely employed due to their versatility, affordability, and compatibility with diverse bioinks. Yet, they continue to face critical challenges in balancing rheological demands with structural fidelity and cell viability. These persistent limitations highlight the need for innovative strategies. As will be elaborated in the following sections, temperature-controlled cryoprinting (TCC) represents a next-generation extrusion-based approach that directly addresses these shortcomings and expands the potential of bioprinting for both biomedical and food applications.

3 Materials Used for 3D Extrusion Bioprinting

A wide variety of biomaterials has been investigated for extrusion-based bioprinting. Extensive studies have reviewed their properties, performance, and suitability for specific applications [29–35]. Table 2 summarizes commonly used naturally derived bioinks, highlighting their sources, strengths, and limitations.

4 Temperature-Controlled Cryoprinting (TCC)

4.1 Principle of TCC

Extrusion-based bioprinting faces critical challenges. Bioinks must have sufficiently low viscosity to facilitate extrusion through the nozzle and, when cells are involved, to minimize shear-induced cellular

Table 2
Commonly used naturally derived bioinks for extrusion bioprinting

Material	Source	Strengths	Limitations	Reference
Collagen	A highly abundant, naturally occurring protein composed of self-assembling polypeptide chains stabilized by hydrogen and covalent bonds	Excellent biocompatibility and natural cell adhesion properties, among the most widely used biomaterials in bioprinting	Relatively low mechanical strength	[30]
Hyaluronic acid	The most abundant glycosaminoglycan in the extracellular matrix (ECM)	Outstanding biocompatibility; mimics the ECM of soft tissues such as brain, skin, and cartilage	Inherently low mechanical strength; requires chemical modification for structural applications	[31]
Gelatin	Obtained through partial hydrolysis of collagen	High solubility in water; biocompatible; thermoreversible gelation	Requires chemical modification for stable structural use; gelatin methacryloyl (GelMA) was developed to address these limitations	[32]
Chitosan	A natural polymer derived from the alkali deacetylation of chitin found in crustacean shells and fungal cell walls	Biocompatible; antibacterial properties; versatile structural behavior	Weak mechanical properties, limited cell adhesion, and poor solubility	[33]
Agarose	A natural, water-soluble polysaccharide extracted from seaweed	Thermoresponsive, biocompatible, and valued for its gelation behavior and structural stability	Poor cell adhesion, limited bioactivity, and challenging chemical modification	[34]
Alginate	A water-soluble polysaccharide mainly extracted from brown seaweed	Tunable hydrogel properties; compressive modulus ranges from a few kPa to several hundred kPa depending on crosslinking density	Limited cell adhesion and bioactivity; unstable ionic crosslinking	[35]

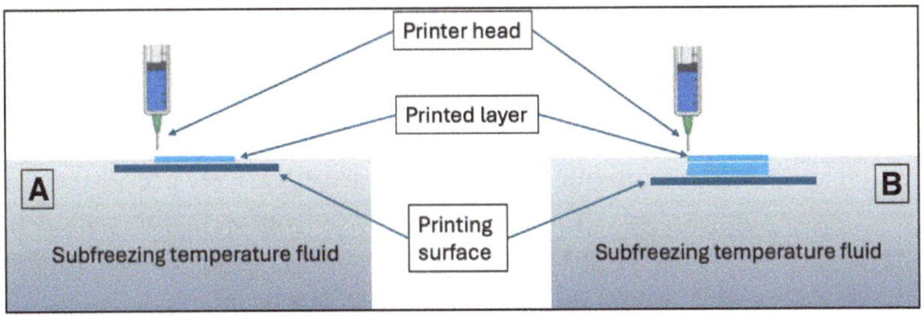

Fig. 1 A schematic illustrating the temperature-controlled cryoprinting (TCC) process. Bioink is delivered at room temperature from a temperature-controlled nozzle (printer head). (**A**) The first layer is deposited onto a printing surface, in contact with a subfreezing temperature liquid. (**B**) As new layers are printed, the printing surface is lowered into the subfreezing temperature fluid, so that at all times the last printed layer is at the same distance from the printer head, while all the previously printed layers are in the subfreezing temperature fluid

damage. While low-viscosity inks improve fluidity and reduce shear stress, they often lack the mechanical strength necessary to maintain structural integrity after deposition, resulting in deformation or collapse. This limitation also restricts the ability to print large, stable structures. Conversely, high-viscosity inks provide better shape fidelity but increase shear stresses that can compromise cell viability. Achieving an optimal balance between printability, mechanical stability, and biological compatibility, therefore, remains one of the major challenges for extrusion-based bioprinting. In addition, the relatively slow speed of extrusion and the prolonged exposure of biological matter during printing further limit the fabrication of large, viable constructs.

Temperature-controlled cryoprinting (TCC) was developed to overcome these inherent limitations (Fig. 1). The key feature of TCC is the precise control of freezing at the voxel scale. This is achieved through a three-pronged strategy: (i) Continuous regulation of bioink temperature during extrusion through the nozzle, (ii) Active control of the deposition surface temperature to ensure uniform freezing upon contact, (iii) Immediate immersion of the construct in a subfreezing fluid bath, which maintains each deposited layer at a constant subzero temperature for the entire duration of the printing process. Together, these measures ensure that each voxel is solidified under reproducible thermal conditions, and the construct remains stably frozen throughout fabrication.

4.2 Advantages Over Conventional Extrusion Printing

4.2.1 Mechanical Rigidity

Freezing provides structural stiffness to the printed material, enabling the fabrication of complex geometries such as overhangs, hollow cavities, and unsupported features. This allows the use of low-viscosity inks, including delicate biomaterials or food formulations that would otherwise collapse under their own weight.

4.2.2 Preservation During Printing

Since the construct remains frozen throughout the process, deterioration and microbial growth are effectively prevented. This feature is highly advantageous in food applications, where shelf-life extension is critical, and in tissue engineering, where maintaining cell viability during extended fabrication is essential.

4.2.3 Control of Microstructure

Voxel-level freezing allows precise regulation of ice crystal size by tuning cooling rates and temperature gradients during solidification. This results in uniform and reproducible microstructures, which are crucial for food preservation and for tissue engineering, as well as for predictable mass transport properties in engineered tissues.

4.2.4 Direct Production of Frozen Products

Often, cellular constructs or food products fabricated by conventional bioprinting require an additional freezing step for long-term preservation. However, traditional freezing methods are problematic: in large biological constructs, they commonly lead to extensive cell death, while in food products, they can compromise texture, flavor, and overall quality. In contrast, temperature-controlled cryoprinting (TCC) eliminates the need for this secondary, often damaging, freezing step by producing constructs that are already frozen at the end of the printing process. This integrated approach not only streamlines manufacturing but also provides clear advantages for both domains—enabling the direct fabrication of consumer-ready frozen foods and facilitating the cryopreservation of large, complex biological structures. Thus, TCC effectively merges fabrication and preservation into a single step, representing a paradigm shift compared with conventional bioprinting and food-processing workflows.

4.3 Chronological Development of Temperature-Controlled Cryoprinting (TCC)

Following is a chronological account of our work that has led to the development of the temperature-controlled cryoprinting (TCC) technology, tracing the trajectory from its conceptual origins to its experimental realization and subsequent applications. Beginning with the initial idea of using cryogenic conditions to impart structural fidelity during extrusion printing, this body of work progressively refined the methodology, introduced novel engineering solutions, and expanded the scope of applications across tissue engineering, cryobiology, and food preservation. Each paper represents a step in the intellectual and technological evolution of TCC, from foundational principles through to practical demonstrations of its advantages and versatility (Fig. 2).

4.3.1 Directional Solidification

The conceptual basis of temperature-controlled cryoprinting (TCC) is rooted in the fundamental crystallographic properties of ice. Ice possesses a tightly ordered lattice structure that excludes solutes during solidification. As aqueous solutions freeze, solutes are rejected into the unfrozen regions, generating characteristic microscale patterns of

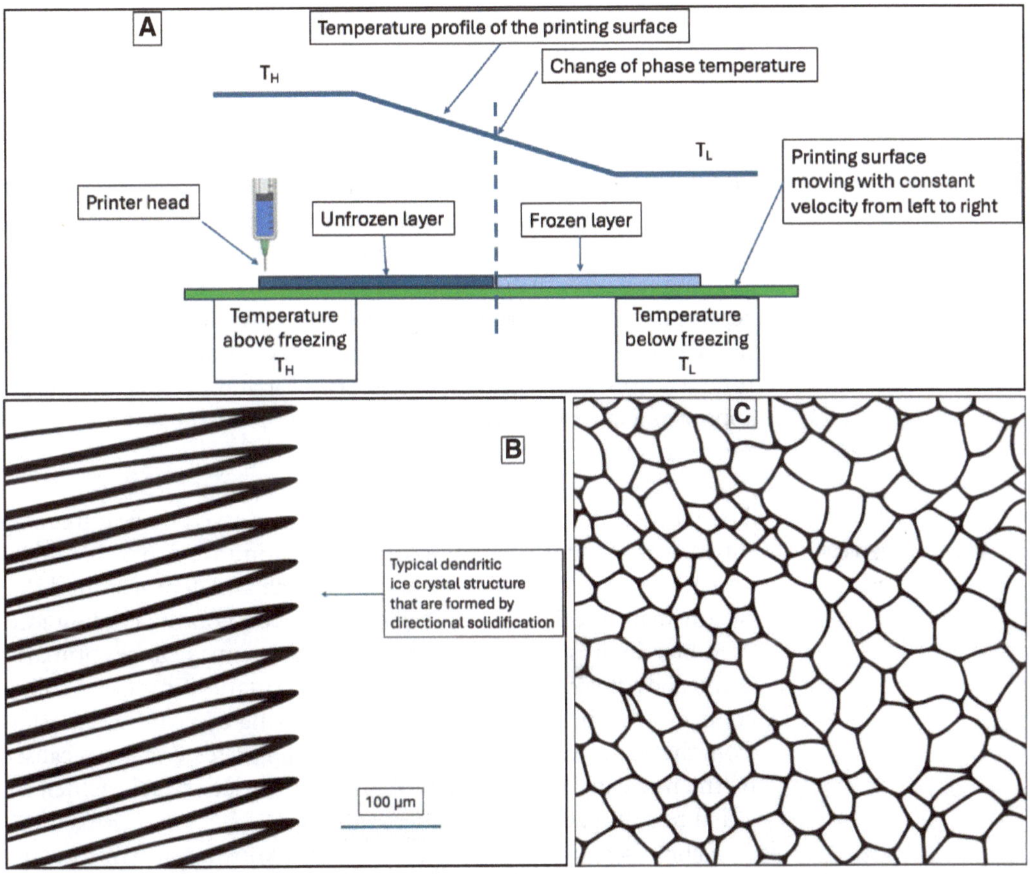

Fig. 2 Research that led to the Temperature Controlled Cryoprinting concept. (**A**) Schematic of a directional solidification stage. (**B**) Schematic of micron-scale ice crystals generated by directional solidification freezing of physiological saline. (**C**) Schematic of an electron microscope image of a tissue scaffold generated by directional solidification of alginate

ice crystals. In gels, these microstructures are preserved upon thawing, with their dimensions determined by both the cooling rate and the orientation of the thermal gradient applied during freezing [36].

Our early cryobiology research provided critical insights that paved the way for voxel-level freezing control in TCC. In particular, we developed a cryomicroscope capable of freezing biological specimens under one-dimensional temperature control [37] This system enabled direct observation and experimental analysis of ice formation dynamics with high spatial precision, establishing laboratory methods for precise, layer-wise freezing that can yield controlled microscale structures [38], These pioneering studies represent an intellectual and technological precursor to cryoprinting, where similar principles of controlled temperature gradients are applied at the voxel scale to guide freezing during 3D printing.

4.3.2 Generating Tissue Scaffolds by Directional Solidification

Studies such as those published in [36–38] established the foundation for using directional freezing of single layers of crosslinked alginate gels to generate tissue scaffolds with controlled microscale architecture. By manipulating the magnitude and orientation of temperature gradients during freezing, it became possible to tune pore size and alignment within the scaffold. An early proposal extended this concept to tissue engineering, suggesting that directional freezing could be exploited to fabricate porous scaffolds with defined microstructural features, thereby planting the seed for additive manufacturing approaches in which controlled ice formation is harnessed to engineer functional microarchitectures [39].

4.3.3 The Potential of TCC for Cryopreservation

The study of how cells are damaged or preserved during freezing is the central focus of cryobiology. Pioneering work by Mazur systematically summarized decades of research and demonstrated that two primary mechanisms of cell injury depend on the cooling rate. At suboptimal (slow) cooling rates, cellular damage arises primarily from dehydration, as extracellular ice formation increases solute concentration in the unfrozen fraction, driving excessive water efflux from the cells. At supraoptimal (rapid) cooling rates, damage is caused by the nucleation and growth of intracellular ice crystals, which are lethal to cells [40, 41].

The discovery that certain chemicals, known as cryoprotective agents(CPAs), can mitigate these effects represented a major breakthrough in cryobiology. CPAs act by reducing ice formation, limiting solute effects, and stabilizing cellular structures during freezing and thawing, thereby enabling the long-term preservation of many isolated cell types [42, 43]. Our own directional freezing studies further elucidated the relationship between cooling rate, ice formation, and cell survival in various cell populations [44–48]. These experiments provided mechanistic insights into the pathways of cell death and survival under controlled freezing conditions, thereby linking fundamental cryobiology with practical strategies for cryopreservation.

Despite the success of cell preservation, extending cryopreservation to tissues and organs has remained a major challenge. The main obstacles include: (i) the difficulty of introducing CPAs uniformly into all cells of a complex tissue via perfusion through the vasculature [49, 50]; (ii) the destructive effects of freezing on the vasculature itself [51]; and (iii) the loss of structural and functional integrity of the organ as a whole [52].

These limitations motivated the development of temperature-controlled cryoprinting (TCC). By enabling the voxel-by-voxel construction of large frozen biological tissues, TCC circumvents many of the traditional problems of bulk freezing, providing a pathway to fabricate large-scale constructs that are structurally stable and protected from freezing-induced damage.

4.3.4 Freeze Casting of Gel Solutions

The process of freeze casting of gel solutions bears some resemblance to temperature-controlled cryoprinting (TCC), though the two approaches are fundamentally distinct. Freeze casting, also referred to as ice templating or freeze gelation, is a well-established method for fabricating porous materials, including gels, across a wide range of compositions and processing conditions. In this technique, a solution, suspension, sol, or gel is subjected to controlled solidification, after which the frozen solvent (most commonly water) is removed through sublimation under reduced pressure. The resulting porous structure is essentially a negative replica of the solvent crystal network.

One of the earliest references to this technology can be traced back to Lottermoser's 1908 paper "Über das Ausfrieren von Hydrosolen" (On the Freezing of Hydrosols) [53]. While freeze casting has since become widely applied in industrial settings, particularly for producing porous ceramics, its introduction into tissue engineering can be credited to the pioneering work of I.V. Yannas and colleagues in the mid-1970s, who employed freeze-drying to generate porous skin substitutes [54]. Subsequent applications have included the fabrication of alginate sponges for cell culture and related biomedical uses [55].

The critical distinction between freeze casting and TCC lies in scale and control. Freeze casting operates on the bulk level, freezing and structuring an entire cast volume at once. In contrast, TCC achieves control at the voxel scale, enabling precise regulation of freezing during the additive assembly of three-dimensional constructs. This voxel-wise strategy imparts unique advantages in terms of structural fidelity, microarchitectural control, and the integration of cryobiological principles into the printing process.

4.3.5 Introducing Temperature-controlled 3D Cryoprinting Technology

The 2015 paper by Adamkiewicz & Rubinsky, "Multi-Layer Cryolithography for Additive Manufacturing" [56], was the pioneering study that first articulated the use of controlled freezing as an enabling mechanism for additive manufacturing. At the time, extrusion-based bioprinting and 3D printing with soft biomaterials faced severe limitations: printed objects often collapsed under their own weight, exhibited poor structural fidelity, and deteriorated during prolonged printing. This study proposed an entirely new paradigm, using controlled solidification of the voxels in each layer to impart immediate mechanical strength and stability, thereby enabling the fabrication of complex and delicate geometries.

The approach involved depositing a voxel of soft material (such as a hydrogel or a cell-laden medium) and then immediately freezing it under precisely regulated thermal conditions. The printer extrudes hydrogel while the entire build is immersed in liquid nitrogen; the liquid's free surface is actively kept level with the highest printed layer, so each new voxel is frozen under similar conditions. This process was repeated layer by layer until a complete three-dimensional structure was assembled. Unlike conventional methods, this strategy provided structural integrity without requiring chemical crosslinking or high-viscosity inks.

The methodology drew heavily from cryobiology, incorporating principles such as the crystallographic properties of ice, solute exclusion during freezing, and the impact of cooling rates on microstructure. By harnessing these principles, the authors envisioned producing constructs with tunable microscale architectures. They further suggested that precise control over cooling rates could, in principle, enable voxel-specific optimization of cryopreservation protocols, allowing survival of embedded cells throughout the entire volume of a large organ.

Using agar gels as a model system, the study successfully demonstrated the feasibility of voxel-scale, layer-wise freezing as a manufacturing tool. It demonstrates complex frozen hydrogel geometries with high fidelity, establishing the feasibility of immersion-cryogenic printing as a platform for later biofabrication and food-texture work. The printed constructs retained shape fidelity and incorporated complex features that were not achievable with conventional extrusion (Fig. 3A). Importantly, the study showed that freezing not only stabilized geometry but also preserved biochemical composition during fabrication.

This 2015 paper laid the intellectual foundation for what later became temperature-controlled cryoprinting (TCC). It was the first published articulation of the concept that freezing itself could serve as the stabilizing "curing" step in bioprinting and food printing. By bridging classical cryobiology with modern additive manufacturing, the work introduced the revolutionary idea that fabrication could occur within a frozen environment. The study marks the conceptual birth of TCC. It directly seeded subsequent advances in nozzle and bioink temperature regulation, controlled deposition surface cooling, and immersion in subfreezing media, features that matured into the full TCC platform in later years.

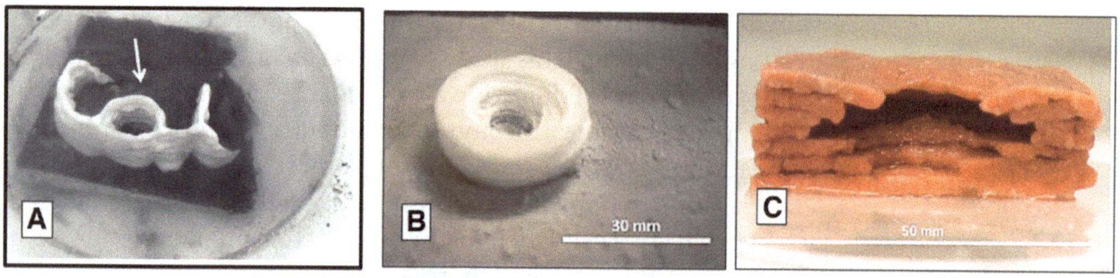

Fig. 3 **(A)** Structure made with temperature-controlled cryoprinting (TCC) [56]. **(B, C)** Structures in which each layer was made with TCC and assembled with MLCL [57]

4.4 "Cryoprinting": Patent Application (US20180304537A1, 2018)

With the proof of concept for cryogenic printing established, there arose a clear need for a comprehensive framework to protect and formalize the systems, methods, and workflows required for printing under controlled subfreezing conditions. The 2018 patent application addressed this need by detailing the apparatus and processes that would later form the foundation of temperature-controlled cryoprinting (TCC).

The application set forth methods for extruding bioink while the nozzle, substrate, and surrounding environment are actively temperature-regulated. The construct is kept subfreezing throughout the fabrication process, either by immersion in a liquid bath or within a controlled cryogenic chamber, ensuring immediate and consistent solidification. In addition, the patent integrated provisions for post-processing, such as crosslinking upon thaw, to guarantee the structural fidelity of the final product.

Among its most significant contributions, the filing claimed voxel-wise freezing control through staged temperature regulation, laying the groundwork for the precise layer-by-layer solidification that characterizes TCC. It also detailed hardware architectures, including temperature-controlled nozzles, descending build stages, and subfreezing baths, and explicitly envisioned applications spanning tissue engineering and food fabrication.

The patent provided exemplars and embodiments that outlined complete operational sequences for cryoprinting, including identifying task-specific temperature windows and strategies for reliable material handling. In doing so, it established the intellectual property framework that would underpin subsequent TCC hardware and workflows, supporting both the technology's scientific maturation and its eventual translation and commercialization.

4.4.1 Introducing the Concept Of Multi-layer Cryolithography in TCC

The 2018 paper by Zawada, Ukpai, Powell-Palm, and Rubinsky, "Multi-layer cryolithography for additive manufacturing" [57], introduced a significant new concept in the evolution of temperature-controlled cryoprinting (TCC). Following the initial proof-of-principle of cryogenic 3D printing [56], a central challenge remained: extrusion-based printing is inherently serial and therefore slow, limiting throughput and scalability, especially for large and complex constructs. To address this bottleneck, the authors proposed a strategy that decoupled material deposition from layer-by-layer extrusion by leveraging parallelized fabrication.

The approach, termed multi-layer cryolithography (MLCL), was inspired by lithographic processes. The workflow involved using TCC to form multiple 2D frozen hydrogel layers in parallel on specialized carrier substrates with hydrophilic/hydrophobic patterning. Freezing each layer separately preserved shape, stiffness, and mechanical integrity, thereby facilitating manipulation of each layer without deformation. This enabled the stacking, alignment, and bonding of the frozen layers into integral 3D constructs via cryogenic adhesion followed by chemical crosslinking. This strategy transformed additive manufacturing from serial voxel-wise deposition into batchwise layer assembly, providing a new dimension of scalability.

Key contributions of this study include conceptual advances: the first demonstration of cryogenic layer assembly as a complement to layer-by-layer printing. It also demonstrated a carrier-based handling system, in which hydrophilic/hydrophobic carriers facilitated reliable transport and positioning of delicate frozen hydrogel sheets, and it introduced a frozen-layer bonding strategy in which freezing served as a reversible "fixture" to stabilize layers during stacking, while crosslinking permanently integrated them into coherent 3D constructs.

The study showed that complex 3D hydrogel structures could be rapidly fabricated by stacking pre-frozen 2D sheets. Interfaces bonded seamlessly, producing constructs that preserved geometric fidelity and structural coherence. Photographic and structural analyses confirmed reproducibility and highlighted the throughput advantages of parallelized layer fabrication compared to traditional extrusion (Fig. 3B, C).

This work marked an expansion of cryogenic fabrication strategies. By showing that freezing could function not only as an in situ stabilizer but also as a manufacturing enabler for parallelized assembly, the study positioned MLCL as a complementary branch to filament-based cryoprinting. The approach broadened the cryogenic toolbox for additive manufacturing, establishing freezing as both a fabrication and assembly mechanism. It foreshadowed later developments such as freezing-modulated crosslinking (FMC) and fully matured temperature-controlled cryoprinting (TCC), where freezing is formalized as an active, controllable step in the construction of complex biological and food-based materials.

4.4.2 A Parallel Multiple Layer Cryolithography Device for the Robotic Manufacture of Biological Materials

By 2019, the concept of multi-layer cryolithography (MLCL) had been introduced as a strategy to overcome the limitations of serial extrusion in cryoprinting. However, until this point, the idea had remained largely theoretical, demonstrated only in laboratory prototypes. The 2019 study by Ukpai and colleagues marked an important step forward by presenting a practical device that implemented MLCL principles in a reproducible, engineering-focused manner [58]. The motivation was to demonstrate that cryogenic fabrication could be scaled beyond proof of concept and translated into hardware capable of consistent, parallelized layer production and assembly.

The device described in this work was designed to fabricate multiple frozen two-dimensional hydrogel layers in parallel, each deposited on specialized carriers. Once frozen, the layers could be transported, aligned, and bonded into three-dimensional structures under cryogenic conditions. Key to the system was its ability to maintain the mechanical stability of each sheet in the frozen state, which not only prevented deformation under gravity but also allowed precise handling and registration during stacking. The frozen interfaces were then permanently integrated through chemical crosslinking, resulting in coherent 3D constructs (Fig. 4).

The study's major contributions lie in transforming MLCL from a conceptual framework into an operational platform. It demonstrated that parallelized production of frozen hydrogel layers could be achieved reliably and that cryogenic bonding between layers was robust, yielding seamless integration without structural defects. The system further introduced modularity into the design, enabling reproducible alignment and stacking across multiple layers, and provided quantitative evidence that cryogenic fixation could serve as an effective manufacturing tool rather than merely a stabilizing mechanism.

Results showed that large and complex hydrogel constructs could be assembled with high geometric fidelity, overcoming the throughput and scalability constraints that had plagued serial extrusion-based cryoprinting. Photographs and structural evaluations confirmed that the interfaces between layers remained coherent and that the frozen state effectively countered gravitational collapse during assembly.

The significance of this paper lies in its demonstration that cryogenic principles could be embodied in a practical device, bridging the gap between laboratory exploration and manufacturable technology. By showing that MLCL could reliably produce integral multi-layer constructs, the study expanded the cryoprinting toolbox and laid the

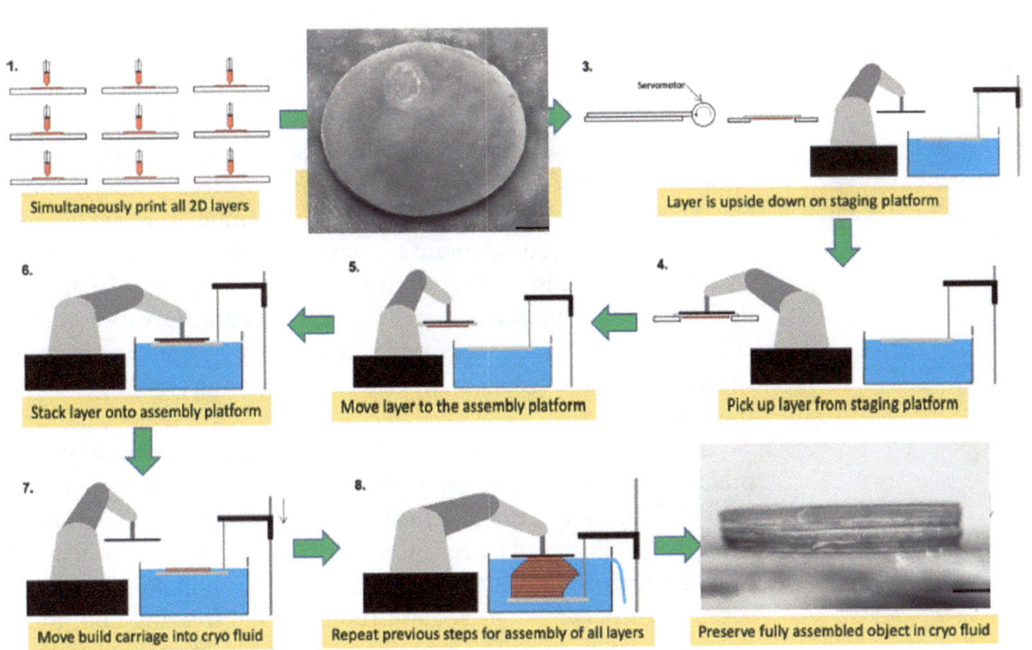

Fig. 4 Schematic of the MLCL process with examples of the single sheet and assembly of multiple sheets. (Modified from [58])

foundation for further advances in temperature-controlled cryoprinting systems. It underscored the value of freezing not only as a preservation tool but also as an active enabler of scalable biofabrication.

4.4.3 Mathematical and Experimental Framework for TCC

In 2020, Ukpai and Rubinsky presented two studies that provided the analytical and experimental foundation for understanding freezing phenomena in temperature-controlled cryoprinting (TCC). These works addressed both the mathematical description of solidification processes and the experimental validation of models, illustrating how theoretical principles could be applied to guide and refine cryoprinting practice.

The first study [59], published in the ASME Journal of Heat Transfer, expanded on the original studies in [37, 39] and examined directional solidification in aqueous solutions and biological matter. The authors developed an analytical framework to describe the propagation of the solid–liquid interface under imposed thermal gradients. The model accounted for how cooling rates and gradient orientation influence the stability of the ice front, as well as the scale and morphology of the resulting microstructure. The analysis showed that one-dimensional approximations, often used in classical freezing studies, may not fully capture the dynamics relevant to cryoprinting geometries, in which two- and three-dimensional effects become important. By extending the theory to higher dimensions and validating predictions against controlled experiments, the study provided a means of linking thermal boundary conditions to structural outcomes in frozen constructs.

The second paper [60], published in Bioprinting, focused on developing a three-dimensional transient model of heat transfer and phase change during cryoprinting. The model incorporated the coupled processes of layer deposition, conduction, and solidification, and simulated how parameters such as nozzle temperature, substrate temperature, print speed, and construct height affect freezing dynamics. A key finding was that without active temperature regulation, thermal conditions deteriorate as construct height increases, leading to variable freezing rates and potential loss of fidelity. To address this, the study proposed a framework in which nozzle, substrate, and environmental temperatures are controlled in a coordinated manner, thereby maintaining consistent voxel-scale freezing conditions across the entire build.

Experiments were conducted to validate the computational predictions. These included measurements of freezing dynamics in hydrogel layers and evaluations of construct fidelity under varying thermal conditions. The results supported the model's conclusion that temperature regulation of both the deposition source and the substrate is necessary to achieve reproducible freezing conditions across multiple layers. The experimental data also demonstrated that directional control of cooling rates could influence microstructural uniformity in a way consistent with theoretical expectations.

Taken together, these two papers illustrate how mathematical analysis and experimental validation can be integrated to form the basis of TCC. The analytical work provided insights into the role of thermal gradients and cooling rates in defining ice-front behavior, while the computational and experimental study demonstrated how these principles could be operationalized through controlled printing conditions. This combination of approaches established a coherent framework for TCC, linking physical models of solidification with practical strategies for fabricating multilayer constructs under reproducible freezing conditions.

4.4.4 Workflow and System Development in TCC

By 2022, the conceptual and mathematical foundations of temperature-controlled cryoprinting (TCC) had been articulated, and attention turned to developing practical workflows and hardware systems to integrate freezing, crosslinking, and structural preservation into reproducible manufacturing sequences. Two papers from this period, focusing respectively on freezing-modulated crosslinking (FMC) and modular cryoprinting systems, illustrate this stage in the evolution of TCC.

The first study introduced freezing-modulated crosslinking (FMC) as a strategy to stabilize cryoprinted constructs during the critical thawing step [61]. A recurring challenge in cryoprinting was that frozen constructs, although geometrically accurate, often lost fidelity when thawed for post-processing. FMC addressed this problem by immersing the frozen construct in a warm crosslinker solution, allowing crosslinking to proceed progressively at the advancing melt front. In this way, each voxel was stabilized as it transitioned from the frozen to the liquid state. Thermal and mass-transport analyses were performed to define safe process windows, and experimental results demonstrated that complex geometries could be preserved during the conversion of frozen constructs into stable, crosslinked scaffolds. This approach formalized freezing not just as a temporary stabilizer but as an active element of the manufacturing sequence, tightly integrated with chemical stabilization.

The second paper [62] presented the design of a modular cryoprinting system that brought together the multiple steps of deposition, freezing, and post-processing into a flexible hardware platform. The system incorporated modules for nozzle and substrate temperature regulation, as well as controlled environments for freezing and crosslinking. Using this platform, the authors studied how variations in cooling rates and crosslinking sequences influenced the microstructure and mechanical properties of porous alginate scaffolds. The modular design enabled reproducible experiments that linked process parameters to structural outcomes, thereby establishing a testbed for systematic exploration of TCC workflows.

Together, these two contributions represented a shift in TCC research from foundational modeling and feasibility demonstrations toward integrated process engineering. FMC provided a robust method for preserving structure during thawing, while the modular system offered a platform for studying the interplay between deposition, freezing, and stabilization. In combination, they demonstrated how TCC could be organized into a reproducible workflow in which each step, extrusion, voxel-level freezing, and post-processing, was coordinated to maintain fidelity and functionality.

4.4.5 Material Development and Transport Properties in TCC

An important stage in the development of temperature-controlled cryoprinting (TCC) has been the formulation of inks that combine ease of extrusion with stability during freezing and post-processing, as well as the characterization of their transport properties after fabrication. Two consecutive studies by Lou and Rubinsky, published in Gels [63] and Bioprinting [64], addressed these questions through systematic analysis of alginate–agar mixtures as cryoprinting inks.

The 2023 paper introduced a composite ink composed of sodium alginate and agar, designed to leverage the complementary properties of the two polymers. Agar provides thermo-reversible gelation, which imparts immediate structural stability upon cooling, while alginate undergoes ionic crosslinking, offering long-term mechanical integrity after post-processing. The study examined how the balance between agar and alginate concentrations affected the ink's printability, its response to freezing, and its compatibility with the freezing-modulated crosslinking (FMC) process. Results showed that the composite ink supported extrusion at relatively low viscosity, enabling smooth deposition, and that frozen constructs maintained shape fidelity prior to permanent stabilization upon thawing in a crosslinker solution. By harmonizing agar's reversible gelation with alginate's crosslinking, this study demonstrated a practical materials approach to improving reproducibility and fidelity in TCC workflows.

The 2024 follow-up paper focused on the diffusion properties of scaffolds produced with the same alginate–agar inks under different freezing histories. Since mass transport of solutes, nutrients, or drugs is a critical factor in the function of bioprinted scaffolds, the authors measured effective diffusion coefficients in constructs fabricated through unfrozen casting, rapid freezing, and TCC printing. The study found that freezing history significantly influenced diffusion behavior, with TCC scaffolds exhibiting anisotropic transport consistent with their ice-templated microstructure. These findings underscore the importance of considering thermal history not only for mechanical fidelity but also for functional mass transport properties. The data provided quantitative guidance for designing scaffolds in which controlled microstructural anisotropy can be used to tailor diffusion for specific applications, including tissue culture and drug delivery.

Together, these two studies represent the progression from material formulation to functional characterization within the TCC framework. The 2023 work established alginate–agar mixtures as inks well-suited for temperature-controlled freezing and crosslinking, while the 2024 study demonstrated how freezing conditions impart predictable transport properties to the resulting constructs. In combination, they show how material selection and thermal history interact in TCC to define not only the structural integrity of printed scaffolds but also their biological and functional performance.

4.4.6 Applications of TCC in Tissue Engineering

The application of temperature-controlled cryoprinting (TCC) to tissue engineering has been explored in two studies by Warburton and Rubinsky, which together illustrate how cryoprinting can be used not only to fabricate three-dimensional constructs but also to preserve them for later use.

The first study, published in Gels in 2023, focused on the cryopreservation of 3D bioprinted scaffolds [65]. In conventional cryoprinting, one limitation was the deterioration of freezing conditions as construct height increased, leading to variability in freezing dynamics and, consequently, reduced uniformity in construct integrity and cell survival. The TCC method addressed this by actively controlling nozzle and substrate temperatures and by immersing the growing construct in a subfreezing bath, thereby ensuring that each layer was frozen under similar thermal conditions regardless of build height. Using hydrogel-based scaffolds containing cells, the authors demonstrated that constructs fabricated with TCC retained geometric fidelity across multiple layers and supported appreciable cell viability after thawing. The study confirmed that voxel-scale freezing control is achievable in practice and critical for generating reproducible scaffolds suitable for tissue engineering applications.

The second paper, published in Bioprinting in 2025 [66], extended this line of work to the cryopreservation of vascularizable tissues. The study explored whether TCC could be used to fabricate large constructs that, after thawing, support the development of vascular networks, an essential requirement for functional engineered tissues. Constructs were printed with architectures designed to facilitate vascular ingrowth and were fabricated under TCC conditions that maintained uniform freezing dynamics throughout the build. After thawing and culturing, the authors observed the formation of vascular structures by endothelial and stromal cells within the constructs, indicating that the frozen, thawed scaffolds provided an appropriate environment for vascularization. The findings suggested that TCC can bridge the fabrication of large engineered tissues and their long-term preservation, addressing two major bottlenecks in tissue engineering: structural fidelity during printing and functional recovery after cryopreservation.

Taken together, these studies highlight the dual role of TCC in tissue engineering: as a fabrication method that enables reproducible printing of delicate, cell-laden constructs and as a preservation strategy that stabilizes those constructs for storage and later use. By integrating printing and cryopreservation into a single process, TCC offers an approach that addresses both the manufacturing and storage challenges associated with engineered tissues.

4.4.7 Applications of TCC in Food Engineering

In parallel with tissue engineering, temperature-controlled cryoprinting (TCC) has also been applied to food engineering, where freezing is central to both preservation and texture modification. Two studies by Lou, Bilbao-Sainz and Rubinsky illustrate how TCC can be adapted to design foods with tailored microstructures and textures, with particular emphasis on personalized nutrition.

The first study, published in Innovative Food Science and Emerging Technologies in 2023 [67], addressed the challenge of producing safe and palatable foods for patients with dysphagia, a condition that requires soft and carefully structured foods to reduce the risk of choking. Traditional food-processing methods often lack the precision needed to engineer textures that are simultaneously safe and appealing. TCC was used to print food formulations that were immediately frozen under controlled thermal conditions, thereby allowing the size and orientation of ice crystals to be tuned by regulating cooling rates and gradients. Upon thawing, the imprinted ice microstructures translated into reproducible textures. The authors demonstrated that food constructs produced in this way could be designed to meet established texture standards for dysphagia diets while also offering improved consistency and sensory properties. This work highlighted how TCC could directly fabricate frozen, consumer-ready foods with targeted textural attributes.

The second paper, published in ACS Food Science and Technology in 2025 [68], expanded the application of TCC to the design of plant-based foods with customized macro- and microstructures. In this study, a coaxial cryoprinting method was developed to create core–shell geometries using plant-derived ingredients such as pea protein, alginate, and agar. By combining composition control with regulated freezing conditions, the authors showed that both large-scale structural features (e.g., marbling patterns) and microscale textures could be tailored in a single process. Texture profile analysis confirmed that the printed foods could be tuned across a range of mechanical properties relevant to consumer preferences and nutritional needs. In particular, the ability to fabricate frozen vegetarian products with reproducible textures highlighted the potential of TCC to create personalized and sustainable food options.

Together, these two studies illustrate the versatility of TCC in food engineering. The 2023 work demonstrated its role in addressing specific dietary and clinical needs, such as producing safe textures for individuals with swallowing disorders, while the 2025 study extended the technology toward broader consumer markets, emphasizing sustainability and customization through plant-based formulations. In both cases, TCC integrated fabrication and preservation into a single workflow, enabling the production of foods that are both structurally tailored and directly suitable for frozen storage and distribution.

4.4.8 Intellectual Property Development in Temperature-Controlled Cryoprinting (TCC)

Alongside the scientific and engineering studies on TCC, a sequence of patent applications and grants has established an intellectual property framework for the technology. These filings documented the conceptual foundations, formalized process elements, and described hardware architectures for cryogenic additive manufacturing, ensuring that the core ideas could be protected and translated toward practical use.

The first filing, published in 2018 (US20180304537A1) [69], captured the essential concept of extruding bioinks or soft materials under controlled subfreezing conditions. It described systems in which the nozzle, substrate, and surrounding environment are temperature-controlled to keep constructs frozen throughout fabrication. The application also outlined workflows that integrated post-processing steps such as crosslinking upon thaw. By doing so, the 2018 filing established the principle of voxel-wise freezing control and envisioned applications in both tissue engineering and food fabrication. This patent was issued in 2021 [70].

In 2019, a related family of applications and subsequent device patents elaborated on multi-layer cryolithography (MLCL), a complementary approach in which two-dimensional frozen hydrogel layers are produced in parallel, transported, and assembled into three-dimensional structures [71]. These filings protected methods of carrier-based handling, cryogenic bonding between layers, and hybrid freezing-plus-crosslinking strategies. Together with the 2018 application, they marked the consolidation of freezing not only as a stabilizing step but also as a central element of the additive manufacturing process.

The intellectual property portfolio expanded further in 2023 with the grant of US11,584,066B2 ("3D Cryoprinting"), which formalized the mature features of TCC. This patent claimed apparatus and methods involving temperature-controlled nozzles, descending build stages immersed in subfreezing baths, and integrated thaw-crosslinking workflows. The issued claims described how coordinated temperature regulation can produce frozen constructs of consistent quality across varying build heights, thereby addressing one of the principal limitations of early cryoprinting methods.

Complementing these, a more recent application by Rubinsky, Rubinsky, and Ukpai (US20230284670, "Systems and methods for additive manufacturing of materials with controllable microscale textures") [72] emphasized the ability to regulate microstructure during fabrication. This filing details systems in which a movable build platform is sequentially exposed to heating and cooling modules, generating controlled thermal gradients and directional solidification. By framing freezing as a tool to produce anisotropic microscale textures, this application aligned directly with the goals of TCC in both biomedical and food contexts, where reproducible microstructural features are critical.

Taken together, these patents represent a structured effort to define and protect the key features of TCC. They span the initial articulation of cryoprinting as a new approach (2018), the extension to parallel layer assembly (2019), the codification of mature hardware and process-control strategies (2023), and the refinement of methods for tailoring microstructure (2023 application). This intellectual property activity runs in parallel with the scientific literature, ensuring that the core technical advances in TCC are supported by an implementable and transferable framework for future development and application.

5 Summary

The paper provides a comprehensive review of the scientific, technical, and application-driven development of TCC. It begins by situating TCC within the broader field of 3D bioprinting and food printing, noting persistent limitations in extrusion-based methods that motivated the search for new approaches. The conceptual foundation is linked to cryobiology research on ice crystallography, solute exclusion, and freezing-induced cell damage, which suggested freezing could be harnessed not only for preservation but also as a stabilizing tool during fabrication.

The chronological development is then detailed: the 2015 proof-of-concept of cryogenic 3D printing, the 2018 introduction of multi-layer cryolithography for scalable assembly, and subsequent device-level demonstrations. The 2020 studies of Ukpai and Rubinsky formalized the mathematical and experimental framework for voxel-level freezing. By 2022, system integration and workflow refinement were pursued through freezing-modulated crosslinking and modular cryoprinting platforms. Later studies introduced new bioinks, particularly alginate–agar mixtures, and quantified their transport properties under different freezing histories.

The review then turns to applications. In tissue engineering, TCC has been shown to maintain fidelity and cell viability in 3D scaffolds and to enable cryopreservation of large constructs that support vascularization after thawing. In food engineering, it was applied to fabricate frozen foods with controlled textures for patients with dysphagia and to create plant-based foods with tunable macro- and microstructures.

Finally, the paper reviews the intellectual property landscape, including patents that formalized cryoprinting methods, devices, and microstructure control. Across these sections, the paper shows how TCC has developed from theoretical and experimental origins into a framework for uniting additive manufacturing with preservation in both biomedical and food contexts.

Acknowledgments

This study was partially funded by the NSF Engineering Research Center for Advanced Technologies for Preservation of Biological Systems (ATP-Bio) NSF EEC #1941543. The authors used OpenAI's ChatGPT to assist in improving the manuscript's grammar, clarity, and style. All content generated or suggested by the model was critically reviewed and independently verified by the authors. The authors take full responsibility for the integrity and accuracy of all statements presented in this work.

References

1. Hull CW (1986) Apparatus for production of three dimensional objects by stereolitography. US Patent US4575330A
2. Murphy SV, Atala A (2014) 3D bioprinting of tissues and organs. Nat Biotechnol 32:773–785. https://doi.org/10.1038/nbt.2958
3. Kang HW, Kengla C, Lee SJ, Yoo JJ, Atala A (2014) 3-D organ printing technologies for tissue engineering applications. In: Narayan R (ed) Rapid prototyping of biomaterials: principles and applications. Woodhead Publishing, Philadelphia, pp 236–253
4. Xu T, Zhao W, Zhu JM, Albanna MZ, Yoo JJ, Atala A (2013) Complex heterogeneous tissue constructs containing multiple cell types prepared by inkjet printing technology. Biomaterials 34:130–139. https://doi.org/10.1016/j.biomaterials.2012.09.035
5. Peloso A, Dhal A, Zambon JP, Li P, Orlando G, Atala A, Soker S (2015) Current achievements and future perspectives in whole-organ bioengineering. Stem Cell Res Ther 6:107. https://doi.org/10.1186/s13287-015-0089-y
6. Kang HW, Lee SJ, Ko IK, Kengla C, Yoo JJ, Atala A (2016) A 3D bioprinting system to produce human-scale tissue constructs with structural integrity. Nat Biotechnol 34:312–319. https://doi.org/10.1038/nbt.3413
7. Moroni L, Boland T, Burdick JA, De Maria C, Derby B, Forgacs G, Groll J, Li Q, Malda J, Mironov VA, Mota C, Nakamura M, Shu W, Takeuchi S, Woodfield TBF, Xu T, Yoo JJ, Vozzi G (2018) Biofabrication: a guide to technology and terminology. Trends Biotechnol 36:384–402. https://doi.org/10.1016/j.tibtech.2017.10.015
8. Gungor-Ozkerim PS, Inci I, Zhang YS, Khademhosseini A, Dokmeci MR (2018) Bioinks for 3D bioprinting: an overview. Biomater Sci 6:915–946. https://doi.org/10.1039/c7bm00765e
9. Heinrich MA, Bansal R, Lammers T, Zhang YS, Michel Schiffelers R, Prakash J (2019) 3D-bioprinted mini-brain: a glioblastoma model to study cellular interactions and therapeutics. Adv Mater 31:e1806590. https://doi.org/10.1002/adma.201806590
10. Liu Z, Zhang M, Bhandari B, Wang Y (2017) 3D printing: printing precision and application in food sector. Trends Food Sci Technol 69:83–94. https://doi.org/10.1016/j.tifs.2017.08.018

11. Wong GHC, Pant A, Zhang Y, Chua CK, Hashimoto M, Leo CH, Tan UX (2022) 3D food printing-sustainability through food waste upcycling. Mater Today Proc 70:627–630. https://doi.org/10.1016/j.matpr.2022.08.565
12. Mantihal S, Kobun R, Lee BB (2022) 3D food printing of as the new way of preparing food: a review. Int J Gastron Food Sci 22:100260. https://doi.org/10.1016/j.ijgfs.2020.100260
13. Sun J, Peng Z, Yan L, Fuh J, Hong GS (2015) 3D food printing-an innovative way of mass customization in food fabrication. Int J Bioprint 1:27–38. https://doi.org/10.18063/IJB.2015.01.006
14. Yang F, Zhang M, Bhandari B (2017) Recent development in 3D food printing. Crit Rev Food Sci Nutr 57:3145–3153. https://doi.org/10.1080/10408398.2015.1094732
15. Nachal N, Moses JA, Karthik P, Anandharamakrishnan C (2019) Applications of 3D printing in food processing. Food Eng Rev 11:123–141.https://doi.org/10.1007/s12393-019-09199-8
16. Lorenz T, Iskandar MM, Baeghbali V, Ngadi MO, Kubow S (2022) 3D food printing applications related to Dysphagia: a narrative review. Foods 11:1789. https://doi.org/10.3390/foods11121789
17. Sun J, Zhou W, Huang D, Fuh JYH, Hong GS (2015) An overview of 3D printing technologies for food fabrication. Food Bioproc Tech 8:1605–1615. https://doi.org/10.1007/s11947-015-1528-6
18. Sun J, Peng Z, Zhou W, Fuh JYH, Hong GS, Chiu A (2015) A review on 3D printing for customized food fabrication. Procedia Manuf 1:308–319. https://doi.org/10.1016/j.promfg.2015.09.057
19. Zhang B, Montgomery M, Chamberlain MD, Ogawa S, Korolj A, Pahnke A, Wells LA, Massé S, Kim J, Reis L, Momen A, Nunes SS, Wheeler AR, Nanthakumar K, Keller G, Sefton MV, Radisic M (2016) Biodegradable scaffold with built-in vasculature for organ-on-a-chip engineering and direct surgical anastomosis. Nat Mater 15:669–678. https://doi.org/10.1038/nmat4570
20. Munialo CD (2024) A review of alternative plant protein sources, their extraction, functional characterisation, application, nutritional value and pinch points to being the solution to sustainable food production. Int J Food Sci Technol 59:462–472. https://doi.org/10.1111/ijfs.16467
21. Derakhshanfar S, Mbeleck R, Xu K, Zhang X, Zhong W, Xing M (2018) 3D bioprinting for biomedical devices and tissue engineering: a review of recent trends and advances. Bioact Mater 3:144–156. https://doi.org/10.1016/j.bioactmat.2017.11.008
22. Kyle S, Jessop ZM, Al-Sabah A, Whitaker IS (2017) 'Printability' of candidate biomaterials for extrusion based 3D printing: state-of-the-art. Adv Healthc Mater 6:1700264. https://doi.org/10.1002/adhm.201700264
23. Gudapati H, Dey M, Ozbolat I (2016) A comprehensive review on droplet-based bioprinting: past, present and future. Biomaterials 102:20–42. https://doi.org/10.1016/j.biomaterials.2016.06.012
24. Peltola SM, Melchels FP, Grijpma DW, Kellomäki M (2008) A review of rapid prototyping techniques for tissue engineering purposes. Ann Med 40:268–280. https://doi.org/10.1080/07853890701881788
25. Melchels FPW, Feijen J, Grijpma DW (2010) A review on stereolithography and its applications in biomedical engineering. Biomaterials 31:6121–6130. https://doi.org/10.1016/j.biomaterials.2010.04.050
26. Guillotin B, Souquet A, Catros S, Duocastella M, Pippenger B, Bellance S, Bareille R, Rémy M, Bordenave L, Amédée J, Guillemot F (2010) Laser assisted bioprinting of engineered tissue with high cell density and microscale organization. Biomaterials 31:7250–7256. https://doi.org/10.1016/j.biomaterials.2010.05.055
27. Guillemot F, Souquet A, Catros S, Guillotin B, Lopez J, Faucon M, Pippenger B, Bareille R, Rémy M, Bellance S, Chabassier P, Fricain JC, Amédée J (2010) High-throughput laser printing of cells and biomaterials for tissue engineering. Acta Biomater 6:2494–2500. https://doi.org/10.1016/j.actbio.2009.09.029
28. Ovsianikov A, Gruene M, Pflaum M, Koch L, Maiorana F, Wilhelmi M, Haverich A, Chichkov B (2010) Laser printing of cells into 3D scaffolds. Biofabrication 2:014104. https://doi.org/10.1088/1758-5082/2/1/014104
29. Vanaei S, Parizi MS, Vanaei S, Salemizadehparizi F, Vanaei HR (2021) An overview on materials and techniques in 3D bioprinting toward biomedical application. Eng Regen 2:1–18. https://doi.org/10.1016/j.engreg.2020.12.001
30. Osidak EO, Kozhukhov VI, Osidak MS, Domogatsky SP (2020) Collagen as bioink for bioprinting: a comprehensive review. Int J Bioprint 6:270. https://doi.org/10.18063/ijb.v6i3.270
31. Petta D, D'Amora U, Ambrosio L, Grijpma DW, Eglin D, D'Este M (2020) Hyaluronic acid as a bioink for extrusion-based 3D printing. Biofabrication 12:032001. https://doi.org/10.1088/1758-5090/ab8752

32. Ahmadi Soufivand A, Faber J, Hinrichsen J, Budday S (2023) Multilayer 3D bioprinting and complex mechanical properties of alginate-gelatin mesostructures. Sci Rep 13: 11253. https://doi.org/10.1038/s41598-023-38323-2
33. Kumi M, Wang T, Ejeromedoghene O, Wang J, Li P, Huang W (2024) Potentials of chitin and chitosan-based bioinks for 3D-printing of flexible electronics: the future of sustainable bioelectronics. Small Methods 8:e2301341. https://doi.org/10.1002/smtd.202301341
34. Wenger L, Radtke CP, Gerisch E, Kollmann M, Niemeyer CM, Rabe KS, Hubbuch J (2022) Systematic evaluation of agarose- and agar-based bioinks for extrusion-based bioprinting of enzymatically active hydrogels. Front Bioeng Biotechnol 10:928878. https://doi.org/10.3389/fbioe.2022.928878
35. Gonzalez-Fernandez T, Tenorio AJ, Campbell KT, Silva EA, Leach JK (2021) Alginate-based bioinks for 3D bioprinting and fabrication of anatomically accurate bone grafts. Tissue Eng Part A 27:1168–1181. https://doi.org/10.1089/ten.tea.2020.0305
36. Kurz W, Fisher DJ, Rappaz M (2023) Fundamentals of solidification. Trans Tech Publications Limited, Switzerland, p 372
37. Rubinsky B, Ikeda M (1985) A cryomicroscope using directional solidification for the controlled freezing of biological material. Cryobiology 22:55–68. https://doi.org/10.1016/0011-2240(85)90008-2
38. Rubinsky B, Lee C, Chow M (1993) Experimental observations and theoretical studies on solidification processes in saline solutions. Exp Therm Fluid Sci 6:157–167. https://doi.org/10.1016/0894-1777(93)90025-E
39. Preciado JA, Skandakumaran P, Cohen S, Rubinsky B (2003) Utilization of directional freezing for the construction of tissue engineering scaffolds. In: ASME 2003 International Mechanical Engineering Congress and Exposition. Paper no. IMECE2003-42067. pp 439–442. https://doi.org/10.1115/IMECE2003-42067
40. Mazur P (1984) Freezing of living cells—mechanisms and implications. Am J Physiol 247:C125–C142. https://doi.org/10.1152/ajpcell.1984.247.3.C125
41. Mazur P (1970) Cryobiology: the freezing of biological systems. Science 168:939–949. https://doi.org/10.1126/science.168.3934.939
42. Smith AU, Polge C (1950) Survival of spermatozoa at low temperatures. Nature 166:668–669. https://doi.org/10.1038/166668a0
43. Polge C, Smith AU, Parkes AS (1949) Revival of spermatozoa after vitrification and dehydration at low temperatures. Nature 164:666. https://doi.org/10.1038/164666a0
44. Ishiguro H, Rubinsky B (1994) Microscopic behavior of ice crystals and biological cells during directional solidification of solutions with cells. Trans Jpn Soc Mech Eng 60:572B
45. Ishiguro H, Rubinsky B (1994) Mechanical interactions between ice crystals and red blood cells during directional solidification. Cryobiology 31:483–500. https://doi.org/10.1006/cryo.1994.1059
46. Tatsutani K, Rubinsky B (1998) A method to study intracellular ice nucleation. J Biomech Eng 120:27–31. https://doi.org/10.1115/1.2834302
47. Tatsutani K, Rubinsky B (1997) A technique to study the effects of cellular dehydration. Cryobiology 35:377
48. Rui J, Tatsutani KN, Dahiya R, Rubinsky B (1999) Effect of thermal variables on human breast cancer in cryosurgery. Breast Cancer Res Treat 53:185–192. https://doi.org/10.1023/A:1006182618414
49. Rubinsky B, Cravalho EG (1982) Transient mass transfer processes during the perfusion of a biological organ with a cryophylactic agent solution. Cryobiology 19:70–82. https://doi.org/10.1016/0011-2240(82)90126-2
50. Rubinsky B, Pegg DE, Diaper MP, Lee C (1984) Analysis of cryophylactic agents introduction and removal in rabbit kidneys using a Krogh cylinder model. Cryobiology 21:705
51. Rubinsky B, Pegg DE (1988) A mathematical model for the freezing process in biological tissue. Proc R Soc B 234:343–358. https://doi.org/10.1098/rspb.1988.0053
52. Rubinsky B, Cravalho EG, Mikic B (1980) Thermal stresses in frozen organs. Cryobiology 17:66–73. https://doi.org/10.1016/0011-2240(80)90009-7
53. Lottermoser A (1908) Über das Ausfrieren von Hydrosolen. Ber Dtsch Chem Ges 41:3976–3979
54. Dagalakis N, Stsikelis FP, Burhke F, Yannas IV (1980) Design of an artificial skin. Part III. Control of pore structure. J Biomed Mater Res 14:511–528. https://doi.org/10.1002/jbm.820140417
55. Shapiro L, Cohen S (1997) Novel alginate sponges for cell culture and transplantation. Biomaterials 18:583–590. https://doi.org/10.1016/S0142-9612(96)00181-0
56. Adamkiewicz M, Rubinsky B (2015) Cryogenic 3D printing for tissue engineering. Cryobiology 71:518–521. https://doi.org/10.1016/j.cryobiol.2015.10.152
57. Zawada B, Ukpai G, Powell-Palm MJ, Rubinsky B (2018) Multi-layer cryolithography for additive manufacturing. Prog Add Manufact

3:245–255. https://doi.org/10.1007/s40964-018-0045-3
58. Ukpai G, Sahyoun J, Stuart R, Wang S, Xiao Z, Rubinsky B (2019) A parallel multiple layer cryolithography device for the manufacture of biological material for tissue engineering. J Med Devices 13:035001. https://doi.org/10.1115/1.4043080
59. Ukpai G, Rubinsky B (2020) A mathematical analysis of directional solidification of aqueous solutions. J Heat Transfer 142:022401. https://doi.org/10.1115/1.4045312
60. Ukpai G, Rubinsky B (2020) A three-dimensional model for analysis and control of phase change phenomena during 3D printing of biological tissue. Bioprinting 18:e00077. https://doi.org/10.1016/j.bprint.2020.e00077
61. Warburton L, Rubinsky B (2022) Freezing modulated crosslinking: a crosslinking approach for 3D cryoprinting. Bioprinting 27:e00225. https://doi.org/10.1016/j.bprint.2022.e00225
62. Warburton L, Lou L, Rubinsky B (2021) A modular 3D bioprinter for printing porous scaffolds for tissue engineering. J Heat Transfer 144:031205. https://doi.org/10.1115/1.4053198
63. Lou L, Rubinsky B (2023) Temperature-controlled 3D cryoprinting inks made of mixtures of alginate and agar. Gels 9:689. https://doi.org/10.3390/gels9090689
64. Lou L, Rubinsky B (2024) Diffusion coefficients in scaffolds made with temperature controlled cryoprinting and an ink made of sodium alginate and agar. Bioprinting 41:e00348. https://doi.org/10.1016/j.bprint.2024.e00348
65. Warburton L, Rubinsky B (2023) Cryopreservation of 3D bioprinted scaffolds with temperature-controlled-cryoprinting. Gels 9:502. https://doi.org/10.3390/gels9060502
66. Warburton L, Cheng A, Rubinsky B (2025) Cryopreservation of vascularizable tissue with temperature-controlled-cryoprinting. Bioprinting 48:e00411. https://doi.org/10.1016/j.bprint.2025.e0041
67. Lou L, Bilbao-Sainz C, Wood D, Rubinsky B (2023) Temperature controlled cryoprinting of food for dysphagia patients. Innov Food Sci Emerg Technol 86:103362. https://doi.org/10.1016/j.ifset.2023.103362
68. Lou L, Bilbao-Sainz C, Rubinsky B (2025) Temperature-controlled cryoprinting vegetarian food with modified macro- and microstructures for Dysphagia. ACS Food Sci Technol 5:3219–3228. https://doi.org/10.1021/acsfoodscitech.5c00556
69. Rubinsky B, Adamkiewicz M, Shaked Z (2018) Systems, apparatus and methods for cryogenic 3D printing. US20180304537A1
70. Rubinsky B, Adamkievicz M, Shaked Z (2021) Systems, apparatus and methods for cryogenic 3D printing. US20180304537A1
71. Rubinsky D, Rubinsky B, Shaked Z, Ukpai G (2021) Parallel-additive manufacturing of objects made of aqueous and/or organic materials. US20210137153
72. Rubinsky, B. Rubinsky D, Ukpai G (2023) Systems and methods for additive manufacturing of materials with controllable microscale textures. US20230284670A1

Part III

Quality Assessment of Stored Biologics and Logistics

Chapter 9

Standard Operating Procedures in Biorepositories and Biobanks

Robyn J. Osborne

Abstract

Whole blood represents one of the most practical and frequently collected human biospecimens. Its accessibility and minimally invasive collection make it a cornerstone for both clinical studies and laboratory research. Once obtained, whole blood can be fractionated into multiple components that provide valuable material for diverse experimental purposes. Plasma, serum, and peripheral blood mononuclear cells (PBMCs) are among the most widely used derivatives, each offering distinct analytical advantages, from biomarker discovery to immunological profiling. Data reproducibility and clinical translation depend heavily on biospecimen quality; maintaining the integrity of whole blood samples throughout processing, storage, and handling is of paramount importance. Variability introduced during these steps can compromise cell viability, distort biomarker profiles, and reduce confidence in study outcomes. Standardized operating procedures (SOPs) serve as a critical safeguard by minimizing inconsistencies, ensuring regulatory compliance, and reinforcing trust in biobanking practices. This chapter details four well-validated SOPs for processing whole blood collected in blood collection tubes (BCT). These SOPs describe methods for preparing Double-Spun Plasma and Serum, and for isolating PBMCs using two complementary approaches, with particular emphasis on strategies that preserve sample quality and integrity across all stages of processing.

Keywords Biobank SOP, PBMC, Plasma, Serum, Standard

1 Introduction

Biobanking of human blood samples underpins critical discovery and translational research, driving the development of novel therapies and diagnostics, and yielding deeper insights into human disease. A biobank is responsible for recording, monitoring, and safeguarding the quality and integrity of biospecimens entrusted to its care. This responsibility begins with a robust quality management system that incorporates meticulous recordkeeping, validated operating protocols, comprehensive training programs, and ongoing competency assessments.

Willem F. Wolkers and Harriëtte Oldenhof (eds.), *Cryopreservation and Freeze-Drying Protocols*, Methods in Molecular Biology, vol. 3060, https://doi.org/10.1007/978-1-0716-5416-3_9,

Standard operating procedures (SOPs) provide clear, standardized guidelines to minimize inconsistencies, reduce errors, and ensure compliance with regulatory requirements. Adherence to well-defined SOPs not only strengthens operational efficiency but also upholds ethical and legal obligations, fostering trust within the research community. Equally important is the design and selection of fit-for-purpose SOPs, which requires an understanding of the intended downstream applications of samples and a consistent focus on preserving their quality and integrity. Failure to maintain these standards can compromise data reproducibility, scientific validity, and the translational value of research.

Aligned with the International Society for Biological and Environmental Repositories (ISBER) Best Practices [1] and Good Clinical Laboratory Practice (GCLP) Guidelines [2], an SOP is a concise, step-by-step document that typically includes a summary or description, safety considerations, a list of required equipment and reagents, the detailed procedure, and relevant attachments. This structured approach ensures SOPs remain clear, accessible, and easily reproducible by any appropriately trained individual. This chapter presents validated SOPs that define sample quality expectations and adhere to established best practices for the processing of human blood-derived specimens, with an unwavering emphasis on preserving sample quality and integrity at every stage.

2 Materials

2.1 Cryopreservation of Double-Spun Plasma from Whole Blood (see Notes 1 and 2)

2.1.1 Specimen Type

1. Whole blood collected in Blood Collection Tubes (BCT) containing anticoagulant (*see* **Note 3**).

2.1.2 Equipment

1. Centrifuge for BCT (e.g., Thermo Scientific).
2. Single Channel Manual Pipettes (e.g., Costar).
3. Automated Serological Pipettor (e.g., Drummond).
4. Biological Safety Cabinet, Class II (e.g., ThermoFisher).
5. Freezer −80°C.

2.1.3 Disposables

1. 2 mL Polypropylene Cryovials (e.g., Thermo Scientific) (*see* **Notes 4 and 5**).
2. Serological Pipets (e.g., Falcon).
3. Pipet Tips (e.g., Falcon).

4. Polypropylene centrifuge conical bottom tubes (e.g., Falcon) (*see* **Note 4**).

2.2 Cryopreservation of Serum from Whole Blood Collected in Blood Collection Tubes With or Without a Gel Barrier (see Notes 6 and 7)

2.2.1 Specimen Type

1. Whole blood was collected in blood collection tubes (BCT) containing a clot activator or no additive. Some BCT may also contain a gel barrier, which separates the clot from the serum after centrifugation.

2.2.2 Equipment

1. Centrifuge for BCT (e.g., Thermo Scientific).
2. Single Channel Manual Pipettes (e.g., Costar).
3. Automated Serological Pipettor (e.g., Drummond).
4. Biological Safety Cabinet, Class II (e.g., ThermoFisher).
5. Freezer −80°C.

2.2.3 Disposables

1. 2 mL Polypropylene Cryovials (e.g., Thermo Scientific) (*see* **Notes 8 and 9**).
2. Serological Pipets (e.g., Falcon).
3. Pipet Tips (e.g., Falcon).

2.3 Isolation and Cryopreservation of Peripheral Blood Mononuclear cells (PBMC) from Whole Blood Using Leucosep™ Tubes Containing Density Gradient Separation Media (see Notes 10 and 11)

2.3.1 Specimen Type

1. Whole blood is collected in blood collection tubes that contain anti-coagulant, typically acid citrate dextrose (ACD) and sodium heparin (Na Hep).

2.3.2 Equipment

1. Centrifuge for BCT (e.g., Thermo Scientific).
2. Single Channel Manual Pipettes (e.g., Costar).
3. Automated Serological Pipettor (e.g., Drummond).

4. CoolCell Cell Freezing Container (e.g., Biocision).
5. Biological Safety Cabinet, Class II (e.g., ThermoFisher).
6. Freezer −80°C.
7. LN_2 Freezer.

2.3.3 Reagents

1. Lymphocyte Separation Medium (LSM) (e.g., Corning).
2. Dulbecco's Phosphate Buffered Saline (DPBS) (e.g., Gibco).
3. Dimethyl Sulfoxide (DMSO) (e.g., Sigma).
4. Fetal Bovine Serum (FBS) (e.g., Gemini).

2.3.4 Disposables

1. 2 mL Polypropylene Cryovials (e.g., Thermo Scientific) (*see* **Notes 12 and 13**).
2. Serological Pipets (e.g., Falcon).
3. Pipet Tips (e.g., Falcon).
4. Polypropylene centrifuge conical bottom tubes (e.g., Falcon) (*see* **Note 12**).
5. Leucosep™ tubes (12 mL and/or 50 mL) (e.g., Greiner Bio-One).

2.3.5 Reagent Preparation

1. Leucosep™ Tube Preparation (*see* **Note 14**).
   (a) For the 12 mL tubes, load 3.3 mL LSM above the frit.
   (b) For the 50 mL tubes, load 15.5 mL LSM above the frit.
   (c) Centrifuge: (i) Temperature = 18–25°C, (ii) Speed = 800–1000 RCF, (iii) Time = 1 min, (iv) Brake: ON or set to Maximum Deceleration.
   (d) Remove tubes from centrifuge, inspect for a gap (or a large bubble) between the frit and LSM (which will have now moved below the frit). If large bubbles are present, add additional LSM and re-centrifuge the tubes.
   (e) Tubes can be prepared in batches, labeled, and stored for 1 month (30 days) at room temperature.
2. Freeze Media—90% hiFBS and 10% DMSO—If a different freeze medium is used, refer to the instructions specific to that product.
   (a) Make fresh daily.
   (b) Freshly made freezing media must be allowed to cool on wet ice for a minimum of 10–15 min prior to use.
   (c) Keep on wet ice until used.
   (d) Discard any remaining media at the end of the day.

2.4 Isolation and Cryopreservation of Peripheral Blood Mononuclear Cells (PBMC) from BD Vacutainer® Cell Preparation Tubes (CPT™) (see Notes 15 and 16)

2.4.1 Specimen Type

1. Whole blood collected in BD Vacutainer® Cell Preparation Tubes (CPT™) that contain the anti-coagulant sodium citrate or sodium heparin.

2.4.2 Equipment

1. Centrifuge for BCT (e.g., Thermo Scientific) (*see* **Note 17**).
2. Single Channel Manual Pipettes (e.g., Costar).
3. Automated Serological Pipettor (e.g., Drummond).
4. CoolCell Cell Freezing Container (e.g., Biocision).
5. Biological Safety Cabinet, Class II (e.g., ThermoFisher).
6. Freezer −80°C.
7. LN2 Freezer.

2.4.3 Reagents

1. Dulbecco's Phosphate Buffered Saline (DPBS) (e.g., Gibco).
2. Dimethyl Sulfoxide (DMSO) (e.g., Sigma).
3. Fetal Bovine Serum (FBS) (e.g., Gemini).
4. High Yield Lyse (Manufacturer: Life Tech).

2.4.4 Disposables

1. 2 mL Polypropylene Cryovials (e.g., Thermo Scientific) (*see* **Notes 18 and 19**).
2. Serological Pipets (e.g., Falcon).
3. Pipet Tips (e.g., Falcon).
4. Polypropylene centrifuge conical bottom tubes (e.g., Falcon) (*see* **Note 18**).

2.4.5 Reagent Preparation

1. High-Yield Lyse [3].
   (a) Lyse may be aliquoted and kept at room temperature.
   (b) Place aliquot(s) of High Yield Lyse into the 37°C water bath each day to warm for a minimum of 10 min prior to processing whole blood samples for PBMC.
2. Freeze Media—90% hiFBS and 10% DMSO—If a different freeze medium is used, refer to the instructions specific to that product.

(a) Make fresh daily.

(b) Freshly made freezing media must be allowed to cool on wet ice for a minimum of 10–15 min prior to use.

(c) Keep on wet ice until used.

(d) Discard any remaining media at the end of the day.

3 Methods

3.1 Cryopreservation of Double-Spun Plasma from Whole Blood

Plasma is a derivative of whole blood that is collected in BCT that contains an anticoagulant (*see* **Notes 20 and 21**).

1. Gently invert the BCT containing freshly collected whole blood approximately five times to ensure the anticoagulant is properly mixed.
2. Centrifuge under the following conditions: (i) Temperature = 4–25°C (*see* **Note 22**), (ii) Speed = 2500 RCF, (iii) Time = 15 min, (iv) Brake: ON or set to Maximum Deceleration.
3. Remove plasma and transfer to a clean, conical-bottom tube.
4. Centrifuge a second time as in Step 2 above.
5. Immediately, upon completion of the second centrifugation, aliquot plasma (*see* **Note 23**) into appropriately labeled cryovials and place in −80°C freezer for long-term storage.

3.2 Cryopreservation of Serum from Whole Blood Collected in Blood Collection Tubes With or Without a Gel Barrier

Serum is a derivative of whole blood that was collected in a BCT that contained either a clot activator or no additive and formed a clot (*see* **Note 24**). Processing of BCT for serum should begin as soon as possible after the clot is fully formed. (see **Note 25**).

1. Centrifuge under the following conditions:

   (a) Temperature = 4–25°C (*see* **Note 26**).

   (b) For BCT containing a gel barrier: Centrifuge Speed should be set between 1000–1300 RCF. Higher speeds may dislodge the gel [4]. (i) Centrifuge for 10 min with the brake set to ON or Maximum Deceleration. (ii) After centrifugation, the serum may remain in the BCT or be aliquoted into cryovials.

   (c) For BCT with no gel barrier: Centrifuge Speed should be set at ≤1300 RCF [4] (*see* **Note 27**). (i) Centrifuge for 10–15 min with the brake set to ON or Maximum Deceleration. (ii) Aliquot immediately after centrifugation.

   (d) Aliquoted serum should be frozen immediately in a low-temperature freezer for long-term storage.

   (e) (*see* **Notes 28 and 29**).

3.3 Isolation and Cryopreservation of Peripheral Blood Mononuclear Cells (PBMC) from Whole Blood Using Leucosep™ Tubes Containing Density Gradient Separation Media

This method [5, 6] employs density gradient centrifugation to isolate peripheral blood mononuclear cells (PBMC from anticoagulated whole blood (*see* **Notes 30 and 31**).

1. Gently invert each BCT approximately three to four times to ensure the anticoagulant is properly mixed.
2. Use a serological pipette to transfer the whole blood from each BCT into LSM-loaded Leucosep™ tube(s) (*see* **Note 32**).
   (a) For blood volumes ≥15 mL, use a 50 mL LeucoSep™ tube (*see* **Note 33**).
   (b) For blood volumes <15 mL, use the 12 mL Leucosep™ tube (*see* **Note 34**).
3. Centrifuge under the following conditions: (swing-out head, horizontal rotor): (i) Temperature = 18–5°C, (ii) Speed = 800 RCF, (iii) Time = 20 min (*see* **Note 35**), (iv) Brake: OFF or set to Minimum Deceleration (*see* **Note 36**).
4. Gently remove the samples from the centrifuge to avoid disturbing the layers. After centrifugation, the whole blood will have separated into the following distinct layers starting at the top of the tube and moving down: (i) Plasma (*see* **Notes 37 and 38**), (ii) Interphase containing peripheral blood mononuclear cells (PBMC), (iii) Density gradient separation medium (LSM), (iv) Porous Disk, (v) Pellet containing erythrocytes and granulocytes.
5. Carefully transfer the interphase layer containing the PBMC to a sterile, properly labeled, 50 mL conical tube. Use one 50 mL conical tube per Leucosep™ tube, regardless of the size of the Leucosep™. Do NOT combine PBMC layers from multiple Leucosep™ tubes (from a single donor) at this step.
6. Add DPBS up to approximately 20–45 mL total volume to the PBMC suspension.
7. Centrifuge the PBMC suspension: (i) Temperature = 18–25°C, (ii) Speed = 300 RCF, (iii) Time = 10 min, (iv) Brake: ON or set to Maximum Deceleration.
8. Confirm the cell pellet is visible and then carefully pour off the supernatant (*see* **Note 39**).
9. Mechanically disrupt the pellet by flicking the tube with your fingers or other means. Do not vortex.
10. Resuspend the cells in a total volume of 20–45 mL DPBS, depending on the size of the cell pellet.
11. At this stage, if multiple Leucosep™ tubes from the same donor are being processed, the resulting cell pellets may be combined. Do not combine pellets before this point. To combine, resuspend the first pellet in the single-pellet volume

(e.g., 35 mL of DPBS), reserving 5–10 mL of DPBS. Use the same 35 mL suspension to resuspend each additional pellet. Then, with a separate pipette, use the reserved 5–10 mL of DPBS to rinse residual cells from each tube and add them to the combined suspension. This process should result in a single 50 mL conical containing 40–45 mL of cells resuspended in DPBS.

12. Centrifuge the PBMC suspension: (i) Temperature = 18–25°C, (ii) Speed = 300 RCF, (iii) Time = 10 min, (iv) Brake: ON or set to Maximum Deceleration.
13. Confirm the cell pellet is visible and then carefully pour off the supernatant.
14. Mechanically disrupt the pellet by flicking the tube with your fingers or other means. Do not vortex.
15. Resuspend the PBMCs in a volume appropriate for the counting method utilized (*see* **Note 40**).
16. After counting is completed, centrifuge the resuspended cells as follows: (*see* **Note 41**): (i) Temperature = 18–25°C, (ii) Speed = 300 RCF, (iii) Time = 10 min, (iv) Brake: ON or set to Maximum Deceleration.
17. Confirm cell pellet is visible and then carefully pour off the supernatant, removing as much as possible, without disturbing the pellet (*see* **Note 42**).
18. Mechanically disrupt the pellet by flicking the tube with your fingers or other means. Do not vortex.
19. Ensure labeled cryovials are prepped and under the biosafety cabinet.
20. Very slowly add ice-cold freezing media (*see* **Note 43**) dropwise to the cells while simultaneously swirling the tube (*see* **Note 44**).
21. Pipette up and down gently to mix, without causing bubbles, and then aliquot 1 mL into each of the labeled 2 mL cryovials.
22. Transfer aliquots to a CoolCell (*see* **Notes 45 and 46**).
23. Place the CoolCell in a low-temperature freezer (–70 to –80°C) overnight (*see* **Note 47**).
24. Aliquots must be transferred to a –150°C or LN2 freezer for long-term storage within 3 weeks from the date placed in –80°C, optimally the transfer should take place the next working day [5, 7].

3.4 Isolation and Cryopreservation of Peripheral Blood Mononuclear Cells (PBMC) from BD Vacutainer® Cell Preparation Tubes (CPT™)

This method employs density gradient centrifugation and a gel barrier in a self-contained evacuated tube to isolate peripheral blood mononuclear cells (PBMCs) from anticoagulated whole blood (*see* **Notes 48–50**).

1. Gently invert each CPT™ approximately 8–10 times to ensure the anticoagulant is properly mixed.
2. Centrifuge under the following conditions (swing-out head, horizontal rotor): (i) Temperature = 18–25°C, (ii) Speed = 1500–1800 RCF (*see* **Note 51**), (iii) Time = 20 min, (iv) Brake: OFF or set to Minimum Deceleration.
3. Gently remove the samples from the centrifuge so as not to disturb the layers. After centrifugation, the whole blood will have separated into the following distinct layers starting at the top of the tube and moving down: (i) Plasma (*see* **Notes 52 and 53**), (ii) Interphase containing peripheral blood mononuclear cells (PBMC), (iii) Density gradient separation medium, (iv) Polyester Gel, (v) Erythrocytes and Granulocytes.
4. Carefully, remove the rubber top of each CTP™ and place them in a tube rack.
5. Using a transfer pipette, collect the interphase layer and transfer to a 50 mL conical bottom centrifuge tube (*see* **Note 54**).
6. Use a transfer pipette to gently rinse the surface of the gel in the CPT™ with 5–7 mL of DPBS. Pipette up and down gently a few times (*see* **Note 55**).
7. Transfer the DPBS used for rinsing the CPTs into the 50 mL conical tube containing the PBMCs.
8. Add additional DPBS up to approximately 20–45 mL total volume to the PBMC suspension.
9. Centrifuge the PBMC suspension: (i) Temperature = 18–25 °C, (ii) Speed = 300 RCF, (iii) Time = 10 min, (iv) Brake = ON or set to Maximum Deceleration.
10. Confirm the cell pellet is visible and then carefully pour off the supernatant.
11. Mechanically disrupt the pellet by flicking the tube with your fingers or other means. Do not vortex.
12. Re-suspend the pellet in pre-warmed (37°C) High Yield Lyse (*see* **Note 56**).
    (a) High-Yield Lyse should be added in a volume that is approximately equivalent to 10 times the volume of the pellet (*see* **Note 57**).
    (b) Mix the cell pellet and lyse gently.
    (c) Incubate for 10 min at room temperature.
    (d) After 10 min, bring volume up to 20–45 mL using DPBS.

13. Centrifuge conical tube under the following conditions: (i) Temperature = 18–25°C, (ii) Speed = 300 RCF, (iii) Time = 10 min, (iv) Brake = ON or set to Maximum Deceleration.
14. Confirm the cell pellet is visible and then carefully pour off the supernatant.
15. Mechanically disrupt the pellet by flicking the tube with your fingers or other means. Do not vortex.
16. Resuspend the PBMCs in a volume appropriate for the counting method utilized (*see* **Note 58**).
17. After counting is completed, centrifuge the resuspended cells as follows (see **Note 59**): (i) Temperature = 18–25°C, (ii) Speed = 300 RCF, (iii) Time = 10 min, (iv) Brake: ON or set to Maximum Deceleration.
18. Confirm that the cell pellet is visible, then carefully pour off the supernatant, removing as much as possible without disturbing the pellet (*see* **Note 60**).
19. Mechanically disrupt the pellet by flicking the tube with your fingers or other means. Do not vortex.
20. Ensure labeled cryovials are prepped and under the biosafety cabinet.
21. Very slowly add ice-cold freezing media (*see* **Note 61**) dropwise to the cells while simultaneously swirling the tube (*see* **Note 62**).
22. Pipette up and down gently to mix, without causing bubbles, and then aliquot 1 mL into each of the labeled 2 mL cryovials.
23. Transfer aliquots to a CoolCell (*see* **Notes 63 and 64**).
24. Place the CoolCell in a low-temperature freezer (−70 to −80°C) overnight (*see* **Note 65**).
25. Aliquots must be transferred to a −150°C or LN2 freezer for long-term storage within 3 weeks from the date placed in −80°C, optimally the transfer should take place the next working day [5, 7].

4 Notes

1. Advantages: This protocol outlines a basic procedure for the cryopreservation of plasma that is spun or centrifuged twice, i.e., double-spun. The benefit of spinning the plasma twice (once for the initial separation from whole blood and once for further purification) results in a "cleaner" sample with reduced platelets and cellular contamination [8, 9].

Additionally, higher speed spins, to further clarify the plasma, have also been recommended [10].

2. Potential Assay Utilization: Identification of blood-based biomarkers such as inflammatory, angiogenic and tumor-related growth factors or cell-free DNA [11–16]. Common assays include the following: (i) Multiplex Immunoassays, (ii) Enzyme-linked immunosorbent (ELISA) assays, (iii) Genomic, and (iv) Mass Spectrometry.
3. EDTA, heparin, citrate, and lithium heparin [13, 17–19] are common anticoagulants utilized for blood collections intended for plasma. Depending on the intended downstream application, other anticoagulants may also be suitable. However, consistency within each experimental group is critical; use the same anticoagulant across all samples within an experimental group to prevent variation in experimental conditions [20].
4. Polypropylene is recommended for storing plasma, serum, and other biological samples due to its nonreactive surface that prevents it from binding proteins.
5. Cryovials with barcodes embedded on the bottom are valuable for efficient inventory management and sample tracking in biobanks. Including human-readable text alongside the barcode adds convenience and confidence for both the technician retrieving the samples and the customer receiving them at handoff.
6. Advantages: This protocol outlines a basic procedure for the cryopreservation of serum collected from BCT with or without a gel barrier. The advantages of serum tubes with a gel barrier, otherwise known as serum separator tubes (SST™), are increased safety (limited or no aerosolization), higher yields, allowing for transport in the primary container by removing the necessity of aliquoting, and preventing the clot from dissolving back into the serum [17, 21].
7. Potential Assay Utilization: Identification of blood-based biomarkers such as inflammatory or injury markers, circulating nucleic acids, proteins, and hormones [11–14]. Common assays include the following: (i) Multiplex Immunoassays, (ii) Enzyme-linked immunosorbent (ELISA) assays, (iii) Genomic, and (iv) Mass Spectrometry.
8. Polypropylene is recommended for storing serum, plasma, and other biological samples due to its nonreactive surface that prevents it from binding proteins.
9. Cryovials with barcodes embedded on the bottom are valuable for efficient inventory management and sample tracking in biobanks. Including human-readable text alongside the

barcode adds convenience and confidence for both the technician retrieving the samples and the customer receiving them at handoff.

10. Advantages: PBMC isolation using Leucosep™ tubes offers the convenience of adding undiluted whole blood directly into the tube pre-filled with density gradient medium. This approach is advantageous as it minimizes the need for technical proficiency and allows for both the separation of PBMC and the preservation of plasma in its original, undiluted form.
11. Potential assay utilization: PBMCs are a readily accessible source of primary immune cells that can be used immediately or cryopreserved for later use. These cells serve as a snapshot of the current immune condition at the time of collection and are used for therapeutic treatments, vaccine discovery, and disease modeling. Common assays include the following [22–24]: (i) Flow Cytometry, (ii) ELISpot, (iii) Cytotoxicity, (iv) Proliferation, and (v) Immunogenicity.
12. Polypropylene is recommended for storing PBMC, serum, plasma, and other biological samples due to its nonreactive surface that prevents it from binding proteins.
13. Cryovials with barcodes embedded on the bottom are valuable for efficient inventory management and sample tracking in biobanks. Including human-readable text alongside the barcode adds convenience and confidence for both the technician retrieving the samples and the customer receiving them at handoff.
14. Per the manufacturer, the volume of separation media to add to the 50 mL Leucosep™ is 15.0 and 3.0 mL for the 12 mL tube, we have found that if we add 15.5 mL to the 50 mL Leucosep™ tube and 3.3 mL to the 12 mL one there is less of an issue with a gap between the separation media and the frit.
15. Advantages: PBMC isolation using CPT™ offers the convenience of separating PBMCs in a single centrifugation step and the ability to transport them overnight without removing them from the original container. This approach is advantageous as it minimizes the need for technical proficiency and allows for both the separation of PBMC and the preservation of plasma in its original, undiluted form.
16. Potential assay utilization: PBMCs are a readily accessible source of primary immune cells that can be used immediately or cryopreserved for later use. These cells serve as a snapshot of the current immune condition at the time of collection and are used for therapeutic treatments, vaccine discovery, and disease modeling. Common assays include the following [22–24]: (i) Flow Cytometry, (ii) ELISpot, (iii) Cytotoxicity, (iv) Proliferation, and (v) Immunogenicity.

17. CPT™ tubes are available in two draw volumes: 4 mL and 8 mL. The 8 mL tube measures 16×125 mm, making it 25 mm longer than standard 8.5 mL ACD and 10 mL Red-Top serum tubes. Due to their increased length, the 8 mL CPT™ tubes require taller centrifuge buckets than those typically found on standard centrifuges.
18. Polypropylene is recommended for storing PBMC,serum, plasma, and other biological samples due to its nonreactive surface that prevents it from binding proteins.
19. Cryovials with barcodes embedded on the bottom are valuable for efficient inventory management and sample tracking in biobanks. Including human-readable text alongside the barcode adds convenience and confidence for both the technician retrieving the samples and the customer receiving them at handoff.
20. BCT should be filled properly, per manufacturer recommendations, to ensure the proper sample to anticoagulant ratio [17].
21. It is strongly recommended that whole blood intended for plasma separation be centrifuged within 2 h of collection—or sooner—if required by the stability of specific analytes, depending on downstream applications. While older blood samples may still be usable, analyte integrity can be compromised [13]. To preserve analyte stability, whole blood should be processed as promptly as possible [25, 26]. At a minimum, ensure that all samples within an experimental group are processed within consistent time frames to minimize intra-group variability [20].
22. While spinning samples at 4°C is preferred, the duration of the centrifugation is generally not long enough to significantly impact results (verification of specific analytes is recommended). Additionally, access to a refrigerated centrifuge may not always be available.
23. It is recommended to freeze multiple smaller, single-use aliquots rather than a few large ones to minimize freeze-thaw cycles, which can compromise sample integrity [20].
24. BCT with gel separators should be filled appropriately per manufacturer recommendations [4].
25. Whole blood collected in BCT intended for serum must be allowed to clot 30–60 min prior to centrifugation. Centrifugation is generally recommended within 2 h of clot formation to prevent analyte degradation, as substances like glucose and potassium are particularly sensitive to delays between blood collection and serum separation [27–29].

26. While spinning samples at 4°C is preferred, the duration of the centrifugation is generally considered not long enough to significantly impact results (verification of specific analytes is recommended). Additionally, access to a refrigerated centrifuge may not always be available.
27. BCT without gel separators is often spun at higher RCF (2500) with the plasma BCT due to time and space constraints, with no issue at our laboratory.
28. It is recommended to freeze multiple smaller, single-use aliquots rather than a few large ones to minimize freeze-thaw cycles, which can compromise sample integrity [20].
29. The remaining blood clot can serve as a source of DNA for genotyping and other DNA-based analyses [30, 31].
30. Whole blood intended for PBMC should be processed within 8 h to maintain optimal sample integrity [5, 7]. Delayed processing has been shown to impact the frequency of some populations, cytokine and signaling pathways, but ensuring consistency by planning for consistent processing times within each experimental group will help control for possible variation [32].
31. Sample acceptance criteria: (i) BCT should be filled properly, per manufacturer recommendations. (ii) Whole blood to be processed for PBMC should be received in ample time to allow for processing to be completed in the allotted time frame defined in the study parameters, noting that processing completion within 8 h or less is considered optimal [5].
32. The blood may be poured directly into the Leucosep™ tube or transferred by pipette. Pipetting allows for a precise measurement of the volume of whole blood collected, a required variable used for calculating initial cell yield, a recommended pre-analytical quality metric [5].
33. A minimum of 15 mL to a maximum of 30 mL of whole blood is required per 50 mL LeucoSep™ tube. Do not exceed 30 mL of blood per 50 mL LeucoSep™ tube or go under 15 mL. Volumes below the required minimum may cause the interphase layer to form within or beneath the frit.
34. A minimum of 3 mL to a maximum of 8 mL of whole blood is required per 12 mL LeucoSep™ tube. Do not exceed 8 mL of blood per small LeucoSep™ tube or go under 3 mL. Volumes below the required minimum may cause the interphase layer to form within or beneath the frit.
35. Our laboratory standardly centrifuges the Leucosep™ tubes for 20 min, which is different than the 15 min required in the manufacturer's recommendation [6]. It provides a cleaner PBMC interphase.

36. The centrifuge brake must be set to off, or the deceleration set to "0" for this step. Failure to do so could result in the layers being disrupted and difficult to distinguish.
37. Plasma Collection and Cryopreservation: Due to the unique nature of the Leucosep™ tube, plasma remains undiluted and can be retained. If retaining the plasma, after the initial separation is completed, use a transfer pipette to collect the plasma layer and transfer it to a 50 mL conical tube. Be careful not to disturb the interphase layer containing the PBMC, leaving behind a small amount of plasma as a buffer next to the layer. Follow the plasma processing SOP, "Cryopreservation of Double-Spun Plasma from Whole Blood," to complete the processing of the plasma by starting at the 2nd centrifugation step.
38. Plasma collected from whole blood samples intended for PBMC isolation is often "older" than plasma obtained from blood collected specifically for plasma preparation. Additionally, this SOP requires whole blood for PBMC isolation to be centrifuged at room temperature, whereas blood collected for plasma is typically centrifuged at 4°C. It is crucial to document the timing and method of plasma collection, as these factors may impact downstream data analysis.
39. Troubleshooting: Issue—PBMC cell-pellet is grossly contaminated with Erythrocytes or red blood cells (RBCs) [3]. RBC contamination can result in a loose cell pellet that is easily lost during wash buffer removal, interfere with accurate cell counts, and cause cells to become sticky and clumpy after thawing. The following "High Yield Lyse Method for Removal of Red Blood Cells from PBMC" may be followed to remove the RBC contamination, leaving behind a clean PBMC pellet. (i) Without disturbing the pellet, use a pipette to carefully remove as much of the wash buffer as possible. (ii) Add pre-warmed, 37°C, High Yield Lyse (Manufacturer: Life Tech) to the conical tube containing the PBMC cell pellet at a volume equal to approximately 10× the volume of the pellet after the wash buffer has been removed (minimum 10 mL). For example, if the volume remaining is approximately 2 mL, add 20 mL of High Yield Lyse. (iii) If there are multiple tubes from the same donor, treat each tube separately; do not combine the tubes. (iv) Mix the cell pellet and lyse mixture gently by pipetting up and down. (v) Incubate at room temperature or in a water bath at 37°C (for severely contaminated samples) for 10 min. (vi) Centrifuge conical tube containing the PBMC and lyse mixture under the following conditions: 18–25°C, 300 RCF, 10 min, Brake: ON/set to Maximum Deceleration. (vii) Confirm the cell pellet is visible and then

carefully pour off the supernatant. (viii) Proceed to the second wash in the procedure, Step "10."

40. When resuspending PBMC for counting, it is important not to overmix. Do not allow bubbles to form; be gentle, but mix thoroughly. If counting multiple times, remix between counts, PBMC will settle quickly.

41. We recommend preparing and labeling the PBMC cryovials during the final centrifugation step. The number of cryovials required should be based on the cell count obtained immediately prior.

    i. Typically, PBMC are cryopreserved at a concentration of 5–20 million cells per mL in 1 mL aliquots.

42. We have found that decanting the fluid, followed by gently inverting the conical tube and tapping it onto a clean, dry paper towel inside the biosafety cabinet, is effective for removing residual liquid. However, if absolute sterility is required, this step should be carefully evaluated. In such cases, sterile towels must be used, or the step should be omitted entirely. Do not perform this step if there is any concern that the cell pellet has been disturbed. A clean, RBC-free pellet will remain intact, but a disturbed pellet may pour off with the liquid.

43. Ensure freezing media was allowed to cool on wet ice for a minimum of 10–15 min prior to use.

44. It is extremely important that the freezing media be added one drop at a time to the cells. Any faster can cause the cells to rupture and die.

45. Never add cells to a CoolCell that is already in a freezer.

46. Freeze media contains DMSO, which is toxic to cells, and its toxicity increases with temperature. Therefore, once the freeze media has been added, it is essential to work quickly. We recommend transferring the cells to the freezer within 5 min of adding the media.

47. We recommend a minimum of 12–14 h before removing the PBMC from the CoolCell.

48. Processing of CPT™ for PBMC should begin as soon as possible from the time of collection. Delays greater than 2 h from the time of collection in completing the first centrifugation can result in a reduction of overall cell recovery.

49. CPT™ is designed to allow for a partial delay in processing. (i) Centrifuge under the following conditions: 18–25°C, 1500–1800 RCF, 20 min, Brake: OFF/Minimum Deceleration. (ii) Remix the cells with the plasma by inverting the CPT™ gently 5–10 times. (iii) Store or transport the sample at room temperature for up to 24 h after centrifugation. (iv) To

complete processing, proceed to Step 5. Note that the entire portion above the gel will be collected. To separate and retain the plasma, an additional centrifugation step above what is described in this SOP is required.

50. Sample acceptance criteria: (i) CPT™ should be filled properly, per manufacturer recommendations. (ii) Whole blood to be processed for PBMC should be received in ample time to allow for processing to be completed in the allotted time frame defined.
51. Our laboratory standardly centrifuges CPT™ for 20 min at 1800 RCF.
52. Plasma collection and cryopreservation: Due to the nature of the CPT™, plasma remains undiluted and can be retained. If retaining the plasma, after the initial separation is completed, use a transfer pipette to collect the plasma layer and transfer it to a 50 mL conical tube. Be careful not to disturb the interphase layer containing the PBMC, leaving behind a small amount of plasma as a buffer next to the layer. Follow the plasma processing SOP, "Cryopreservation of Double-Spun Plasma from Whole Blood," to complete the processing of the plasma by starting at the second centrifugation, Step 4.
53. Plasma collected from whole blood samples intended for PBMC isolation is often "older" than plasma obtained from blood collected specifically for plasma preparation. Additionally, this SOP requires whole blood for PBMC isolation to be centrifuged at room temperature, whereas blood collected for plasma is typically centrifuged at 4°C. It is crucial to document the timing and method of plasma collection, as these factors may impact downstream data analysis.
54. PBMCs from up to 4 CPTs™ from a single donor can be collected into a single 50 mL conical tube.
55. CPT™ from the same donor can be rinsed sequentially using the same DPBS by transferring from tube to tube. Rinsing should remove most, if not all, of the reddish coloring from the gel matrix (the wash will also be reddish in color). Be careful not to touch or damage the gel matrix with the transfer pipette.
56. High-Yield Lyse (HYL) is used to remove any residual red blood cell contamination.
57. When adding HYL to the PBMC cell pellet, the volume of the pellet should be estimated. A minimum of 10 mL should be used when pellets are very small.
58. When resuspending PBMC for counting, it is important not to overmix. Do not allow bubbles to form; be gentle, but

mix thoroughly. If counting multiple times, remix between counts, PBMC will settle quickly.

59. We recommend preparing and labeling the PBMC cryovials during the final centrifugation step. The number of cryovials required should be based on the cell count obtained immediately prior. Typically, PBMC are cryopreserved at a concentration of 5–20 million cells per mL in 1 mL aliquots.

60. We have found that decanting the fluid, then gently inverting the conical tube and tapping it onto a clean, dry paper towel inside the biosafety cabinet, is effective at removing residual liquid. However, if absolute sterility is required, this step should be carefully evaluated. In such cases, sterile towels must be used, or the step should be omitted entirely. Do not perform this step if there is any concern that the cell pellet has been disturbed. A clean, RBC-free pellet will remain intact, but a disturbed pellet may pour off with the liquid.

61. Ensure freezing media was allowed to cool on wet ice for a minimum of 10–15 min prior to use.

62. It is extremely important that the freezing media be added one drop at a time to the cells. Any faster can cause the cells to rupture and die.

63. Never add cells to a CoolCell that is already in a freezer.

64. Freeze media contains DMSO, which is toxic to cells, and its toxicity increases with temperature. Therefore, once the freeze media has been added, it is essential to work quickly. We recommend transferring the cells to the freezer within 5 min of adding the media.

65. We recommend a minimum of 12–14 h before removing the PBMC from the CoolCell.

References

1. Gehrie EA, Young PP, Basavaraju SV, Bracey AW, Cap AP, Culler L, Dunbar NM, Homer M, Isufi I, Macedo R, Petraszko T, Ramsey G, Tormey CA, Kaufman RM, Snyder EL (2024) Addressing platelet insecurity—a national call to action. Transfusion 64:2001–2013. Snapes E (2023) ISBER best practices-5th edn
2. National Institute of Allergy and Infectious Diseases (2021) Good Clinical Laboratory Practice (GCLP) Guidelines, Version 4.1. Division of AIDS, National Institutes of Health, Bethesda, Maryland
3. The International Society for Biological and Environmental Repositories (2017) Due North: aligning biobanking practice with evolving evidence and innovation. Biopreserv Biobank 15. https://doi.org/10.1089/bio.2017.29021.abstracts
4. PAS_BC_Vacutainer-blood-collection-system. Becton, Dickinson and Company. BD Vacutainer® Evacuated Blood Collection System – Instructions for Use. Reference 018. BD Life Sciences – Preanalytical Systems.
5. Sarzotti-Kelsoe M, Needham LK, Rountree W, Bainbridge J, Gray CM, Fiscus SA, Ferrari G, Stevens WS, Stager SL, Binz W, Louzao R, Long KO, Mokgotho P, Moodley N, Mackay M, Kerkau M, McMillion T, Kirchherr J, Soderberg KA, Haynes BF, Denny TN (2014) The Center for HIV/AIDS Vaccine Immunology (CHAVI) multi-site quality assurance program for cryopreserved human peripheral blood mononuclear

cells. J Immunol Methods 409:21–30. https://doi.org/10.1016/j.jim.2014.05.013
6. Greiner Bio-One (2016) Instruction manual Leucosep™preparation. Lit. No. 071609W. VWR International
7. Bull M, Lee D, Stucky J, Chiu YL, Rubin A, Horton H, McElrath MJ (2007) Defining blood processing parameters for optimal detection of cryopreserved antigen-specific responses for HIV vaccine trials. J Immunol Methods 322:57–69. https://doi.org/10.1016/j.jim.2007.02.003
8. Arraud N, Linares R, Tan S, Gounou C, Pasquet JM, Mornet S, Brisson AR (2014) Extracellular vesicles from blood plasma: determination of their morphology, size, phenotype and concentration. J Thromb Haemost 12:614–627. https://doi.org/10.1111/jth.12554
9. Lacroix R, Judicone C, Poncelet P, Robert S, Arnaud L, Sampol J, Dignat-George F (2012) Impact of pre-analytical parameters on the measurement of circulating microparticles: towards standardization of protocol. J Thromb Haemost 10:437–446. https://doi.org/10.1111/j.1538-7836.2011.04610.x
10. Nielsen MH, Beck-Nielsen H, Andersen MN, Handberg A (2014) A flow cytometric method for characterization of circulating cell-derived microparticles in plasma. J Extracell Vesicles 2014:3. https://doi.org/10.3402/jev.v3.20795
11. Zhang T, Zhu J, George DJ, Nixon AB (2016) Metastatic clear cell renal cell carcinoma: circulating biomarkers to guide antiangiogenic and immune therapies. Urol Oncol 34:510–518. https://doi.org/10.1016/j.urolonc.2016.06.020
12. Liu Y, Starr MD, Bulusu A, Pang H, Wong NS, Honeycutt W, Amara A, Hurwitz HI, Nixon AB (2013) Correlation of angiogenic biomarker signatures with clinical outcomes in metastatic colorectal cancer patients receiving capecitabine, oxaliplatin, and bevacizumab. Cancer Med 2:234–242. https://doi.org/10.1002/cam4.71
13. Plebani M, Banfi G, Bernardini S, Bondanini F, Conti L, Dorizzi R, Ferrara FE, Mancini R, Trenti T (2020) Serum or plasma? An old question looking for new answers. Clin Chem Lab Med 58(2):178–187. https://doi.org/10.1515/cclm-2019-0719
14. Psychogios N, Hau DD, Peng J, Guo AC, Mandal R, Bouatra S, Sinelnikov I, Krishnamurthy R, Eisner R, Gautam B, Young N, Xia J, Knox C, Dong E, Huang P, Hollander Z, Pedersen TL, Smith SR, Bamforth F, Greiner R, McManus B, Newman JW, Goodfriend T, Wishart DS (2011) The human serum metabolome. PLoS One 6:e16957. https://doi.org/10.1371/journal.pone.0016957
15. Boddy JL, Gal S, Malone PR, Harris AL, Wainscoat JS (2005) Prospective study of quantitation of plasma DNA levels in the diagnosis of malignant versus benign prostate disease. Clin Cancer Res 11:1394–1399. https://doi.org/10.1158/1078-0432.ccr-04-1237
16. Greytak SR, Engel KB, Parpart-Li S, Murtaza M, Bronkhorst AJ, Pertile MD, Moore HM (2020) Harmonizing cell-free DNA collection and processing practices through evidence-based guidance. Clin Cancer Res 26:3104–3109. https://doi.org/10.1158/1078-0432.Ccr-19-3015
17. Bowen RA, Remaley AT (2014) Interferences from blood collection tube components on clinical chemistry assays. Biochem Med 24:31–44. https://doi.org/10.11613/BM.2014.006
18. Khadka M, Todor A, Maner-Smith KM, Colucci JK, Tran V, Gaul DA, Anderson EJ, Natrajan MS, Rouphael N, Mulligan MJ, McDonald CE, Suthar M, Li S, Ortlund EA (2019) The effect of anticoagulants, temperature, and time on the human plasma metabolome and lipidome from healthy donors as determined by liquid chromatography-mass spectrometry. Biomolecules 9:200. https://doi.org/10.3390/biom9050200
19. Mohamadkhani A, Poustchi H (2015) Repository of human blood derivative biospecimens in biobank: technical implications. Middle East J Dig Dis 7:61–68
20. Tuck MK, Chan DW, Chia D, Godwin AK, Grizzle WE, Krueger KE, Rom W, Sanda M, Sorbara L, Stass S, Wang W, Brenner DE (2009) Standard operating procedures for serum and plasma collection: early detection research network consensus statement standard operating procedure integration working group. J Proteome Res 8:113–117. https://doi.org/10.1021/pr800545q
21. Bowen RA, Hortin GL, Csako G, Otañez OH, Remaley AT (2010) Impact of blood collection devices on clinical chemistry assays. Clin Biochem 43:4–25. https://doi.org/10.1016/j.clinbiochem.2009.10.001
22. Afonso G, Scotto M, Renand A, Arvastsson J, Vassilieff D, Cilio CM, Mallone R (2010) Critical parameters in blood processing for T-cell assays: validation on ELISpot and tetramer platforms. J Immunol Methods 359:28–36. https://doi.org/10.1016/j.jim.2010.05.005
23. Sanchez AM, Rountree W, Berrong M, Garcia A, Schuetz A, Cox J, Frahm N, Manak M, Sarzotti-Kelsoe M, D'Souza MP, Denny T, Ferrari G (2014) The External Quality Assurance Oversight Laboratory (EQAPOL) proficiency program for IFN-gamma enzyme-linked immunospot (IFN-gamma ELISpot)

assay. J Immunol Methods 409:31–43. https://doi.org/10.1016/j.jim.2014.03.017
24. Weinberg A, Zhang L, Brown D, Erice A, Polsky B, Hirsch MS, Owens S, Lamb K (2000) Viability and functional activity of cryopreserved mononuclear cells. Clin Diagn Lab Immunol 7:714–716. https://doi.org/10.1128/cdli.7.4.714-716.2000
25. Lippi G, Betsou F, Cadamuro J, Cornes M, Fleischhacker M, Fruekilde P, Neumaier M, Nybo M, Padoan A, Plebani M, Sciacovelli L, Vermeersch P, von Meyer A, Simundic AM, Working Group for Preanalytical Phase EFoCC, Laboratory M (2019) Preanalytical challenges—time for solutions. Clin Chem Lab Med 57:974–981. https://doi.org/10.1515/cclm-2018-1334
26. Malmodin D, Bay Nord A, Zafar H, Paulson L, Karlsson BG, Naluai AT (2024) Preanalytical (mis) handling of plasma investigated by (1)H NMR metabolomics. ACS Omega 9:48727–48737. https://doi.org/10.1021/acsomega.4c08215
27. Tanner M, Kent N, Smith B, Fletcher S, Lewer M (2008) Stability of common biochemical analytes in serum gel tubes subjected to various storage temperatures and times pre-centrifugation. Ann Clin Biochem 45:375–379. https://doi.org/10.1258/acb.2007.007183
28. Dash P, Tiwari R, Nayak S, Mangaraj M (2022) Impact of time delay in the analysis of serum ionized calcium, sodium, and potassium. J Lab Physicians 14:373–376. https://doi.org/10.1055/s-0042-1757228
29. Mokhsin A, Subramaniam P, Sivaneson S, Nheu N, Ramaloo G, Hanifah AS, Mahathevan SB, Nadarajah M, Sampasivam G, Mohd Ismail A, Abdul Rahman T (2025) Assessing the stability of uncentrifuged serum and plasma analytes at various post-collection intervals. J Lab Med 49:21–28. https://doi.org/10.1515/labmed-2024-0062
30. Stanzick KJ, Simon J, Zimmermann ME, Schachtner M, Peterhoff D, Niller HH, Uberla K, Wagner R, Heid IM, Stark KJ (2023) DNA extraction from clotted blood in genotyping quality. Biotechniques 74:23–29. https://doi.org/10.2144/btn-2022-0061
31. Zhang S, Cheng X, Yang G, Peng H, Zou C, Yao D, Qian K (2024) Evaluation of DNA extracted from residual blood clots after serological testing. Cell Biochem Funct 42:e4023. https://doi.org/10.1002/cbf.4023
32. Posevitz-Fejfar A, Posevitz V, Gross CC, Bhatia U, Kurth F, Schutte V, Bar-Or A, Meuth SG, Wiendl H (2014) Effects of blood transportation on human peripheral mononuclear cell yield, phenotype and function: implications for immune cell biobanking. PLoS One 9:e115920. https://doi.org/10.1371/journal.pone.0115920

Chapter 10

Vaccine Distribution Logistics

Eric R. James

Abstract

Lyophilized vaccines are typically stored in cold rooms and refrigerators and transported in cold boxes through a 2–8°C cold chain, with a few exceptions, such as those containing *Varicella* that may be transported at −50°C to −15°C. Vaccines stored as liquid formulations also require a 2–8°C cold chain, while frozen vaccines are stored in freezers at around −20°C or ultralow −80°C freezers. These require cold chains, with cold boxes containing frozen gel packs or preconditioned phase-change materials (PCMs), or insulated boxes with dry ice (e.g., BioNTech's COVID-19 mRNA vaccine). Cryopreserved vaccines require liquid nitrogen (LN2) vapor phase (LNVP) freezers and cold chains that use Dewars with LN_2 or LNVP dry shippers. Storage and distribution methods for all vaccines follow strict national and international guidelines and regulations, maintain an end-to-end chain of custody, and are monitored by temperature data loggers and other devices. Different distribution models are employed for different products and by different countries, ranging from the hub-and-spoke model for LNVP distribution to models with several intermediate stops. Storage and distribution of lyophilized, frozen, and cryopreserved vaccines are described.

Keywords Vaccine storage, Vaccine distribution, Vaccine cold chains

1 Introduction

1.1 General Introduction

Live-attenuated vaccines are stabilized by one of four main methods, i.e., as a: (i) liquid formulation, (ii) lyophilized (freeze-dried) product, (iii) frozen liquid (−15°C to −90°C), or (iv) cryopreserved liquid. The focus of this chapter is on lyophilized, frozen and cryopreserved vaccines. The distribution logistics of licensed vaccines are generally handled by a third-party logistics (3PL) provider in partnership with the manufacturer under national health authority oversight and/or World Health Organization (WHO) guidance and following Good Distribution Practice (GDP). Companies or institutes with vaccines in early-stage clinical development generally need to be more directly involved in establishing and running their own logistics infrastructure to handle shipments at a smaller scale, and

Willem F. Wolkers and Harriëtte Oldenhof (eds.), *Cryopreservation and Freeze-Drying Protocols*, Methods in Molecular Biology, vol. 3060, https://doi.org/10.1007/978-1-0716-5416-3_10,

this section is designed to provide information on these, pre-clinical and phase 1/2 level-supporting logistics operations for lyophilized, frozen, and cryopreserved vaccines.

Liquid formulated and lyophilized vaccines are mostly distributed through 2–8°C cold chains in glass vials, with elastomer stoppers that also serve as an access septum, and with crimped aluminum seals and protective caps that complete the container closure (CC), providing integrity (CCI), and tamper evidence. The vials may be single or multidose (e.g., the measles, mumps and rubella (MMR), and COVID-19 vaccines). At the immunization center where vaccine is administered, for liquid formulations (removed from the 2–8°C refrigerator, or for frozen product thawed on the benchtop to ambient temperature) after the protective cap is removed, a syringe with appropriately sized needle (for intramuscular, subcutaneous, or intradermal etc. injection) is used to extract the vaccine dose from the vial by piercing the septum and aspirating the required volume from the inverted vial [1, 2].

1.2 Lyophilized Vaccines

Lyophilized vaccines, whether distributed and stored at 2–8°C or in freezers, require the addition of diluent, usually water, 0.9% saline, or phosphate-buffered saline, that may include an adjuvant (e.g., AS01, aluminum phosphate, or similar) to reconstitute the freeze-dried cake into its (usually) original pre-lyophilization volume. Diluent is provided in separate glass vials, packaged similar to liquid formulation vaccines and typically stored and shipped at 2–8°C (never frozen [3, 4]) or controlled ambient (+15°C to +25°C) in temperature-controlled environments and/or in passive packaging with insulation and gel packs or phase change material (PCM) panels (gel/PCM). A kit of ancillary supplies, which includes syringes and needles both for aspiration of diluent and for dose preparation and administration, may also be shipped with the vaccine and diluent.

Most human (and veterinary) vaccines against infectious agents fit into a target product profile that requires shipment at 2–8°C. This includes most vaccines in the US childhood immunization program (VFC) [3] run by a 3PL (McKesson) under contract from the Centers for Disease Control and Prevention (CDC) [5], and also the Expanded Program on Immunization (EPI) organized through the WHO [6] and administered by member nations. The distribution networks for these vaccines utilize 2–8°C cold rooms at intermediate depots/nodes in the cold chain and refrigerators powered by electricity, or, as in many locations in Africa, by propane or solar. Distribution networks are varied and range from the direct-to-site (from the vaccine manufacturer or a central warehouse to the point of use) or hub-and-spoke from central depots (as in the US VFC), to hierarchical expanding networks of depots/nodes of decreasing size from national to regional to district to clinic/dispensary, as is typical for countries participating in the EPI. Vaccine distribution systems in other countries utilize a range of models from those more similar to the VFC centralized system, as in the United Kingdom, France, and Germany [7, 8] to networks similar to the extended tiered system of the EPI, as in China [9].

Cold chains are tightly monitored for temperature using data loggers, or with temperature tags that change color to report accumulated temperature excursions and their duration. The temperature range for 2–8°C cold chains is particularly tight and, in consequence, more difficult to maintain, and combined with the often multiple handling steps down through the cold chain, leads to high levels of vaccine wastage [10, 11]. Freezing of vaccines in the 2–8°C cold chain is a more frequent cause of vaccine loss than temperature excursions above +8°C [12].

1.3 Frozen and Cryopreserved Vaccines

The liquid formulation vaccines for chicken pox (Varicella: VAR) and measles, mumps, rubella and Varicella (MMRV) are stored and shipped at between −50°C and −15°C [13] while the Moderna and BioNTech COVID-19 vaccines are stored long-term between −50°C and −15°C [14] and between −90°C and −60°C [15], respectively. And the Ebola vaccine is stored and shipped at between −80°C and −60°C [16]. The cold chain for these vaccines utilizes freezers, or ultralow temperature freezers supplemented with cold boxes incorporating gel/PCM, or dry ice for last mile delivery. Cryopreserved vaccines are stored and shipped below −150°C [17], and in the clinic are thawed to 0 °C or ambient and may then be extracted from the cryovial using syringe and needle, or diluted to reduce the concentration of cryoprotective additive (CPA) before loading syringes for injection [18, 19]. If distributed in standard screw-cap cryovials, dilution and syringe preparation must occur in an ISO 5 aseptic environment (e.g., biological safety cabinet (BSC), laminar flow cabinet (LFC), or isolator). Vaccines produced by closed vial filling (e.g., Aseptic Technologies (AT) vials; Synflorix/PCV10 [20]) or in vials through sterile crimp-sealed tubing (e.g., CellSeal vials) can be prepared for injection in the dispensary without a BSC/LFC. Standard glass vials and screw-cap cryovials do not retain CCI below around −90°C due to the stoppers or o-rings/gaskets, respectively, becoming hard and separating from the vial body due to differential thermal contraction. The vaccine against East Coast Fever (ECF) [21] of cattle is produced and shipped locally by road in sealed straws in liquid nitrogen (LN_2) Dewars, whereas cryopreserved human vaccines are stored in LN_2 vapor phase (LNVP) or below −150°C freezers and shipped in LNVP dry shippers. Liquid, frozen, and cryopreserved vaccines are shipped in triple packaging [22]: primary = vials, secondary = carton/tray/box or canister, and tertiary = reinforced box or LNVP dry shipper. Boxes with vaccine in dry ice, as well as LN_2 Dewars and LNVP dry shippers, can remain on-site at the immunization clinic for many weeks (30-day limit for the BioNTech COVID-19 vaccine [15]) if replenished with dry ice or LN2, thus acting as local vaccine stores. Dry shippers consume no electricity and may typically provide hold times of 10 or more days (30 days for IC Biomedical TW47-6), making them ideal for cryopreserved vaccine distribution to remote locations [19, 23] and for travel medicine clinics cycling through a 2–3-week recharging and/or replenishment schedule [24].

2 Materials

2.1 Materials and Equipment for Distribution of Lyophilized Vaccines

1. Boxes for holding vaccine vials: one vial per box, together with the package insert.
2. Secondary outer box/carton for bulk packaging several units of vaccine.
3. Tertiary reinforced box (overpack) to consolidate vaccine bulk-packed boxes (may be Styrofoam with gel/PCM), or a large actively-chilled/heated temperature container (e.g., Envirotainer), to maintain 2–8°C or (for diluent) a controlled ambient environment.
4. Boxes for holding diluent vials: one vial per box, or multiple vials in a gridded box.
5. Boxes for holding immunization kits, including syringes and needles for diluent and vaccine.
6. Secondary packaging carton for bulk packaging several units of diluent.
7. Tertiary reinforced box (overpack) to consolidate diluent bulk-packed boxes (may be Styrofoam with gel/PCM if shipping at 2–8°C).
8. Temperature data logger, freeze indicators, or temperature-sensitive tags/labels to monitor the temperature of vaccine shipment through the complete cold chain. Temperature monitoring devices are also used for diluent shipment.
9. Other separate or integrated monitoring devices that may include a tilt monitor, a shock/vibration monitor, and a global positioning system (GPS) monitor.
10. Cold rooms for storage, pre-distribution, or at intermediate depots/nodes. Also, to precondition the gel/PCM. Controlled ambient rooms for diluent.
11. Transportation using a large actively chilled vehicle or temperature container (e.g., Envirotainer) maintaining controlled 2–8°C and ambient environments for distribution.

12. At national, regional, district or local stores/waypoints, 2–8°C cold room/refrigerators for vaccine and a controlled ambient environment for diluent.
13. Last mile distribution uses insulated cold boxes or vaccine carriers with gel/PCM holding sufficient vaccine for the day or immunization event, and insulated containers with gel/PCM for diluent.
14. Clinic/immunization center: 2–8°C refrigerator storage of vaccine in cartons or boxes, and an air-conditioned controlled ambient environment for diluent.

2.2 Materials and Equipment for Distribution of Cryopreserved Vaccines

1. For vaccine storage: LN_2 tanks (for veterinary vaccines in straws), LNVP freezers (for human vaccines in standard or custom cryovials), or stirling/mechanical freezers maintaining −150°C or below.
2. Transportation/distribution for veterinary vaccines: LN_2 Dewars or LNVP dry shippers.
3. Transportation/distribution of human vaccines: LNVP dry shippers maintaining −150°C or below with validated static hold times, and sized for vaccine dose distribution and in-clinic use. Available dry shippers range in capacity from units holding a few straws or cryovials (e.g., IC Biomedical CRX100 with a hold time of 16 days, or MVE Bio SC4/3V with 26-day hold time) through intermediate shippers holding several hundred cryovials or several thousand straws (e.g., IC Biomedical TW47 with 30-day hold time, or MVE Bio IATA with 14-day hold time) to large shippers holding many thousands of cryovials and straws (e.g., MVE 1536PD).
4. Protective enclosure/casing for LNVP dry shipper; various sizes.
5. Canisters, goblets, cryoboxes, racks, or canes to hold straws or cryovials. Vaccine doses may be supplied as a single dose per box or grouped by lot or batch into goblets or boxes (usually up to 25, 81, 96 (SBS format), or 100, depending on primary container size and number).
6. Adsorbent pads for shipping diluent and "socks" padding or similar to protect vaccine boxes during transit in LNVP dry shippers.
7. A data logger with a cryoprobe usually reads from at least ambient temperature down to −200°C.
8. Other separate or integrated monitoring devices that may include a tilt monitor, a shock/vibration monitor, and a global positioning system (GPS) monitor.

3 Methods

3.1 Method of Distribution of Lyophilized Vaccines

1. Follow Good Manufacturing Practice (GMP) and GDP requirements and guidelines, and develop SOPs for shipping lyophilized product, diluent, and supplies.
2. Work with the selected 3PL or carrier to arrange the best shipping date, avoiding inclement weather and holidays that might interfere with transportation—particularly important for international shipments. Note that certain 3PLs, couriers, or airlines are preferred for specific routes and destinations—information that is usually gained through experience and certain ports of entry may be specified.
3. Pre-condition the appropriate gel/PCM in a refrigerator or freezer at the required temperature for the vaccine, and at 2–8°C or +15 to +25°C for the diluent.
4. For vaccine shipments at 2–8°C or −15°C, assemble a certified passive insulated shipping container (e.g., Styrofoam, or Crēdo Cube (Peli Bio Thermal)) with the gel/PCM enclosing cartons of lyophilized vaccine, or pack cartons into an active shipping container (e.g., Envirotainer).
5. For shipments at −80°C (similar to BioNTech's COVID-19 and Ervebo for Ebola), pack trays or cartons of boxed vaccine into a Styrofoam or custom-fit-sized-for-purpose cold box (e.g., Softbox, AeroSafe Global) with dry ice.
6. Pack diluent boxes into insulated shipping containers, along with sufficient absorbent pads to capture the liquid contents of all vials, should they be damaged during transit.
7. Activate the temperature data loggers (and any additional devices) for vaccine and diluent.
8. Complete documentation: Waybill for domestic shipments; Air Waybill (AWB), customs or pro-forma invoice (listing each item with its Harmonized Tariff Schedule (HTS) code, specifying desired incoterms), export permit and/or import permit (3PL, courier or consignee can/may assist with these), packing list and delivery form. Sign up for text or email updates on shipments while in transit.
9. Importation into the USA will require a USDA Statement indicating material is noninfectious to livestock and poultry, a US FDA End Use Letter, and, only if the product is not attenuated, a CDC permit. Many other countries require import permits, and the lead time for obtaining them can be long.
10. Deliver to the courier or call for pickup, activate temperature/GPS data loggers for vaccine and diluent and seal boxes.

11. Continuously monitor the progress of the shipment via GPS loggers and/or courier/3PL/airline online tracking sites. Be prepared to call/text/email the courier/3PL/airline immediately to resolve any "on-hold" issues or if a temperature excursion appears imminent. Many couriers/3PLs can perform in-transit dry ice replenishment.
12. Monitor shipment through arrival and last-mile delivery. For international shipments, prepare to handle customs duty/tax payments as needed.
13. Receipt by the consignee should be acknowledged by text/email, and the delivery form completed, noting temperature history and any damage to the packaging that may compromise the vaccine or diluent.
14. Transfer vaccine to appropriate storage: 2–8°C refrigerator, −15°C to −20°C freezer, or ultralow freezer, or remain in dry ice-replenished shipping box. Maintain continuous temperature monitoring.
15. Store diluent and immunization kits at appropriate temperature (2–8°C, controlled ambient or ambient) with continuous temperature monitoring.

3.2 Methods for Distribution of Cryopreserved Vaccines

1. Follow GMP and GDP requirements and guidelines and develop SOPs for shipping cryopreserved product, diluent, and supplies.
2. Cryopreserved vaccines stored and transported in liquid phase LN_2 (e.g., the ECF vaccine) are limited to local distribution by road using Dewars with LN_2.
3. Cryopreserved human vaccines are stored in LNVP freezers and can be transported by all modes, including by air in LNVP dry shippers.
4. Work with the selected 3PL or carrier to arrange the best shipping date, avoiding inclement weather and holidays that might interfere with transportation—particularly important for international shipments. Note that certain 3PLs, couriers, or airlines are preferred for specific routes and destinations—information that is usually gained through experience and certain ports of entry may be specified.
5. Fully charge the dry shipper with LN_2 according to the manufacturer's instructions (usually overnight).
6. For the accompanying diluent, precondition the appropriate gel/PCM in a refrigerator or freezer at the required temperature, 2–8°C or +15 to +25°C.
7. Complete documentation: Waybill for domestic shipments; For international shipments: AWB, customs or proforma invoice (listing each item with its Harmonized Tariff

Schedule (HTS) code, desired incoterms), export permit and/or import permit (3PL, courier or consignee can/may assist with these), packing list and delivery form. Sign up for text or email updates on shipments while in transit.

8. Pack the diluent in a certified passive insulated shipping container (e.g., Styrofoam or Crēdo Cube, etc.) with the gel/PCM at the required temperature, or pack cartons into an active shipping container (e.g., Envirotainer).
9. If the 3PL/courier requires an inversion test to demonstrate the dry shipper is "dry," perform it when the dry shipper is fully charged and immediately before loading product. Complete an inversion test certification form (on company/institute letterhead).
10. Pack Vaccine cryovials in appropriate box(es) (*see* Sect. 2.2.5), or on canes, and place into a canister for small dry shippers or, for larger numbers/boxes of vaccine cryovials, into the IATA canister of a large dry shipper.
11. Add padding materials (e.g., socks) to fill the void space in and above the vaccine and beneath the neck plug of small dry shippers, and in the IATA canister of large dry shippers.
12. Activate the temperature data loggers (and any additional devices) for vaccine and diluent, and close the dry shipper with the lid/neck plug.
13. Secure documents for the consignee (e.g., MSDS, Packing List, Delivery Form) in an envelope inside the dry shipper protective case.
14. Place shipping documents (AWB, customs/pro-forma invoice, all permits in triplicate) in an adhesive pouch attached to the top of the dry shipper protective case.
15. Activate temperature/GPS data loggers for vaccine and diluent, and seal diluent boxes.
16. Latch the dry shipper protective casing and secure with zip ties.
17. Call for pickup and/or delivery to 3PL/courier.
18. Continuously monitor the progress of the shipment via GPS loggers and/or courier/3PL/airline online tracking sites. Be prepared to call/text/email the courier/3PL/airline immediately to resolve any "on-hold" issues or if a temperature excursion appears imminent. Some couriers/3PLs can perform in-transit LN_2 recharging.
19. Monitor the shipment from arrival to last-mile delivery. For international shipments, prepare to handle customs duty/tax payments as needed.

20. Receipt by consignee should be acknowledged by text/email, and the delivery form completed, noting temperature history and any damage to the packaging that may compromise the vaccine or diluent.
21. Keep vaccine in LN_2 dewar or LNVP dry shipper, replenishing with LN_2 as needed. Maintain continuous temperature monitoring.
22. Store diluent and immunization kits at appropriate temperature (2–8°C, controlled ambient or ambient) with continuous temperature monitoring.

References

1. Hunt D (2017) US Pharmacopoea. https://www.uspnf.com/sites/default/files/usp_pdf/EN/USPNF/revisions/659_rb_notice.pdf
2. World Health Organization (2026). World Health Organization Essential Programme on Immunization. https://www.who.int/teams/immunization-vaccines-and-biologicals/essential-programme-on-immunization/supply-chain/guidance
3. CDC (2024) CDC Vaccine Storage and Handling Toolkit 2024. https://www.cdc.gov/vaccines/hcp/downloads/storage-handling-toolkit.pdf
4. World Health Organization (2015) How to monitor temperatures in the vaccine supply chain. https://iris.who.int/server/api/core/bitstreams/87e73054-8634-4c32-b7a4-783ef105be19/content
5. World Health Organization (2009) Effective Vaccine Management (EVM) initiative (current hub) WHO Vaccine Handling 2015–2020. https://www.who.int/teams/immunization-vaccines-and-biologicals/essential-programme-on-immunization/supply-chain/effective-vaccine-management-%28evm%29
6. CDC (2024) CDC VFC & Centralized Distribution. https://www.cdc.gov/vaccines-for-children/media/pdfs/2024/08/vfc-ops-guide_version-4.0_july-2024_low-res-508-rev-2.pdf
7. World Health Organization (2015) WHO EPI/Immunization in Practice. https://apps.who.int/iris/bitstream/handle/10665/193412/9789241549097_eng.pdf
8. UKHSA ImmForm/Movianto (2022). https://www.england.nhs.uk/south/wp-content/uploads/sites/6/2022/09/ImmForm-Manual-Input-Guidance-2223.pdf
9. China Vaccine Administration Law 2019(2019). https://english.nmpa.gov.cn/2019-07/29/c_773013.htm
10. World Health Organization (2006) Temperature sensitivity of vaccines. World Health Organization, Geneva. https://apps.who.int/iris/handle/10665/69387
11. PATH (2003) Effects of freezing on vaccine potency. Literature review. https://media.path.org/documents/TS_cc_effects.pdf
12. Das MK, Arora NK, Mathew T, Vyas B, Devi SK, Yadav A (2020) Temperature integrity and exposure of vaccines to suboptimal temperatures in cold chain devices at different levels in three states of India. Trop Dis Travel Med Vacc 6:2–9. https://doi.org/10.1186/s40794-020-00109-z
13. Merck Sharp and Dohme LLC (2024) Merck VARIMAX proQUAD package inserts. ver 11/2024. https://www.fda.gov/media/76008/download
14. Moderna TX Inc (2025) Moderna Spikevax package insert. ver 08/2025. https://www.fda.gov/media/155675/download
15. BioNTech Manufacturing GmbH (2025) BioNTech Comirnaty package insert. ver 08/2025. https://www.fda.gov/media/151707/download
16. Jusu MO, Glauser G, Seward JF, Bawoh M, Tempel J, Friend M, Littlefield D, Lahai M, Jalloh HM, Sesay AB, Caulker AF, Samai M, Thomas V, Farrell N, Widdowson MA (2018) Rapid establishment of a cold chain capacity of -60 °C or colder for the STRIVE Ebola Vaccine trial during the Ebola outbreak in Sierra Leone. J Infect Dis 18:S48–S55. https://doi.org/10.1093/infdis/jix336
17. Natalie McKinne (2018) USP <1044> Cryopreservation of Cells (open PDF). https://www.usp.org/sites/default/files/usp/document/our-work/biologics/resources/gc-1044-cryopreservation-of-cells.pdf
18. Mordmüller B, Surat G, Lagler H, Chakravarty S, Ishizuka AS, Lalremruata A, Gmeiner M,

Campo JJ, Esen M, Ruben AJ, Held J, Calle CL, Mengue JB, Gebru T, Ibáñez J, Sulyok M, James ER, Billingsley PF, Natasha KC, Manoj A, Murshedkar T, Gunasekera A, Eappen AG, Li T, Stafford RE, Li M, Felgner PL, Seder RA, Richie TL, Sim BKL, Hoffman SL, Kremsner PG (2017) Sterile protection against human malaria by chemoattenuated PfSPZ vaccine. Nature 542:445–449. https://doi.org/10.1038/nature21060

19. Oneko M, Steinhardt LC, Yego R, Wiegand RE, Swanson PA, Kc N, Akach D, Sang T, Gutman JR, Nzuu EL, Dungani A, Kim Lee Sim B, Oloo PN, Otieno K, Bii DK, Billingsley PF, James ER, Kariuki S, Samuels AM, Jongo S, Chebore W, Abdulla S, Daubenberger C, Mpina M, Styers D, Potter GE, Abarbanell G, Richie TL, Hoffman SL, Seder RA (2021) Safety, immunogenicity and efficacy of PfSPZ Vaccine against malaria in infants in western Kenya: a double-blind, randomized, placebo-controlled phase 2 trial. Nat Med 27:1636–1645. https://doi.org/10.1038/s41591-021-01470-y
20. Wyeth Pharmacruticals LLC (2021) PREVNAR 20 package insert. ver 6/2021. https://fda.report/media/149987/Package+Insert+-+PREVNAR+20.pdf
21. Patel E, Mwaura N, Di Giulio G, Cook EAJ, Lynen G, Toye P (2019) Infection and treatment method (ITM) vaccine against East Coast fever: reducing the number of doses per straw for use in smallholder dairy herds by thawing, diluting and refreezing already packaged vaccine. BMC Vet Res 15:46. https://hdl.handle.net/10568/99527
22. National Archives (2020) US Government 49 Code of Federal Regulations Part 173-Shippers-General requirements for shipments and packagings. ver 11/2020. https://www.ecfr.gov/current/title-49/subtitle-B/chapter-I/subchapter-Cpart-173
23. Diawara H, Healy SA, Mwakingwe-Omari A, Issiaka D, Diallo A, Traore S, Soumbounou IH, Gaoussou S, Zaidi I, Mahamar A, Attaher O, Fried M, Wylie BJ, Mohan R, Doan V, Doritchamou JYA, Dolo A, Morrison RD, Wang J, Hu Z, Rausch KM, Zeguime A, Murshedkar T, Natasha KC, Sim BKL, Billingsley PF, Richie TL, Hoffman SL, Dicko A, Duffy PE, PfSPZ Vaccine Study Team (2024) Safety and efficacy of PfSPZ Vaccine against malaria in healthy adults and women anticipating pregnancy in Mali: two randomised, double-blind, placebo-controlled, phase 1 and 2 trials. Lancet Infect Dis 24:1366–1382. https://doi.org/10.1016/S1473-3099(24)00360-8
24. James ER, Church LWP, Hoffman SL, Richie TL, Robertson BD, Hickey PW, Schwartz DJ, Logan PT, Asare TD, Jones ML, Bay JL, Roschel AK, Pfeiffer JL, Acosta RW, Schiavi E, Acosta AM, Noble M, Henkle T, Young CJ (2024) Piloting delivery of PfSPZ vaccines for malaria through a cryogenic vaccine cold chain to travel and military medicine clinics. Travel Med 31:taae007. https://doi.org/10.1093/jtm/taae007

Part IV

Cryopreservation and Freeze-Drying Protocols

Chapter 11

Cryopreservation of Bovine, Equine, and Porcine Sperm

Harriëtte Oldenhof, Willem F. Wolkers, Anna Tönißen and Harald Sieme

Abstract

In livestock breeding, cryopreserved semen is routinely used for artificial insemination. Sperm cryopreservation enables long-term storage and transport of insemination doses, thereby securing genetic material for later use. Although specifics may vary, commonly used sperm cryopreservation protocols for domestic species feature similar processing steps as outlined in this chapter. We here describe common procedures for cryopreservation of bull, stallion and boar sperm. This includes preparation and use of the following diluents: TRIS egg yolk extenders (TEY) for bull sperm, ordinary and freezing skim milk extenders (INRA-82) for stallion sperm, and Beltsville thawing solution (BTS) and lactose egg yolk extenders (LEY/GO) for use with boar sperm. Details and schematic presentations are given for dilution with cryoprotective agents, using a one-step versus a two-step approach. Furthermore, simple methods are presented for packaging of insemination doses in straws, slow cooling of sperm samples to 5°C, controlled freezing in a polystyrene box with liquid nitrogen vapor, and rapid warming at 37°C.

Keywords Boar, Bull, Cryopreservation, Lactose-egg yolk extender, Skim milk extender, Semen, Spermatozoa, Stallion, TRIS-egg yolk extender

1 Introduction

Cryopreservation of sperm facilitates long-term storage and long-distance transport. It allows artificial insemination to be performed at a desired time point, considering factors like location, logistics, and the availability of both the male donor and female recipient. Cattle breeding almost entirely relies on the use of cryopreserved semen. In horse breeding, diluted semen of 5°C is predominantly used; however, there is growing interest in using cryopreserved semen. Pig insemination is done with diluted semen, which is stored and shipped at ~16–20°C. In all cases, sperm cryopreservation is used to secure valuable genetic resources [1, 2].

Sperm cryopreservation protocols vary amongst animal species, due to differences in sperm cell membrane composition and other intrinsic factors, as well as differences in the historical development of specific protocols for sperm from different species. Processing sperm

Willem F. Wolkers and Harriëtte Oldenhof (eds.), *Cryopreservation and Freeze-Drying Protocols*, Methods in Molecular Biology, vol. 3060, https://doi.org/10.1007/978-1-0716-5416-3_11,

from different species typically involves the use of specific diluents, i.e., extenders with different compositions, for cooling and freezing. In general, sperm freezing extenders include non-permeating cryoprotectants (milk and/or egg yolk), a permeating cryoprotectant (typically glycerol), a buffering compound (e.g., TRIS or HEPES), sugars (e.g., glucose, lactose, fructose), nutrients and salts, as well as antibiotics. There is ongoing interest in the further development of sperm diluents, to enhance sperm viability and functionality during storage at ambient, refrigerated, and cryogenic temperatures [3, 4].

Although specifics may vary among species, the main processing steps for cryopreservation of sperm from domestic species are very similar. After collection, semen is directly diluted with the primary extender at 37°C, followed by cooling to room temperature. After dilution and cooling, sperm samples can be subjected to centrifugation to remove part of the seminal plasma and non-viable sperm and to concentrate the sample. Thereafter, the sperm concentration can be adjusted, and freezing extender is added. When freezing extender is directly added to the resuspended sperm pellet, this is referred to as one-step dilution, whereas mixing defined volumes of freezing extender and diluted sperm is referred to as two-step dilution. The primary diluent contains salts, sugars, proteins, and lipids, while the freezing extender also contains a penetrating protectant like glycerol (2–6% final concentrations). The required sperm concentration varies depending on the specific needs of the artificial insemination procedure. Sperm diluted in freezing extender needs to be slowly cooled (~0.1°C/min) from room temperature to 5°C, after which they can be packaged in plastic straws (of 0.25 or 0.5 mL). Freezing is done by placing straws in racks, in a horizontal position. The racks can be placed either in a controlled-rate freezer or at a defined distance above liquid nitrogen in a closed polystyrene box. Sperm samples are typically frozen at cooling rates ranging from 10–100°C/min [5]. Straws can be thawed by plunging them in a 37°C water bath. This results in a warming at ~2000°C/min [6].

In this chapter, we describe cryopreservation protocols for bovine, equine, and porcine sperm. This includes preparation and use of TRIS egg yolk extenders (TEY) for bull sperm, INRA-82-based diluents for stallion sperm, and lactose egg yolk extenders (LEY/GO) for boar sperm. Both one-step and two-step dilution approaches are discussed, i.e., for dilution with cryoprotective agents. Furthermore, simple methods are presented for packaging of insemination doses in straws, slow cooling of sperm samples to 5°C, controlled freezing in a polystyrene box with liquid nitrogen vapor, and rapid warming at 37°C.

2 Materials

2.1 Materials for Processing of Raw Semen and Dilution

1. Standard laboratory equipment and supplies.
2. Glassware and/or tubes for diluting and handling of sperm; e.g., ordinary 15- or 50-mL polypropylene conical tubes.
3. Phase-contrast microscope (with heated stage), microscope glass slides and coverslips, for inspecting sperm characteristics like motility.
4. Hemocytometer ("Neubauer improved" counting chamber) for determining the sperm concentration. Alternatively, use a photometer or nucleocounter (*see* **Note 1**).
5. Water bath set at 37°C, for pre-warming of diluents.
6. Centrifuge for processing of sperm samples, i.e., a device that allows for centrifugation at 600×*g* (stallion) or 2400×*g* (boar). For boar sperm, a temperature-controlled device is preferred for centrifugation at 17°C.
7. For preparation of clarified egg yolk, tubes and a temperature-controlled centrifuge are needed that can be used at 10,000×*g* and 4°C.
8. Device for removing the supernatant after centrifugation; e.g., water vacuum pump or syringe with extended tip.
9. Diluents (so-called extenders) for diluting and storage of semen. Details on preparing the following diluents are given below: INRA-82 (primary and freezing) extenders for equine semen, TRIS-egg yolk (TEY-1 and -2) extenders for bovine semen, Beltsville Thawing Solution (BTS) and lactose-egg yolk (LEY cooling and freezing) extenders for porcine semen (*see* **Note 2**).
10. For use with egg yolk: Equex STM paste (e.g., from Minitube).
11. Cushion fluid (e.g., Optiprep) can be used with high-speed centrifugation of diluted semen.

2.2 Materials and Equipment for Cooling, Cryopreservation, and Thawing of Sperm Samples

1. Cryoprotective gloves, goggles, and tweezers/forceps for safe handling of liquid nitrogen and straws.
2. Straws made of polypropylene or polyvinyl chloride (e.g., from Minitube or IMV-Technologies). Typically, 0.25-mL straws are used for bull sperm and 0.5-mL straws for stallion and boar sperm.
3. A device for labeling straws or a permanent marker.
4. Device(s) for (semi-)automated filling and sealing of straws (e.g., from IMV-Technologies or Minitube). Alternatively, for filling straws manually, a 1-mL syringe can be used with a small

piece of silicone tubing attached to the opening to fit in a straw (plugged site). Sealing can also be done using heat, sealing powder, or balls.

5. Cooling cabinet or refrigerator/cold room, for handling of materials at ~5°C.
6. Thermocouple, preferably with a data logger, for measuring cooling rates.
7. Controlled rate freezer (e.g., model "Icecube" from Minitube, or "Minidigitcool" from IMV-technologies), with accompanying racks for holding straws. Alternatively, a polystyrene box with liquid nitrogen can be used, with the rack/straws being placed at a fixed position above the liquid nitrogen level in the vapor phase. Racks can be stapled, with the lowest rack being placed in liquid nitrogen, or a rack with (polystyrene) floaters can be placed directly on the liquid nitrogen level. Use racks in which straws can be held horizontally in separate slots. Adjust the height at which the straws are positioned above the liquid nitrogen level to attain a specific cooling rate (monitor using a thermocouple positioned in a mock straw).
8. Cryogenic container and goblets for storage of straws in liquid nitrogen (vapor).
9. Water bath set at 37°C for thawing of straws.

3 Methods

3.1 Preparation of (Clarified) Egg Yolk: For Use in Semen Extenders

1. Separate egg yolks from egg whites, then carefully roll them on filter paper to remove any remaining yolk. Cut the surrounding membrane using a scalpel.
2. Recover the yolks only, and collect in a cylinder (*see* **Note 3**). This egg yolk solution (~100% solution) can be directly used for preparing TRIS- or lactose-egg yolk extenders (bull, boar).
3. To prepare clarified egg yolk, add an equal volume of distilled water, which gives a ~50% egg yolk solution. Mix well, and centrifuge at 10,000 × *g* for 20 min at 4°C.
4. Recover the clear water-soluble fraction, and discard the pellet (as well as lipid material floating at the surface). Filter through a folded/fluted paper filter. Use directly or freeze as aliquots for later use (*see* **Note 4**). Use clarified egg yolk for preparing INRA-82 freezing extender (stallion).

3.2 Preparation of INRA-82 Skim Milk and Freezing Extender: For Use with Equine Semen

For stallion sperm, (modified) INRA-82 skim milk extender can be used as a primary diluent [7, and references therein]. INRA-82 supplemented with clarified egg yolk and glycerol can be used as a freezing extender. Final concentrations for cryopreservation are typically 2% (v/v) clarified egg yolk and 2.5% (v/v) glycerol. Components for preparations are listed in Table 1.

1. Prepare buffered glucose saline solution (GS1) by dissolving the following chemicals in 500 mL distilled water: 25 g glucose (anhydrous), 1.5 g lactose monohydrate, 1.5 g raffinose pentahydrate, 0.25 g sodium citrate dihydrate, 0.41 g potassium citrate monohydrate, 4.76 g HEPES, and antibiotics (*see*

Table 1
Preparation of INRA-82 diluent and freezing extender, for use with stallion sperm

Medium	Component	MW (g/mol)	Amount
Glucose-saline solution	Glucose; anhydrous, or monohydrate	180.16, or 198.17	25 g, or 27.50 g
	Lactose monohydrate	360.32	1.5 g
	Raffinose pentahydrate	594.50	1.5 g
	Trisodium citrate dihydrate	294.10	0.25 g
	Potassium citrate monohydrate	324.41	0.41 g
	HEPES	238.30	4.76 g
	Antibiotics		(*see* **Note 6**)
	Water		to 500 mL (GS1) or variable (GS2)
	Adjust to pH 6.8–7.0, osmolality 300–330 mOsm/kg		
INRA-82	Glucose-saline solution (GS1)		500 mL
	UHT milk, 0.3%		500 mL
INRA-82, 5% EY	Glucose-saline solution (GS2)		400 mL
	UHT milk, 0.3%		500 mL
	Clarified egg yolk, 50%		100 mL
2× INRA-82 freezing extender	Glucose-saline solution (GS2)		370 mL
	UHT milk, 0.3%		500 mL
	Clarified egg yolk, 50%		80 mL
	Glycerol, 100%		50 mL
1× INRA/EY/GLY	INRA-82 (primary diluent)		50 mL
	2× INRA-82 freezing extender		50 mL

Note 5). Mix well (use a magnetic stirrer) until all components are dissolved. The pH should be adjusted to 7.2 using KOH, and osmolality to ~330 mOsm/kg by adding water or glucose if needed (*see* **Note 6**). Filter-sterilize the solution by passing through a 0.2 μm filter, and store at 4°C for use within 1–2 weeks. This solution can be frozen as aliquots (e.g., of 50 mL) for later use.

2. INRA-82, for primary dilution and cooled storage, is prepared by mixing equal volumes of the above-described buffered glucose saline solution (GS1) and commercial 0.3% ultra-heat-treated skim milk. Typically, mix 50 or 500 mL each. This should give a pH of 6.8–7.0 and osmolality of 300–330 mOsm/kg. Prepare fresh, prior to use. If needed, store at 4°C for use within 3 days.
3. 2× INRA-82-freezing extender, containing two fold the desired final concentration of cryoprotective agents, is prepared using modified buffered saline solution (GS2), taking into account volumes needed for adding egg yolk and glycerol (i.e., aqueous solutions). Prepare a modified buffered saline solution by adding the above-listed amounts of chemicals in 370 mL (instead of 500 mL as done for ordinary INRA-82). Then, prepare 1 L 2× INRA-82-freezing extender by adding: 500 mL commercial 0.3% ultra-heat-treated skim milk, 80 mL clarified egg yolk (~50% solution, prepared as described above) and 50 mL glycerol (~100% solution). Mix well, and if needed, freeze as aliquots for later use.
4. In case of performing a "one-step" dilution with freezing extender (i.e., direct addition to a sperm pellet resuspended in a little volume as remaining after centrifugation), prepare 1× INRA-82 freezing extender, containing 2% (v/v) egg yolk and 2.5% (v/v) glycerol. This can be done by mixing equal volumes of INRA-82 and 2× INRA-82-freezing extender. Alternatively, prepare INRA-82 supplemented with 5% egg yolk by dissolving glucose saline solution components in 400 mL water, add 500 mL 0.3% ultra-heat-treated skim milk, and 100 mL ~50% clarified egg yolk; and add glycerol just before use (see below).

3.3 Preparation of TEY-1 and -2: For Use with Bovine Semen

For bull sperm, TRIS-egg yolk (TEY) extenders can be used [8, and references therein]. Both TEY-1 and -2 contain egg yolk, whereas TEY-2, which is added for cryopreservation, also includes glycerol. Final concentrations for cryopreservation are 20% (v/v) egg yolk and 6.4% (v/v) glycerol. Components for preparations are listed in Table 2.

1. Prepare TRIS stock solution by dissolving the following chemicals in 1 L distilled water: 36.05 g TRIS, 20.24 g citric acid monohydrate, 14.88 g fructose, and antibiotics (*see* **Note 5**).

Table 2
Preparation of TEY-1 and-2, for dilution and cryopreservation of bull sperm

Medium	Component	MW (g/mol)	Amount
TRIS stock solution	TRIS	121.14	36.05 g
	Citric acid monohydrate	210.14	20.24 g
	Fructose	180.16	14.88 g
	Antibiotics		(*see* **Note 6**)
	Water		to 1000 mL
	Adjust to pH 7.0		
TEY-1	TRIS stock solution		672 mL
	Water		128 mL
	Egg yolk, 100%		200 mL
TEY-2	TRIS stock solution		672 mL
	Egg yolk, 100%		200 mL
	Glycerol, 100%		128 mL
1× TEY freezing extender	TRIS stock solution		672 mL
	Water		64 mL
	Egg yolk, 100%		200 mL
	Glycerol, 100%		64 mL
1× TEY/EY/GLY	TEY-1 (primary diluent)		50 mL
	TEY-2 (freezing diluent		50 mL

The pH should be adjusted to 7.0, after which the solution should be filter-sterilized by passing through a 0.2 μm filter. If needed, the solution can be stored at 4°C for use within 1–2 weeks.

2. For 1 L TEY-1, add together: 672 mL TRIS stock solution, 128 mL distilled water, and 200 mL egg yolk (~100% solution).
3. For 1 L TEY-2, add together: 672 mL TRIS stock solution, 200 mL egg yolk (~100% solution), and 128 mL glycerol (~100% solution).
4. TEY containing the desired final concentration of cryoprotective agents is prepared by adding together: 672 mL TRIS stock solution, 64 mL water, 200 mL egg yolk (~100% solution), and 64 mL glycerol (~100% solution). These final concentrations are also obtained by mixing equal volumes of TEY-1 and TEY-2.

3.4 Preparation of BTS, LEY, and LEYGO for use with Porcine Semen

For boar sperm, Beltsville Thawing Solution (BTS) can be used for primary dilution [9], and lactose-egg yolk (LEY) cooling and freezing extenders for cryopreservation [10, 11]. Final concentrations for cryopreservation are typically 20% (v/v) egg yolk, 0.5% Equex paste and 2% (v/v) glycerol. Components for preparations are listed in Table 3.

1. Prepare BTS by dissolving the following chemicals in 1 L distilled water: 6.0 g trisodium citrate dihydrate, 1.25 g sodium hydrogen carbonate, 0.75 g potassium chloride, 40.5 g glucose monohydrate, 1.25 g ethylenediaminetetraacetic acid disodium salt dihydrate, and antibiotics (*see* **Note 5**). The pH should be adjusted to 7.2, and the osmolality should be 290–300 mOsm/kg. Filter-sterilize via passing the solution through

Table 3
Preparation of BTS, LEY cooling and LEYGO freezing extenders for dilution and cryopreservation of boar sperm

Medium	Component	MW (g/mol)	Amount
BTS	Glucose; anhydrous, or monohydrate	180.16, or 198.17	37 g, or 40.70 g
	Potassium chloride	74.55	0.75 g
	Sodium bicarbonate	84.01	1.25 g
	Trisodium citrate dihydrate	294.10	6.0 g
	Ethylenediaminetetraacetic acid disodium salt dihydrate	372.24	1.25 g
	Antibiotics		(*see* **Note 6**)
	Water		to 1000 mL
	Adjust to pH 7.2		
	Osmolality 290–300 mOsm/kg		
Lactose solution	Lactose	342.30	11 g
	Antibiotics		(*see* **Note 6**)
	Water		to 100 mL
LEY cooling extender	Lactose solution		80 mL
	Egg yolk, 100%		20 mL
LEYGO freezing extender	LEY cooling extender		92.5 mL
	Equex paste		1.5 mL
	Glycerol, 100%		6 mL
1× LEY/GO	LEY cooling extender		100 mL
	LEYGO freezing extender		50 mL

a 0.2 μm filter. If needed, the solution can be stored at 4°C for use within 1–2 weeks.

2. Prepare lactose solution (310 mM) by dissolving 11 g lactose in 100 mL distilled water.
3. Prepare LEY-cooling extender by combining 80 mL lactose solution and 20 mL egg yolk (~100% solution). Filter the solution through a folded/fluted paper filter.
4. Prepare LEY-freezing extender (also referred to as LEYGO: LEY-glycerol-Orvus-ES-paste), by supplementing 92.5 mL LEY-cooling extender with 1.5 mL Equex paste (*see* **Note 7**) and 6 mL glycerol (~100% solution).
5. The final cryopreservation solution is prepared by adding 1 volume of LEY-freezing extender to 2 volumes of LEY-cooling extender.

3.5 Semen Collection and Dilution with Primary Extender

Use standard methods and approaches for collecting semen from domestic species.

1. Have all materials, including the primary extender, pre-warmed at 37°C.
2. In case of collecting boar semen, collect the sperm rich fraction only. For stallion and bull, use the whole ejaculate. Directly after semen collection, determine the ejaculate (gel-free) volume, appearance/consistency, and (approximate) sperm concentration.
3. Dilute raw semen by slowly adding the primary extender; i.e., INRA-82 for equine semen, TEY-1 for bovine semen, and BTS for porcine semen. Add at least an equal volume of extender or dilute to, e.g., $\sim 100 \times 10^6$ sperm/mL. This is referred to as diluted semen.
4. Evaluate sperm motility using a phase contrast microscope with a heated stage. In addition, inspect further characteristics like sperm morphology and membrane intactness (*see* **Note 8**).
5. Let the diluted semen sample cool down to room temperature, i.e., ~22°C, for further handling. After attaining room temperature, cool diluted boar semen to 17°C.

3.6 Centrifugation Processing for the Removal of Seminal Plasma and to Concentrate the Sample

Stallion and boar sperm samples are generally subjected to centrifugation processing prior to cryopreservation, whereas bull sperm samples are not. To increase the recovery rate while preventing sperm packing in a dense pellet, a higher centrifugation speed can be applied by using cushion fluid below the sperm sample.

1. When performing ordinary centrifugation, with stallion or boar sperm:

(a) Transfer the diluted sperm sample to a tube (e.g., 50 mL) for centrifugation. Centrifuge stallion sperm samples for 10 min at 600 × *g*, at room temperature. Centrifuge boar sperm for 3 min at 2400 × *g*, while maintaining at 17°C.

(b) Carefully remove the supernatant; using a Pasteur pipette connected to a vacuum pump or a syringe with extension, leaving a little volume (~5 mL in case of 50 mL original volume) for re-suspension of the sperm pellet (*see* **Note 9**). Gently re-suspend, e.g., using a shaker.

2. When performing high-speed cushioned centrifugation, with stallion sperm:

   (a) Add diluted sperm to a centrifugation tube. Underlay a small amount of cushion fluid in the bottom of the centrifugation tube, such that it covers the full diameter of the tube (i.e., ~4 mL in case of a 50 mL tube). Centrifuge for 20 min at 1000 × *g*, at ambient temperature.

   (b) Carefully remove the supernatant and cushion fluid. Cushion fluid can be aspirated from below the layer with sperm (*see* **Note 10**). Alternatively, recover and transfer the sperm layer to a new tube.

3. Determine the sperm concentration (and number) in the resuspended pellet or recovered layer, using a hemocytometer.

4. After centrifugation, either a "one-" or "two-step" dilution can be used to add freezing extender.

3.7 One-Step Approach for Diluting Sperm with Freezing Extender

When performing a one-step dilution, freezing extender is directly added to a concentrated sperm sample. This approach is followed at many equine breeding stations and involves the addition of INRA-82 supplemented with 2%–10% clarified egg yolk (freeze-thawed aliquots) to which glycerol is added just prior to use, such that ~400 × 10⁶ sperm/mL (*see* **Note 11**) are cryopreserved in INRA-82 supplemented with (~5%) egg yolk and 2.5% glycerol. This procedure is depicted in Fig. 1A–C.

1. After initial dilution (in INRA-82) and centrifugation (10 min, 600 × *g*, at 22°C), resuspend the obtained sperm pellet in a little volume (i.e., remaining after removal of the supernatant) and determine the sperm concentration, recovered volume and total sperm number.

2. Determine the amount of freezing extender (i.e., volume) needed for diluting to the desired final sperm concentration (400 × 10⁶ sperm/mL).

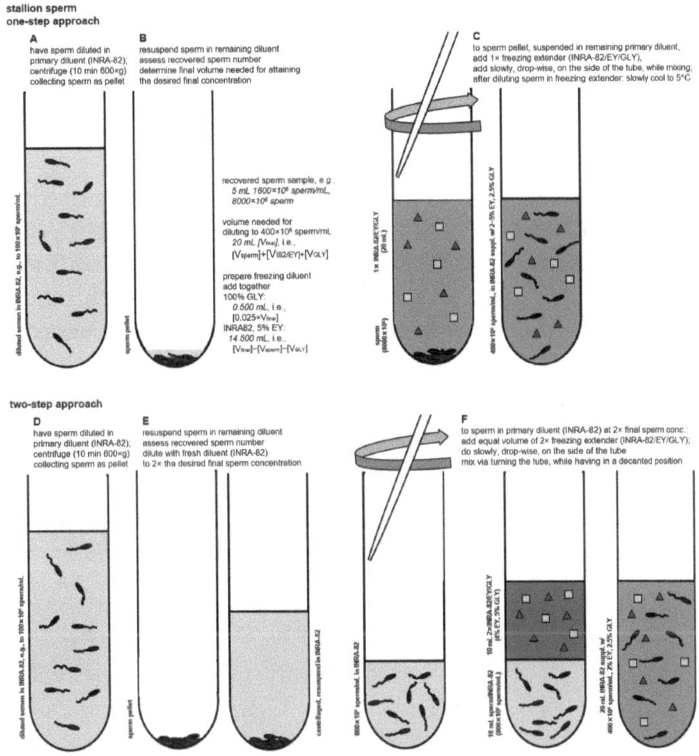

Fig. 1 Schematic presentation of one-step (**A**–**C**) and two-step dilution (**D**–**F**) approach, for diluting stallion sperm with freezing extender; i.e., eventually resulting in 400×10^6 sperm/mL in INRA-82 supplemented with 2% clarified egg yolk (yellow squares) and 2.5% glycerol (purple triangles). In both cases, ordinary diluted sperm samples, of 100×10^6 sperm/mL in INRA-82 (**A**, **D**) are subjected to centrifugation for recovering sperm in a pellet. Then, with the one-step procedure, the sperm number and remaining volume are determined for calculating the needed final volume for dilution, and thus the needed amounts of INRA-82 supplemented with 5%–10% egg yolk and 100% glycerol (**B**). In this case, after preparing the needed volume, (1×) INRA-82 freezing extender is directly added to the suspended sperm pellet (**C**). Addition is done slowly, drop-wise on the side of the tube, while turning to ensure mixing. In case of a two-step dilution protocol, after centrifugation, sperm are diluted to 800×10^6 sperm/mL in INRA-82, i.e., twofold the desired final concentration (**E**). To such centrifuged/diluted sperm sample, and an equal volume of 2× INRA-82 freezing extender is added, i.e., INRA-82 supplemented with 4% egg yolk and 5% glycerol, which is prepared in advance (**F**). Again, addition needs to be done slowly. Stallion sperm is cooled to 5°C after adding freezing extender

3. Supplement INRA-82 containing egg yolk (2%–10%) with glycerol (~100% solution) such that the final glycerol concentration is 2.5%, i.e., in the sample subjected to cryopreservation (taking into account the volume in which the sperm pellet is suspended) (*see* **Note 12**).

4. Slowly add 1× INRA-82 freezing extender, i.e., diluent containing the final concentrations of cryoprotective agents. Add dropwise to the side of the tube containing the sperm sample, while gently turning the tube to ensure mixing.

3.8 Two-Step Approach for Diluting Sperm with Freezing Extender

In case of a two-step dilution, a sperm sample is first diluted with a primary extender which does not contain glycerol; this is done to a concentration two-fold (stallion and bull) or 3/2-fold (boar) the desired final sperm concentration. Then, for cryopreservation, freezing extender is added, which contains two-fold (stallion and bull) or three-fold (boar) the desired final concentrations of cryoprotective agents. This procedure is depicted in Fig. 1D–F (stallion), Fig. 2A–C (bull) and Fig. 2D–F (boar).

1. Stallion sperm:
   (a) Use INRA-82 for initial dilution and centrifugation (10 min, 600 × *g*, at 22°C).
   (b) After centrifugation, dilute sperm in fresh INRA-82 to 200–800 × 10^6 sperm/mL.
   (c) Thereafter, add an equal volume (e.g., 10 mL to 10 mL) of 2× INRA-82 freezing extender (i.e., INRA-82 supplemented with 4% egg yolk and 5% glycerol). Add slowly, drop-wise on the side of the tube with the sperm sample while gently turning to ensure mixing.
   (d) This results in 100–400 × 10^6 sperm/mL, 2% egg yolk and 2.5% glycerol.
2. Bull sperm:
   (a) Use TEY-1 (contains 20% egg yolk) for diluting raw semen to 120 × 10^6 sperm/mL.
   (b) Thereafter, add an equal volume of TEY-2 (containing 20% egg yolk and 12.8% glycerol). Add slowly, drop-wise on the side of the tube while gently turning to ensure mixing.
   (c) This results in 60 × 10^6 sperm/mL, 20% egg yolk and 6.4% glycerol.
3. Boar sperm:
   (a) Use BTS for initial dilution, cooling the sperm sample to 17°C and centrifugation (3 min, 2400 × *g*, at 17°C).
   (b) After centrifugation, use LEY cooling extender (contains 20% egg yolk) for diluting to 1500–750 × 10^6 sperm/mL.
   (c) Slowly cool the sperm sample to 5°C.

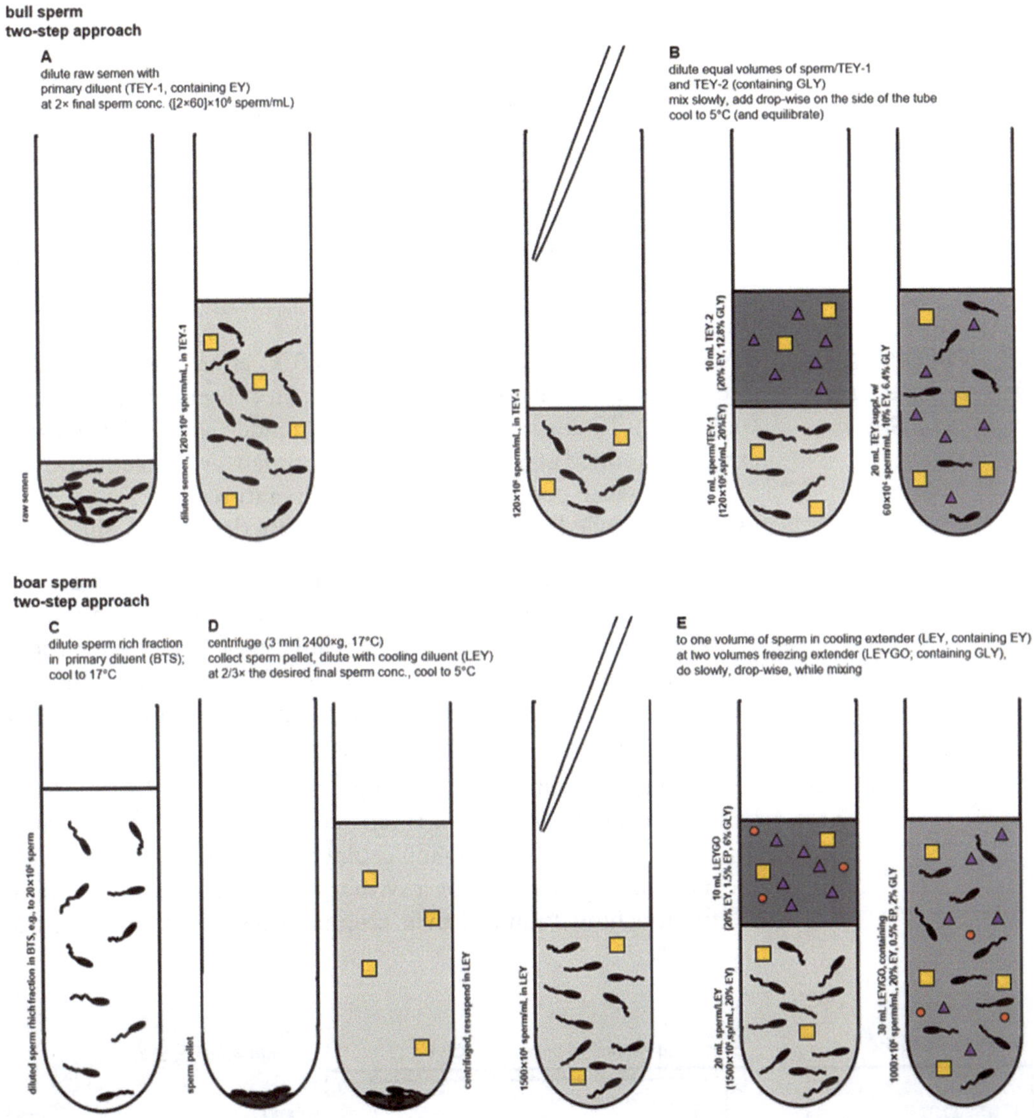

Fig. 2 Schematic presentation of the two-step approach for diluting bull sperm (**A**, **B**) as well as boar sperm (**C**–**E**) with freezing extender. Raw bull semen is diluted to 120×10^6 sperm/mL with TEY-1, i.e., twofold the desired final concentration (**A**). Then, an equal volume of TEY-2 is added; which in addition to ordinary egg yolk (yellow squares) contains glycerol (purple triangles). The final sample is composed of 60×10^6 sperm/mL in TEY supplemented with 20% egg yolk and 6.4% glycerol, and will be equilibrated at 5°C prior to cooling further. In case of boar semen, the sperm rich fraction is diluted with BTS, e.g., to $20–100 \times 10^6$ sperm/mL (**C**). After cooling to 17°C and centrifugation, sperm are diluted with LEY cooling extender to $750–1500 \times 10^6$ sperm/mL (**C**) followed by cooling to 5°C. Then, one volume of LEYGO freezing extender is added to two volumes of boar sperm in LEY. Both LEY and LEYGO contain ordinary egg yolk (yellow squares), whereas LEYGO also contains Equex paste (red circles) and glycerol (purple squares). LEYGO needs to be added slowly, and the final concentrations for cryopreservation are: $500–1500 \times 10^6$ in sperm/mL, 20% egg yolk, 0.5% Equex paste and 2% glycerol

(d) While maintaining at this temperature, dilute to 1000–500 × 10^6 sperm/mL with LEY freezing extender (containing 20% egg yolk, 1.5% Equex paste and 6% glycerol). This is done at a 1/2, v/v, ratio; e.g., via adding 10 mL LEY freezing extender to 20 mL sperm in LEY cooling extender. Add slowly, drop-wise on the side of the tube while gently turning.

(e) This results in 1000–500 × 10^6 sperm/mL, 20% egg yolk, 0.5% Equex paste and 2% glycerol.

3.9 Slow Cooling and Freezing of Sperm Packaged in Straws

Stallion and bull sperm are cooled to 5°C after being diluted in freezing extender at room temperature. In contrast, boar sperm is cooled from 17 down to 5°C in cooling extender, after which freezing extender is added at 5°C. In all cases, specimens are cooled slowly to 5°C, after which cooling to subzero temperatures/freezing is done using a controlled rate. These procedures are depicted in Fig. 3A,B.

1. Slowly cool sperm down to 5°C, at ~0.1°C/min. Such a cooling rate is reached when placing a tube with a sperm sample of ambient temperature (e.g., 50 mL) in a beaker with room temperature water (e.g., 250 mL) at 5°C for 2–4 h. With bull sperm, equilibration periods at 5°C up to 18 h can be used (*see* **Note 13**).
2. Load sperm samples in freezing extender into 0.25-mL (bull) or 0.5-mL (stallion, boar) labeled straws, while maintaining samples at 5°C (in a cooling cabinet/cold room). Use an automatic filling machine, or a syringe with a little piece of tubing for attaching straws to their plugged sites. Make sure that the

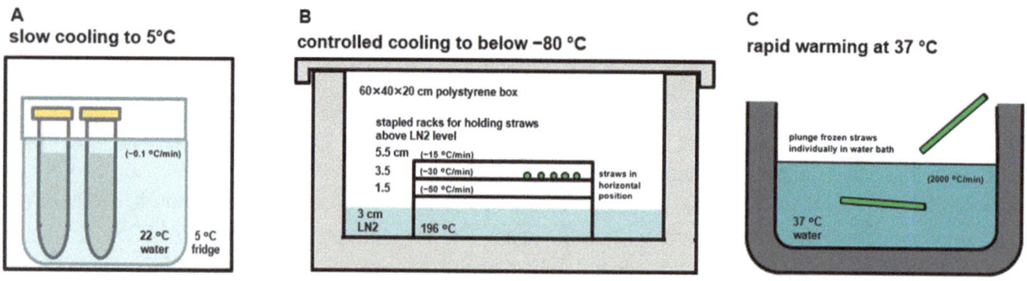

Fig. 3 Schematic presentation of simple procedures for slow cooling of sperm samples to 5°C in a fridge (**A**), controlled freezing in the liquid nitrogen vapor phase in a polystyrene box (**B**), and rapid warming in a 37°C water bath (**C**). Slow cooling of sperm samples to 5°C is feasible by placing room-temperature samples, packaged in tubes, in a container with room-temperature water. When placed in the fridge, the water and samples slowly cool from ~22 to 5°C during 2–4 h. Controlled cooling to temperatures below −80°C can be done using a polystyrene box with defined dimensions filled with a defined level of liquid nitrogen (e.g., 60 × 40 × 20 cm polystyrene box, 3 cm liquid nitrogen). Then, stapled racks can be used for holding straws at a defined height above the liquid nitrogen level, which results in a defined cooling profile (e.g., 30°C/min at 3.5 cm). Rapid warming of straws is simply done by plunging single frozen straws in a water bath set at 37°C

sealing powder on the plugged site is moistened upon filling, to close the straw on that site. Typically, straws are labeled with the date of semen collection, donor and manufacturer.

3. Seal the straws further using a sealing machine, sealing powder or balls.
4. Place straws on the racks used for freezing, while maintaining at 5°C.
5. For cooling down to subzero temperatures (i.e., freezing), use a controlled-rate freezer connected to a liquid nitrogen tank or place straws in the vapor phase of liquid nitrogen built up in a polystyrene box (*see* **Note 14**).
   (a) When using a controlled rate freezer: place racks with straws in the freezing chamber of the device held at 5°C, and cool down to −120°C, at 10–60°C/min).
   (b) Alternatively, straws can be frozen in the vapor phase above liquid nitrogen in a polystyrene box. In case of a box with 60×40×20 cm dimensions, fill with liquid nitrogen up to 3 cm height; for reaching a cooling rate of ~30°C/min when straws are placed in horizontal position ~3 cm above the surface of liquid nitrogen. Then, temperatures below −80°C are attained within 15 min.
6. After reaching −80 to −120°C, plunge the straws in liquid nitrogen. Thereafter, for storage, straws can be transferred into goblets and liquid nitrogen containers. Store for a minimum of 24 h in liquid nitrogen; after which, post-thaw sperm motility can be assessed.

3.10 Thawing of Cryopreserved Samples

This procedure is depicted in Fig. 3C.

1. Thaw straws by plunging and incubating in a water bath for 30 s at 37°C (*see* **Note 15**).
2. Determine post-thaw sperm motility and further characteristics. When needed, the sperm sample can be diluted; however, do so slowly when using an ordinary diluent to minimize osmotic stress (i.e., due to permeating cryoprotective agents still being present in the thawed sample).

4 Notes

1. Microscopic counting of sperm using a hemocytometer allows for actually inspecting sperm. When measuring the optical density as a measure for the sperm concentration, using a photometer, one cannot discriminate between cell types and

debris, and extender components like milk and egg yolk particles will interfere. Determining cell concentrations and percentages of membrane-intact sperm by using a nucleocounter eliminates the latter issue, since a DNA-intercalating fluorescent dye is used. This approach is based on comparing fluorescence intensities with a calibration curve, both before and after full membrane permeabilization.

2. In addition to diluents prepared in the laboratory by yourselves, many species-specific commercial extenders are available. Manufacturers of extenders commonly used in the breeding industry of domestic species include Minitube and IMV-Technologies. Furthermore, instead of "egg yolk" and "skim milk" alternative components (of non-animal origin) can be used, like liposomes prepared from defined or plant-derived lipids, low-density proteins, and phosphocaseinates.
3. Approximately ~150 mL will be recovered from ten medium-sized eggs.
4. If the obtained fraction is not clear, the centrifugation duration can be increased, e.g., to 30 min, or the obtained fraction can be subjected to a second centrifugation round using the same conditions.
5. Antibiotics can be added to cryopreserved semen, taking into account effective regulations. Recommendations as formulated by the European Council Directive can be found online, i.e., Commission Delegated Regulation EU 2020/686 of 17 December 2019 (see: https://eur-lex.europa.eu/eli/reg_del/2020/686/oj/eng), last modified on 10 April 2023. When adding antibiotic(s) to semen, their name(s) and concentration(s) should be listed in the accompanying animal health certificate. Where necessary, antibiotics or mixtures of antibiotics with a bactericidal activity at least equivalent to that of the following antibiotics or their mixtures in each mL of semen, may be added to semen or contained in semen diluents:

   - in the case of semen of bovine and porcine animals, a mixture of lincomycin-spectinomycin (150/300 μg), penicillin (500 IU), and streptomycin (500 μg); or
   - in the case of semen of ovine and caprine animals, gentamicin (250 μg) or a mixture of penicillin (500 IU) and streptomycin (500 μg); or
   - a mixture of gentamicin (250 μg), tylosin (50 μg), lincomycin-spectinomycin (150/300 μg), penicillin (500 IU), and streptomycin (500 μg); or
   - a mixture of amikacin (75 μg) and divekacin (25 μg).

6. The osmolality of a solution can be measured using a freezing-point depression osmometer. When the osmolality of a solution is too high, add water while taking into account how much to dilute the sample. When the osmolality is too low: calculate the number of osmoles missing (i.e., difference from e.g., 300 or 330 mOsm/kg) and the amount that needs to be added for the prepared solution volume (taking into account that 1 kg represents ~1 L water). In the case of INRA-82, the major component is glucose; thus, glucose can be added to reach the desired osmolality. Glucose does not dissociate in solution; the number of osmoles missing in the solution equals the number of moles to be added. If correcting an osmolality by adding NaCl, note that it dissociates in solution, and the number of osmoles missing should be divided by 2 to obtain the number of moles to be added. Calculate the amount of grams to be added to the solution volume by multiplying the number of moles needed by the molecular weight of the compound.
7. Equex STM and Orvus ES paste contain the detergent/surfactant sodium dodecyl sulfate. Addition likely results in solubilizing egg yolk lipids, which have been implicated in increasing protective properties.
8. Use a 10×20 magnification for counting a minimum of 250 cells. Sperm motility can be observed using a 10×20–40 magnification. For evaluation of morphological abnormalities, use immobilized/fixed specimens and a 10×100 magnification with an oil immersion objective. When available, computer-assisted sperm analysis (CASA) can be used to assess sperm motility characteristics. Furthermore, specific fluorescent dyes can be used with fluorescence microscopy or flow cytometric analysis, e.g., for determining plasma and acrosomal membrane integrity.
9. Presence of large amounts of seminal plasma in preserved semen has been shown to negatively affect sperm quality. Seminal plasma contents of ~5%, however, have been reported to result in higher post-thaw motility.
10. For cryopreservation, 1–2% (v/v) cushion fluid/iodixanol can remain in sperm samples. Iodixanol has been described to have cryoprotective properties.
11. For commercial purposes, take into account regulations on minimum values for numbers and/or percentages of motile and morphological sperm that need to be present in a dose for use in artificial insemination. Generally, the total number of sperm per straw or insemination dose is calculated. For bulls, one 0.25-mL straw containing ~20×10^6 sperm typically results in ~10×10^6 progressively motile sperm post-thaw,

when taking into account a 50% survival after cryopreservation. For stallions, 4-to-8 straws of 0.5 mL containing ~50–100×10^6 sperm typically result in ~250×10^6 progressively motile sperm post-thaw.

12. Calculate the number of recovered sperm after centrifugation [(total sperm count, i.e., sperm number): (concentration, in sperm/mL) × (V_{sperm}, in mL)]. Determine the final volume for diluting to 400×10^6 sperm/mL [(V_{final}, in mL): (# × 10^6 sperm)/(400×10^6 sperm/mL)]. Calculate the amount of glycerol (100%) that needs to be added to attain a 2.5% final concentration [(V_{GLY}, in mL): 0.025 × (V_{final})]. Prepare freezing extender for adding to the resuspended sperm pellet, by mixing INRA-82 supplemented with 2–10% egg yolk [($V_{I82/EY}$, in mL): (V_{final}) − (V_{sperm}) − (V_{GLY})] with 100% glycerol [(V_{GLY}, in mL)].
13. Bull and equine sperm samples can be filled in straws at ambient temperature, after which cooling down to 5°C can be programmed in a controlled-rate freezer. With bull sperm, prolonged equilibration periods in freezing extender at 5°C have been described as beneficial, likely due to longer times for exchange of protective extender components while the sperm metabolism is slowed.
14. If desired, specimens can be held at a particular subzero temperature for (mechanically) inducing ice nucleation. When using a controlled-rate freezer, cooling rates can be held constant and differently programmed for different temperature ranges. When using a polystyrene box, allow time for building up the vapor phase by keeping the box closed to freeze straws under similar conditions. Note that cooling rates dramatically decrease when placing straws further away from the liquid nitrogen level, and cooling rates may not be linear anymore. The optimal cooling rate for cryopreserving sperm is dependent on the type and concentration of cryoprotective agents used in the freezing extender.
15. Faster thawing, achieved by using higher incubation temperatures for a shorter period, has been shown to result in higher post-thaw sperm motility. This has been attributed to a reduced incidence of re-crystallization during thawing.

Acknowledgments

Employees of the Unit for Reproductive Medicine of the University of Veterinary Medicine Hannover (Hannover, Germany), National Stud of Lower Saxony (Celle, Germany), and Masterrind GmbH (Verden, Germany) are acknowledged for sharing protocols.

References

1. Wiebke M, Hensel B, Nitsche-Melkus E, Jung M, Schulze M (2022) Cooled storage of semen from livestock animals (part I): boar, bull, and stallion. Anim Reprod Sci 246:106822. https://doi.org/10.1016/j.anireprosci.2021
2. Yánez-Ortiz I, Catalán J, Rodríguez-Gil JE, Miró J, Yeste M (2022) Advances in sperm cryopreservation in farm animals: cattle, horse, pig and sheep. Anim Reprod Sci 246:106904. https://doi.org/10.1016/j.anireprosci.2021.106904
3. Sieme H, Oldenhof H, Wolkers WF (2016) Mode of action of cryoprotectants for sperm preservation. Anim Reprod Sci 169:2–5. https://doi.org/10.1016/j.anireprosci.2016.02.004
4. Hungerford A, Bakos HW, Aitken RJ (2023) Sperm cryopreservation: current status and future developments. Reprod Fertil Dev 35:265–281. https://doi.org/10.1071/RD22219
5. Oldenhof H, Bigalk J, Hettel C, de Oliveira Barros L, Sydykov B, Bajcsy ÁC, Sieme H, Wolkers WF (2017) Stallion sperm cryopreservation using various permeating agents: interplay between concentration and cooling rate. Biopreserv Biobank 15:422–431. https://doi.org/10.1089/bio.2017.0061
6. Uhlmannsiek L, Shen H, Eylers H, Martinsson G, Sieme H, Wolkers WF, Oldenhof H (2025) Preserving frozen stallion sperm on dry ice using polymers that modulate ice crystalization kinetics. Cryobiology 114:104852. https://doi.org/10.1016/j.cryobiol.2024.104852
7. Vidament M, Ecot P, Noue P, Bourgeois C, Magistrini M, Palmer E (2000) Centrifugation and addition of glycerol at 22°C instead of 4°C improve post-thaw motility and fertility of stallion spermatozoa. Theriogenology 54:907–919. https://doi.org/10.1016/s0093-691x(00)00401-5
8. Wall RJ, Foote RH (1999) Fertility of bull sperm frozen and stored in clarified egg yolk-tris-glycerol extender. J Dairy Sci 82:817–821. https://doi.org/10.3168/jds.S0022-0302(99)75301-4
9. Pursel VG, Johnson LA (1975) Freezing of boar spermatozoa: fertilizing capacity with concentrated semen and a new thawing procedure. J Anim Sci 40:99–102. https://doi.org/10.2527/jas1975.40199x
10. Westendorf P, Richter L, Treu H (1975) Zur Tiefgefrierung von Ebersperma. Labor-und Besamungsergebnisse mit dem Hülsenberger Pailetten-Verfahren. Dtsch Tierärztl Wschr 82:261–267
11. Parrilla I, Gil MA, Cuello C, Martinez EA (2025) Sperm cryopreservation in boars. Methods Mol Biol 2897:183–192. https://doi.org/10.1007/978-1-0716-4406-5_13

Standardized Cryopreservation Protocols for European Perch (*Perca fluviatilis*) and Pikeperch (*Sander lucioperca*) Semen Enabling Hatchery-Scale Fertilization

Sylwia Wałdowska, Joanna Nynca, Daniel Żarski, Katarzyna Palińska-Żarska, Mariola A. Dietrich and Andrzej Ciereszko

Abstract

Cryopreservation of fish semen ensures the year-round availability of gametes from species with seasonal spawning, supports genetic resource banking, and underpins selective breeding programs. However, percid species such as European perch and pikeperch pose specific challenges, notably the spontaneous activation of sperm motility in seminal plasma, which can reduce sperm motility reserves during cryopreservation. Building on systematic studies, here we present standardized protocols for semen collection, sperm quality evaluation, extender preparation, freezing, thawing, and post-thaw handling. These protocols support reliable commercial cryobanking and enable consistent, hatchery-scale fertilization in percid aquaculture.

Keywords Cryopreservation, European perch, Fish semen, Glucose-methanol extender, Pikeperch, Sperm motility, Standardized protocol

1 Introduction

European perch (*Perca fluviatilis*) and pikeperch (*Sander lucioperca*) are key freshwater species of high ecological, economic, and scientific importance [1]. Both are employed as model organisms in studies spanning reproductive physiology, environmental monitoring, ecotoxicology, developmental biology, and domestication [2–5]. Both species are used as models for scientific purposes aimed at reproduction and developmental biology—covering gametogenesis, photothermal control of spawning, and egg-quality management—while also playing important roles in trophic dynamics and biomanipulation projects to regulate prey populations [6–8]. European perch and pikeperch are both commercially important species for recirculating aquaculture systems (RAS), where intensive rearing conditions can affect reproductive physiology, leading to seasonal spawning patterns and variable sperm quality, which hinder consistent

Willem F. Wolkers and Harriëtte Oldenhof (eds.), *Cryopreservation and Freeze-Drying Protocols*, Methods in Molecular Biology, vol. 3060, https://doi.org/10.1007/978-1-0716-5416-3_12,

production of high-quality semen [1]. Therefore, cryopreservation of sperm offers a practical solution to these constraints by enabling off-season fertilization, standardizing sperm availability and quality, and supporting selective breeding and genetic resource conservation efforts essential for the sustainable development of European perch and pikeperch aquaculture [9–12].

Reproductive biology differs between these species: European perch spawn eggs as a long, cylindro-conical gelatinous ribbon, with individual eggs embedded in a thick zona radiata externa, forming strands up to 2 m long, facilitating estimation of fecundity and requiring specific fertilization handling [13, 14]. Pikeperch eggs are released as batches of single, adhesive eggs that require removal of adhesiveness in hatchery conditions before incubation [8]. These differences influence fertilization protocols and highlight species-specific considerations in hatchery operations. Moreover, unpredictable and variable reproductive performance remains a major bottleneck for reliable, year-round juvenile production and therefore efficient broodstock management [1, 15].

It is well-known that fish spermatozoa are commonly immotile in the seminal plasma [16]. However, in percids, spontaneous sperm motility activation in seminal plasma occurs, which complicates the cryopreservation process and further affects post-thaw sperm motility [17, 18]. Only a handful of studies concerning European perch [12, 19–22] and pikeperch [10, 11] sperm cryopreservation have been published, and they relied on different compositions of extenders at variable sperm-to-extender dilution ratios. It is well-established that spermatozoa concentration varies among species, stocks, and even within samples from the same fish, depending on the collection period during the reproductive season [23]. Therefore, the spermatozoa concentration in the straws after dilution with an extender at a constant ratio further differs among individuals, which could cause differences in post-thaw sperm motility and conditions of fertilization with respect to the sperm-to-egg ratio. For this reason, the standardization of the cryopreservation protocols by establishing an optimal (which is species-specific and necessitates experimental validation) and constant spermatozoa concentration in straws at a constant cryoprotectant concentration is essential. The standardization of spermatozoa concentration in the straws simplifies the fertilization procedure, since the final spermatozoa concentration in the straw is already known and is similar across each straw from all individuals. The use of these procedures enables to obtain post-thaw sperm motility at the level of 70%–80% for European perch and 50%–60% for pikeperch [17, 18]. Fixing both sperm density and cryoprotectant composition in every 0.5-mL straw, which resulted in the fact that each straw contains a known sperm dose at a constant cryoprotectant level—lets hatcheries apply precise, biologically optimal sperm:egg ratios, one of the most important determinants of fertilization success in percid eggs [24].

2 Materials

Prepare all solutions and stocks for the extender using ultrapure water and analytical grade reagents right before usage. Store all solutions at 4°C and use within 7 days.

2.1 Solutions and Stocks for Extender

Sperm motility activation requires using a two-step procedure for European perch (applying dilution of sperm in immobilizing solution and then activation solution) or a one-step procedure for pikeperch (applying activation solution).

1. Immobilizing solution (European perch): 150 mM NaCl, 5 mM KCl, 1 mM $MgSO_4$, 1 mM $CaCl_2$, 20 mM Tris–HCl, pH 8.0 [20, 25].
2. Activating solution (European perch and pikeperch): 75 mM NaCl, 2 mM KCl, 1 mM $MgSO_4$, 1 mM $CaCl_2$, 20 mM Tris–HCl, 0.5% (w/v) BSA, pH 8.0 [20, 25].
3. Stocks for extender for cryopreservation of European perch semen: 1 M glucose, 1 M KCl, 100% methanol, ddH_2O for obtaining final concentrations of extender components at 0.30 M glucose, 7.5% (v/v) methanol, 25 mM KCl (*see* Table 1) [18].
4. Stocks for extender for cryopreservation of pikeperch semen: 1 M glucose, 100% methanol, ddH_2O for obtaining final concentrations of extender components at 0.13 M glucose, 7.5% (v/v) methanol (*see* Table 2) [17].

2.2 Materials for Semen Collection and Sperm Motility Analysis

1. Anesthesia solution: hatchery water containing 150 mg/L MS-222.
2. Soft catheters for semen collection. Recommended flexible infant feeding tubes of size CH6 cut into pieces of length between 3 cm (for European perch) and 5 cm (for pikeperch).
3. Computer-Assisted Sperm Analysis (CASA) system (e.g., Hamilton-Thorne CEROS II system) equipped with negative-phase 20× objective.
4. Immobilizing (European perch) and activation solutions (European perch and pikeperch).
5. Cooling system for Eppendorf tubes (e.g., Eppendorf ThermoStat C) enabling keeping polypropylene tubes with semen, immobilizing and activation solutions at 4°C.

Table 1

Volumes of semen and extender components (1 M glucose, 1 M KCl, methanol, and ddH_2O) were calculated for cryopreservation of European perch sperm based on initial sperm concentrations. For each European perch male, semen was diluted to achieve a final sperm concentration of 3×10^9spermatozoa/mL, 0.3 M glucose, 25 mM KCl, and 7.5% (v/v) methanol in a final volume of 20 mL. The bold values in table cells show the calculated volumes of semen and each extender component required to reach these final concentrations, whereas unbold cells contain basic information for calculations of these volumes

Male no.	Sperm concentration ($\times 10^9$ spz/mL)	Final sperm concentration ($\times 10^9$ spz/mL)	Final glucose concentration (M)	Dilution of semen ratio	Semen (mL)	1 M glucose (mL)	1 M KCl (mL)	100% methanol (mL)	ddH20 (mL)	Final volume (mL)
1	30	3	0.3	10.0	**2.000**	**6.006**	**0.500**	**1.500**	**9.994**	20
2	32	3	0.3	10.7	**1.875**	**6.006**	**0.500**	**1.500**	**10.119**	20
3	35	3	0.3	11.7	**1.714**	**6.006**	**0.500**	**1.500**	**10.280**	20

Table 2

Volumes of semen and extender components (1 M glucose, 100% methanol, and ddH_2O) were calculated for cryopreservation of pikeperch sperm based on initial sperm concentrations. For each pikeperch male, semen was diluted to achieve a final sperm concentration of 3×10^9spermatozoa/mL, 0.13 M glucose, and 7.5% (v/v) methanol in a final volume of 20 mL. The bold values in table cells show the calculated volumes of semen and each extender component required to reach these final concentrations, whereas unbold cells contain basic information for calculations of these volumes

Male no	Sperm concentration (× 10^9 spz/mL)	Final sperm concentration (× 10^9 spz/mL)	Final glucose concentration (M)	Dilution of semen ratio	Semen (mL)	1 M glucose (mL)	100% methanol (mL)	ddH20 (mL)	Final volume (mL)
1	15	3	0.13	5.0	**4.000**	**2.601**	**1.500**	**11.899**	20
2	18	3	0.13	6.0	**3.333**	**2.601**	**1.500**	**12.566**	20
3	20	3	0.13	6.7	**3.000**	**2.601**	**1.500**	**12.899**	20

2.3 Materials for Sperm Concentration Evaluation

1. NucleoCounter SP-100 computer-aided fluorescence microscope (Chemometec) or equivalent cell counter with the necessary consumables (*see* **Note 5**).
2. PBS solution.

2.4 Materials for Sperm Cryopreservation

1. 0.5 mL plastic straws.
2. Liquid nitrogen (~20 L).
3. Styrofoam box with a tightly fitting cover, an inner stainless container and a floating rack (enabling positioning of the straws 3 cm above the LN_2 surface).
4. Water bath (ensuring proper circulation of the water, ensuring equal distribution of the temperature inside the bath) for thawing the straws at 40°C.
5. Cryopreservation security gear (gloves, forceps/tweezers, and protective eyewear).
6. Timer.

3 Methods

3.1 Cryopreservation of European Perch Semen

1. Anesthetize fish in hatchery water containing 150 mg/L MS-222 (Fig. 1A; *see* **Note 1**).
2. Collect semen using a catheter by inserting it gently into the sperm duct and providing a massage to the abdomen directly into a dry 1.5 mL polypropylene tube (Fig. 1B). Stripping the sperm through the catheter allows for avoiding contamination with urine or feces.
3. Keep collected semen at 4°C and process within 30 min. Record fish ID, date, and sperm volume for each sample.

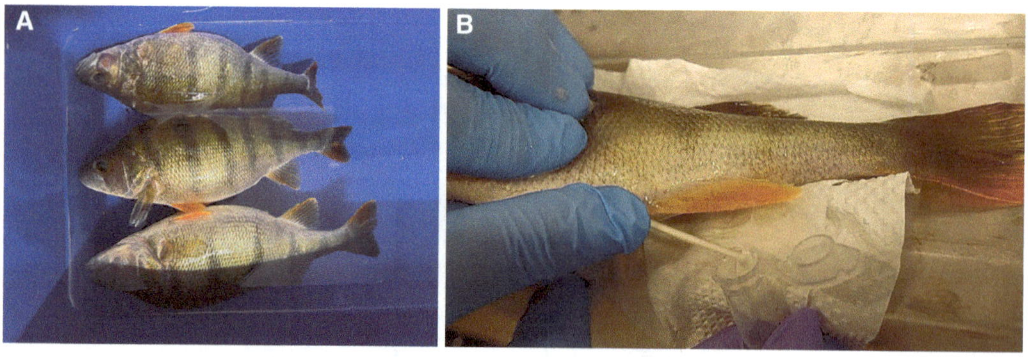

Fig. 1 Anesthetized fish (**A**) and collection of semen using the catheter (**B**)

4. Assess baseline sperm motility using CASA applying two-step procedure: dilute 1 µL semen in 49 µL immobilizing solution, then activate 1 µL of pre-diluted semen with 19 µL of activating solution; record motility parameters within 10 s (Fig. 2a; *see* **Note 2 and 3**).
5. Measure sperm concentration with a NucleoCounter SP-100 (Fig. 2b). Dilute semen 100 times with PBS, then 101 times with Reagent S100 (Chemometec) and subsequently load to a disposable cassette [26] (*see* **Note 4**). Enlarge the pipette tip opening to collect an appropriate volume of sperm (*see* **Note 5**).
6. Fill in the data in the spreadsheet containing formulas for calculating the volume of semen and extender components (*see* Table 1). Pipette into the glass beaker the calculated amounts of extender components and semen based on the spreadsheet (Table 1). It is recommended to pipette them in the following order: glucose, KCl, methanol, ddH_2O, and semen. Mix them thoroughly.
7. Fill the 0.5 mL straws and put them on a floating rack 3 cm height (*see* **Note 6**).
8. Cryopreservation is carried out by placing the floating rack on a liquid nitrogen surface in a tightly covered styrofoam box with an inner stainless container for 5 min before plunging the samples into liquid nitrogen for storage (Fig. 3).
9. Thaw by immersion of each straw in a controlled temperature water bath (40°C) for 10 s. Transfer thawed semen into dry polypropylene Eppendorf tube and immediately place in 4°C before further procedures.
10. Assess post-thaw sperm quality using the CASA system. Thawed sperm can then be used for fertilization of eggs or for other applications (*see* **Note 7**).

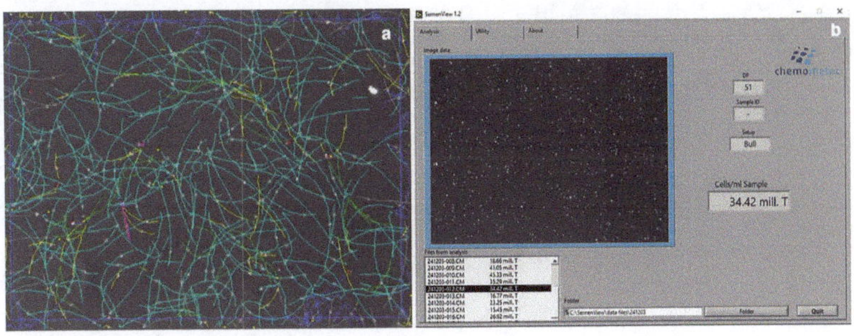

Fig. 2 Tracks of the motile spermatozoa applied by the CASA system (**a**) and view from SemenView software with results of sperm concentration evaluated using Nucleocounter (**b**)

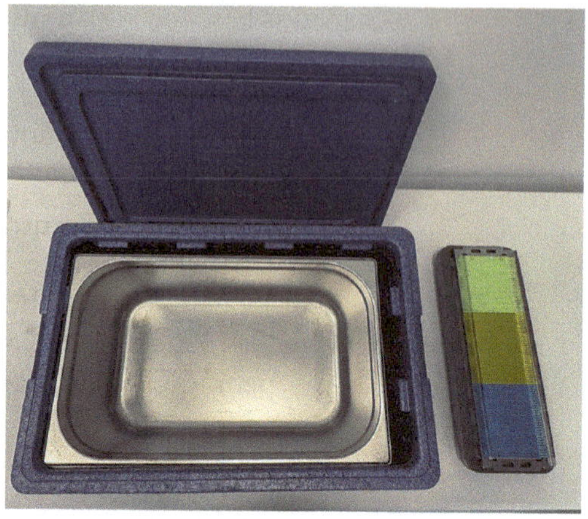

Fig. 3 Equipment used for cryopreservation in liquid nitrogen vapor: tightly covered styrofoam box with an inner stainless container and 3 cm height floating rack with 90 pieces of 0.5 mL straws

3.2 Cryopreservation of Pikeperch Semen

1. Anesthetize fish in hatchery water containing 150 mg/L MS-222.
2. Collect semen using a catheter by inserting it gently into the sperm duct and providing a massage to the abdomen directly into a dry 1.5 mL polypropylene tube. Stripping the sperm through a catheter allows for avoiding contamination with urine or feces.
3. Keep collected semen at 4°C and process within 30 min. Record fish ID, date, and sperm volume for each sample.
4. Assess baseline sperm motility using CASA: dilute 1 μL semen in 199 μL of activating solution; record motility parameters within 10 s (Fig. 2a, *see* **Note 4**).
5. Measure sperm concentration with a NucleoCounter SP-100 (Fig. 2b, *see* **Note 4**).
6. Fill in the data in the spreadsheet containing formulas for calculating the volume of semen and extender components (*see* Table 2).
7. Pipette into the glass beaker the calculated amounts of extender components and semen based on the spreadsheet (Table 2). It is recommended to pipette them in the following order: glucose, methanol, ddH_2O, and semen. Mix them thoroughly.
8. Fill the 0.5 mL straws and put them on a floating rack 3 cm high.
9. Cryopreservation is carried out by placing the floating rack on a liquid nitrogen surface in a tightly covered styrofoam

box with an inner stainless container for 5 min before plunging the samples into liquid nitrogen for storage (Fig. 3).

10. Thaw by immersion of each straw in a controlled temperature water bath (40°C) for 10 s. Transfer thawed semen into dry polypropylene Eppendorf tube and immediately placed in 4°C before further procedures.
11. Assess post-thaw sperm quality using the CASA system. Thawed sperm can then be used to fertilize eggs or for other applications.

4 Notes

1. Semen should be obtained using a catheter after 5 days of hormonal treatment. Hormonal stimulation of males should be performed using salmon gonadoliberine analog (sGnRHa; at a dose of 50 μg/kg in a single injection), mammalian GnRHa (mGnRHa; at a dose of 100 μg/kg), or with human chorionic gonadotropin (hCG; at a dose of 250 IU/kg^{-1}) [3, 27].
2. Duration of sperm motility in European perch and pikeperch after activation lasts about 30 s; however, after activation of sperm motility, velocities rapidly decreased, therefore CASA analysis should be performed as fast as possible (within 10 s [25]).
3. Spontaneous sperm motility can be caused by lower than 300 mOsm/kg (reference value for seminal and blood plasma osmolality), seminal plasma osmolality, which can be the effect of stress caused by fish handling.
4. Depending on the device you have, sperm concentration can be measured using a Makler or Bűrker chamber, spectrophotometer, or flow cytometer, etc. [12, 28].
5. European perch semen is characterized by high density (more than 30×10^9spermatozoa/mL). Cutting the end of the pipette tip to enlarge the opening is highly recommended.
6. Due to small amounts of semen collected from European perch and pikeperch (about 5 mL) using an automated filling and sealing machine, a controlled-rate freezer is not recommended.
7. Due to a time-dependent decrease in sperm motility after thawing, quality assessment and fertilization of eggs should be performed immediately after thawing of straws [9, 18].

Acknowledgments

This work was supported by funds from the National Science Center, granted for research project No. 2020/38/E/NZ9/00394 entitled "Transcriptomic and zootechnical exploration of parental contribution to progeny quality in Eurasian perch, *Perca fluviatilis*" and research project No. 2023/51/B/NZ9/01877 entitled "Blood circulating small noncoding RNAs (c-sncRNAs) as nonlethal biomarkers of reproductive status and gamete quality in pikeperch (Akronym: sAnDeRNA)."

References

1. Policar T, Schaefer FJ, Panana E, Meyer S, Teerlinck S, Toner D, Żarski D (2019) Recent progress in European percid fish culture production technology—tackling bottlenecks. Aquac Int 27:1151–1174. https://doi.org/10.1007/s10499-019-00433-y
2. Kupren K, Palińska-Żarska K, Krejszeff S, Żarski D (2019) Early development and allometric growth in hatchery-reared Eurasian perch, *Perca fluviatilis* L. Aquac Res 50:2528–2536. https://doi.org/10.1111/are.14208
3. Żarski D, Fontaine P, Roche J, Alix M, Blecha M, Broquard C, Król J, Milla S (2019) Time of response to hormonal treatment but not the type of a spawning agent affects the reproductive effectiveness in domesticated pikeperch, *Sander lucioperca*. Aquaculture 503:527–536. https://doi.org/10.1016/j.aquaculture.2019.01.042
4. Żarski D, Le Cam A, Nynca J, Klopp C, Ciesielski S, Sarosiek B, Montfort J, Król J, Fontaine P, Ciereszko A, Bobe J (2020) Domestication modulates the expression of genes involved in neurogenesis in high-quality eggs of *Sander lucioperca*. Mol Reprod Dev 87:934–951. https://doi.org/10.1002/mrd.23414
5. Fontaine P, Teletchea F (2019) Domestication of the Eurasian Perch (*Perca fluviatilis*). In: Animal domestication. IntechOpen. pp 137–159
6. Żarski D, Ben Ammar I, Bernáth G, Baekelandt S, Bokor Z, Palińska-Żarska K, Fontaine P, Horváth Á, Kestemont P, Mandiki SNM (2020) Repeated hormonal induction of spermiation affects the stress but not the immune response in pikeperch (*Sander lucioperca*). Fish Shellfish Immunol 101:143–151. https://doi.org/10.1016/j.fsi.2020.03.057
7. Żarski D, Krejszeff S, Palińska K, Targońska K, Kupren K, Fontaine P, Kestemont P, Kucharczyk D (2012) Cortical reaction as an egg quality indicator in artificial reproduction of pikeperch, *Sander lucioperca*. Reprod Fertil Dev 24:843. https://doi.org/10.1071/RD11264
8. Żarski D, Krejszeff S, Kucharczyk D, Palińska-Żarska K, Targońska K, Kupren K, Fontaine P, Kestemont P (2015) The application of tannic acid to the elimination of egg stickiness at varied moments of the egg swelling process in pikeperch, *Sander lucioperca* (L.). Aquac Res 46:324–334. https://doi.org/10.1111/are.12183
9. Judycka S, Żarski D, Dietrich MA et al (2019) Standardized cryopreservation protocol of European perch (*Perca fluviatilis*) semen allows to obtain high fertilization rates with the use of frozen/thawed semen. Aquaculture 498:208–216. https://doi.org/10.1016/j.aquaculture.2018.08.059
10. Bokor Z, Horvath A, Horvath L, Urbany B (2008) Cryopreservation of pike perch sperm in hatchery conditions. Isr J Aquac 60:166–169. https://doi.org/10.46989/001c.20493
11. Bokor Z, Müller T, Bercsényi M, Horváth L, Urbányi B, Horváth A (2007)) Cryopreservation of sperm of two European percid species, the pikeperch (*Sander lucioperca*) and the Volga pikeperch (*S. volgensis*). Acta Biol Hung 58:199–207. https://doi.org/10.1556/ABiol.58.2007.2.6
12. Bernáth G, Bokor Z, Żarski D, Várkonyi L, Hegyi Á, Staszny Á, Urbányi B, Radóczi Ifj J, Horváth Á (2016) Commercial-scale out-of-season cryopreservation of Eurasian perch (*Perca fluviatilis*) sperm and its application for fertilization. Anim Reprod Sci 170:170–177. https://doi.org/10.1016/j.anireprosci.2016.05.005
13. Treasurer JW, Holliday FGT (1981) Some aspects of the reproductive biology of perch *Perca fluviatilis* L. A histological description of the reproductive cycle. J Fish Biol 18:359–376. https://doi.org/10.1111/j.1095-8649.1981.tb03778.x

14. Żarski D, Horváth Á, Kotrik L, Targońska K, Palińska K, Krejszeff S, Bokor Z, Urbányi B, Kucharczyk D (2012) Effect of different activating solutions on the fertilization ability of Eurasian perch, *Perca fluviatilis* L., eggs. J Appl Ichthyol 28:967–972. https://doi.org/10.1111/jai.12098
15. Teletchea F, Gardeur J-N, Psenicka M, Kaspar V, Le Doré Y, Linhart O, Fontaine P (2009) Effects of four factors on the quality of male reproductive cycle in pikeperch *Sander lucioperca*. Aquaculture 291:217–223. https://doi.org/10.1016/j.aquaculture.2009.03.029
16. Dzyuba B, Boryshpolets S, Rodina M, Gela D, Linhart O (2010) Spontaneous activation of spermatozoa motility by routine freeze-thawing in different fish species. J Appl Ichthyol 26:720–725. https://doi.org/10.1111/j.1439-0426.2010.01553.x
17. Judycka S, Dietrich MA, Żarski D, Karol H, Palińska-Żarska K, Błażejewski M, Ciereszko A (2021) Towards standardization of the cryopreservation procedure of cultured pikeperch (*Sander lucioperca*) semen. Aquaculture 538:736539. https://doi.org/10.1016/j.aquaculture.2021.736539
18. Judycka S, Żarski D, Dietrich MA et al (2022) Toward commercialization: improvement of a semen cryopreservation protocol for European perch enables its implementation for commercial-scale fertilization. Aquaculture 549:737790. https://doi.org/10.1016/j.aquaculture.2021.737790
19. Bernáth G, Milla S, Várkonyi L, Ledoré Y, Griffitts JD, Fontaine P, Urbányi B, Bokor Z (2022) The effect of two different experimental rearing temperatures on the quality and the large-scale cryopreservation of Eurasian perch (*Perca fluviatilis*) sperm. Theriogenology 185:127–133. https://doi.org/10.1016/j.theriogenology.2022.03.033
20. Bernáth G, Żarski D, Krejszeff S, Palińska-Żarska K, Bokor Z, Król J, Kollár T, Kucharczyk D, Urbányi B, Horváth Á (2015) Optimization of conditions for the cryopreservation of Eurasian perch (*Perca fluviatilis* Linnaeus, 1758) sperm. J Appl Ichthyol 31:94–98. https://doi.org/10.1111/jai.12740
21. Ciereszko A, Ramseyer L, Dabrowski K (1993) Cryopreservation of yellow perch semen. Prog Fish-Cult 55:261–264
22. Rodina M, Policar T, Linhart O, Rougeot C (2008) Sperm motility and fertilizing ability of frozen spermatozoa of males (XY) and neomales (XX) of perch (*Perca fluviatilis*). J Appl Ichthyol 24:438–442. https://doi.org/10.1111/j.1439-0426.2008.01137.x
23. Kowalski RK, Cejko BI (2019) Sperm quality in fish: determinants and affecting factors. Theriogenology 135:94–108. https://doi.org/10.1016/j.theriogenology.2019.06.009
24. Kristan J, Zarski D, Blecha M, Policar T, Malinovskyi O, Samarin AM, Palinska-Zarska K, Nowosad J, Krejszeff S, Kucharczyk D (2018) Fertilizing ability of gametes at different post-activation times and the sperm-oocyte ratio in the artificial reproduction of pikeperch *Sander lucioperca*. Aquac Res 49:1383–1388. https://doi.org/10.1111/are.13570
25. Lahnsteiner F (2011) Spermatozoa of the teleost fish *Perca fluviatilis* (perch) have the ability to swim for more than two hours in saline solutions. Aquaculture 314:221–224. https://doi.org/10.1016/j.aquaculture.2011.02.024
26. Nynca J, Ciereszko A (2009) Measurement of concentration and viability of brook trout (Salvelinus fontinalis)spermatozoa using computer-aided fluorescent microscopy. Aquaculture 292:256–258. https://doi.org/10.1016/j.aquaculture.2009.04.020
27. Żarski D, Horváth Á, Bernáth G, Krejszeff S, Radóczi J, Palinska-Zarska K, Bokor Z, Kupren K (2017) Controlled reproduction of wild Eurasian perch. Springer International Publishing, Cham
28. Nynca J, Dietrich GJ, Liszewska E, Judycka S, Karol H, Dobosz S, Krom J, Ciereszko A (2016) Usefulness of a portable flow cytometer for sperm concentration and viability measurements of rainbow trout spermatozoa. Aquaculture 451:353–356. https://doi.org/10.1016/j.aquaculture.2015.09.027

Check for updates

Chapter 13

Oocyte and Embryo Vitrification in Horses

Daniel Angel-Velez, Elisa Serrano-Revuelta, Ann Van Soom and Katrien Smits

Abstract

Vitrification has emerged as a decisive technique in assisted reproductive technology, offering superior cryopreservation outcomes for both oocytes and embryos compared with traditional slow freezing. This chapter provides a comprehensive overview of the vitrification of equine oocytes and embryos, emphasizing the influence of technical factors, including cryoprotectant selection, exposure time, temperature control, and device volume, on vitrification outcomes. Biological factors such as the oocyte maturation stage, the presence of cumulus cells, and embryo size are also explored, highlighting species-specific responses to cryodamage. Detailed protocols for oocyte and embryo vitrification and warming are presented, including media preparation, handling conditions, and critical timing steps. Additionally, storage logistics, labeling strategies, and cryo-room design to prevent sample mix-ups and ensure regulatory compliance are discussed.

Keywords Assisted reproduction, Cryopreservation, Embryos, Fertility preservation, Horse, Oocytes, Vitrification

1 Introduction

Oocyte and embryocryopreservation has emerged as a revolutionary method in equine assisted reproductive medicine and a significant component of animal breeding programs [1–6]. Equine in vitro embryo production (IVP) has grown rapidly, evolving from an experimental technique to a commercial reality. However, transferable embryos are produced between days 6 and 10 after intracytoplasmic sperm injection (ICSI), while optimal pregnancy rates require recipient mares to be at day 4 post-ovulation [7]. Embryo freezing allows embryos to be preserved and transferred to the recipient mare at the optimal time [5, 8–10]. Likewise, embryocryopreservation facilitates the reproductive management of donor mares by enabling breeding outside the reproductive season [11–13], allowing broader geographical distribution of valuable genetics, supporting genetic

Willem F. Wolkers and Harriëtte Oldenhof (eds.), *Cryopreservation and Freeze-Drying Protocols*, Methods in Molecular Biology, vol. 3060, https://doi.org/10.1007/978-1-0716-5416-3_13,

banking, and preserving embryos after biopsy for pre-implantation genetic testing [5, 10, 12, 14, 15].

Oocyte vitrification serves as a valuable adjunct in assisted reproductive technology due to its capacity to: (i) defer decisions regarding the selection of the fertilizing stallion for ICSI; (ii) enable the recovery of oocytes from recently deceased animals in regions lacking IVP facilities; and (iii) preserve high-quality germ lines from young mares prior to their participation in competitions. This is particularly important because older mares typically exhibit a reduced number of follicles suitable for ovum pick-up (OPU), thereby diminishing the chances of successful embryo development following IVP. Additionally, horseoocyte banking would facilitate the global dissemination of valuable female genetics for commercial applications and for breeding programs targeting endangered equid breeds and species [16, 17]. Finally, storing abattoir-derived oocytes could provide a sustained source of material for research and cloning in nations with limited oocyte availability [16, 18].

Two main cryopreservation methods are used to preserve biological samples: slow freezing and vitrification. In slow freezing, low concentrations (5%–10%) of cryoprotective agents (CPAs) allow extracellular ice to form gradually, promoting water efflux from cells and minimizing damaging intracellular ice formation, provided that cooling is performed slowly. In contrast, vitrification employs much higher CPA concentrations (30%–40%) and extremely rapid cooling rates (e.g., 70,000°C/min), which altogether prevent ice formation inside or outside the cells [19]. Although slow freezing showed promise initially, particularly in mice and humans, vitrification has since become the preferred method due to its superior outcomes in oocyte and embryo survival rates and embryo development across species [20, 21]. In horses, embryos can be successfully cryopreserved by either slow freezing or vitrification; however, only vitrified oocytes have effectively supported embryo development, making vitrification the method of focus in this chapter [16, 18].

Vitrification is the process of converting an aqueous solution into a glass-like solid state while avoiding ice crystal formation. This can be achieved either by employing high concentrations of solutes (stable vitrification) or by applying ultra-rapid cooling rates that prevent ice nucleation (metastable vitrification) [22, 23]. High concentrations of cryoprotective agents, including membrane-permeating compounds such as dimethyl sulfoxide (DMSO), glycerol (also known as 1,2,3-propanetriol), ethylene glycol (EG), and propylene glycol (PG; also known as 1,2-propanediol), together with nonpermeating solutes like sugars, make vitrification possible. Once vitrified, oocytes can be safely stored at −196°C in liquid nitrogen, effectively halting all molecular motion [24, 25].

One of the most widely used and successful cryoprotectant combinations for equine embryo vitrification is DMSO and EG, typically both at concentrations of 7.5%–15%, together with a nonpermeating agent [26]. In mature human oocytes, EG exhibits lower membrane permeability than PG and DMSO; nevertheless, it is widely used because of its low toxicity [20]. In equine oocytes, EG has primarily been used in combination with nonpermeating CPAs to maintain oocyte viability. However, blastocyst development from vitrified equine oocytes was not achieved until combinations such as EG–DMSO–sucrose and PG–EG–galactose were employed [27, 28]. Using a comparative approach, our group evaluated various combinations of permeating and nonpermeating CPAs for immature equine oocyte vitrification. The combination of PG, EG, and galactose yielded the best results, with blastocyst development rates comparable to those obtained from fresh oocytes [28]. Therefore, this chapter describes the optimized method for equine oocyte vitrification based on these findings.

Extracellular nonpermeating CPAs—such as sugars, macromolecules, and synthetic polymers like ice blockers—help reduce the required concentration of toxic permeating CPAs and mitigate osmotic damage during warming. By increasing solution viscosity, they influence ice nucleation and improve vitrification efficiency at lower permeating CPA levels [29,30]. Different protocols using sugars—such as sucrose, trehalose, and galactose—in combination with hydroxypropyl cellulose as a synthetic polymer have yielded successful results in embryo vitrification [6, 26]. In oocyte vitrification, sugars are also commonly used as nonpermeating CPAs; however, the highest vitrification efficiency has been achieved with media containing trehalose or galactose [27, 28, 31].

The exposure time in the vitrification solution is critical in balancing cryoprotection and CPA toxicity. While equine embryos tolerate traditional long protocols (10–15 min of equilibration [6, 26, 32], prolonged CPA exposure in oocytes has been associated with toxicity, negatively impacting embryo development [27, 33]. In contrast, shorter exposure times still permit significant CPA permeation into the oocyte, and intracellular osmolality rapidly matches that of the extracellular vitrification solution—often within seconds [34]. Consequently, the most effective protocols for immature equine oocytes involve brief exposure (less than 2 min of equilibration) [27, 28, 33, 35]. Further research is needed on CPA concentrations and combinations in equine oocytes, as much of the current knowledge has been extrapolated from other species. This chapter describes our routine protocols, specifying both a commercial protocol for embryo vitrification and a short protocol for equine oocyte vitrification.

Removing CPAs after warming is crucial, as the greatest osmotic damage often occurs during this phase. Stepwise warming through different solutions with progressively lower osmolality has traditionally been the gold standard for CPA removal. However, a recent study in equine embryos showed that a single-step warming using commercial embryo-holding medium without cryoprotectants can achieve high pregnancy rates with IVP equine embryos [4]. Similarly, in humans, recent studies demonstrate that ultra-fast single-step warming produces embryological and clinical outcomes comparable to those of standard protocols [36, 37]. For both human and equine oocytes, single-step warming has been as effective as traditional multi-step protocols with decreasing CPA concentrations [31, 38]. This chapter describes the single-step warming method for both embryos and oocytes.

Rapid cooling and warming rates are essential for successful vitrification, as they allow the process to occur at lower CPA concentrations and help prevent chilling injury. Additionally, high warming rates are fundamental for inhibiting the growth of intracellular ice and preventing the associated cellular damage during warming [30, 39, 40]. The rates of cooling and warming depend on: (i) the volume and surface area of the sample and (ii) the efficiency of heat transfer between the sample and the cooling or warming system [30]. Consequently, minimum-volume cryodevices have been developed to achieve very high cooling rates, thereby enhancing vitrification performance compared with larger-volume devices [41]. Hence, this chapter describes a protocol that uses a minimal-volume device.

In humans, the standard protocol involves vitrification of in vivo matured oocytes, achieving blastocyst development and pregnancy rates comparable to those obtained with fresh oocytes. Similarly, these outcomes have been observed in horses [42, 43]. Notwithstanding, the inability to effectively induce superovulation in mares significantly restricts the availability of in vivo matured oocytes for vitrification [44]. Consequently, equine oocyte cryopreservation usually involves either immature oocytes or those matured in vitro. The optimal meiotic stage for vitrification remains poorly defined, with most research predominantly targeting immature oocytes [27, 28, 33, 45]. Our group compared the development and transcriptome of both meiotic stages, finding similar survival rates after vitrification in both study groups. Nevertheless, mature oocytes exhibited similar efficiency to fresh oocytes, whereas vitrification before in vitro maturation (IVM) resulted in a more disrupted transcriptome [46, 47]. This chapter describes a vitrification methodology applicable to both meiotic stages. However, further investigation is indispensable to optimize vitrification procedures, particularly for mature oocytes and denuded in vitro matured equine oocytes.

The role of cumulus cells during oocyte cryopreservation appears to be species-specific and dependent on the maturation stage [33, 48]. Excessive cumulus layers may hinder water and CPA exchange, thereby decreasing protection against cryodamage. Reducing cumulus layers and minimizing the volume of medium used when loading oocytes onto cryodevices also increases cooling and warming rates. In horses, where ICSI is optimized, and cumulus cells are not required for fertilization, partial or total denudation of oocytes seems preferable due to more homogeneous permeability, lower media volume, faster cooling and warming rates, and easier evaluation of oocyte survival and meiotic stage. Consequently, this chapter recommends partial denudation for immature oocyte vitrification and complete denudation for mature oocytes.

Equine in vitro embryos can be cryopreserved—either by slow freezing or vitrification—with pregnancy rates comparable to those of fresh IVP embryos. This is largely due to the absence or rudimentary development of the glycoprotein capsule, which otherwise serves as a major diffusion barrier for CPAs. Additionally, their small size (< 200 μm) and limited blastocoelic fluid increase the surface-to-volume ratio, enabling more efficient cooling and warming—both essential for cryosurvival [5, 6]. Small in vivo-derived embryos (< 300 μm) can also be effectively cryopreserved using either approach, with post-thaw pregnancy rates ranging from 45% to 67% [49, 50]. However, obtaining an embryo of this size remains challenging. For those exceeding 300 μm in diameter, perforation of the capsule and trophectoderm—combined with blastocoelic fluid aspiration—is required before vitrification [6, 32, 51]. This section presents a protocol applicable to all embryo types discussed; however, it does not include procedures for blastocoelic fluid aspiration by puncture.

This chapter outlines a comprehensive methodology for vitrifying equine oocytes and embryos. It begins with a concise list of required materials, followed by streamlined protocols. Oocyte procedures include slaughterhouse ovary processing, IVM, and a rapid two-step vitrification using a minimal-volume device, followed by single-step warming in 0.3 M galactose. Embryo cryopreservation relies on the commercial Equine VIT-Kit®, involving sequential equilibration and a single-step warming in embryo-holding medium. Detailed identification and storage procedures ensure traceability and sample integrity. Finally, extensive notes justify the choice of media, stripper-tip diameters, temperature control, CPA exposure times, and storage conditions, emphasizing that strict adherence to these parameters is essential for achieving post-warming survival and developmental rates comparable to fresh oocytes and embryos.

2 Materials

2.1 Oocyte Vitrification

2.1.1 Post-Mortem Oocyte Collection

Oocytes can be obtained via ovum pick-up or post-mortem collection. This chapter focuses on slaughterhouse-derived oocytes.

1. Oocyte collection medium supplemented with heparin (e.g., EquiPlus OPU Recovery Medium) (*see* **Note 1**).
2. Sterile glass bottles for fluid collection.

3. Vacuum pump with adjustable aspiration pressure.
4. Sterile Medical Grade Teflon (PTFE) aspiration tubing.
5. One 10 mL syringe without a rubber plunger.
6. One 18-gauge needle and three 16-gauge needles.
7. One liter of sterile NaCl solution.
8. Two small 35/10 mm Petri dishes for oocyte selection.
9. One large 100/20 mm Petri dish.
10. Pipette controller.
11. Serological 10 mL pipettes.
12. Permanent marker.
13. Two 50 mL centrifuge tubes.
14. Container for fluid waste disposal.
15. Oocyte handling solution: Tissue Culture Medium 199 (TCM-199) with Hank's salts (e.g., Gibco) supplemented with 10% (v/v) heat-inactivated fetal bovine serum (FBS, e.g., Gibco) and 50 μg/mL gentamicin (e.g., Sigma) (*see* **Note 2**).
16. Adjustable-volume pipettes (range: 0.5–10 μL, 10–100 μL, and 100–1000 μL) with corresponding sterile tips.
17. Syringe filters 0.22 μm, PSE, Puradisc™ (Whatman™).
18. Scalpel number 4 handle fitted with a No. 20–24 blade.
19. Stereomicroscope for oocyte handling.

2.1.2 In Vitro Maturation (IVM)

Oocytes can be vitrified at the immature stage or after IVM. The following materials and conditions are required for oocyte maturation:

1. Maturation medium: TCM-199 with Earle's salts (Gibco) supplemented with 10% (v/v) fetal bovine serum (FBS, Gibco), 9.4 μg/mL follicle-stimulating hormone (FSH), 1.88 μg/mL luteinizing hormone (LH) (Stimufol, Reprobiol, Ouffet, Belgium), and 50 μg/mL gentamicin (Sigma G1264) (*see* **Note 3**).
2. Four-well dishes, for oocyte maturation.
3. Sterile paraffin oil (Cooper Surgical, Venlo, The Netherlands).
4. Incubator set at 38.2°C in atmospheric air.
5. Adjustable-volume pipettes (range: 0.5–10 μL, 10–100 μL, and 100–1000 μL) with corresponding sterile tips.
6. Syringe filters 0.22 μm, PSE, Puradisc™ (Whatman™).
7. One 10 mL syringe without a rubber plunger.
8. Two 5 mL round sterile tubes (Falcon).
9. Stereomicroscope for oocyte handling.

2.1.3 Preparation of the Oocyte for Vitrification

Partial denudation is advisable for vitrification of immature oocytes, while mature oocytes require complete denudation. The materials needed for each procedure are listed below:

1. Partial denudation (immature oocytes):
   a. Oocyte handling solution, specified in Sect. 2.1.1.
   b. Two 35/10 mm Petri dishes for storing oocytes and performing partial denudation.
   c. Stripper device with 200 μm stripper tips for handling and partially denuding immature oocytes.
   d. Adjustable-volume pipettes (range: 0.5–10 μL, 10–100 μL, and 100–1000 μL) with corresponding sterile tips.
   e. Syringe filters 0.22 μm, PSE, Puradisc™ (Whatman™).
   f. One 10 mL syringe without a rubber plunger.
   g. Two 5 mL round sterile tubes (Falcon).
   h. Stereomicroscope for oocyte handling.
2. Complete denudation (mature oocytes):
   a. Oocyte handling solution, specified in Sect. 2.1.1.
   b. 60/10 mm Petri dish for denudation.
   c. Stripper device with 135 μm and 175 μm stripper tips for complete denudation of mature oocytes (*see* **Note 4**).
   d. Hyaluronidase (0.1%, Sigma-Aldrich H3757) diluted in oocyte handling solution.
   e. Paraffin oil to overlay the denudation dish.
   f. Adjustable-volume pipettes (0.5–10 μL, 10–100 μL, and 100–1000 μL) with sterile pipette tips corresponding to each pipette size.
   g. Syringe filters 0.22 μm, PSE, Puradisc™ (Whatman™).
   h. One 10 mL syringe without a rubber plunger.
   i. Two 5 mL round sterile tubes (Falcon).
   j. Stereomicroscope for oocyte handling.

2.1.4 Preparation of Vitrification and Warming Solutions

The following materials are required for preparing vitrification and handling solutions:

1. Four sterile tubes, one of 15 mL for warming solution, one of 50 mL for base solution, and two more of 5 mL for each vitrification solution.
2. Syringe filters 0.22 μm, PSE, Puradisc™ (Whatman™).
3. One 20 mL syringe without a rubber plunger.

4. Base solution: TCM-199 with Hank's salts (Gibco) supplemented with 0.4% (w/v) bovine serum albumin (BSA, A6003) and 50 μg/mL gentamicin (Sigma G1264) (*see* **Note 5**).
5. Vitrification Solution 1: base solution supplemented with 10% (v/v) ethylene glycol (EG, Sigma #102466), 10% (v/v) propylene glycol (PG, Sigma #P4347), and 50 μg/mL gentamicin (Sigma G1264).
6. Vitrification Solution 2: base solution supplemented with 20% (v/v) ethylene glycol (EG, Sigma #102466), 20% (v/v) propylene glycol (PG, Sigma #P4347), 0.5 M galactose (Sigma G5388), and 50 μg/mL gentamicin (Sigma G1264).
7. Warming solution: base solution supplemented with 0.3 M galactose (Sigma G5388) and 50 μg/mL gentamicin (Sigma G1264).
8. Adjustable-volume pipettes (0.5–10 μL, 10–100 μL, and 100–1000 μL) with sterile pipette tips corresponding to each pipette size.
9. An analytical balance accurate to 0.001 g.

2.1.5 Vitrification

During the vitrification process, the following materials are required:

1. Base and vitrification solutions (Vitrification 1 and 2).
2. Stripper device with 200 μm stripper tips for immature oocytes or 150 μm stripper tips for mature oocytes.
3. 100/10 mm Petri dish for vitrification droplets.
4. Adjustable-volume pipette (10–100 μL range) with sterile pipette tips.
5. Minimal volume vitrification device (<1 μL), such as Cryolock (*see* **Note 6**).
6. Thermal plate maintained at 38°C (*see* **Note 7**).
7. 10 mm goblet for loading and storing vitrified samples (*see* Sect. 3.5 and **Note 8**).
8. Kelly clamp or fine-tip tweezers for handling vitrification devices.
9. Vitrification box (e.g., Styrofoam or isolated box with at least 2.5 L of liquid nitrogen).
10. Stereomicroscope for monitoring oocyte handling steps.
11. Printer-compatible labels for sample identification.
12. Cryo-resistant labels compatible with ultra-low temperatures (suitable for use in liquid nitrogen).

2.1.6 Warming

During the warming process, the following materials are required:

1. Warming solution: base solution supplemented with 0.3 M galactose (Sigma G5388) and 50 µg/mL gentamicin (Sigma G1264).
2. Stripper device with 200 µm stripper tips for immature oocytes or 150 µm stripper tips for mature oocytes.
3. 35/10 mm Petri dish or 60 mm center-well dish for warming and rinsing oocytes.
4. Adjustable-volume pipette (100–1000 µL range) with sterile pipette tips.
5. Thermal plate maintained at 38°C.
6. Kelly clamp or fine-tip tweezers for handling vitrification devices.
7. Vitrification box.
8. Stereomicroscope for monitoring oocyte handling steps.

2.2 Embryo Vitrification

In this chapter, we describe the use of a commercial vitrification medium (Equine VIT Kit; Tiefenbach, Germany), which is based on the formulation originally developed by Eldridge-Panuska et al. [50].

2.2.1 Vitrification

The following materials are required when using the Equine VIT-Kit® for embryo vitrification:

1. Commercial embryo handling solution (EmCare™, ICPbio Reproduction) (*see* **Note 9**).
2. Equine VIT-Kit® vitrification solutions (VS1, VS2, VS3).
3. Stereomicroscope for monitoring embryo handling steps.
4. Thermal plate set at 38°C for taking the embryo out of culture and at 23°C during vitrification.
5. Four-well dish or equivalent for sequential exposure to vitrification solutions (*see* **Note 10**).
6. 100/20 mm Petri dish for general handling and preparation of media droplets.
7. Minimal volume vitrification device (<1 µL), such as Cryolock (*see* **Note 6**).
8. Stripper device with 175 µm stripper tips for embryo loading.
9. Adjustable-volume pipette (100–1000 µL range) with sterile pipette tips.
10. 10 mm goblet for loading and storing vitrified samples (*see* Sect. 3.5 and **Note 8**).
11. Kelly clamp or fine-tip tweezers for handling vitrification devices.
12. Vitrification box.

13. Printer-compatible labels for sample identification.
14. Cryo-resistant labels compatible with ultra-low temperatures (suitable for use in liquid nitrogen).
15. Color code identification rod (*see* **Note 11**).

2.2.2 Warming

During the warming process, the following materials are required:

1. Commercial embryo handling solution (EmCare™, ICPbio Reproduction) (*see* **Note 9**).
2. Oocyte/Embryo manipulation system with 10–100 μL tip (BRAND micro-classic pipette controller).
3. 35/10 mm Petri dish or 60 mm center-well dish for warming the embryos.
4. Four-well dish for rinsing the embryo.
5. Adjustable-volume pipette (100–1000 μL range) with sterile pipette tips.
6. Thermal plate maintained at 38°C.
7. Kelly clamp or fine-tip tweezers for handling vitrification devices.
8. Vitrification box.
9. Stereomicroscope for monitoring embryo handling steps.

3 Methods

3.1 Oocyte Collection

This procedure can be applied to ovaries obtained from slaughterhouse animals, post-mortem cases, or mares that must be euthanized.

3.1.1 Preparation of Oocyte Handling Solution

1. Under the laminar flow, add the following to a sterile 10 mL tube: 9 mL of TCM-199 with Hank's salts, 1 mL of FBS, and 10 μL of gentamicin stock solution (stock concentration: 5 mg/mL).
2. Mix well and filter the solution using a sterile 0.22 μm syringe filter into another sterile 10 mL tube.
3. Store at 4°C for a maximum of 2 days.

3.1.2 Oocyte Collection

1. At least 1 h before the ovaries arrive, warm 1 L of sterile NaCl solution, 10 mL of handling solution, and 200 mL of EquiPlus OPU recovery medium in a warming unit to 38°C.
2. Perform the procedure in a clean environment where the room temperature is approximately 23°C.

3. Use up to two collection bottles per vacuum pump, with the aspiration pressure set between 100- and 120-mm Hg (prepare two sterile tubes or glass bottles per operator). Each bottle should have two lines: one connected to the vacuum pump, and the other fitted with a 16-G white needle.
4. Retrieve the pre-warmed flush medium (EquiPlus) from the warming unit.
5. Aspirate follicles ranging from 5 to 30 mm in diameter. Insert the needle directly into each follicle and scrape vigorously along the follicle walls. Optionally, refill the follicle with flush medium using a syringe without a rubber plunger and an 18-G needle, and aspirate again to increase recovery yield.
6. Once the surface follicles have been aspirated, bisect the ovary using a scalpel. Further slice each half into 2–3 mm sections. Aspirate and flush any visible internal follicles in these slices.
7. Rinse the collection tube with flush medium every two to three follicle aspirations and again upon completing the procedure for each ovary.
8. Take the bottle and move to a laminar airflow cabinet, and ensure the warming plates are set to room temperature (23°C).
9. Use a fine-point permanent marker to draw a grid of squares on the bottom of a large petri dish (100/20 mm).
10. Dispense 5 mL of oocyte handling medium into a small petri dish (35/10 mm) and keep the dish covered throughout the oocyte selection and collection process.
11. Use a sterile 10 mL serological pipette to aspirate follicular fluid from the bottom of the collection bottle. Use the pipette controller to aspirate the fluid.
12. Transfer 10–20 mL of follicular fluid into the large grid-marked petri dish. Adjust the volume based on the level of cell debris—less fluid should be used if debris is abundant. Begin scanning for COCs systematically, starting from the top-left square and proceeding toward the bottom-right.
13. Identify and isolate COCs under a stereomicroscope.
14. Gently swirl the dish to resuspend any debris, then repeat the search to maximize recovery.
15. Discard the examined follicular fluid and repeat the process until no oocytes are found in three consecutive aspirates, or until the collection bottle is empty.
16. After completing the search, dispense approximately 2 mL of handling medium into the lid of the 35 mm dish to create a "wash dish."

17. Wash the identified COCs in the wash dish, grouping them in sets of no more than 25.
18. At this stage, perform an initial quality selection: retain COCs with intact cumulus investment or partial denudation up to the corona radiata. Slightly expanded COCs may also be included. Discard fully denuded oocytes and extremely expanded ones. Avoid transferring any somatic cells—COCs only.
19. Store the oocytes in embryo-holding medium or move to a pre-equilibrated four-well plate with IVM medium.

3.2 Oocyte In Vitro Maturation (IVM)

1. In a sterile 15 mL conical tube, add 9 mL of TCM-199 with Earle's salts, 1 mL of FBS, and 10 μL of gentamicin stock solution (stock concentration: 5 mg/mL).
2. Mix thoroughly and filter the solution through a sterile 0.22 μ m filter into a labeled sterile tube.
3. Using a 1000 μL pipette, dispense 500 μL of base maturation medium into each well of the four-well dish.
4. Gently add 500 μL of oil to each well, carefully covering the medium. Add 10 μL of Stimufol® Reprobiol stock solution (0.5 mg pFSH; 0.1 mg pLH) to each well.
5. Add 1000 μL TCM-199 with Earle's salts (Gibco) to the center of the dish using a sterile glass pipette.
6. Label dishes based on treatment and place the dish in an incubator at 38.2°C, 5% CO_2 in air.
7. After equilibration, take one IVM plate and wash the COCs two times (in the left upper and lower wells) before putting them in the final well (maximum: 25 oocytes per 500 μL of maturation medium).
8. Label the dish. Work fast and try to minimize the volume of medium transferred along with the COCs.
9. Put the four-well plate back in the long-term CO_2 incubator (38.2°C, 5% CO2 in air) for approximately 30 h.

3.3 Oocyte Vitrification Procedure

3.3.1 Preparation of Solutions

The base solution can be stored refrigerated for up to one week. Conversely, vitrification and warming solutions must be made freshly for each use. It is advisable to make at least 2.5 mL of each vitrification solution and 5 mL of warming solution. Consequently, a minimum of 20 mL of total volume is necessary for each vitrification session. This chapter delineates the preparation method for a singular session.

1. Base Solution. Preparation of 20 mL of Base solution:
   (a) Weigh 0.08 g (80 mg) of bovine serum albumin (BSA).
   (b) Transfer the BSA powder into a sterile 50 mL conical tube.
   (c) Add TCM-199 with Hank's salts to a final volume of 20 mL.
   (d) Allow the BSA to dissolve completely at room temperature.
   (e) Filter the solution using a sterile 0.22 μm syringe filter.
   (f) Store the filtered solution at 4°C for up to 1 week.
2. Vitrification Solution 1. Preparation of 2.5 mL of Vitrification Solution 1:
   (a) Using a 1000 μL pipette, add 250 μL of ethylene glycol (EG) to a sterile 5 mL conical tube.
   (b) Add 250 μL of propylene glycol (PG).
   (c) Add 2 mL of the prepared base solution to the tube (*see* **Note 12**).
   (d) Mix the solution thoroughly and warm it to 38°C before use.
3. Vitrification Solution 2. Preparation of 2.5 mL of Vitrification Solution 2:
   (a) Weigh 0.225 g (225 mg) of galactose (Sigma G5388).
   (b) Transfer the galactose into a sterile 5 mL conical tube.
   (c) Using a 1000 μL pipette, add 500 μL of ethylene glycol (EG).
   (d) Add 500 μL of propylene glycol (PG).
   (e) Add base solution to bring the final volume to 2.5 mL.
   (f) Mix thoroughly and warm the solution to 38°C before use.
4. Warming Solution. Preparation of 5 mL of Warming Solution:
   (a) Weigh 0.270 g (270 mg) of galactose (Sigma G5388).
   (b) Transfer the galactose into a sterile 15 mL conical tube.
   (c) Add base solution to bring the final volume to 5 mL.
   (d) Mix thoroughly and warm the solution to 38°C before use.

3.3.2 Vitrification

1. Warm the base and vitrification solutions at 38°C for at least 1 h.
2. Turn on the thermal plate at 38°C.
3. Prepare a 35/10 mm petri dish with 4 mL of pre-warmed (38°C) base solution.

4. Prepare the vitrification box with liquid nitrogen for handling oocytes under liquid nitrogen (*see* **Note 13**).
5. Prepare the goblet and the identification rod for labeling, and subsequently immerse them in liquid nitrogen, as described in Sect. 3.5.
6. Select the COCs or denuded oocytes designated for vitrification (maximum of four oocytes per device) (*see* **Note 14**).
7. Maintain all selected COCs or denuded oocytes in a 35/10 mm Petri dish containing 4 mL of base solution at 38°C until vitrification.
8. Place a maximum of four COCs or denuded oocytes together in a 100 μL droplet of base solution on a 100/10 mm Petri dish pre-warmed to 38°C.
9. Transfer the oocytes through two successive 100 μL droplets of vitrification solution 1 (VS1) for 25 s at 38°C (*see* **Note 15**).
10. Immediately transfer the oocytes through two successive 100 μL droplets of vitrification solution 2 (VS2) for a maximum of 15 s.
11. Load the oocytes onto a minimal volume device (e.g., Cryolock), ensuring a final volume of <1 μL.
12. Remove excess vitrification medium by capillarity using a pipette to ensure minimal volume remains.
13. Plunge the device directly into liquid nitrogen within 30–45 s after the oocytes are placed in VS2 (*see* **Note 16**).
14. If the device includes a protective cap, secure it using a Kelly clamp or tweezers while keeping the device submerged in liquid nitrogen.
15. Insert the vitrification device into a 10 mm goblet, ensuring it always remains submerged in liquid nitrogen.
16. Transfer the goblet into the liquid nitrogen storage tank (*see* **Note 8**).
17. Store vitrified oocytes in liquid nitrogen, maintaining them fully submerged. Do not remove them from liquid nitrogen until the warming procedure begins (*see* Sect. 3.5).

3.3.3 Warming

1. Place the warming solution in a warming unit (e.g., stove or incubator) for at least 1 h at 38°C (*see* **Note 17**).
2. Remove the vitrification device from the storage tank and transfer it to a vitrification box, ensuring it remains submerged at all times.
3. Position the vitrification box as close as possible to the working area to minimize exposure to ambient temperature once warming begins.

4. Dispense 4 mL of warming solution into a 35/10 mm Petri dish.
5. If the vitrification device is capped, carefully remove the cap while it is under liquid nitrogen using clamps or tweezers.
6. Immediately immerse the device in 4 mL of pre-warmed solution at 38°C.
7. Start the timer and keep the device in the warming solution until the oocytes are released.
8. After 5 min, pass the oocytes through two 100 µL droplets of handling solution and hold them for a maximum of 10 min, or transfer them directly to maturation medium, 30 h for immature oocytes and 2 h for mature oocytes.

3.4 Embryo Vitrification Procedure

3.4.1 Vitrification

1. Place the commercial embryo holding medium in a warming unit (e.g., stove or incubator) for at least 1 h at 38°C (*see* **Note 17**).
2. Place the Equine VIT-Kit® solutions at least 30 min at room temperature (~23°C).
3. Prepare the vitrification box for handling embryos under liquid nitrogen.
4. Prepare the goblet and the identification rod for labeling, and subsequently immerse them in liquid nitrogen, as described in Sect. 3.5 "Identification and storage of oocytes and embryos."
5. Prepare two clean work areas if possible: one with a thermal plate at 38°C and one at room temperature (~23°C). If only one work area is available, set it to 38°C initially.
6. Prepare a four-well dish or equivalent with 500 µL of pre-warmed (38°C) embryo holding medium in one well.
7. Remove the culture dish from the incubator and gently aspirate the selected embryo for vitrification with a stripper device with a 175 µm stripper tip. Transfer the embryo to the first well with embryo-holding medium.
8. If using two stereomicroscopes, move the dish to the second work area (room temperature). If only one is available, turn off the thermal plate and allow the area to cool until room temperature (~23°C). Subsequent steps are performed at room temperature.
9. Dispense 300 µL of VS1 solution into the second well.
10. Transfer the embryo into the VS1 solution with a minimal volume of medium using the stripper tip of 175 µm and incubate for 5 min. Embryo shrinkage during this step is a positive indication of effective equilibration.

11. Dispense 300 μL of VS2 solution into the third well.
12. After the 5-min equilibration in VS1, transfer the embryo into VS2 using minimal volume with the stripper tip of 175 μ m and incubate for 5 min.
13. While the embryo is in VS2, prepare three 100 μL droplets of VS3 solution on the lid of a 100/20 mm Petri dish (or equivalent), and prepare a timer for 40 s.
14. Ensure the vitrification box is within reach. Place the vitrification device nearby with its identification mark facing upward and the cap removed.
15. After the 5-min VS2 incubation, aspirate the embryo and place it in the first droplet of VS3 solution. Immediately start the 40-s timer.
16. Quickly move the embryo through each of the three VS3 droplets to ensure proper exposure and washing.
17. Aspirate the embryo and load it onto the vitrification device (Cryolock) with a minimal volume of medium.
18. Remove excess medium by capillarity using the stripper tip, ensuring a final volume of <1 μL.
19. Plunge the Cryolock directly into liquid nitrogen within 40 s of first contact with VS3.
20. As the device has a cap, secure it while submerged in liquid nitrogen using a clamp or tweezers.
21. Store the vitrified embryo in a nitrogen tank (−196°C) until warming.

3.4.2 Warming

1. Place the commercial embryo holding medium in a warming unit (e.g., stove or incubator) for at least 1 h at 38°C.
2. Remove the vitrification device from the storage tank and transfer it to the vitrification box containing liquid nitrogen, ensuring that the device always remains submerged.
3. Position the vitrification box as close as possible to the working area to minimize exposure time at ambient temperature once thawing begins.
4. Dispense 4 mL of embryo holding medium into a 35/10 mm Petri dish.
5. As the vitrification device is capped, carefully remove the cap under liquid nitrogen using clamps or tweezers.
6. Immediately and swiftly transfer the vitrification device into the pre-warmed holding medium.
7. Start a timer and maintain the device in the warming solution until the embryos are spontaneously released from the device.

8. After 5 min, transfer the embryos through two 100 μL droplets of embryo-holding medium. Prepare them for either transfer or transport as appropriate.

3.5 Identification and Storage of Oocytes and Embryos

Accurate identification and proper labeling of vitrification and storage devices are critical steps in assisted reproductive procedures. These practices ensure the traceability, safety, and proper allocation of biological material. To minimize the risk of sample mix-ups, it is recommended to use a combination of physical labeling, color coding, and double witnessing.

1. Use a cryo-resistant label printer (e.g., Brady) to create three identical labels.
2. Each label must include the identification number (OPU cycle number+ embryo/oocyte number, owner, stallion, mare, date of vitrification (YYYY-MM-DD), and lab code (*see* **Note 11**).
3. Choose a color-coded 10 mm goblet for easy identification, and apply the first label to its outer surface.
4. Choose a color-coded identification rod—preferably a different color than the goblet—and apply the second label to the ID rod.
5. Take the Cryolock and apply the third label, making sure it does not interfere with device handling or storage.
6. Insert the information into the database and have another qualified staff member verify the identity and correct labeling of each item (double witnessing).
7. Place the labeled ID rod inside the labeled goblet.
8. Insert the goblet within the vitrification box.
9. Once vitrification is performed and the Cryolock is capped, insert the labeled Cryolock into the goblet.
10. Transfer the goblet into the liquid nitrogen tank for storage (*see* **Note 8**).
11. Ensure the device is fully submerged in liquid nitrogen and remains so until it warms.

4 Notes

1. In transvaginal follicle aspiration (ovum pick-up or OPU), the choice of collection media is critical for preserving the viability and developmental potential of cumulus-oocyte complexes (COCs). Various collection media are available for

flushing follicles during OPU, with the most used being commercial embryo flushing solutions in which heparin should be added to avoid coagulation [11, 12]. Some commercial media specifically designed for oocyte collection in horses already contain heparin. However, for those requiring heparin supplementation, special attention is necessary, as the components of the flushing medium can significantly influence outcomes, such as the blastocyst rate [52].

2. During oocyte handling procedures in ambient air, such as searching, denuding, or ICSI, the selection of media prioritizes pH stability, nutrient support, and compatibility with atmospheric conditions. Common formulations include inorganic salts, amino acids, glucose, and antibiotics, often supplemented with serum or albumin to provide essential proteins and maintain osmotic balance. Widely used options include HEPES-buffered media such as Dulbecco's phosphate-buffered saline (DPBS), ORIGIO Handling Medium, and TCM-199 with HEPES. In equine reproduction, although various media have been used for oocyte handling, no comparative studies have assessed their effects on embryo development. However, Swain et al. demonstrated in humans that using bi- or tri-buffered media—containing combinations of HEPES, MOPS, or DIPSO—offers advantages over media containing only HEPES or MOPS [53].

3. Equine oocyte in vitro maturation (IVM) depends on various culture media, whose compositions significantly influence both maturation outcomes and developmental competence [54]. Commonly used base media include TCM-199 or DMEM/F-12, typically supplemented with combinations of serum (fetal calf serum or estrogenic mare serum), hormones such as follicle-stimulating hormone (FSH), luteinizing hormone (LH), or equine chorionic gonadotropin (eCG), and additional factors like epidermal growth factor (EGF) or insulin–transferrin–selenium (ITS) [11, 12, 55–57].

4. The internal diameter of the pipette (stripper) tip must be carefully matched to the size and morphological characteristics of the structure being vitrified, as it directly influences handling precision and volumetric control during transfer. A tip that is too narrow can increase resistance when loading the oocyte or embryo, resulting in prolonged manipulation and an excessive expulsion of medium during placement onto the vitrification device. Conversely, an excessively wide tip compromises precision, leading to imprecise loading, increased handling time, and residual volumes that may compromise the efficiency of vitrification. For this reason, a tip diameter of approximately 135 μm is recommended for vitrifying

denuded mature oocytes, 175 μm for embryos, and 200 μm for oocytes surrounded by the corona radiata.

5. In equine oocyte vitrification, both the base medium and the type of protein supplement are critical for ensuring post-warming survival and developmental competence. Common base media include TCM-199, DMEM/F-12, α-MEM, and occasionally PBS, each serving as a vehicle for cryoprotectants like DMSO and EG [28, 33, 42, 45, 58]. Among these, TCM-199 is the most widely used in horses due to its balanced composition and compatibility with oocyte physiology. However, the protein supplement in the vitrification medium plays a pivotal role in mitigating osmotic stress and protecting cellular structures. Fetal bovine serum (FBS) is rich in growth factors and proteins that stabilize membranes, but its undefined nature introduces batch-to-batch variability and potential immunogenic concerns. Bovine serum albumin (BSA), by contrast, provides a more consistent and defined alternative, though it lacks the broader protective profile of serum. Hydroxypropyl cellulose (HPC), a synthetic, protein-free polymer, is used in some protocols to reduce biological variability and contamination risks, but its use in equine vitrification remains limited. Ultimately, the combination of a physiologically supportive base medium and an appropriate protein source, balancing protection, consistency, and safety, is essential for optimizing equine oocyte cryopreservation outcomes [59, 60].

6. Various vitrification devices have been developed for the cryopreservation of gametes and embryos. These are broadly categorized into open and closed systems, each with specific implications for cooling rates, sterility, and practical application. Open systems, such as Cryotop®, Cryoloop, and Cryolock, allow direct exposure of the sample to liquid nitrogen, enabling ultra-rapid cooling rates (>20,000°C/min), thereby minimizing ice crystal formation [61]. However, open systems carry an inherent biosecurity risk due to potential contamination from direct contact with liquid nitrogen. Closed systems, like sealed straws (e.g., CBS High Security Straws), Cryotip, or cryovials, mitigate this risk by isolating the sample from liquid nitrogen. However, they often result in lower cooling rates, which may compromise post-warming survival, particularly in large or lipid-rich cells like equine oocytes. In equine reproduction, open systems such as Cryotop and Cryolock have demonstrated higher post-warming survival and developmental competence, especially for embryos, due to their sensitivity to suboptimal cooling rates [6, 27, 28, 33]. In humans, some centers have adopted closed systems to meet regulatory or biosafety requirements. Finally,

device selection depends on the balance between biosecurity, cooling efficiency, sample type (oocyte vs. embryo), and institutional protocol.

7. Vitrification procedures for equine oocytes and embryos are typically performed either at physiological temperatures (~37°C) or at room temperature (~23°C), based on workflow and biological context. Performing vitrification at physiological temperature aligns with the natural cellular environment and maintains membrane fluidity and cytoskeletal stability, particularly important for sensitive structures like the meiotic spindle in oocytes [62]. However, this approach also accelerates the evaporation of vitrification solutions, especially when using small droplets (<100 μL) in open systems. Evaporation increases the concentration of cryoprotectants, which raise osmolarity and cytotoxicity, leading to cellular damage if not strictly controlled [63]. This requires rapid handling, minimized exposure times, and environmental control measures such as oil overlays and closed dishes. In contrast, vitrification at room temperature reduces evaporation risk, offering more stable conditions for longer handling when using long protocols (>10 min). However, cryoprotectant permeation is less efficient, requiring adjusted equilibration times [64, 65]. The choice of temperature, therefore, must balance biological preservation with practical manipulation considerations, and should be adapted to the specific stage (oocyte vs. embryo), species, and lab infrastructure.

8. In the cryopreservation of oocytes and embryos, the selection and configuration of liquid nitrogen storage systems are crucial in ensuring sample integrity, safety, and traceability. Liquid nitrogen storage containers range from basic static tanks, requiring manual monitoring of liquid levels, to advanced, digitally controlled tanks equipped with real-time temperature monitoring, automated refill mechanisms, and alarm systems. While basic dewars offer a cost-effective solution, they pose higher risks due to reliance on manual oversight. For highly sensitive biological samples like vitrified oocytes and embryos, which are stored in minimal volumes and are highly susceptible to devitrification, even minor temperature deviations during handling can compromise viability [62, 63]. Thus, vapor-phase storage tanks with temperature monitoring and rapid-access configurations are preferred to reduce thermal fluctuations. Additionally, using multiple smaller tanks, rather than a few large ones, helps minimize temperature shifts during routine access and limits potential contamination zones. Separation of oocyte and embryostorage not only improves traceability but also reduces the risk of misidentification or cross-contamination. Moreover,

regulatory requirements often mandate quarantine protocols, necessitating the allocation of dedicated tanks or isolated storage canisters for patient samples under infectious disease surveillance [66]. These practices reflect current best standards in reproductive cryobiology, combining technological safeguards with procedural rigor to ensure sample security and patient safety.

9. Several commercial embryo-holding media are widely used in equine embryo transfer and cryopreservation procedures, including ViGro Holding Plus® (Bioniche), Emcare Holding Medium® (ICPbio Reproduction), and BotuEmbryo® (Botypharma), among others. These media are designed to maintain embryo viability during short-term handling or transport, typically containing energy substrates, pH stabilizers (like HEPES), and proteins such as BSA or synthetic substitutes to support osmotic balance. Although these commercial media are generally similar in composition, subtle differences exist in osmolarity, protein source, and buffering systems, which may impact embryo development depending on exposure time and temperature. Comparative studies in equine reproduction remain limited, but empirical use suggests that no single medium consistently outperforms others for embryo holding or warming embryos, since all provide basic quality criteria. In earlier years, laboratories prepared in-house holding solutions based on modified Dulbecco's phosphate-buffered saline (DPBS) or TCM-199 supplemented with 0.4% BSA and 25 mM HEPES, which offered cost-effective alternatives with customizable [67, 68] composition. However, the use of commercial media has become more common due to batch consistency and ease of use.

10. The methodology for equine embryo and oocyte vitrification offers flexibility in the handling of equilibration and vitrification solutions, with common approaches including the use of larger individual droplets (ranging from approximately 50 to 300 μL) placed in a standard Petri dish or employing specialized commercially available vitrification dishes, such as those by Cryotech or Kitazato, which often feature designated wells designed for specific volumes. Utilizing smaller solution volumes in open droplets within a Petri dish can potentially lead to faster temperature fluctuations and increased rates of evaporation, which may alter cryoprotectant concentrations if not carefully managed, whereas specialized dishes with defined wells may offer more stability in solution temperature and osmolarity and can streamline the handling process. Irrespective of the dish system, crucial elements for success include maintaining precise temperatures of the solutions, ensuring accurate timing of exposure to each cryoprotectant

step to prevent toxicity or insufficient dehydration, minimizing mechanical stress during oocyte/embryo transfer between droplets or wells, and operator proficiency in managing the chosen system to achieve consistent outcomes [62, 64, 69].

11. The identification and storage of oocytes and embryos represents a critical component, ensuring both the safety of biological material and the accurate matching with the intended recipient. At a minimum, each device must be labeled with a unique patient code or identification number and the date of vitrification. Additional recommended identifiers include the owner's name, donor/recipient details, the number and type of cells stored (e.g., oocytes or embryos), and a lab or case code. Labeling must be legible, resistant to cryogenic temperatures, and applied directly on the cryodevice (e.g., Cryolock, straw) as well as on associated storage items such as ID rods, goblets, visotubes, or canes. Labeling options include printed cryo-resistant labels (e.g., via Brady printers), color-coded components for quick visual identification, and digital systems using barcodes or RFID chips. To further minimize errors, many laboratories employ double-witnessing protocols or electronic witnessing systems that confirm sample identity at every critical handling step.
12. When preparing vitrification and warming solutions, it is essential to strictly adhere to the specified cryoprotectant proportions to maintain efficacy and ensure reproducibility. For optimal accuracy, the cryoprotectant should be measured and added first, followed by the appropriate volume of handling solution to achieve the final desired concentration and volume.
13. In procedures involving the vitrification or warming of oocytes and embryos, it is critical to ensure that the selected container, typically a Styrofoam box or a designated liquid nitrogen vessel, is filled with liquid nitrogen at the start of the process. Inadequate nitrogen levels can create a vapor-phase interface near the top of the container, temporarily exposing the vitrification device to temperatures above –150°C. This transitional zone slows the cooling or warming rate, potentially compromising vitrification efficiency and increasing the risk of devitrification.
14. Cumulus cells play a critical role in equine oocyte physiology and cryopreservation, mediating metabolic and meiotic support through transzonal projections and gap junctions. In immature oocytes, these cells are essential for meiotic competence, and their presence during vitrification, particularly the inner corona radiata layer, helps preserve chromatin integrity and mitigate cryodamage. However, excessive cumulus

investment can hinder cryoprotectant permeation and cooling efficiency, negatively impacting outcomes [33, 70, 71]. For this reason, partial denudation (i.e., removing outer layers while retaining the corona radiata) is the recommended approach for immature oocytes. In contrast, mature oocytes no longer rely on cumulus cell support and should be fully denuded prior to vitrification. This not only facilitates optimal CPA diffusion and ultra-rapid cooling but also enables direct assessment of nuclear maturity via first polar body visualization and evaluation of oolemma integrity. Such selection is critical for excluding compromised oocytes and improving cryosurvival and developmental outcomes, particularly in ICSI-based programs.

15. Vitrification protocols for embryos and oocytes commonly employ a two-step or multi-step approach involving a gradual increase in cryoprotectant agent (CPA) concentrations. This strategy has proven effective for equine embryos as well as in vivo–matured human and equine oocytes [26, 28, 42, 72]. The stepwise method facilitates cellular dehydration and CPA influx while minimizing osmotic stress. However, the duration of CPA exposure is critical. Traditional or "long" protocols involve equilibration and vitrification steps that may extend over several minutes (typically 10–15 min), cumulatively exposing cells to high concentrations of potentially toxic CPAs. Prolonged exposure has been clearly linked to detrimental effects on subsequent embryo development, particularly in equine oocytes, where CPA toxicity can compromise viability [27, 33]. In contrast, "short" protocols drastically reduce exposure times, often to just a few minutes or even less than a minute in the final vitrification solution, aiming to minimize toxic effects by leveraging very rapid cooling rates. Recent research has turned attention to "ultra-fast" vitrification techniques. The rationale behind these methods is that extremely rapid cooling and warming rates can effectively achieve vitrification at lower CPA concentrations or with significantly shorter exposure times. This may improve post-warming survival and developmental competence by further reducing CPA toxicity and osmotic injury [36, 38].

16. Upon immersion in liquid nitrogen, vitrification devices should be inserted vertically to maximize the cooling rate. There is no risk of the oocyte or embryo dislodging during this step if the device has been properly loaded. Gentle agitation during immersion is recommended to prevent the formation of CO_2 bubbles on the sample surface, a phenomenon explained by Henry's Law, as adhered gas bubbles can interfere with sample recovery during warming. To ensure rapid and safe handling, the vitrification box should be positioned

close to the working area but must remain stable to avoid accidental collisions with the device, which could compromise both operator safety and sample viability. Additionally, it is imperative that minimal-volume devices remain fully submerged in liquid nitrogen at all times. Transfers between the vitrification container and the storage tank must be performed using tools or systems, such as goblets or cassettes filled with liquid nitrogen, that ensure continuous immersion and minimize the risk of transient warming and devitrification.

17. The warming solution plays a critical role in the successful recovery of vitrified oocytes and embryos, and it is recommended to maintain the solution at 38°C to achieve the highest possible warming, ensuring optimal effectiveness. Rapid warming is essential to prevent the formation of intracellular ice crystals, which can develop if warming is too slow, leading to irreversible cellular damage and reduced viability. To achieve this, the solution should be thoroughly pre-equilibrated to the target temperature and dispensed into the dish immediately before use to minimize heat loss to the surrounding environment. Maintaining the solution at 38°C ensures a steep thermal gradient during the critical initial seconds of warming, facilitating ultra-rapid rehydration and the transition from the vitrified to liquid state, thereby preserving the structural and functional integrity of the cells.

References

1. Taylor MJ, Weegman BP, Baicu SC, Giwa SE (2019) New approaches to cryopreservation of cells, tissues, and organs. Transfus Med Hemother 46:197–215. https://doi.org/10.1159/000499453
2. Oldenhof H, Wolkers WF, Sieme H (2021) Cryopreservation of semen from domestic livestock: bovine, equine, and porcine sperm. Methods Mol Biol 2180:365–377. https://doi.org/10.1007/978-1-0716-0783-1_15
3. Tharasanit T, Thuwanut P (2021) Oocyte cryopreservation in domestic animals and humans: principles, techniques and updated outcomes. Animals 11:2949. https://doi.org/10.3390/ani11102949
4. Canesin HS, Ortiz I, Rocha Filho AN, Salgado RM, Brom-de-Luna JG, Hinrichs K (2020) Effect of warming method on embryo quality in a simplified equine embryo vitrification system. Theriogenology 151:151–158. https://doi.org/10.1016/j.theriogenology.2020.03.012
5. Lazzari G, Colleoni S, Crotti G, Turini P, Fiorini G, Barandalla M, Landriscina L, Dolci G, Benedetti M, Duchi R, Galli C (2020) Laboratory production of equine embryos. J Equine Vet Sci 89:103097. https://doi.org/10.1016/j.jevs.2020.103097
6. Choi YH, Hinrichs K (2017) Vitrification of in vitro-produced and in vivo-recovered equine blastocysts in a clinical program. Theriogenology 87:48–54. https://doi.org/10.1016/j.theriogenology.2016.08.005
7. Cuervo-Arango J, Claes AN, Stout TA (2019) The recipient's day after ovulation and the number of corpora lutea influence the likelihood of pregnancy in mares following transfer of ICSI frozen embryos. Theriogenology 135:181–188. https://doi.org/10.1016/j.theriogenology.2019.06.020
8. Stout TAE (2020) Clinical application of in vitro embryo production in the horse. J Equine Vet Sci 89:103011. https://doi.org/10.1016/j.jevs.2020.103011
9. Cuervo-Arango J, Claes AN, Stout TA (2018) Effect of embryo-recipient synchrony on post-ET survival of in vivo and in vitro-produced equine embryos. J Equine Vet Sci

66:163–164. https://doi.org/10.1016/j.jevs.2018.05.058

10. Claes A, Stout TAE (2022) Success rate in a clinical equine in vitro embryo production program. Theriogenology 187:215–218. https://doi.org/10.1016/j.theriogenology.2022.04.019
11. Lazzari G, Colleoni S, Crotti G, Turini P, Fiorini G, Barandalla M, Landriscina L, Dolci G, Benedetti M, Duchi R, Galli C (2020) Laboratory production of equine embryos. J Equine Vet Sci 89:103097. https://doi.org/10.1016/j.jevs.2020.103097
12. Fonte JS, Alonso MA, Junior MPM, Gonçalves MA, Pontes JH, Bordignon V, Fleury PDC, Fernandes CB (2024) Successful equine in vitro embryo production by ICSI—effect of season, mares' age, breed, and phase of the estrous cycle on embryo production. Theriogenology 223:47–52. https://doi.org/10.1016/j.theriogenology.2024.04.007
13. Cuervo-Arango J, Claes AN, Stout TA (2019) A retrospective comparison of the efficiency of different assisted reproductive techniques in the horse, emphasizing the impact of maternal age. Theriogenology 132:36–44. https://doi.org/10.1016/j.theriogenology.2019.04.010
14. Herrera C (2016) Clinical applications of preimplantation genetic testing in equine, bovine, and human embryos. J Equine Vet Sci 41:29–34. https://doi.org/10.1016/j.jevs.2016.04.002
15. Choi YH, Gustafson-Seabury A, Velez IC, Hartman DL, Bliss S, Riera FL, Roldán JE, Chowdhary B, Hinrichs K (2010) Viability of equine embryos after puncture of the capsule and biopsy for preimplantation genetic diagnosis. Reproduction 140:893–902. https://doi.org/10.1530/REP-10-0141
16. De Coster T, Angel-Velez D, Soom AV, Woelders H, Smits K (2020) Cryopreservation of equine oocytes: looking into the crystal ball. Reprod Fertil Dev 32:453–467. https://doi.org/10.1071/RD19229
17. Smits K, Hoogewijs M, Woelders H, Daels P, Van Soom A (2012) Breeding or assisted reproduction? Relevance of the horse model applied to the conservation of endangered equids. Reprod Domest Anim 47 Suppl 4:239–248. https://doi.org/10.1111/j.1439-0531.2012.02082.x
18. Hinrichs K (2020) Advances in holding and cryopreservation of equine oocytes and embryos. J Equine Vet Sci 89:102990. https://doi.org/10.1016/j.jevs.2020.102990
19. Paredes E, Mazur P (2013) The survival of mouse oocytes shows little or no correlation with the vitrification or freezing of the external medium, but the ability of the medium to vitrify is affected by its solute concentration and by the cooling rate. Cryobiology 67:386–390. https://doi.org/10.1016/j.cryobiol.2013.09.003
20. Edgar DH, Gook DA (2012) A critical appraisal of cryopreservation (slow cooling versus vitrification) of human oocytes and embryos. Hum Reprod Update 18:536–554. https://doi.org/10.1093/humupd/dms016
21. Rienzi L, Gracia C, Maggiulli R, LaBarbera AR, Kaser DJ, Ubaldi FM, Vanderpoel S, Racowsky C (2017) Oocyte, embryo and blastocyst cryopreservation in ART: systematic review and meta-analysis comparing slow-freezing versus vitrification to produce evidence for the development of global guidance. Hum Reprod Update 23:139–155. https://doi.org/10.1093/humupd/dmw038
22. Chen T, Fowler A, Toner M (2000) Literature review: supplemented phase diagram of the trehalose–water binary mixture. Cryobiology 40:277–282. https://doi.org/10.1006/cryo.2000.2244
23. Fahy GM, Levy DI, Ali SE (1987) Some emerging principles underlying the physical properties, biological actions, and utility of vitrification solutions. Cryobiology 24:196–213. https://doi.org/10.1016/0011-2240(87)90023-X
24. Pegg DE (2007) Principles of cryopreservation. Methods Mol Biol 368:39–57. https://doi.org/10.1007/978-1-59745-362-2_3
25. Mazur P (1984) Freezing of living cells: mechanisms and implications. Am J Physiol 247:C125–C142. https://doi.org/10.1152/ajpcell.1984.247.3.C125
26. Kovacsy S, Ismer A, Funes J, Hoogewijs M, Wilsher S (2024) Successful vitrification of equine embryos >300 microns without puncture or aspiration. Equine Vet J 56:815–822. https://doi.org/10.1111/evj.14081
27. Canesin HS, Brom-de-Luna JG, Choi YH, Ortiz I, Diaw M, Hinrichs K (2017) Blastocyst development after intracytoplasmic sperm injection of equine oocytes vitrified at the germinal-vesicle stage. Cryobiology 75:52–59. https://doi.org/10.1016/j.cryobiol.2017.02.004
28. Angel-Velez D, De Coster T, Azari-Dolatabad N, Fernandez-Montoro A, Benedetti C, Bogado Pascottini O, Woelders H, Van Soom A, Smits K (2021) New alternative mixtures of cryoprotectants for equine immature oocyte vitrification. Animals 11:3077. https://doi.org/10.3390/ani11113077
29. Kasai M (1997) Vitrification: refined strategy for the cryopreservation of mammalian embryos. J Mamm Ova Res 14:17–28
30. Orief Y, Schultze-Mosgau A, Dafopoulos K, Al-Hasani S (2005) Vitrification: will it replace the conventional gamete cryopreservation techniques?. Middle East Fertil Soc J 10:171–184

31. Canesin HS, Brom-de-Luna JG, Choi YH, Pereira AM, Macedo GG, Hinrichs K (2018) Vitrification of germinal-vesicle stage equine oocytes: effect of cryoprotectant exposure time on in-vitro embryo production. Cryobiology 81:185–191. https://doi.org/10.1016/j.cryobiol.2018.01.001
32. Wilsher S, Rigali F, Couto G, Camargo S, Allen WR (2019) Vitrification of equine expanded blastocysts following puncture with or without aspiration of the blastocoele fluid. Equine Vet J 51:500–505. https://doi.org/10.1111/evj.13039
33. Ortiz-Escribano N, Bogado Pascottini O, Woelders H, Vandenberghe L, De Schauwer C, Govaere J, Van den Abbeel E, Vullers T, Ververs C, Roels K, Van De Velde M, Van Soom A, Smits K (2018) An improved vitrification protocol for equine immature oocytes, resulting in a first live foal. Equine Vet J 50:391–397. https://doi.org/10.1111/evj.12747
34. Woelders H, Guignot F, Ortiz-Escribano Van Soom A, Smits K (2018) Simulations of osmotic events in vitrification of equine oocytes and porcine embryos. Cryobiology 85:154–155. https://doi.org/10.1016/j.cryobiol.2018.10.136
35. Clérico G, Rodríguez MB, Taminelli G, Butteri A, Veronesi JC, Fernández S, Salamone D, Sansinena M (2018) Vitrification of immature oocytes for the production of equine embryos by ICSI: protocol effect on maturation, embryo development, mitochondrial distribution and functionality. J Equine Vet Sci 66:192–193. https://doi.org/10.1016/j.jevs.2018.05.083
36. Lammers J, Reignier A, Loubersac S, Chaillot M, Freour T (2025) Ultra-fast warming procedure of vitrified blastocysts results in maintained embryology and clinical outcomes. Reprod Sci 32:495–501. https://doi.org/10.1007/s43032-024-01762-x
37. Liebermann J, Hrvojevic K, Hirshfeld-Cytron J, Brohammer R, Wagner Y, Susralski A, Jasulaitis S, Chan S, Takhsh E, Uhler M (2024) Fast and furious: pregnancy outcome with one-step rehydration in the warming protocol for human blastocysts. Reprod Biomed Online 48:103731. https://doi.org/10.1016/j.rbmo.2023.103731
38. Schiewe MC, Reichelderfer R, Wozniak K, De Romana C, Nordbak M, Baek K, Chung K (2024) Ultra-fast vitrification and rapid elution of human oocytes: Part I. germinal vesicle model validation. Reprod Biomed Online 49:104691. https://doi.org/10.1016/j.rbmo.2024.104691
39. Seki S, Mazur P (2009) The dominance of warming rate over cooling rate in the survival of mouse oocytes subjected to a vitrification procedure. Cryobiology 59:75–82. https://doi.org/10.1016/j.cryobiol.2009.04.012
40. Mazur P, Paredes E (2016) Roles of intracellular ice formation, vitrification of cell water, and recrystallisation of intracellular ice on the survival of mouse embryos and oocytes. Reprod Fertil Dev 28:1088. https://doi.org/10.1071/RD16021
41. Li W, Zhou X, Wang H, Liu B (2012) Numerical analysis to determine the performance of different oocyte vitrification devices for cryopreservation. Cryo Lett 33:144–150
42. Maclellan LJ, Albertini DF, Stokes JE, Carnevale EM (2023) Use of confocal microscopy and intracytoplasmic sperm injection (ICSI) to assess viability of equine oocytes from young and old mares after vitrification. J Assist Reprod Genet 40:2565–2576. https://doi.org/10.1007/s10815-023-02935-4
43. Maclellan LJ, Stokes JE, Preis KA, McCue PM, Carnevale EM (2010) Vitrification, warming, ICSI and transfer of equine oocytes matured in vivo. Anim Reprod Sci 121:260–261
44. Maclellan LJ, Carnevale EM, Coutinho da Silva MA, Scoggin CF, Bruemmer JE, Squires EL (2002) Pregnancies from vitrified equine oocytes collected from super-stimulated and non-stimulated mares. Theriogenology 58:911–919. https://doi.org/10.1016/s0093-691x(02)00920-2
45. Clérico G, Taminelli G, Veronesi JC, Polola J, Pagura N, Pinto C, Sansinena M (2021) Mitochondrial function, blastocyst development, and live foals born after ICSI of immature vitrified/warmed equine oocytes matured with or without melatonin. Theriogenology 160:40–49. https://doi.org/10.1016/j.theriogenology.2020.10.036
46. Angel-Velez D, Meese T, Hedia M, Fernandez-Montoro A, De Coster T, Pascottini OB, Van Nieuwerburgh F, Govaere J, Van Soom A, Pavani K, Smits K (2023) Transcriptomics reveal molecular differences in equine oocytes vitrified before and after in vitro maturation. Int J Mol Sci 24:6915. https://doi.org/10.3390/ijms24086915
47. Angel D, Canesin HS, Brom-de-Luna JG, Morado S, Dalvit G, Gomez D, Posada N, Pascottini OB, Urrego R, Hinrichs K, Velez IC (2020) Embryo development after vitrification of immature and in vitro-matured equine oocytes. Cryobiology 92:251–254. https://doi.org/10.1016/j.cryobiol.2020.01.014
48. Fujihira T, Nagai H, Fukui Y (2005) Relationship between equilibration times and the presence of cumulus cells, and effect of taxol treatment for vitrification of in vitro matured porcine oocytes. Cryobiology 51:339–343. https://doi.org/10.1016/j.cryobiol.2005.08.002

49. Hochi S, Maruyama K, Oguri N (1996) Direct transfer of equine blastocysts frozen-thawed in the presence of ethylene glycol and sucrose. Theriogenology 46:1217–1224. https://doi.org/10.1016/S0093-691X(96)00292-0
50. Eldridge-Panuska WD, di Brienza VC, Seidel GE, Squires EL, Carnevale EM (2005) Establishment of pregnancies after serial dilution or direct transfer by vitrified equine embryos. Theriogenology 63:1308–1319. https://doi.org/10.1016/j.theriogenology.2004.06.015
51. Diaz F, Bondiolli K, Paccamonti D, Gentry GT (2016) Cryopreservation of day 8 equine embryos after blastocyst micromanipulation and vitrification. Theriogenology 85:894–903. https://doi.org/10.1016/j.theriogenology.2015.10.039
52. Cuervo-Arango J, Claes AN, Beitsma M, Stout TAE (2019) The effect of different flushing media used to aspirate follicles on the outcome of a commercial ovum pickup-ICSI program in mares. J Equine Vet Sci 75:74–77. https://doi.org/10.1016/j.jevs.2019.01.015
53. Swain JE, Pool TB (2009) New pH-buffering system for media utilized during gamete and embryo manipulations for assisted reproduction. Biomed Online 18:799–810. https://doi.org/10.1016/S1472-6483(10)60029-6
54. Fernández-Hernández P, Sánchez-Calabuig MJ, García-Marín LJ, Bragado MJ, Gutiérrez-Adán A, Millet Ó, Bruzzone C, González-Fernández L, Macías-García B (2020) Study of the metabolomics of equine preovulatory follicular fluid: a way to improve current in vitro maturation media. Animals 10:883. https://doi.org/10.3390/ani10050883
55. Papas M, Govaere J, Peere S, Gerits I, Van de Velde M, Angel-Velez D, De Coster T, Van Soom A, Smits K (2021) Anti-Müllerian hormone and OPU-ICSI outcome in the mare. Animals 11:2004. https://doi.org/10.3390/ani11072004
56. Foss R, Ortis H, Hinrichs K (2013) Effect of potential oocyte transport protocols on blastocyst rates after intracytoplasmic sperm injection in the horse. Equine Vet J 45:39–43. https://doi.org/10.1111/evj.12159
57. Lewis N, Canesin H, Choi YH, Foss R, Felix M, Rader K, Hinrichs K (2023) Equine in vitro produced blastocysts: relationship of embryo morphology, stage and speed of development to foaling rate. Reprod Fertil Dev 35:338–351. https://doi.org/10.1071/RD22224
58. Hochi S, Kozawa M, Fujimoto T (1996) In vitro maturation and transmission electron microscopic observation of horse oocytes after vitrification. Cryobiology 33:300–310. https://doi.org/10.1006/cryo.1996.0030
59. Checura CM, Seidel GE (2007) Effect of macromolecules in solutions for vitrification of mature bovine oocytes. Theriogenology 67:919–930. https://doi.org/10.1016/j.theriogenology.2006.09.044
60. Gallardo M, Hebles M, Migueles B, Dorado M, Aguilera L, González M, Piqueras P, Lucas A, Montero L, Sánchez-Martín P, Sánchez-Martín F, Risco R (2017) Hydroxypropyl cellulose supplementation in vitrification solutions: a prospective study with donor oocytes. J Assist Reprod Genet 34:417. https://doi.org/10.1007/s10815-016-0841-y
61. Vajta VG, Nagy ZP (2006) Are programmable freezers still needed in the embryo laboratory? Review on vitrification. Reprod Biomed Online 12:779–796. https://doi.org/10.1016/s1472-6483(10)61091-7
62. Vajta G, Kuwayama M (2006) Improving cryopreservation systems. Theriogenology 65:236–244. https://doi.org/10.1016/j.theriogenology.2005.09.026
63. Lane M, Schoolcraft WB, Gardner DK (1999) Vitrification of mouse and human blastocysts using a novel cryoloop container-less technique. Fertil Steril 72:1073–1078. https://doi.org/10.1016/s0015-0282(99)00418-5
64. Vajta G (2013) Vitrification in human and domestic animal embryology: work in progress. Reprod Fertil Dev 25:719–727. https://doi.org/10.1071/RD12118
65. Vajta G (2000) Vitrification of the oocytes and embryos of domestic animals. Anim Reprod Sci 60–61:357–364
66. Practice Committees of the American Society for Reproductive Medicine, Society for Reproductive Biologists and Technologists, and Society for Assisted Reproductive Technology (2020) Cryostorage of reproductive tissues in the in vitro fertilization laboratory: a committee opinion. Fert Steril 114:486–490. https://doi.org/10.1016/j.fertnstert.2020.06.019
67. Hinrichs K, Choi YH (2005) Assisted reproductive techniques in the horse. Clin Tech Equine Pract 4:210–218. https://doi.org/10.1053/j.ctep.2005.07.002
68. McKinnon AO, Squires EL (1988) Equine embryo transfer. Vet Clin North Am Equine Pract 4:305–333. https://doi.org/10.1016/S0749-0739(17)30643-0
69. Rall WF, Fahy GM (1985) Ice-free cryopreservation of mouse embryos at −196°C by vitrification. Nature 313:573–575. https://doi.org/10.1038/313573a0
70. Tharasanit T, Colleoni S, Galli C, Colenbrander B, Stout TA (2009) Protective effects of the cumulus-corona radiata complex during

vitrification of horse oocytes. Reproduction 137:391–401. https://doi.org/10.1530/REP-08-0333
71. Tharasanit T, Colleoni S, Lazzari G, Colenbrander B, Galli C, Stout TA (2006) Effect of cumulus morphology and maturation stage on the cryopreservability of equine oocytes. Reproduction 132:759–769. https://doi.org/10.1530/rep.1.01156
72. Kuwayama M, Vajta G, Kato O, Leibo SP (2005) Highly efficient vitrification method for cryopreservation of human oocytes. Reprod Biomed Online 11:300–308. https://doi.org/10.1016/S1472-6483(10)60837-1

Chapter 14

Cryopreservation of Bovine Blastocysts and Blastocoel Fluid Aspiration for Large Embryos

Iris Martínez-Rodero and Teresa Mogas

Abstract

The aspiration of blastocoele fluid (BF) from bovine blastocysts is a minimally invasive technique that allows artificial collapse of large embryos before cryopreservation, and, optionally, the collection of the BF from the blastocoelic cavity for its further analysis. The procedure aims to remove most of the fluid from the blastocoelic cavity to mitigate the severity of osmotic changes when embryos are exposed to cryoprotectants and to decrease the chance of intracellular ice crystal formation during cryopreservation. This chapter provides a step-by-step methodology for BF aspiration prior to cryopreserve in vitro-produced (IVP) bovine blastocysts by vitrification. Vitrification using the Cryotop ® method and following a short equilibration, as well as warming and step-wise dilution, are also described.

Keywords Artificial blastocoel collapse, Cell-free DNA, Non-invasive biomarkers, In vitro production, Micromanipulation, Vitrification

1 Introduction

Cryopreservation is becoming an essential procedure in assisted reproductive technologies, particularly in the context of in vitro-produced (IVP) bovine embryos, where it facilitates genetic preservation, simplifies logistics, and enhances reproductive efficiency on a global scale. Among the cryopreservation methods, vitrification has become the method of choice over slow freezing due to its ultrarapid cooling rates, which effectively prevent ice crystal formation and reduce membrane chilling injury, thereby improving post-warming survival (reviewed in [1, 2]). The fully expanded blastocyst, typically attained by Day 7 of in vitro embryo development, is generally considered as the optimal stage for cryopreservation of IVP bovine embryos [3]. At this stage, embryo exhibit higher post-warming survival rates than those preserved at earlier developmental phases. However, a significant challenge associated with expanded blastocysts is the presence of a large blastocoel cavity, which contains a considerable volume

Willem F. Wolkers and Harriëtte Oldenhof (eds.), *Cryopreservation and Freeze-Drying Protocols*, Methods in Molecular Biology, vol. 3060, https://doi.org/10.1007/978-1-0716-5416-3_14,

of blastocoele fluid (BF). This fluid increases the risk of intracellular ice formation and intensifies osmotic stress during exposure to cryoprotective agents and throughout the vitrification process [4]. This can result in mechanical and osmotic damage, disruption of cellular organelles, and loss of cellular integrity, which can potentially compromise embryo survival [5].

To mitigate these risks, artificial collapse of the blastocoel prior to vitrification has been employed. Numerous studies across various species—including bovine [6–8], human [4, 9], equine [10], murine [11], feline [12], and buffalo [13]—have demonstrated that reducing the blastocoelic volume through artificial collapse techniques significantly enhances cryotolerance and embryo viability following warming compared to their non-collapsed counterparts.

Artificial collapse of the blastocoel can be achieved using either laser pulses [13] or by puncture with or without aspiration of the BF. The aspiration approach involves the active removal of BF using an ICSI pipette, ensuring that nearly all the fluid is extracted until the inner cell mass (ICM) contacts the opposite trophectoderm layer. This technique is well established in both equine [10, 14, 15] and bovine species [8, 16], where puncture or aspiration prior to vitrification has been shown to markedly improve the cryotolerance of large embryos. In addition to reducing the blastocoelic volume and thereby enhancing the embryo's resistance to cryoinjury, aspiration also provides access to the BF, which contains valuable information for diagnosis [8, 17] or research purposes [18].

Recent comparative studies in humans have evaluated the efficacy of pipette aspiration versus laser-induced blastocyst collapse before vitrification [19]. While laser-treated blastocysts re-expanded significantly faster after warming than those collapsed by pipette aspiration, no statistically significant differences were observed in survival, pregnancy, or live birth rates between the two methods. This indicates that although laser-induced collapse may accelerate re-expansion kinetics, pipette aspiration remains at least as effective in terms of key clinical outcomes, with the added advantage of enabling BF collection for further genetic analysis. The BF contains cell-free DNA (cfDNA), which can be utilized, in bovine or equine species, for preimplantation genetic testing (PGT) without the need for more invasive procedures like trophectoderm (TE) biopsy. Studies have shown that cfDNA obtained from BF is sufficient for accurate sex determination and genetic assessments with comparable efficiency to traditional biopsy methods [8, 10].

The dual benefits of BF aspiration—enhancing cryotolerance and enabling non-invasive genetic analysis—underscore its value in bovine embryocryopreservation protocols. As the field advances, integrating BF aspiration into standard practices could lead to improved outcomes in both embryo survival and genetic screening, thereby optimizing the efficiency of bovine reproductive technologies.

2 Materials

Prepare all solutions in sterile conditions (under laminar flow hood) and filter them before their use (0.22 μm filters). Keep them at room temperature during preparation and store them at 4°C until their use. Except where otherwise noted, all chemicals and reagents were supplied from Sigma-Aldrich (Merck, MA, USA).

2.1 Micromanipulation Dish

1. Holding medium (HM): TCM199 with HEPES and 20% fetal bovine serum (FBS). Add 8 mL of TCM199-HEPES and 2 mL of FBS (kept frozen and heat-inactivated) in a 10 mL-falcon tube. Mix, filter, seal with parafilm, and store at 4°C.
2. DEPC-treated water for RT-PCR.
3. Mineral oil: embryo-tested. Kept at room temperature. Aliquot 4 mL of mineral oil in a 5 mL round tube for each dish of micromanipulation planned to be prepared. Close tightly and store at 4°C. Pre-heat at 38.5°C before its use (1–2 h).
4. Tips used to prepare drops should be DNAse and RNAse free, with a filter, sterile, and discarded after use.

2.2 Micromanipulator Calibration

1. Micromanipulation system: inverted microscope with heated stage (38.5°C), hydraulic or electronic micromanipulators, oil- or air-based microinjectors.
2. Microtools: Holding pipette with 15–20 μm internal diameter, 90 μm outer diameter of angle and ICSI pipette with 5–5.5 μm internal diameter and beveled. Both must have a tip bend angle that matches the angle in the micromanipulation arms (30 or 35°, *see* **Note 1**).

2.3 Collection and Storing of Blastocoel Fluid

1. Collection tubes: 0.2 mL sterile RNase-free PCR tubes.
2. Tips used to handle BF and TE biopsy samples should be DNAse and RNAse-free, with a filter, sterile, and discarded after use.
3. Liquid nitrogen or dry ice for snap freezing of BF samples.
4. WGA kit: REPLI-g Single Cell Kit (Qiagen).
5. Sexing PCR reagents:
    a. 2× PCR Master Mix (Thermo Fisher).
    b. BRY4a primers (Y-specific; 300 bp) [20] (*see* **Note 2**): Fw: CTCAGCAAAGCACACCAGAC, Rv: GAACTTTCA AGCAGCTGAGGC
    c. SAT1 primers (bovine marker; 216 bp) [20] (*see* **Note 2**): Fw: TGGAAGCAAAGAACCCCGCT, Rv: TCGTGAGAAA CCGCACACTG

6. Thermocycler.
7. Electrophoresis gel of 2% agarose with 10 µL SafeView™ Plus.

2.4 Vitrification and Warming Procedures

1. Equilibration solution (ES): HM with 7.50% (v/v) ethylene glycol (EG) and 7.50% (v/v) dimethyl sulfoxide (DMSO). Mix 6.5 mL of TCM199-HEPES, 2 mL of FBS, 0.75 mL EG, and 0.75 mL DMSO in a 10 mL-falcon tube. Mix, filter, seal with parafilm, and store at 4°C.
2. Vitrification solution (VS): HM containing 15% (v/v) EG, 15% (v/v) DMSO, and 0.50 M sucrose. Add 5 mL of TCM199-HEPES, 2 mL of FBS, 1.50 mL EG, 1.50 mL DMSO, and 1.70 g of sucrose in a 10 mL-falcon tube. Mix, filter, seal with parafilm, and store at 4°C.
3. Cryotop® or similar open vitrification device.
4. Liquid nitrogen and porexpan box.
5. Warming solution 1 (WS1): HM with 1 M sucrose. Mix 8 mL of TCM199-HEPES with 2 mL FBS and 3.40 g of sucrose. Mix, filter, seal with parafilm, and store at 4°C.
6. Warming solution 2 (WS2): HM with 0.5 M sucrose. Mix 8 mL of TCM199-HEPES with 2 mL FBS and 1.70 g of sucrose. Mix, filter, seal with parafilm, and store at 4°C.

3 Methods

IVP embryos can be produced under each lab condition (37.80°C, 5% CO_2, 5% O_2, and humidified atmosphere), culture system (i.e., microdroplet, 4-well dish, microwell), and culture media (i.e., in-house or commercially prepared). Blastocysts management should be done with a pulled Pasteur pipette or an embryo handling microcapillary pipette such as Stripper® or EZ® (Cooper Surgical). Work under sterile conditions (laminar flow hood) and using a stereomicroscope with a heated stage set at 38.50°C. Make pertinent checks for the real temperature of the embryo when they are on top of the heated surface using a thermocouple.

3.1 Preparation

1. Pre-warm mineral oil to 38.50°C for 1–2 h, ES and VS at room temperature.
2. Micromanipulation dish: prepare two lines of 5–6 microdroplets of 25 µL HM next to 1 µL DEPC-water onto a 60 mm Petri dish. Rapidly, cover the drops by pouring the aliquot of pre-heated mineral oil. Leave it at least 15 min before its use (Fig. 1).

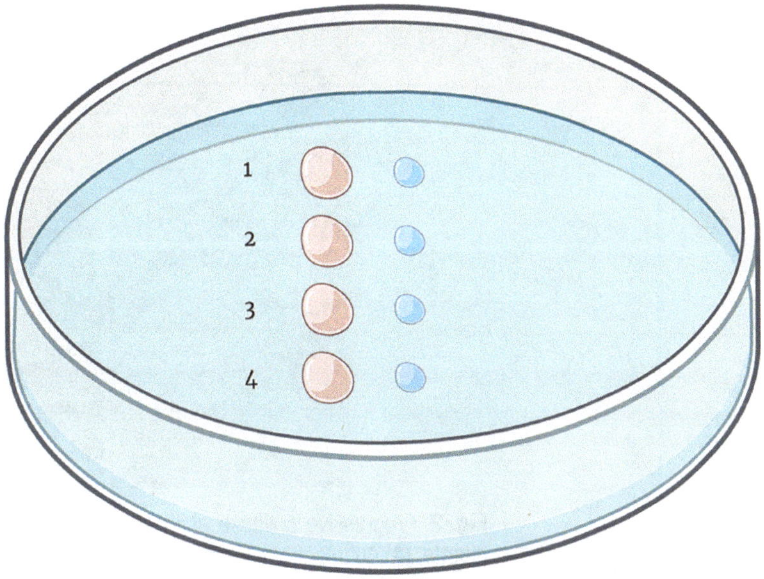

Fig. 1 Schematic representation of the micromanipulation dish layout used for blastocoel-fluid collection and downstream RT-PCR. Four rows (labeled with permanent marker underneath the dish as 1–4) of 25 μL microdroplets paired with 1 μL microdroplets are arranged in a 60 mm dish and overlaid with pre-heated (38.50°C) mineral oil. In each row, the left (pink) droplet contains Holding Medium (HM: TCM 199-HEPES supplemented with 20% heat-inactivated FBS), and the right (blue) droplet contains DEPC-treated water for RT-PCR. Mineral oil (embryo-tested) maintains temperature and prevents evaporation during embryo handling

3. Micromanipulator equilibration: Mount the holding and ICSI pipettes onto their respective arms. Align them so they are on the same focal plane and facing each other correctly. Slightly raise both pipettes to prevent them from hitting the dish when placing them on the stage.
4. Select those Day 5 or Day 6 good quality (Grade 1 or 2; IETS [21]) blastocysts to cryopreserve.

3.2 Blastocoel Fluid Aspiration

1. Transfer a single blastocyst to the HM drop in the micromanipulation dish.
2. Label the micromanipulation dish with a unique ID per embryo (to maintain traceability with its corresponding BF).
3. Under the inverted microscope, orient the embryo so the ICM is positioned opposite (9′) or lateral (12′) to the aspiration site.
4. Gently insert the aspiration pipette through the trophectoderm away from the ICM.
5. Slightly aspirate blastocoel fluid from the blastocoelic cavity until the blastocoel is collapsed and before any cell content can be accidentally suctioned (*see* **Note 3**, Fig. 2).

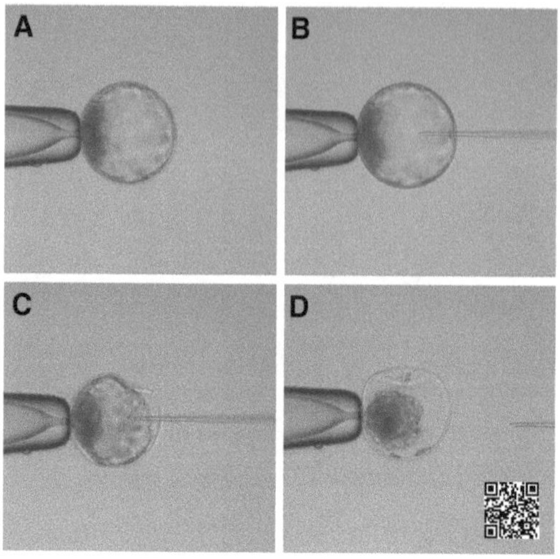

Fig. 2 Progressive collapse of the blastocoel via fluid aspiration with an ICSI pipette. (**A**) An intact blastocyst is stabilized by the holding pipette, with the inner cell mass (ICM) oriented at the 9 o'clock position. (**B**) The ICSI pipette is inserted through the zona pellucida, carefully avoiding the ICM. (**C**) Partial deflation of the blastocoel cavity following gentle withdrawal of blastocoelic fluid. (**D**) Fully collapsed blastocyst after complete aspiration of the fluid. Scan the QR code to view a video of the entire procedure

6. Slowly retract the pipette. In case of analyzing BF, expel all the content of the ICSI pipette in a 1 μL drop of DEPC-treated water until seeing an air bubble (*see* **Note 4**).
7. Leave temporarily the embryo/s collapsed onto the heated stage of the laminar flow hood while preparing the vitrification dish.

3.3 Vitrification of Collapsed Blastocyst

1. Prepare the equilibration dish by placing 100 μL of pre-heated ES (38.50°C, 1–2 h) in a 35 mm petri dish, and identify the dish with the embryo ID.
2. Label the Cryotop ® with at least the embryo ID. Other data, such as vitrification date or blastocysts quality, can be additionally written.
3. Transfer with the microcapillary pipette the collapsed embryo to the ES solution. Set a timer to 3 min.
4. During this time, prepare vitrification dish with two drops of 80 μL VS.
5. After 3 min in ES, the embryo remains collapsed, and no apparent volumetric changes have occurred (*see* **Note 5**). Transfer it to the first VS drop. Immediately, move it to the second VS drop and leave it for 30 s.
6. Then, take the embryo with the microcapillary pipette and load it onto the Cryotop ® strip in a minimal volume drop

(*see* **Note 6**). If the drop exceeds a thin layer of solution on top of the blastocysts, remove some VS with the pipette.

7. The Cryotop® is immediately plunged into liquid nitrogen and inserted into the straw cap.
8. The entire procedure from initial VS contact to immersion in liquid nitrogen takes less than 1 min.
9. Loaded Cryotops® are stored in liquid nitrogen, properly identified in a plastic gobelet, until further use.

3.4 Warming of Collapsed Blastocyst

1. Prepare post-warming culture dishes the day before.
2. Prepare a warming dish by placing 2 mL of WS1 in a 35 mm Petri dish and keep it on a heated stage or an incubator (38.50°C, 1–2 h). Leave at room temperature WS2 and HM media.
3. Take the specimens to warm from its storage and place them in the porexpan box filled with liquid nitrogen (*see* **Note 7**).
4. Prepare the dilution dish by placing one 80 µL drop of WS2 and two 80 µL drops of HM. No oil is used here.
5. Remove the cap from the device prior to its warming. Place it in the box corner closest to the heated stage where the warming dish will be.
6. Prepare the embryo handling tool and set a timer for 1 min.
7. Take the dish out of the incubator and quickly place it on the heated stage. Immediately immerse the tip of the Cryotop® in the WS1 media and press the timer (*see* **Note 8**).
8. After 1 min, take the embryo from WS1 and aspirate some WS1 after the embryo, so the tip of the pipette is filled with a 1 cm column of media. Then, transfer the embryo to WS2. First, a layer of WS1 will appear and then the embryo on top (*see* **Note 9**). Set the timer to 3 min.
9. After that, take the embryo from WS2 again carrying 1 cm column of media after the embryo. Transfer it to the first drop of HM and incubate for 5 min.
10. Wash the embryo for 1 min in the second drop of HM and then transfer it to the culture dish, again washing the embryo before leaving it for 2–3 h of post-warming culture.
11. Assess the survival of the embryo at 0 and 3 h after warming and use it for the intended purpose.

3.5 Blastocoel Fluid Analysis for Blastocysts Sexing

3.5.1 Whole-Genome Amplification (WGA)

1. Thaw BF tubes at room temperature and then do a spin to collect all the content (centrifuge at 8000×*g* for 30 s) (*see* **Note 10**).

2. Add 2.50 μL of REPLI-g Denaturation Buffer, incubate 3 min at RT, and add 0.50 μL REPLI-g Neutralization Buffer.
3. Add 40 μL of REPLI-g Master Mix, incubate 16 h at 30°C, and then heat-inactivate at 65°C for 3 min to stop DNA amplification (*see* **Note 11**).
4. Store the 0.20 mL tubes with the WGA product properly identified with ID of the corresponding embryo at 4°C until its analysis by conventional PCR.

3.5.2 Conventional PCR

1. Prepare the PCR reaction mix (50 μL total):
   - 25 μL of 2× PCR Master Mix.
   - 2 μL BRY4a primers (Rv + Fw) (1 μM).
   - 2 μL SAT1 primers (Rv + Fw) (1 μM).
   - 16 μL DEPC-treated water.
   - 1 μL DNA template (WGA product).

3.5.3 Sex Assignment

1. Interpretation of the PCR products to assign the sex to the embryo from the sexing genes that have been amplified from its corresponding BF is:
   (a) Male: both 300 bp and 216 bp bands present.
   (b) Female: only 216 bp band.
   (c) Non-amplified: no bands observed.
2. When setting up the technique, a further validation step can be done to check the proper amplification of sexing genes and the absence of contamination. For that, the final products remaining after conventional PCR with sexing genes can be analyzed on a bioanalyzer using a high-sensitivity DNA assay to confirm amplicon identity and molecular weight.

4 Notes

1. The bend angle of both the ICSI and holding pipettes must match the angle of the micromanipulator arms to ensure precise control and alignment during micromanipulation. When the angles are matched, the pipette tips remain parallel to the focal plane, allowing accurate visualization and manipulation under the microscope. Misalignment can lead to poor focus, reduced precision, and an increased risk of damaging the embryo or pipette during injection. Consistency in angles also minimizes mechanical stress on the pipettes, improving handling stability and reproducibility.
2. The primers provided here were based on bibliography [20], as they were validated in this and other numerous works.

Primers can also be designed using a tool by the NCBI (available at: https://www.ncbi.nlm.nih.gov/tools/primer-blast/), but the authors do not recommend it based on their own experience.

3. Occasionally, the elastic plasma membrane of trophectoderm cells sticks to the puncture opening, causing the blastocoel to partially or totally collapse. In some instances, puncturing needs to be repeated at a separate trophectoderm site until the blastocoel collapses.
4. To ensure that all the content expelled in the DEPC-water drop is BF aspirated from the blastocysts, the ICSI pipette has to be initially purged from any media or oil that has entered. It must be ensured that the ICSI pipette enters the blastocysts without any content charged rather than air [8].
5. When the embryo is already collapsed at the initial incubation in ES, no apparent volumetric changes are occurring, and the operator may be alarmed as intact blastocysts embryo shrink is normally noticed. However, cryoprotectant influx and water efflux are happening during the 3 min of incubation in ES of a collapsed blastocyst, although it is hardly noticeable to the naked eye [22].
6. The key to pipetting the embryo onto the Cryotop® in a minimal volume drop of VS is the size of the microcapillary used. Diameters of 260–300 μm, which are close to the size of the embryo, are recommended.
7. To fully fill the porexpan box is important so the transfer time of the specimen from the liquid nitrogen (−196°C) to the warming dish (38.5°C) is minimized and the warming rate is high enough to avoid ice crystal formation within the glass transition temperature.
8. To achieve the highest warming rates, maintaining the warming dish at 38.5°C remains fundamental. For that reason, it is recommended to take out the warming dish from the incubator only when the rest of the material is ready.
9. To increase the gradual change of osmotic pressure of the embryo during warming, to carry over media from one step to the following is recommended.
10. Attempting to measure cell-free DNA in a BF sample using a Nanodrop before amplification is futile: yields are well below detection limits, and any processing for quantification would consume the entire sample.
11. A WGA kit (e.g., REPLI-g Single Cell Kit) uses phi29 DNA polymerase and random primers to perform isothermal amplification of minute DNA inputs into microgram-scale yields. Heat-inactivation (e.g., 65°C for 3 min) is required to terminate polymerase activity and prevent nonspecific

over-amplification or branched by-products. Quantifying WGA products on Nanodrop gives misleading results, since the highly branched, size-heterogeneous fragments migrate and absorb unpredictably.

References

1. Ferré LB, Kjelland ME, Taiyeb AM, Campos-Chillon F, Ross PJ (2020) Recent progress in bovine in vitro-derived embryo cryotolerance: impact of in vitro culture systems, advances in cryopreservation and future considerations. Reprod Domest Anim 5:659–676. https://doi.org/10.1111/rda.13667
2. Mogas T, García-Martínez T, Martínez-Rodero I (2024) Methodological approaches in vitrification: enhancing viability of bovine oocytes and in vitro-produced embryos. Reprod Domest Anim 59(Suppl 3):e14623. https://doi.org/10.1111/rda.14623
3. Morato R, Izquierdo D, Paramio MT, Mogas T (2010) Survival and apoptosis rates after vitrification in cryotop devices of in vitro-produced calf and cow blastocysts at different developmental stages. Reprod Fertil Dev 22:1141–1147. https://doi.org/10.1071/RD10013
4. Kovačič B, Taborin M, Vlaisavljević V, Reljič M, Knez J (2022) To collapse or not to collapse blastocysts before vitrification? a matched case-control study on single vitrified-warmed blastocyst transfers. Reprod Biomed Online 45:669–678. https://doi.org/10.1016/j.rbmo.2022.03.030
5. Hendriks WK, Roelen BA, Colenbrander B, Stout TA (2015) Cellular damage suffered by equine embryos after exposure to cryoprotectants or cryopreservation by slow-freezing or vitrification. Equine Vet J 47:701–707. https://doi.org/10.1111/evj.12341
6. Min SH, Kim JW, Lee YH, Park SY, Jeong PS, Yeon JY, Park H, Chang KT, Koo DB (2014) Forced collapse of the blastocoel cavity improves developmental potential in cryopreserved bovine blastocysts by slow-rate freezing and vitrification. Reprod Domest Anim 49:684–692. https://doi.org/10.1111/rda.12354
7. Marques TC, Santos ECDS, Diesel TO, Martins CF, Cumpa HCB, Leme LO, Dode MAN, Alves BG, Costa FPH, Oliveira EB, Gambarini ML (2021) Blastocoel fluid removal and melatonin supplementation in the culture medium improve the viability of vitrified bovine embryos. Theriogenology 160:134–141. https://doi.org/10.1016/j.theriogenology.2020.10.028
8. Martínez-Rodero I, Salas-Huetos A, Diaz-Muñoz J, Ordóñez-León EA, García-Martínez T, Yeste M, Olegario Hidalgo C, Mogas T (2024) Blastocoel fluid aspiration improves vitrification outcomes and produces similar sexing results of in vitro-produced cattle embryos compared to microblade biopsy. Theriogenology 218:142–152. https://doi.org/10.1016/j.theriogenology.2024.01.042
9. Kovačič B, Taborin M, Vlaisavljević V (2018) Artificial blastocoel collapse of human blastocysts before vitrification and its effect on re-expansion after warming - a prospective observational study using time-lapse microscopy. Reprod Biomed Online 36:121–129. https://doi.org/10.1016/j.rbmo.2017.10.111
10. Herrera C (2021) Vitrification of equine in vivo-derived embryos after blastocoel aspiration. In: Wolkers WF, Oldenhof H (eds) Cryopreservation and freeze-drying protocols. Humana, New York, NY, pp 517–522. https://doi.org/10.1007/978-1-0716-0783-1_25
11. Frank LA, Rose RD, Anastasi MR, Tan TCY, Barry MF, Thompson JG, Brown HM (2019) Artificial blastocyst collapse prior to vitrification significantly improves Na^+/K^+-ATPase-dependent post-warming blastocoel re-expansion kinetics without inducing endoplasmic reticulum stress gene expression in the mouse. Reprod Fertil Dev 31:294–305. https://doi.org/10.1071/RD17500
12. Ochota M, Wojtasik B, Niżański W (2017) Survival rate after vitrification of various stages of cat embryos and blastocyst with and without artificially collapsed blastocoel cavity. Reprod Domest Anim 52:281–287. https://doi.org/10.1111/rda.12826
13. Yang C, Zheng H, Moussa M, Amin A, Huang J, El-Sayed A, Shang J, Liu Q (2020) Effects of laser zona thinning and artificial blastocoel collapse on the cryosurviving and hatching of buffalo (*Bubalus bulalis*) blastocysts of different ages. Theriogenology 147:197–201. https://doi.org/10.1016/j.theriogenology.2019.11.015
14. Wilsher S, Rigali F, Couto G, Camargo S, Allen WR (2019) Vitrification of equine expanded blastocysts following puncture with or without aspiration of the blastocoele fluid. Equine Vet J 51:500–505. https://doi.org/10.1111/evj.13039
15. Wilsher S, Rigali F, Kovacsy S, Allen WT (2021) Successful vitrification of manually punctured

equine embryos. Equine Vet J 53:1227–1233. https://doi.org/10.1111/evj.13400
16. Min SH, Lee E, Son HH, Yeon JY, Koo DB (2013) Forced collapse of the blastocoel enhances survival of cryotop vitrified bovine hatching/hatched blastocysts derived from in vitro fertilization and somatic cell nuclear transfer. Cryobiology 66:195–199. https://doi.org/10.1016/j.cryobiol.2013.01.004
17. Herrera C, Morikawa MI, Castex CB, Pinto MR, Ortega N, Fanti T, Garaguso R, Franco MJ, Castañares M, Castañeira C, Losinno L, Miragaya MH, Mutto AA (2015) Blastocele fluid from in vitro- and in vivo-produced equine embryos contains nuclear DNA. Theriogenology 83:415–420. https://doi.org/10.1016/j.theriogenology.2014.10.006
18. Jensen PL, Grøndahl ML, Beck HC, Petersen J, Stroebech L, Christensen ST, Yding Andersen C (2014) Proteomic analysis of bovine blastocoel fluid and blastocyst cells. Syst Biol Reprod Med 60:127–135. https://doi.org/10.3109/19396368.2014.894152
19. Martínez-Rodero I, Kovačič B, Shebl O, Enengl S, Lastinger J, Ebner T (2025) The effect of pipette- and laser-induced blastocyst collapse before vitrification on their re-expansion and clinical outcome after warming. Reprod Biomed Online 50:104476. https://doi.org/10.1016/j.rbmo.2024.104476
20. Bermejo-Alvarez P, Rizos D, Rath D, Lonergan P, Gutiérrez-Adán A (2008) Can bovine in vitro-matured oocytes selectively process X- or Y-sorted sperm differentially?. Biol Reprod 79:594–597. https://doi.org/10.1095/biolreprod.108.070169
21. Bó G, Mapletoft R (2013) Evaluation and classification of bovine embryos. Anim Reprod 10:344–348
22. Martinez-Rodero I, Diaz-Muñoz J, Higgins AZ, López Béjar M, Mogas T, García-Martínez T (2024) In silico-designed vitrification protocols: an approach to improve survival of in vitro produced-bovine embryos. Reproduction 167:e24003. https://doi.org/10.1530/REP-24-0036

Chapter 15

Cryopreservation of Seeds and Seed Embryos in Orthodox, Intermediate, and Recalcitrant Seeded Species

Daniel Ballesteros, Natalia Fanega-Sleziak, Sershen, Jayanthi Nadarajan and Rachael M. Davies

Abstract

Seeds are one of the preferred and most used sources of germplasm for the ex situ preservation of plant genetic resources. They are generally stored dry at −20°C in seed banks following international standards. However, some seeds do not tolerate drying and/or storage at −20°C, or present short lifespans at these conditions. For them, cryopreservation is indicated for long-term preservation. When seeds tolerate desiccation (i.e., orthodox seeds), they can be dried at about 32% ± 3% relative humidity at 18°C and stored in the vapour phase of liquid nitrogen. This is the method followed in the Millennium Seed Bank of the Royal Botanic Garden`s, Kew for wild species with short life spans in the standard conditions of seeds banks. When seeds do not tolerate desiccation (i.e., recalcitrant seeds) or their tolerance to desiccation and/or −20°C storage is limited (i.e., intermediate seeds), drying and cooling procedures must be adjusted, and often, antioxidants and/or cryoprotection are required to avoid physical and metabolic lesions. Some methods are detailed for diverse species of temperate and tropical origin.

Keywords Seed embryonic axes, Drying, Antioxidants, Cathodic water, *Aesculus* sp., *Quercus* sp., *Fagus sylvatica* , *Citrus* sp., *Coffea* sp., Oxidative stress, Vacuum infiltration vitrification

1 Introduction

Seeds are one of the most used sources of germplasm for the ex situ preservation of genetic resources for agriculture, horticulture, forestry, and biodiversity conservation [1]. For their storage behavior in ex situ preservation facilities, seeds are classified as orthodox, intermediate, and recalcitrant based on their tolerance to drying and exposure to low temperatures. Orthodox seeds tolerate drying to low moisture contents (e.g., less than about 0.07 g H_2O/g dry weight) and also tolerate storage at sub-zero temperatures [2]. Recalcitrant seeds do not tolerate drying to water contents less than about 0.20 g H_2O/g dry weight and are killed by exposure to sub-zero temperatures due to massive ice crystal formation [3–5]. In the intermediate seed category, we find a heterogeneous group that

Willem F. Wolkers and Harriëtte Oldenhof (eds.), *Cryopreservation and Freeze-Drying Protocols*, Methods in Molecular Biology, vol. 3060, https://doi.org/10.1007/978-1-0716-5416-3_15,

encloses seeds that may: tolerate some drying but not to the same levels as orthodox seeds, exhibit anomalous longevity responses to temperatures between +10°C and −30°C, or have short lifespans no matter how they are dried or cooled (reviewed in [3]).

Following international standards of gene banking [6, 7], orthodox seeds are routinely stored dry (after drying in a controlled environment of 10%–25% relative humidity [RH] at 5°C–20°C) in hermetically sealed containers at −18°C ± 3°C [6]. For example, in the Millennium Seed Bank (MSB), Royal Botanic Gardens (RBG), Kew, seeds are routinely stored at −20°C ± 3°C after drying at 15% RH ± 3% at 15°C ± 3% [8]. This standard storage method is potentially valid for seeds of most plant species as it has been predicted that about 90% of them may produce orthodox seeds [9]. However, some orthodox seeds, when stored at the standard seed storage conditions, present a short lifespan and deteriorate faster than expected, dying in just a few years [10–12]. In addition, intermediate seeds store poorly or do not tolerate storage at −18°C ± 3°C (even if they can be dried to levels at which water does not freeze) and recalcitrant seeds do not tolerate the drying needed for safe storage at sub-zero temperatures. For all these species, cryopreservation is recommended for their long-term preservation [4].

In the MSB, sub-samples of the seed collections of orthodox-seeded species/accessions that are relatively short-lived at standard seed storage conditions (e.g., *Salix, Populus, Primula, Anemone*, and *Corylus*; [11, 12]) are stored dry at ~−185°C to −190°C [i.e., in the vapour phase of liquid nitrogen (LN)] [12]. The drying conditions for LN storage of orthodox seeds are slightly different to those used for standard storage at −20°C. This consideration has been taken to avoid over-drying problems that may arise when seeds are dried to low moisture contents at ambient temperature and then stored at LN temperatures [13]. In addition, the MSB also stores intermediate seeds of some species at LN temperatures. The drying levels for LN storage of intermediate seeds are higher than those used for standard storage at −20°C to avoid drying damage (intermediate seeds tend to present tolerances to desiccation between orthodox and recalcitrant seeds [14, 15]), but lower than the water contents at which water can freeze [14, 15].

For recalcitrant seeds, the cryopreservation of the whole seed is not usually possible due to their large size (>1 cm). Large seeds tend to be dried, cryoprotected, and cooled slower than desired to reach the maximum survival rates [16]. In this case, the embryonic axes of the seeds are excised and cryopreserved. The drying, cryoprotection, and cooling procedures must be adjusted, and are often species-specific [4, 6]. But in general, fast drying and cooling procedures tend to increase survival and regeneration of the cryopreserved embryonic axes [17]. In addition, it has been shown that the use of cryoprotectants, such as glycerol, that have inherent antioxidant properties [18] and/or vacuum infiltration to speed up the intake of cryoprotectants could improve post-cryopreservation regeneration rates in oily or large embryo/embryonic axes by reducing stress [19].

The success of orthodox seed storage at sub-zero temperatures (−20°C and LN) is mostly based on the natural seed tolerance to low water contents (up to 3%–7% moisture content) without viability loss [3]. At these low water contents, the water molecules in the seed tissues are strongly bonded to macromolecules [20] and do not freeze (i.e., do not form ice crystals) when seeds are stored at sub-zero temperatures [21]. This water fraction is known as “unfreezable” or “unfrozen water [22]. Knowing the window of water contents at which water is “unfreezable” or “unfrozen” can be important for the development of seed cryopreservation protocols for orthodox and intermediate seeds, but also for different tissues of recalcitrant-seeded species [14, 15, 23, 24]. Differential Scanning Calorimetry is commonly used to determine the maximum moisture level of the optimal hydration window.

Additionally, the success of cryopreservation of recalcitrant seed tissues relies on the ability to reduce oxidative stress during excision, the cryoprotective steps, and the early embryo regeneration [18]. For instance, early studies preferred rapid nonequilibrium rewarming, which typically involves immersing cryotubes or naked explants into a rewarming medium at 40°C–45°C [5, 18, 25–29]. However, rewarming and rehydration are characterized by significant levels of uncontrolled free radical production, and more recent studies have tried to mitigate this oxidative stress by rewarming and rehydrating explants in cathodic water(CW) prepared from a dilute solution Ca^{2+} and Mg^{2+} cations [18]. The method used to prepare the CW and its application during rewarming and rehydration are described below.

In this chapter, the methods for cryopreservation of orthodox seeds and some intermediate seeds (e.g., *Fagus sylvatica*) followed in the MSB are detailed. Methods for other intermediate seeds (e.g., *Citrus* sp., *Coffea* sp.) have been extracted from the literature [14, 15, 30]. In addition, a fast (flash) drying method successfully used for the cryopreservation of the embryonic axes of diverse temperate (e.g., *Aesculus* sp. and *Quercus* sp.) and tropical [17] species at different institutions is described. Furthermore, we describe the vacuum infiltration vitrification (VIV) method, which was successfully used for speeding up the intake of cryoprotective agents and increasing embryonic axes survival and regeneration [19]. Finally, the preparation and use of a strongly reducing, high pH, and cathodic water produced by electrolysis of a solution containing calcium and magnesium chloride is described due to its high antioxidant capacity and success in the cryopreservation of the embryonic axes for some tropical species [31].

2 Materials

Please note that the equipment mentioned in this chapter is used by the authors at their respective institutions and has been chosen carefully using our many years’ experience. The list of suppliers is for guidance only and does not represent an endorsement by the authors’ institutions. The manufacturer's instructions must be followed when using any of the equipment referred to in this chapter.

2.1 Drying of Orthodox Seeds for Cryopreservation

1. Humidity and temperature-controlled hydration cabinet set up at 32% RH and 18°C (e.g., Climate chamber Memmert HPP110, Memmert GmBh) (*see* **Note 1**).
2. Water activity analyzer and probe for measuring seed moisture (henceforth called as "Rotronic", e.g., Hygropalm 3 portable water activity analyzer with an AW-DIO station probe, Rotronic UK).

2.2 Drying of Intermediate Seeds for Cryopreservation

1. Humidity and temperature-controlled hydration cabinets set up at a range of 50%–75% RH and 20°C (e.g., Climate chamber Memmert HPP110, Memmert GmBh) (*see* **Note 2**).
2. "Rotronic" (e.g., Hygropalm 3 portable water activity analyzer with an AW-DIO station probe, Rotronic UK).

2.3 Cryopreservation and Warming of Orthodox and Intermediate Seeds

1. Cryovials (2, 5, or 10 mL depending on the sample size, e.g., L-Brady) and boxes (e.g., Nalgene Cryobox 81 × 5 mL Vials, Nalgene Cryobox 81 × 2 mL Vials, Scientific Laboratory Supplies Ltd, UK) or canes, depending on the cryo-dewar used.
2. Foil bags (e.g., 85 × 120 mm, Moore and Buckle Ltd, UK). Foil bags can be placed inside a large vial box, with the vial insert removed. One stack will fit 11 × 2 mL vial boxes or 7 × 5 mL vial/foil bag boxes. 30–34 stacks will fit a large dewar.
3. Heat sealer for foil bags (e.g., Heat Sealer HM305CTD, Hulme Martin).
4. Cryo label printer (e.g., Brady BBP11, Brady, UK); Brady vial labels (e.g., Admiral Labels & Print, Ltd, UK), 2500 per roll.
5. Cryo-dewar large/small (e.g., Archive Dewar, Biosystem 36 and Biorack 6000, Statebourne.com).
6. Cold-protective gloves (cryo gloves) and protective glasses.
7. Controlled rate or programmable freezer (e.g., Minicool LC 40, Air Liquide) or freezing container (e.g., Nalgene® Mr. Frosty) (for intermediate seeds only).
8. Ultralow temperature mechanical freezer set at −80°C (e.g., U410, New Brunswick) (for Mr. Frosty).
9. Water-bath set up at 37°C–40°C (for intermediate seeds only).

2.4 Excision of Embryonic Axes of Recalcitrant Seeds

1. Freshly harvested mature seeds (*see* **Note 3**).
2. Freshly prepared "antioxidant solution" (*see* **Note 4**); e.g., 50 mg/L ascorbic acid prepared in water or liquid culture medium (*see* **Note 5**) or cathodic water (see Sects. 2.5 and 3.10).
3. Sterile blot paper.

4. Sterile tweezers.
5. Sterile scalpel.
6. 9 mm diameter Petri dishes.

2.5 Cathodic Water Protection

1. Ca/Mg solution: 0.5 μM $CaCl_2.2H_20$ and 0.5 mM $MgCl_2.6H_2O$ [32].
2. Two 250 mL glass beakers.
3. An agar-based salt bridge (30% KCl and 3% agar bacteriological) prepared within a curved glass tube (*ca.* 8 mm outer diameter and 5 cm in length).
4. Constant voltage generator (e.g., Bio-Rad PowerPac HC Electrophoresis Power Supply).
5. Two platinum electrodes (rectangular: 2 cm by 0.5 cm by *ca.* 1.5–2 mm thick) to connect to each terminal of the constant voltage generator.

2.6 Partial and Rapid ("Flash") Drying of Embryonic Axes of Recalcitrant Seeds

1. "Flash" drier (see Figs. 1 and 2G) (*see* **Note 6**).
2. Activated silica-gel cooled down to room temperature for 15 min (better if oven dried [activated] overnight before is used in the morning) (*see* **Note 7**).
3. Forceps.

2.7 Moisture Content Determination

1. Micro-balance (e.g., XP6U/XP2U Ultra Micro Balance, Mettler Toledo).
2. Drying oven set up at 103°C.
3. Small (about 1 cm diameter) aluminum pans.

2.8 Cooling and Cryopreservation of Recalcitrant Seeds

1. Vacuum pump to generate nitrogen slush (*see* **Note 8**).
2. Polystyrene foam cup to make nitrogen slush.
3. Aluminum foil strips (15 × 35 mm).
4. 2 mL cryovials (e.g., L-Brady) and boxes (e.g., Nalgene Cryobox 81 × 2 mL Vials, Scientific Laboratory Supplies Ltd, UK) or canes, depending on the cryo-dewar used.
5. Cryo-dewar large/small (e.g., Archive Dewar, Biosystem 36 and Biorack 6000, Statebourne.com).
6. Cold-protective gloves (cryo gloves) and protective glasses.
7. Liquid nitrogen.

2.9 Vacuum Infiltration Vitrification (VIV) of Embryonic Axes of Recalcitrant Seeds

1. A single-stage vacuum pump (e.g., model VP130) attached to a 5 L stainless steel vacuum chamber (Fig. 3A).
2. Cryoprotective solutions (*see* **Note 9**).
3. 2 mL cryovials (e.g., L-Brady).
4. Ice bath (small beaker containing crushed ice).

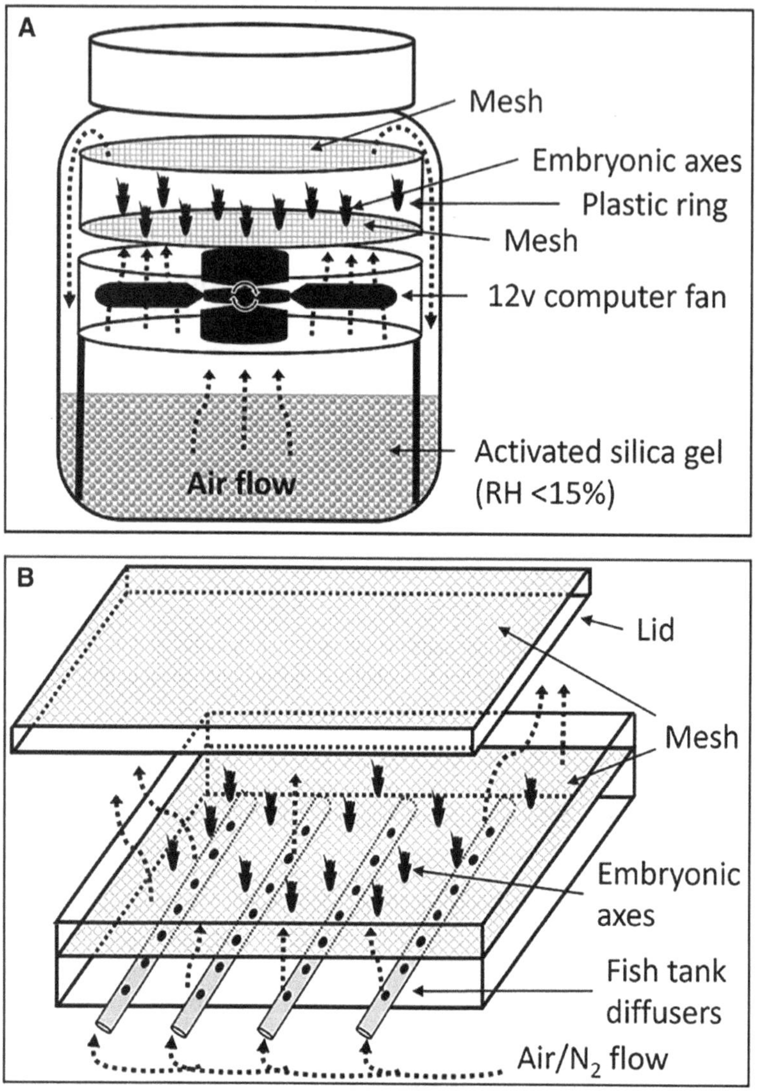

Fig. 1 "Flash" drying devices used for the rapid and partial desiccation of embryonic axes of recalcitrant seeds. (**A**) "Flash" drier used in Royal Botanic Gardens, Kew based in the design from [5]. In this design, air at room temperature (about 20°C) is moved at 2 m/s using a 12 v computer fan situated in the middle of a 2 L jar with 450 mL of activated silica gel on the bottom. Excised embryonic axes are placed in an 8 cm diameter × 1.5 cm height ring with a fine-mesh nylon attached to both ends on top of the fan. (**B**) "Flash" drier designed by [42]. In this design, air or N_2 gas at room temperature (about 20°C) is passed at 9 L/min through four fish-tank air diffusers in parallel, evenly spaced on the bottom of a plastic box (10 cm long, 10 cm wide, and 4 cm deep). Excised embryonic axes are placed on a fine-mesh nylon supported 2 cm above the air stones. A lid with a fine-mesh nylon on top can also be used

5. Cryo-dewar large/small (e.g., Archive Dewar, Biosystem 36 and Biorack 6000, Statebourne.com).
6. Cold-protective gloves (cryo gloves) and protective glasses.
7. Liquid nitrogen.

2.10 Warming and Rehydration of Embryonic Axes of Recalcitrant Seeds

1. Ca/Mg solution for rehydration (see Sect. 2.5).
2. Water bath set up at 45°C.
3. Thermometer.
4. Beaker or polystyrene foam cup.

2.11 Sterilization of Embryonic Axes and In Vitro Culture

1. Sterilization solution: 1% sodium hypochlorite (prepared by adding 200 mL of NaClO and 1 mL Tween 20 to 800 mL of H_2O). It can be stored in the fridge for up to 5 days.
2. Culture medium for embryonic axes. McCown Woody Plant Medium (WPM) [33] for *Aesculus* and *Quercus* sp. is used at RBG Kew. It can be prepared using 2.3 g/L of commercially available WPM salts (e.g., Sigma), 30 g/L of sucrose, and MS vitamins (e.g., Sigma). The pH is adjusted to 5.6 ± 0.1 before autoclaving, and phytagel (2.5 g/L) is used as a gelation agent. Some authors also add 0.3% charcoal to the medium when working with *Quercus* sp. [25].
3. Sterile tubes (e.g., 2 cm diameter × 15 cm length) and lids for plant in vitro culture or sterile Petri dishes (5–6 cm diameter).
4. Long sterile forceps.

3 Methods

3.1 Drying Orthodox Seeds for Cryopreservation

1. Place the seeds in an appropriately sized "Rotronic" measurement pan or another suitable open container. Seeds should be spread and fill no more than 50% of the dish or container to allow efficient drying.
2. Label all samples adequately, such as by placing an empty pre-labelled cryovial in the container.
3. Orchids and other small seeds, which fill less than 1 mL of a 2 mL vial, can be equilibrated in a storage cryovial or a folded paper packet prior to storage in a foil bag.
4. Measure and record the initial eRH of the seeds using a "Rotronic" water activity sensor (see Sect. 3.2) or another suitable hygrometer [34]. Where sample volumes are too small to measure, a surrogate sample of seeds may be used.
5. Place orthodox seeds at 32% RH and 18°C or 24% RH and 5°C (*see* **Note 1**). This drying will be equivalent to 5%–9% MC (*see* **Note 10** and Tables 1 and 2).
6. Measure and record eRH of the seeds during drying using a "Rotronic" sensor (see Sect. 3.2) or another suitable hygrometer [34]. Alternatively, other low-cost monitors of seed moisture status can be used [35]. Most samples should be tested every 2 weeks and very short-lived samples (e.g., *Salix* sp., *Populus* sp., orchids and ferns) every 2 days.

Table 1

Comparison of moisture content at 18°C and 30% relative humidity (RH) for species with a range of oil contents

Species	Oil content (%)	MC % (fwb) (18°C and 30% RH)
Pisum sativum	1	8.7
Ranunculus sceleratus	25.9	6.7
Helianthus annuus	37.9	5.6

Table 2

Comparison of moisture content under different temperature and humidity conditions for *Ranunculus sceleratus*

Drying conditions	Drying temp (°C)	Drying RH (%)	Storage MC (% fwb)
MSB drying room for conventional storage	18	15	4.6
MSB drying cabinet for cryo-storage	18	30–32	6.7–6.9
NLGRP drying room for cryo and conventional	5	24	6.6
CPC for cryo and conventional storage	5	32	7.7
CPC for cryo and conventional storage	15–25	35	7.5–6.9

7. Finish drying when seed eRH has not changed in 24 h, or a flat curve is given when measuring with a computer-linked "Rotronic" However, total drying time should not exceed 4 weeks for most samples or 3–14 days for very short-lived samples (*Salix and Populus* sp.).

3.2 Use of a "Rotronic" Water Activity Sensor

1. Place the seed sample into the "Rotronic" measurement chamber (*see* **Note 11**). The headspace should be minimized; ideally, the seed sample should fill at least 20%–50% of the space (including headspace of the rotronic chamber [39]. Use the appropriately sized "Rotronic" measurement pan as necessary (sizes from <3 to 64 cm^3 available).
2. Close the unit containing the seed sample and wait at least 20 min for equilibration [39].
3. Record the eRH and temperature given on the unit for that sensor (*see* **Note 12**). Ideally, multiple measurements should be taken to ensure equilibrium has been reached, or data can be graphically presented using software or exported to Excel (*see* **Note 13**).

3.3 Packing and Cryopreservation of Orthodox Seeds

1. Once equilibrated, pack and seal samples in a suitable container (*see* **Note 14**). Screw caps onto vials or seal foil bags using a heat sealer. This should be done quickly so moisture is maintained once samples are removed from the equilibration chamber.
2. Place vials/bags into a suitable Cryobox.
3. Wearing protective gloves and eye wear, place this box into the appropriate rack within the cryo-storage dewar (*see* **Note 15**) and lower directly into the LN vapour phase (*see* **Note 16**).
4. Where possible, the storage of subsamples (for germination assays or seed interchange) should be included to avoid repeated freeze/thaw cycles on bulk collections.

3.4 Removing Orthodox Seeds from LN

1. Wearing protective gloves and eye wear, carefully lift the appropriate stack and remove the storage box, cane, cryovial, or foil bag required and replace the stack (*see* **Note 15**).
2. Place the cryovial into a temporary storage box containing silica gel for around 20 min on a nearby bench to warm under ambient conditions (~20°C) before moving to the dry room or humidity box as appropriate (*see* **Note 17**).
3. Use samples after containers have been at ambient temperature for about 24 h to ensure all seeds have been thawed.

3.5 Cryopreservation of* Fagus sylvatica *Whole Seeds

1. Following collection, place seeds in a 47%–50% RH atmosphere at 18°C for 2–4 weeks (*see* **Note 18**). This drying step should be kept as short as possible as seeds can deteriorate fast in this atmosphere.
2. Monitor moisture during this period with a "Rotronic" sensor.
3. Once equilibrated, seal seeds in foil cryo bags and place into the vapour phase of LN at ~−185°C to −190°C.
4. To thaw, remove foil bags from storage and leave to warm in ambient conditions for 24 h before opening.

3.6 Cryopreservation of Whole Seed of* Coffea *sp. in LN

This protocol is adapted from [14, 30].

1. Remove *Coffea* seed testa (endocarp).
2. Desiccate seeds between 75% and 81% RH for 3 weeks at 25°C (*see* **Note 19**) in a humidity and temperature-controlled hydration cabinet or using salt solutions (*see* **Note 2**).
3. Seal seeds inside 10 mL cryovials or foil bags.
4. For seeds of *C. pseudo-zanguebariae*, *C. racemosa*, and *C. sessiliflora* immerse the cryovials in LN.

5. For seeds of *C. arabica* and *C. eugenioides*, precool the cryovials to −50°C at 1°C/min prior to immersion in LN (*see* **Note 20**). For the precooling step, a controlled rate freezer or a freezing container, such as Mr. Frosty, can be used. Follow the instructions of the instrument or the supplier (*see* **Note 21**).
6. Thaw by plunging the cryotubes in a 40°C water bath for 2 min.
7. Place seeds in a high RH atmosphere for 24 h to allow slow rehydration prior to placing in germination conditions (e.g., 30°C dark). (*see* **Note 22**).

3.7 Cryopreservation of Whole Seed of Citrus *sp.* in LN

This protocol is adapted from [15].

1. Dry seeds at 75%–85% RH and 25°C for a period of 4 weeks (*see* **Note 23**) in a humidity and temperature-controlled hydration cabinet or using salt solutions (*see* **Note 2**).
2. Monitor moisture during this period with a "Rotronic" sensor.
3. Place seeds in 2 mL polypropylene cryotubes and close the screw cap.
4. Immerse directly in LN or in the LN vapour phase.
5. After removal from cryo-storage, place seeds immediately in a 37°C water-bath for 5 min.

3.8 Melting Lipids After Cryo-storage

1. Expose dry seed to a heat pulse of 45°C for at least 1 h prior to hydration (*see* **Note 24**).
2. Place directly into germination conditions.

3.9 Excision of Embryonic Axes (Quercus *sp.* and Aesculus *sp.*)

1. Retrieve the seeds from cold storage and wash under the tap to remove excess dirt.
2. Remove the pericarp and seed coat using a scalpel and forceps and carefully excise the embryonic axes from the cotyledon tissue (see Fig. 2). Dissection of embryonic axes can be done under a laminar flow hood on sterile paper for maximum sterile conditions or, alternatively, on a lab bench previously cleaned with ethanol at 70% or another surface disinfectant.
3. The excised embryonic axes are then placed in a Petri dish on the sterile blot paper soaked with antioxidant solution. Embryonic axes should be used within 2 h after excision.

3.10 Cathodic Water Protection

1. Prepare 1 L of Ca/Mg solution and place it in two 250 mL beakers.
2. Insert one platinum electrode, connected to a constant volt generator, into each beaker and maintain charge balance within the internal circuit using the agar-based salt bridge.
3. Mount the cathodic water generation system as shown in Fig. 3B.
4. Apply a 60 V potential difference and 400 mA for 60 min.

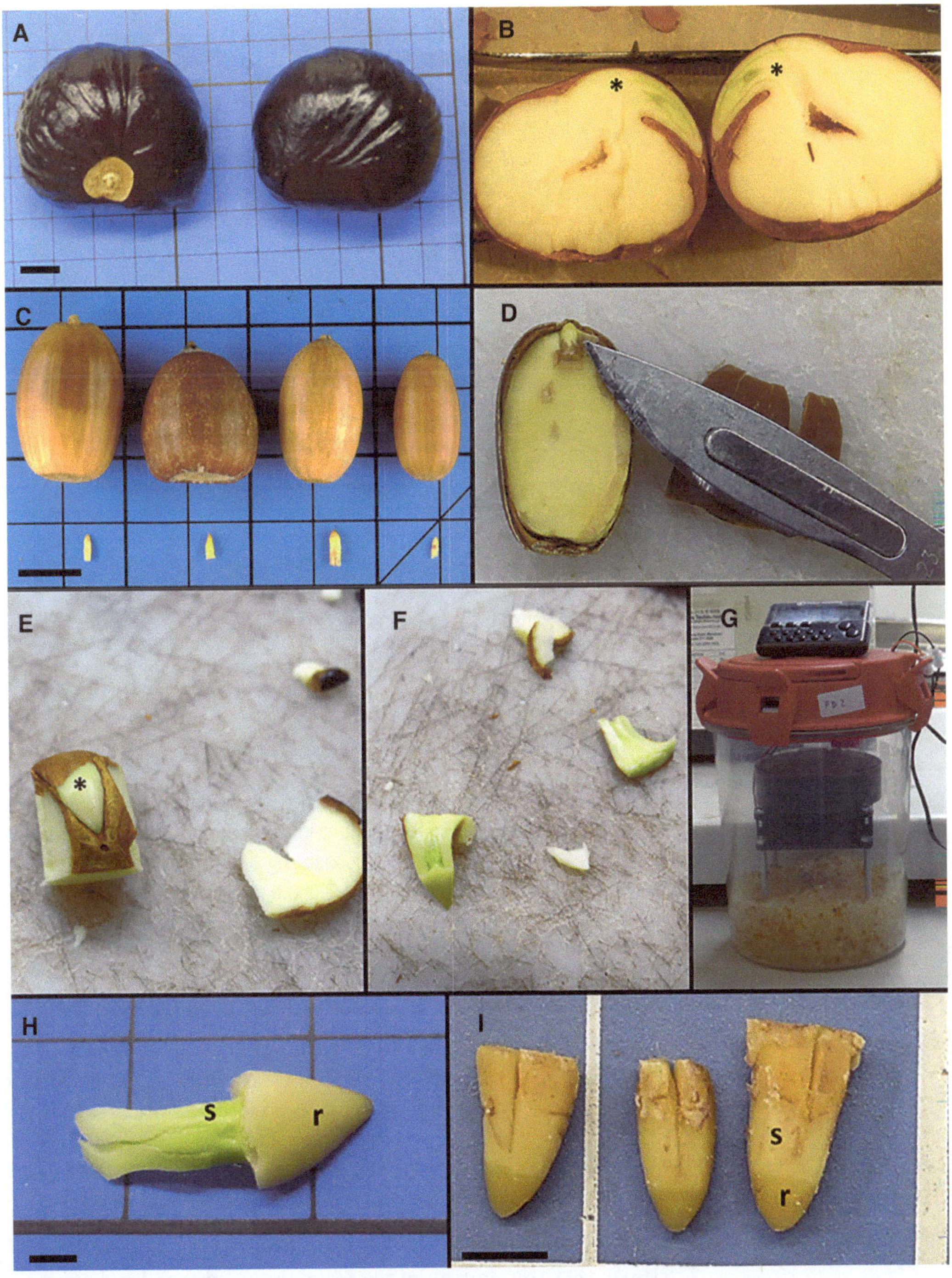

Fig. 2 Excision of *Quercus* sp. and *Aesculus* sp. embryonic axes. (**A**) *Aesculus indica* seeds. (**B**) Location of the embryonic axes within the seed, in a pocket formed by the seed coat (marked with an asterisk). (**C**) *Quercus* sp. acorns (fruit + seed) and embryonic axes (at the bottom of the image). (**D**) Location of the embryonic axes within the acorn (near the tip). (**E**) *Aesculus indica* embryonic axes (marked with an asterisk) after the removal of the seed coat and part of the cotyledons. (**F**) *Aesculus indica* embryonic axes after removing most of the cotyledonary tissue. (**G**) "Flash" drier used in the labs at Royal Botanic Gardens Kew. (**H**) *Aesculus indica* embryonic axes after full excision. (**I**) *Quercus pyrenaica* embryonic axes after full excision. s: shoot tip; r: root tip. Bar represents 1 cm (except in H and I where bar represents 2.5 mm)

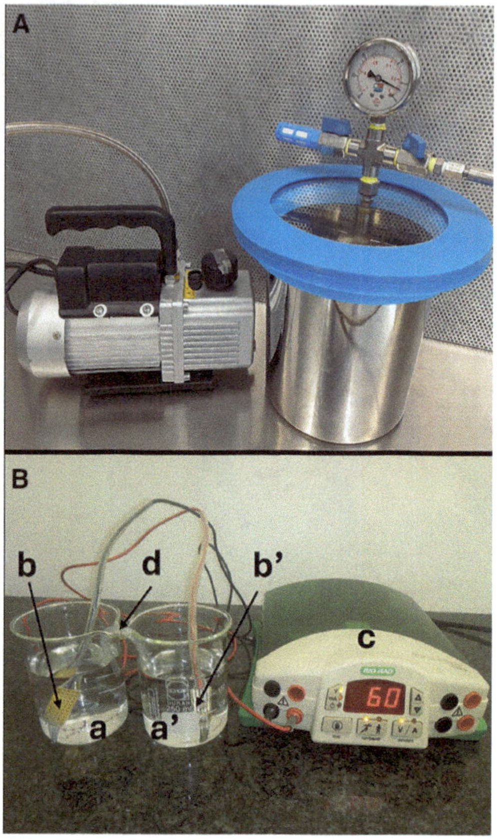

Fig. 3 (**A**) Single-stage vacuum pump (Model VP130) attached to a 5 L stainless steel vacuum chamber used in the vacuum-infiltration-vitrification (VIV) method. (**B**) Cathodic water generation system consisting of two 250 mL beakers, each containing 200 mL of Ca/Mg solution (a, a') and a platinum electrode (b, b') connected to a constant volt generator (c), and an agar-based salt bridge connecting both beakers to complete the circuit (d)

5. Use the cathodic fraction, within 24 h of preparation, to rewarm (at 40°C–45°C for 2 min) and rehydrate (ambient temperature for 30 min) explants recovered from cryostorage.

3.11 Rapid (Flash) Drying of Embryonic Axes of Recalcitrant Seeds

1. Blot-dry the embryonic axes carefully and place in the basket inside the flash dryer to prevent them from getting lost.
2. Add the activated silica gel to the flash drier.
3. Introduce the embryonic axes in the mesh of the flash drier (see Fig. 1A).
4. Dry the embryonic axes in the flash drier for a certain amount of time until water contents are in the 0.25–0.30 g H_2O/g dry weight range; the tested drying times in this lab range from 1 to 8 h, but in most species studied the targeted water content is reached between 2 and 4 h of rapid drying (*see* **Notes 25–27**).

3.12 Cryopreservation of Embryonic Axes of Recalcitrant Seeds

1. Place five (*Aesculus* sp.) to ten (*Quercus* sp.) embryonic axes (see Fig. 2H, I) in each 2 mL cryovial (cryovials should be properly labelled).
2. Plunge cryovial into LN using a canister for 5 min.
3. Store the cryovial in a box for cryovials in the vapor phase of LN. The box should be precooled in LN to avoid temperature fluctuations (*see* **Note 15**).

3.13 Vacuum-Infiltration-Vitrification (VIV) Cryopreservation

1. Prepare CPA solutions (*see* **Note 9**).
2. Place 5–10 embryos into a 1.5 mL cryovial and treat with LS solution (*see* **Note 9**) for 20 min at room temperature.
3. Remove LS from 1.5 mL cryovial and place the cryovial inside the beaker containing crushed ice.
4. Add 1 mL of prechilled PVS2 (*see* **Note 9**) into the 1.5 mL cryovial containing embryos within. Do not close the vial with the cap.
5. Quickly transfer the beaker with open cryovials containing the embryos and PVS2 solution inside the vacuum chamber, seal the chamber, and start applying a vacuum of 381 mm (15 in) Hg (50 kPa) for 1–10 min (*see* **Note 28**).
6. Remove the vacuum very slowly. When it reaches the room atmospheric pressure (100 kPa), unseal the vacuum chamber.
7. Quickly transfer cryovials to the workstation, seal the cryovials with their caps, and immerse them into LN.

3.14 Rewarming of Dried Cryopreserved Embryonic Axes

1. Remove the cryovial from the LN. The box containing cryovials should be kept on LN in a polystyrene foam box to avoid heating of the other samples in the box (*see* **Note 15**).
2. Open the cap of the cryovial and place the cryopreserved embryonic axes in a beaker or polystyrene foam cup filled with Ca/Mg solution preheated to 42°C ± 2°C in a bath.
3. Remove the embryonic axes from the warm bath and gently place into Ca/Mg solution for 2 min. Check with a thermometer that the temperature of the solution is around 40°C.
4. Transfer the thawed embryonic axes to a Petri dish with the Ca/Mg solution for rehydration for 30 min in the dark.

3.15 Rewarming of Vacuum Infiltration Vitrification (VIV) Cryopreserved Embryo

1. Remove cryovial from the LN. The box containing cryovials should be kept on LN in a polystyrene foam box.
2. Replace PVS2 solution in the cryovial with preheated unloading solution (ULS) (*see* **Note 9**) and maintain for 20 min in a bath at 42°C ± 2°C.
3. Transfer the thawed embryos to a Petri dish layered with sterile filter paper and blot dry the embryos.

4. Transfer embryos onto Petri dishes containing regeneration medium with a single layer of sterile filter paper.
5. Seal Petri dishes with sealing film (e.g., Parafilm® Sigma Aldrich).
6. Cover Petri dishes with aluminum foil and maintain them at a standard growth environment for 24 h.
7. After 24 h, transfer embryos onto a new Petri dish containing regeneration medium and keep them in the dark (cover with aluminum foil) at standard growth environment for a further 1–7 days.
8. Remove aluminum foil from the Petri dishes and maintain the embryos at a standard growth environment.

3.16 In-Vitro Culture and Regeneration of Embryonic Axes

1. Working under sterile conditions in the laminar flow; wash the embryonic axes for 2 or 5 min (for *Quercus* sp. and *Aesculus* sp., respectively) with the sterilization medium and rinse three times with sterile water (60 s each rinse).
2. Plant the embryonic axes in test tubes or Petri dishes containing the proper culture medium, making sure the plumule or shoot tip is a few millimeters above the culture medium and the root tip is inserted in the medium.
3. The in-vitro cultures are stored in the dark for 2 weeks in a light incubator at 25°C.
4. Move the cultures into the light after 2 weeks in the dark and check for contamination and growth measurement.

4 Notes

1. The use of saturated salt solutions (e.g., $MgCl_2$ for 33% RH at 15°C [36]) or dissolutions of LiCl [37] can be an alternative to generate an environment of 30% RH in a sealed jar when controlled environment chambers are not available. Tables for the relationship of RH and temperature of saturated salt solutions can be found in [36] and [38].
2. The use of saturated salt solutions (e.g., $Ca(NO_3)_2$ for 51% and 56% RH at 15°C and 25°C, respectively, and NaCl for 75% RH at 5°C–45°C [36]) or dissolutions of LiCl [37] can be an alternative to generate an environment of 50%–75% RH in a hermetically sealed jar or container (e.g., electrical enclosure box) when controlled environment chambers are not available. Saturated salt solutions can also be used to generate RH >75%. Tables for the relationship of RH and temperature of saturated salt solutions can be found in [36] and [38].

3. Seeds must be collected when fully mature. For example, for *Aesculus* sp. and *Quercus* sp., seeds are collected from the ground, right after falling from the tree. Seeds that remain a long time after shedding on the ground can dry and increase the chances of predation or infestation. Cryopreservation of embryonic axes should be done on freshly harvested seeds, but if needed, seeds of *Aesculus* sp. and *Quercus* sp. can be stored for a few months in the dark in cold storage at 5°C in sealed bags to avoid drying. For example, in such conditions, embryonic axes of *Aesculus hippocastanum* can be successfully cryopreserved after 6 months of seed storage, but the optimal response to cryopreservation in *A. hippocastanum* embryonic axes is only maintained for up to 3 months [39]. Indications for short-term storage of *Quercus* sp. acorns can be found in [40]. Sprouting seeds should not be used for cryopreservation of the embryonic axes.
4. Several antioxidant solutions can be found in the literature, e.g.,:
   (a) filter-sterilized ascorbic acid solution (50 mg/L) [26];
   (b) citric and ascorbic acid (0.2 g/L) solution [25];
   (c) a strongly reducing, high pH cathodic water produced by electrolysis of a solution containing calcium and magnesium chloride (0.5 mM $CaCl_2.2H_2O$ and 0.5 mM $MgCl_2.6H_2O$) used with particular success in *Strychnos gerrardii* axes [31].
5. For *Aesculus* sp. and *Quercus* sp., liquid medium containing 1/2 the macronutrients and micronutrients of MS [41] or the macronutrients and micronutrients of the Woody Plant Medium have been used [33].
6. Rapid ("flash") drying using a "flash drier" has proved to be far more effective for cryopreservation of embryonic axes of recalcitrant seeds than slow drying over silica gel [42]. Mainly, two types of "flash" driers can be found in the literature, one that is built in a jar with a computer fan [27] (see Fig. 1A) and one that uses a stream of air or nitrogen gas through fish tank diffusers [25, 42] (see Fig. 1B). Drying of embryonic axes is also usually done under the air flow or a laminar flow chamber [28].
7. 450 mL of activated silica gel is used in a 2 L container where the flash drier is built (see Fig. 1A). This ratio creates an RH in the container that is <15% during the drying time needed.
8. Rapid cooling of the samples can be done following several methods, each method providing different cooling rates [25]. For embryonic axes of recalcitrant seeds, a common method is to enclose five to ten embryonic axes in a cryovial and submerge the cryovial in LN. This method allows cooling

Quercus sp. embryonic axes to about 3–8°C/s [25]. Another method used to provide a faster cooling rate is to enclose two to three axes in lightweight aluminum foil packets (e.g., 2 cm × 3 cm) that are submerged with forceps in N_2 slush (formed by exposing to vacuum LN, about –210°C). This method allows cooling *Quercus* sp. embryonic axes to about 30–80°C/s [25].

9. Cryoprotectant solutions used in the vitrification protocol included Loading Solution (LS), Plant vitrification Solution 2 (PVS2), and unloading solution (ULS). The LS consisted of MS Medium [27] supplemented with 2.0 M glycerol and 0.4 M sucrose. PVS2 consisted of MS medium supplemented with 0.4 M sucrose, 30% w/v glycerol, 15% w/v ethylene glycol, and 15% w/v dimethyl sulfoxide (DMSO). The ULS consisted of MS medium supplemented with 0.75 M sucrose.

10. As storage temperature decreases, the critical or optimal moisture content for seed storage increases, so moisture requirements for cryo-storage are typically above those recommended for conventional storage [13]. Prior to cryo-storage, seeds must be dried at above-zero temperatures to a suitable moisture. To reach 5%–12% moisture content (fresh weight basis) or 0.05–0.14 g H_2O/g dry weight (suggested moisture content for cryo-stored seeds [7]), orthodox seeds can be dried at 32% RH and 18°C (as performed in the MSB, RBG Kew, UK), at 24% RH and 5°C (as performed in the National Laboratory for Genetic Resources Preservation [NLGRP], Fort Collins, USA), or at 32% RH and 5°C [as suggested by the Center for Plant Conservation (CPC)]. Moisture content at 30%–32% RH and 18°C will generally range from 5% to 9% depending on oil content (see Table 1). Drying at 24% RH and 5°C or 35% RH and 25°C will give a similar moisture range. The NLGRP and the CPC use these drying conditions for both conventional and cryo storage. These moisture conditions are compared for *Ranunculus sceleratus* using the SID (http://data.kew.org/sid/viability/mc1.jsp) equations (see Table 2). However, it could be possible that the optimal moisture contents for seed cryo-storage are higher than the ones indicated in this chapter. For example, [13] suggested that the optimal water contents for cryo-storage (0.17 g H_2O/g dry weight) would be obtained after equilibration at 75% RH at room temperature.

11. "Rotronic" water activity sensors provide a nondestructive, quick method for measuring sample moisture status, making it ideal for small collections (e.g., those of rare and threatened species). The method measures equilibrium RH (eRH); that is, the RH of air at equilibrium with seeds held in a sealed chamber. RH is the ratio of actual water vapour pressure to the saturation water vapour pressure at the same temperature.

12. Always note the exact instrument used and use this again for the same sample. The temperature of the seed sample and "Rotronic" chamber must be equal for an accurate measurement and should always be recorded with the eRH. Calibrations at 20°C are recommended every 6 months. Workstations and display units should be calibrated as pairs. More information about the theoretical basis and the use of a "Rotronic" water activity sensor for measuring seed moisture can be found in [37] and [43].
13. Software for "Rotronic" workstation units (e.g., HW4 "Rotronic" monitoring system, www.rotronic.com) can be logged by connecting the display unit to a "Roline" software loaded on a PC, tablet, or smartphone. This allows controlled logging and data manipulation via Excel. Although these units can be powered by battery, routinely mains connection is preferred.
14. The choice of container depends on the volume of the sample and should be as full as possible, ensuring air gaps are kept to a minimum. Multiple containers per sample are preferable, allowing for samples to be removed for testing without the need to remove the whole collection. Any quarantine material must be foil bagged, due to risk of leakage from vials. Seeds over 10 mm in diameter should not be backed up to cryo-storage due to risk of the seed coat cracking and therefore becoming damaged during freezing/rewarming procedures. Any seeds with sharp or spiny edges, which may damage foil bags, should be prepacked in a protective packet, such as card or paper. A suitable database for assigning and storing bank locations is recommended.
15. Ideally, samples stored in the cryo-dewar should not be exposed to room temperature to avoid temperature fluctuations. These fluctuations can increase the temperature of the sample and could produce negative effects in the samples, either by ice crystallization (in the case of samples stored at high water contents, such as embryonic axes) or by lipid melting and re-crystallization cycles (in intermediate seeds, orthodox oily seeds, and fern spores). However, as moving racks, canisters, boxes, or canes from the cryo-dewar may be unavoidable when collections are inserted or removed from LN, some recommendations must be followed to avoid damage to other collections already stored in the cryo-dewar. These recommendations follow those suggested in the Best Practices for Repositories of the International Society for Biological and Environmental Repositories (ISBER) [44]:
    (a) If a new box or cane is introduced in the cryo-dewar, the rack or the canister containing other boxes or canes should be kept just above the cryo-dewar only for the

brief time needed to insert the box in an empty space. In samples going into large cryo-dewars, two people may be needed, one holding the rack and the other inserting or removing the box. This needs to be done quickly and usually takes less than 20 s. After the box is inserted or removed, the rack should be placed back inside the cryo-dewar. It is recommended to insert just one box per rack at a time to avoid warming up the tank or the adjacent boxes.

(b) If a sample is added or removed from a box or a cane that already holds samples, these containers should remain in LN or in the vapour phase of LN. This can be accomplished by adding a 3–4 cm of LN into a polystyrene foam box, in which the box or cane is moved from the rack or canister. When cryovials are added or removed, the other samples will remain in contact with LN. The use of forceps is recommended to avoid the heat of fingers warming up adjacent cryovials.

16. No pre-freeze step at −20°C is necessary as this may be harmful to some seeds, such as those with high oil contents or freeze-sensitive properties [45].

17. If samples are in vials, ice may have formed on the outside. This step will prevent excess water from entering the vial during rewarming.

18. Desiccation sensitivity, and therefore storage behavior, can vary considerably across accessions for *Fagus sylvatica*. This may be due to maturity at collection, provenance, post-harvest handling, genetic or other factors [46]. Some accessions can maintain viability of 70%–80% after desiccation to equilibrium to as low as 15% RH at 15°C–18°C, though typically around 50% viability is lost after desiccation below ~47% RH (MSB unpublished data). Therefore, drying conditions of at least 47% RH are recommended to maintain maximum viability across multiple collections.

19. The optimal hydration level for cryopreservation of *Coffea* seed will vary by species but is typically between 75% and 81% RH. This is higher than that for conventional storage at −20°C, which varies between 43% and 50% RH. Critical water contents for cryo-storage will increase with decreasing temperature and may not correspond to the critical water content for desiccation sensitivity at room temperature. Some useful tables can be found in [14, 30].

20. *Coffea* species vary considerably in response to cryopreservation. Species with high cryo survival include C. *pseudozanguebariae*, *C.racemosa*, and *C. sessiflora*, some survival is reported for *C. arabica* and *C. eugenioides*, and no survival for *C.*

brevipes, *C. canephora*, *C. liberica*, and *C. stenophylla* [14]. Survival of *C. arabica* and *C. eugenioides* seed is enhanced by slow cooling (1°C per minute to −50°C) before immersion in LN.

21. The use of a Mr. Frosty is convenient when a programmable freezer is not available. Nalgene Mr. Frosty containers are inexpensive, reusable, and convenient devices for freezing biological samples at a controlled cooling rate of −1°C/min. To use a Mr. Frosty:
    (a) Remove the tube holder and foam insert from the Mr. Frosty.
    (b) Add 100% isopropyl alcohol to the fill line.
    (c) Replace tube holder.
    (d) Place cryovials containing the seed samples into holes in the tube holder.
    (e) Place the Mr Frosty in the bottom of a −80°C mechanical freezer.
    (f) Leave undisturbed for 3–4 h to reach the desired temperature (−50°C).
    (g) Remove frozen cryovials from the Mr. Frosty and submerge in LN.
22. For some species (e.g., *C. liberica* and *C. stenophylla*), freezing injury was restricted to the endosperm only, with embryos still recoverable by extraction and tissue culture (when above a critical water content of 0.27 and 0.25 g H_2O/g dry weight) [14].
23. The optimal hydration level for cryopreservation of non-orthodox oily seeds will vary by species but is generally found to correspond to the upper limit of the seed's unfrozen water content. For species within *Citrus*, this can be obtained by equilibrating seeds at 75%–85% RH and 25°C [15].
24. Seeds of some species, including *Cuphea* sp., neem (*Azadirachta indica*) and papaya (*Carica papaya*), may require a short period of high temperatures to melt lipids after storage [47, 48]. This is required to prevent imbibition damage and promote germination. In some cases, the method of slow rehydration in a high RH atmosphere (as used for *Coffea* sp.) may damage seed [47].
25. A drying curve must be done in advance for each species/accession to determine the appropriate drying time to reach the target water content. Target water contents found in the literature are species-specific (range between 0.25 and 0.40 g H_2O/g dry weight) (see Table 3).

Table 3

Guide on drying methods and moisture contents of embryo axes of some recalcitrant seeds for successful cryopreservation

Family	Species	Drying method	Moisture content (fwb)	Moisture content (g H_2O/g dw)	Survival	Normal plant growth	Reference
Sapindaceae	*Aesculus hippocastanum*	"Flash drier" (Fig. 1A)	20%–29%	0.25–0.40	100%	100%	[27]
	Acer saccharinum	"Flash drier" (Fig. 1A)	20%	0.25	90%	75%	[5]
Fagaceae	*Castanea sativa*	Airflow of a laminar flow chamber	20%	0.25	100%	63%	[29]
	Castanea sativa	Airflow of a laminar flow chamber	24%	0.32	93%	61%	[29]
	Quercus faginea	Airflow of a laminar flow chamber	21%	0.27	75%	60%	[28]
	Quercus rubra	Stream of nitrogen gas (Fig. 1B)	23%	0.30	90%	60%	[25]

fwb fresh weight basis, *dw* dry weight

26. Water content determination: The water content is determined at each drying time to find the appropriate drying time to reach the desired water content. The fresh weight is measured with a microbalance before the partial desiccation in the flash drier. After drying, the embryonic axes are then placed in the drying oven for 16–17 h at 103°C. The dry weight is measured on the microbalance after cooling the embryos for 10 min to room temperature in a dry box containing activated silica gel.
27. The drying method showed in this protocol works well for diverse species that have been researched at Kew and other species found in the literature (e.g., *Aesculus hippocastanum*, *A. californica*, *Aesculus* × *arnoldiana*, *Quercus pyrenaica*, *Q. rubra*). However, for some other species, the use of cryoprotectants, such as glycerol and sucrose, before the drying step is needed (e.g., *Amaryllis belladonna* [49]). Another cryopreservation method using vacuum infiltration of the cryoprotectant Plant Vitrification Solution 2 (PVS2) has been developed for embryonic axes of oily seeds, including the recalcitrant seeds of *Laurus nobilis* [19]. While the development of protocols using cryoprotectants may be necessary for diverse species, partial drying and fast cooling as exposed in this chapter is recommended as the first method to try in a recalcitrant-seeded species with unknown response to cryopreservation.
28. A weaker or "soft" vacuum (381 mm [15 in] of Hg, which is equivalent to ~50 kPa) pressure is recommended for plant tissues based on previous application in gene transformation studies of 1-week old tobacco seedlings [50]. The application can vary from 1 min to several minutes based on the plant materials used, i.e., until vitrification is achieved. For smaller embryonic axes, such as *Passiflora edulis*, 1 min VIV treatment at 0°C was sufficient for good viability and regrowth (above 55%) following cryopreservation [19]. The application of vacuum infiltration vitrification can be carried out at room temperature (i.e., without the use of ice bucket) [19].

Acknowledgments

We thank diverse researchers and technicians from RBG Kew and the MSB that have supported and revised the development of these protocols, particularly John Adams for helping to build the "Flash drier" used at RBG Kew and Frances Stanley and Kirstine Manger for critically reviewing the manuscript. RBG Kew received a grant-in-aid from Defra.

References

1. Li DZ, Pritchard HW (2009) The science and economics of ex situ plant conservation. Trends Plant Sci 14:614–621. https://doi.org/10.1016/j.tplants.2009.09.005
2. Roberts EH (1973) Predicting the storage life of seeds. Seed Sci Technol 1:499–514
3. Walters C (2015) Orthodoxy, recalcitrance and in-between: describing variation in seed storage characteristics using threshold responses to water loss. Planta 242:397–406. https://doi.org/10.1007/s00425-015-2312-6
4. Walters C, Berjak P, Pammenter N, Kennedy K, Raven P (2013) Preservation of recalcitrant seeds. Science 339:915–916. https://doi.org/10.1126/science.1230935
5. Wesley-Smith J, Berjak P, Pammenter NW, Walters C (2014) Intracellular ice and cell survival in cryo-exposed embryonic axes of recalcitrant seeds of *Acer saccharinum*: an ultrastructural study of factors affecting cell and ice structures. Ann Bot 113:695–709. https://doi.org/10.1093/aob/mct284
6. FAO, 2014. Genebank standards for plant genetic resources for food and agriculture, Rev. FAO, Rome
7. Center for Plant Conservation, 2019. CPC Best plant conservation practices to support species survival in the wild. Center for Plant Conservation, Escondido, CA
8. Millennium Seed Bank (MSB) Partnership (2015) Seed conservation standards for MSB partnership collections. Royal Botanic Gardens, Kew, UK
9. Wyse SV, Dickie JB (2017) Predicting the global incidence of seed desiccation sensitivity. J Ecol 105:1082–1093. https://doi.org/10.1111/1365-2745.12725
10. Walters C, Wheeler LM, Grotenhuis JM (2005) Longevity of seeds stored in a genebank: species characteristics. Seed Sci Res 15:1–20. https://doi.org/10.1079/SSR2004195
11. Ballesteros D, Pence VC (2017) Survival and death of seeds during LN storage: a case study on seeds with short lifespans. CryoLetters 38:278–289
12. Davies RM, Dickie JB, Ballesteros D (2018) Evaluation of short-lived seeds' cryopreservation as alternative to conventional seed banking. Cryobiology 85:140–141. https://doi.org/10.1016/j.cryobiol.2018.10.088
13. Vertucci CW, Roos EE, Crane J (1994) Theoretical basis of protocols for seed storage III. Optimum moisture contents for pea seeds stored at different temperatures. Ann Bot 74:531–540. https://doi.org/10.1006/anbo.1994.1151
14. Dussert S, Chabrillange N, Rocquelin G, Engelmann F, Lopez M, Hammon S (2001) Tolerance of coffee (*Coffea* spp.) seeds to ultra-low temperature exposure in relation to calorimetric properties of tissue water, lipid composition, and cooling procedure. Phys Plant 112:495–504. https://doi.org/10.1034/j.1399-3054.2001.1120406.x
15. Hor YL, Kim YJ, Ugap A, Chabrillange N, Sinniah UR, Engelmann F, Dussert S (2005) Optimal hydration status for cryopreservation of intermediate oily seeds: citrus as a case study. Ann Bot 95:1153–1161. https://doi.org/10.1093/aob/mci126
16. Berjak P, Pammenter NW (2007) From *Avicennia* to *Zizania*: seed recalcitrance in perspective. Ann Bot 101:213–228. https://doi.org/10.1093/aob/mcm168
17. Sershen, Pammenter NW, Berjak P, Wesley-Smith J (2007) Cryopreservation of embryonic axes of selected amaryllid species. CryoLetters 28:387–399
18. Sershen S, Varghese B, Pammenter NW, Berjak P (2012) Cryo-tolerance of zygotic embryos from recalcitrant seeds in relation to oxidative stress-a case study on two amaryllid species. J Plant Physiol 169:999–1011. https://doi.org/10.1016/j.jplph.2012.02.016
19. Nadarajan J, Pritchard HW (2014) Biophysical characteristics of successful oilseed embryo cryoprotection and cryopreservation using vacuum infiltration vitrification: an innovation in plant cell preservation. PLoS One 9:e96169. https://doi.org/10.1371/journal.pone.0096169
20. Vertucci CW, Leopold CA (1987) The relationship between water binding and desiccation tolerance in tissues. Plant Physiol 85:232–238. https://doi.org/10.1104/pp.85.1.232
21. Vertucci CW (1990) Calorimetric studies of the state of water in seed tissues. Biophys J 58:1463–1471. https://doi.org/10.1016/S0006-3495(90)82491-7
22. Wolfe J, Bryant G, Koster KL (2002) What is 'unfreezable water', how unfreezable is it and how much is there?. CryoLetters 23:157–166
23. Pritchard HW (1995) Cryopreservation of seeds. In: Day JG, Pennington MW (eds.) Cryopreservation and freeze-drying protocols. Humana Press, Totowa, NJ, 133–144
24. Nadarajan J, Mansor M, Krishnapillay B, Staines HJ, Benson EE, Harding K (2008) Applications of differential scanning calorimetry in developing cryopreservation strategies for *Parkia speciosa*, a tropical tree producing recalcitrant seeds. CryoLetters 29:95–110

25. Xia K, Hill LM, Li DZ, Walters C (2014) Factors affecting stress tolerance in recalcitrant embryonic axes from seeds of four *Quercus* (Fagaceae) species native to the USA or China. Ann Bot 114:1747–1759. https://doi.org/10.1093/aob/mcu193
26. Gonzalez-Benito ME, Prieto RM, Herradón E, Martín C (2002) Cryopreservation of *Quercus suber* and *Quercus ilex* embryonic axes: in vitro culture, desiccation and cooling factors. CryoLetters 23:283–290
27. Wesley-Smith J, Walters C, Pammenter NW, Berjak P (2001) Interactions among water content, rapid (nonequilibrium) cooling to -196°C, and survival of embryonic axes of *Aesculus hippocastanum* L. seeds. Cryobiology 42:196–206. https://doi.org/10.1006/cryo.2001.2323
28. Gonzalez-Benito ME, Perez-Ruiz C (1992) Cryopreservation of *Quercus faginea* embryonic axes. Cryobiology 29:685–690. https://doi.org/10.1016/0011-2240(92)90072-A
29. Corredoira E, San-José MC, Ballester A, Vieitez AM (2004) Cryopreservation of zygotic embryo axes and somatic embryos of European chestnut. CryoLetters 25:33–42
30. Eira MTS, Amaral da Silva AE, De Castro RD, Dussert S, Walters C, Bewley JD, Hilhorst HWM (2006) Coffee seed physiology. Braz J Plant Physiol 18:149–163. https://doi.org/10.1590/S1677-04202006000100011
31. Berjak P, Varghese B, Pammenter NW (2011) Cathodic amelioration of the adverse effects of oxidative stress accompanying procedures necessary for cryopreservation of embryonic axes of recalcitrant-seeded species. Seed Sci Res 21:187–203. https://doi.org/10.1017/S0960258511000110
32. Mycock D (1999) Addition of calcium and magnesium to a glycerol and sucrose cryoprotectant solution improves the quality of plant embryo recovery from cryostorage. CryoLetters 20:77–82
33. McCown BH, Lloyd G (1981) Woody plant medium (WPM) - a mineral nutrient formulation for microculture of woody plant species. Hort Sci 16:453–453
34. Gold, K, Manger, K, 2014. Technical Information Sheet_05: measuring seed moisture status using a hygrometer. Board of Trustees of the Royal Botanic Gardens, Kew, http://brahmsonline.kew.org/Content/Projects/msbp/resources/Training/05-eRH-moisture-measurement.pdf. Accessed 31 July 2019
35. Sutcliffe, V, Adams, J, 2014. Technical Information Sheet_07: low-cost monitors of seed moisture status. Board of Trustees of the Royal Botanic Gardens, Kew, http://brahmsonline.kew.org/Content/Projects/msbp/resources/Training/07-Low-cost-moisture-monitors.pdf, Accessed 31 July 2019
36. Vertucci CW, Roos EE (1993) Theoretical basis of protocols for seed storage II. The influence of temperature on optimal moisture levels. Seed Sci Res 3:201–213. https://doi.org/10.1017/S0960258500001793
37. Gold, K, Hay, F, 2014. Technical Information Sheet_09: equilibrating seeds to specific moisture levels. Board of Trustees of the Royal Botanic Gardens, Kew, http://brahmsonline.kew.org/Content/Projects/msbp/resources/Training/09-Seed-air-moisture-relations.pdf, Accessed 28 July 2019
38. Sun WQ (2002) Methods for the study of water relations under desiccation stress. In: Black M, Pritchard HW (Eds.)Eds. Desiccation and survival in plants: drying without dying. CABI, Wallingford, UK, 47–91
39. Fanega-Sleziak N, Pritchard HW, Ballesteros D (2018) Contribution of embryo size and age to the successful cryopreservation of *Aesculus* species. Cryobiology 85:140. https://doi.org/10.1016/j.cryobiol.2018.10.087
40. Bonner FT, Karrfalt RP (2008) The woody plant seed manual. Agriculture handbook, vol 727, USDA-FS
41. Murashige T, Skoog F (1962) A revised medium for rapid growth and bio assays with tobacco tissue cultures. Physiol Plant 15:473–497. https://doi.org/10.1111/j.1399-3054.1962.tb08052.x
42. Pammenter NW, Vertucci CW, Berjak P (1991) Homeohydrous (recalcitrant) seeds: dehydration, the state of water and viability characteristics in *Landoiphia kirkii*. Plant Physiol 96:1093–1098. https://doi.org/10.1104/pp.96.4.1093
43. Probert R, Manger K, Adams J (2003) Non-destructive measurement of seed moisture. In: Smith RD, Dickie JB, Linington SH, Pritchard HW, Probert RJ (Eds.)Eds. Seed conservation: turning science into practice. The Royal Botanic Gardens, Kew, London, 367–387
44. Campbell LD, Astrin JJ, DeSouza Y, Giri, J, Patel AA, Rawley-Payne M, Rush A, Sieffert N (2018) The 2018 revision of the ISBER best practices: summary of changes and the editorial team's development process. Biopreserv Biobank 16:3–6. https://doi.org/10.1089/bio.2018.0001
45. Chau MM, Chambers T, Weisenberger L, Keir M, Kroessig TI, Wolkis D, Kam R, Yoshinaga AY (2019) Seed freeze sensitivity and ex situ longevity of 295 species in the native Hawaiian flora. Am J Bot 106:1248–1270. https://doi.org/10.1002/ajb2.1351

46. Kalemba EM, Janowiak F, Pukacka S (2009) Desiccation tolerance acquisition in developing beech (*Fagus sylvatica* L.) seeds: the contribution of dehydrin-like protein. Trees 23:305–315. https://doi.org/10.1007/s00468-008-0278-8
47. Crane J, Miller AL, Van Roekel JW, Walters C (2003) Triacylglycerols determine the unusual storage physiology of *Cuphea* seed. Planta 217:699–708. https://doi.org/10.1007/s00425-003-1036-1
48. Neya O, Golovina EA, Nijsse J, Hoekstra FA (2004) Ageing increases the sensitivity of neem (*Azadirachta indica*) seeds to imbibitional stress. Seed Sci Res 14:205–217. https://doi.org/10.1079/SSR2004170
49. Sershen, Berjak P, Pammenter NW, Wesley-Smith J (2012) The effects of various parameters during processing for cryopreservation on the ultrastructure and viability of recalcitrant zygotic embryos of *Amaryllis belladonna*. Protoplasma 249:155–169. https://doi.org/10.1007/s00709-011-0274-5
50. Dulk-Ras A, Regensburg-Tuink TJG, Hooykaas PJJ (2003) VirD4-independent transformation by CloDF13 evidences an unknown factor required for the genetic colonization of plants via Agrobacterium. Mol Microbiol 47:891–901. https://doi.org/10.1046/j.1365-2958.2003.03328.x

Chapter 16

Cryopreservation of Potato Shoot Tips Using PVS3 Droplet-Vitrification

Marion Grübe and Manuela Nagel

Abstract

Efficient cryopreservation protocols ensuring high regrowth of shoot tips after rewarming are crucial for the secure long-term conservation of the more than 60,000 globally maintained potato accessions. At Leibniz Institute of Plant Genetics and Crop Plant Research (IPK), over 2,150 accessions have been securely cryopreserved over the last 30 years. To ensure high regrowth potential and genetic stability after rewarming, the cryopreservation method was changed from dimethyl sulfoxide (DMSO) droplet freezing to plant vitrification solution 3 (PVS3) droplet vitrification and further improved and refined through key modifications. The last modifications included the adoption of microtubers for in vitro plant propagation, the preculture of in vitro plantlets and excised shoot tips in the dark, the direct cooling of PVS3-treated shoot tips and the rewarming using lower sugar concentration. As a result of these substantial improvements, the average regrowth potential of the complete secure cryo-collection has increased to 70%. However, further modifications and experiments will be conducted in future studies to ensure secure long-term conservation of the world's most important noncereal food crop.

Keywords Cryoprotection, Long-term storage, PVS3, Plant genetic resources, Tissue culture, Ultra-low temperature

1 Introduction

Potato (*Solanum tuberosum* L.) is the world's most important noncereal food crop with a global production of 370 million tons produced on 17.3 million hectare [1]. It was domesticated in the South American Andes about 8000–10,000 years ago, utilizing the genetically complex genepool of di-, tri-, tetra-, penta-, and hexaploid wild species that reproduce both sexually and clonally [2, 3]. To this day, clonal reproduction remains the primary source for field production of potato varieties. Consequently, of the 82,293 accessions maintained in 93 genetic resources collections, about 80% of the material is conserved clonally as landraces, improved varieties or breeding lines and requires field or in vitro maintenance [1]. However, long-term secure conservation can only be achieved through cryopreservation.

Willem F. Wolkers and Harriëtte Oldenhof (eds.), *Cryopreservation and Freeze-Drying Protocols*, Methods in Molecular Biology, vol. 3060, https://doi.org/10.1007/978-1-0716-5416-3_16,

Plant cryopreservation involves the preparation and storage of viable plant material under ultralow temperature conditions, typically below −130°C. To maintain viability and prevent lethal ice crystallization, the vegetative plant tissue of herbaceous species often requires a complex procedure involving preculture, shoot tip excision, dehydration, cryoprotection, and rewarming, washing, and regeneration [4]. The first successful experiments on potato cryopreservation using dimethyl sulfoxide (DMSO) freezing were conducted by Bajaj [5]. However, it was not until the mid-1990s that researchers from the predecessor institution of the Leibniz Institute DSMZ (German Collection of Microorganisms and Cell Cultures) refined and improved the method [6, 7]. Subsequently, DMSO droplet freezing was further developed and used at the Leibniz Institute of Plant Genetics and Crop Research (IPK) to preserve around 1500 accessions [8] (Fig. 1). Nevertheless, concerns arose regarding the potential impact of DMSO on DNA methylation in various cell types, e.g., post-thaw functionality was affected in bovine embryo [9]. As a result, DMSO droplet freezing was replaced by droplet vitrification using plant vitrification solution 3 (PVS3) by the end of 2016 [10] (Fig. 1).

Nowadays, the routine PVS3 droplet vitrification procedure at IPK involves several key steps (Fig. 2). First, potato in vitro plantlets are propagated from microtubers and grown under optimum growth conditions for 2–5 weeks. For an additional 1–3 weeks, the plantlets are acclimated in the dark at 5°C. The acclimated plants are used to excise the apical shoot tips that are precultured in a 10% sucrose solution in the dark overnight. The next day, the shoot tips undergo dehydration and cryoprotection using a Loading and PVS3

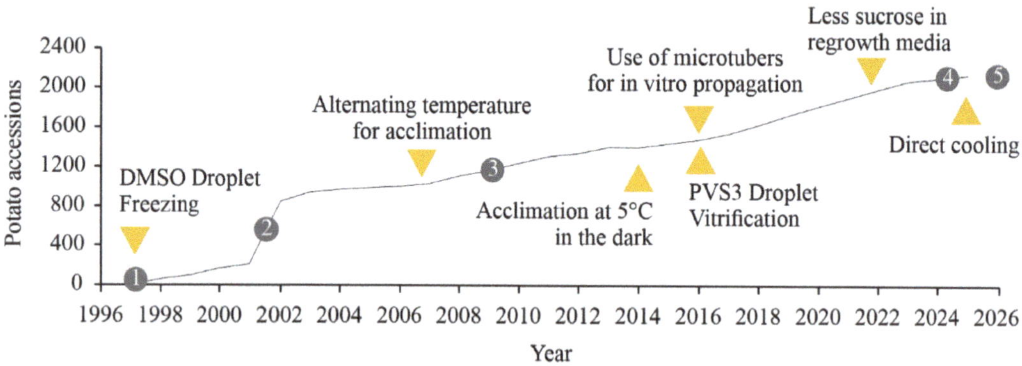

Fig. 1 Development of the IPK potato cryo-collection including significant changes in the cryopreservation procedure and collection management. **(1)** Potato cryopreservation was initiated in 1997 using the DMSO droplet freezing. **(2)** In 2002, the collection was expanded by transferring approximately 600 cryo-accessions from the former West German genebank to IPK [16]. **(3)** Since 2009, a duplicate system was established at the DSMZ in Braunschweig. **(4)** In 2024, the duplicate system was upgraded to a triplicate system, utilizing QR and barcodes for improved tracking. **(5)** Additionally, at the backup storage, passport and protocol data are cryopreserved along the material providing an additional layer of security and redundancy

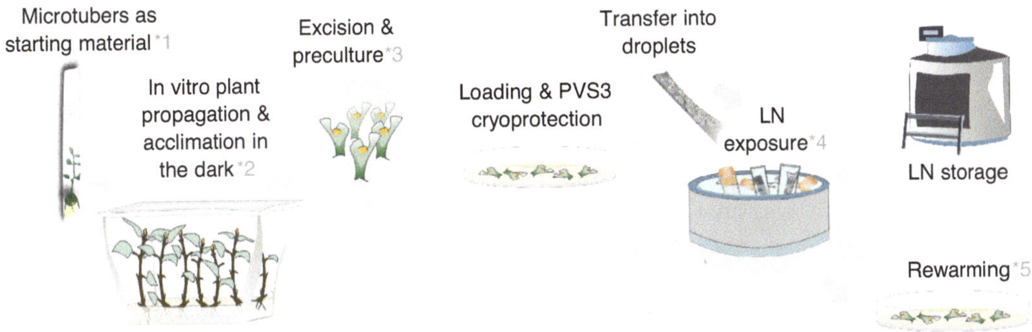

Fig. 2 Routine PVS3 droplet vitrification procedure for potato used at IPK. Green stars (*) indicate key modifications that have led to substantial increase in regrowth after cryopreservation. PVS3, plant vitrification solution 3

solution for approximately 2–2.5 h. Afterwards, the shoot tips are transferred into fresh PVS3 droplets and exposed to liquid nitrogen for cryopreservation. The cryopreserved potatoes can then be either stored or rewarmed for control purposes.

Several key modifications have been implemented to improve the regrowth potential of material cryopreserved by PVS3 droplet vitrification. These modifications include: (i) The adoption of microtuber-based in vitro plant propagation (Fig. 2, *1), which has been shown to reduce the frequency of endophytic contaminations during regrowth [10]. The number of subcultivations has a significant impact on the success of cryopreservation, as endophytes are assumed to accumulate with increasing rounds of subcultivation. The use of microtubers, which are initiated through dehydration and cold treatment, may restrict endophyte growth and support regrowth. (ii) The preculture of in vitro plantlets (*2) and excised shoot tips (*3) in the dark involves the combination of cold and dark (*2) and osmotic stress and dark (*3). These combinations are hypothesized to down-regulate electronic transport, reduce the availability of reactive oxygen species, and positively affect the chlorophyll biosynthesis and degradation [11], and leading to improved regrowth. (iii) The direct cooling of PVS3-treated shoot tips (*4) increase the cooling rate, vitrify the tissue and prevent lethal ice crystallization [12]. The previous simplified method involved transferring aluminum strips holding PVS3 droplets and shoot tips into cryo vials, closing the vial and plunging it into liquid nitrogen. This process created an air isolation reducing the cooling rate, and potentially leading to ice crystal formation. (iv) Rewarming using lower sugar concentration (*5) has been found to be beneficial for potato accessions cryopreserved with PVS2 [13]. High sucrose concentrations are considered to support rehydration and stimulate recrystallization of small ice crystals during rewarming. However, a study involving 86 potato accessions revealed that using a lower sucrose concentration (half of the original concentration) resulted in higher regrowth rates for 99% of the accessions [13]. This finding was confirmed at IPK using PVS3 droplet vitrification and implemented.

Following a series of key modifications, the average regrowth potential of the complete secure cryo-collection has increased to 70%. However, despite these improvements, some cases of endophytic contaminations and incomplete plant development have still been observed. To support healthy plant development, increase regrowth, and ensure secure conservation, further modifications will be tested in future studies.

2 Materials

2.1 *In Vitro* Culture Establishment, Maintenance, and Propagation

2.1.1 General Equipment

1. Laminar flow bench.
2. Incubators (CLF Plant Climatics GmbH, Wertingen, Germany) to adjust light and temperatures between 5°C and 25°C.
3. Dissection binocular (Stemi 305; Carl Zeiss Microscopy GmbH, Jena, Germany).
4. Heat sterilizer (250°C).
5. Autoclave.
6. UV light.
7. Gas safety burner to flame culture tubes.
8. Glass bead sterilizer to sterilize instruments.
9. Magnetic stirrer with heating.
10. pH meter.
11. Balance.
12. Benchtop open air shaker.
13. Sterile instruments: scalpel, forceps, pipette, counter, syringe, and needle.

2.1.2 Surface Sterilization of Plant Material Derived from the Field or Greenhouse

1. Tap water.
2. Ethanol 70% (v/v).
3. Hand disinfection solution.
4. Sodium hypochlorite solution [NaClO] 12% Cl, diluted 1:4 (v:v) with deionized water.
5. Surface detergent Tween 20.

6. Autoclaved deionized water.
7. Autoclaved round filter papers, grade 598, 90 mm for preparation (Whatman GmbH, Dassel, Germany).
8. Erlenmeyer flasks, 100 mL or bigger.

2.1.3 In Vitro Maintenance and Propagation

1. In vitro cultures of the former subculture as sources of explants.
2. Autoclaved round filter paper, grade 595, 125 mm (Whatman GmbH, Dassel, Germany).
3. Glass Petri dishes (150 mm).
4. Culture tubes 3 cm, 10 cm long, glass vials (750 mL), oval culture boxes with "breathing" hermetic cover (Duchefa Biochemie BV, Haarlem, The Netherlands).
5. Aluminum foil for caps, thickness 50 μm.
6. Murashige and Skoog (MS) [14] basal medium, standardized powder containing macroelements and microelements (Duchefa Biochemie BV, Haarlem, The Netherlands) (*see* **Note 1**).
7. Agar Agar Kobe I (SERVA Electrophoresis GmbH, Heidelberg, Germany).
8. Potato culture medium; MS medium with 0.058 M sucrose, 1% agar at pH 5.8.

2.2 Cryopreservation

2.2.1 Equipment for Cryopreservation

1. Cryotank.
2. Boxes for cryovials.
3. Dewar vessel for liquid nitrogen.
4. Label printer.

2.2.2 Plant Preparation and Cryopreservation Treatment

1. Dissection binocular.
2. Sterile instruments: scalpel, forceps, pipette, counter, syringe, and needle.
3. Cryovials 1.8 mL NUNC CryoTubes™ with internal thread and starfoot (Fisher Scientific GmbH Schwerte, Germany).
4. Labels resistant to liquid nitrogen (Laser CRYO-BABIES, Diversified Biotech, Dedham, USA or Cryo Barcode Labels, size 44×16+Ø 9.5 mm, LabId Technologies B.V., The Netherlands).
5. Petri dishes (60 mm).
6. Round filter paper, grad 391, 45 mm.
7. Parafilm.
8. Ethanol 70% (v/v) for surface sterilization.

9. Liquid nitrogen.

2.2.3 Media and Solutions for Cryopreservation

1. Murashige and Skoog [14] vitamin medium (MS with vitamin), standardized powder containing macroelements and microelements and vitamins (*see* **Note 1**).
2. Preculture solution: MS medium with vitamin, 0.3 M sucrose, at pH 5.8 (liquid).
3. Loading solution A: MS medium with vitamin, 0.4 M sucrose, 2 M glycerol (liquid), pH 5.8.
4. Plant Vitrification Solution 3 (PVS3): MS medium with vitamin, 1, 5 M sucrose, 5.4 M glycerol [15] (liquid), pH 5.8.
5. Washing solution: MS with vitamin, 0.6 M sucrose in medium (liquid).
6. Regeneration medium for potato: MS medium with vitamin, 0.5 mg/L trans-zeatin riboside, 0.5 mg/L IAA, 0.2 mg/L Gibberellic acid (GA3), 0.088 M sucrose, 0.8%–1% agar at pH 5.8 (*see* **Note 2**).

3 Methods

3.1 Preparation and Sterilization of the Culture Media

1. Potato culture medium (1000 mL). Dissolve 4.302 g MS and 20 g sucrose in 400 mL water to create a stock solution. Adjust the pH of the media to 5.7–5.8 using 1 M HCl or 1 M NaOH. Weigh 9–10 g agar (for a semisolid medium) and dissolve in 100 mL water. Heat up 500 mL of water and pour it into dissolved agar (100 mL), then boil. After boiling, add 400 mL stock solution to create a volume of 1000 mL. Divide the culture medium into 50–90 mL jars with minimum 1 cm height. Autoclave at 121°C for 20 min (118 kPa steam pressure). If plastic boxes are used, autoclave the medium in bottles and then pour it into plastic containers under the laminar flow hood. Store the jars with culture medium at 6°C for a maximum of 2 months. Before using, equilibrate the medium to room temperature.
2. MS stock solution (10-times concentrated) (100 mL). Weigh 4.405 g MS with vitamins und fill up with water to a total volume of 100 mL. Distribute 10 mL of the stock solution in tubes, label (e.g., MS 10×) and store them at −20°C.
3. Preculture solution (MS10, 100 mL). Weigh 10.26 g sucrose, mix with 10 mL MS 10× und fill up with water to a total volume of 100 mL. Adjust the pH of the solution to 5.7–5.8 using 1 M HCl or 1 M NaOH, label (e.g., MS 10) and autoclave at 121°C for 20 min (118 kPa steam pressure).

4. Loading solution (LS, Cryoprotectant, 100 mL). Weigh 13.7 g sucrose and 18.42 g glycerol mix with 10 mL MS 10×. Fill up with water to a total volume of 100 mL. Adjust the pH of the solution to 5.7–5.8 using 1 M HCl or 1 M NaOH, divide solution into two bottles and label (e.g., LS) and autoclave at 121°C for 20 min (118 kPa steam pressure).
5. PVS3 solution (PVS3, Cryoprotectant, 100 mL). Weigh 50 g sucrose. Mix 10 mL MS 10× with 50 g glycerol. Slowly add sucrose while warming and stirring (with a heated magnet stirrer) until everything is solved. Fill up with water to a total volume of 100 mL and cool down at room temperature. At 23°C, adjust the pH of the solution to 5.7–5.8 using 1 M HCl or 1 M NaOH, and autoclave at 121°C for 20 min (118 kPa steam pressure).
6. Rewarming and unloading solution (MS21, 100 mL). Weigh 21 g sucrose and mix with 10 mL MS 10×. Fill up with water to a total volume of 100 mL. Adjust the pH of the solution to 5.7–5.8 using 1 M HCl or 1 M NaOH, divide solution into two bottles and label (e.g., MS21) and autoclave at 121°C for 20 min (118 kPa steam pressure).
7. MS3 solution (1000 mL). Weigh 4.405 g MS and 30.0 g sucrose and fill up with water to a total volume of 1000 mL. Adjust the pH of the solution to 5.7–5.8 using 1 M HCl or 1 M NaOH. Distribute in bottles, label (e.g., MS3) and autoclave at 121°C for 20 min (118 kPa steam pressure).
8. Regeneration media solution (MS3m, 1000 mL). Prepare stock solutions of phytohormones (1 mg/mL). Accurately weigh 10 mg GA3 and 10 mg IAA and add each 10 mL ethanol (96% (v/v)) and store at 4°C. If IAA turns red, prepare solution again. For zeatin riboside, dissolve 10 mg zeatin riboside in 10 mL methanol (use agitator), disperse solution in 1 mL Eppendorf tubes and store at –20°C. Before use, mix zeatin riboside well using vortex, blow air with pipette through solution. Mix 975 mL water with 30 g sucrose and 4.405 g MS with vitamins (MS3 modified). Adjust the pH of the solution to 5.7–5.8 using 1 M HCl or 1 M NaOH. Add 1.6 g agar into each of five bottles and add 195 mL of MS3 modified in each of the five bottles and shake. Autoclave at 121°C for 20 min (118 kPa steam pressure) and store at 4°C–10°C. Before use, rewarm MS3 modified (195 mL) until liquid and cool down to 60°C. Pour 5 mL water into a small beaker and add phytohormones (GA3, IAA, Zeatin riboside) and filter-sterilize (0.2 μm filter) into the bottle with MS3 modified and mix. Pour the solution into sterile Petri dishes, wait until it solidifies and seal with Parafilm. Store at 4°C in the dark for a maximum of 3 weeks.

3.2 Potato Material for Establishment of In Vitro Cultures

1. Sprout potato tubers in the darkness at room temperature to achieve long nodal segments.
2. Cut potato shoots, remove leaves, cut nodal segments into pieces of 2–3 cm and continue with surface sterilization.

3.3 Surface Sterilization

1. Prepare sodium hypochlorite solution (see Sect. 2.1.2) and add a drop of Tween 20.
2. Begin surface sterilization by shaking the explants in 70% ethanol for 10–20 s.
3. Transfer to sodium hypochlorite solution and shake for 20 min.
4. Wash explants with sterile water four to five times.
5. Transfer explants on the potato culture medium and cultivate them at 20°C–25°C and 16 h illumination of 60–80 µmol/ m^2/s (*see* **Note 3**).

3.4 Culturing of Potato In Vitro Plants and Cold-hardening

1. For micropropagation, apices and nodal segments of 2–3 cm were used (*see* **Note 4**, Fig. 3A, B).
2. Transfer to fresh culture medium (see Sect. 2.1.3) and expose explants to a 16 h photoperiod (60 µmol/m^2/s) at 22°C for 3–5 weeks (*see* **Note 3**).
3. Prior cryopreservation, in vitro plants are exposed to cold hardening either at 5°C in the dark or at 22/8°C for 8/16 h day/night cycle for 1–3 weeks.

3.5 Potato Shoot Tip Preparation

1. For potato, only the tips of in vitro plants are used for the preparation of explants for the cryopreservation.
2. Use apical shoot tips of well-developed in vitro plants and excise explants (0.5–1.0 mm thick and 2–3 mm long) under the dissection microscope (Fig. 3C).
3. Preculture shoot tip explants in 2 mL preculture medium (MS10, liquid). Seal Petri dishes with Parafilm and keep at 25°C in the dark overnight (Fig. 3D).

3.6 Cryopreservation

1. Loading phase: transfer explants into Petri dish containing 2 mL loading solution. Incubate at room temperature for 20 min.
2. Cryoprotection: transfer explants into 2 mL PVS3 solution. Shoot tips of potato are incubated at room temperature for 1.5–2 h. In the meantime, fill the box with liquid nitrogen, move cryo vials into the box, open vials, fill vials with liquid nitrogen and; prepare aluminum stripes (5 × 20 mm) and pipette ten droplets of 2 µL fresh PVS3 solution on the stripes (Fig. 3E). Precaution: Make sure that after 2 h, shoot tips are cooled immediately in liquid nitrogen.

Fig. 3 Plant material used and steps of PVS3 droplet vitrification. (**A**) Microtubers used for micropropagation of potato shoot tips. (**B**) Final in vitro propagation of 6 × 50 in vitro plantlets in plastic boxes. (**C**) Apical shoots before excision, and (**D**) excised shoot tips kept in preculture solution overnight. (**E**) Prepared aluminum stripe with fresh PVS3 droplets and (**F**) with excised shoot tips. (**G**) Direct cooling of shoot tips held on aluminum strips in cryo vials filled with liquid nitrogen and (**H**) continuous storage in liquid nitrogen afterwards. (**I, J**) Cryo boxes used in back up with labeled cryo vials and space for a microfilm holding passport and protocol data. (**K**) To estimate regrowth, shoot tips are rewarmed and survival is evaluated after 3–6 weeks and regrowth after 6–12 weeks. (**L**) If regrowth is ≥30%, accessions can be considered as safely preserved. (M) If the regrowth is <30%, cryopreservation procedure should be repeated (*see* **Note 6**)

3. After 1.5–2 h cryoprotection, immediately transfer shoot tips into fresh PVS3 droplets, one shoot tip for each droplet (Fig. 3F). Refill the cryo vials with liquid nitrogen. Transfer the aluminum stripe into the cryo vials, one stripe per vial (Fig. 3G). Keep stripes for few seconds below liquid nitrogen to adjust for temperature. Refill the cryo vials with liquid nitrogen.
4. Close the vials and plunge them into a Dewar vessel containing liquid nitrogen (Fig. 3H). Afterwards, keep the vials above or in liquid nitrogen and avoid any temperature increase!
5. Arrange cryovials for storage in the liquid nitrogen tank (Fig. 3I).

3.7 Rewarming and Regeneration

1. Rewarming and washing: prepare box filled with liquid nitrogen, and sterile beaker with 50 mL ambient washing solution. Transfer the cryovials from the cryotank or Dewar vessel into the box. Open the vials and refill with liquid nitrogen. Move aluminum foil with forceps into the warming solution and shake it so that the shoot tips will rewarm very quickly. Keep at room temperature and in the dark for 20 min and shake from time to time.
2. Regeneration: place shoot tips on regeneration medium. For regrowth of samples Petri dishes are put at 22°C for 1 week in the darkness and then with a 16 h photoperiod in lowlight.

3.8 Explant Regeneration

1. Count the survival of the explants 2–4 weeks after rewarming and calculate percentage of survival (*see* **Note 5**, Fig. 3K).
2. Count the regrowth of the explants 8–12 weeks after rewarming (*see* **Note 6**, Fig. 3L, M).

4 Notes

1. Murashige and Skoog medium can be prepared from single micronutrients and macronutrients and vitamins. For convenience and standardization, powdered medium sourced from a commercial supplier is recommended (Duchefa Biochemie BV, Haarlem, The Netherlands).
2. To support plant growth and propagation, a combination of 0.1 mg/L NAA (1-Naphthalenc acetic acid) and 0.5 mg/L 2iP(6-(y-y-Dimethylallylamino)purine) can be added to the media. Also, other hormone combinations were published and proved to be suitable under our conditions, such as 0.1 mg/L kinetin and 0.1 mg/L IAA (Indole-3-acetic acid) or 2 mg/L BAP(6-Benzylaminopurine) and 0.2 mg/L NAA.

3. Plants can be colonized by microbial endophytes, mainly bacteria, which cannot be removed by surface sterilization. The outbreak of these microorganisms is visible by white or yellowish exudates or as a more or less permanent tiny veil surrounding the basis of the explant. It can be either tolerated by the plants or contribute to its death. During subcultivation, microbial endophytes can accumulate and restrict the number of propagation cycles. Culture vessels showing high endophytic contaminations need to be removed.
4. To avoid high number of subcultivations, and hence, endophytic contamination, in vitro plantlets may derive from microtubers grown under low temperature and osmotic conditions. When microtubers start to grow after dormancy phase, micropropagate nodal segments (see Sect. 3.4).
5. Survival of the explants is characterized by yellowish-green color. Sometimes explants may swell or stop growing. Regrowth is characterized by the development of regularly formed shoots. Roots develop often after transfer to potato culture medium.
6. For secure storage, following criteria are applied at IPK. Overall, 310 explants are cryopreserved. Of these 210 explants are stored as triplicates (3×70 shoot tips) in cryo storage tanks. Additional 100 explants are used for regrowth control. If the control explants are fully contaminated after regrowth and/or regrowth is between 0% and <5%, another one to two vials are tested for confirmation; if the results are confirmed, the entire batch is removed from storage, and the accession is cryopreserved again using new and healthy in vitro plants. If the control explants show regrowth between 5% and <30%, it is considered as "preserved in cryo" (one star), but the accession is cryopreserved again using new and healthy in vitro plants until 30% regrowth is achieved. If the control explants show a minimum of 30% regrowth, it is considered as "save in cryo" (two stars). In this case, triplicates are divided into two storage tanks at IPK and one storage tank at the safety storage at DSMZ Braunschweig; at DSMZ passport data and protocols are printed on films and are stored beside cryomaterial. When the number of vials at IPK storage drops below 60 explants, cryopreservation needs to be repeated to add another 210 explants.

Acknowledgments

We like to acknowledge Angelika Senula, Anton Peterson, Rainer Vollmer, and Bart Panis for helpful comments and suggestions during the improvement of the method.

References

1. Nagel M, Dulloo E, Bissessur P, Gavrilenko T, Bamberg J, Ellis D, Giovannini P (2022) Global strategy for the conservation of potato. Global Crop Diversity Trust, Bonn, Germany. https://doi.org/10.5447/ipk/2022/29
2. Ovchinnikova A, Krylova E, Gavrilenko T, Smekalova T, Zhuk M, Knapp S, Spooner DM (2011) Taxonomy of cultivated potatoes (Solanum section Petota: Solanaceae). Bot J Linn Soc 165:107–155. https://doi.org/10.1111/j.1095-8339.2010.01107.x
3. Spooner DM, Ghislain M, Simon R, Jansky SH, Gavrilenko T (2014) Systematics, diversity, genetics, and evolution of wild and cultivated potatoes. Bot Rev 80:283–383. https://doi.org/10.1007/s12229-014-9146-y
4. Nagel M, Pence V, Ballesteros D, Lambardi M, Popova E, Panis B (2024) Plant cryopreservation: principles, applications, and challenges of banking plant diversity at ultralow temperatures. Annu Rev Plant Biol 75:797–824. https://doi.org/10.1146/annurev-arplant-070623-103551
5. Bajaj YPS (1978) Tuberization in potato plants regenerated from freeze-preserved meristems. Crop Improv 5:137–141
6. Schäfer-Menuhr A, Müller E, Mix-Wagner G (1996) Cryopreservation: an alternative for the long-term storage of old potato varieties. Potato Res 39:507–513. https://doi.org/10.1007/Bf02358469
7. Schäfer-Menuhr A, Schumacher HM, Gunda MW (1994) Langzeitlagerung alter Kartoffelsorten durch Kryokonservierung der Meristeme in flüssigen Stickstoff. Landbauforsch Volkenrode 44:301–313. https://www.openagrar.de/receive/timport_mods_00031848
8. Keller ERJ, Senula A, Grübe M, Diekmann K, Dehmer KJ (2014) Fifteen years of cryopreservation in the IPK genebank—experience, conclusions and outlook. Acta Hortic 1039:249–263. https://doi.org/10.17660/ActaHortic.2014.1039.32
9. Shida M, Ito J, Inoue Y, Hara S, Shirasuna K, Iwata H (2025) Dimethyl sulfoxide-induced DNA demethylation during vitrification of early cleavage-stage embryos and possible countermeasures. J Assist Reprod Genetics 42:1275–1286. https://doi.org/10.1007/s10815-025-03415-7
10. Köpnick C, Grübe M, Stock J, Senula A, Mock H-P, Nagel M (2018) Changes of soluble sugars and ATP content during DMSO droplet freezing and PVS3 droplet vitrification of potato shoot tips. Cryobiology 85:79–86. https://doi.org/10.1016/j.cryobiol.2018.09.005
11. Li X, Zhang W, Niu D, Liu X (2024) Effects of abiotic stress on chlorophyll metabolism. Plant Sci 342:112030. https://doi.org/10.1016/j.plantsci.2024.112030
12. Teixeira AS, Faltus M, Zamecnik J, Gonzalez-Benito ME, Molina-Garcia AD (2014) Glass transition and heat capacity behaviors of plant vitrification solutions. Thermochim Acta 593:43–49. https://doi.org/10.1016/j.tca.2014.08.015
13. Vollmer R, Villagaray R, Castro M, Anglin NL, Ellis DJPC (2019) Cryopreserved potato shoot tips showed genotype-specific response to sucrose concentration in rewarming solution (RS). Plant Cell Tiss Organ Cult 136:353–363. https://doi.org/10.1007/s11240-018-1520-8
14. Murashige T, Skoog F (1962) A revised medium for rapid growth and bio assays with tobacco tissue cultures. Physiol Plant 15:473–497. https://doi.org/10.1111/j.1399-3054.1962.tb08052.x
15. Senula A, Nagel M (2021) Cryopreservation of plant shoot tips of potato, mint, garlic, and shallot using plant vitrification solution 3. In: Wolkers WF, Oldenhof H (eds) Cryopreservation and freeze-drying protocols. Methods in molecular biology, vol 2180. Springer, pp 647–661. https://doi.org/10.1007/978-1-0716-0783-1_35
16. Weise S, Blattner FR, Börner A, Dehmer KJ, Grübe M, Harpke D, Lohwasser U, Oppermann M, Stein N, Willner E, Nagel M (2025) The German federal ex situ genebank for agricultural and horticultural crops—conservation, exploitation and steps towards a bio-digital resource centre. Genet Resour 2025(S2):91–105. https://doi.org/10.46265/genresj.GYDY5145

Check for updates

Chapter 17

Cryopreservation by Vitrification for the Aquatic Plant Family Lemnaceae

Anton Peterson, Olena Kishchenko, Ingo Schubert and Manuela Nagel

Abstract

Cryopreservation provides the conditions necessary for the long-term storage of plant germplasm, ensuring its viability and genetic stability over potentially limitless time periods. To enable regrowth of the propagated vegetatively plant germplasm, the plant material must undergo cryoprotection treatment before cryo-cooling and rewarming, followed by washing and revitalization procedures. Dehydration of plant tissue during cryoprotection often results in a substantial decrease or complete inhibition of regrowth after cryopreservation. Here, we describe a vitrification-based cryopreservation protocol for the genera *Spirodela*, *Landoltia*, *Lemna*, and *Wolffia*, which are mostly vegetatively propagating flowering plants belonging to the Lemnaceae family (duckweeds), and serve as a model for aquatic plants. Our protocol involves cryo-cooling whole duckweed plants and overcomes their inherent high desiccation sensitivity and exceptional buoyancy, which are characteristics of their aquatic lifestyle. Additionally, we describe the introduction and maintenance of duckweed in an aseptic in vitro culture. Overall, this protocol has been successfully applied to over 25 duckweed accessions from four genera.

Keywords Cryopreservation by vitrification, Dehydration, Duckweed, Lemnaceae, Light-induced damages, Pulsed illumination, Regrowth

1 Introduction

Long-term conservation of clonally propagating plant genetic resources still remains challenging due to genetic instability, the risk of loss during long-term field and in vitro propagation, and high workloads [1]. Cryopreservation, the introduction and storage of material at ultra-low (<−130°C) temperatures, overcomes these challenges and is considered to be the most suitable approach for long-term storage of clonal plants. Under ultra-low temperatures, specifically in the liquid or vapor phase of liquid nitrogen (<−196°C), the rate of biochemical processes tends to zero, and hence the plant germplasm can be stored for unlimited time periods. Two most common principles of plant cryopreservation demonstrated their efficiency for different types of plant materials: slow cooling and

Willem F. Wolkers and Harriëtte Oldenhof (eds.), *Cryopreservation and Freeze-Drying Protocols*, Methods in Molecular Biology, vol. 3060, https://doi.org/10.1007/978-1-0716-5416-3_17,

vitrification. For organized plant tissue, cryopreservation by vitrification has become the most often used technique nowadays due to high efficiency and independence from expensive equipment [2].

Vitrification is a process that allows inter and intracellular liquid phase transition to the glassy state, avoiding lethal ice crystal formation during cryo-cooling. Ultra-fast rewarming avoids ice recrystallization in the plant tissue after releasing vitrified plant material from cryo-storage.

Cryopreservation by vitrification requires the use of cryoprotectants to avoid ice crystal formation in plant tissues during cryo-cooling. This includes substances that dehydrate the vegetative plant material before cryo-cooling and increase the cellular solute concentration. Several plant vitrification solutions were proposed [3], and plant vitrification solution 3 (PVS3; 50% (w/v) sucrose, 50% (w/v) glycerol) [4] has become one of the most commonly used [5]. However, the water content of PVS3 is about 22%, whereas the critical relative water content for desiccation-sensitive vegetative tissue is often above 30% [6]. Thus, cryoprotection of desiccation-sensitive plant cells using PVS3 leads to substantial alteration of subcellular structures that could be irreversible and lead to loss of viability after cryopreservation [7]. Therefore, cryoprotection before cryopreservation requires additional protective measures to ensure the viability of plant vegetative germplasm after rewarming and rehydration.

Duckweed represents the smallest flowering plants, characterized by a reduced plant architecture. In nature, these free-floating freshwater plants flower very rarely and commonly propagate clonally. Duckweed species are considered to be a model for aquatic plants due to their simple in vitro cultivation, low demand for volume of cultivation, and illumination [8]. Using duckweed as a research model allows for substantial acceleration of experimentation through its high growth rate, as well as providing high interlaboratory reproducibility and traceability. The relatively small size and reduced architecture of duckweeds allow the use of whole plants for cryopreservation.

The following protocol for cryopreservation by vitrification of duckweed is based on a previously published method [9] and includes further developments increasing its efficiency, robustness, and versatility (Fig. 1). The protocol utilizes whole duckweed fronds or colonies and does not require time-consuming isolation of any specific plant parts or tissues. Further, the protocol includes special steps intended to compensate for the following: (i) the unpredictable relative water content resulting from the growth in liquid nutrition media that affect vitrification, (ii) the buoyancy of the free floating fonds that prevent complete immersion in the cryoprotectant solution, and (iii) the high desiccation-sensitivity of duckweed somatic tissue which causes dehydration-induced damage and inhibits regrowth of new plants from reproductive tissue of the cryopreserved plant.

The described protocol utilizes rich nutrition media (including glucose, protein hydrolysate, and vitamins). This bears the risk of

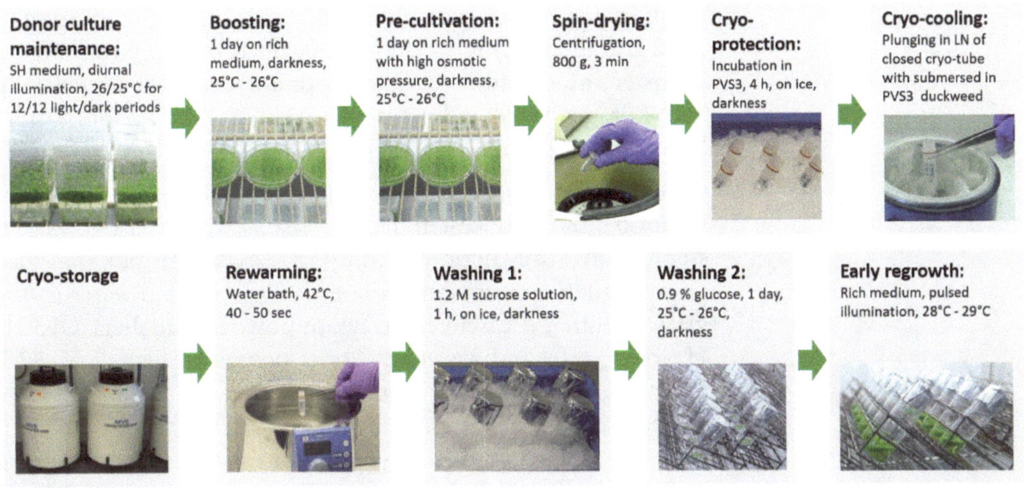

Fig. 1 Graphic presentation of the cryopreservation protocol for Lemnaceae (duckweeds). SH Schenk and Hildebrand nutrition medium, PVS3 plant vitrification solution 3, LN liquid nitrogen, Pulsed illumination (see Sect. 3.6 for details)

overgrowth by hidden endophytes, which do not reveal their presence during duckweed in vitro cultivation in conventional nutrition media, and may strongly inhibit or completely block regrowth of new plants after cryopreservation. Therefore, it is important to use as start material for cryopreservation confirmed aseptic in vitro cultures of duckweed. To obtain aseptic in vitro culture, protocols for surface sterilization of duckweed fronds, as well as cultural testing to prove the absence of interfering contaminant microbes, are included.

2 Materials

2.1 Stock Solutions

1. KOH solution for pH adjusting (A-solution): 1 M KOH. Put the magnetic bar in a 0.2 L glass. Add about 80 mL purified water to the glass. Weigh 5.6 g KOH and add it to the glass with the water. Dissolve the powder by stirring on a magnetic stirrer. Transfer the solution to a 0.1 L graduate cylinder and add purified water up to 100 mL. Transfer the solution to an appropriate plastic vessel with a hermetic lid for storage. Store at room temperature.
2. Glycerol/sucrose stock solution (3× GS-solution): 15% (v/v) glycerol, 410 g/L sucrose, sterile. Put a magnetic bar in a 1 L glass and put the glass on a magnetic stirrer. Add about 300 mL purified water to the glass. Weigh 205 g of sucrose and add it to the glass with the water. Dissolve the sucrose by stirring on a magnetic stirrer. Add 75 mL glycerol to the sucrose

solution and mix on a magnetic stirrer. Transfer the solution to a 0.5 L graduate cylinder and add purified water up to 500 mL. Transfer the solution to an appropriate vessel for autoclaving and storage. Sterilize by autoclavation, at 121°C for 20 min. Store at room temperature.

3. Casein hydrolysate/yeast extract stock solution (20× Pep-solution): 20 g/L casein hydrolysate, 20 g/L yeast extract, sterile. Put a magnetic bar in a 0.5 L glass and put the glass on a magnetic stirrer. Add about 150 mL purified water to the glass. Weigh 4 g casein hydrolysate powder (Duchefa C1301, Haarlem, NL) and 4 g yeast extract powder (Duchefa Y1333, Haarlem, NL) and add to the glass with the water. Dissolve the powder by stirring on a magnetic stirrer. Transfer the solution to a 0.25 L graduate cylinder and add purified water up to 200 mL. Transfer the solution to the appropriate vessel for autoclaving and storage. Sterilize by autoclaving, at 121°C for 20 min. Store at room temperature.
4. Glucose stock solution (20× Gluc-solution, 90 g/L glucose, sterile): 90 g/L glucose, sterile. Put a magnetic bar in a 0.5 L glass and put the glass on a magnetic stirrer. Add about 150 mL purified water 18 g glucose and add to the glass with the water. Dissolve the glucose by stirring on a magnetic stirrer. Transfer the solution to a 0.25 L graduate cylinder and add purified water up to 200 mL. Transfer the solution to an appropriate vessel for autoclaving and storage. Sterilize by autoclaving, at 121°C for 20 min. Store at room temperature.
5. Sugars/antioxidant buffered stock solution (2× SAB-solution): 10 g/L sucrose, 10 g/L glucose, 2 g/L fructose, 1 g/L polyvinylpyrrolidone (PVP 10,000), 2 g/L 2-(N-morpholino) ethanesulfonic acid (MES), pH 7.3–7.4 before autoclavation, sterile (*see* **Note 1**). Put a magnetic bar in a 1 L glass and put the glass on a magnetic stirrer. Add about 450 mL purified water to the glass. Weigh 5 g sucrose, 5 g glucose, 1 g fructose, 0.5 g PVP 10,000, 1 g MES, and add to the glass with the water. Dissolve the powder by stirring on a magnetic stirrer. Using an A-solution and a pH-meter, adjust pH to 7.3–7.4. Transfer the solution to a 0.5 L graduate cylinder and add purified water up to 500 mL. Transfer the solution to the appropriate vessel for autoclaving and storage. Sterilize by autoclaving, at 121°C for 20 min. Store at room temperature.
6. Liquid 2× concentrate of nutrition medium (2× SH-solution): 6.36 g/L Schenk and Hildebrand basal salt medium powder (Duchefa S0225, Haarlem, NL), 2020 mg/L Schenk and Hildebrand vitamin mixture powder (Duchefa S0411, Haarlem, NL), sterile [10]. Put a magnetic bar in a 1 L glass and put the glass on a magnetic stirrer. Add about 450 mL purified water to the glass. Weigh 3.18 g of the medium powder,

1010 mg of the vitamins powder, and add to the glass with the water. Dissolve by stirring on a magnetic stirrer. Transfer the solution to a 0.5 L graduate cylinder and add purified water up to 500 mL. Transfer the solution to an appropriate vessel for autoclaving and storage. Sterilize by autoclaving, at 121°C for 20 min. Store at room temperature.

7. Solid 2× concentrate of Schenk and Hildebrand nutrition medium (2× SHS): 6.36 g/L Schenk and Hildebrand basal salt medium powder (Duchefa S0225, Haarlem, NL), 2020 mg/L Schenk and Hildebrand vitamin mixture powder (Duchefa S0411, Haarlem, NL), 20 g/L agar, sterile. Put a magnetic bar in a 1 L glass and put the glass on a magnetic stirrer. Add about 450 mL purified water to the glass. Weigh 3.18 g of the medium powder, 1010 mg of the vitamins powder, and add to the glass with the water. Dissolve by stirring on a magnetic stirrer. Transfer the solution to a 0.5 L graduate cylinder and add purified water up to 500 mL. Transfer the solution to an appropriate vessel for autoclaving and storage. Weigh 10 g of agar and add it to the vessel with the medium. Sterilize by autoclaving at 121°C for 20 min. Mix vigorously (when hot) by hand shaking to dissolve the agar homogeneously. Store at room temperature.
8. 1.1× concentrate of Schenk and Hildebrand nutrition medium (SH5C-solution): 3.5 g/L Schenk and Hildebrand basal salt medium powder (Duchefa S0225, Haarlem, NL), 1111 mg/L Schenk and Hildebrand vitamin mixture powder (Duchefa S0411, Haarlem, NL), 5.56 g/L sucrose, pH 5.7–5.8 before autoclaving, sterile. Put a magnetic bar in a 1 L glass and put the glass on a magnetic stirrer. Add about 850 mL purified water to the glass. Weigh 3.18 g of the medium powder, 1111 mg of the vitamins powder, and 5 g of sucrose, and add to the glass with the water. Dissolve by stirring on a magnetic stirrer. Using an A-solution and a pH-meter, adjust pH to 5.7–5.8. Transfer the solution to a 1 L graduate cylinder and add purified water up to 900 mL. Transfer the solution to an appropriate vessel for autoclaving and storage. Sterilize by autoclaving, at 121°C for 20 min. Store at room temperature.

2.2 Working Solutions

1. Purified water, sterile. Aliquot purified water to appropriate vessels (usually 1 L glass laboratory bottle with screw cup type GL45) for autoclaving and storage. Sterilize by autoclaving, at 121°C for 20 min. Store at room temperature.
2. Surface sterilization solution for duckweed fronds (S-solution), non-sterile: 0.24% NaClO, ≈ 0.01% Tween 20. Add 1 mL NaClO solution (12% Cl, Art.-Nr. 90062.3, Carl Roth GmbH, Germany) to 50 mL sterile purified water and add one drop of Tween 20. Mix briefly manually. Use always freshly prepared.

3. Plant vitrification solution 3 (PVS3): 500 g glycerol, 500 g sucrose, 3.18 g Schenk and Hildebrand basal salt medium powder (Duchefa S0225, Haarlem, NL, USA), 290 mL purified water, pH 7.1–7.2 before autoclaving, sterile. Put a magnetic bar in a 1 L glass, and put the glass on a magnetic stirrer. Add 145 mL purified water to the glass. Weigh 1.59 g of the medium powder and add it to the glass with the water. Dissolve by stirring on a magnetic stirrer. Adjust pH to 7.1–7.2 with A-solution using a pH-meter. Weigh 250 g of sucrose and add it to the glass with the medium solution. Start dissolving sucrose by intensive stirring on magnetic stirrer for 3–5 min. Heat the suspension of sucrose crystals in a water bath at 75–90°C. Mix vigorously manually or use stirring on magnetic stirrer under heating for 1–2 min. Add 250 g of glycerol and continue vigorous mixing by hand shaking with periodic heating on a water bath to keep the temperature 75–90°C or use stirring on a magnetic stirrer with heating to dissolve sucrose crystals completely (10–15 min). Transfer the solution to an appropriate vessel for autoclaving and storage. Sterilize by autoclaving at 121°C for 20 min (*see* **Note 2**). Store at room temperature.
4. Washing solution 1 (W1-solution): 410 g/L sucrose, sterile. Put a magnetic bar in a 1.5 or 2 L glass and put the glass on a magnetic stirrer. Add about 600 mL purified water to the glass. Weigh 410 g of sugar and add it to the glass with the water. Dissolve on a magnetic stirrer. Transfer the solution to a 1 L graduate cylinder and add purified water up to 1 L. Divide aliquots of 10 mL into 50 mL glass tubes. Cover the tubes with foil cups and sterilize by autoclaving, at 121°C for 20 min. Store at room temperature.
5. Washing solution 2 (W2-solution): 9 g/L glucose, 1 g/L MES, 0.5 g/L PVP 10.000, pH 7.1–7.2 before autoclavation, sterile. Put a magnetic bar in a 1.5 or 2 L glass and put the glass on a magnetic stirrer. Add about 950 mL purified water, 9 g glucose, 0.5 g PVP 10,000, 1 g MES, and add to the glass with the water. Dissolve on a magnetic stirrer. Adjust the pH to 7.1–7.2 with A-solution using a pH-meter. Transfer the solution to a 1 L graduate cylinder and add purified water up to 1 L. Transfer the solution to an appropriate vessel for autoclaving and storage. Sterilize by autoclaving, at 121°C for 20 min. Store at room temperature.

2.3 Nutrition Media

1. Liquid nutrition medium [10] for routine in vitro duckweed culture maintenance (SH5L-medium): 3.18 g/L Schenk and Hildebrand basal salt medium powder (Duchefa S0225, Haarlem, NL), 1010 mg/L Schenk and Hildebrand vitamin mixture powder (Duchefa S0411, Haarlem, NL), 5 g/L sucrose, pH 5.7–5.8 before autoclavation. Put a magnetic bar

in a 1.5 or 2 L glass and put the glass on a magnetic stirrer. Add about 950 mL purified water to the glass. Weigh 3.18 g of the medium powder, 1010 mg of the vitamins powder, and add to the glass with the water. Weigh 5 g of sucrose and add it to the glass with the water and the medium powder. Dissolve by stirring on a magnetic stirrer. Adjust pH to 5.7–5.8 with A-solution using a pH-meter. Transfer the solution to a 1 L graduate cylinder and add purified water up to 1 L. Transfer the solution to the appropriate vessel for autoclaving and storage. Sterilize by autoclaving, at 121°C for 20 min. Store at room temperature.

2. Solid nutrition medium for duckweed growth after surface sterilization of the fronds (SH5S-medium): 3.18 g/L Schenk and Hildebrand basal salt medium powder (Duchefa S0225, Haarlem, NL), 1010 mg/L Schenk and Hildebrand vitamin mixture powder (Duchefa S0411, Haarlem, NL), 5 g/L sucrose, 10 g/L agar, pH 5.7–5.8 before autoclavation. Put a magnetic bar in a 1.5 or 2 L glass and put the glass on a magnetic stirrer. Add about 950 mL purified water to the glass. Weigh 3.18 g of medium powder, 1010 mg of the vitamins powder, and add to the glass with the water. Weigh 5 g of sucrose and add it to the glass with the water and the medium powder. Dissolve by stirring on a magnetic stirrer. Adjust pH to 5.7–5.8 with A-solution using a pH-meter. Transfer the solution to a 1 L graduate cylinder and add purified water up to 1 L. Transfer the solution to an appropriate vessel for autoclaving. Weigh 10 g of agar and add it to the vessel with the medium. Sterilize by autoclaving, at 121°C for 20 min. Cool down to 50–45°C, mix by manual shaking, and transfer 20 mL each to sterile Petri dishes of 9–10 cm diameter aseptically using a laminar cabinet. Seal the dishes with Parafilm. Store at room temperature.

3. Solid nutrition medium for checking absence of contaminants in aseptic duckweed in vitro culture (CASO–medium): 3 g/L CASO–Bouillon nutrition medium powder (X938.1, Carl Roth GmbH + Co. KG, Germany), 10 g/L agar, sterile. Put a magnetic bar in a 1.5 or 2 L glass and put the glass on a magnetic stirrer. Add about 950 mL purified water to the glass. Weigh 3 g of CASO–Bouillon nutrition medium powder and add to the glass with the water. Dissolve by stirring on a magnetic stirrer. Transfer the solution to a 1 L graduate cylinder and add purified water up to 1 L. Transfer the solution to an appropriate vessel for autoclaving. Weigh 10 g of agar and add it to the vessel with the medium. Sterilize by autoclaving, at 121°C for 20 min. Cool down to 45–50°C, mix manually, and transfer 20–25 mL each into sterile plastic Petri dishes of 9–10 cm diameter aseptically using a laminar cabinet. Seal the dishes with Parafilm. Store at room temperature.

4. Liquid nutrition medium for growth boosting of in vitro duckweed culture before cryopreservation (BL-medium): 3.18 g/L Schenk and Hildebrand basal salt medium powder (Duchefa S0225, Haarlem, NL), 1010 mg/L Schenk and Hildebrand vitamin mixture powder (Duchefa S0411, Haarlem, NL), 5 g/L sucrose, 4.5 g/L glucose, 1 g/L casein hydrolysate, 1 g/L yeast extract, sterile. This medium is prepared from sterile stock solutions aseptically, using a laminar cabinet. To a sterilized 1 L glass laboratory bottle with screw cup type GL45, add 900 mL SH5C-solution, 50 mL 20× Pep-solution, and 50 mL 20× Gluc-solution. Mix by manual shaking by hand. Store at room temperature.
5. Liquid nutrition medium for regrowth of duckweed after cryopreservation (RL-medium): 3.18 g/L Schenk and Hildebrand basal salt medium powder (Duchefa S0225, Haarlem, NL), 1010 mg/L Schenk and Hildebrand vitamin mixture powder (Duchefa S0411, Haarlem, NL), 5 g/L sucrose, 5 g/L glucose, 1 g/L fructose, 0.5 g/L PVP 10.000, 1 g/L MES, sterile. This medium is prepared from sterile stock solutions aseptically, using a laminar cabinet. To a sterilized 1 L glass laboratory bottle with screw cup type GL45, add 500 mL 2× SH-solution and 500 mL 2× SAB-solution. Mix by manual shaking and store at room temperature.
6. Solid nutrition medium for regrowth of duckweed after cryopreservation (RS-medium): 3.18 g/L Schenk and Hildebrand basal salt medium powder (Duchefa S0225, Haarlem, NL), 1010 mg/L Schenk and Hildebrand vitamin mixture powder (Duchefa S0411, Haarlem, NL), 5 g/L sucrose, 5 g/L glucose, 1 g/L fructose, 0.5 g/L PVP 10.000, 1 g/L MES, 10 g/L agar, sterile. This medium is prepared from sterile stock solutions aseptically, using a laminar cabinet. To a sterilized 1 L glass laboratory bottle with screw cup type GL45, add 500 mL of melted (in microwave oven) 2× SHS and 500 mL of 2× SAB-solution preheated to 50–45°C. Mix by manual shaking and transfer 20 mL each to sterile plastic Petri dishes of 9–10 cm diameter aseptically in a laminar cabinet. Seal the dishes with Parafilm. Store at room temperature.

2.4 Special Consumables

1. Single-use foil inserts, made in-house, sterile (*see* **Note 3** and graphic instruction on Fig. 2). Use aluminum foil of 0.03–0.05 mm thickness. Make a marking (with a needle or the sharp and thin end of small tweezers) on a piece of the foil to allow us to cut out the necessary amount of foil strips with ≈8 mm width and ≈ 65 mm length. Mark the central part of the piece ≈ 1 cm wide (Fig. 2A). Make perforations in the marked central part of the foil piece using a needle or the sharp and thin end

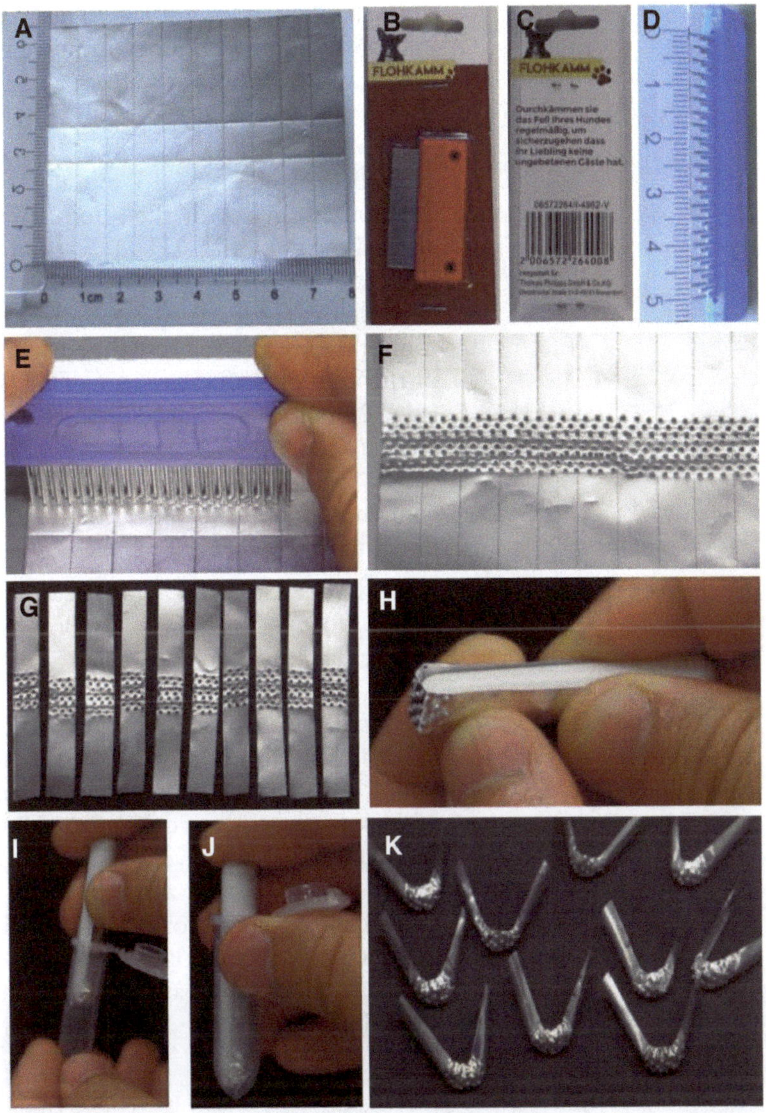

Fig. 2 Graphic instruction for in-house preparation of the single-use foil insert. (**A**) A piece of the foil with the marking made by a needle or the sharp and thin end of small tweezers. (**B**, **C**) Comb for animals in manufacturer's packaging, used in our laboratory to make the many-needled tool. (**D**) The result of making the teeth of the comb sharp and straightening them in slightly different directions. (**E**) Making the perforations in the foil using the many-needled tool. (**F**) The result of the perforation. (**G**) The strips with the perforations are cut according to the marking. (**H**) The magnetic stirrer bar is wrapped by the foil strip. (**I**) Insertion of the bar into the centrifugal test-tube. (**J**) Pressing the bar into the bottom of the centrifugal test-tube. (**K**) The single-use inserts are ready for sterilization and further use in cryopreservation

of small tweezers, to get 15–20 perforations per strip. To accelerate and simplify this operation, we made a special many-needled tool for this purpose from a comb for animals. We made the teeth of the comb sharp and straightened them in slightly different directions (Fig. 2B–F). Cut the strips using premade marked lines (Fig. 2G). Wrap a magnetic stirrer bar (7 mm) or another stick with the same diameter and U-shaped end around the strip (Fig. 2H). Insert the bar wrapped with the foil strips into the plastic centrifugal test-tube with a round bottom (Eppendorf type, 2.0 mL). Gently press the bar into the bottom of the centrifugal test-tube to obtain the "U-shape" form of the central perforated part of the strip inside the test-tube (Fig. 2I, J). Extract the preformed insert from the test-tube and pull the top edges apart to get the angle in between them of 20°–30° (Fig. 2K). Transfer the preformed foil insert to the container for autoclavation. Sterilize by autoclavation at 121°C for 20 min.

2. Swabs of 6–8 mm diameter made from autoclavable fibers, sterile (*see* **Note 4** and Fig. 3). Sterilize by autoclavation at 121°C for 20 min.

2.5 Plant Material and In Vitro Culture Maintenance

1. To obtain aseptic duckweed in vitro cultures, two subsequent rounds of sterilization within 2 days are needed (*see* **Note 5**). For all duckweed genera, except *Wolffia*, transfer several tenth of duckweed fronds or colonies to 50 mL of S-solution in sterilized 100 mL Erlenmeyer flasks aseptically in a laminar cabinet. Cover the flasks with a sterilized lid (made from aluminum foil). Shake manually or on an orbital shaker for 5 min. Wash with 50 mL portions of sterile purified water three times in a laminar cabinet. Incubate aseptically in 50 mL portions of sterile purified water for 2 days at 22°C in the dark.

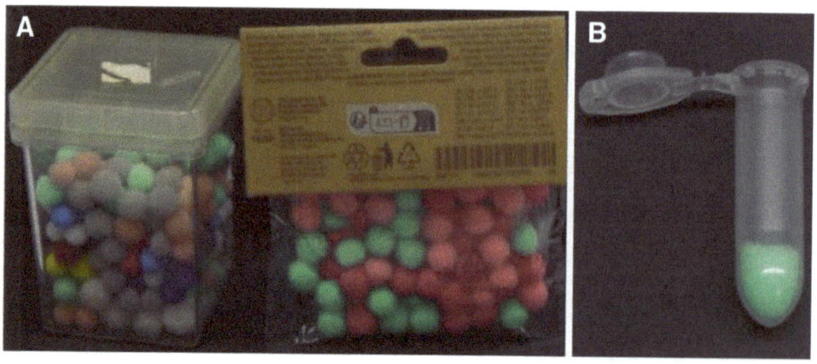

Fig. 3 Used in our laboratory, swabs for retaining liquids during the spin-drying of the duckweed fronds. (**A**) Example of swabs used in our laboratory (left: in a container after sterilization by autoclavation, right: in the manufacturer's package). (**B**) The swab in the centrifugal test tube

2. For genus *Wolffia*, transfer the fronds (in a laminar cabinet) to sterilized 2 mL plastic test-tubes (Eppendorf type) containing 1 mL of S-solution. Shake manually or on an orbital shaker for 5 min. Wash three times in a laminar cabinet with 1 mL portions of sterile purified water and incubate aseptically in 1 mL portions of sterile purified water for 2 days at 22°C in the dark.
3. After 2 days, perform the second round of sterilization using the same solution and conditions as for the first round. Blot duckweed fronds with sterile filter paper in a laminar cabinet. Transfer (in a laminar cabinet) duckweed fronds to Petri dishes containing solid SH5S-medium and seal the dishes with Parafilm. Incubate for 3 days at 22°C in the dark. Transfer the Petri dishes with the duckweed to the cultivation area with a 12/12 h light/dark illumination regime and a photosynthetic photon flux density (PPFD) of approximately 60 μmol $m^{-2} s^{-1}$ at 26/25°C for the light/dark periods. Cultivate until the accumulation of 5–10 normal and healthy fronds without any visible signs of microbial growth.
4. Transfer (in a laminar cabinet) the fronds or colonies to sterilized transparent plastic or glass cultivation containers containing approximately 150 mL SH5L-medium. Cover the containers with a sterilized lid that allows aseptic gas exchange. Transfer the containers to the cultivation area with a 12/12 h light/dark illumination regime and a PPFD of approximately 60 μmol $m^{-2} s^{-1}$, and 26/25°C for the light/dark periods. Cultivate the plants for 1 month until the next sub-cultivation passage.
5. For one sub-cultivation passage of aseptic duckweed in vitro culture, transfer (in a laminar cabinet) several tens of duckweed fronds or colonies to sterilized transparent plastic or glass cultivation containers containing approximately 150 mL SH5L-medium and continue as described in Step 4.
6. To check for the absence of endophytic biome in aseptic duckweed in vitro culture, transfer (in a laminar cabinet) several tens of duckweed fronds or colonies to 3–5 Petri dishes containing CASO–medium, and distribute them on the medium surface in a way that allows free space in between fronds or colonies. Seal the dishes with Parafilm, and transfer them to the cultivation area with a 12/12 h light/dark illumination regime and a PPFD of approximately 60 μmol $m^{-2} s^{-1}$, and 26/25°C for the light/dark periods. Cultivate for one week, and make sure that under a stereo microscope, there is no visible microbial growth.

2.6 Equipment

1. Dewar tanks and containers for liquid nitrogen handling and cryogenic storage.

2. Laminar cabinet.
3. Steam sterilizer.
4. Water bath.
5. Magnetic stirrer.
6. pH-meter.
7. Scales.
8. Centrifuge with rotor for 2 mL plastic test-tubes with round bottom (Eppendorf type).
9. Stereo microscope.
10. Cultivation chambers with suitable temperatures and illumination regimes. The illumination systems of the climate chambers, in addition to the diurnal regime, should be able to support two pulsed round-the-clock illumination regimes: (i) 15 min light impulses separated by 45 min dark periods and (ii) 30 min light impulses separated by 30 min dark periods. For both pulsed regimes, the temperature should be set at 28–29°C.

3 Methods

3.1 Growth Boosting of In Vitro Duckweed Culture Before Cryopreservation

1. Transfer (in laminar cabinet) 14 mL BL-medium to sterile plastic Petri dishes of 9–10 cm diameter.
2. Transfer a portion of duckweed fronds (or colonies) from the aseptic in vitro donor culture (*see* **Note 6**) to the Petri dish. Use enough fronds to cover the medium by 1–2 layers of fronds.
3. Seal the dishes with Parafilm.
4. Transfer the Petri dishes with the duckweeds to the incubation area at 25–26°C in the dark (*see* **Note 7**).
5. Incubate for 20–24 h.

3.2 Pre-Cultivation (Induction of Cryo-Tolerance)

1. After the boost is completed, transfer the Petri dishes with the duckweed to the laminar cabinet.
2. Add aseptically 7 mL 3× GS-solution to each dish and mix the medium in the dish by pipetting. During the manipulations, protect duckweeds as much as possible against visible light (*see* **Note 8**).
3. Immediately seal the dishes with Parafilm and transfer them to darkness.
4. Transfer the Petri dishes with duckweeds to an incubation area at 25–26 °C in the dark.
5. Incubate one day (or at least overnight).

3.3 Dehydration and Cooling

During the manipulations, use a laminar flow cabinet and protect duckweeds as much as possible against visible light as during the previous step.

1. Fill 1 mL sterile PVS3 into each sterile 2 mL-cryo-tube. Close, and cool down on ice.
2. Insert autoclaved swabs into the sterile plastic centrifuge tubes (2 mL-Eppendorf type) using sterile forceps. Push the swabs down to the bottom of the tubes.
3. Collect a portion of duckweed fronds from the donor culture (after boosting and pre-cultivation) into the sterile tubes, supplemented with a swab (*see* **Note 4** and Fig. 4), using autoclaved foil inserts (as a single-use sampler) and sterile forceps (*see* **Notes 3 and 9**).
4. Slightly roll and clamp the edges of the foil insert at the top of the tubes with sterile forceps to form the foil envelope with the fronds inside the tubes (*see* **Note 10** and Fig. 5).
5. Close the test-tubes, transfer them to the centrifuge, and provide the correct orientation of the tubes containing envelopes with duckweed fonds in the rotor to prevent a loss of fronds from the envelope during centrifugation (*see* Fig. 6).

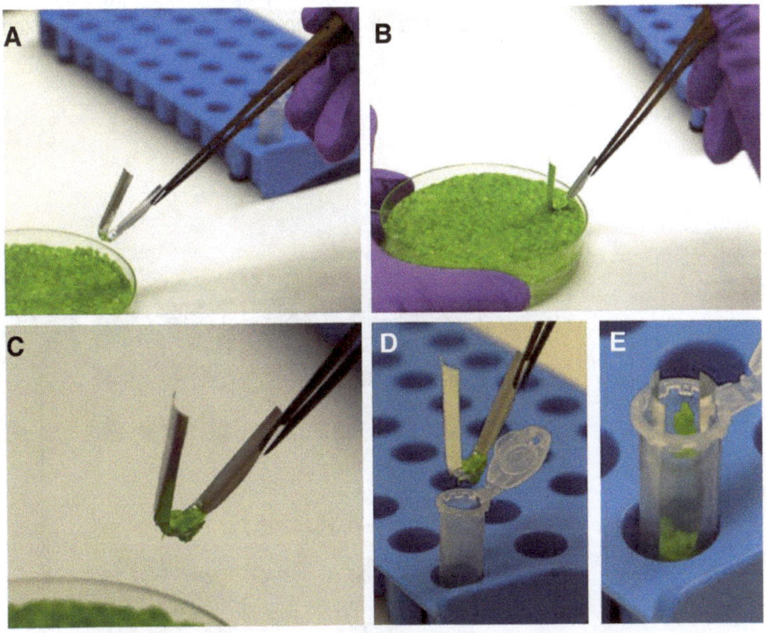

Fig. 4 Sampling of the duckweed fronds from donor culture for cryopreservation by a single-use foil insert. (**A**) Starting position, (**B**) sampling of the fronds from a Petri dish with the donor culture, (**C**) insert with a collected portion of fronds, (**D**) transferring of the sample into a centrifuge tube (with inserted swab), and (**E**) centrifuge tube with the insert and duckweed fronds

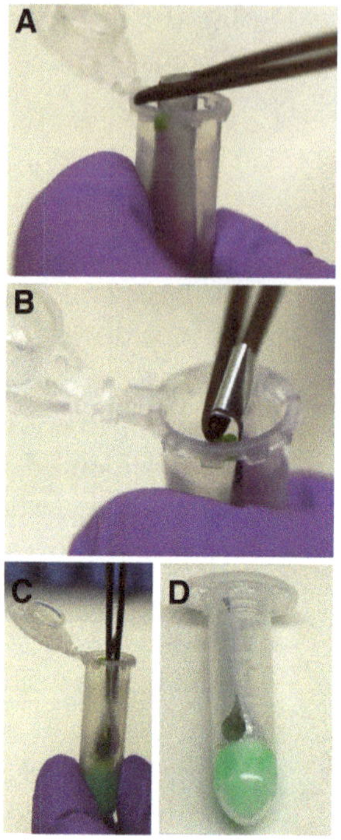

Fig. 5 Forming the foil envelope with the duckweed fronds inside the centrifuge tube. (**A**) Start of screwing the edges of the foil insert with fronds, (**B**) result of the screwing, (**C**) clamping the upper parts of the sides of the foil insert with the fronds to form the envelope, and (**D**) fronds packed in the foil envelope ready for spin-drying

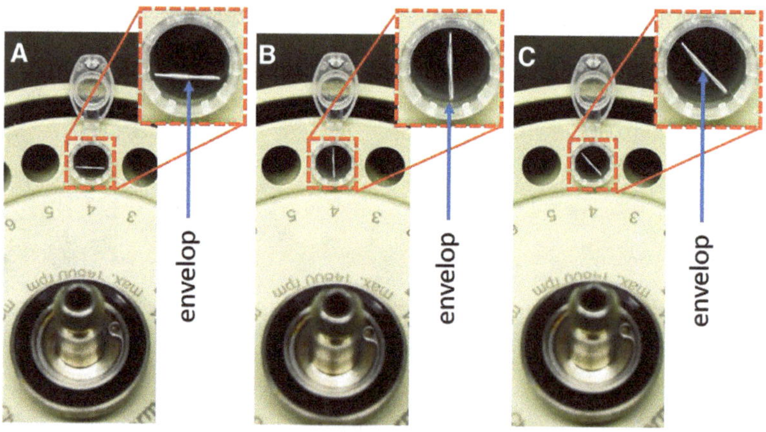

Fig. 6 Correct orientation of the envelope (inside the centrifuge tube) in the centrifuge rotor. (**A**) The correct orientation, (**B**, **C**) examples of incorrect orientations. The tubes are open to visualize the positions of the envelop inside the tube

6. Spin-dry the duckweed fronds by centrifugation at room temperature for 3 min and 800 × *g*.
7. Transfer the foil envelope with spin-dried fronds to precooled ice cryo-tubes filled with 1 mL of PVS3 (*see* Fig. 7).
8. Immediately close the cryo-tubes, and transfer it on ice in the dark.
9. Incubate for 4 h on ice in the dark.
10. Transfer the cryo-tubes with the duckweed to liquid nitrogen. Store the sample in the liquid or vapor phase of liquid nitrogen.

3.4 Rewarming, Washing, and Unloading

During all the manipulations, protect duckweed as much as possible against visible light, as before.

1. Precool and keep on ice a glass test-tube filled with 10 mL W1-solution.
2. Precool the W2-solution at 10°C.
3. Preheat a water bath to 42°C (*see* **Note 11**).
4. Take out the cryo-tube with the duckweed from liquid nitrogen, and immediately rewarm it by incubation in a water bath at 42°C for 40–50 s, keeping the lower part of the tube submersed in the bath (*see* **Note 12**).
5. Immediately after incubation at 42°C, briefly rinse the cryo-tube with 70% ethanol and open the cryo-tube in a laminar cabinet. Take out the foil envelope with plant material from the cryo-tube and disclose it by pulling the halves apart with two sterile forceps.

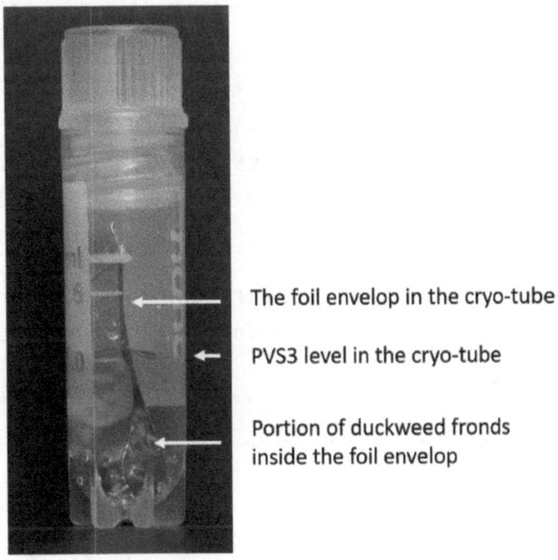

Fig. 7 The cryo-tube with the foil envelope containing a portion of duckweed fronds ready for the cryoprotection, with fronds completely immersed in PVS3

6. Immediately transfer the open foil envelope with duckweed to a pre-cooled (on ice) glass tube filled with W1-solution, and close the glass tubes with their cap (*see* **Note 13**).
7. Immediately transfer the glass tube to ice in the dark.
8. Incubate for 1 h on ice in the dark. During this time, most of the duckweed fronds should unstick from the surface of the foil and float on the surface of the solution.
9. Under the laminar flow cabinet, remove the W1-solution from the glass tube, leaving the duckweed fronds inside the tube, and add 10 mL of W2-solution (*see* **Note 14**).
10. Transfer the glass tubes with the fronds to an incubation area of 25–26°C in the dark.
11. Incubate there for 1 day in darkness.

3.5 Early Regrowth in the Dark

During all the manipulations, protect duckweeds as much as possible against visible light, as in Sects. 3.3 and 3.4

1. In a laminar cabinet, remove the W2-solution from the glass tube, leaving the fronds in the tube, and add 10 mL of RL-medium (*see* **Note 15**).
2. Transfer the tube to an incubation area at 28–29°C in the dark.
3. Incubate for 3 days.

3.6 Early Regrowth Under Pulsed Illumination

1. Transfer the tube with the duckweeds to an incubation area at 28–29°C under a pulsed illumination with PPFD during the light impulses of approximately 20 μmol m^{-2} s^{-1} (*see* **Note 16**). Duration of light impulses –15 min. Duration of dark intervals between light impulses –45 min ("15–45" regime).
2. Incubate for 3 days (*see* **Note 17**).
3. Transfer the tube to an incubation area that provides 28–29°C under a pulsed illumination with PPFD during the light impulses of approximately 20 μmol m^{-2} s^{-1}. Duration of light impulses—30 min. Duration of dark intervals between light impulses—30 min ("30–30" regime).
4. Incubate for 3 days.
5. Regrown plants can be maintained under normal cultivation conditions (*see* Step 5 in Sect. 2.5).

4 Notes

1. The mentioned pH range is necessary to reach a final pH diapason in RL-medium of 7.1–7.2, which is chosen for all media used during duckweed cryopreservation and early

regrowth. In nature, duckweeds live in nearly neutral pH environments. The pH value also affects the photosynthesis rate. A higher pH decreases the rate of photosynthesis [11]. An imbalance caused by cryopreservation of photosynthetic activity is harmful through increased ROS generation as a by-product. Thus, we assume, that decreasing photosynthesis through increasing the pH (typical plant culture media have pH 5.6–5.8) during cryopreservation and early regrowth may be helpful to reduce photosynthetic activity and thus to attenuate the stress.

2. Occasionally, PVS3 became light yellow to light brown after autoclavation. However, this coloration did not interfere with cryopreservation results and might result from a combination of certain lots of the medium powder, sucrose, and glycerol.
3. The water-facing surface of duckweed has hydrophilic properties and can transfer substantial amounts of liquid and dilute the PVS3 solution. This, in turn, could impair vitrification and should be avoided as far as possible. Therefore, it is necessary to remove the remnants of liquid nutrition medium from the frond surface before transferring it to PVS3 for dehydration. For this aim, we designed a single-use foil insert that (i) samples duckweed fronds from the donor culture, (ii) packages them for spin-drying to remove liquids, (iii) keeps duckweed fronds completely submersed in PVS3 preventing them from floating on the surface.
4. During spin-drying of fronds, it is necessary to keep the foil insert holding the fronds away from the bottom of the tube where the liquid is collected during centrifugation. To meet this requirement, we put swabs of 6–8 mm diameter made from autoclavable fibers at the bottom of the sterilized 2 mL plastic tubes. The liquid is sampled in these swabs during centrifugation, thus preventing the liquid from returning to the surface of the fronds.
5. Green tissue is very sensitive to sterilizing agents such as NaClO. Thus, a low concentration of NaClO (0.24%) was used for surface sterilization of duckweed fronds. A 2-day interval between sterilization rounds allows germination of spores that were not eliminated during the first sterilization round. The second round of sterilization ensures complete decontamination.
6. The duckweed fronds prepared for cryopreservation should be free of microbes. In most cases, surface sterilization (Steps 1–4 in Sect. 2.5) is sufficient. However, some microorganisms survive in the intercellular space, being partially protected against the sterilizing agent. Transfer of such contaminated duckweed culture on rich nutrition media containing glucose, amino acids, vitamins, and other growth factors (as used

in the regrow medium) may lead to accelerated growth of the endophytic microorganisms. This overgrowth interferes with the revitalization of the duckweeds after cryopreservation and inhibits the regrowth of the new fronds. Therefore, it is important to ensure the absence of endophytic organisms in the donor culture before cryopreservation. Step 6 of Sect. 2.5 provides the protocol for contamination evaluation. If contamination is still detectable, repeat sterilization according to Steps 1–4 in Sect. 2.5 to solve contamination problems.

7. Starting from the boosting stage, the plant material must be protected against the visible or low-energy light (even LEDs or on/off indicators) during and after cryopreservation. For this purpose, black plastic boxes with tight lids were used to light-protect plant material immediately after finishing every single operation and for all transportation.
8. During the manipulations, it is important to limit the exposure of the plant material to visible light (especially blue and red parts of the spectrum; direct sunlight or any kind of bright illumination) as much as possible. If applicable, substantially attenuate artificial illumination and use green monochrome illumination (from an LED source).
9. When sampling the duckweed fronds from the donor culture by single-use inserts, the fronds in the sample are not counted or weighted. Due to the small size and large number of fronds in the Petri dish with the donor culture, the sampling portion is in the range of 10–50 fronds. Some donor cultures with long roots or larger frond sizes (such as *Spirodela*) may require assistance by forceps during loading the fronds into the insert. When transferring duckweeds from Petri dishes with the donor culture to the centrifuge tubes, some fronds could stick to the outer surface of the insert. Also, during insertion into the tubes, some fronds could stick to the wall of the tube. These situations are acceptable, and no additional measures need to be taken against them.
10. During clamping of the upper sides of the insert (when forming the envelope), it is necessary to leave free access for PVS3 to the duckweed fronds inside the envelope, and the fronds should not be damaged mechanically by the clamping. Thus, the clamping should not involve the lower part of the insert, and the edges of the lower parts of the insert should not be tightened in a way that they block the aperture between the lower parts of the edges. The size of the free aperture between the edges should be adjusted according to the size of the fronds to avoid fronds floating out of the envelope during incubation in PVS3.
11. It is recommended to use a water bath in combination with a magnetic stirrer or with an integrated stirring option. The

heat exchange directly depends on the temperature difference inside and outside the cryo-tube and defines the speed of warming. Mechanical stirring of water increases the warming speed, which is very important to achieve a higher viability through omitting ice crystallization, and standardizes the warming process to reduce lot-to-lot variation in regrowth after cryopreservation.

12. Using a water bath without stirring option requires manual shaking of the cryo-tubes with duckweed fronds during rewarming. The shaking should be neither too slow nor too fast. Fast shaking may generate air gaps between the water and the tube, decreasing the heat exchange. Correct shaking increases the warming speed about twofold in comparison to warming without shaking.
13. Efficient unloading and washing of remnants of PVS3 is important. In some cases, fronds stick tightly to each other and/or the surface of the envelope. In this case, the unloading and washing are inefficient. To overcome this situation, it is recommended to take care visually that the majority of the fronds are separated from the envelope and shake the tube gently after transferring the envelope to the W1-solution, removing the fronds from the envelope.
14. Before substituting the W1-solution with the W2-solution, it is necessary to ensure that the fronds are separated from the surface of the envelope. When some fronds still remain on the foil surface, wash them away from the foil by gently pipetting the W1-solution in the tube.
15. The condition for early regrowth after cryopreservation is crucial for the success of cryopreservation. As duckweeds are free-floating, liquid media provide optimal regrowth conditions. Nevertheless, for revitalization and regrowth after cryopreservation the choice between liquid and solid media is species- or even accession-specific. In case of low regrowth or for accessions of *Spirodela* and *Wolffia* genera, parallel testing of regrowth on liquid RL-medium and solid RS-medium is recommended.
16. Initially, white fluorescent lamps were used. However, some experiments using white LED-based light sources revealed decreased regrowth after cryopreservation in comparison to white fluorescent lamps. Addition of PVP 10,000 to the W2-solution and the RL-medium (and the RS-medium) partially compensated for the mentioned decrease.
17. During revitalization and early regrowth, the combination of increased temperature (28–29°C) and pulsed illumination was designed to counteract the light-induced degradation of chloroplasts and the light-induced initiation of programmed cell death of cryopreserved fronds. This was necessary to

provide enough time for meristematic tissue to generate viable daughter fronds. The efficient duration of the treatments is species- and accession-specific. A period of 3 days was appropriate for most tested accessions. However, for certain accessions belonging to the genus *Wolffia*, the period of cultivation at this stage may need to be extended. The optimal duration should be tested out visually. The transfer of the material from the pulsed illumination "15–45" regime to "30–30" is reasonable when newly developed daughter fronds sprout from reproductive pockets of cryopreserved fronds as visible under the stereo microscope. New daughter fronds, developed under pulsed illumination "15–45" regime, typically look creamy to fresh creamy green and turn into light green, while the cryopreserved mother fronds appear deep green to gray–green, up to completely gray.

Acknowledgments

Philip Schwartz Initiative for researchers at risk by Alexander von Humboldt Foundation, grant numbers 0901323 and 0901324.

References

1. Panis B, Nagel M, Van den Houwe I (2020) Challenges and prospects for the conservation of crop genetic resources in field genebanks, in in vitro collections and/or in liquid nitrogen. Plants 9:1634–1656. https://doi.org/10.3390/plants9121634
2. Nagel M, Pence V, Ballesteros D, Lambardi M, Popova E, Panis B (2024) Plant cryopreservation: principles, applications, and challenges of banking plant diversity at ultralow temperatures. Annu Rev Plant Biol 75:797–824. https://doi.org/10.1146/annurev-arplant-070623-103551
3. Zamecnik J, Faltus M, Bilavcik A (2021) Vitrification solutions for plant cryopreservation: modification and properties. Plants 10:2623. https://doi.org/10.3390/plants10122623
4. Nishizawa S, Sakai A, Amano Y, Matsuzawa T (1993) Cryopreservation of asparagus (*Asparagus officinalis* L.) embryogenic suspension cells and subsequent plant regeneration by vitrification. Plant Sci 91:67–73. https://doi.org/10.1016/0168-9452(93)90189-7
5. Senula A, Nagel M (2021) Cryopreservation of plant shoot tips of potato, mint, garlic, and shallot using plant vitrification solution 3. In: Wolkers WF, Oldenhof H (eds) Cryopreservation and freeze-drying protocols. Methods in molecular biology 2180. Springer US, New York, NY, pp 647–661. https://doi.org/10.1007/978-1-0716-0783-1_35
6. Oliver MJ, Farrant JM, Hilhorst HWM, Mundree S, Williams B, Bewley JD (2020) Desiccation tolerance: avoiding cellular damage during drying and rehydration. Annu Rev Plant Biol 71:435–460. https://doi.org/10.1146/annurev-arplant-071219-105542
7. Stock J, Bräutigam A, Melzer M, Bienert GP, Bunk B, Nagel M, Overmann J, Keller ERJ, Mock HP (2020) The transcription factor WRKY22 is required during cryo-stress acclimation in Arabidopsis shoot tips. J Exp Bot 71:4993–5009. https://doi.org/10.1093/jxb/eraa224
8. Acosta K, Appenroth KJ, Borisjuk L, Edelman M, Heinig U, Jansen MAK, Oyama T, Pasaribu B, Schubert I, Sorrels S, Sree KS, Xu S, Michael TP, Lam E (2021) Return of the Lemnaceae: duckweed as a model plant system in the genomics and post-genomics era. Plant Cell 33:3207–3234. https://doi.org/10.1093/plcell/koab18
9. Peterson A, Kishchenko O, Kuhlmann M, Tschiersch H, Fuchs J, Tikhenko N, Schubert I, Nagel M (2023) Cryopreservation of duckweed

genetic diversity as model for long-term preservation of aquatic flowering plants. Plants 12:3302–3327. https://doi.org/10.3390/plants12183302
10. Schenk RU, Hildebrandt AC (1972) Medium and techniques for induction and growth of monocotyledonous and dicotyledonous plant cell cultures. Can J Bot 50:199–204. https://doi.org/10.1139/b72-026
11. Spilling K (2007) Dense sub-ice bloom of dinoflagellates in the Baltic Sea, potentially limited by high pH. J Plankton Res 29:895–901. https://doi.org/10.1093/plankt/fbm067

Check for updates

Chapter 18

Isochoric Freezing of Food Products

Alan Lenon Maida and Boris Rubinsky

Abstract

Isochoric freezing is an emerging preservation method that leverages the combined effects of constant-volume thermodynamics and the expansion of ice-Ih relative to liquid water to create a uniquely coupled set of physical conditions. This protocol provides a step-by-step approach for applying isochoric freezing to food products. The described method includes chamber preparation, sample loading, and monitoring of pressure–temperature conditions. Applications extend to food safety and quality preservation, with potential to reduce spoilage and extend shelf life.

Keywords Cryopreservation, Food preservation, Isochoric freezing

1 Introduction

Food preservation methods must balance reduced metabolism, microbial safety with retention of nutritional and organoleptic quality. Conventional refrigeration and freezing are widely used for food preservation; however, these methods often compromise food quality through microbial growth, enzymatic activity, and physical damage caused by ice crystal formation [1].

Isochoric freezing, in contrast, confines water and food samples within a rigid container of fixed volume. As ice nucleates and expands, the constant volume constraint elevates pressure, which depresses the freezing temperature and maintains a two-phase equilibrium of ice and liquid water [2, 3]. In a pressure–temperature phase diagram, isochoric freezing follows the liquidus line, where decreasing temperature corresponds to increased pressure, up to the triple point of water (Fig. 1). The fraction of ice formed under equilibrium conditions as temperature decreases below 0°C. Under isochoric conditions only a portion of the chamber volume crystallizes, leaving

Willem F. Wolkers and Harriëtte Oldenhof (eds.), *Cryopreservation and Freeze-Drying Protocols*, Methods in Molecular Biology, vol. 3060, https://doi.org/10.1007/978-1-0716-5416-3_18,

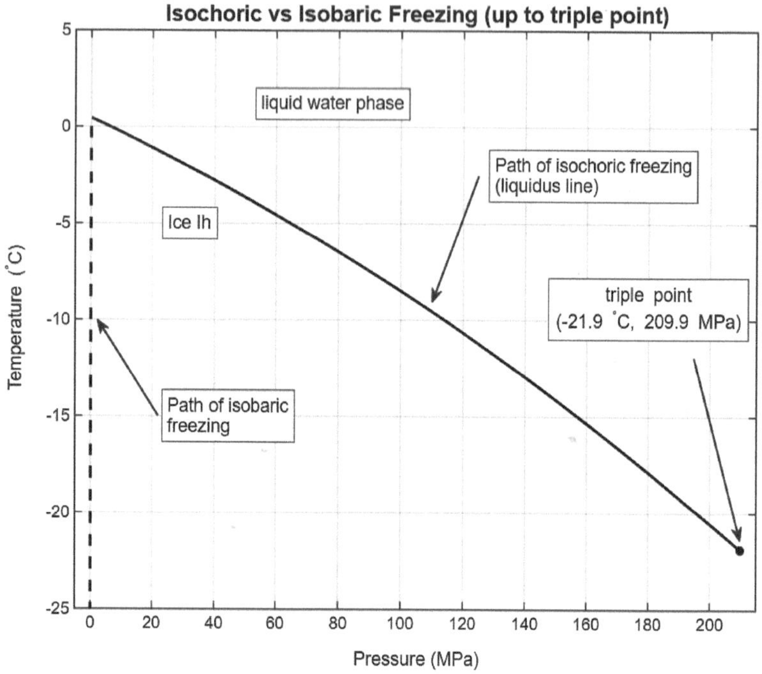

Fig. 1 Pressure–temperature phase diagram for isochoric versus isobaric freezing. Comparison of freezing under constant pressure (isobaric, dashed line) and constant volume (isochoric, solid line). Isochoric freezing follows the liquidus line, where decreasing temperature corresponds to increased pressure, up to the triple point of water (−21.9°C, 209.9 MPa)

sufficient liquid phase to store and protect food products (Fig. 2). This allows for food to be preserved in the unfrozen portion of the system at subfreezing temperatures and elevated pressure without suffering ice-related texture damage for sensitive fresh food products. Isochoric conditions have been successfully applied to meat, fruits, vegetables, and juices, maintaining better physicochemical and sensory properties than conventional freezing [4–12]. A rigid isochoric pressure vessel equipped with a digital pressure transducer can be used for isochoric freezing (Fig. 3). As temperature decreases, nucleation occurs at a designated site, producing ice that elevates internal pressure under constant volume conditions. The remaining unfrozen liquid maintains the food sample at subfreezing temperatures without ice formation.

This chapter provides a practical guide for setting up and performing isochoric freezing experiments on food products. It includes materials, stepwise procedures, and considerations for maintaining reproducibility.

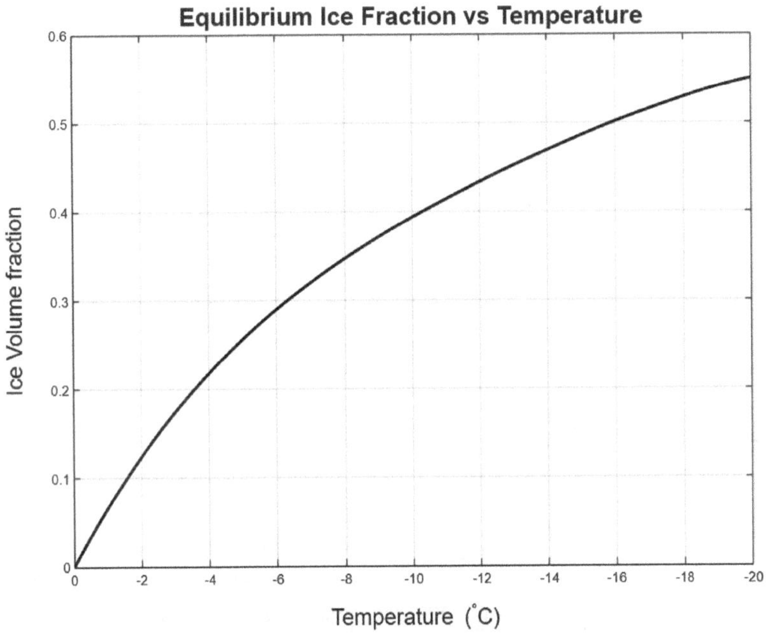

Fig. 2 Equilibrium ice volume fraction as a function of temperature in isochoric freezing. Graph showing the fraction of ice formed under equilibrium conditions as temperature decreases below 0°C. Ice fraction increases progressively, reaching ~0.55 at −20°C. Significance: Demonstrates that only a portion of the chamber volume crystallizes under isochoric conditions, leaving sufficient liquid phase to store and protect food products from structural damage associated with bulk ice formation

2 Materials

2.1 Equipment and Containers

1. Isochoric chamber: Rigid pressure vessel capable of withstanding the elevated pressures at the preservation temperatures. The liquidus line in a temperature pressure phase diagram for ice Ih and water provides the pressure for each storage temperature.
2. Temperature control system: Cooling bath or refrigerated circulator to maintain subzero conditions.
3. Pressure transducer: For continuous monitoring of chamber pressure.
4. Thermocouples/data logger: For monitoring internal and bath temperatures.
5. Temperature control: Refrigerated bath circulator or programmable cooling unit.
6. Safety: Pressure relief valve (rated <5 MPa).
7. Sterile sampling bags: Heat-sealed, food-grade bags for larger-volume isochoric freezing.

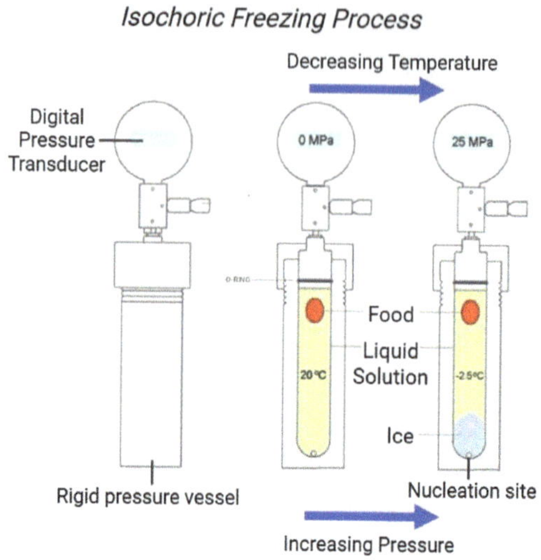

Fig. 3 Schematic of the isochoric freezing process. Illustration of a rigid isochoric pressure vessel equipped with a digital pressure transducer. As temperature decreases, nucleation occurs at a designated site, producing ice that elevates internal pressure under constant volume conditions. The remaining unfrozen liquid maintains the food sample at subfreezing temperatures without ice formation. Significance: Demonstrates how isochoric freezing stabilizes food products by combining low temperature and elevated pressure while minimizing/eliminating ice-induced damage

2.2 Food Samples

1. Raw food material (e.g., chicken breast, orange juice, cut potatoes).
2. Food-safe bag (e.g., polyethylene or barrier pouch) [optional for containment within chamber].

2.3 Solutions and Media

1. Water (H_2O) or tap water—used as a base or control medium in many isochoric freezing experiments.
2. Preservation solution-with similar osmolarity as food product:

   (a) Isotonic saline solution (e.g., 2.5%–3% NaCl w/v)—recommended for preserving meat or fish products to minimize osmotic imbalance.

   (b) Sucrose solution (e.g., 10%–15% w/v)—optional for preserving fruits, vegetables, or tissue with high sugar content to maintain osmotic balance.

2.4 Cleaning and Sterilization

1. 70% ethanol or bleach solution.
2. Autoclave (for reusable parts).
3. Filtered air drying rack.

3 Methods

3.1 Sample Preparation

1. Trim and weigh the food sample to fit into the volume of the isochoric chamber.
2. If using food bags, heat seal the sample in a food-grade pouch with no headspace.
3. For solid foods, vacuum seal in a food-grade pouch with no headspace.

3.2 Filling and Sealing the Isochoric Chamber

1. Fill the chamber completely with water or designated preservation solution (*see* **Note 1**), ensuring no air bubbles remain (*see* **Note 2**).
2. Optional—facilitating nucleation. To promote controlled ice nucleation, place one of the following nucleating agents at the bottom of the chamber, underneath or adjacent to the food product (*see* **Note 3**):
   (a) Small screw or bolt.
   (b) A tuft of steel wool.
   (c) A trace amount of silver iodide (AgI) (<0.01% w/v) dissolved into the fill solution.
3. Insert the food sample (directly or within a sealed pouch) into the chamber.
4. Gently tap and top off the chamber to ensure complete filling.
5. Seal the chamber using the lid.
6. Connect pressure and temperature sensors.
7. Inspect for leaks and tighten fittings as needed.

3.3 Isochoric Freezing and Storage

1. Place the sealed chamber in the temperature-controlled bath or cooling unit.
2. Decrease temperature to desired setpoint (e.g., −5°C, −10°C, or −15°C).
3. Monitor pressure rise due to ice formation. Record steady-state pressure and temperature (*see* **Note 4**).
4. Maintain isochoric conditions for the target duration (e.g., 1–7 days).

3.4 Post-Processing

1. After completion, slowly return the bath to ~5°C to allow gradual pressure release.
2. After reaching atmospheric pressures, disconnect sensors and unseal the chamber cautiously (*see* **Note 5**).
3. Remove the food product and inspect for structural integrity, color, and fluid separation.

4. Store samples at 4°C for short-term use or conduct physicochemical, microbial, or sensory testing (*see* **Note 6**).

4 Notes

1. Osmotic balance: When food is stored in direct contact with the preservation solution (not sealed), it is important to match the solution's osmolarity to that of the food product. For meats, this may involve using a mild saline solution (e.g., 2.5%–3% NaCl); for juices or plant tissues, sucrose or citric acid buffers may be used. Failure to match osmolarity can result in structural or chemical degradation due to water loss/gain or solute exchange [9–12].
2. Air removal is critical: Even small air bubbles can destabilize supercooling and reduce isochoric pressure buildup [13].
3. Facilitating ice nucleation: In some cases, it may be desirable to trigger ice formation at specific temperatures to prevent spontaneous supercooling. Positioning the nucleation aid at the bottom allows ice to initiate in a localized, predictable zone, reducing chance of ice damage to food products [2, 4].
4. Choice of temperature and time: Lower temperatures enhance microbial inhibition but increase pressure, ensure chamber rating is adequate. For sensitive food products, higher pressures might lead to changes in sensory properties, especially when stored for long periods of time.
5. Safety: Always use pressure relief valves when working above to prevent over pressurization.
6. Sample validation: For reproducibility, perform microbial plate counts, water activity, drip loss, or vitamin content post-treatment.

References

1. Maida AL, Bilbao-Sainz C, Karman A et al (2025) Low-pressure isochoric freezing as a scalable technique for fresh food preservation. Transl Food Sci 1:vxaf004
2. Rubinsky B, Perez PA, Carlson ME (2005) The thermodynamic principles of isochoric cryopreservation. Cryobiology 50:121–138. https://doi.org/10.1016/j.cryobiol.2004.12.002
3. Powell-Palm MJ, Aruda J, Rubinsky B (2019) Thermodynamic theory and experimental validation of a multiphase isochoric freezing process. J Biomech Eng 141:081011. https://doi.org/10.1115/1.4043521
4. Bilbao-Sainz C, Zhao Y, Takeoka G, Williams T, Wood D, Chiou BS, Powell-Palm MJ, Wu VCH, Rubinsky B, McHugh TH (2020) Effect of isochoric freezing on quality aspects of minimally processed potatoes. J Food Sci 85:2656–2664. https://doi.org/10.1111/1750-3841.15377
5. Bilbao-Sainz C, Olsen C, Chiou BS, Rubinsky B, Wu VCH, McHugh TH (2024) Benefits of isochoric freezing for carrot juice preservation. J Food Sci 89:1324–1336. https://doi.org/10.1111/1750-3841.16963
6. Lou L, Takeoka G, Rubinsky B, Bilbao-Sainz C (2024) Isochoric freezing to extend the shelf life

of pomegranate juice. J Food Sci 89:1347–1360. https://doi.org/10.1111/1750-3841.16941

7. Atci S, McGraw VS, Takeoka G, Vu VCH, McHugh TH, Rubinsky B, Bilbao-Sainz C (2024) Assessing the impact of isochoric freezing as a preservation method on the quality attributes of orange juice. J Food Sci 89:3167–3182. https://doi.org/10.1111/1750-3841.17071
8. Maida AL, Bilbao-Sainz C, Karman A, Takeoka G, Powell-Palm MJ, Rubinsky B (2023) Effects of isochoric freezing on the quality characteristics of raw bovine milk. Foods 12:4150. https://doi.org/10.3390/foods12224150
9. Bilbao-Sainz C, Millé A, Chiou BS, Takeoka GR, Rubinsky B, McHugh TH (2024) Calcium impregnation during isochoric cold storage to improve postharvest preservation of fresh blueberries. Postharvest Biol Technol 211:112841. https://doi.org/10.1016/j.postharvbio.2024.112841
10. Rinwi TG, Sun DW, Ma J, Wang QJ (2023) Effects of isochoric freezing on freezing process and quality attributes of chicken breast meat. Food Chem 405:134732. https://doi.org/10.1016/j.foodchem.2022.134732
11. Rinwi TG, Sun DW, Ma J, Wang QJ (2024) Enhancing frozen chicken meat quality by isochoric storage as compared with conventional frozen storage. LWT 208:116608. https://doi.org/10.1016/j.lwt.2024.116608
12. Bilbao-Sainz C, Sinrod AJ, Dao L, Takeoka GR, Williams TG, Wood DF, Ukpai G, Aruda J, Bridges DF, Wu VCH, Rubinsky B, McHugh TH (2018) Preservation of sweet cherry by isochoric (constant volume) freezing. Innov Food Sci Emerg Technol 52:108–115. https://doi.org/10.1016/j.ifset.2018.10.016
13. Perez PA, Preciado J, Carlson G, DeLonzor R, Rubinsky B (2016) The effect of undissolved air on isochoric freezing. Cryobiology 72:225–231. https://doi.org/10.1016/j.cryobiol.2016.04.002

Chapter 19

Cryopreservation of Rodent Ovarian Tissues and Oocytes

Yuksel Agca and Cansu Agca

Abstract

Germplasm cryobanking of rodents is a valuable tool for protecting scientifically important human disease models from genetic drift, genetic contamination, and loss of breeding colonies due to disease or catastrophic disasters to the housing facilities, as well as avoiding stress associated with domestic and international live animal shipment. Furthermore, germplasm cryopreservation enhances management efficiency by saving vivarium space, reducing workload for staff, financial burden of maintaining live animals, reducing the number of animals used to maintain a breeding colony, facilitating transportation of genetics by allowing distribution of frozen germplasm rather than live animals, which also reduces the risk of transfer of pathogens between facilities. In addition to spermatozoa and preimplantation embryos, effective long-term preservation of female germplasm within the ovarian tissues containing primordial follicles and mature oocytes via superovulation is critical for future reconstitution of scientifically important rodent strains for biomedical research. They also serve as a model to investigate and develop optimized oocyte and ovarian tissue cryopreservation and transplantation strategies for human infertility patients and family planning.

Keywords Cryopreservation, Mouse, rat, Oocytes, ovary, Vitrification

1 Introduction

Rodent genetic modification in combination with assisted reproductive technologies (ART) has opened many biomedical research opportunities as well as treatment options for solving the human infertility problem. Cryopreservation has become an integral part of these procedures, thereby providing significant logistical support to these research efforts [1, 2]. Since sperm is abundant and easy to collect, it has become the preferred method of cryopreservation for strains with single-engineered mutations on commonly used inbred strain backgrounds [3–5].

Embryo cryopreservation, nonetheless, is a required mode for preserving mutant rodent models that cannot be backcrossed due to complex genetic backgrounds or multiple engineered alleles [6, 7]. Female germplasm can be cryopreserved by either ovarian tissue

Willem F. Wolkers and Harriëtte Oldenhof (eds.), *Cryopreservation and Freeze-Drying Protocols*, Methods in Molecular Biology, vol. 3060, https://doi.org/10.1007/978-1-0716-5416-3_19,

(OT) containing primordial follicles or fully matured oocytes after superovulation by administering gonadotropins. The concept of female germplasm preservation via ovarian tissuecryopreservation and transplantation (OTCT) has a long history. The production of the first live offspring after orthotopic transplantation of fresh mouse ovaries was done [8]. This achievement, in combination with the discovery of glycerol as a cryoprotective agent(CPA), paved the way for transplantation of frozen–thawed OT in rodents [9]. Parkes and Smith [10] cryopreserved rat OT at ~80°C in glycerol and were able to restore the reproductive cycle.

A few other studies during the 1950s investigated the longevity of transplanted ovaries as well as the incidence of maintaining a regular reproductive cycle after exposure of rat OT graft to various glycerol concentrations and duration, and cooling rates [11, 12]. This research effort was complemented by the delivery of live offspring after ovarian bursa orthotopic transplantation of slowly frozen mice and hamster ovaries using glycerol [13, 14]. OTCT technologies in rodents had not received much attention for several decades. In the 1990s, dimethyl sulfoxide (DMSO), ethylene glycol (EG), and propylene glycol (PG) were found to have cryoprotective properties. Among these, DMSO as well as EG, and sometimes in combination with sucrose have been the most commonly used CPAs to cryopreserve rodent OT due to their higher permeation rates as compared to glycerol. OT has been suggested as an additional approach for preserving the female germplasm of GM rodent strains in circumstances such as sudden death, diseases, accidents, poor breeding, or reproductive performance.

Orthotopic transplantation of frozen-thawed OT was successfully used to rescue GM mice [15–20] and rats [21–23] by using syngeneic or athymic models [24, 25]. While the rodent OTCT approach provides a rederivation option for GM rodent strains, they also serve as models to investigate OT cryopreservation as a potential alternative for human infertility patients, for the development of improved OTCT strategies. Furthermore, rodent models are extremely useful for determining post-thaw OT graft sites as well as potential medical interventions to expedite angiogenesis, inhibit inflammatory/immune response, OT longevity, and follicular integrity [26, 27]. Although thousands of primordial follicles (PF) are present in the ovary, it is technically very challenging to recover them successfully and efficiently. Furthermore, the requirements for in vitro maturation conditions for the development of these PF to maturity are challenging. Thus, OTCT can offer a simple and straightforward approach for fertility restoration. Rodent modeling of OTCT is critical for the studies involving general and reproductive health since such experiments in humans have ethical concerns associated with experimenting on human oocytes and embryo and time constraints [28–32].

The achievements made in rodent models have already led to successful autologous orthotopic (pelvic cavity) transplantation of frozen-thawed OT and led to pregnancies and live births in humans via natural conception [33–36]. Similarly, there are other ARTs by which laboratory rodents serve as a model to restore endocrine and reproductive function; these ARTs include in vitro growth and maturation of ovarian follicles followed by IVF, microinsemination, enabling embryo production to transfer [37–39]. Heterotopic OT grafts are highly dependent on the determination of the optimal anatomical site, which would provide an angiogenic advantage for rapid vascularization [40]. Posttransplantation follicle viability of ovarian cortex is highly dependent on improving the revascularization with angiogenic and antiapoptotic factors, which can be studied in the use of rodent models [41].

Furthermore, the choice of site must also provide convenient access for monitoring or follicle isolation when needed, and also have to be transferable to human reproductive medicine. To this end, various sites (e.g., subcutaneous, omentum, renal capsule, neck muscle, intraperitoneal, and abdominal wall) have been experimented on rodents [39, 42–49]. Among these sites, SC transplantation requires minimal technical effort and may allow monitoring follicular development using currently available imaging techniques such as ultrasound. In addition, it would be possible to retrieve oocytes from the grafts, fertilize them in vitro, and transfer the resulting embryo back to the patient who received the OT autograft to restore fertility. Rodent studies have been a useful model to investigate premature ovarian failure [50], estrogen function in longevity [51], chemo-induced ovarian function [52], GnRH analog and chemo [53]. Pharmacological intervention (rapamycin) of the mTOR pathway, which involves follicle activation [26, 27, 40]. In addition, OT cryopreservation for future transplantation was used to investigate menopause or age-related metabolic and cognitive disorders [31, 54–56]. Rodents as model animals are invaluable to distinguish the impact of reproductive function or reproductive aging from the natural aging process.

OT cryopreservation from oncology patients before cancer treatment, and subsequent autotransplantation after remission, may provide future fertility options and prevention of conditions such as osteoporosis, obesity, and heart disease caused by menopause [36, 41, 57–63]. The safety, effectiveness, and longevity of cryopreserved OT for maintaining the endocrine and oocyte development posttransplantation are important factors. Thus, it is imperative that patients with hematological diseases, such as leukemias, be treated with caution because of the potential presence of malignant cells in the OT [64]. These parameters are difficult to assess directly in humans due to ethical and safety problems associated with these potential reproductive therapeutic approaches. Thus, experimental rodent studies can be designed to address this concern and obtain valuable information to determine the safety of OT transplantation as an option for human patients with gonadal dysfunction. A series of immunohistochemistry and molecular bioassays can be performed on the OT after xenotransplantation into immune-compromised mouse models, and clinically detect the presence of residual malignant cells in the OT. Another approach before cryopreservation and transplantation is "ovarian purging," by which treatment of OT with inhibitors that affect the malignant cells primarily, without damaging the follicles and stromal compartments [65, 66]. Few studies on leukemia patients demonstrated that in vitro treatment of the OT with tumor-specific inhibitors effectively reduced cancer cells, suggesting a potential method for minimizing the risk of reintroducing tumor cells [66–69].

One of the limitations of the use of fresh oocytes for IVF cryo-recovery (IVF-CR) is that it requires superovulation procedures and involves weekend/holiday injections. Cryopreservation via vitrification is the ideal method for preserving Metaphase II (MII) rodent oocytes since it is very effective and easy [70–72]. Oocyte cryopreservation significantly reduces operating costs, animal shipping costs, and other logistical difficulties. For instance, IVF-rederivation can begin immediately without delay due to lengthy and costly animal ordering or maintaining in-house colonies for oocyte production. Thus, oocyte cryobanking provides a long-term supply for IVF-CR procedures with existing cryopreserved mutant rodent sperm in repositories, institutional, and corporate transgenic core facilities. It would also eliminate the need for repeated live animal shipments and housing costs, and allow IVF-CR without any prior scheduling or ordering of donor females. In addition, it would enable testing of each batch of oocytes for developmental competency for successful live births and reduce turnover time and operating expenses. It is especially valuable because cryopreserved sperm and oocytes can be used for instant IVF-CR to reestablish a breeding colony and can be an integral part of the overall success of genome resources. It can be applied to cryobanking of mutant rodent models, which would significantly shorten the time necessary for the generation of double-transgenic rodents from cryopreserved gametes. It would allow the resuscitation of two distinct models in a single IVF attempt, eliminating the need for first-generation crosses and shortening the breeding period. One can also generate homozygous transgenic rodents that can be used to regenerate homozygous knockout mice, as well as the combination of different transgenes. This can also be applied to the Cre-Lox system and other instances where two different transgenes in cryopreserved sperm and oocytes can generate double-transgenic rodents without the need for mating mature animals, significantly shortening the required time for generations. Recently, a novel anti-inhibin monoclonal antibody (AIS) ultra-superovulation method allowed large-scale rodent oocyte and embryo cryobanking [70, 72]. Superovulation efforts using AIS on major inbred (C57BL/6) mouse strains resulted in up to two to three times more oocytes than equine chorionic gonadotropin (eCG)

[73–79]. It has also been reported for wild strains of subspecies of *Mus musculus*, that the number of oocytes obtained from AIS superovulation is significantly higher than eCG [80, 81]. Thus, AIS superovulation, cryopreservation, and subsequent in vitro embryo production from these unique rodent strains would provide wider research opportunities. AIS superovulation also found application in rats as well [82–85]. The AIS superovulation approach would also fulfill one of the 3Rs (Replacement, Reduction, and Refinement) of live animal use [86]. This chapter provides a detailed protocol for mouse and rat oocyte vitrification as well as OTCT for these species.

2 Materials

2.1 Equipment, Materials

1. Laminar flow hood.
2. Protective equipment (nitrile gloves, masks, and lab coats).
3. Cryoprotective gear: cryo-apron, cryo gloves, and safety goggles.
4. Cryo-label printer, cryo-canes, cryo-cane sleeves, and cryo-cane goblets.
5. Programmable freezer.
6. 0.25 cc and 0.5 cc freezing straws.
7. Petri dishes (35 mm).
8. Humidified CO_2 incubator (37°C, 5% CO_2, 95% air).
9. Stereomicroscope with fiber-optic light source.
10. Slide warmer adjusted to 37°C.
11. Water bath.
12. Precision balance.
13. Appropriately sized Griffin beaker and volumetric flask, stir bar, weigh paper, and spatula.
14. Osmometer and pH meter.
15. Heat sealer.
16. Timer.
17. VersiDry Lab Soaker for benchtop and Kimwipes.
18. Pipetman and pipette tips (20, 200, and 1000 μL capacity).
19. Micropipette.
20. Mouth pipette.
21. Glass capillaries for embryo handling.
22. Filter unit (0.22 μm pore size).
23. Precision microscissors and forceps.
24. Precision tweezers.

25. Small straight scissors.
26. Curved-tip tweezers.
27. Curved, serrated forceps, hemostats.
28. 1 cc syringe with a Luer-slip connection and 27 g needle.
29. Benchtop dewar flasks (2 L capacity).
30. VersiDry Lab Soaker for benchtop.
31. 70% Ethanol in a spray bottle.
32. Kimwipes.
33. Microcentrifuge tubes (1.5 mL).
34. Microscope slides and fire-polished Pasteur pipettes.
35. Methylene blue.

2.2 Reagents

1. Milli-Q water.
2. Saline solution.
3. Dulbecco's Phosphate Buffered Saline (DPBS without calcium chloride and magnesium chloride).
4. Heat-treated fetal calf serum.
5. Liquid mineral oil.
6. Hyaluronidase.
7. Cryoprotectants: dimethyl sulfoxide (DMSO) and ethylene glycol (EG), and sucrose.
8. Anti-inhibin Serum (Kyudo Co., Ltd., CARD HyperOva®).
9. Equine chorionic gonadotropin (eCG).
10. Human chorionic gonadotropin (hCG).

2.3 Surgical Materials for Ovarian Tissue Transplantation

1. Sterile surgical gloves and nitrile gloves.
2. Ketamine HCI and Xylazine.
3. Flunixin meglumine.
4. Ophthalmic ointment.
5. Epinephrine.
6. Electric shaver.
7. Surgical scissors, spring-type microscissors.
8. 2 × 2 inches cotton gauze and eye spears.
9. Betadine solution.
10. Curved, serrated forceps, fine forceps, and a bulldog clamp.
11. Beed sterilizer.
12. Scalpel holder and blade.
13. Absorbable suture materials (Vicryl® 4-0 and 8-0)
14. Castroviejo needle holder.
15. Autoclip applier and remover (9 mm).

3 Methods

3.1 Preparation of Oocyte Handling, Washing, and Holding Medium

FHM medium is modified KSOM, where part of the bicarbonate is replaced with HEPES buffer. It is often used for washing and handling mouse embryo on the lab bench [87]. Similarly, mR1ECM–HEPES is a modified version of the R1ECM medium, which also contains HEPES for handling rat oocytes and one-cell embryo on the bench [88].

1. FHM Flushing and holding medium for mice material (see Table 1):
   (a) Bring an appropriate size Griffin beaker and pour embryo-tested Milli-Q water (~80% of the total volume of media), and then add the components as listed in Table 1. All the recipes are calculated based on a 1-L final volume, so adjust the volume of the media for your needs (*see* **Note 1**).
   (b) After all components are completely dissolved, transfer the contents to the appropriate volumetric flask and bring to

Table 1
Composition of flushing and holding medium

Components	FW (g)	Amount per L	Concentration [mM]
$CaCl_2$–$2H_2O$	147.02	0.251 g	1.71
KCl	74.55	0.1864 g	2.5
NaCl	58.44	5.5518 g	95
$NaHCO_3$	84.01	0.42 g	4
D-glucose	180.2	0.036 g	0.2
Sodium lactate (60% syrup)	186.8	1.740 mL	11.77
Sodium pyruvate	110	0.022 g	0.2
Penicillin G potassium salt	372.2	0.06 g	n/a
Streptomycin sulfate	1457	0.0125 g	n/a
KH_2PO_4	136.08	0.0476 g	0.35
$MgSO_4$–$7H_2O$	246.47	0.0493 g	0.2
EDTA	292.24	0.0037 g	0.01
HEPES	238.31	4.7662 g	20
Phenol red (0.5%)	354.38	0.66 mL	n/a
GlutaMAX-1		5 mL	1.0
MEM EAA 50×		10 mL	n/a
MEM NEAA 100×		5 mL	n/a

the desired volume by rinsing out the Griffin beaker with embryo-tested Milli-Q water and adding it to the volumetric flask (*see* **Note 2**).

(c) Add 4.0 g/L Bovine Serum Albumin Fraction V into the stock solution and let it dissolve slowly (*see* **Note 3**).

(d) Measure the osmolality (270–290 mOsm) and pH (~7.2), and record and filter into an appropriately sized sterile bottle using 0.22 μm pore size filters.

(e) Aliquot into desired volume, label the tubes with make and expiration date (3 months from make date), initials, and batch number.

(f) Store in −20°C freezer for up to 3 months.

2. mR1ECM-HEPES stock solution (see Table 2A):

(a) Fill an appropriate-size Griffin Beaker with ~800 mL (~80% of the final volume) of embryo-tested, ultrapure culture-grade water, and then add the components as listed in Table 2A.

(b) After all components are dissolved, transfer the contents to a 1 L volumetric flask and bring to the desired volume by rinsing the Griffin Beaker with embryo-tested water and adding it to the volumetric flask. Measure the osmolality, record, and filter into an appropriately sized sterile bottle using a 0.22 μm filter.

(c) Label the bottle with the date made and the expiration date.

(d) Store at 4°C for about 4 weeks ± 1 week.

3. mR1ECM–HEPES for rat oocyte holding and washing (see Table 2B):

(a) Fill an appropriate-size Griffin Beaker with ~500 mL (~50% of the final volume) of embryo-tested water, and then add the components as listed in Table 2B.

(b) After all components are dissolved, check to ensure the pH of the solution is 7.4 ± 0.2 and adjust if needed. If any precipitates form in the solution, discard the solution and start over.

(c) Transfer the contents to a 1 L volumetric flask and bring to the desired volume by rinsing the Griffin beaker with embryo-tested water and adding it to the volumetric flask.

(d) Check the osmolality of the solution and remake if it does not fall between 290 and 320 mOsm.

(e) Sterilize the solution using a 0.22 μm filter unit.

(f) Label the bottle with expiration date, pH, osmolality, initials, and batch number.

Table 2
(A) Composition of stock solution for mR1ECM–HEPES. (B) Composition of mR1ECM–HEPES for rat oocyte holding and washing

Components	FW (g)	Amount per L	Concentration [mM]
(A)			
NaCl	58.44	46.7520	80
KCl	74.55	2.3856	3.2
D-Glucose	180.2	13.5150	7.5
Penicillin G K Salt	372.2	0.7500	100 μg/mL
Streptomycin Sulfate	1457	0.5000	50 μg/mL
Sodium Lactate (60% syrup)	186.8	25.2740	13.53
$CaCl_2$–$2H_2O$	147.02	2.9404	2
$MgCl_2$–$6H_2O$	203.31	1.0166	0.5
(B)			
$NaHCO_3$	84.01	0.4200 g	5
HEPES	238.31	5.2428 g	22
Sodium Pyruvate	110	0.0552 g	0.5
NaCl	58.44	1.7532 g	80
MEM NEAA 100×		10 mL	0.5×
MEM EAA 50×		20 mL	0.5×
GlutaMAX-1		0.500 mL	0.1
Rat Stock solution		100 mL	

(g) Store at 4°C for up to 2 weeks.

(h) Prepare mR1ECM–HEPES working solution by adding 4 mg/mL fatty acid-free BSA to an aliquot/the amount needed. Place the tube in a 37°C water bath to allow the BSA to dissolve. Sterilize using a 0.22 μm filter before use.

3.2 Oocyte Vitrification Solution

For the oocyte vitrification and rewarming procedure, base solution, pre-vitrification solution, vitrification solution, and dilution solution need to be prepared (see Table 3).

1. Base solution for oocyte vitrification procedure: DPBS + 20% heat-treated FCS
2. Oocyte pre-vitrification solution: 7.5% ethylene glycol (EG) + 7.5% DMSO in DPBS base solution

Table 3
Composition of oocyte pre-equilibration and vitrification solution

Solutions	Components	Amount per 100 mL base solution
Base solution	DPBS + FCS	80 mL + 20 mL
Pre-vitrification solution	DMSO + EG	7.5 mL + 7.5 mL
Vitrification solution	DMSO + EG + sucrose	15 mL + 15 mL + 17 g
Dilution solution	Sucrose	34 g

DPBS: without Ca^{2+} and Mg^{2+}

3. Oocyte vitrification solution: 15% ethylene glycol (EG) + 15% DMSO + 0.5 M sucrose in DPBS base solution
4. Oocyte post-warming dilution solution: 1 M sucrose in DPBS base solution without Ca^{2+} and Mg^{2+}

3.3 Superovulation, Recovery of Oocyte–Cumulus Complexes

1. Inject 3–4-week-old female donors with equine chorionic gonadotropin (eCG, 5 IU for mice and 25 IU for rats IP). Alternatively, donors can be ultra-superovulated via 100 μL AIS, followed by IP injection of human chorionic gonadotropin (hCG, 5 IU for mice and 25 IU for rats) 48 h post-eCG or AIS injection using a 27-g needle.
2. At 14–15 h post-hCG injections, humanely sacrifice female mice or rats using cervical dislocation only, as euthanasia via CO_2 inhalation has adverse effects on the fertility of oocytes.
3. Place scissors and fine tweezers on the bench, and always wear gloves.
4. Apply 70% ethanol to the abdomen and cut the abdominal skin and muscle of the mice using scissors.
5. Pull both edges of the cut skin and expose the peritoneum.
6. Find the uterine horns and oviducts, and ovaries by shifting the position of the gut and internal organs.
7. Gently grab the utero-tubal junction with a pair of fine tweezers, cut the utero-tubal and isthmic–ampullary junctions, and dissect the oviducts from the uterus.
8. Eliminate the oviducts from the fat tissue and blood and place them into a 1 mL petri dish containing warm mineral oil and keep them at 37°C (*see* **Note 4**).
9. Using fine forceps, remove each oviduct from the petri dish and transfer the edges of 100 μL prewarmed flushing and holding medium (FHM, mice) or HEPES-buffered rat one-cell embryo culture (R1ECM–HEPES, rats) drops containing 1 mg/mL hyaluronidase under mineral oil.

10. Using fine forceps, place each oviduct next to a hyaluronidase drop while depressing the oviduct, tear each oviduct to release clutches of oocytes using a 27-g needle and drag them into the hyaluronidase drop and hold COCs until cumulus cells are detached from the oocytes (*see* **Note 5**, Fig. 1).
11. Pipette the cumulus-free oocytes using a fine glass pipette, and wash in FHM (mice) or R1ECM-HEPES (rats) medium three times before equilibrating the oocytes in pre-vitrification solution (Fig. 1). Count the number of oocytes from each donor after removal of morphologically abnormal oocytes, as they are distinguished by cytoplasmic leakage and fragmentation.

3.4 Oocyte Vitrification Procedure

1. Using a fine glass pipette, transfer 20–30 oocytes into 35 mm petri dishes containing 3 mL pre-equilibration solution for 5 min at room temperature (*see* **Note 6**).
2. During the last minute of pre-equilibration, use a 1 cc syringe with a Luer-slip connection and load a pre-labeled 0.25 mL French straw with ~65 mm column of 1 M sucrose solution first, then sandwich ~25 mm vitrification solution between ~7 mm air, and finally 5 mm vitrification solution. Leave a ~5 mm air space at the tip of the straw for heat-sealing (*see* **Note 7**, Fig. 2A).
3. Load the pre-equilibrated oocytes into a fine glass pipette and carefully insert the tip into the open end of the straws. Then gently deposit the oocytes into the 25 mm vitrification solution column, quickly heat-seal, hold for 30 s, and immediately plunge the straw into LN2-containing dewar flasks (*see* **Note 8**, Fig. 2B).
4. Once all the oocytes are vitrified, transfer an empty cryo cane with plastic goblets and hold for a few minutes until the goblet is filled with LN2.

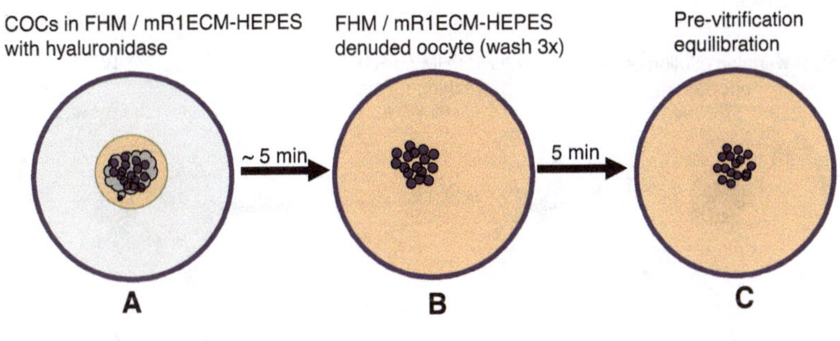

Fig. 1 Removal of cumulus cells from the oocytes with hyaluronidase treatment (**A**), washing (**B**), and equilibration of oocytes in pre-vitrification solution (**C**)

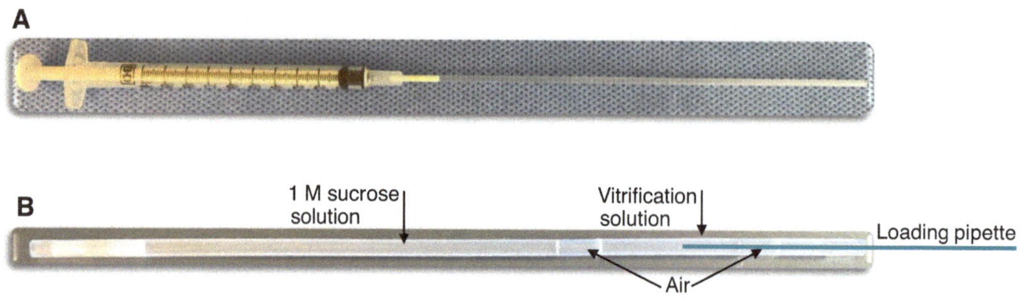

Fig. 2 A 1 cc syringe attached to a 0.25 cc vitrification straw (**A**) and the configurations of the vitrification solutions (**B**)

5. Grab the cotton end of the straw with forceps and transfer each straw into the plastic goblets submerged in LN2.
6. Place the cryo-canes having the straws into the cryo-sleeve to protect the straws from accidentally falling from the cryo-goblets, and then into a designated LN2 dewar for long-term storage.

3.5 Oocyte Thawing Procedure

1. Remove the straws from the LN2 storage tanks and immediately transfer them into a small LN2 dewar.
2. Remove the straw and immediately plunge it into a 37°C water bath and hold until the ice crystals disappear (~10 s) and dry it using a Kimwipe.
3. First, cut the sealed end of the straw and then the cotton top, expel the straw's contents onto a 35 mm Petri dish, then expel the remaining solution using a syringe and hold for 5 min (see Fig. 3).
4. Transfer the oocytes having intact oolemma into a 35 mm Petri dish containing 3 mL FHM (for mouse oocytes) or mR1ECM–HEPES (for rat oocytes), hold for another 5 min, and wash a few more times prior to IVF procedure (*see* **Note 9**).

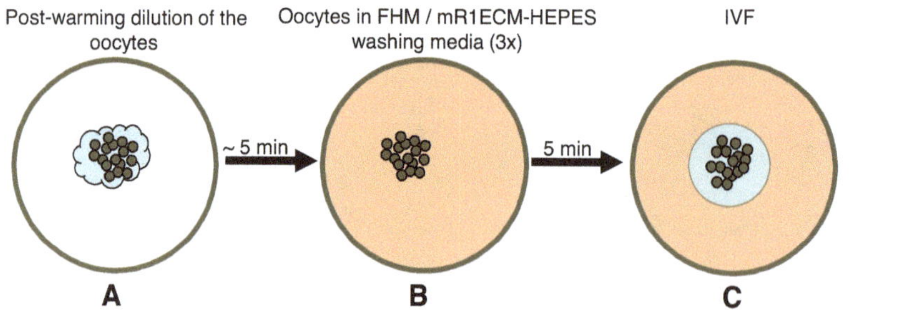

Fig. 3 A diagrammatic representation of post-warming dilution (**A**), washing (**B**), and IVF (**C**) procedures

3.6 Preparation of HEPES-Buffered Tyrode's Albumin, Lactate, and Pyruvate Medium (TALP-HEPES)

TALP-HEPES medium is modified from Tyrode's-Lactate medium [89], and part of the bicarbonate is replaced with HEPES. TL-HEPES is used for washing OT, holding before cryopreservation, base solution for making freezing solution, removing cryoprotectant after thawing, and dilution (*see* **Note 10**).

1. Add embryo-tested Milli-Q water (~80% of the final volume) to an appropriate size Griffin Beaker and then add the components in grams as listed in Table 4. All the recipes are for a 1-L final volume, so adjust the volume of the media for your needs.
2. Solubilize the $CaCl_2 \cdot 2H_2O$ and $MgCl_2 \cdot 6H_2O$ separately from the other reagents before combining the solutions.
3. Add 4.0 g/L bovine serum albumin Fraction V into the stock solution and let it dissolve slowly without using a stir bar.
4. After all components are completely dissolved, transfer the contents to an appropriately sized volumetric flask and bring to 95% of the desired volume by rinsing out the Griffin Beaker with embryo-tested Milli-Q water and adding it to the volumetric flask.
5. Adjust the pH to 7.4 before bringing to the final volume.
6. Check osmolarity (275–290 mOsm) and record. Filter into a sterile bottle using a 0.22 μm pore size filter. If the osmolarity is significantly different from 275 to 290 mOsm or precipitates form in the solution, remake the solution.

Table 4
Composition of TALP-HEPES media

Components	FW (g)	Amount per L	Concentration [mM]
[a]$CaCl_2$–$2H_2O$	147.02	0.300 g	2.0
[a]$MgCl_2$–$6H_2O$	203.30	0.100 g	0.5
KCl	74.55	0.240 g	3.2
NaCl	58.44	6.66 g	114.0
NaHCO3	84.01	0.168 g	2.0
NaH_2PO_4–$2H_2O$	177.99	0.056 g	0.4
Sodium lactate (60% syrup)		1.740 mL	11.77
Sodium pyruvate	110	0.036 g	0.2
Penicillin G potassium salt	372.2	0.06 g	n/a
Streptomycin sulfate	1457	0.0125 g	n/a
HEPES	238.31	2.400 g	20
Phenol red (0.5%)	354.38	0.66 mL	n/a

[a]Add last after all other reagents are in solution; it may precipitate

7. Aliquot into the desired volume, label the tubes, and include the expiration date, initials, and batch number. TL-HEPES can be stored for a maximum of 4 weeks at 4°C or for up to 3 months in −20°C.

3.7 Preparation of Ovary Freezing solution

Cryopreservation solution consists of 1.5 M DMSO in TL-HEPES, supplemented with 10% FCS and 0.5 M sucrose. Components of the OT cryopreservation solution are presented in Table 5.

1. Add 50 mL TL-HEPES solution to a 100 mL capacity graduated cylinder.
2. Add 12.9 mL DMSO and 17.1 g sucrose and stir until sucrose is completely dissolved.
3. Add 10 mL FCS and then bring the final volume to 100 mL by adding TL-HEPES solution.

3.8 Ovarian Tissue Donors and Recipients

It is preferable to isolate whole ovaries from sexually immature young females (3–4 weeks). The choice of recipient strain depends on the strain of the OT. Ideally, syngeneic female mice or rats (6–8 weeks) can be used (*see* **Note 11**). Alternatively, if the resources allow, immune incompetent athymic nude mice or rats can be used to avoid immune rejection (*see* **Notes 12 and 13**).

3.9 Isolation of Donor Ovaries for Cryopreservation

1. Euthanize the ovarian tissue donor females by institutionally approved animal care and use method/s (e.g., anesthesia, cervical dislocation, and CO_2 inhalation).
2. Spray 70% ethanol on the ventral body side and cut open the abdominal cavity to reach the reproductive track. Dissect out the ovaries with all the surrounding tissues aseptically and place them into a Petri dish containing 3–4 mL of room-temperature TL-HEPES solution.
3. Carefully dissect out the fat pad, ovarian bursa, and oviduct tissues under a stereo dissecting microscope using fine forceps and microscissors.

Table 5
Composition of ovarian tissue freezing solution

Components	Amount per 100 mL TL-HEPES
Dimethyl sulfoxide	12.9 mL
Sucrose	17.1 g
Fetal calf serum	10 mL

4. Transfer the cleaned ovaries into a Petri dish containing fresh 3–4 mL of room-temperature TL-HEPES solution (see Fig. 4A).
5. For rats, first cut each ovary into half using a scalpel, and then cut each half into three equal-sized pieces using microscissors (total of 12 pieces per donor (see Fig. 4B). Mouse ovaries do not require cutting since they are often small enough to load into 0.5 cc French straws.

3.10 Ovary Cryopreservation Procedure

1. Pre-label the 0.5 cc French straws with strain identification information.
2. Transfer the washed and cut OT pieces into a Petri dish containing 3–4 mL freezing solution TL-HEPES with 1.5 M DMSO, 10% FCS, and 0.5 M sucrose (Table 4) at room temperature and hold for 20 min (see Fig. 4C).
3. Load the OT into a 0.5 cc sterile French straw by using a 3 cc syringe (2 pieces per straw) along with freezing solution (see Fig. 5).
4. Transfer the straws into a controlled freezing machine at 4°C and reduce the temperature to −7°C with a cooling rate of 1°C/min.
5. Induce ice formation by seeding by touching the straws with a cotton-tipped swab presoaked with LN2 at −196 °C and holding the straw at −7°C for 10 min.
6. Reduce the temperature to −55°C with a cooling rate of 0.5°C/min.
7. Rapidly plunge the straws into a 2-L capacity dewar filled with LN2.
8. Place a cryo-canister with cryo-goblets and insert the straws into cryo-goblets, and place the cryo-canister into a cryo-sleeve to protect the straws from accidentally falling from the cryo-goblets.

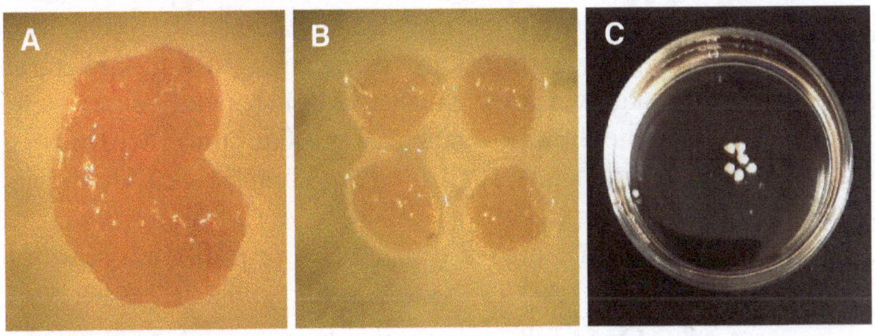

Fig. 4 Three-week-old rat ovary after removal of fat pad, ovarian bursa, and oviduct tissues (**A**), ovarian tissue pieces (**B**) before cryopreservation in a petri dish containing freezing solution (**C**)

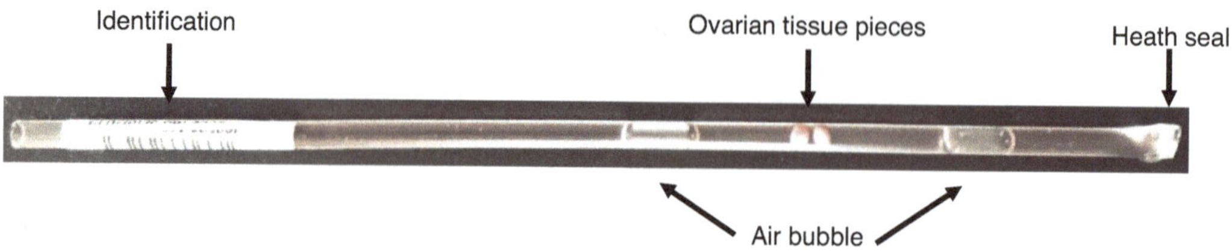

Fig. 5 A 0.5 cc French straw containing ovarian tissue pieces in freezing solution

9. Place the cryo-canister enclosed with cryo-sleeve into an LN2 tank for long-term storage.

3.11 Ovary Thawing

1. Remove the cryo-cane containing the OT samples from the large LN2 dewar and quickly transfer it into a 2-L capacity dewar filled with LN2.
2. Remove a single straw by holding from the cotton tip end, followed by holding at room temperature until the ice totally melts.
3. Wipe the straw with 70% Ethanol and Kimwipes and cut the sealed end first, point the cut end to a Petri dish containing 3–4 mL TL-HEPES solution, expel the contents including the OT pieces, and hold for 10 min at room temperature to allow removal of DMSO.
4. Transfer the OT into a fresh TL-HEPES solution by using fine forceps, hold for another 10 min, and finally transfer into another fresh TL-HEPES solution.
5. At this stage, the OT can either be grafted orthotopically, a desired site heterotopically, or fixed in 10% neutral buffered formalin for histological analysis.

3.12 Ovarian Tissue Transplantation

1. Set up a standard surgical area in a laminar flow clean bench having a stereo dissecting microscope with a fiber-optic light source, a heating pad, and all the necessary autoclaved surgical equipment before the operation.
2. After weighing, anesthetize the recipient females by IP injection of a ketamine and xylazine cocktail (rats: 80 mg/kg ketamine + 9 mg/kg xylazine; mice: 90 mg/kg ketamine + 10 mg/kg xylazine). Inject (SC) postoperative pain reliever, flunixin meglumine (rats: 2.5 mg/kg; mice: 2.0 mg/kg), once the vital signs, such as the toe pinch reflex, disappear. Apply ophthalmic ointment to the eye to protect the cornea. Place the animal into the heated cage until the anesthesia has taken effect.
3. Once fully anesthetized, remove the animal from the cage, place her on a heated sterile surgical field in a prone position. Shave the dorsal skin using a clipper and gently wipe

the shaved area with 2×2-in. cotton gauze presoaked in 70% ethanol, and then apply betadine to the shaved site using a cotton-tipped swab presoaked with betadine solution. Repeat these steps two times before starting the operation.

4. To access the abdominal muscle, make a single sagittal midline dorsal incision on the skin across the lumbar area. This incision across the lumbar area will allow access to both ovaries for orthotopic transplantation into the ovarian bursa after bilateral ovariectomy, as well as simple SC or under the renal capsule OT grafting.
5. Separate the skin from the muscle by using scissors SC from the incision, both the left and right flank regions of the animal. Bluntly dissect the connective tissue and access the peritoneal cavity by making a paralumbar muscle incision (3–4 mm in mice and 5–6 mm in rats) away from the lumbar vertebra, a little below the kidney, identify the ovary, and surround it with fat tissue.
6. Using blunt curved-serrated forceps, gently hold the peri-ovarian fat tissue and pull out the ovary through the incision and place it on a 2×2-in. cotton gauze placed on the animal and secure with a bulldog clamp. The ovarian fat pad appears little whitish than the surrounding.
7. Place the animal under the dissecting stereomicroscope using a fiber-optic light source, and focus the microscope on the ovarian bursa. Apply a few drops of epinephrine onto the bursal surface with a syringe and wait a few minutes. This would minimize the bleeding during the incision of the ovarian bursa.
8. To expose the ovary, gently pinch the bursa with fine forceps and incise the bursa by spring-type microscissors adjacent to the ovarian fat pad for the mouse and between the two blood vessels for the rats due to high vascularization (see Fig. 6A). Gently expose and cut the ovary from the ovarian bursa using a 27 g needle tip and discard while clamping at the ovarian hilum with fine forceps to prevent bleeding (*see* **Note 14**, Fig. 6C).
9. After removal of the ovary from the recipient, apply a few drops of epinephrine (50 mg/mL) and wait a few minutes, and touch with hemostatic eye spears on the empty bursa to control bleeding and aid in blood clotting (*see* **Note 15**).
10. Locate the frozen-thawed donor ovary in the Petri dish containing TL-HEPES solution, and gently place the OT by the empty ovarian bursa using fine forceps. Gently pinch one edge of the ovarian bursa incision and place the donor OT (orthotopically) under the bursa (see Fig. 6D). If the OT pieces do not fit inside the recipient bursa, trim them to fit by using spring-type microscissors under the microscope.

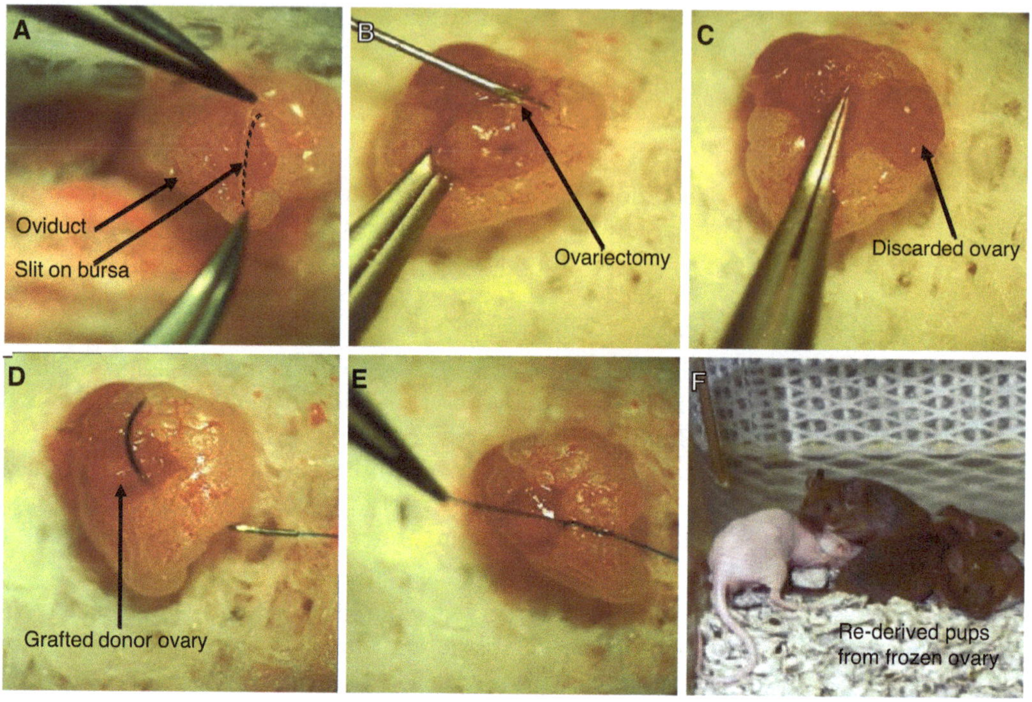

Fig. 6 The major steps of orthotopic ovarian tissue transplantation in athymic nude mice: incision on ovarian bursa (**A**), ovariectomy (**B**, **C**), grafting donor OT and suturing the ovarian bursa with the fat pad (**D**, **E**), and the production of live pups after natural mating (**F**)

11. Locate one edge of the ovarian bursa incision and secure the transplanted OT by suturing the edge of the bursa to the fat pad using a Castroviejo needle driver with an 8-0 Vicryl absorbable suture needle (*see* **Note 16**, Fig. 6D, E).
12. Using sterile physiological saline solution, irrigate the suture sites, remove the fat-pad clamp, and gently return the ovary/fat pad to the abdominal cavity. Remove the drape and close the abdominal wall with a 4-0 Vicryl absorbable suture. Repeat the procedure at the contralateral site as described above and close the skin using 9 mm wound clips (*see* **Note 17**).
13. Place the animal from the heat pad to the recovery cage on the slide warmer away from the light and monitor the animal frequently until recovery from anesthesia. If needed, an additional dose of Banamine may be administered postoperatively. Remove the wound clips approximately 7–9 days after surgery.

3.13 Determination of Reproductive Cycle via Vaginal Cytology

Although orthotopic, intramuscular, and renal capsule transplantation provide the fastest neovascularization, the other sites would also provide revascularization within less than a week of post-grafting. However, it must be noted that the establishment of a proper hypothalamic/pituitary/gonadal/adrenal axis signaling network may take several months, depending on the OT donor species [28]. For laboratory rodents, restoration of the reproductive cycle can be monitored for 3–4 weeks post-grafting.

3.14 Vaginal Smears

1. Transfer 10 mL warm physiological saline solution into a 15 mL test tube and prepare fire-polished Pasteur pipettes for each animal.
2. Draw approximately 30–50 μL saline solution on the tip of the Pasteur pipette using a rubber bulb and restrain the animal by holding it from the base of the tail.
3. Gently insert the Pasteur pipette into the vagina and wash a few times while still inside, and finally suck the content and expel onto a microscope slide and cover with a cover slip for immediate evaluation under a 10× objective (*see* **Notes 18 and 19**).
4. Alternatively, air dry the slides completely, plunge into a jar containing methanol for 5 min to fix before staining with methylene blue for 2 min, and wash the slide twice in saline solution. After completely drying, examine the slides immediately or later.

3.15 Vaginal Cytology and Corresponding Estrus Cycle Stages

The reestablishment of the reproductive cycle of the recipients can be determined by monitoring vaginal cytology as early as 4 weeks after transplantation surgery. To produce live pups after orthotopic OT grafting, the OT recipients displaying a regular reproductive cycle (4–5 days) can be mated with an appropriate proven breeder male. It is expected that the recipients will become pregnant within 3 months after transplantation and produce live pups. Depending on the strain, the OT recipients are able to produce 3–6 live pups per litter (see Fig. 6F).

The complete estrous cycle in laboratory mice and rats occurs over 4 or 5 days and can be divided into four stages: proestrus, estrus, metestrus, and diestrus. The duration of the proestrus phase is about 0.5 days and corresponds to the preovulatory stage with an increase in serum estradiol level before the preovulatory LH surge and subsequent ovulation (see Fig. 7A). The microscope field is predominantly covered with nucleated epithelial cells, individually or in clusters. Occasionally, a few cornified cells may be observed. The duration of the estrus phase is also about 0.5 days and corresponds with the presence of cornified squamous epithelial cells, mostly in clusters. These cells have irregular edges and have no visible nucleus (see Fig. 7B). Serum estradiol remains high throughout the morning and declines back to basal levels in the afternoon. The duration of the metestrus phase is about 1 day and is marked by the presence of a mixture of cell types, including mostly leukocytes, a few nucleated and/or cornified squamous epithelial cells (see Fig. 7C). The serum estradiol concentration is low in metestrus. The duration of the diestrus phase is 2–3 days and is marked by the presence of predominantly leukocytes, some round or oval epithelial cells, and thick mucus (see Fig. 7D). The serum estradiol levels begin to increase in diestrus.

4 Notes

1. If any precipitates form in the solution, discard the solution and start over. Use glassware that has been rinsed at least ten times with Milli-Q water, which has been properly dried.
2. Solubilize the $CaCl_2 \cdot 2H_2O$ separately from the other reagents before combining the solutions.
3. Do not shake it to dissolve.
4. This would prevent evaporation and keep the oocytes for future processing. Holding time should not exceed 10–15 min after removal from the body.
5. It usually takes about 5 min to eliminate all cumulus cells surrounding the oocytes.
6. The oocytes will be dispersed in the pre-vitrification solution, so group the oocytes using a fine glass pipette before transferring them into the vitrification solution.
7. Gently wipe off the straw's open ends to get rid of excess sucrose solution before loading the vitrification solution and heat-sealing the straws.
8. Plunge the straws halfway into LN2 for 10 s, then submerge the rest up to the cotton plug and hold another 20 s and then drop it into LN2. This will prevent the potential cracking problem of the plastic straw.
9. Please refer to sperm freezing papers for rat IVF [90].
10. For all the protocols described, it is imperative to use sterile materials, solutions, and tools. All procedures must be

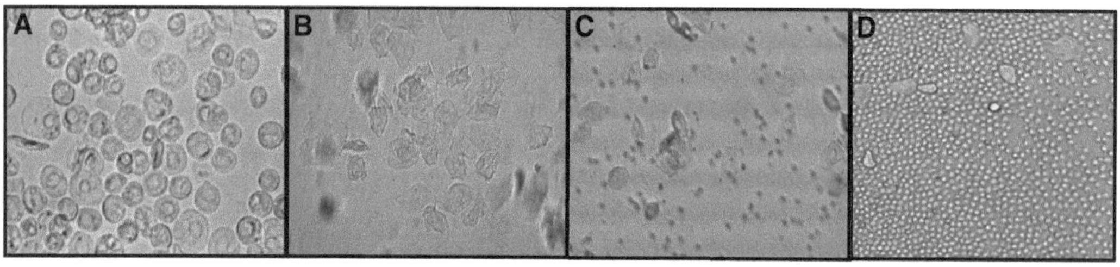

Fig. 7 Representative images of unstained vaginal cells. Proestrus with the presence of predominantly nucleated epithelial cells (**A**), estrus with the presence of anucleated cornified cells (**B**), metestrus, with the presence of a mixture of cell types, including mostly leucocytes, a few nucleated and/or cornified squamous epithelial cells (**C**), and diestrus with the presence of predominantly leukocytes (**D**)

performed under a clean workbench to prevent potential infections.

11. Syngeneic younger, nulliparous females provide the best reproductive performance, but older breeders can also be used as OT recipients. If a syngeneic recipient is not available, athymic nude mice or rats must be used.
12. Autoclaved water, bedding, and food must be provided to maintain athymic nude mice or rats.
13. The OT transplantation procedure is most likely to fail if a female does not become pregnant within 4 months posttransplantation in the presence of a proven breeder male.
14. One must make sure about complete ovariectomy since remaining endogenous recipient OT may develop oocytes and result in undesired genetic background pups in the case of orthotopic grafting. Incomplete ovariectomy may also confound the study in the case of heterotopic grafting by producing ovarian hormones.
15. Bleeding must be under control at this stage since clotting may block the infundibular opening and thus transport of the developing embryos into the uterine horn.
16. If the Castroviejo needle driver is not available, suturing can alternatively be performed with a watchmaker or other fine forceps.
17. Working under the laminar flow hood for an extended time period may dry the cornea of the technician. Thus, it is important to protect the personnel from dry eyes.
18. This chapter largely describes OT cryopreservation and orthotopic transplantation for laboratory mice and rats with the intention of producing offspring. However, the cryopreservation procedures described here can be used to perform heterotopic transplantation, such as under the renal capsule (Fig. 8A–C), subcutaneous (Fig. 9A–C), intramuscular, or

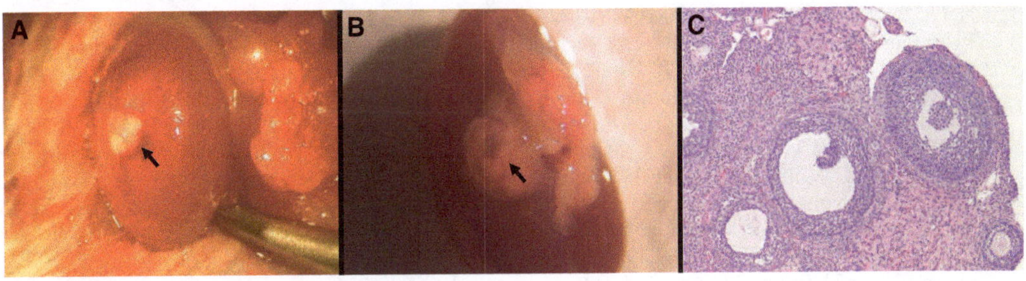

Fig. 8 Frozen-thaw rat ovarian tissue xeno-grafting beneath the renal capsule of athymic nude mice following bilateral ovariectomy (**A**), gross (**B**), and histologic cross-section of ovarian tissue graft containing preovulatory follicles (**C**)

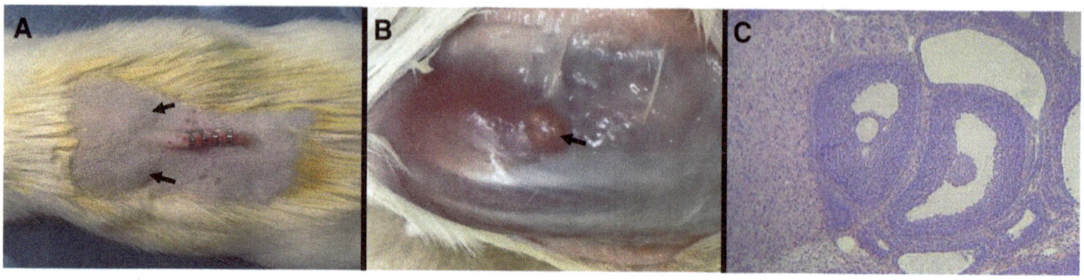

Fig. 9 Subcutaneous frozen–thaw rat ovarian tissue auto-grafting (arrows) following bilateral ovariectomy (**A**), gross (**B**), and histologic cross-section of ovarian tissue graft containing preovulatory follicles (**C**)

other areas with some modifications in the standard surgical procedures to access the desired graft site [43–47, 49].

19. Animals may be initially stressed if they are not properly handled. To avoid this, it is important to practice the procedure a few times before handling the experimental animals. It is also important not to touch deep into the vagina and stimulate the cervix since this would induce pseudopregnancy in the female.

Acknowledgments

The University of Missouri Mutant Mouse Resource and Research Center (NIH) U42 OD010918-20.

References

1. Agca Y (2012) Genome resource banking of biomedically important laboratory animals. Theriogenology 78:1653–1665. https://doi.org/10.1016/j.theriogenology.2012.08.012
2. Agca Y, Amos-Landgraf J, Araiza R, Brennan J, Ciavatta D, Clary D, Korf I, Lutz C, Magnuson T, Manuel de Villena F, Mirochnitchenko O, Patel S, Port D, Reinholdt L, Lloyd KC (2024) The mutant mouse resource and research center (MMRRC) consortium: the US-based public mouse repository system. Mamm Genome 35:524–536. https://doi.org/10.1007/s00335-024-10070-3
3. Ostermeier GC, Wiles MV, Farley JS, Taft RA (2008) Conserving, distributing and managing genetically modified mouse lines by sperm cryopreservation. PLoS ONE 30:e2792. https://doi.org/10.1371/journal.pone.0002792
4. Sztein JM, Takeo T, Nakagata N (2018) History of cryobiology, with special emphasis in evolution of mouse sperm cryopreservation. Cryobiology 82:57–63. https://doi.org/10.1016/j.cryobiol.2018.04.008
5. Takeo T, Sztein J, Nakagata N (2019) The CARD method for mouse sperm cryopreservation and in vitro fertilization using frozen-thawed sperm. Methods Mol Biol 1874:243–256. https://doi.org/10.1007/978-1-4939-8831-0_14
6. Leibo SP, Sztein JM (2019) Cryopreservation of mammalian embryos: derivation of a method. Cryobiology 86:1–9. https://doi.org/10.1016/j.cryobiol.2019.01.007
7. Rall WF, Schmidt PM, Lin X, Brown SS, Ward AC, Hansen CT (2000) Factors affecting the efficiency of embryo cryopreservation and rederivation of rat and mouse models. ILAR J 41:221–227. https://doi.org/10.1093/ilar.41.4.221
8. Robertson GAG (1940) Ovarian transplantation in the house mouse. Proc Soc Exp Biol Med 44:302–304. https://doi.org/10.3181/00379727-44-11438

9. Polge C, Smith AU, Parkes AS (1949) Revival of spermatozoa after vitrification and dehydration at low temperatures. Nature 164:666. https://doi.org/10.1038/164666a0
10. Parkes AS, Smith AU (1954) Preservation of ovarian tissue at –79°C for transplantation. Acta Endoc 17:313–320
11. Deanesly R (1957) Egg survival in immature rat ovaries grafted after freezing and thawing. Proc R Soc Lond Ser B 147:412–421. (Donnell TC, General Endocrinology, W.B. Saunders, Philadelphia)
12. Green SH, Smith AU, Zuckerman S (1956) The number of oocytes in ovarian autografts after freezing and thawing. J Endocrinol 13:330–334. https://doi.org/10.1677/joe.0.0130330
13. Parrott DM (1959) Orthotopic ovarian grafts in the golden hamster. J Endocrinol 19:126–138. https://doi.org/10.1677/joe.0.0190126
14. Parrott DMV (1960) The fertility of mice with orthotopic ovarian grafts derived from frozen tissue. J Reprod Fertil 1:230–241. https://doi.org/10.1530/jrf.0.0010230
15. Sztein JM, McGregor TE, Bedigian HJ, Mobraaten LE (1999) Transgenic mouse strain rescue by frozen ovaries. Lab Anim Sci 49:99–100
16. Agca Y (2000) Cryopreservation of oocytes and ovarian tissue. ILAR J 41:207–220. https://doi.org/10.1093/ilar.41.4.207
17. Takahashi E, Miyoshi I, Nagasu T (2001) Rescue of a transgenic mouse line by transplantation of a frozen-thawed ovary obtained postmortem. Contemp Top Lab Anim Sci 40:28–31
18. Shaw JM, Trounson AO (2002) Ovarian tissue transplantation and cryopreservation. Application to maintenance and recovery of transgenic and inbred mouse lines. Meth Mol Biol 180:229–251. https://doi.org/10.1385/1-59259-178-7:229
19. Dawes J, Liu B, Mars W, Michalopoulos G, Khillan JS (2010) Multiple ovarian transplants to rescue a transgenic line of mice. Lab Anim 39:191–193. https://doi.org/10.1038/laban0610-191
20. Sztein JM (2011) Ovarian transplant for transgenic rescue. Lab Anim 40:9. https://doi.org/10.1038/laban0111-9a
21. Dorsch M, Wedekind D, Kamino K, Hedrich HJ (2004) Orthotopic transplantation of rat ovaries as a tool for strain rescue. Lab Anim 38:307–312. https://doi.org/10.1258/002367704323133691
22. Dorsch M, Wedekind D, Kamino K, Hedrich HJ (2007) Cryopreservation and orthotopic transplantation of rat ovaries as a means of gamete banking. Lab Anim 41:247–254. https://doi.org/10.1258/002367707780378195
23. Dorsch M, Wedekind D (2010) Cryopreservation and orthotopic transplantation of rat ovaries. Methods Mol Biol 597:301–310. https://doi.org/10.1007/978-1-60327-389-3_21
24. Ding Y, Shao JL, Li JW, Zhang Y, Hong KH, Hua KQ, Wang X (2018) Successful fertility following optimized perfusion and cryopreservation of whole ovary and allotransplantation in a premature ovarian insufficiency rat model. J Ovarian Res 11: 35. https://doi.org/10.1186/s13048-018-0401-4
25. Ding Y, Shao J, Hua K, Wang X (2018) Modified model of whole ovary transplantation in rats. Transplant Proc 50:3881–3886. https://doi.org/10.1016/j.transproceed.2018.06.038
26. Bindels J, Squatrito M, Bernet L, Nisolle M, Henry L, Munaut C (2023) The mTOR inhibitor rapamycin counteracts follicle activation induced by ovarian cryopreservation in murine transplantation models. Medicina 16:1474. https://doi.org/10.3390/medicina59081474
27. Bindels J, Squatrito M, Bernet L, Nisolle M, Munaut C (2025) Ovarian cryopreservation with rapamycin improves fertility restoration in a murine orthotopic transplantation model. Sci Rep 15: 9441. https://doi.org/10.1038/s41598-025-94588-9
28. Cargill SL, Medrano JF, Anderson GB (1999) Infertility in a line of mice with the high growth mutation is due to luteal insufficiency resulting from disruption at the hypothalamic-pituitary axis. Biol Reprod 61:283–287. https://doi.org/10.1095/biolreprod61.1.283
29. Torrents E, Boiso I, Barri PN, Veiga A (2003) Applications of ovarian tissue transplantation in experimental biology and medicine. Hum Reprod Update 9:471–481. https://doi.org/10.1093/humupd/dmg036
30. Parker WH, Broder MS, Chang E, Feskanich D, Farquhar C, Liu Z, Shoupe D, Berek JS, Hankinson S, Manson JE (2009) Ovarian conservation at the time of hysterectomy and long-term health outcomes in the Nurses' Health Study. Obstet Gynecol 113:1027–1037. https://doi.org/10.1097/AOG.0b013e3181a11c64
31. Peterson RL, Parkinson KC, Mason JB (2017) Restoration of immune and renal function in aged females by re-establishment of active ovarian function. Reprod Fertil Dev 29:2052–2059. https://doi.org/10.1071/RD16333
32. Anderson RA, Baird DT (2019) The development of ovarian tissue cryopreservation in Edinburgh: translation from a rodent model through validation in a large mammal and then into clinical practice. Acta Obstet Gynecol Scand 98:545–549. https://doi.org/10.1111/aogs.13560
33. Silber S (2016) Ovarian tissue cryopreservation and transplantation: scientific implications. J

Assist Reprod Genet 33:1595–1603. https://doi.org/10.1007/s10815-016-0814-1

34. Oktay K, Taylan E, Sugishita Y, Goldberg GM (2017) Robot-assisted laparoscopic transplantation of frozen-thawed ovarian tissue. J Minim Invasive Gynecol 24:897–898. https://doi.org/10.1016/j.jmig.2017.02.021
35. Jensen AK, Macklon KT, Fedder J, Ernst E, Humaidan P, Andersen CY (2017) 86 successful births and 9 ongoing pregnancies worldwide in women transplanted with frozen-thawed ovarian tissue: focus on birth and perinatal outcome in 40 of these children. J Assist Reprod Genet 34:325–336. https://doi.org/10.1007/s10815-016-0843-9
36. Fabbri R, Vicenti R, Magnani V, Paradisi R, De Meis L, Raimondo D, Arena A, Venturoli S, Raffone A, Raspollini A, Seracchioli R (2024) Ovarian tissue transplantation: 10 years of experience at the Bologna University. Front Endocrinol 15:1332673. https://doi.org/10.3389/fendo.2024.1332673
37. Eppig JJ, O'Brien MJ (1996) Development in vitro of mouse oocytes from primordial follicles. Biol Reprod 54:197–207. https://doi.org/10.1095/biolreprod54.1.197
38. Sztein JM, O'Brien MJ, Farley JS, Mobraaten LE, Eppig JJ (2000) Rescue of oocytes from antral follicles of cryopreserved mouse ovaries: competence to undergo maturation, embryogenesis, and development to term. Hum Reprod 15:567–571. https://doi.org/10.1093/humrep/15.3.567
39. Liu F, Lin Z, Huang K (2025) Developing mouse models for ovarian tissue transplantation and xenotransplantation: a review. Med Sci Monit 31:e946386. https://doi.org/10.12659/MSM.946386
40. Terren C, Bindels J, Nisolle M, Noël A, Munaut C (2022) Evaluation of an alternative heterotopic transplantation model for ovarian tissue to test pharmaceuticals improvements for fertility restoration. Reprod Biol Endocrinol 20: 35. https://doi.org/10.1186/s12958-022-00910-9
41. Donnez J, Dolmans MM (2017) Fertility preservation in women. N Engl J Med 377:1657–1665. https://doi.org/10.1056/NEJMra1614676
42. Deng X, Zheng H, Yu X, Yu H, Zhang C, Chao L, Li R, Liu W (2009) Cryopreserved ovarian tissues can maintain a long-term function after heterotopic autotransplantation in rats. Reproduction 138:519–525. https://doi.org/10.1530/REP-09-0151
43. Agca C, Lucy M, Agca Y (2009) Gene expression profile of rat ovarian tissue following xeno transplantation into immune-deficient mice. Reproduction 137:957–967. https://doi.org/10.1530/REP-09-0048
44. Demeestere I, Simon P, Emiliani S, Delbaere A, Englert Y (2009) Orthotopic and heterotopic ovarian tissue transplantation. Hum Reprod Update 15:649–665. https://doi.org/10.1016/j.bpobgyn.2009.09.002
45. Youm HY, Lee JR, Lee J, Jee BC, Suh CS, Kim SH (2015) Transplantation of mouse ovarian tissue: comparison of the transplantation sites. Theriogenology 83:854–861. https://doi.org/10.1016/j.theriogenology.2014.11.026
46. Tavana S, Rezazadeh Valojerdi M, Eimani H, Abtahi NS, Fathi R (2017) Auto-transplantation of whole rat ovary in different transplantation sites. Vet Res Forum 8:275–280
47. Wietcovsky L, Til D, Salvador RA, Amaral NLL, Senn AP, Amaral VLL (2018) Ovarian tissue vitrification and heterotopic autologous transplantation in prepubertal Wistar rats. JBRA Assist Reprod 22:116–122. https://doi.org/10.5935/1518-0557.20180019
48. Yan J, Cai Y, Dai S, He X, Sun Y, Cao Y, Qiao J, Chian R (2022) Offspring from oocytes grown in frozen-thawed ovarian tissues transplanted to male and female bodies. Reprod Dev Med 6:13–19. https://doi.org/10.1097/RD9.0000000000000005
49. Taketsuru H, Hirayama R, Nakatsukasa E, Natsume R, Takao K, Abe M, Sakimura K (2024) Generation of rat offspring from ovarian oocytes by xenotransplantation. Sci Rep 14:20109. https://doi.org/10.1038/s41598-024-71030-0
50. Hao L, Jia Q, Tian N, Tang H, Liu G, Li Y, Geng M, Chen W, Chen J, Cai Z (2025) Allogeneic ovarian transplantation promotes the recovery of ovarian function in a rat model of cisplatin-induced premature ovarian failure. Sci Rep 15:9500. https://doi.org/10.1038/s41598-025-94456-6
51. Ruhliada NN (2024) Ovarian tissue autotransplantation improves longevity in mice. Front Physiol 15:1443494. https://doi.org/10.3389/fphys.2024.1443494
52. Li Q, Chen Y, Yang L, Sun Y, Wang Y, Zhu J (2013) Ovarian function reconstruction by orthotopic transplantation of novel vitrification cryopreserved ovaries in chemotherapy-induced ovary damage rat model. Zhongguo Xiu Fu Chong Jian Wai Ke Za Zhi 27:1116–1121
53. Li Q, Szatmary P, Liu Y, Ding Z, Zhou J, Sun Y, Luo F, Wang Y, Zhu J (2015) Orthotopic transplantation of cryopreserved mouse ovaries and gonadotrophin releasing hormone analogues in the restoration of function following chemotherapy-induced ovarian damage. PLoS ONE 10:e0120736. https://doi.org/10.1371/journal.pone.0120736
54. Mason JB, Cargill SL, Anderson GB, Carey JR (2009) Transplantation of young ovaries to old

mice increased life span in transplant recipients. J Gerontol A Biol Sci Med Sci 64:1207–1211. https://doi.org/10.1093/gerona/glp134
55. Mason JB, Cargill SL, Griffey SM, Reader JR, Anderson GB, Carey JR (2011) Transplantation of young ovaries restored cardioprotective influence in postreproductive-aged mice. Aging Cell 10:448–456. https://doi.org/10.1111/j.1474-9726.2011.00691.x
56. Mason JB, Terry BC, Merchant SS, Mason HM, Nazokkarmaher M (2015) Manipulation of ovarian function significantly influenced trabecular and cortical bone volume, architecture and density in mice at death. PLoS ONE 10:e0145821. https://doi.org/10.1371/journal.pone.0145821
57. Salama M, Isachenko E, Ludwig S, Einzmann T, Rahimi G, Mallmann P, Isachenko VA (2019) Successful multidisciplinary approach for treatment and for preserving the reproductive potential in a rare case of acute lymphocytic leukemia during pregnancy. Gynecol Endocrinol 35:115–118. https://doi.org/10.1080/09513590.2018.1498833
58. Telfer EE, Zelinski MB (2021) Innovation in ovarian tissue cryopreservation and transplantation. Front Endocrinol 12:784938
59. Johnson J, Lawley SD, Emerson JW, Oktay KH (2024) Modeling delay of age at natural menopause with planned tissue cryopreservation and autologous transplantation. Am J Obstet Gynecol 230:426.e1–426.e8. https://doi.org/10.1016/j.ajog.2023.12.037
60. Erden M, Gayete-Lafuente S, Vural NA, Oktay KH (2024) Utility and outcomes of ovarian tissue cryopreservation and transplantation for gynecologic cancers. Obstet Gynecol 144:481–492. https://doi.org/10.1097/AOG.0000000000005708
61. Finkelstein T, Zhang Y, Vollenhoven B, Rolnik D, Horta F (2024) Successful pregnancy rates amongst patients undergoing ovarian tissue cryopreservation for non-malignant indications: a systematic review and meta-analysis. Eur J Obstet Gynecol Reprod Biol 2:30–39. https://doi.org/10.1016/j.ejogrb.2023.11.004
62. Erden M, Oktay KH (2025) Does gonadotoxic chemotherapy deplete the ovarian reserve through activation of primordial follicles. Hum Reprod 40:571–579. https://doi.org/10.1093/humrep/deaf024
63. Finkelstein T, Rolnik DL, Agresta F, Rozen G, Stern K, Vollenhoven B, Catt S, Horta F (2025) Pregnancy outcomes following ovarian tissue cryopreservation: an Australian cohort Study. Am J Obstet Gynecol 233:108.e1–108.e11. https://doi.org/10.1016/j.ajog.2025.02.007
64. Eijkenboom L, Saedt E, Zietse C, Braat D, Beerendonk C, Peek R (2022) Strategies to safely use cryopreserved ovarian tissue to restore fertility after cancer: a systematic review. Reprod Biomed Online 45:763–778. https://doi.org/10.1016/j.rbmo.2022.05.020
65. Peek R, Bastings L, Westphal JR, Massuger LFAG, Braat DDM, Beerendonk CCM (2015) Preliminary study on a new model system to evaluate tumor-detection and tumor-purging protocols in ovarian cortex tissue intended for fertility preservation. Hum Reprod 30:870–876. https://doi.org/10.1093/humrep/dev013
66. Peek R, Eijkenboom LL, Braat DDM, Beerendonk CCM (2021) Complete purging of Ewing sarcoma metastases from human ovarian cortex tissue fragments by inhibiting the mtorc1 signaling pathway. J Clin Med 10:4362. https://doi.org/10.3390/jcm10194362
67. Mulder CL, Eijkenboom LL, Beerendonk CCM, Braat DDM, Peek R (2019) Enhancing the safety of ovarian cortex autotransplantation: cancer cells are purged completely from human ovarian tissue fragments by pharmacological inhibition of YAP/TAZ oncoproteins. Hum Reprod 34:506–518. https://doi.org/10.1093/humrep/dey384.
68. Eijkenboom L, Mulder C, Van Der Reijden B, Van Mello N, Van Leersum J, Koorenhof-Scheele T, Braat D, Beerendonk C, Peek R (2021) Purging human ovarian cortex of contaminating leukaemic cells by targeting the mitotic catastrophe signalling pathway. J Assist Reprod Genet 38:1571–1588. https://doi.org/10.1007/s10815-021-02081-9
69. Eijkenboom L, Palacio-Castañeda V, Groenman F, Braat D, Beerendonk C, Brock R, Verdurmen W, Peek R (2021) Assessing the use of tumor-specific DARPin-toxin fusion proteins for ex vivo purging of cancer metastases from human ovarian cortex before autotransplantation. F S Sci 2:330–344. https://doi.org/10.1016/j.xfss.2021.09.004
70. Nakagata N, Takeo T, Fukumoto K, Kondo T, Haruguchi Y, Takeshita Y, Nakamuta Y, Matsunaga H, Tsuchiyama S, Ishizuka Y, Araki K (2013) Applications of cryopreserved unfertilized mouse oocytes for in vitro fertilization. Cryobiology 67:188–192. https://doi.org/10.1016/j.cryobiol.2013.06.011
71. Kohaya N, Fujiwara K, Ito J, Kashiwazaki N (2013) Generation of live offspring from vitrified mouse oocytes of C57BL/6J strain. PLoS ONE 8:e58063. https://doi.org/10.1371/journal.pone.0058063
72. Nakagata N, Sztein J, Takeo T (2019) The CARD method for simple vitrification of mouse oocytes: advantages and applications. Methods Mol Biol 1874:229–242. https://doi.org/10.1007/978-1-4939-8831-0_13

73. Wang H, Herath CB, Xia G, Watanabe G, Taya K (2001) Superovulation, fertilization and in vitro embryo development in mice after administration of an inhibin-neutralizing antiserum. Reproduction 122:809–816. https://doi.org/10.1530/rep.0.1220809
74. Hasegawa A, Mochida K, Matoba S, Yonezawa K, Ohta A, Watanabe G, Taya K, Ogura A (2012) Efficient production of offspring from Japanese wild-derived strains of mice (*Mus musculus* molossinus) by improved assisted reproductive technologies. Biol Reprod 86:167. https://doi.org/10.1095/biolreprod.111.098491
75. Takeo T, Nakagata N (2015) Superovulation using the combined administration of inhibin antiserum and equine chorionic gonadotropin increases the number of ovulated oocytes in C57BL/6 female mice. PLoS ONE 10:e0128330. https://doi.org/10.1371/journal.pone.0128330
76. Takeo T, Nakagata N (2016) Immunotherapy using inhibin antiserum enhanced the efficacy of equine chorionic gonadotropin on superovulation in major inbred and outbred mice strains. Theriogenology 86:1341–1346. https://doi.org/10.1016/j.theriogenology.2016.04.076
77. Hasegawa A, Mochida K, Inoue H, Noda Y, Endo T, Watanabe G, Ogura A (2016) High-yield superovulation in adult mice by anti-inhibin serum treatment combined with estrous cycle synchronization. Biol Reprod 94:21. https://doi.org/10.1038/10.1095/biolreprod.115.134023
78. Wuri L, Agca C, Agca Y (2020) Morphometric, subcellular, in-vitro fertilization and embryonic developmental assessment of mouse oocytes produced by anti-inhibin serum or PMSG superovulation. Reprod Fertili Dev 32:474–483. https://doi.org/10.1071/RD19131
79. Hasegawa A, Mochida K, Nakamura A, Miyagasako R, Ohtsuka M, Hatakeyama M, Ogura A (2022) Use of anti-inhibin monoclonal antibody for increasing the litter size of mouse strains and its application to in vivo-genome editing technology. Biol Reprod 107:605–618. https://doi.org/10.1093/biolre/ioac068
80. Meguro K, Komatsu K, Ohdaira T, Nakagata N, Nakata A, Fukumoto M, Miura T, Yamashiro H (2019) Induction of superovulation using inhibin antiserum and competence of embryo development in wild large Japanese field mice (Apodemus speciosus. Reprod Domest Anim 54:1637–1642. https://doi.org/10.1111/rda.13573
81. Hasegawa A, Mochida K, Matoba S, Inoue K, Hama D, Kadota M, Hiraiwa N, Yoshiki A, Ogura A (2021) Development of assisted reproductive technologies for *Mus spretus*. Biol Reprod 104:234–243. https://doi.org/10.1093/biolre/ioaa177
82. Ishigame H, Medan MS, Watanabe G, Shi Z, Kishi H, Arai KY, Taya KA (2004) A new alternative method for superovulation using passive immunization against inhibin in adult rats. Biol Reprod 71:236–243. https://doi.org/10.1095/biolreprod.104.027789
83. Ishigame H, Medan MS, Kawaguchi M, Fukuda A, Watanabe G, Arai KY, Taya KJ (2005) Induction of superovulation by immunoneutralization of endogenous inhibin in immature rats. Reprod Dev 51:559–566. https://doi.org/10.1262/jrd.17020
84. Honda A, Tachibana R, Hamada K, Morita K, Mizuno N, Morita K, Asano M (2019) Efficient derivation of knock-out and knock-in rats using embryos obtained by in vitro fertilization. Sci Rep 9: 11571. https://doi.org/10.1038/s41598-019-47964-1
85. Mochida K, Morita K, Sasaoka Y, Morita K, Endo H, Hasegawa A, Asano M, Ogura A (2024) Superovulation with an anti-inhibin monoclonal antibody improves the reproductive performance of rat strains by increasing the pregnancy rate and the litter size. Sci Rep 14:8294. https://doi.org/10.1038/s41598-024-58611-9
86. Combrisson H (2017) Animal experiment, can we replace?. Transfus Clin Biol 24:93–95. https://doi.org/10.1016/j.tracli.2017.06.005
87. Lawitts JA, Biggers JD (1993) Culture of preimplantation embryos. Methods Enzymol 225:153–164. https://doi.org/10.1016/0076-6879(93)25012-q
88. Oh SH, Miyoshi K, Funahashi H (1998) Rat oocytes fertilized in modified rat 1-cell embryo culture medium containing a high sodium chloride concentration and bovine serum albumin maintain developmental ability to the blastocyst stage. Biol Reprod 59:884–889. https://doi.org/10.1095/biolreprod59.4.884
89. Bavister BD, Leibfried ML, Lieberman G (1983) Development of preimplantation embryos of the golden hamster in a defined culture medium. Biol Reprod 28:235–247. https://doi.org/10.1095/biolreprod28.1.235
90. Takeo T, Nakao S, Mikoda N, Yamaga K, Maeda R, Tsuchiyama S, Nakatsukasa E, Nakagata N (2022) Optimized protocols for sperm cryopreservation and in vitro fertilization in the rat. Lab Anim 51:256–274. https://doi.org/10.1038/s41684-022-01053-5

Check for updates

Chapter 20

Cryopreservation of Marine Invertebrates: From Sperm to Organismal Cryopreservation

Estefanía Paredes, Pablo Heres, Sara Campos, Alba Lago, Catarina Anjos and Elsa Cabrita

Abstract

Marine invertebrates constitute the vast majority of marine biodiversity, and they are extremely diverse and play a key role in marine ecosystems, thus playing an important role at the socioeconomic level. Some invertebrates, such as sea urchins, ascidians, and horseshoe crabs, are well-known model organisms for research and bio-compound discovery. In this chapter, we revisit the importance of cryopreservation for conservation and rational use in research, fisheries management, or aquaculture, and provide comprehensive protocols for the cryopreservation of sperm, embryos, larvae, and juveniles.

Keywords Cryopreservation, Echinoderms, Juveniles, Marine invertebrates, Molluscs

1 Introduction

1.1 General Introduction

Marine invertebrates represent the vast majority of marine biodiversity, including corals, worms, mollusks, sea urchins, starfish, crustaceans, sea cucumbers, sea slugs, and jellyfish. They are extremely diverse in size, complexity, form, and living/feeding/reproducing mechanisms. Filter feeders, grazers, or browsers play a key role in marine ecosystems. Many marine invertebrates are detritus-eaters/sediment processers and thus play important roles in the recycling of organic matter in the ocean, impacting water fluxes, nutrient release, pore water chemistry, and the bacterial characteristics of sediments [1].

Some marine invertebrates, such as the Pacific oyster (*Crassostrea gigas*), blue and green-shell mussels (*Mytilus galloprovincialis* and *Perna canaliculus*), and crustaceans such as the king crab (*Paralithodes camtschaticu*), are important at the socioeconomic level. The well-being of populations of these organisms affects thousands of

Willem F. Wolkers and Harriëtte Oldenhof (eds.), *Cryopreservation and Freeze-Drying Protocols*, Methods in Molecular Biology, vol. 3060, https://doi.org/10.1007/978-1-0716-5416-3_20,

local communities and accounts for hundreds of millions of dollars in the global market [2].

Other invertebrates, such as sea urchins, ascidians, and horseshoe crabs, are very well-known model organisms for biomedical research and biocompound discovery and are used to study intracellular signaling, fertilization mechanisms, cell division, drug development, and endotoxic detection [3, 4]. Overall, marine invertebrates play crucial ecological roles in ocean health, and their welfare influences the structure of the food chain, directly or indirectly affecting local coastal economies and fisheries management at the global scale. The large-scale degradation of the natural environment in combination with poor management can lead to a combined impact on certain species and drastic population changes, in many cases favoring the proliferation of nonnative species [5, 6]. This is the case, for example, for the decline of populations (and consequently fisheries) of the Portuguese native oyster, *C. angulata*, following a common sequence of events in many places worldwide [7, 8]. Typically, the extensive harvest of wild oyster populations results in the loss of reef structure, which exacerbates the impact of additional stresses from anoxia, sedimentation, and disease [6–13]. Populations of sea urchins, where fisheries have been active for decades, are also declining quickly, leading to decreasing captures and, in some regions, temporary closures of fisheries. The annual capture of sea urchins worldwide continues to increase annually [14, 15]. Sea urchins are grazers and therefore play an active role in their local communities. Overfishing and ecological imbalance due to the disappearance of natural predators create notable imbalances in communities, such as the disappearance of kelp forests due to the overproliferation of sea urchins along the U.S. Pacific coast.

Blue mussel culture (*Mytilus galloprovincialis*), mostly based out of the NW coast of Spain, has been showing a steady decrease in harvesting [16–21]. The juvenile natural recruitment on ropes and of coastal rocky areas is the only source of seedlings for the business [22], and producers have been struggling to obtain the right amount of seed due to a decline in mussel production, controlled rock harvesting areas, and conflicts about access to the resource.

The aim of this chapter is to provide standardized protocols for the cryopreservation of several cell types and organisms from different marine invertebrates.

1.2 Cryopreservation of Marine Invertebrates, Why Is It Useful?

By extending ex situ access to marine biological resources, cryopreservation will increase the use of marine species as model species for research, increase the use of sperm and embryo-larval bioassays, and promote applications in the fields of aquaculture, fisheries management, and the conservation of endangered species. The development of cryopreservation protocols will allow the storage of marine invertebrates in biobanks and, consequently, provide year-round access to high-quality material regardless of the reproductive season or meteorological and environmental conditions. The rapid decline in biodiversity [16] has made the role of cryopreservation even more crucial in the context of progress in scientific knowledge, the conservation of marine biodiversity, efficient fisheries management, and aquaculture practices.

Knowledge about the cryopreservation of marine invertebrates has increased in recent decades [17], with several protocols having been published for gametes and larvae of different species [17–21, 23]. In fact, most of these studies focused on the cryopreservation of sperm from species with high commercial value [17, 24]. The storage of sperm has several advantages in aquaculture production, especially for hatcheries, through the synchronization of gamete availability, allowing the provision of sperm outside the reproductive season, permitting simplified broodstock establishment, and facilitating sperm transport between facilities [24, 25]. Cryopreservation also makes it possible to create genetic resource banks [26], which are repositories for selected strains and endangered species, where the material is protected from secure diseases, natural disasters, and contamination [23, 24, 27]. This sperm, which has specific characteristics, can subsequently be used in genetic programs and preservation strategies.

Although the cryopreservation of oyster sperm can help production and conservation activities, the maternal genome clearly needs to be preserved as well. Eggs are more difficult to cryopreserve, and mussels, oysters, and sea urchins are not an exception [28, 29]. An alternative is to cryopreserve embryos or larvae, which have been demonstrated to produce viable offspring in the case of the Pacific oyster (*Crassostrea gigas*) [30].

1.3 Target Species

Paracentrotus lividus is a regular sea urchin that is widespread in the Mediterranean Sea and eastern Atlantic intertidal and subtidal coasts from Scotland to Morocco. It is one of the main edible species in Europe, although it is not the only one [31].

M. galloprovincialis is the most widespread mussel species. In addition to their important ecological role in filtering seawater, mussels are among the most farmed mollusks, and the annual harvest is close to 2 million tonnes per year [2].

Crassostrea angulata (Portuguese oyster) is an example of a species for which genetic resources can disappear. Portuguese oyster natural populations have collapsed due to the onset of pathologies and poor remediation and management measures [32]. Creating mechanisms that reduce the loss of biodiversity is crucial. One tool that allows preservation of the genetic profile and improves reproductive technologies is cryopreservation.

2 Materials

2.1 Expendables

1. Polystyrene rack and styrofoam box with a lid (capacity 1–2 L).
2. Cryopreservation straws (0.25 or 0.5 mL; from Minitube or IMV) and either sealing powder or a heat sealer or cryovials (2 mL).
3. Liquid nitrogen.
4. Cryoprotecting solutions.
5. Filtered and sterilized Sea Water or Artificial Sea Water.

2.2 Equipment

1. Portable programmable freezer (e.g., from Asymptote Grant EF600, UK or Cryologic Australia Ltd.).
2. Computer-Assisted Sperm Analysis (CASA) system and accompanying software (e.g., from ISAS; Proiser, Valencia, Spain) and a Makler chamber.
3. Microscope with a 10× negative-phase contrast objective coupled with a digital camera.
4. Fluorescence microscope.
5. Flow cytometer.

2.3 Lab Safety Equipment

1. Cryopreservation security gear, such as gloves, forceps/tweezers, protective eyewear, or a face mask.

3 Methods

3.1 Cryopreservation of Sea Urchin (Paracentrotus lividus) Sperm

1. Sperm is obtained via direct collection of pure concentrated sperm from the gonad after dissection (Fig. 2). The sperm can be stored at room temperature for short periods of time.
2. Prepare the equipment (Fig. 1), adding liquid nitrogen to the polystyrene box (1.2 L capacity) and a 7.8 cm thick, floating polystyrene rack (for cryopreservation at an approximate cooling rate of 11 °C/min). The volume of liquid nitrogen should be sufficient to ensure cooling, but space should be left for the polystyrene rack in which 0.25–0.5 mL straws will be placed. Cryopreservation security gear (gloves, forceps/tweezers, and protective eyewear) and a timer will also be required.
3. Add the cryoprotective agent (a solution of 15% (v/v) dimethyl sulfoxide in sterile natural seawater or artificial seawater filtered through a 0.22 μm filter and/or UVA-treated) to the sperm suspension in a 1:1 proportion in one step and allow 5 min to reach osmotic equilibrium while loading the straws with the solution. The straws are placed in the comb to make

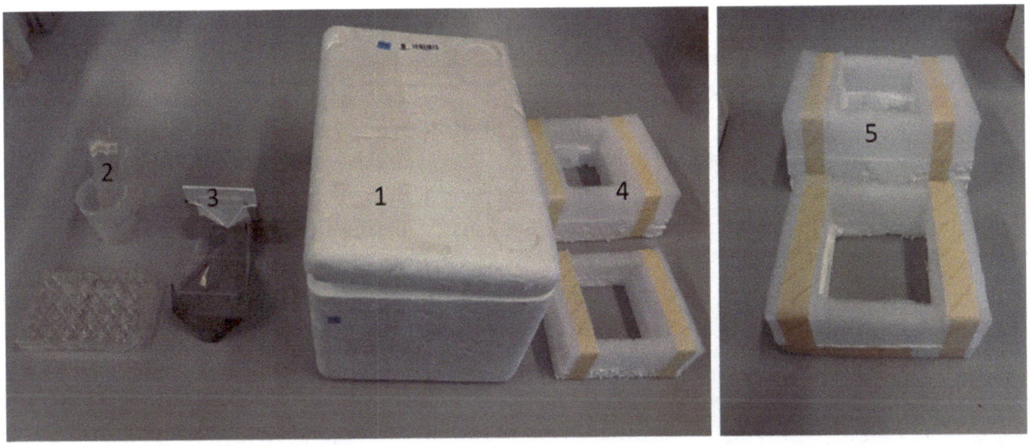

Fig. 1 Equipment used in cryopreservation in liquid nitrogen vapor: (1) styrofoam box, (2) straws, (3) comb to place the straws prior to applying sealing powder, (4) hand-made floating racks, and (5) close up of two floating racks with different distances to the liquid nitrogen surface (i.e., different cooling rates)

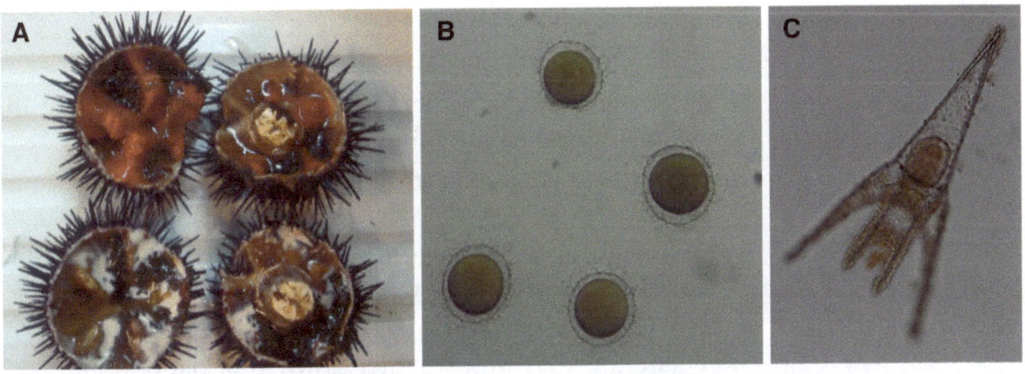

Fig. 2 **A** female (red or orange) and male (yellowish) gonads of adult sea urchins; **B** sea urchin eggs (90 μm) showing the fertilization membrane (excess of sperm can be seen concentrated around the eggs); **C** sea urchin 4-arm pluteus larvae (48 h, 450–500 μm)

room for sealing, and the straws are sealed with PVC powder. The powder is allowed to polymerize by immersion of the tip of the straw in a water bath for a few seconds, and the straw is carefully dried.

4. After equilibration, put the straws immediately on the polystyrene rack and cover the box. The mixture is allowed to cool for 15 min, after which it is plunged into liquid nitrogen. The samples can subsequently be stored in liquid nitrogen for as long as necessary.
5. For thawing, a water bath is prepared at 28°C, and the straws are plunged into the water for 10 s. The content of each straw is subsequently transferred to TCD plates and diluted with

seawater and a 1:1 ratio of cryoprotective agent to minimize toxic effects.

6. Post-thaw quality assessment: this protocol results in fewer motile sperm with free movement and slower movement than the control sperm. Fertilization with a 100:1 sperm-to-oocyte ratio leads to 50% production of normal 4-arm pluteus larvae after 48 h. Fertilization occurs in a matter of seconds; therefore, for a shorter period of time, the sperm/oocyte ratio can be increased until 100% fertilization is reached, and then the excess sperm should be removed prior to incubation (*see* **Note 1**).

3.2 Cryopreservation of Mussel (*Mytilus galloprovincialis*) Sperm

1. Spawn the mussel and, when a male is located, separate the individual into a dry beaker and prevent the mussel from closing the valves completely so that the concentrated sperm can be collected (*see* **Note 2**).
2. Dilute the sperm 1:1 with a cryoprotective agent solution (final concentration of 10% (v/v) dimethyl sulfoxide) and suck the mixture into 0.25 mL straws.
3. Allow equilibration for 5 min at room temperature (18°C) while loading the straws.
4. Place the straws into the 5-cm-thick floating rack above the liquid nitrogen level for 8 min in a polystyrene box.
5. Plunge into liquid nitrogen. The samples are stored in liquid nitrogen for as long as needed.
6. Thawing can be performed by immersion of the straws in a controlled-temperature water bath (35°C) until the samples have melted (6 s).
7. Dilute the CPA with seawater slowly to avoid osmotic shock (4% FSW addition in each step) until motility is regained.
8. Post-thaw quality assessment: when a visual examination of thawed sperm is performed, the cells are slow, and there is aggregation of cells with no movement. With cryopreserved sperm, fertilization requires at least a 100:1 sperm ratio to yield approximately 50% fertilization success after a contact time of 15 min. Mussels are sensitive to polyspermy; therefore, cells should be rinsed of the excess sperm before incubation (*see* **Note 3**).

3.3 Cryopreservation of Portuguese Oyster (*Crassostrea angulate*) Sperm

1. The oyster is opened via an oyster knife, and sex is determined microscopically.
2. The sperm is collected via the dry method directly from the gonad via a scalpel to make small cuts in the gonad (Fig. 3). A micropipette is used to collect the sperm in an Eppendorf tube.

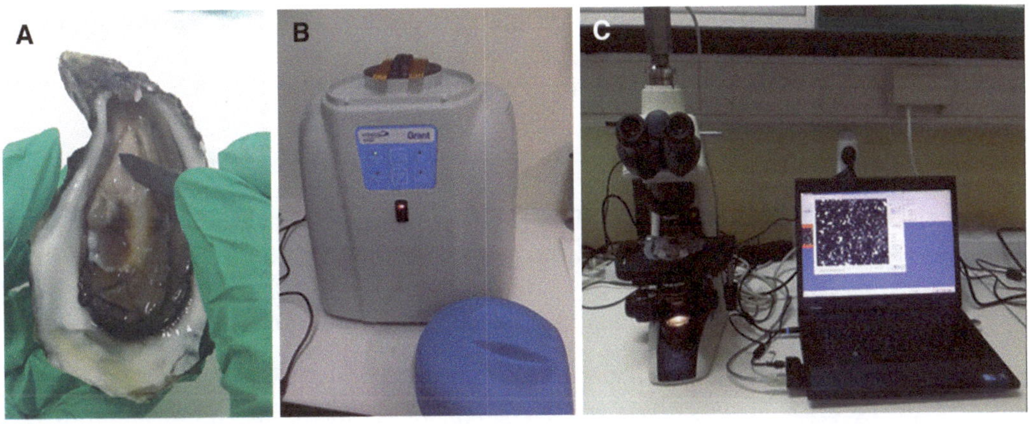

Fig. 3 **A** *Crassostrea angulata* sperm collection. **B** Freezing process and **C** post-thaw motility analysis

3. The sperm is filtered through 100 μm and 20 μm sieves to eliminate debris and other types of cells, and diluted 1:10 in artificial seawater. The cell concentration and total motility are assessed. Only samples with motility higher than 40% and final concentrations between 1 and 2×10^9spermatozoa/mL are cryopreserved.
4. For cryopreservation, dilute the sperm in a 1:1 (v/v) cryoprotectant mixture (20% (v/v) dimethyl sulfoxide in artificial seawater Reef-Salt, TMC) for a 10-min equilibration. The obtained suspension is placed into 0.5 mL plastic straws (Minitube, Germany, or IMV, France). If the straws are to be used for cryobanking purposes, they should be sealed according to the manufacturer's instructions. For laboratory experiments, these materials can be used without sealing.
5. A portable programmable biofreezer programmed with a freezing rate of 6°C/min from 0 to −70°C is used to freeze the straws (biofreezer Asymptote Grant EF600, UK, see Fig. 3). After freezing, the straws are stored directly in a liquid nitrogen container.
6. The straws are thawed in a water bath at 37°C for 10 s.
7. Post-thaw quality assessment: sperm quality is assessed via motility and cell viability. Motility is evaluated via computer-assisted sperm analysis (CASA). Several commercial CASA software programs are available (e.g., ISAS software (ISAS, Proiser, Valencia, Spain)). Analyses are carried out in a Makler chamber under a 10× negative-phase contrast objective coupled with a digital camera set at 25 fps. The samples were diluted 1:10 in artificial seawater prior to analysis. Motility analyses are performed by placing 10 μL of diluted sperm into a Makler chamber. After activation, the sperm move for several days. To

assess sperm viability, two methods can be used: fluorescence microscopy or flow cytometry, both using SYBR green and propidium iodide (PI) as stains. For fluorescence microscope analysis, 15 μL of diluted sperm, 0.5 μL of SYBR Green (final concentration 100 nM), and 1.5 μL of PI are mixed, and the samples are observed in a fluorescence microscope. At least 100 cells, live (SYBR green-positive, green cells) and dead (PI-stained, red cells), are counted. For flow cytometry analysis, 5 μL of sperm is diluted in 500 μL of 1% NaCl buffer. Two microliters of PI at a concentration of 2.4 mM are added to the suspension. The samples are analyzed with a flow cytometer after 5 min of incubation in the dark. After freezing and thawing, the percentage of motile sperm with no agglutination decreases drastically (approximately 2% motility), and the percentage of viable cells is approximately 60%. For fertilization, the sperm is used immediately after thawing, as the viability of spermatozoa decreases with time (*see* **Note 4**).

3.4 Cryopreservation of Sea Urchin (*Paracentrotus lividus*) Embryo

1. *P. lividus* early blastulas are obtained via incubation for 7–8 h at 20°C. For this purpose, male and female sea urchins should be dissected, and gametes should be collected directly from the gonads (Fig. 2). Upon quick examination of gamete quality (spherical oocytes with homogeneous color and motile sperm), fertilization should be conducted, and the percentage of fertilization success should be calculated. Batches with a fertilization rate under 80% should not be used.
2. *P. lividus* cells are incubated for 8 h (20°C) with gentle aeration. The incubation density should be between 40 and 80 fertilized eggs/mL.
3. Filter the embryos carefully with a 40–60 μm mesh filter and concentrate them. The cell density is calculated prior to cryopreservation.
4. The cryoprotective agents(CPAs) used consist of 3 M dimethyl sulfoxide (Me_2SO) + 0.08 M trehalose (TRE). The CPAs are added stepwise. One milliliter of CPA solution is added to 1 mL of embryo suspension in seawater in 15 equimolar steps 1 min apart (1:1 final dilution) at 19 ± 1°C.
5. The controlled rate freezer (Fig. 4) is programmed with a hold at 4°C for 2 min and then cooled at a rate of 1°C/min to −12°C. At this point, cryopreservation vials (2 mL) are seeded for 2 min, followed by cooling at 1°C/min to −80°C. Cryopreservation security gear (gloves, forceps/tweezers, and protective eyewear) and a timer will also be required. Thawing is performed by immersion in a 17 ± 1°C water bath until the ice is melted upon visual inspection.

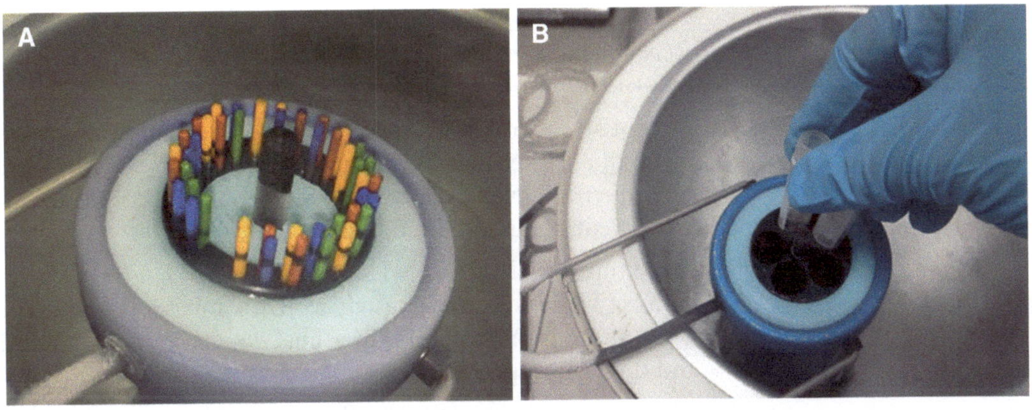

Fig. 4 Portable controlled-rate freezer with straws (**A**) and with 2 mL vials (**B**) (Cryologic Ltd. Australia)

6. CPAs should then be diluted with clean ASW in 12 equimolar steps, 1 min apart at room temperature (19 ± 1°C), and embryos can finally be rinsed with clean ASW.
7. Post-thaw quality assessment: embryos can be incubated for 96 h until they reach the 4-armed pluteus stage. Cryopreserved sea urchin embryos take up to 96 h to develop into 4-armed pluteus, whereas fresh embryos take 48 h (18–20°C). After 96 h, cryopreserved embryos will lead to >90% healthy 4-armed pluteus larvae that are typically half the size of control larvae (Fig. 2) (*see* **Note 5**).

3.5 Cryopreservation of Sea Urchin (*Paracentrotus lividus*) Pluteus Larvae

1. See general equipment needed for controlled rate freezer cryopreservation via straws and straw-specific equipment (Fig. 4, *see* **Note 5**).
2. *P. lividus* pluteus larvae are obtained via incubation for 48 h at 20 ± 1°C. For this purpose, male and female sea urchins should be dissected, and gametes should be collected directly from the gonads (Fig. 2). Upon quick examination of gamete quality (spherical oocytes with homogeneous color and motile sperm), fertilization should be conducted, and the percentage of fertilization success should be calculated. Batches with a fertilization rate under 80% are not used.
3. Following fertilization, the embryos are incubated for 48 h in UV-sterilized filtered seawater (UV-FSW) with gentle aeration, inside an isothermal chamber maintained at 20 ± 1°C. The incubation density should be between 20 and 40 fertilized eggs/mL.
4. Larval feeding is initiated at 24 h post-fertilization (hpf) with an *ad libitum* microalgal diet, adjusted according to the developmental stage-specific feeding demands. The standard hatchery formulation included *Chaetoceros neogracilis*

Van Land., 1968, *Tisochrysis lutea* El M. Bendif & I., 2013, *Rhodomonas lens* Pascher & Ruttner, 1913, *Phaeodactylum tricornutum* Bohlin, 1897.

5. After 48 h, 4-arm pluteus larvae are collected by filtration through a 40 μm mesh. The larvae retained on the mesh are subsequently transferred to fresh FSW for larval counting and assessment of larval development.
6. A mixture of 3 M dimethyl sulfoxide (Me_2SO) + 0.08 M trehalose (TRE) solution is used, which is mixed 1:1 with the larval suspension, resulting in a final concentration of 1.5 M and 0.04 M, respectively.
7. The cryoprotective agents(CPAs) are added stepwise. One milliliter of CPA solution is added to 1 mL of larvae suspension in seawater in 15 equimolar steps 1 min apart (1:1 final dilution) at 20 ± 1°C. Use French straws (0.25 mL).
8. Apply the cooling rate in the speed-controlled freezer. The cooling is started by holding at 4 °C for 2 min followed by cooling at a rate of 1 °C/min to −12°C. At this point, the straws are seeded for 2 min, followed by cooling at 1 °C/min to −35°C. At this stage, the straws should be transferred into liquid nitrogen for cryogenic storage, either for long-term preservation in a biobank or for subsequent thawing and use.
9. Thawing is performed by immersion in a 35 ± 1°C water bath until the ice is melted upon visual inspection (6 s). The contents of the straw are then transferred to a vial, and FSW/ASW is added at a 1:1 ratio in 12 equimolar steps/min to dilute the CPAs at room temperature (20 ± 1°C), and larvae can finally be rinsed with clean FSW or ASW.
10. Post-thaw quality assessment: pluteus larvae survival can be assessed in the short term in several ways, such as by monitoring feeding activity or swimming behavior. Long-term assessment will require the assessment of growth and development during larval rearing and settlement capacity (*see* **Note 5**) [33].

3.6 Cryopreservation of Mussel (*Mytilus galloprovincialis*) D-Larva

1. See general equipment needed for controlled rate freezer cryopreservation via straws and straw-specific equipment (Fig. 4, *see* **Note 6**).
2. The spawned mussels are collected separately. In the case of males, they are placed separately in backers with the umbo facing upwards. Upon quick examination of gamete quality (spherical oocytes with homogeneous color and motile sperm), fertilization should be conducted, and the percentage of fertilization success should be calculated. Batches with a fertilization rate under 80% are not used.

3. Fertilized *M. galloprovincialis* eggs are incubated for 72 h (18 °C ±1°C) with gentle aeration. The incubation density should be between 20 and 40 fertilized eggs/mL.
4. Larval feeding is initiated at 48 h post-fertilization (hpf) with an *ad libitum* microalgal diet, adjusted according to the developmental stage-specific feeding demands. The standard hatchery formulation included *Chaetoceros neogracilis* Van Land., 1968, *Tisochrysis lutea* El M. Bendif & I. Probert, 2013, *Rhodomonas lens* Pascher & Ruttner, 1913, *Phaeodactylum tricornutum* Bohlin, 1897, and *Tetraselmis suecica* Butcher, 1959.
5. Solutions of ethylene glycol 20% (v/v) and trehalose 0.8 M are prepared and mixed in a 1:1 ratio with the larval suspension, resulting in a final concentration of 10% and 0.4 M, respectively, in filtered (0.22 μm + UVA) natural seawater (FSW) or artificial seawater (ASW).
6. The cryoprotective agents are added in one step. One milliliter of CPA solution is added to 1 mL of larvae suspension in seawater, at 18 ± 1°C. Use French straws (0.25 mL).
7. Allow for an equilibration time of 60 min at 18°C while loading into straws and sealing them. After 60 min, the straws are placed in the controlled-rate freezer, ensuring that they are completely dry on the outside.
8. The cryopreservation protocol starts with holding at 4°C, cooling at 1°C/min to −12°C, and then holding for 2 min for seeding. After this, the mixture is cooled at 1°C/min to −35°C. At this stage, the straws should be transferred into liquid nitrogen for cryogenic storage, either for long-term preservation in a biobank or for subsequent thawing and use.
9. Thawing is performed by immersion in a 35 ± 1 °C water bath until the ice is melted upon visual inspection (6 s). The content of the straws is then transferred to a vial, FSW/ASW is added at a 1:1 ratio, and larvae can finally be rinsed with clean FSW or ASW.
10. Post-thaw quality assessment: D-larvae survival can be assessed in the short term in several ways, such as by monitoring feeding activity or swimming behavior. Long-term assessment will require the assessment of growth and development during larval rearing and settlement capacity (*see* **Note 6**) [34, 35].

3.7 Cryopreservation of Mussel (Mytilus galloprovincialis) 72 hpf–20 dpf larvae

1. See general equipment needed for controlled rate freezer cryopreservation via straws and straw-specific equipment (Fig. 4, *see* **Note 6**).

2. The spawned mussels are collected separately. In the case of males, they are placed separately in backers with the umbo facing upwards. Upon quick examination of gamete quality (spherical oocytes with homogeneous color and motile sperm), fertilization should be conducted, and the percentage of fertilization success should be calculated. Batches with a fertilization rate under 80% should not be used.
3. Fertilized *M. galloprovincialis* eggs are incubated until the desired larval stage, from 72 hpf (hours postfertilization) to 20 dpf (days postfertilization) (20 ± 1°C) with gentle aeration. The incubation density should be between 20 and 40 fertilized eggs/mL.
4. Larval feeding is initiated at 48 hpf with an *ad libitum* microalgal diet, adjusted according to the developmental stage-specific feeding demands. The standard hatchery formulation included *Chaetoceros neogracilis* Van Land., 1968, *Tisochrysis lutea* El M. Bendif & I. Probert, 2013, *Rhodomonas lens* Pascher & Ruttner, 1913, *Phaeodactylum tricornutum* Bohlin, 1897, and *Tetraselmis suecica* Butcher, 1959.
5. Solutions of ethylene glycol 20% (v/v) and trehalose 0.8 M are prepared and mixed in a 1:1 ratio with the larval suspension, resulting in a final concentration of 10% and 0.4 M, respectively, in filtered (0.22 μm + =UV) FSW/ASW.
6. The cryoprotective agents are added in one step. One milliliter of CPA solution is added to 1 mL of larvae suspension in seawater, at 18 ± 1°C. Use French straws (0.25 mL).
7. Allow for an equilibration time of 60 min at 18 ± 1°C while loading into straws and sealing them. After 60 min, the straws are placed in the controlled-rate freezer, ensuring that they are completely dry on the outside.
8. The cryopreservation protocol starts with holding at 4°C, cooling at 1°C/min to −12°C, and then holding for 2 min for seeding. After this, the mixture is cooled at 1°C/min to −35°C. At this stage, the straws should be transferred into liquid nitrogen for cryogenic storage, either for long-term preservation in a biobank or for subsequent thawing and use.
9. Thawing is performed by immersion in a 35 ± 1°C water bath until the ice is melted upon visual inspection (6 s). The content of the straws is then transferred to a vial, FSW/ASW is added at a 1:1 ratio, and larvae can finally be rinsed with clean FSW/ASW.
10. Post-thaw quality assessment: The larvae's survival can be assessed in the short term in several ways, such as by monitoring feeding activity or swimming behavior. Long-term assessment will require the assessment of growth and development

during larval rearing and settlement capacity (*see* **Note 7**) [36].

3.8 Cryopreservation of Mussel (M. galloprovincialis) Juvenile (40 dpf–56 dpf)

1. See general equipment needed for controlled rate freezer cryopreservation via straws and straw-specific equipment (Fig. 4, *see* **Note 7**).
2. The spawned mussels are collected separately. In the case of males, they are placed separately in backers with the umbo facing upwards. Upon quick examination of gamete quality (spherical oocytes with homogeneous color and motile sperm), fertilization should be conducted, and the percentage of fertilization success should be calculated. Batches with a fertilization rate under 80% should not be used.
3. Fertilized *M. galloprovincialis* eggs are incubated for incubate until the desired larval stage, from 40 dpf to 56 dpf (20 ± 1°C) with gentle aeration. The incubation density should be between 20 and 40 fertilized eggs/mL.
4. Larval feeding is initiated at 48 h postfertilization (hpf) with an *ad libitum* microalgal diet, adjusted according to the developmental stage-specific feeding demands. The standard hatchery formulation included *Chaetoceros neogracilis* Van Land., 1968, *Tisochrysis lutea* El M. Bendif & I. Probert, 2013, *Rhodomonas lens* Pascher & Ruttner, 1913, *Phaeodactylum tricornutum* Bohlin, 1897, and *Tetraselmis suecica* Butcher, 1959.
5. Solutions of ethylene glycol 24% (v/v) and trehalose 0.8 M are prepared and mixed in a 1:1 ratio with the larval suspension, resulting in a final concentration of 12% and 0.4 M, respectively, in filtered (0.22 μm + UV) FSW/ASW.
6. The cryoprotective agents(CPAs) are added in one step. One milliliter of CPA solution is added to 1 mL of larvae suspension in seawater, at 18 ± 1°C. Use French straws (0.50 mL).
7. Allow for an equilibration time of 60 min at 18 ± 1°C while loading into straws and sealing them. After 60 min, the straws are placed in the controlled-rate freezer, ensuring that they are completely dry on the outside.
8. The cryopreservation protocol starts with holding at 4°C, cooling at 1 °C/min to −12°C, and then holding for 2 min for seeding. After this, the mixture is cooled at 1 °C/min to −35°C. At this stage, the straws should be transferred into liquid nitrogen for cryogenic storage, either for long-term preservation in a biobank or for subsequent thawing and use.
9. Thawing is performed by immersion in a 35 ± 1°C water bath until the ice is melted upon visual inspection (around 10 s). The content of the straws is then transferred to a vial, FSW/

ASW is added at a 1:1 ratio, and larvae can finally be rinsed with clean FSW/ASW.

10. Post-thaw quality assessment: juvenile survival can be assessed in the short term in several ways, such as by monitoring feeding activity or swimming behavior. Long-term assessment will require the assessment of growth and development during larval rearing and settlement capacity (*see* **Note 7**).
11. For damage analysis and the determination of survival percentage, live-dead fluorescence dye can be used (Fig. 5), consisting of propidium iodide 1 mg/1 mL (staining shows dead cells in red) and Hoechst 1 M (live cells with active DNA are shown in blue). Larval damage was analyzed using fluorescence microscopy (*see* **Note 7**).

3.9 Cryopreservation of Portuguese Oyster (Crassostrea angulate) D-Larvae

1. The oysters are opened, and the sex is determined by microscopic observation of gametes obtained by a small incision in the gonad.
2. Egg suspensions are filtered at 100 μm and retained in a 20 μm mesh screen, while spermatozoa are sieved at 20 μm, to reduce potential contaminations (see previous protocol). Both gametes are maintained in filtered and UV-sterilized seawater (FSW), until the fertilization.
3. For quality assessment before fertilization, sperm motility and concentration are checked by the CASA system, while oocyte sphericity and concentration are confirmed at light microscopy after 20 min FSW contact.
4. Fertilization is done in a biker by mixing both gametes at a proportion of 1000:1 (spermatozoon to oocyte ratio).
5. One hour after fertilization, the eggs are filtered at a 20 μm mesh screen to remove the remaining sperm and incubated (100 eggs/mL) at 21 ± 1°C for 24–30 h, with slight aeration until reaching the D-larvae stage.
6. Before the cryopreservation procedure, D-larvae are recovered in a 30 μm mesh screen and kept in FSW to determine their concentration using a Sedgewick Rafter chamber. D-larvae are concentrated at 120,000 larvae/mL in FSW and kept on ice (4°C) for a maximum of 2 h.
7. Dilute D-larvae in the freezing media at 1:1 (larvae suspension: CPAs media) consisting of a final concentration of 10% (v/v) dimethyl sulfoxide (Me_2SO), 1% (w/v) Polyvinylpyrrolidone (PVP-40) 40,000 MW, and 0.2 M Sucrose prepared in mil-li-Q water.
8. Larvae are incubated in the cryoprotectant solution for 3 min (equilibrium time) at 4°C. Higher exposure time may interfere with larval movement patterns. During the equilibrium

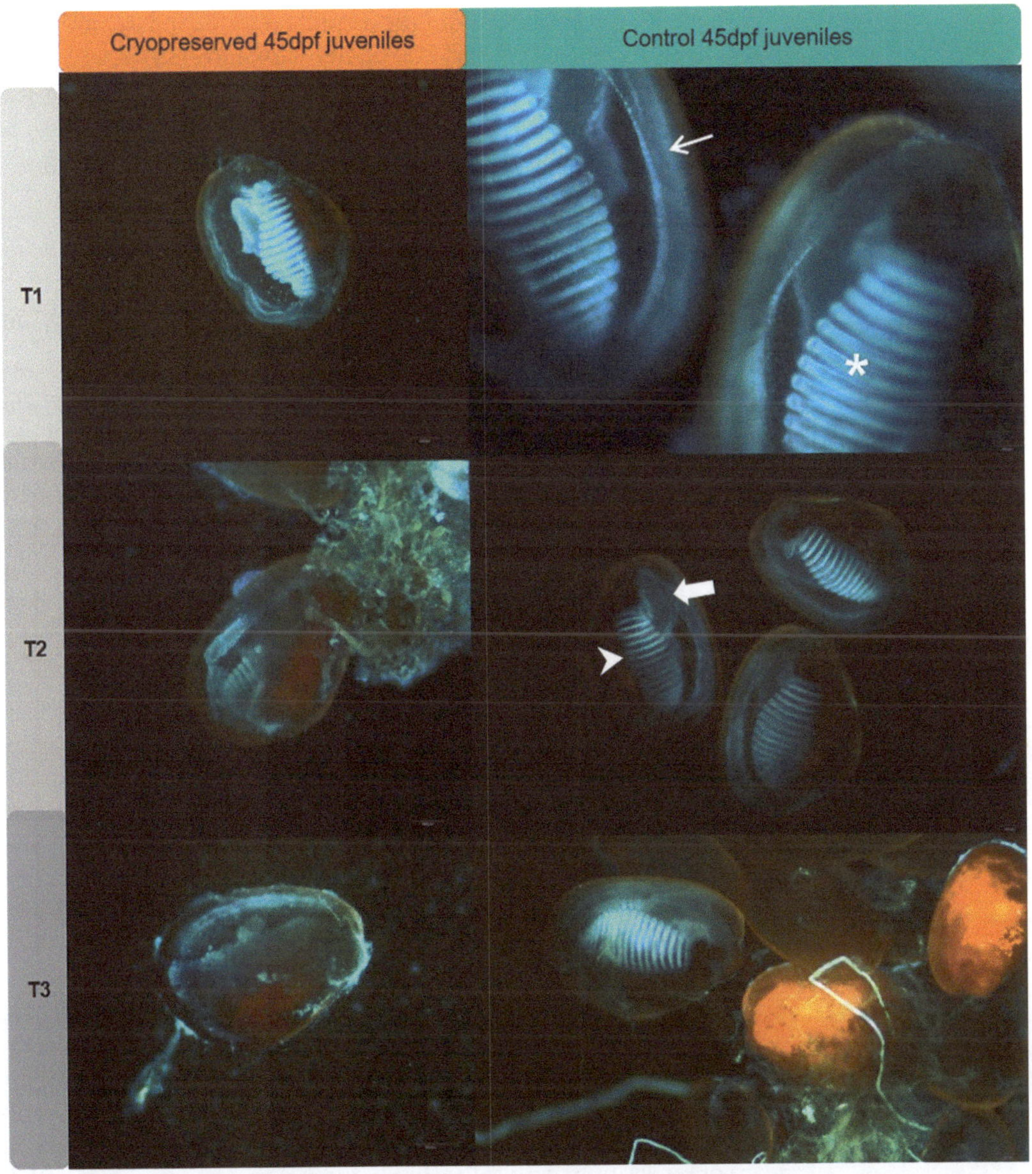

Fig. 5 Images of cryopreserved juveniles with EG12% + TRE 0.4 M at three time points T1, (24 hpt), T2 (48 hpt), and T3 (72 hpt) imaged with Propidium Iodide (red, dead cells) and Hoechst (blue, active DNA) dye (D) The thin arrow points to the edge of the mantle, the thick arrow to the foot, the caret to the digestive tract and the asterisk to the gills

time, D-larvae are loaded into 0.5 mL French straws (30,000 larvae per straw) and maintained at 4°C.

9. Straws are placed in programable biofreezer (e.g., for field use or low number of straws Asymptote Grant EF600, UK is appropriated; for commercial use a benchtop biofreezer is recommended) following the freezing protocol: 2.5°C/min

from 0 to −10°C, hold for 5 min at −10°C, 0.3°C/min from −10°C to −20°C and 2.5°C/min down to −35°C, and finally plunged into liquid nitrogen and stored in a liquid nitrogen container.

10. The straws are thawed in a water bath set at 37°C for 10 s.
11. A recovery bath is prepared by diluting the contents of each straw in 2 L of FSW for an incubation period of 1 h at room temperature, allowing cryoprotectant dissipation.
12. For further use, wash and concentrate D-larvae using a 30 μm mesh screen in 1.5 mL of FSW.
13. D-larvae quality assessment: cryodamage can be assessed using cellular and molecular markers. The candidate genes *adgre3*, *dynein beta* (oyster growth), *mp*, *fbn2*, *myob3b* (mantle and shell formation), *epx*, *hsp70*, *bp10*, *muc19*, *socs5*, *Lectin* (oxidative stress and immune response) are good biomarkers of quality, both validated using RNAseq and qPCR. Motility pattern and velocity can also be a good indicator of larval status [37, 38] (*see* **Note 8**).

3.10 What About Egg Cryopreservation?

Egg cryopreservation in marine invertebrates is a developing field with considerable potential, but also significant biological and technical challenges. It is well-established in mammals and some fish but is still experimental in most marine invertebrates. Eggs have proved to be more challenging to cryopreserve than other cell types, and several reasons have been mentioned as possible causes. Among these reasons, the large size of the eggs and subsequently their low surface area to volume ratio have been to blame for this hardship. The high-water content, their high lipid content, and their low permeability, membrane complexity, and sensitivity to chemicals in water are also frequently mentioned factors that make their cryopreservation difficult. The egg is one of the largest cells in the organism and, for this reason, is characterized by a low surface area-to-volume ratio compared with other cell types, which means that, in order to survive cryopreservation, it must lose more water. Also, their large size makes uniform freezing and thawing difficult. Furthermore, they have high water and lipid content, which makes cellular dehydration necessary to prevent ice formation and consequently damages structures, low permeability, which makes CPA loading difficult, and high chilling sensitivity even to moderate cooling (above freezing), which leads to damage before freezing occurs [39].

Attempts have been made to cryopreserve eggs from starfish, oysters, and mussels with a certain percentage of success, but the results have not been ideal. Starfish (*Asterina miniata*) eggs were cryopreserved using short exposure to a CPA cocktail with dimethyl sulfoxide (Me_2SO), trehalose, and hypotonic artificial seawater (ASW), followed by ultra-rapid cooling. After thawing, CPA removal, and 1 h recovery, they observed a survival greater than 34%, but a small percentage was capable of developing into an embryo [40]. Regarding oyster (*Crassostrea gigas*) eggs cryopreservation, the best results were obtained with 10% ethylene glycol (EG) in Milli-Q water cooling at 1°C/min to a −10°C hold and then 0.3°C/min to −35°C followed by plunging into liquid nitrogen. High fertilization rates and development to eyed larvae and spat were reported [41]. Blue mussels (*Mytilus galloprovincialis*) eggs cryopreservation was done using 7.5% Ficoll (FIC) plus 10% EG, and the post-thaw CPA removal medium consisting of 9% sucrose. After cryopreservation, the developmental capacity of eggs was reduced, resulting in a significantly lower D-larval rate in comparison with controls. However, the subsequent development of resultant D-larvae was not affected by the cryopreservation, as their survival and growth rates to the spat stage were similar to controls [42].

In sea urchin *Paracentrotus lividus*, egg cryopreservation is still not possible nowadays. The blastula embryo (8 h-postfertilization) is the earliest cell that can be cryopreserved [43] and has been proven to be able to complete development to juveniles [44]. In sea urchins, the main limiting factor is neither the cell's shape nor volume, nor membrane impermeability, as has been observed in fish eggs [39]. However, it is worth noting that sea urchin eggs do exhibit low membrane permeability. One of the main challenges in sea urchin egg cryopreservation is the cytotoxic nature of the CPAs [45]. These cells exhibit high sensitivity to CPAs, even at low concentrations, which has complicated the development of effective cryopreservation protocols. Additionally, the susceptibility of these eggs to chilling injury—even at exposure to low but nonfreezing temperatures like 4°C—further complicates preservation efforts. These factors combined have so far prevented the establishment of a reliable cryopreservation or vitrification protocol for sea urchin eggs, highlighting the need for continued research into less toxic cryoprotectants and alternative preservation strategies [46].

4 Conclusions

Whether our ultimate aim is to advance basic research, improve aquaculture practices, or support biodiversity conservation, it is essential to understand key aspects of aquatic organisms' reproductive modes and strategies, spawning induction, gamete collection, fertilization conditions, gamete quality, and successful larval incubation in order to develop cryopreservation protocols for gametes, larvae, and juveniles. The protocols presented here are the latest state-of-the-art for these cell types and species, but they will continue to evolve as new techniques are implemented, lower toxicity CPAs are designed, and more tools are developed to diagnose cryodamage. Low toxicity CPAs are one of the most crucial factors that the cryobiology community can focus on in order not only to develop new protocols but also to improve the survival rates of the existing ones in the near future. Marine invertebrates are well known for being very successful and sensitive marine pollution bioindicators, and that characteristic that makes them exceptionally valuable in marine ecotoxicology makes them more complicated to cryopreserve. Marine invertebrate cryopreservation and aquatic cryopreservation in general have developed significantly since the 70s, and more and improved protocols are being produced every year.

5 Notes

1. A method for non-detrimental sea urchin sex differentiation can be found in [47].
2. There is always sperm activation due to intervalvular seawater, but the concentration is maximized.
3. Related information: Spawning can be conducted by thermal cycling. Mussels can be placed on a PVC container, and spawning can be triggered by changes in the seawater temperature from 20°C to 4°C over several cycles.
4. More information on Portuguese oyster sperm cryopreservation was provided by Riesco et al. [23, 48] and Anjos et al. [49].
5. More information on *P. lividus*embryocryopreservation can be found in [44]. For information on the existing protocols for other species of sea urchins, see Paredes et al. [44, 50].
6. More information on mussel larvae cryopreservation is available in [17, 36, 50, 51].
7. More information on mussel juvenile cryopreservation is available in [36].
8. More information on Portuguese oyster D-larvae quality assessment can be found in [37, 38].

Acknowledgments

Sea urchin work was supported by OCIMER and OCIMER+ projects funded by Fundación Biodiversidad—Pleamar Grants. Mussel work was funded by EU H2020 Assemble plus and recent advances by Xunta de Galicia with NextGenerationEU and the European Maritime Fisheries and Aquaculture Funds. Oyster work received Portuguese national funds from FCT—Foundation for Science and Technology through contracts UID/04326/2025, UID/PRR/04326/2025, and LA/P/0101/2020 (DOI:10.54499/LA/P/0101/2020). E.P. holds a Ramon y Cajal grant by MICIU/AEI/10.13039/501100011033 and by "European Union Next-GenerationEU/PRTR". A.L.'s salary is funded through this grant.

References

1. Mermillod-Blondin F, François-Carcaillet F, Rosenberg R (2005) Biodiversity of benthic invertebrates and organic matter processing in shallow marine sediments: an experimental study. J Exp Marine Biol Ecol 315:187–209. https://doi.org/10.1016/j.jembe.2004.09.013
2. FAO Yearbook (n.d.) Fishery and Aquaculture Statistics 2017/FAO annuaire. Statistiques des pêches et de l'aquaculture 2017/ FAO Yearbook. Fishery and Aquaculture Statistics 2017. Retrieved August 27, 2025, from https://openknowledge.fao.org/items/7cac7d45-83e9-49ec-901b-1c7ad5e4d44c
3. Krisfalusi-Gannon J, Ali W, Dellinger K, Robertson L, Brady TE, Goddard MKM, Tinker-Kulberg R, Kepley CL, Dellinger AL (2018) The role of horseshoe crabs in the biomedical industry and recent trends impacting species sustainability. Front Mar Sci 5:185. https://doi.org/10.3389/fmars.2018.00185
4. Poon J, Fries A, Wessel GM, Yajima M (2019) Evolutionary modification of AGS protein contributes to formation of micromeres in sea urchins. Nat Commun 10: 3779. https://doi.org/10.1038/s41467-019-11560-8
5. Clark GF, Johnston EL (2011) Temporal change in the diversity-invasibility relationship in the presence of a disturbance regime. Ecol Lett 14:52–57. https://doi.org/10.1111/j.1461-0248.2010.01550.x
6. Lawrence JM (2006) Edible sea urchins: biology and ecology, Volume 38, 2nd Edition. https://www.elsevier.com/books/edible-sea-urchins-biology-and-ecology/lawrence/978-0-444-52940-4
7. MacKenzie CL (1997) The history, present condition, and future of the molluscan fisheries of North and Central America and Europe (Vol. 127). US Department of Commerce
8. Council NR (2004) Studies D on E and L., & Board, O. S. Nonnative oysters in the Chesapeake Bay. National Academies Press
9. Lenihan HS, Peterson CH (1998) How habitat degradation through fishery disturbance enhances impacts of hypoxia on oyster reefs. Ecol Appl 8:128–140. https://doi.org/10.1890/1051-0761(1998)008[0571:WICPTS]2.0.CO;2
10. Lenihan HS, Peterson CH (2004) Conserving oyster reef habitat by switching from dredging and tonging to diver-harvesting. Fish Bull 102:298–306
11. Lenihan HS (1999) Physical–biological coupling on oyster reefs: how habitat structure influences individual performance. Ecol Monogr 69:251–275. https://doi.org/10.1890/0012-9615(1999)069[0251:PBCOOR]2.0.CO;2
12. Lenihan HS, Micheli F, Shelton SW, Peterson CH (1999) The influence of multiple environmental stressors on susceptibility to parasites: an experimental determination with oysters. Limnol Oceanogr 44:910–924. https://doi.org/10.4319/lo.1999.44.3_part_2.0910
13. Comps M, Bonami JR, Vago C (1976) Une virose de l'huître portugaise (*Crassostrea angulata* Lmk). C R l'Acad Sci 282:139–143
14. Andrew NL, Agatsuma Y, Ballesteros E, Bazhin AG, Creaser EP, Barnes DK, Botsford LW, Bradbury A, Campbell A, Dixon JD (2002) Status and management of world sea urchin fisheries. Oceanogr Mar Biol. 351–438
15. Stefánsson G, Kristinsson H, Ziemer N, Hannon C, James P (2017) Markets for sea urchins: a review of global supply and markets. Skýrsla Matís 45:10–17
16. Intergovernmental Science-Policy Platform on Biodiversity and Ecosystem Services (IPBES) 2019 https://www.ipbes.net/assessment-reports
17. Paredes E (2015) Exploring the evolution of marine invertebrate cryopreservation—landmarks, state of the art and future lines of research. Cryobiology 71:198–209. https://doi.org/10.1016/j.cryobiol.2015.08.011
18. Fatihah SN, Jasmani S, Abol-Munafi AB, Noorbaiduri S, Muhd-Farouk H, Ikhwanuddin M (2016) Development of a sperm cryopreservation protocol for the mud spiny lobster, *Panulirus polyphagus*. Aquaculture 462:56–63. https://doi.org/10.1016/j.aquaculture.2016.04.025
19. Heres P, Rodriguez-Riveiro R, Troncoso J, Paredes E (2019) Toxicity tests of cryoprotecting agents for *Mytilus galloprovincialis* (Lamark, 1819) early developmental stages. Cryobiology 86:40–46. https://doi.org/10.1016/j.cryobiol.2019.01.001

20. Ribeiro RC, da Silva Veronez AC, Tovar TT, Adams S, Bartolomeu DA, Peronico C, Furley TH (2018) Cryopreservation: extending the viability of biological material from sea urchin (*Echinometra lucunter*) in ecotoxicity tests. Cryobiology 80:139–143. https://doi.org/10.1016/j.cryobiol.2017.10.002
21. Ribeiro MB, Furley T, Spago FR, Paredes E (2018) First steps towards *Echinometra lucunter* embryo cryopreservation. Cryobiology 80:51–54. https://doi.org/10.1016/j.cryobiol.2017.12.005
22. FAO FAO Pesca Mytilus galloprovincialis. (s. f.). Recuperado 19 de julio de 2024, de https://firms.fao.org/fi/website/FIRetrieveAction.do?dom=culturespecies&xml=Mytilus_galloprovincialis.xml
23. Riesco MF, Félix F, Matias D, Joaquim S, Suquet M, Cabrita E (2017) First study in cryopreserved *Crassostrea angulata* sperm. Gen Comp Endocrinol 245:108–115. https://doi.org/10.1016/j.ygcen.2016.05.003
24. Viyakarn V, Chavanich S, Chong G, Tsai S, Lin C (2018) Cryopreservation of sperm from the coral *Acropora humilis*. Cryobiology 80:130–138. https://doi.org/10.1016/j.cryobiol.2017.10.007
25. Hassan MM, Qin JG, Li X (2015) Sperm cryopreservation in oysters: a review of its current status and potentials for future application in aquaculture. Aquaculture 438:24–32. https://doi.org/10.1016/j.aquaculture.2014.12.037
26. Tiersch TR, Green CC, World Aquaculture Society (2011) Cryopreservation in aquatic species: a comprehensive overview of current practices, programmatic development and future directions for cryopreservation of gametes embryos and larvae of aquatic species
27. Martínez-Páramo S, Horváth Á, Labbé C, Zhang T, Robles V, Herráez P, Suquet M, Adams S, Viveiros A, Tiersch TR, Cabrita E (2017) Cryobanking of aquatic species. Aquaculture 472:156–177. https://doi.org/10.1016/j.aquaculture.2016.05.042
28. Adams SL, Smith JF, Roberts RD, Janke AR, King NG, Tervit HR, Webb SC (2008) Application of sperm cryopreservation in selective breeding of the Pacific oyster, *Crassostrea gigas* (Thunberg). Aquacult Res 39:1434–1442. https://doi.org/10.1111/j.1365-2109.2008.02013.x
29. Adams S, Tervit HR, Salinas-Flores L, Smith JF, Mcgowan L, Roberts RD, Janke A, King N, Webb S, Gale S (2011) Cryopreservation of Pacific oyster oocytes. Cryopreserv Aquat Species 616–623
30. Suquet M, Labbé C, Puyo S, Mingant C, Quittet B, Boulais M, Queau I, Ratiskol D, Diss B, Haffray P (2014) Survival, growth and reproduction of cryopreserved larvae from a marine invertebrate, the Pacific oyster (*Crassostrea gigas*). PLoS ONE 9:e93486. https://doi.org/10.1371/journal.pone.0093486
31. Boudouresque C, Verlaque M (2007) Chapter 13 Ecology of Paracentrotus lividus. Developments in Aquaculture and Fisheries Science 37:243–285. https://doi.org/10.1016/S0167-9309(07)80077-9
32. Batista F, Leitao A, Huvet A, Lapegue S, Heurtebise S, Boudry P (2005) The taxonomic status and origin of the Portuguese oyster *Crassostea angulata* (Lamark, 1819). En 1st International Oyster Symposium, Tokyo, Japan, July 13–14. https://archimer.ifremer.fr/doc/00000/3433/
33. Lago A, Paredes E (2023) Modulation of stress factors for cryopreservation of *Paracentrotus lividus* (Lamarck 1816) larvae. Cryobiology 110:8–17. https://doi.org/10.1016/j.cryobiol.2023.01.006
34. Heres P (2021) Mollusk cryopreservation and variability along development and species (Http://purl.org/dc/dcmitype/Text, Universidade de Vigo). https://dialnet.unirioja.es/servlet/tesis?codigo=303462
35. Heres P, Troncoso J, Paredes E (2022) Long-term study on survival and development of successive generations of *Mytilus galloprovincialis* cryopreserved larvae. Sci Rep 12:1–14. https://doi.org/10.1038/s41598-022-17935-0
36. Lago A, Troncoso J, Paredes E (2025) Cryopreservation of juvenile *Mytilus galloprovincialis* to safeguard mollusk biodiversity and support aquaculture. Sci Rep 15: 25587. https://doi.org/10.1038/s41598-025-11439-3
37. Anjos C, Duarte D, Diogo P, Matias D, Cabrita E (2022) Assessment of larval quality of two bivalve species, *Crassostrea angulata* and *Chamelea gallina*, exposed and cryopreserved with different cryoprotectant solutions. Cryobiology 106:24–31. https://doi.org/10.1016/j.cryobiol.2022.04.007
38. Anjos C, Duarte D, Fatsini E, Matias D, Cabrita E (2024) Comparative transcriptome analysis reveals molecular damage associated with cryopreservation in *Crassostrea angulata* D-larvae rather than to cryoprotectant exposure. BMC Genomics 25:591. https://doi.org/10.1186/s12864-024-10473-1
39. Campos S, Troncoso J, Paredes E (2021) Major challenges in cryopreservation of sea urchin eggs. Cryobiology 98:1–4. https://doi.org/10.1016/j.cryobiol.2020.11.008
40. Hamaratoglu F, Eroglu A, Toner M, Sadler KC (2005) Cryopreservation of starfish oocytes. Cryobiology 50:38–47. https://doi.org/10.1016/j.cryobiol.2004.09.008
41. Tervit HR, Adams SL, Roberts RD, McGowan LT, Pugh PA, Smith JF, Janke AR (2005) Successful cryopreservation of Pacific oyster

(*Crassostrea gigas*) oocytes. Cryobiology 51:142–151. https://doi.org/10.1016/j.cryobiol.2005.06.001
42. Liu Y, Li X (2015) Successful oocyte cryopreservation in the blue mussel *Mytilus galloprovincialis*. Aquaculture 438:55–58. https://doi.org/10.1016/j.aquaculture.2015.01.002
43. Bellas J, Paredes E (2011) Advances in the cryopreservation of sea-urchin embryos: potential application in marine water quality assessment. Cryobiology 62:174–180. https://doi.org/10.1016/j.cryobiol.2011.02.005
44. Paredes E, Bellas J, Costas D (2015) Sea urchin (*Paracentrotus lividus*) larval rearing—culture from cryopreserved embryos. Aquaculture 437:366–369. https://doi.org/10.1016/j.aquaculture.2014.12.022
45. Olver DJ, Heres P, Paredes E, Benson JD (2023) Rational synthesis of total damage during cryoprotectant equilibration: modelling and experimental validation of osmomechanical, temperature, and cytotoxic damage in sea urchin (*Paracentrotus lividus*) oocytes. PeerJ 11:e15539. https://doi.org/10.7717/peerj.15539
46. Campos S, Rodriguez U, Troncoso J, Paredes E (2025) Understanding osmoregulation, cryoprotectant toxicity, and cold tolerance in sea urchin eggs: implications for cryopreservation. Cryobiology 119:105235. https://doi.org/10.1016/j.cryobiol.2025.105235
47. Paredes E, Costas D (2020) Non-lethal sex identification of sea urchins: method and advantages. Lab Anim (NY) 49:7–8. https://doi.org/10.1038/s41684-019-0439-y
48. Riesco MF, Félix F, Matias D, Joaquim S, Suquet M, Cabrita E (2019) Comparative study on cellular and molecular responses in oyster sperm revealed different susceptibilities to cryopreservation. Aquaculture 498:223–229. https://doi.org/10.1016/j.aquaculture.2018.08.049
49. Anjos C, Santos AL, Duarte D, Matias D, Cabrita E (2021) Effect of trehalose and sucrose in post-thaw quality of crassostrea angulata sperm. Front Physiol 12:749735. https://doi.org/10.3389/fphys.2021.749735
50. Rodriguez-Riveiro R, Heres P, Troncoso J, Paredes E (2019) Long term survival of cryopreserved mussel larvae (*Mytilus galloprovinciallis*). Aquaculture 512:734326. https://doi.org/10.1016/j.aquaculture.2019.734326
51. Wang H, Li X, Wang M, Clarke S, Gluis M, Zhang Z (2011) Effects of larval cryopreservation on subsequent development of the blue mussels, *Mytilus galloprovincialis* Lamarck: mussel larval cryopreservation. Aquacult Res 42:1816–1823. https://doi.org/10.1111/j.1365-2109.2010.02782.x

Chapter 21

Preservation of Microorganisms for Storage in Culture Collections

Felizitas Bajerski, Anna Barheine, Lisa Belke and Silke Pradella

Abstract

Microbial culture collection centers are pivotal for preserving and providing microorganisms for scientific and industrial applications. Proper preservation methods, like freezing or drying, are crucial. Cryostorage in the vapor or liquid nitrogen (LN) phase at temperatures below −150°C is used to ensure that the biological material is stored genetically unaltered. In addition, microorganisms are preserved as dried lyophilizates using appropriate freeze–drying protocols, where applicable. Maintaining the quality of source material through optimal growth conditions is also essential, and can be found in species descriptions, in culture collections catalogs, or online databases. This applies in particular to microorganisms that are difficult to cultivate and is of crucial importance for preservation, but is often a challenge due to the laborious cultivation process. The protocols described in this chapter represent examples of how microorganisms in culture collections are preserved at the Leibniz Institute DSMZ-German Collection of Microorganisms and Cell Cultures GmbH.

Keywords Aerobe/anaerobe/microaerophile microorganisms, Biological resource centers, Cyanobacteria, Freezing, Freeze–drying, Fungi, Lyophilization, Long-term storage, Yeast

1 Introduction

Microbial culture collections are crucial for the conservation and provision of diverse biological resources for future advances in science, public health, and the bioeconomy [1]. Several microbial culture collection centers have established common laboratory guidelines ("CABRI" guidelines: Common Access to Biological Resources and Information) to harmonize maintenance and preservation procedures [2–4]. Microorganisms are usually preserved by freezing or drying. Microorganisms are frozen and stored in ampoules, straws, polypropylene tubes, or in glass capillaries in the vapor or liquid nitrogen (LN) phase, as well as in freezers at ≤−70°C. Cryogenic containers are filled with a mixture of cells in an appropriate suspension medium, together with a cryoprotectant such as glycerol or dimethyl sulfoxide (DSMO) for cryogenic storage, or a combination

Willem F. Wolkers and Harriëtte Oldenhof (eds.), *Cryopreservation and Freeze-Drying Protocols*, Methods in Molecular Biology, vol. 3060, https://doi.org/10.1007/978-1-0716-5416-3_21,

of skimmed milk and sucrose for lyophilization [5]. The process may include controlled rate freezing protocols or preincubation steps, e.g., in the vapor-nitrogen phase or in a pre-cooling device such as CoolCell LX (Corning) or Mr. Frosty (Nalgene) at −70°C before reaching the final storage temperature. Microorganisms can be dried for long-term preservation by freeze–drying, liquid drying, or vacuum drying using protective formulations and suitable equipment.

The quality of the source material is the most important prerequisite for successful conservation. The microbial cultures should therefore be cultivated according to their optimum conditions, which may, however, vary due to the diversity of microorganisms. Information on various media and growth conditions can usually be found in species descriptions of type strains, which can be found in the literature or in databases such as MediaDB (https://mediadb.systemsbiology.net), KOMODO (https://komodo.modelseed.org), BacDive (https://bacdive.dsmz.de) [6], or MediaDive (https://mediadive.dsmz.de) [7]. Culture collection centers usually provide information on media and growth conditions in their online catalogs as well.

As this is a protocol manual, a practical example is demonstrated for the aerobe, psychrophilic bacterium *Chryseobacterium frigidisoli* isolated from sandy dry permafrost on R2A media (0.05% Proteose peptone, 0.05% Casamino acids, 0.05% yeast extract, 0.05% dextrose, 0.05% soluble starch, 0.03% dipotassium phosphate, 0.005% magnesium sulfate 7× H_2O, 0.03% sodium pyruvate (w/v); pH 7.2) [8] after incubation at 10°C for three weeks (Table 1) [9]. The type strain PB4T is deposited at the Leibniz Institute DSMZ-German Collection of Microorganisms and Cell Cultures (DSMZ) as DSM 26000 (https://www.dsmz.de/collection/catalogue/details/culture/DSM-26000) and at the Belgian Coordinated Collections of Microorganisms (BCCM) as LMG 27025 (https://bccm.belspo.be/page/lmg-catalogue-display/fields/name/LMG%2027025). Thus, cultures of *Chryseobacterium frigidisoli* could be obtained by incubation on R2A or Nutrient medium at 20–25°C (Table 1).

A stable and healthy culture is particularly important for the preservation of microorganisms that are difficult to cultivate, such as fastidious anaerobes or nitrite-oxidizing bacteria (NOB), which are obligate or facultative chemolithoautotrophic bacteria. The laborious cultivation procedures often result in a lack of stable preservation methods. For nitrite-oxidizing bacteria, a generally applicable cryopreservation method using 10% DMSO as the CPA with optional tenfold diluted trypticase soy broth (https://mediadive.dsmz.de/medium/535) as preservation medium and storage in liquid nitrogen was established by Vekemann et al. [10]. However, it has to be considered that DMSO may act as a cytotoxin in some microorganisms [5]. Usually, information and protocols on preservation of microorganisms are not included in strain-specific literature, but reported in standard textbooks or reviews describing preservation protocols of

Table 1
Where can I find information on cultivation conditions for certain microorganisms? Information on the cultivation of individual species of microorganisms can be found in initial descriptions of type strains, databases, or in the catalogs of culture collections, which is illustrated here using the example of the bacterium *Chryseobacterium frigidisoli*

Source	Strain ID	Media for growth	Temperature for growth	Growth conditions
Species description [9]	$PB4^T$	R2A media (https://mediadive.dsmz.de/medium/830)	18°C, optimum of 20°C, range of 0–25°C	pH 6.5, 0.5% NaCl
BacDive, https://doi.org/10.13145/bacdive24035.20250331.9.3	24035	Nutrient agar (1.5% Agar, 0.5% Peptone, 0.3% Meat extract (w/v); https://mediadive.dsmz.de/medium/1)	20°C, optimum of 20°C, range of 0–25°C	pH 6.5, range pH 5.5–8.5, 0.5% NaCl
Culture collection (DSMZ)	DSM 26000	DSMZ Medium 1 (Nutrient agar)	20°C	
Culture collection (BCCM)	LMG 27025	LMG medium 193 (R2A agar)	25°C	

certain taxonomic groups, such as thermophiles [11] or general principles on vacuum-drying and cryopreservation of prokaryotes [12].

2 Materials

The material and all steps of the preservation procedure should be handled or processed under sterile conditions. Specific equipment (e.g., laminar flow cabinet with HEPA filters) may be required to handle microorganisms according to their risk assessment. The organism needs to be cultivated under appropriate conditions (see Sect. 1). The cells for cryopreservation should be harvested at the end of the exponential growth phase or at the beginning of the stationary phase (i.e., the cultures are densely overgrown but healthy). The sample volume needs to be adapted to the preservation device/container that is used and the number of samples that are produced. It also depends on the quantity of cells that can be obtained in a specific volume.

2.1 Materials for Freeze Drying (Lyophilization) and Vacuum Concentration

1. Constrictor (for glass ampoules).
2. Freeze Drying machine (Christ Alpha 1–4 LD).
3. Vacuum centrifuge (Christ RVC2-25 CDplus), for use with anaerobic microorganisms.
4. Gas station/anaerobic chamber (for anaerobic microorganisms only).
5. For aerobic and microaerophilic microorganisms and fungi/yeasts: skimmed milk containing vials (see Sect. 3.1); using skim milk powder (from, e.g., Sigma-Aldrich).
6. For fungi/yeasts: Tween tubes, 0.1% (w/v, dissolved in distilled water).
7. Hungate tubes or Hungate bottles.
8. Anaerobic medium such as M78 (https://mediadive.dsmz.de/medium/78) supplemented with 12% (w/v) sucrose and 5% (w/v) FeS.

2.2 Materials for Freezing Aerobic/Microaerophilic Microorganisms

1. 20% (w/v) glycerol solution (dissolved in distilled water).
2. Sterile glass capillaries (or alternative cryovial).
3. Micro-horizontal burner.

2.3 Materials for Freezing Anaerobic Microorganisms

1. Sterile glass capillaries (or alternative cryovial).
2. Dimethyl sulfoxide (DMSO) solution: cultivation medium supplemented with 5% v/v DMSO (final concentration).
3. Capillary adapter for micro pipette controller.
4. Micro pipette controller.
5. Micro horizontal burner.

2.4 Materials for Freezing Fungi

1. Short sterile straws.
2. Plate with material to be preserved: medium consisting of malt supplemented with 5% v/v glycerol.

2.5 Materials for Freezing Yeasts

1. Hydroxyethyl starch (HES) solution, 10% HES w/v dissolved in distilled water.
2. Half straws (pressed closed on one side): cut straws to approx. 1 cm length, dip one side in acetone, and then melt with a foil sealer. Then autoclave.

2.6 Materials for Freezing Cyanobacteria

1. Sterile pipettes and 5 mL pipette tips (if necessary, cutoff pipette tips to gently pipette clumpy cultures).
2. Ice bath for precooling.
3. Sterile filter 100% DMSO (Duchefa Biochemie D1370) with a Teflon filter (e.g., Membrane Solutions (MS), Sterile Syringe Filter, PTFE-B, 0.22 μm, SFPTFE025022SBI). Sterile-filtered DSMO is stored in a light-protected bottle.
4. Precool a CoolCell LX (Corning) in the refrigerator at 4°C overnight.
5. Label 22 sterile cryotubes (Thermo Scientific Nunc Cat. No. 377267) per culture, e.g., with a colored lid insert.
6. Suitable tray for preincubation of the samples in liquid nitrogen.

3 Methods

3.1 Preparation of Skimmed Milk Containing Vials (for Drop-on Method)

1. Prepare a 20% skimmed milk solution. For microaerophiles, 10% skim milk may be used as an alternative method. Here, the cell mass is directly added to the liquid milk powder suspension (see Sect. 3.2.1 Step 3b).
2. Fill each 500 μL into empty flat-bottomed glasses (3 mL).
3. Plug with a cotton roll/dental roll.
4. Autoclave at 115°C for 13 min (do not autoclave too long, too hot because the milk may caramelize).
5. Freeze sterile milk tubes at −20°C overnight.
6. Freeze–drying of the frozen milk tubes for 3 days.
7. Check the sterility: dissolve three tubes of the series in a complex/nutrient-rich medium such as Trypticase Soy Yeast Extract medium (3% Trypticase soy broth, 0.3% Yeast extract, 1.5% Agar, if necessary (w/v), DSMZ M92, https://mediadive.dsmz.de/medium/92) and incubate one tube each at 28, 37, and 55°C for 3 days.

8. Drop 1 drop of each suspension onto a plate of DSMZ M92 (or other complex/nutrient-rich medium) and incubate at the same temperature as before for 7 days. No growth = sterility test passed.

3.2 Freeze–Drying (Lyophilization)

In the method described, the cell suspension is first frozen in a suitable suspension liquid (e.g., growth medium with or without cryoprotectant, skimmed milk) and then quickly placed under vacuum before the vials thaw. The ice contained in the samples is then removed under vacuum by sublimation. The protocol used at the DSMZ is based on a 20% (w/v, drop-on method) or 10% (w/v, direct-mix method) sterile skimmed milk suspension (*see* **Note 1**).

3.2.1 Freeze–Drying of Aerobic/Microaerophilic Microorganisms

1. Cultivate organism under appropriate conditions (see Sect. 1): solid medium (at least one plate, up to three agar plates or ten slant agar tubes, depending on growth) or liquid medium (high turbidity is advantageous, several aliquots recommended).
2. The described information refers to a 5 mL culture suspension to be preserved (*see* **Note 2**).
   (a) Solid: completely harvest the plate(s) with an inoculation loop and resuspend the biomass in liquid medium (approximately 5 mL).
   (b) Liquid: combine culture tubes and centrifuge (angle rotor, 5700 rpm, 10 min). (i) Remove and discard the supernatant until approximately 5 mL remain (do not discard the pellet). (ii) Resuspend the pellet in the remaining 5 mL supernatant until a homogeneous suspension is obtained. (iii) Mix 1 mL of the suspension with 1 mL of sterile glycerol 20% in a 4 mL vial with a screw cap and store for preservation in liquid nitrogen (see Sect. 3.3).
3. Prepare 15 sterile milk tubes (see Sect. 3.1).
4. Drop-on-method: drop 25 μL of the suspension onto the surface of the milk pellets and close again with cotton plugs/dental rolls (see Fig. 1A).
5. Direct mix-method (alternative for microaerophiles): add the cell mass directly to a 10% liquid skimmed milk suspension. (i) Pre-freeze the pellets at −20°C. (ii) Dry under the bonnet on the freeze–drying machine overnight.
6. Cut off the protruding part of the cotton plugs and place the milk tubes in (e.g., special round-bow) glass tubes with silica gel used as a color indicator for humidity.
7. Plug the glass tubes with fireproof glass wool. Take appropriate safety precautions: may cause mechanical irritation to the eyes, respiratory tract, and skin.

Fig. 1 Freeze–drying (Lyophilization) of microorganism using glass ampoules. (**A**) Double glass ampoules are used at the DSMZ at different stages of the process. Flat-bottom-glass with cotton plug before cutting off the protruding part of the cotton plugs, glass vial containing the flat-bottom-glass with cutoff cotton plug, constricted ampoules containing the microbial suspension on/mixed with the prepared skimmed milk vials, ready to use ampoule. (**B**) Customized manifold for constricting ampoules used at the DSMZ. (**B**, **C**) Ampoule during the sealing "constricting" process. The ampoules are closed by forming a small spot for subsequent melting, but are not completely sealed. **D** Ampoules attached to the freeze–drying machine during the preservation process. After completion, the ampoules are sealed by melting and removed in one step

8. Place the glass tubes on the constrictor (see Fig. 1B). Then, using a hot flame, melt the glass at a point in the middle behind the glass wool, creating a capillary between the two parts (see Fig. 1C).
9. Place the constricted tubes on the manifold "tree" of a precooled freeze–drying system (following the manufacturer´s instructions).
10. Dry under vacuum overnight (3 h for the direct-mix method).

11. Cut off the ampoules the next day under vacuum using a gas torch (see Fig. 1D). Important: check whether the silica gel has changed color and whether the ampoules are really sealed.
12. Store the ampoules cool and dark.

3.2.2 Freeze–Drying of Anaerobic Microorganisms

1. Cultivate organism under appropriate conditions in Hungate tubes or Hungate bottles (several aliquots recommended to achieve a higher concentration of organism; https://www.dsmz.de/fileadmin/Bereiche/Microbiology/Dateien/Kultivierungshinweise/Kultivierungshinweise_neu_CD/engl_Anaerob_update.pdf).
2. Centrifuge aliquots (swing-out rotor, 3000 rpm, 20 min).
3. Open aliquots under anaerobic conditions (fumigation, anaerobic chamber, etc.) and remove and discard the supernatant (dump or suck off, depending on the strength of the pellet).
4. Resuspend the pellets in complex anaerobic medium such as M78 (https://mediadive.dsmz.de/medium/78) supplemented with 12% sucrose and 5% FeS to form a homogeneous suspension (approximately 4 mL final volume).
5. Add 200 μL of the suspension into sterile, empty flat-bottom glass tubes (same as used for the skimmed milk containing tubes) with stoppers.
6. Place the tubes onto a rotor of a rotary vacuum concentrator (Christ RVC2-25 CDplus) and centrifuge at 800 rpm and 10°C for approximately 2.5–3 h.
7. Continue from freeze–drying aerobic/microaerophilic (Sect. 3.2.1, Step 4).

3.2.3 Freeze–Drying of Fungi/Yeasts

1. Cultivate organism under appropriate conditions (one plate per strain).
2. Completely harvest the plate with an inoculation loop and resuspend the biomass in 3.5 mL Tween80 (0.1%).
   (a) Fungi: take up some Tween solution with a sterile Pasteur pipette and rinse the plate several times and, if necessary, scrape off a little until the Tween solution becomes cloudy (important: beware of spores! Open the plate only as far as necessary). Transfer the cloudy Tween solution to the original vial used for resuspension.
   (b) Yeasts: suspend the biomass of approximately one inoculation loop into the Tween solution.
3. Continue from freeze–drying aerobic/microaerophilic (Sect. 3.2.1, Step 4).

3.3 Preservation in Liquid Nitrogen

Freezing in glass capillaries is effective, both in terms of freezing success and space saving. The small volume ratio enables rapid freezing and thawing, and the small size means that many parallel capillaries/batches can be stored for one strain, which is particularly important for a large culture collection such as the DSMZ. Furthermore, it reduces the risk of contamination, because a single capillary is used for each new inoculation. Using hermetically sealed samples also eliminates the risk for cross-contamination during cryostorage. Although the risk for cross-contamination was evaluated as low, it is recommended to use sterile conditions throughout the whole sample processing, perform contaminant screening, and clean the LN storage tanks at regular intervals, if possible [13]. In the 1970s, the capillary preservation for anaerobes [14] was adapted and developed further at the DSMZ. Instead of capillaries and storage in liquid nitrogen, the described cell-mixture can be used for freezing in, e.g., standard cryovials at −80°C (*see* **Note 3**, and see freeze–drying aerobic/microaerophilic (Sect. 3.2.1, Step 2b, iii)).

3.3.1 Cryopreservation of Aerobic/Microaerophilic Microorganisms

1. Mix 1 mL of suspension with 1 mL of sterile glycerol 20% in a 4 mL vial with a screw cap.
2. Immerse sterile capillary in the suspension, the suspension is drawn into the capillary by capillary forces, fill capillary approximately halfway. Remove the capillary, hold it horizontally so that a gap remains at both ends (the liquid should not be held directly in the flame).
3. Melt and seal the capillaries on both ends using a micro-horizontal burner (avoid air bubbles in the capillaries; see Fig. 2A). The sealed capillaries may be immersed in cold alcohol to make the outside sterile.

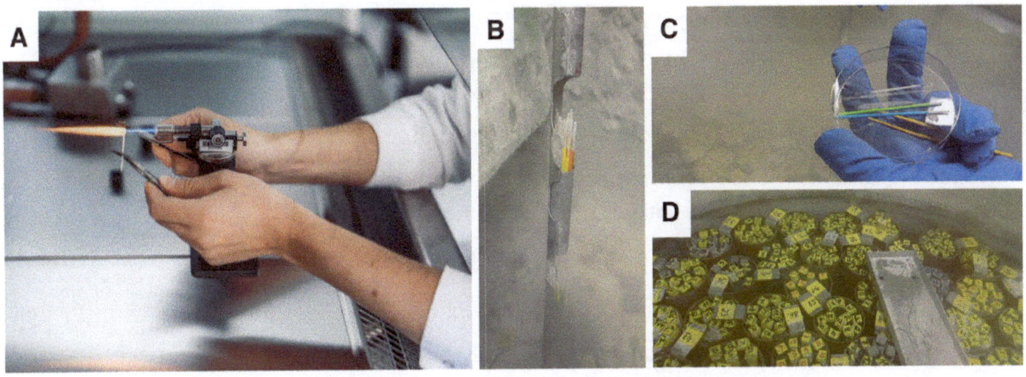

Fig. 2 Liquid nitrogen preservation of microorganisms in glass capillaries at the DSMZ. (**A**) Sealing of the glass capillaries by melting with a micro-horizontal burner after the microbial culture has been drawn into the capillary by the capillary forces. (**B**) Sealed and frozen glass capillaries in the container of a holder in the liquid nitrogen tank. Different colored sleeves may be used for different strains. (**C**) Glass capillaries in a petri dish. (**D**) Arrangement of the metal holders in metal cans in the liquid nitrogen tanks

4. First, put the capillaries in a Petri dish and place them in the gas phase of liquid nitrogen for at least 20 min to prevent the capillaries from bursting.
5. Storage in gaseous or liquid nitrogen phase (see Fig. 2B, C) (*see* **Notes 4 and 5**).

3.3.2 Cryopreservation of Anaerobic Microorganisms

1. Add 5% (DMSO) to a Hungate tube containing the cultivation medium appropriate for the strain.
2. Preparation of the culture: see freeze–drying anaerobic (Sect. 3.2.2, Steps 1–3).
3. To prepare 10 capillaries, add 500 µL of the prepared medium from Step 1 to the pellet and resuspend it.
4. Transfer suspension to a sterile, empty flat-bottom tube with a stopper and sparge medium with 100% N_2 gas to make it anoxic.
5. Attach the sterile capillary adapter to the micro pipetting aid and pick up the sterile capillary with the pipetting aid.
6. Draw up the cell suspension with the pipetting aid (start with a little gas phase, then draw up the cell suspension, then finish with a little gas phase again).
7. Melt and seal the capillaries on both ends with a micro-horizontal burner (note: avoid air bubbles in the capillaries). Check proper sealing under the binocular microscope, if possible.
8. First, store the capillaries in the gas phase of liquid nitrogen for at least 20 min to prevent the capillaries from bursting.
9. Storage in gaseous or liquid nitrogen phase (*see* **Note 4**).

3.3.3 Cryopreservation of Fungi

1. Cultivate the organism on an agar plate containing the cultivation medium appropriate for the strain, supplemented with glycerol (5% final concentration on the plates).
2. Cut out the agar and biomass with approx. 1 cm long, sterile straws until the straw is filled with agar.
3. Transfer the straws into an empty, sterile cryotube.
4. First, store the cryotube in the gas phase of liquid nitrogen for at least 20 min.
5. Final storage in gaseous or liquid nitrogen phase (*see* **Note 4**).

3.3.4 Cryopreservation of Yeasts

1. Cultivate organism under appropriate conditions (one plate per strain, see Sect. 1).
2. Resuspend the biomass from one inoculation loop in 600 µL 10% hydroxyethyl starch (HES) solution.
3. Using a sterile syringe, inject the suspension into approximately 1 cm long, sterile half-open straws (one side closed/

welded shut). Preparation of the straws: dip one side of the straws in acetone and seal them with a foil sealer; autoclave the straws.

4. Transfer the straws into an empty, sterile cryogenic tube (open side of the straws up).
5. First, store the cryogenic tube in the gas phase of liquid nitrogen for at least 20 min.
6. Storage in gaseous or liquid nitrogen phase (*see* **Note 4**).

3.3.5 Cryopreservation of Cyanobacteria

1. Grow the cyanobacterial strains in a volume of 20 mL in cell culture flasks (TTP 90026) containing the cultivation medium appropriate for the strain. The protocol includes the preparation of 22 cryopreserves in 22 CryoTubeTM vials, Thermo Scientific Nunc Cat. No. 377267) each.
   (a) For 22 cryopreserves, two cell culture flasks containing 20 mL of culture each are generally required per well-growing strain.
   (b) For strains producing small amounts of biomass, more than two cell culture bottles must be inoculated, which are concentrated for cryopreservation by suitable methods (gentle centrifugation or decanting of the supernatant).
2. Homogenize the cells depending on the culture so that the cell material is evenly distributed when filling the cryotubes.
   (a) If cell adhesions are present, these are carefully scraped off with a cell scraper (see Fig. 3A).

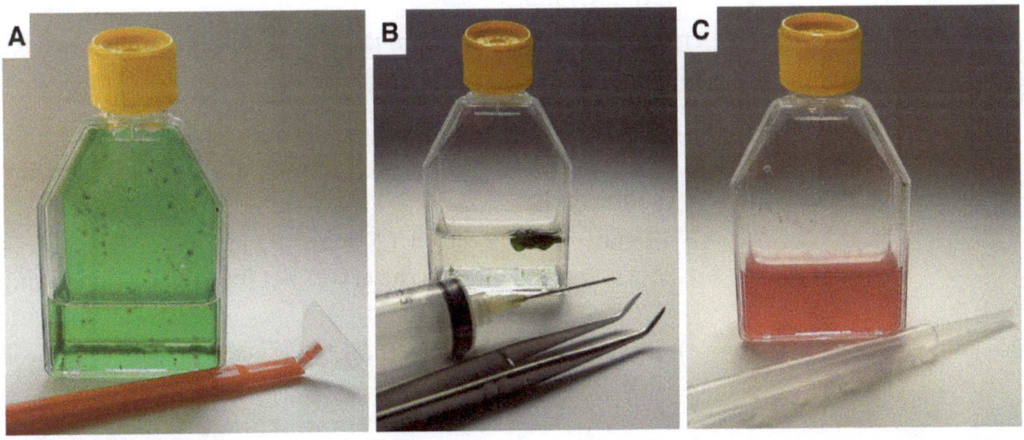

Fig. 3 Homogenization of cyanobacterial cells in preparation for the preservation process. Depending on the culture, different methods are used to ensure that the cell material is evenly distributed when the cryotubes are filled. (**A**) If cell adhesions are present, these are carefully scraped off with a cell scraper. (**B**) Some cultures require vigorous pipetting (up and down) or cutting to break up cell clumps. (**C**) Other cultures need to be carefully mixed and gently pipetted. ©Steph Seufert (Cyanobacteria and Protists working group, DSMZ)

(b) Some cultures require vigorous pipetting (up and down) or cutting to break up cell clumps (see Fig. 3B).

(c) Other cultures need to be carefully mixed and gently pipetted (see Fig. 3C).

3. Fill up each homogenized cyanobacterium culture up to 20 mL with the corresponding medium according to the marking on the cell culture bottle.
4. Place each cell culture bottle on ice for 10 min and swirl occasionally.
5. Combine the contents of the two cell culture bottles in one bottle for cryopreservation of a strain (see above).
6. Pipette 4 mL of 100% DMSO (sterile-filtered, room temperature) into 40 mL of the combined culture (≈9% DMSO) while swirling on ice. Mix gently but thoroughly. Caution: as heat is generated during the addition of DMSO, it is important that the cultures are kept on ice and the DSMO is pipetted to the culture gradually and with swirling.
7. Incubate the culture/DMSO mixture in an ice bath for 20 min to allow the DMSO to penetrate the cells and fill the 22 cryogenic tubes during the 20-min incubation period.

   (a) Fill 1.5 mL of the culture/DMSO mixture into each cryotube and place immediately on ice (stop the time).

   (b) During this time, no direct light should hit the cells, and the light of the safety cabinet should be switched off.

8. Transfer the 22 cryotubes to a CoolCell LX and freeze for 4 h in a −80°C freezer (cooling rate in the CoolCell LX: approximately 1°C/min).
9. Transfer the cryotubes from the −80°C freezer to the nitrogen tank.

   (a) Place the final storing cryobox in a tray filled with LN so that the cryotubes are well immersed in the LN, but not submerged.

   (b) Remove the CoolCell LX with the cryotubes from the −80°C freezer and put the cryotubes into the cryobox using tweezers.

   (c) Place the cryobox into the LN tank according to its specified position.

3.4 Thawing

3.4.1 Thawing of Cryopreserved Samples (Capillaries of Aerobic/ Microaerophilic/Anaerobic Microorganisms)

The cryo device can be thawed in a sterile water bath at a defined temperature or at room temperature.

3.4.2 Thawing of Cyanobacteria

1. Remove the samples from the LN tank and directly transfer them to a Dewar vessel filled with LN for transport.
   (a) Avoid strong light irradiation during thawing (e.g., safety cabinet without lighting).
   (b) Place samples in the dark immediately after thawing.
2. Place one cryotube per strain in a thawing float and thaw in the sterile water bath at a temperature of exactly 37°C (see above).
   (a) The thawing behavior of each culture is slightly different, so the culture should be observed closely. It is processed further as soon as no more ice crystals are visible.
   (b) Alternatively, the cryotube can be thawed in a sterile water bath at 30–35°C or at room temperature.
3. Wipe the cryotube with 70% ethanol and pipette the contents of the cryotube into a culture flask containing the provided 20 mL medium.
4. Incubate the culture bottle hours in the dark at ≈20°C (or at the temperature at which it is usually kept) for 24 and then continue culturing under the specified light regime.
   (a) It may be necessary to gradually expose the culture to light for 2–3 days (or longer).
   (b) Various light filter films are used to shade the cultures. Less light is often advantageous after thawing.
   (c) Grow the culture for the appropriate cultivation condition time (4–16 weeks).
   (d) Check growth macroscopically and microscopically.

4 Notes

1. The quality of the source material is most important for successful conservation.
2. The lower the storage temperature, the better/longer the stability and viability.
3. In general, the growth medium with or without cryoprotectant or skimmed milk may be used as a suspension medium.

4. Viability tests on a regular basis are recommended during long-term storage.
5. Identification/authenticity of the material is recommended prior to storage.

Acknowledgments

I would like to thank the current and former technical and scientific staff of the DSMZ who have been and continue to be actively involved in the development of the methods and protocols described. Thanks go to the Cyanobacteria and Protists working group for the protocols for the preservation of cyanobacteria and especially to Steph Seufert for providing the photos (Fig. 3). Furthermore, I thank Dr. Rüdiger Pukall for critical proofreading and valuable comments.

References

1. Overmann J (2015) Significance and future role of microbial resource centers. Syst Appl Microbiol 38:258–265
2. CABRI—Common Access to Biological Resources and Information: guidelines see Part 3: guidelines for maintaining deposits: flow diagrams, procedures and appendixes 1999–2023 [cited 2025 07.05.2025]. Available from http://cabri.org/guidelines/micro-organisms/MCover1.html
3. Romano P, Kracht M, Manniello MA, Stegehuis G, Fritze D (2005) The role of informatics in the coordinated management of biological resources collections. Appl Bioinform 4:175–186. https://doi.org/10.2165/00822942-200594030-00002
4. Romano P, Dawyndt P, Piersigilli F, Swings J (2005) Improving interoperability between microbial information and sequence databases. BMC Bioinform 6(Suppl 4):S23. https://doi.org/10.1186/1471-2105-6-S4-S23
5. Hubalek Z (2003) Protectants used in the cryopreservation of microorganisms. Cryobiology 46:205–229. https://doi.org/10.1016/s0011-2240(03)00046-4
6. Schober I, Koblitz J, Sardà Carbasse J, Ebeling C, Schmidt ML, Podstawka A, Gupta R, Ilangovan V, Chamanara J, Overmann J, Reimer LC (2025) BacDive in 2025: the core database for prokaryotic strain data. Nucleic Acids Res 53(D1):D748–D756. https://doi.org/10.1093/nar/gkae959
7. Koblitz J, Halama P, Spring S, Thiel V, Baschien C, Hahnke RL, Pester M, Overmann J, Reimer LC (2023) MediaDive: the expert-curated cultivation media database. Nucleic Acids Res 51(D1):D1531–D1538. https://doi.org/10.1093/nar/gkac803
8. Reasoner DJ, Geldreich EE (1985) A new medium for the enumeration and subculture of bacteria from potable water. Appl Environ Microbiol 49:1–7. https://doi.org/10.1128/aem.49.1.1-7.1985
9. Bajerski F, Ganzert L, Mangelsdorf K, Padur L, Lipski A, Wagner D (2013) Chryseobacterium frigidisoli sp. nov., a psychrotolerant species of the family Flavobacteriaceae isolated from sandy permafrost from a glacier forefield. Int J Syst Evol Microbiol 63:2666–2671. https://doi.org/10.1099/ijs.0.046904-0
10. Vekeman B, Hoefman S, De Vos P, Spieck E, Heylen K (2013) A generally applicable cryopreservation method for nitrite-oxidizing bacteria. Syst Appl Microbiol 36:579–584.https://doi.org/10.1016/j.syapm.2013.07.002
11. Spring S (2006) Preservation of thermophilic microorganisms. In: Methods in microbiology. Academic Press, pp 349–368
12. Tindall B (2007) Vacuum-drying and cryopreservation of prokaryotes. In: Day J, Stacey G (eds) Cryopreservation and freeze-drying protocols. Humana Press, pp 73–97
13. Bajerski F, Nagel M, Overmann J (2021) Microbial occurrence in liquid nitrogen storage tanks: a challenge for cryobanking. Appl Microbiol Biotechnol 105:7635–7650. https://doi.org/10.1007/s00253-021-11531-4
14. Hippe HH (1991) Maintenance of methanogenic bacteria. In: ADBE Kirsop (ed) Maintenance of microorganisms and cell cultures. Academic Press, London, UK, pp 101–114

Chapter 22

Cryopreservation of Mammalian Oocytes: Slow Cooling and Vitrification Approaches

Victoria Keros and Barry J. Fuller

Abstract

Two basic methods for the laboratory-focused cryopreservation of mammalian oocytes are discussed, based initially on work with murine oocytes and subsequently transferred to human infertility services. One method uses a relatively low concentration of the cryoprotectant propanediol plus sucrose and requires controlled-rate cooling equipment to achieve a slow cooling rate. This method has also produced live births from cryopreserved human oocytes. The second method, which is described here, employs a high concentration of the cryoprotectant dimethyl sulfoxide plus a low concentration of polyethylene glycol. This is a vitrification method, which involves ultra-rapid cooling by plunging standard straws into liquid nitrogen vapor, hence avoiding the need for specialized equipment, but requires technical ability to manipulate the oocytes quickly in the highly concentrated cryoprotectant solutions. Murine oocytes that have been vitrified using this technique have resulted in live births. Vitrification using other cryoprotectant mixtures is now a clinically accepted method of choice for cryobanking of human oocytes.

Keywords Human, Murine, Oocyte, Slow-cooling, Vitrification

1 Introduction

Successful cryopreservation of mature oocytes has been one area of reproductive cryobiology, which has challenged our understanding of cryobiology over the past 30 years, in ways unforeseen at the beginning of the studies. Much had to be learned about the specific ultrastructural and physiological characteristics of these large, unique cells, in order to be able to recover functional oocytes from cryogenic storage, and these details have been published in many previous reviews [1–5]. With improved outcomes, in the past decade, there was a renaissance of interest in applying oocyte cryopreservation [6, 7], and the technology has moved from a largely research-focused activity into the mainstream of clinical reproductive medicine, resulting in the recent designation from the ASRM as a justified therapy [8].

Storage of unfertilized oocytes has numerous applications [9]. Originally proposed as a potential option for women undergoing

Willem F. Wolkers and Harriëtte Oldenhof (eds.), *Cryopreservation and Freeze-Drying Protocols*, Methods in Molecular Biology, vol. 3060, https://doi.org/10.1007/978-1-0716-5416-3_22,

anti-cancer treatment which may render the women infertile [10, 11], it may now have a wider remit, including application in banking of donated oocytes, premature ovarian failure, ethical objections to embryocryopreservation, or in women seeking to defer their reproductive potential until later in life but who do not have a current partner [4] and for fertility preservation in trans masculine individuals [12]. The method can also be used to cryo-bank excess oocytes produced within a particular cycle during in vitro fertilization.

Applications in animal management include the preservation of important genetic lineages; the ability to restock following outbreaks of diseases such as foot and mouth; the preservation of genetically modified strains, thereby reducing the cost of continuous breeding and avoiding problems of genetic drift; and the preservation of endangered species [13–15].

Oocyte cryopreservation has also developed alongside a greater understanding of what is needed for successful recovery of function after cryogenic storage. The need for optimized cryogenic dehydration of the cells, coupled with the introduction of specific non-toxic stabilizing solutes called cryoprotectants (CPA) is a central tenet of cryobiology [2, 16, 17]. The term cryopreservation itself has come to mean different things in applied cryobiology. For true long-term cryogenic preservation, cells and their supporting medium need to be stored in this optimally cryo-dehydrated state below the temperature range at which any residual water molecules undergo a "glassy transition" [2, 18]. However, there are two different approaches to achieve this optimally dehydrated state, which can be termed controlled rate slow cooling (CRSC), or vitrification (VS—literally changing into a glassy state). In CRSC, CPAs are essential to support cell survival, but the dehydration is produced by the extracellular water within the sample being changed into ever-increasing ice fraction as temperatures progress slowly (at rates <−1°C/min) down from 0°C to −60°C and beyond. In VS, CPAs are equally essential, but they are applied in very high concentrations to produce optimal cryo-dehydration and suppress ice crystal formation before cooling begins [2–4, 18]. Because these high CPA concentrations are potentially toxic, the samples must be cooled very rapidly (at rates >200°C/min) to below the glassy transition range (about −120°C, depending on the CPA mixture selected). VS, as this is widely used, can only be practically achieved using small volume samples (usually <10 μL) in special containers to achieve these high cooling rates. Lastly, for both CRSC and VS, the rate of rewarming back from cryogenic storage is important to control, and should be fast, to avoid the propensity for free water molecules to form ice crystals as the temperature profile climbs (the so-called event of freezing during warming) [19].

Oocyte cryopreservation has been at the heart of this debate within applied cryobiology. Originally, the early reports of oocyte cryopreservation in both human and animal oocytes were made using

CRSC [20–27]. Since then, pregnancies and live births resulting from cryopreserved oocytes have been reported in a number of species [28–31] and humans [32]. Clinical programs were developed in the early 2000s using CRSC with optimized CPA mixtures [33, 34]. At the same time, VS techniques were becoming optimized [4, 35]. It became apparent that experienced practitioners could recover higher percentages of oocytes from cryogenic storage using VS methods [36–38].

Historically, oocytes have been considered to be difficult cells to cryopreserve for a number of reasons. The oocyte is a large single cell with relatively low permeability to water and cryoprotective agents. This means that oocytes have a tendency to retain water as ice starts to form in the suspending medium during cooling and, if this leads to the formation of intracellular ice as cooling proceeds, damage to the cell results. Permeability of oocytes varies between species, strains, and the maturational status of the oocyte. The mature oocyte is also a short-lived cell that must undergo timely fertilization in order for it to continue to survive and develop. For fertilization to occur naturally, the oocyte must retain the integrity of a number of its unique structural features. These include the surrounding zona pellucida, the cortical granules, and the microtubular spindle on which the chromosomes are organized. The zona pellucida is a glycoprotein coat surrounding the oocyte. Changes to this layer, triggered by the action of a single sperm binding to its receptor, induce the cortical granules in the periphery of the ooplasm to release their contents. The enzymes released from the cortical granules act to crosslink the glycoproteins of the zona pellucida, thus rendering it impenetrable to further sperm [39]. Cryopreservation has been shown to result in the premature release of the contents of cortical granules, thus creating a block to sperm penetration [40]. CPAs themselves can cause a transient increase in intracellular calcium similar to the initial increase caused by sperm entry and subsequent induction of zona hardening [41]. Conversely, cryopreservation can lead to physical damage to the zona pellucida, resulting in multiple sperm entry. Both of these problems are overcome by the application of the intracytoplasmic sperm injection (ICSI) technique, whereby a single sperm is injected into the oocyte to achieve fertilization. The microtubular spindle is the structure upon which the condensed chromosomes are aligned in mature oocytes and is responsible for the movement of chromosomes during cell division. Damage to this structure can lead to aneuploidy after fertilization. During cooling, the microtubular spindle has been shown to disassemble, although there is growing evidence that the spindle is capable of repair on warming and post-thaw culture [42–44]. One way of avoiding the potential for damage to the microtubular spindle is to cryopreserve the immature oocyte before the spindle is formed, when the chromosomes are contained within the germinal vesicle and the oocyte is termed immature [45].

However, success following cryopreservation of immature oocytes is less than that following cryopreservation of mature oocytes [46]. If protocols for hormonal stimulation of ovaries to induce the release of numerous mature oocytes are not available or not easily applicable, for example, in patients with polycystic ovary syndrome or in cancer patients, when harvesting of ovarian tissue is the only option for fertility preservation, then only immature oocytes, contained within the ovaries, may be available for cryo-storage. The major problem with cryopreserving immature oocytes is that they must be matured in vitro in order to become fertilizable. This maturation involves communication with the cumulus cells that surround the oocyte [47]. The cumulus cells are much smaller than the oocyte and are connected with it by numerous gap junctions. Development of cryopreservation protocols that allow survival of the oocyte, the cumulus cells, and connections between the two is difficult [48]. However, it was shown that preservation of this cell-to-cell connection can be achieved even within such a complex ovarian tissue when the method of VS was applied [49]. Also, techniques for maturation in vitro require further refinement in most species. In vitro maturation of oocytes prior to cryopreservation may produce poor-quality oocytes that are more prone to damage during cryopreservation [50]. Both IVM and vitrification can have an impact on the spindle morphology of oocytes and impact fertilization and embryo development [51, 52], while it has been shown that vitrification of human oocytes before and after IVM does not impair maturation ability but induces meiotic spindle alterations [53]. As well as these general problems, the oocytes of some species have further characteristics that make them still more difficult to freeze; for example, porcine oocytes have cytoplasmic high lipid content and are sensitive to chilling [54].

The protocols used for the cryopreservation of unfertilized oocytes have largely been adopted from earlier embryocryopreservation techniques. As with embryocryopreservation, both CRSC and VS techniques have been successfully applied. In the 1980s, murine oocytes were successfully cryopreserved using a range of CPA, such as glycerol or dimethyl sulfoxide, using slow cooling controlled rate protocols, and early work in human oocytes followed slow cooling in the presence of propanediol and sucrose. This technique, combined with ICSI, first yielded a live birth from cryopreserved human oocytes in 1997 [32]. Over the following years, by the 2010s, the number of human live births increased steadily, escalating into hundreds [43, 55]. VS has also produced an accumulating number of human live births [56, 57], with more than 250 recorded by 2009 [58] and coming to more than a thousand babies born already in 2016–2017 [59–62], following significant improvement of oocyte VS by the Cryotop method [63, 64]. Clinical outcomes of infertility treatment using vitrified-warmed autologous oocytes are comparable to treatment cycles using fresh autologous oocytes, as well as to treatment using donor fresh and vitrified oocytes [65, 66]. Combination of the (i) stepwise equilibration at the room temperature in increasing concentrations of permeating CPA with addition of non-permeating agents to the VS solution (minimizing the toxic effect by using a mixture of CPAs and equilibration at room temperature); (ii) placing an oocyte on the surface of the fine propylene strip attached to a plastic handle in the minimum (˂1 μL) volume of VS solution (reaching a high cooling rate (>20,000°C/min), to minimize the toxic effect of the CPAs and associated osmotic injury, also preventing any accidental ice formation, when a lower concentration of CPA is used, and avoiding any temperature-related fracture of the vitrified oocyte or surrounding solution); (iii) direct plunging into liquid nitrogen in an open system and shaking thus reaching a high cooling rate and avoiding a bubbles from the strip. To prevent contamination of the sample, liquid nitrogen can be sterilized by filtration or exposure to UV; (iv) rapid warming at 37°C with stepwise equilibration of oocytes in decreasing concentration of sucrose to prevent the effects of devitrification and osmotic shock resulted in good morphological survival of 95% of vitrified oocytes, high rates of fertilization (91%), in vitro blastocyst formation (50%) and subsequent delivery of healthy babies (see Fig. 1). To achieve the best result at fertilization of the warmed oocytes and subsequent development of the embryo and onset of pregnancy leading to the birth of the child, it is recommended to follow certain time limits when manipulating the oocytes (see Fig. 2). The Cryotop® method using different mixtures of CPAs is the most established and commonly used worldwide for oocyte VS. Detailed explanation of the Cryotop®—Open System method (Kitazato Biopharma, Shizuoka, Japan) as successfully applied worldwide for human oocyte VS can be found on the Kitazato company's homepage.

Different groups throughout the early 2010s applied both CRSC and VS, and initially there was debate about which methods were superior [67–70]. Then gradually there was an emerging body of experience, which demonstrated that fertilization and embryo development were higher in the VS groups [38, 61, 71–73]. Following the scientific publications for oocyte cryopreservation from the PubMed database (*see* Fig. 3), we can easily observe the growing interest in oocyte cryopreservation through the years. With the improvement of the methods for in vitro fertilization and assisted reproductive techniques (ART), developing the method for oocyte cryopreservation was necessary to be able to preserve any surplus of oocytes harvested and used for infertility treatment or especially for fertility preservation. While it is clear that at the beginning of infertility services, the slow programmed freezing was the method of choice, the focus was changed to adjust the method for vitrification for preserving such complicated cells as oocytes. Now, vitrification is by far the dominant method of choice. However, some scientists

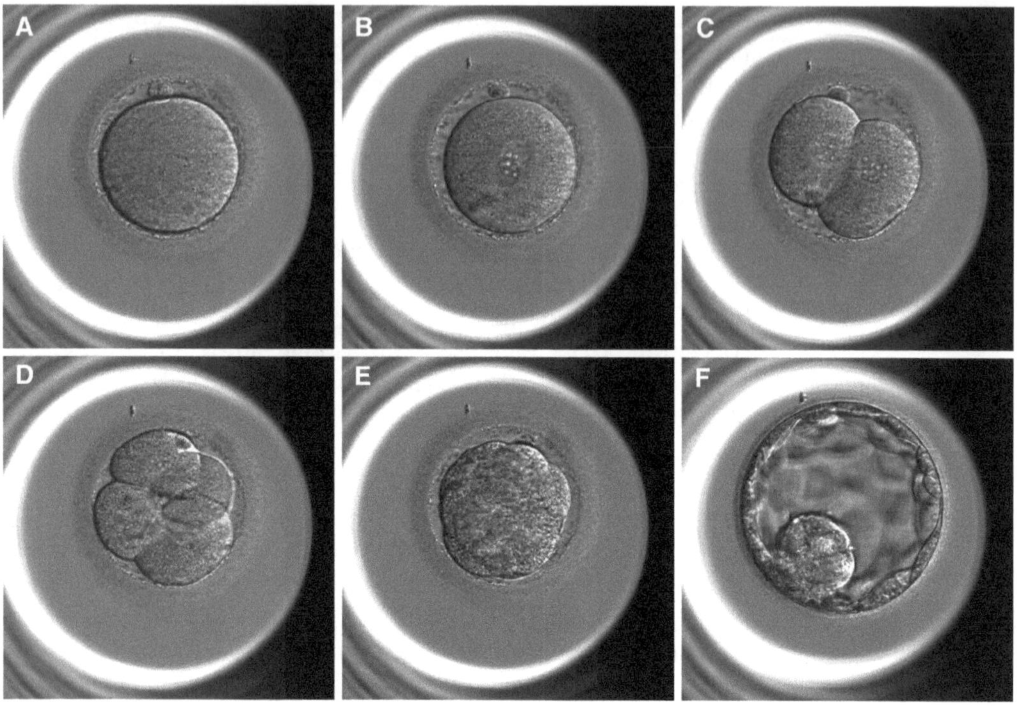

Fig. 1 A human embryo undergoing development from an oocyte vitrified using the Kitazato Cryotop®—Open System method and fertilized by intracytoplasmic sperm injection. (**A**) Matured MII oocyte, recovered from a vitrification after warming and 3 h incubation in the IVF medium (Vitrolife) at 5% O_2 and 6% CO_2. After ICSI, the oocyte retains good refractive cytoplasm with intact plasma membrane, typical first polar body, and intact, homogenous zona pellucida. (**B**) Zygote 16 h after insemination. The presence of two juxtaposed, centrally located pronuclei (PN) within a spherical oocyte and the appearance of a second polar body (visible behind the oocyte) are clear evidence of fertilization after ICSI. (**C**) Cleavage stage embryo 27 h after insemination with two blastomers, each containing one nucleus. (**D**) Four-cell embryo 44 h after insemination. (**E**) Morula with finilized blastomere compaction 94 h after insemination. (**F**) Blastocyst 114 h after insemination with a large compacted inner cell mass (ICM) and many trophectoderm cells (TE) lining the blastocoel cavity, a greater volume than original volume of the embryo and thinned ZP. The blastocyst was transferred, implanted and resulted in a clinical pregnancy followed by live birth. 'Images reproduced here with permission from Springer Nature'

are still interested in the details and principles of slow cooling for animal oocyte cryo-banking.

Slow cooling may remain a useful technique for banking non-human oocytes in circumstances where operators have traditionally used controlled rate cooling devices or large numbers of oocytes may need to be cryopreserved at one time, but there has not been a significant alteration in slow cooling protocols for many years. The slow controlled rate cooling method uses a mixture of the permeating cryoprotectant PrOH and non-permeating sucrose. It has been shown that 0.3 M sucrose yields better survival than 0.2 M sucrose in human oocytes [74, 75]. This is thought to be largely due to greater dehydration of the cells prior to freezing. Optimal exposure time to PrOH plus 0.2 M sucrose is 5–10 min whereas optimal exposure time to PrOH plus 0.3 M sucrose is yet to be determined. However,

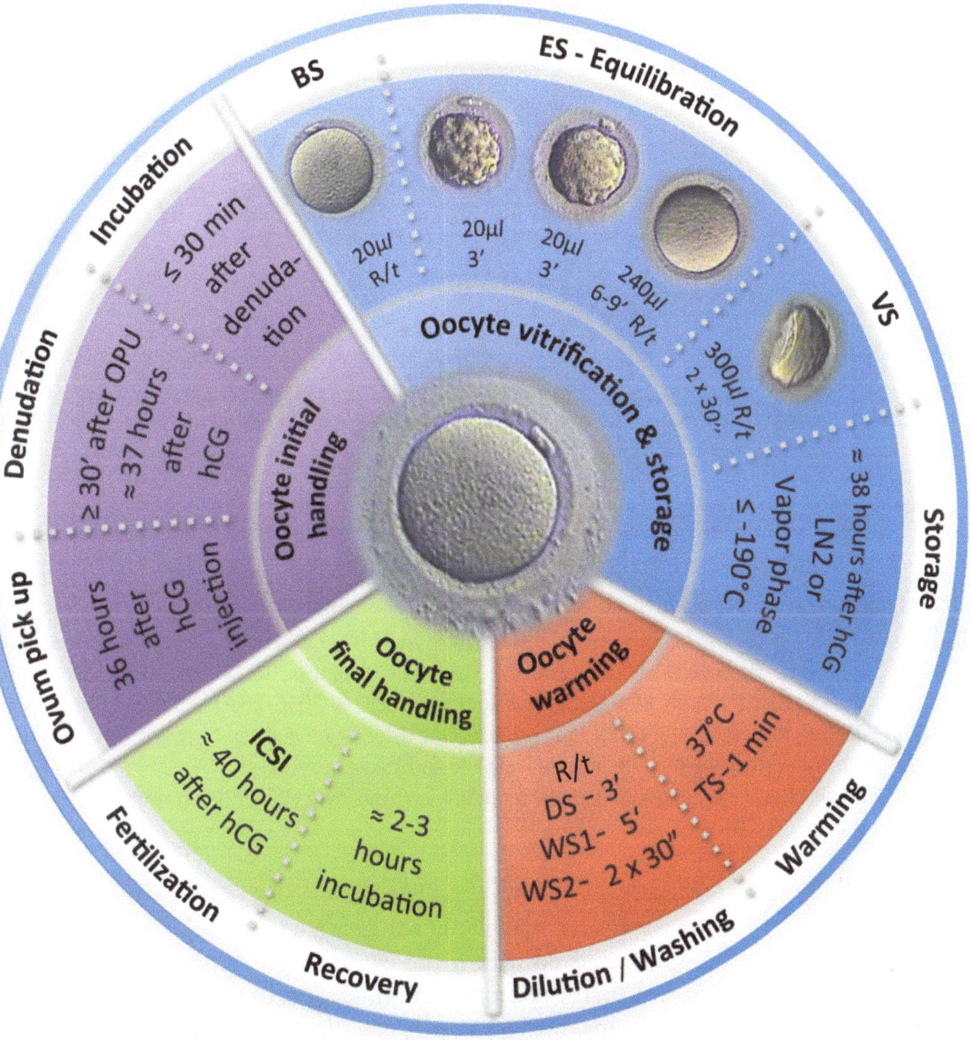

Fig. 2 A circular diagram outlines the recommended time frames for the most critical steps in human oocyte processing for vitrification and subsequent in vitro fertilization, linked to injection of the hCG trigger as the starting point, viewed clockwise from Ovum Pick Up. The timeline divided into four color-coded sections representing different stages: OPU and initial handling (purple), vitrification and storage (blue), warming (red), recovery and final handling (light green). The center image symbolizing the main focus of the diagram and expected quality of the oocyte at the start and final point. The initial handling section details timing after ovulation trigger injection, emphasizing quick processing with intervals around 30 minutes between different stages. The vitrification and storage section shows small images of oocytes at various stages with specific volumes and times for equilibration and vitrification solutions, plus storage conditions in liquid nitrogen or vapor phase at or below minus 190°C. The warming section lists times and temperatures for oocyte warming starting at 37°C with subsequent dilution and washing steps at room temperature. The final handling and recovery sections note incubation times of about two to three hours incubation in the IVF culture medium at 37°C, 5% O_2 and 6% CO_2 and fertilization by intracytoplasmic sperm injection approximately 40 hours after hormone trigger. This diagram is important for guiding precise timing, solution volumes, temperatures, and handling steps to ensure successful vitrification, storage and thawing of oocytes for fertility treatments. Ovum pick up (OPU); human chorionic gonadotropin (hCG); room temperature (R/t); basic solution (BS); equilibration solution (ES); vitrification solution (VS). Note! Equilibration of oocytes in VS at this stage is to be performed in two steps, V1 and V2, each within 300 μL and 30″; liquid nitrogen (LN2); warming solution (TS); dilution solution (DL); washing solution (WS); intra cytoplasmic sperm injection (ICSI)

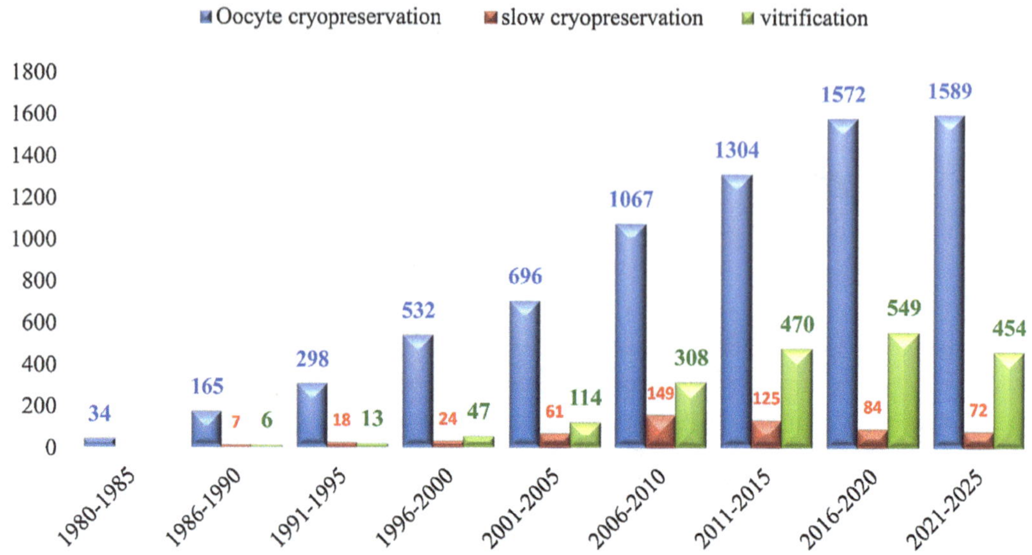

Fig. 3 The chart demonstrates a chronological illustration of the publications on oocyte cryopreservation and compares the number of articles focused on two methods of oocyte preservation over different time periods from 1980 to 2025. The blue bars show interest to oocyte cryopreservation, which steadily increases from 34 cases in 1980-1985 to 1,589 cases in 2021-2025, indicating growing use. The green bars represent vitrification, starting low but rising sharply after 2000, reaching 549 cases in 2016-2020 before slightly dropping to 454 in 2021-2025 as the successful vitrification protocol was established and clinically approved. The red bars show slow cryopreservation, which remains relatively low and stable, peaking at 149 cases in 2006-2010 and then declining. This chart highlights the increasing preference for oocyte vitrification over slow cryopreservation in recent years

an exposure time of 2 min will give a level of dehydration equivalent to that achieved with 5 min exposure to PrOH plus 0.2 M sucrose [55]. A further potential modification to the method is the choice of media in which to dilute the CPAs. Studies have also shown that a medium in which some of the sodium has been replaced by choline is preferable to PBS [44, 76, 77]. For example, we reported a successful method for slow cooling murine oocytes applying 1,2 propanediol as CPA, linked to sucrose as a secondary non-permeating CPA, and details of this can be found in Keros and Fuller [78].

On the other hand, many human oocytes, cryopreserved in earlier years using slow-programmed protocol, currently remain stored in cryobanks and, on request, will be used for future infertility treatment. It is well known that thawing can significantly influence the survival of the biological system, and results can be improved by optimizing the warming procedure [79]. To achieve outcomes equivalent to vitrified oocytes, the modified rehydration protocol can be applied [80]. Beyond the choice of CPAs, many ancillary factors, such as oocyte quality and selection, hormonal stimulation regime, and in vitro manipulation conditions, can impact cryopreservation outcomes [60, 81, 82]. As far as can be assessed from current information, oocyte cryopreservation does not have a negative impact on birth abnormalities [58, 60]. Health, physical, and mental development of children born after oocyte vitrification is comparable to that of those who were naturally conceived and born after use of fresh oocytes [83, 84].

Oocyte cryopreservation has moved away from assignment as an experimental procedure, towards recognition of the true value of this powerful technique with far-reaching potential in reproductive technologies. The current conventional vitrification protocols, consisting of cooling using various mixtures of penetrating CPAs such as DMSO, EG, PrOH, and non-penetrating sugars such as sucrose, trehalose, in combination with heating protocols designed for stepwise dilution and dehydration of vitrified oocytes after storage, consisting mainly of sucrose in decreasing concentrations, are widely used in clinical practice and are the standard clinical procedure at present. All types of media are mixed in stock solutions of varying composition and supplemented with buffers and antibiotics for stability when handling sensitive oocytes. The whole cycle of vitrification (≈15 min) and warming (≈10 min) requires approximately 25 min for one set of oocytes. As the number of patients seeking oocyte cryopreservation is growing continuously, it's lead to the time-consuming overload for busy ART laboratories is becoming clear. Semi-automated vitrification of oocytes, proposed as an attempt to optimize the procedure and make it reproducible, gives similar clinical results [52, 85, 86]; it was found to be less convenient and not preferred by laboratories experienced in manual vitrification staff [87].

Recently, on the basis of the biophysical model, shortening the vitrification and dehydration protocol was validated by evaluation of the osmotic behavior of human matured-M2 oocytes in response to exposure to CPAs in a short fast-vitrified/fast-warmed and standard protocols, and resulted in a comparable survival rate to the expected survival rate of 95% [88]. Similar results were obtained in a preclinical validation of a fast protocol on discarded human oocytes and supported by an animal study, with survival rates of 90%–100% of mouse and rabbit metaphase II oocytes and comparable to fertilization rate after ICSI using fresh oocytes, although implantation and live-birth rates in mice were decreased but not significantly, compared to the slow-vitrified/slow-warmed and slow-vitrified/fast-warmed oocytes (38.7% vs. 47.8% vs. 43.2% accordingly) [89]. Reduced exposure to CPAs at ultra-fast (UF) vitrification of mouse oocytes caused minimal alterations in cell organelles and no significant changes in the meiotic spindle and chromosomes [90]. Human immature oocytes successfully survived UF vitrification/warming, matured in vitro, and retained the ability to blastocyst development [91–93]. Summarizing laboratory and clinical outcomes [94, 95], Kitazato IVF company announced UF warming protocols, and in the year 2024, the new Kitazato medium for UF warming of oocytes, vitrified by the conventional Kitazato Cryotop® method, was CE marked

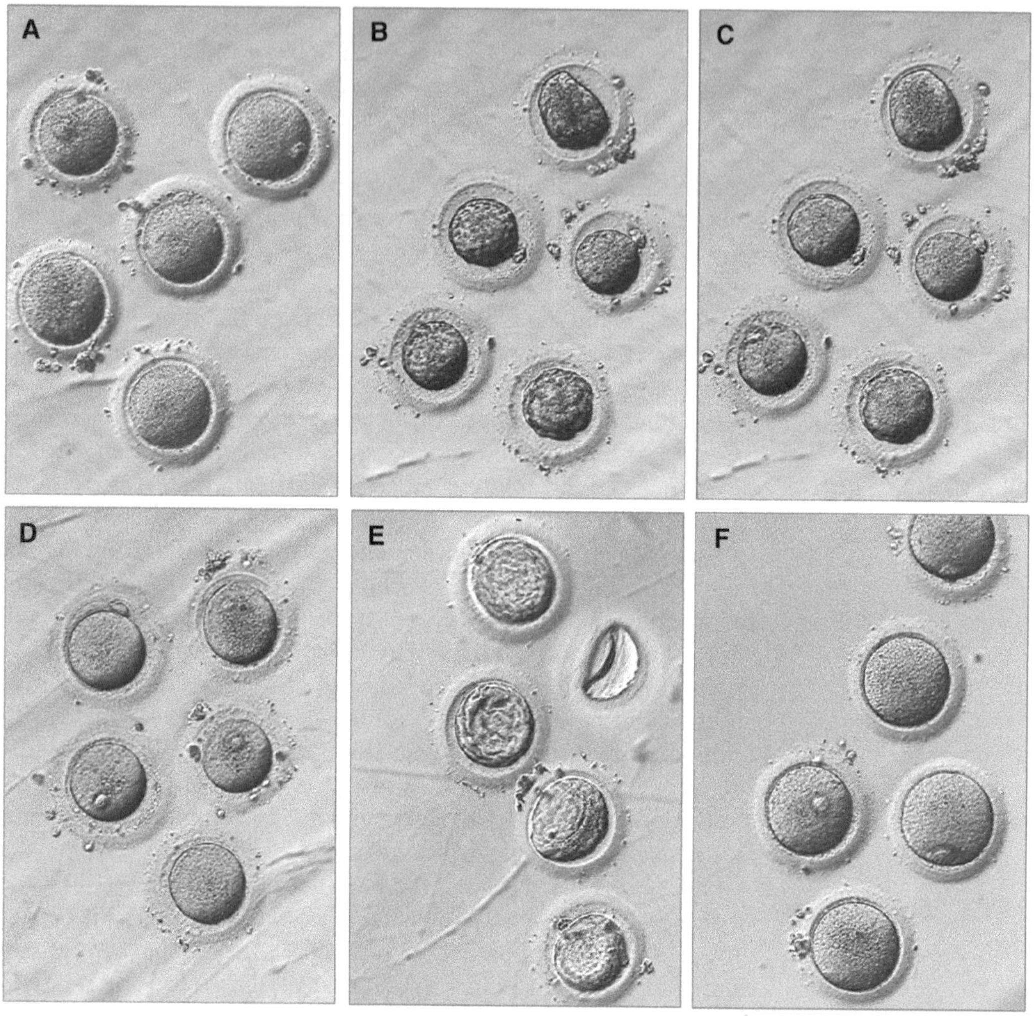

Fig. 4 Dynamical changes in the shape of immature metaphase I (M1) and in vitro matured metaphase II (M2) human oocytes during equilibration in CPA solutions before vitrification according to the Kitazato Cryotop®—Open System vitrification protocol and subsequent ultra-fast (UF) warming with Kitazato UF Warm solutions. (**A**) M1 and M2 oocytes in the basic solution (BS) before addition of equilibration solution (ES). (**B**) Reduced volume of oolemma and concurrent increased perivitelline space in the shrunken oocytes after addition of ES containing CPAs. (**C**) Expanding oocytes immersed in ES. (**D**) Completely expanded oocytes after immersing in ES, showing the width of the perivitelline space equal to that of the non-immersed oocyte. (**E**) Dehydrated shrunken oocyte in the vitrification solution just before transferring to the Cryotop and plunging into liquid nitrogen. (**F**) 100% survived vitrified-warmed oocytes showing intact morphological shape and structure after 2 h recovery in the IVF medium (37.0°C, 6.0% CO_2, 5% O_2) following UF warming

and accepted for clinical use. The warming procedure takes 2 min including 1-min exposure in TS at 37°C, and washing in three drops of oocyte culture medium, excluding dilution step, with subsequent incubation in the IVF medium for recovery before fertilization (see Fig. 4). Recently, the same company announced the UF vitrification protocol where the whole vitrification procedure takes only 2 min and consists of exposure of oocytes during 1 min in equilibration solution (ES) and 1 min in vitrification solution (VS) with subsequent plunging into LN2. The simplified time-saving protocol for vitrification and warming takes just 3 min, which is almost ten times less than the current Cryotop® method, thereby also reducing the exposure of sensitive oocytes to the high concentrations of CPAs and damaging temperature lower than 37°C. However only limited preliminary data show high up to 98% human M2 oocyte survival rate, 0.8% post-ICSI degeneration, 58% high-quality day 5 blastocysts as well as no differences in fertilization, cleavage, euploidy, or live birth ($p > 0.05$) compared with standard warming vs. UF-warming with clinical pregnancy rate per single blastocyst transfer at 65% for UF vs. 54.3% for current protocol [95], while long-term follow-up effect on epigenetic modifications and health of children born after application of UF methods for oocyte vitrification/warming is as yet lacking and more research is required [96].

As we approach the mid-2020s, there is a highly significant switch to VS for those centers performing oocyte cryopreservation, such that, for example, in the UK and Scandinavia, none are now using CRSC (Fuller, Keros personal communication 2025) as far as we are aware. Therefore, in this chapter, we shall concentrate on providing technical details of the VS method outlined below.

The robust method of oocyte vitrification is described below, with the necessary technical details to introduce this into a standard laboratory setting. Both the method for slow controlled rate cooling [2, 78] and the VS method, which have been developed for murine oocytes, can yield reliable, good outcomes, but (as discussed above) there may still be opportunities for further optimization in the various steps, should investigators be interested in pursuing those.

2 Materials

1. Plastic straws (IMV, L'Aigle, France) (*see* **Note 1**).
2. Plugs for straws (IMV, L'Aigle, France) or sealing powder or heat sealer (*see* **Note 2**).
3. Pulled glass pipettes, denudation glass, or plastic pipettes.
4. Tissue culture dishes (Falcon, Becton and Dickinson Co., USA).
5. Dissecting microscope.
6. Thermocouple.
7. Device to hold a straw horizontally above liquid nitrogen vapor without covering the area of the straw containing oocytes (*see* **Note 3**).

8. Hotplate set at 37°C.
9. Liquid nitrogen dewars, preferably at least two storage dewars (*see* **Note 4**), plus a container capable of holding liquid nitrogen and accommodating a straw held horizontally.
10. Scalpel.
11. 3× 1 mL syringe and 2× needle.
12. Heated water bath set at 20°C.
13. Safety equipment, e.g., cryo-gloves, face shield, and oxygen depletion monitor.
14. Heated gas incubator.
15. Oocyte culture medium, e.g., for human oocytes Fertilization medium (Cook IVF, Brisbane, Australia), for murine oocytes, Tyrode's medium (Invitrogen, Paisley, UK).
16. Dulbecco's phosphate-buffered solution (PBS; Invitrogen, Paisley, UK) for human oocytes supplemented with 30% protein supplement, e.g., plasmaprotein supplement (Baxter AG, Vienna, Austria) or serumprotein supplement (Pacific Andrology, CGA/Diasint, Florence, Italy); for mouse oocytes, supplementation with 5% heat-inactivated fetal bovine serum (Invitrogen, Paisley, UK).
17. VS solutions: 6 M dimethyl sulfoxide (Me_2SO) plus 1 mg/mL polyethylene glycol (PEG; MW 8000) added to 4× strength PBS medium without $CaCl_2$ (*see* **Note 5**). The solution (for a volume of 10 mL) should be made up by adding 1 mL of distilled water to 2.5 mL of 4× PBS prior to adding the Me_2SO (4.69 mL) and PEG (1 mL of 10 mg/mL aqueous stock). Then add protein supplement (e.g., 0.5 mL FBS) and make up almost to volume with water prior to adding $CaCl_2$ (100 μL of aqueous stock solution of 15.9 g/L). Finally, add distilled water to make up to the final volume (10 mL in this case). Keep at room temperature until required. This solution will be referred to as VSDP (*see* **Note 6**). Make up dilutions of 25% and 65% VSDP using PBS containing protein supplement as required.
18. Dilution solution consisting of 1 M sucrose made up in PBS containing protein supplement.
19. Hyaluronidase (optional, *see* **Note** 7).

3 Methods

Vitrification is advantageous over slow controlled rate cooling in that neither expensive cooling machines nor special laboratory environment is required. However, there is a need for highly skilled practitioners because the need for high concentrations of CPA means that the times stated for CPA exposure must be strictly adhered to, and manipulation of oocytes in these highly viscous solutions is technically demanding. Different CPA mixtures and minimum volume VS have been proven to be effective for oocyte VS. Toxicity of VS solution can be reduced by a combination of decreased concentration of permeating CPA, which can be balanced by increasing concentrations of non-permeating agents like polymers and sugars [63, 97–100]. Accurately performed VS (see Fig. 1) can result in oocytes with similar to unfrozen oocytes post-warming morphological recovery, embryo development, and live birth rates in humans. The VS method described using VSDP has resulted in high blastocyst formation [101] and live births [102] in mice, although the technique is prone to variability [58, 97]. The CPA mixture and cryopreservation vessel could be modified, but the basic techniques described are applicable to all VS protocols.

1. Straws should be prepared by pushing the cotton plug approximately one-third of the way down the straw. Using a needle and syringe, inject a column of sucrose solution, via what was the open end, so that the plug is wetted. Take care not to wet the sides of the straw with sucrose (*see* **Note 8**). The column of sucrose should fill approximately one-third of the straw. Using a separate needle and syringe, inject a small ~0.5 cm column of 100% VSDP, leaving at least 1 cm between the sucrose and the VS solution. Leave prepared straw at room temperature until required.
2. Place one 50 μL droplet each of 25%, 65%, and 100% VSDP in a tissue culture dish. Label each droplet.
3. Place liquid nitrogen to a level equivalent to at least the length of the straw into the vessel capable of accommodating the straw horizontally. Use the thermocouple to determine the point above the liquid nitrogen at which the temperature is −140°C, and temporarily affix the thermocouple at this level.
4. Pipette a maximum of five oocytes (*see* **Note 9**) into 25% VSDP droplet and leave at room temperature for 3–5 min.
5. Transferring as little of the solution as possible (*see* **Note 10**), pipette the oocytes from the 25% VSDP into 65% VSDP droplet.
6. As quickly as possible, move oocytes into a droplet of 100% VSDP.
7. Immediately draw up a small amount of 100% VSDP into a glass pipette and collect the oocytes. Transfer the oocytes into the column of 100% VSDP contained within the prepared straw.

8. Seal the open end of the straw (*see* **Note 2**). Hold the straw using the holder and, wearing suitable protective equipment, position the straw horizontally at the position above the liquid nitrogen at which the temperature is −140°C (*see* **Note 11**). Keep the straw in this position for 3 min and then plunge the straw into the liquid nitrogen.
9. Transfer the straw to a liquid nitrogen storage vessel (*see* **Note 12**).
10. Prior to warming the straw, place 1 mL of 1 M sucrose solution in a culture dish, half-fill a 1 mL syringe with 1 M sucrose, and place 2× 50 μL droplets of sucrose solution and 2× 50 μ L droplets of PBS with protein supplement in a culture dish. Label the droplets.
11. Wearing suitable protective equipment, remove the straw from the liquid nitrogen storage. Hold the straw in the air for 10 s, hold either end of the straw, not the area containing the oocytes, then plunge the straw into the water bath at 20°C for 10 s (*see* **Note 11**).
12. Wipe the straw dry and cut through the straw using a scalpel in the area containing the sucrose. Remove the plug from the other end, if used, or cut with a scalpel. Attach the syringe containing sucrose to one end of the straw and hold the other end over the dish containing 1 mL 1 M sucrose solution. Flush the contents of the straw and syringe into the dish and ensure good mixing.
13. Immediately begin to visually identify the oocytes using the dissecting microscope (*see* **Note 13**). As soon as the oocytes are identified, place them in one of the droplets of 1 M sucrose solution.
14. Immediately move the oocytes into the second droplet of 1 M sucrose solution. When the oocytes have been in contact with 1 M sucrose for a total of 5 min, transfer the oocytes into a droplet of PBS and leave for 10 min at room temperature. Move the oocytes into the second droplet of PBS and leave them for 10 min on the warm plate (*see* **Note 14**).
15. The oocytes should then be placed in oocyte culture medium, within an incubator (2–3 h for human, 30 min for mouse) to achieve post-dilution equilibrium prior to addition of sperm (*see* **Note 15**).

4 Notes

1. Faster cooling/warming rates can be achieved and the risk of ice crystal formation reduced by using straws that have been pulled to reduce the wall thickness, or by the use of such

devices as nylon loops or microscope grids. However, such devices are even more susceptible to temperature change, and some of these systems are "open," i.e., the sample is in direct contact with liquid nitrogen (*see* **Note 12**).

2. Straws can be sealed by heating the ends (and this is required for clinical samples), although care must be taken not to heat the solution containing the oocytes, or by inserting plugs, which should be wet in order to give a good seal, or with sealing powder.
3. A piece of plastic that can be inserted into a straw and is approximately twice the length of a straw is ideal. The plastic is inserted into the end of the straw and then bent through 90°. The straw can then be placed close to the nitrogen liquid surface in the horizontal position, and the holder can be held by hand in the cold nitrogen vapor at a safe distance from the liquid nitrogen. A floating straws carrying device (CryoFloater, Nidacon, Sweden), that provides a constant distance between the straws and the nitrogen surface, thereby allowing freezing in the cold nitrogen vapor, can also be an option.
4. Ideally, oocytes should be stored individually within straws and the straws kept in at least two different locations so that should an accident occur with one straw, then all is not lost. Accurate records should be kept of their location within the storage vessel.
5. The addition of the high concentrations of CPA used in VS solutions directly to the PBS medium will result in a reduction of the salt concentrations within that medium. Using a concentrated PBS medium allows the CPA to be added prior to the addition of water to give the required volume of single-strength PBS medium with appropriate isotonic salt balance.
6. The VS solution described here could be replaced with other combinations of CPA, for example, ethylene glycol together with sucrose, or ethylene glycol plus propanediol plus sucrose often in combination with a macromolecule such as Ficoll or trehalose, which has been used successfully for the preservation of bovine and human oocytes, or in combination with newly applied cryoprotective agent carboxylated ε-poly-L-lysine resulted in production of live offspring in mice [103]. Exposure to this CPA mixture can be performed at 37°C, thereby avoiding any cooling-related damage.
7. Studies have been performed with oocytes that have been denuded of cumulus cells (by means of treatment with hyaluronidase and/or gentle pipetting) prior to freezing or frozen with the cumulus intact. No clear advantage either way is evident. Human oocytes ought to be denudated within

approximately 37 h post-hCG administration for ovulation induction and to be vitrified directly after denudation. Cryopreservation of denuded oocytes is a routine method; it has been shown that mouse oocytes with remaining intact cumulus cells successfully survive VS in calcium-free medium with ethylene glycol as CPA [104].

8. The presence of the sucrose solution on the internal sides of the straw would dilute the concentration of CPA in the VSDP column and, more importantly, may allow the propagation of ice crystals from the sucrose solution into the area within which the oocytes are contained.
9. In order to adhere to the timing of each step of the procedure, oocytes should be processed in small batches.
10. Care must be taken not to dilute the CPA concentration in these small droplets; the use of small droplets aids the location of the oocytes in these viscous solutions.
11. The straw is held at −140°C (below the glass transition temperature range for aqueous CPS mixtures), rather than plunging directly into liquid nitrogen, in order to reduce the occurrence of cracks in the vitrified glass, which, on warming, may form sites of ice nucleation. It is important to measure the temperature as extreme fluctuations in temperature are present across a few centimeters in liquid nitrogen vapor. Higher cooling/warming rates are recommended when lower CPA concentrations are used, and the oocyte is vitrified in the low—less than 1 μL volume—of the VS solution. Immediate warming in a large volume of warming solution at 37°C gives better survival if oocytes are vitrified in a closed system [105].
12. It should be noted that straws allow rapid heat transfer and hence are susceptible to temperature change during handling, for example, on transfer from the cooling machine to liquid nitrogen for storage and during handling for identification purposes prior to thawing. Straws can be stored in the liquid (−196°C) or vapor phase (≤−190°C) of liquid nitrogen. If stored within the liquid phase, care should be taken to prevent leakage of liquid nitrogen into the straw, as this will expand as a gas on warming and may cause the straw to explode or crack. Also, the liquid nitrogen may be contaminated with viruses, which have been shown to survive such temperatures. Straws can also be placed within a second vessel, for example, a second larger straw (straw-in-straw method), to reduce liquid nitrogen entry. One end of the larger straw should be presealed and plunged into liquid nitrogen, then the VS straw containing oocytes should be placed inside of it, and the other end of the covering straw should be accurately sealed, avoiding warming of the sample. However, debate continues

about the impact on cooling and warming rates, which are enclosed in a second straw produce, and whether this has an impact on oocyte survival [6]. If storing in the vapor phase is chosen, the level of liquid nitrogen in the storage dewar should be carefully monitored to ensure stability of storage temperature. Alarms and automatic filling systems are available, but it is recommended that the level of liquid nitrogen still be checked manually at regular intervals.

13. Oocytes are particularly difficult to locate in this solution, and operator practice is important. The oocytes must spend no longer than 5 min in 1 M sucrose solution at this point, and this time can pass quickly. Other groups have developed post-vitrification dilution steps using 0.5 M sucrose solutions [103], which could be applied here, although we have no direct evidence about their use with VSDP protocols.
14. It may be possible to use different media for rewarming (devitrification) of oocytes vitrified with different media compositions and from different companies through the combination of different vitrification/warming kits [106]. The idea of a universal rewarming medium would be highly appreciated by personnel involved in oocyte cryopreservation, especially in case of interruptions in the commercial medium supply and when transporting cryopreserved oocytes between different laboratories around the world [107].
15. There is evidence to suggest that the microtubular spindle is capable of repair during a period of culture post-thaw [52, 89, 108].

Acknowledgments

We would like to gratefully acknowledge the significant role played by Dr. Sharon Paynter, who, as a collaborator over many years, developed the basic practical aspects of the two methods described in this chapter.

References

1. Fuller B, Paynter S (2004) Fundamentals of cryobiology in reproductive medicine. Reprod Biomed Online 9:680–691. https://doi.org/10.1016/s1472-6483(10)61780-4
2. Keros V, Fuller BJ (2015) Cryopreservation of mammalian oocytes. In: Wolkers WF, Oldenhof H (Eds.)Eds. Methods in cryopreservation and freeze-drying, Methods in molecular biology. Springer, 289–304. https://doi.org/10.1007/978-1-4939-2193-5_11
3. Yurchuk T, Petrushko M, Fuller B (2018) Science of cryopreservation in reproductive medicine—embryos and oocytes as exemplars. Early Hum Dev 126:6–9. https://doi.org/10.1016/j.earlhumdev.2018.08.016
4. Iussig B, Maggiulli R, Fabozzi G, Maggiulli R, Fabozzi G, Bertelle S, Vaiarelli A, Cimadomo D, Ubaldi FM, Rienzi L (2019) A brief history of oocyte cryopreservation: arguments and facts. Acta Obstet Gynecol Scand

98:550–558. https://doi.org/10.1111/aogs.13569
5. Gook DA (2011) History of oocyte cryopreservation. Reprod Biomed Online 23:281–289. https://doi.org/10.1016/j.rbmo.2010.10.018
6. Boldt J (2011) Current results with slow freezing and vitrification of the human oocyte. Reprod Biomed Online 23:314–322. https://doi.org/10.1016/j.rbmo.2010.11.019
7. Chang CC, Shapiro DB, Nagy ZP (2022) The effects of vitrification on oocyte quality. Biol Reprod 106:316–327. https://doi.org/10.1093/biolre/ioab239
8. Practice Committees of American Society for Reproductive Medicine; Society for Assisted Reproductive Technology (2013) Mature oocyte cryopreservation: a guideline. Fertil Steril 99:37–43. https://doi.org/10.1016/j.fertnstert.2012.09.028
9. Sciorio R, Pluchino N, Fuller BJ (2023) Review of human oocyte cryopreservation in ART programs: current challenges and opportunities. Cryobiology 113:104590. https://doi.org/10.1016/j.cryobiol.2023.104590
10. Al-Hasani S, Diedrich K, van der Ven H, Reinecke A, Hartje M, Krebs D (1987) Cryopreservation of human oocytes. Hum Reprod 2:695–700. https://doi.org/10.1093/oxfordjournals.humrep.a136616
11. Cimadomo D, Cobo A, Galliano D et al. (2024) Oocyte vitrification for fertility preservation is an evolving practice requiring a new mindset: societal, technical, clinical, and basic science-driven evolutions. Fertil Steril 121:555–561. https://doi.org/10.1016/j.fertnstert.2024.01.003
12. Asseler JD, Knieriem J, Huirne JA, Goddijn M, Verhoeven MO, van Mello NM (2023) Outcomes of oocyte vitrification in trans masculine persons. Reprod Biomed Online 47:94–101. https://doi.org/10.1016/j.rbmo.2023.03.007
13. Kamoshita M, Sugita H, Kageyama A et al. (2024) Recent advances of oocyte/embryo vitrification in mammals from rodents and large animals. Anim Sci J 95(1):e13931. https://doi.org/10.1111/asj.13931
14. Mogas T, García-Martínez T, Martínez-Rodero I (2024) Methodological approaches in vitrification: enhancing viability of bovine oocytes and in vitro-produced embryos. Reprod Domest Anim 59(Suppl 3):e14623. https://doi.org/10.1111/rda.14623
15. Simone R, Čižmári D, Holtze S, Mulot B, Lamglait B, Knauf-Witzens T, Weigold A, Hermes R, Hildebrandt TB (2024) Cryopreservation of okapi (Okapia johnstoni) oocytes following in vitro maturation. Theriogenol Wild 4:100088. https://doi.org/10.1016/j.therwi.2024.100088
16. Leibo SP, Pool TB (2011) The principal variables of cryopreservation: solutions, temperatures and rate changes. Fertil Steril 96:269–276. https://doi.org/10.1016/j.fertnstert.2011.06.065
17. Fuller B, Paynter S (2009) The rational basis for controlled rate slow cooling. In: Borini A, Coticchio G (eds) Preservation of human oocytes, from cryobiology science to clinical applications. Inform Healthcare, pp 25–35
18. Fuller B, Gonzalez-Molina J, Erro E et al. (2017) Applications and optimization of cryopreservation technologies to cellular therapeutics. Cell Gene Ther Insights 3:359–378. https://doi.org/10.1016/10.18609/cgti.2017.038
19. Rall WF, Fahy GM (1985) Ice-free cryopreservation of mouse embryos at −196 degrees C by vitrification. Nature 313:573–575. https://doi.org/10.1038/313573a0
20. Chen C (1986) Pregnancy after human oocyte cryopreservation. Lancet 1:884–886. https://doi.org/10.1016/s0140-6736(86)90989-x
21. Trounson A (1986) Preservation of human eggs and embryos. Fertil Steril 46:1–12
22. Gook DA, Schiewe MC, Osborn SM, Asch RH, Jansen RP, Johnston WI (1995) Intracytoplasmic sperm injection and embryo development of human oocytes cryopreserved using 1,2-propanediol. Hum Reprod 10:2637–2641. https://doi.org/10.1093/oxfordjournals.humrep.a135759
23. Fabbri R, Porcu E, Marsella T, Primavera MR, Seracchioli R, Ciotti PM, Magrini O, Venturoli S, Flamigni C (1998) Oocyte cryopreservation. Hum Reprod 13(Suppl 4):98–108. https://doi.org/10.1093/humrep/13.suppl_4.98
24. Carroll J, Wood MJ, Whittingham DG (1993) Normal fertilization and development of frozen-thawed mouse oocytes: protective action of certain macromolecules. Biol Reprod 48:606–612. https://doi.org/10.1095/biolreprod48.3.606
25. Whittingham DG (1977) Fertilization *in vitro* and development to term of unfertilized mouse oocytes previously stored at −196 degrees C. J Reprod Fertil 49:89–94. https://doi.org/10.1530/jrf.0.0490089
26. Bernard A, Fuller BJ (1996) Cryopreservation of human oocytes: a review of current problems and perspectives. Hum Reprod Update 2:193–207. https://doi.org/10.1093/humupd/2.3.193
27. Leibo SP, McGrath JJ, Cravalho EG (1978) Microscopic observation of

intracellular ice formation in unfertilized mouse ova as a function of cooling rate. Cryobiology 15:257–271. https://doi.org/10.1016/0011-2240(78)90036-6
28. Al-Hasani A, Kirsch J, Diedrich K, Blanke S, van der Ven H, Krebs D (1989) Successful embryo transfer of cryopreserved and *in vitro* fertilised rabbit oocytes. Hum Reprod 4:77–79. https://doi.org/10.1093/oxfordjournals.humrep.a136849
29. Fuku E, Kojima T, Shioya Y, Marcus GJ, Downey BR (1992) In vitro fertilization and development of frozen-thawed bovine oocytes. Cryobiology 29:485–492. https://doi.org/10.1016/0011-2240(92)90051-3
30. MacLellan LJ, Carnevale EM, Coutinho da Silva MA, Scoggin CF, Bruemmer JE, Squires EL (2002) Pregnancies from vitrified equine oocytes collected from super-stimulated and non-stimulated mares. Theriogenology 58:911–919. https://doi.org/10.1016/s0093-691x(02)00920-2
31. Pope CE, Gómez MC, Kagawa N, Kuwayama M, Leibo SP, Dresser BL (2012) In vivo survival of domestic cat oocytes after vitrification, intracytoplasmic sperm injection and embryo transfer. Theriogenology 77:531–538. https://doi.org/10.1016/j.theriogenology.2011.08.028
32. Porcu E, Fabbri R, Seracchioli R et al. (1997) Birth of a healthy female after intracytoplasmic sperm injection of cryopreserved human oocytes. Fertil Steril 68:724–726
33. De Santis L, Cino I, Rabellotti E, Papaleo E, Calzi F, Fusi FM, Brigante C, Ferrari A (2007) Oocyte cryopreservation: clinical outcome of slow-cooling protocols differing in sucrose concentration. Reprod Biomed Online 14:57–63. https://doi.org/10.1016/10.1016/s1472-6483(10)60764-x
34. Borini A, Coticchio G (2009) The efficacy and safety of human oocyte cryopreservation by slow cooling. Semin Reprod Med 27:443–449. https://doi.org/10.1055/s-0029-1241053
35. Rienzi L, Romano S, Albricci L, Maggiulli R, Capalbo A, Baroni E, Colamaria S, Sapienza F, Ubaldi F (2010) Embryo development of fresh 'versus' vitrified metaphase II oocytes after ICSI: a prospective randomized sibling-oocyte study. Hum Reprod 25:66–73. https://doi.org/10.1093/humrep/dep346
36. Antinori M, Licata E, Dani G, Cerusico F, Versaci C, Antinori S (2007) Cryotop vitrification of human oocytes results in high survival rate and healthy deliveries. Reprod Biomed Online 14:72–79. https://doi.org/10.1016/s1472-6483(10)60766-3
37. Cobo A, Kuwayama M, Pérez S, Ruiz A, Pellicer A, Remohí J (2008) Comparison of concomitant outcome achieved with fresh and cryopreserved donor oocytes vitrified by the cryotop method. Fertil Steril 89:1657–1664. https://doi.org/10.1016/j.fertnstert.2007.05.050
38. Levi-Setti PE, Patrizio P, Scaravelli G (2016) Evolution of human oocyte cryopreservation: slow freezing versus vitrification. Curr Opin Endocrinol Diabetes Obes 23:445–450. https://doi.org/10.1016/10.1097/MED.0000000000000289
39. Bleil JD, Wassarman PM (1980) Structure and function of the zona pellucida: identification and characterization of the proteins of the mouse oocyte's zona pellucida. Develop Biol 76:185–202. https://doi.org/10.1016/0012-1606(80)90371-1
40. Carroll J, Depypere H, Matthews CD (1990) Freeze-thaw induced changes of the zona pellucida explains decreased rates of fertilisation in frozen-thawed mouse oocytes. J Reprod Fertil 90:547–553. https://doi.org/10.1530/jrf.0.0900547
41. Larman MG, Sheehan CB, Gardner DK (2006) Calcium-free vitrification reduces cryoprotectant-induced zona pellucida hardening and increases fertilization rates in mouse oocytes. Reproduction 131:53–61. https://doi.org/10.1530/rep.1.00878
42. Bianchi V, Coticchio G, Fava L, Flamigni C, Borini A (2005) Meiotic spindle imaging in human oocytes frozen with a slow freezing procedure involving high sucrose concentration. Hum Reprod 20:1078–1083. https://doi.org/10.1093/humrep/deh736
43. Stachecki JJ, Cohen J (2004) An overview of oocyte cryopreservation. Reprod Biomed Online 9:152–163. https://doi.org/10.1016/s1472-6483(10)62124-4
44. Rienzi L, Martinez F, Ubaldi F, Minasi MG, Iacobelli M, Tesarik J, Greco E (2004) Polscope analysis of meiotic spindle changes in living metaphase II human oocytes during the freezing and thawing procedures. Hum Reprod 19:655–659. https://doi.org/10.1093/humrep/deh101
45. Combelles CM, Ceyhan ST, Wang H, Racowsky C (2011) Maturation outcomes are improved following Cryoleaf vitrification of immature human oocytes when compared tocholine-based slow-freezing. J Assist Reprod Genet 28:1183–1192. https://doi.org/10.1007/s10815-011-9674-x
46. Khalili MA, Shahedi A, Ashourzadeh S, Nottola SA, Macchiarelli G, Palmerini MG (2017) Vitrification of human immature oocytes before and after in vitro maturation: a review. J

Assist Reprod Genet 34:1413–1426. https://doi.org/10.1007/s10815-017-1005-4
47. Brower PT, Schultz RM (1982) Intercellular communication between granulosa cells and mouse oocytes: existence and possible nutritional role during oocyte growth. Dev Biol 90:144–153. https://doi.org/10.1016/0012-1606(82)90219-6
48. Ruppert-Lingham CJ, Paynter SJ, Godfrey J, Fuller BJ, Shaw RW (2003) Developmental potential of murine germinal vesicle stage cumulus-oocyte complexes following exposure to dimethylsulphoxide or cryopreservation: loss of membrane integrity of cumulus cells after thawing. Hum Reprod 18:392–398. https://doi.org/10.1093/humrep/deg071
49. Keros V, Xella S, Hultenby K, Pettersson K, Sheikhi M, Volpe A, Hreinsson J, Hovatta O (2009) Vitrification versus controlled-rate freezing in cryopreservation of human ovarian tissue. Hum Reprod 24:1670–1683. https://doi.org/10.1093/humrep/dep079
50. Shahedi A, Hosseini A, Khalili MA, Norouzian M, Salehi M, Piriaei A, Nottola SA (2013) The effect of vitrification on ultrastructure of human in vitro matured germinal vesicle oocytes. Eur J Obstet Gynecol Reprod Biol 67:69–75. https://doi.org/10.1016/j.ejogrb.2012.11.006
51. Bartolacci A, Albertini DF (2024) The new ice age: the promise and challenges of rapid oocyte warming protocols. J Assist Reprod Genet 41:2969–2971. https://doi.org/10.1007/s10815-024-03315-2
52. Chaput L, Dollet S, Eymard-Pierre E, Pereira B, Lucas C, Grémeau AS, Tchirkov A, Marteil G, Brugnon F (2025) Analysis of maturation dynamics and oocyte nuclear quality after rescue-IVM and semi-automated vitrification. Hum Reprod 7:1344–1356. https://doi.org/10.1093/humrep/deaf078
53. Marteil G, Metchat A, Dollet S, Cugnot C, Chaput L, Pereira B, Gremeau AS, Brugnon F (2024) Vitrification of human oocytes before or after rescue-IVM does not impair maturation kinetics but induces meiotic spindle alterations. Reprod Sci 31:2625–2636. https://doi.org/10.1007/s43032-024-01596-7
54. Isachenko V, Soler C, Isachenko E, Perez-Sanchez F, Grishchenko V (1998) Vitrification of immature porcine oocytes: effects of lipid droplets, temperature, cytoskeleton, and addition and removal of cryoprotectant. Cryobiology 36:250–253. https://doi.org/10.1006/cryo.1998.2079
55. Bianchi V, Lappi M, Bonu MA, Borini A (2012) Oocyte slow freezing using a 0.2–0.3 M sucrose concentration protocol: is it really the time to trash the cryopreservation machine?. Fertil Steril 97:1101–1107. https://doi.org/10.1016/j.fertnstert.2012.01.127
56. Kuleshova L, Gianaroli L, Magli C, Ferraretti A, Trounson A (1999) Birth following vitrification of a small number of human oocytes. Hum Reprod 14:3077–3079. https://doi.org/10.1093/humrep/14.12.3077
57. Yoon TK, Chung HM, Lim JM, Han SY, Ko JJ, Cha KY (2000) Pregnancy and delivery of healthy infants developed from vitrified oocytes in a stimulated *in vitro* fertilization-embryo transfer program. Fertil Steril 74:180–181. https://doi.org/10.1016/s0015-0282(00)00572-0
58. Noyes N, Porcu E, Borini A (2009) Over 900 oocyte cryopreservation babies born with no apparent increase in congenital anomalies. Reprod Biomed Online 18:769–776. https://doi.org/10.1016/s1472-6483(10)60025-9
59. UK Human Fertilisation and Embryology Authority (September, 2018) Egg freezing in fertility treatment. Trends and figures: 2010–2016. https://www.hfea.gov.uk/media/2656/egg-freezing-in-fertility-treatment-trends-and-figures-2010-2016-final.pdf
60. Levi-Setti PE, Borini A, Patrizio P, Bolli S, Vigiliano V, De Luca R, Scaravelli G (2016) ART results with frozen oocytes: data from the Italian ART registry (2005–2013). J Assist Reprod Genet 33:123–128. https://doi.org/10.1007/s10815-015-0629-5
61. Nagy ZP, Anderson RE, Feinberg EC, Hayward B, Mahony MC (2017) The Human Oocyte Preservation Experience (HOPE) Registry: evaluation of cryopreservation techniques and oocyte source on outcomes. Reprod Biol Endocrinol 15:10. https://doi.org/10.1186/s12958-017-0228-7
62. ESHRE Working Group on Oocyte Cryopreservation in Europe, Shenfield F, de Mouzon J, Scaravelli G, Kupka M, Ferraretti AP, Prados FJ, Goossens V (2017) Oocyte and ovarian tissue cryopreservation in European countries: statutory background, practice, storage and use. Hum Reprod Open 2017:hox003. https://doi.org/10.1093/hropen/hox003
63. Kuwayama M, Vajta G, Kato O, Leibo SP (2005) Highly efficient vitrification method for cryopreservation of human oocytes. Reprod Biomed Online 11:300–308. https://doi.org/10.1016/s1472-6483(10)60837-1
64. Kuwayama M (2007) Highly efficient vitrification for cryopreservation of human oocytes and embryos: the Cryotop method. Theriogenology 67:73–80. https://doi.org/10.1016/j.theriogenology.2006.09.014
65. Garratt J, Shah T, Mclaughlin A, Al-Hashimi B, Macklon N, Linara-Demakakou E, Ahuja

KK (2025) Clinical outcomes of vitrified-warmed autologous oocyte cycles with 15-year follow-up at a single UK centre: consistent and predictable results. Reprod Biomed Online 50:104376. https://doi.org/10.1016/j.rbmo.2024.104376

66. Kostoglou K, Michos G, Najdecki R, Tsakiridis I, Tatiana C, Chouliara F, Dagklis TI, Mamopoulos AM, Papanikolaou E (2025) Comparison of cumulative live birth rates between fresh and vitrified donor oocytes. Cureus 17:e82589. https://doi.org/10.7759/cureus.82589
67. Cao YX, Xing Q, Li L, Cong L, Zhang ZG, Wei ZL, Zhou P (2009) Comparison of survival and embryonic development in human oocytes cryopreserved by slow-freezing and vitrification. Fertil Steril 92:1306–1311. https://doi.org/10.1016/j.fertnstert.2008.08.069
68. Fadini R, Brambillasca F, Renzini MM, Merola M, Comi R, De Ponti E, Dal Canto MB (2009) Human oocyte cryopreservation: comparison between slow and ultrarapid methods. Reprod Biomed Online 19:171–180. https://doi.org/10.1016/s1472-6483(10)60069-7
69. Edgar DH, Gook DA (2012) A critical appraisal of cryopreservation (slow cooling versus vitrification) of human oocytes and embryos. Hum Reprod Update 18:536–554. https://doi.org/10.1093/humupd/dms016
70. Noyes N, Knopman J, Labella P, McCaffrey C, Clark-Williams M, Grifo J (2010) Oocyte cryopreservation outcomes including pre-cryopreservation and post-thaw meiotic spindle evaluation following slow cooling and vitrification of human oocytes. Fertil Steril 94:2078–2082. https://doi.org/10.1016/j.fertnstert.2010.01.019
71. Cobo A, Diaz C (2012) Clinical application of oocyte vitrification: a systematic review and meta-analysis of randomized controlled trials. Fertil Steril 96:277–285. https://doi.org/10.1016/j.fertnstert.2011.06.030
72. Cobo A, Coello A, Remohí J, Serrano J, de Los Santos JM, Meseguer M (2017) Effect of oocyte vitrification on embryo quality: time-lapse analysis and morphokinetic evaluation. Fertil Steril 108:491–497. https://doi.org/10.1016/j.fertnstert.2017.06.024
73. Rienzi L, Gracia C, Maggiulli R, LaBarbera AR, Kaser DJ, Ubaldi FM, Vanderpoel S, Racowsky C (2017) Oocyte, embryo and blastocyst cryopreservation in ART: systematic review and meta-analysis comparing slow-freezing versus vitrification to produce evidence for the development of global guidance. Hum Reprod Update 23:139–155. https://doi.org/10.1093/humupd/dmw038
74. Fabbri R, Porcu E, Marsella T, Rocchetta G, Venturoli S, Flamigni C (2001) Human oocyte cryopreservation: new perspectives regarding oocyte survival. Hum Reprod 16:411–416. https://doi.org/10.1093/humrep/16.3.411
75. Paynter SJ, Borini A, Bianchi V, De Santis L, Flamigni C, Coticchio G (2005) Volume changes of mature human oocytes on exposure to cryoprotectant solutions used in slow cooling procedures. Hum Reprod 20:1194–1199. https://doi.org/10.1093/humrep/deh742
76. Stachecki JJ, Cohen J, Willadsen SM (1998) Cryopreservation of unfertilized mouse oocytes: the effect of replacing sodium with choline in the freezing medium. Cryobiology 37:346–354. https://doi.org/10.1006/cryo.1998.2130
77. Quintans CJ, Donaldson MJ, Bertolino MV, Pasqualini RS (2002) Birth of two babies using oocytes that were cryopreserved in a choline-based freezing medium. Hum Reprod 17:3149–3152. https://doi.org/10.1093/humrep/17.12.3149
78. Keros V, Fuller BJ (2021) Cryopreservation of mammalian oocytes: slow cooling and vitrification as successful methods for cryogenic storage. In: Wolkers WF, Oldenhof H (eds) Methods Mol Biol, vol 2180, pp 437–454. https://doi.org/10.1007/978-1-0716-0783-1_20
79. Keros V, Milenkovic M, Hultenby K (2024) The art of cryopreservation—live birth after transplantation of vitrified ovarian tissue. Cryobiology 117:105085. https://doi.org/10.1016/j.cryobiol.2024.105085
80. Papis K, Hardej K, Stachowiak E, Żyżyńska-Galeńska K, Lewandowski P, Kozioł K (2025) Equivalent outcomes of human oocytes after vitrification or slow freezing with a modified rehydration protocol. Reprod Biol Endocrinol 23: 58. https://doi.org/10.1186/s12958-025-01383-2
81. Paynter SJ (2000) Current status of the cryopreservation of human unfertilised oocytes. Hum Reprod Update 6:449–456. https://doi.org/10.1093/humupd/6.5.449
82. Boldt J (2011) Current results with slow freezing and vitrification of the human oocyte. Reprod Biomed Online 23:314–322. https://doi.org/10.1016/j.rbmo.2010.11.019
83. Van Reckem Blockeel C, Bonduelle M, Buysse A, Roelants M, Verheyen G, Tournaye H, Hes F, Belva F (2021) Health of 2-year-old children born after vitrified oocyte donation in comparison with peers born after fresh oocyte donation. Hum Reprod Open;2021(1):hoab002. 10.1093/hropen/hoab002. Erratum in:

Hum Reprod Open 13;2021(2). 10.1093/hropen/hoab013

84. Takeshige Y, Takahashi M, Hashimoto T, Kyono K (2021) Six-year follow-up of children born from vitrified oocytes. Reprod Biomed Online 42(3):564–571. https://doi.org/10.1016/j.rbmo.2020.11.005
85. Brunetti XO, Cawood S, Gaunt M (2021) The first livebirth using warmed oocytes by a semi-automated vitrification procedure. J Reprod Infertil 22:70–72. https://doi.org/10.18502/jri.v22i1.4998
86. Barberet J, Ducreux B, Bruno C, Guilleman M, Simonot R, Lieury N, Guilloteau A, Bourc'his D, Fauque P (2022) Comparison of oocyte vitrification using a semi-automated or a manual closed system in human siblings: survival and transcriptomic analyses. J Ovarian Res 15: 128. https://doi.org/10.1186/s13048-022-01064-3
87. Hajek J, Baron R, Sandi-Monroy N, Schansker S, Schoepper B, Depenbusch M, Schultze-Mosgau A, Neumann K, Gagsteiger F, von Otte S, Griesinger G (2021) A randomised, multi-center, open trial comparing a semi-automated closed vitrification system with a manual open system in women undergoing IVF. Hum Reprod 36:2101–2110. https://doi.org/10.1093/humrep/deab140
88. Gallardo M, Saenz J, Risco R (2019) Human oocytes and zygotes are ready for ultra-fast vitrification after 2 minutes of exposure to standard CPA solutions. Sci Rep 9: 15986. https://doi.org/10.1038/s41598-019-52014-x
89. Costa-Borges N, Matia-Algué Q, Coello A, Mestres E, Acacio M, Flores-Saiffe Farias A, Castello C, Gallardo M, Chavez-Badiola A, Marco-Jiménez F, Cobo A, Cohen J (2025) Preclinical validation of fast oocyte vitrification and warming protocols with comparable efficiencies to a standard method. Hum Reprod 40:1066–1076. https://doi.org/10.1093/humrep/deaf069
90. Cho JR, Yu EH, Lee HJ, Kim IH, Jeong JH, Lee DB, Cho SK, Joo JK (2024) Ultra-fast vitrification: minimizing the toxicity of cryoprotective agents and osmotic stress in mouse oocyte cryopreservation. Int J Mol Sci 25:1884. https://doi.org/10.3390/ijms25031884
91. Liebermann J, Brohammer R, Wagner Y, Smith R, Even K, Hirshfeld-Cytron J, Uhler ML (2024) Fast and furious: successful survival and resumption of meiosis in immature human oocytes vitrified and warmed using a short protocol. Reprod Biomed Online 49:103976. https://doi.org/10.1016/j.rbmo.2024.103976
92. Schiewe MC, Reichelderfer R, Wozniak K, De Romana C, Nordbak M, Baek K, Chung K (2024) Ultra-fast vitrification and rapid elution of human oocytes: part I. germinal vesicle model validation. Reprod Biomed Online 49:104691. https://doi.org/10.1016/j.rbmo.2024.104691
93. Wozniak K, Reichelderfer R, Ghaemi S, Hupp D, Fuzesi P, Ringler G, Marrs RP, Schiewe MC (2024) Ultra-fast vitrification and rapid elution of human oocytes: part II—verification of blastocyst development from mature oocytes. Reprod Biomed Online 49:104690. https://doi.org/10.1016/j.rbmo.2024.104690
94. Aydin B, Hudkova D, Maggiotto G, Kal NS, Osmanllari U, Unsal E, Baltaci V, Aktuna S, Yonar D, Dorofeyeva U, Kotliarova E, Vitushynska M, Misyura Y, Yetim I, Babarikova V (2024) Human oocyte survival, early embryo development, metabolic fingerprinting, and pregnancy outcomes following ultra-rapid or standard vitrification and thawing. Hum Reprod 39(Suppl 1):deae108.351. https://doi.org/10.1093/humrep/deae108.351
95. Kotliarova O, Aydin B, Dorofeyeva U, Hudkova D, Pylypenko N, Vitushynska M, Misyura Y, Rozhak N, Unsal E, Kal NS, Kubaskova TM, Babarikova V (2025) Optimizing morphokinetic embryo development: oocyte and embryo vitrification and thawing via ultra-rapid warming protocol. Hum Reprod 40(Suppl 1):deaf097.121. https://doi.org/10.1093/humrep/deaf097.121
96. Martinez-Rodero I, Gallardo M, Pisaturo V, Scarica C, Conaghan J, Liebermann J, Cuevas-Saiz I (2025) Shorter protocols for vitrification and post-warming dilution of human oocytes and embryos: a narrative review. Reprod Biomed Online 51:104857. https://doi.org/10.1016/j.rbmo.2025.104857
97. Fahy GM, Wowk B, Wu J, Paynter SJ (2004) Improved vitrification solutions based on the predictability of vitrification solution toxicity. Cryobiology 48:22–35. https://doi.org/10.1016/j.cryobiol.2003.11.004
98. Robles V, Valcarce DG, Riesco MF (2019) The use of antifreeze proteins in the cryopreservation of gametes and embryos. Biomolecules 9:181. https://doi.org/10.3390/biom9050181
99. Shaw PW, Bernard AG, Fuller BJ, Hunter JH, Shaw RW (1992) Vitrification of mouse oocytes using short cryoprotectant exposure: effects of varying exposure times on survival. Mol Reprod Dev 33:210–214. https://doi.org/10.1002/mrd.1080330214

100. Correia LFL, Alves BRC, Batista RITP, Mermillod P, Souza-Fabjan JMG (2021) Antifreeze proteins for low-temperature preservation in reproductive medicine: a systematic review over the last three decades. Theriogenology 176:94–103. https://doi.org/10.1016/j.theriogenology.2021.09.025
101. O'Neil L, Paynter SJ, Fuller BJ, Shaw RW (1997) Vitrification of mature mouse oocytes: improved results following addition of polyethylene glycol to a dimethyl sulfoxide solution. Cryobiology 34:295–301. https://doi.org/10.1006/cryo.1997.2007
102. O'Neil L, Paynter SJ, Fuller BJ, Shaw RW (1999) Birth of live young from mouse oocytes vitrified in 6 M dimethyl sulfoxide supplemented with 1 mg/ml polyethylene glycol. Cryobiology 39:284. https://doi.org/10.1006/cryo.1998.2098
103. Watanabe H, Kohaya N, Kamoshita M, Fujiwara K, Matsumura K, Hyon SH, Ito J, Kashiwazaki N (2013) Efficient production of live offspring from mouse oocytes vitrified with a novel cryoprotective agent, carboxylated ε-poly-L-lysine. PLoS ONE 8:e83613. https://doi.org/10.1371/journal.pone.0083613
104. Kohaya N, Fujiwara K, Ito J, Kashiwazaki N (2011) High developmental rates of mouse oocytes cryopreserved by an optimized vitrification protocol: the effects of cryoprotectants, calcium and cumulus cells. J Reprod Dev 57:675–680. https://doi.org/10.1262/jrd.11-066h
105. De Munck N, Verheyen G, Van Landuyt L, Stoop D, Van de Velde H (2013) Survival and post-warming *in vitro* competence of human oocytes after high security closed system vitrification. J Assist Reprod Genet 30:361–359. https://doi.org/10.1007/s10815-013-9930-3
106. Troncoso-Perez P, Gonzalez-Navas C, Coccia ME, Fucci R, Falcone P, Bertocci F, Picone R, Sosa-Rosales DF, Lopez-Perez N, Criado-Scholz E, Vilches-Ferron MA (2025) Application of a single "Universal warming protocol" for vitrified donor oocytes: a multicenter study. J Assist Reprod Genet 42:1331–1341. https://doi.org/10.1007/s10815-025-03423-7
107. Parmegiani L, Minasi MG, Arnone A, Casciani V, Cognigni GE, Viñoles R, Varricchio MT, Quintero LA, Greco E, Filicori M (2025) "Universal Warming" protocol for vitrified oocytes to streamline cell exchange for transnational donation programs: a multi-center study. J Assist Reprod Genet 37:1379–1385. https://doi.org/10.1007/s10815-020-01798-3
108. Ciotti PM, Porcu E, Notarangelo L, Magrini O, Bazzocchi A, Venturoli S (2009) Meiotic spindle recovery is faster in vitrification of human oocytes compared to slow freezing. Fertil Steril 91:2399–2407. https://doi.org/10.1016/j.fertnstert.2008.03.013

Chapter 23

Cryopreservation of Ovarian Cortex Tissue

Andreas Schallmoser, Rebekka Einenkel-Ulrich and Nicole Sänger

Abstract

Ovarian tissue cryopreservation (OTC) is applied as a fertility preservation option for women and girls who are at risk of losing ovarian function due to medical treatments with gonadotoxic side effects, such as chemotherapy or radiotherapy. Especially in prepubertal girls, it is the only option for fertility preservation. Two modes of OTC are practiced: Vitrification and slow freezing. The aim of OTC is the preservation of follicles. Reimplantation of the ovarian tissue provides the patient with a follicle pool, in order to increase the chances of becoming pregnant by serving oocytes and the ovarian hormonal functions. Vitrification of ovarian tissue is a modern cryopreservation technique that rapidly cryopreserves tissue to prevent ice crystal formation, which can damage cells. It is increasingly studied and applied as an alternative to slow freezing. Current meta-analyses indicate that both techniques appear to provide comparable outcomes. Vitrification provides several advantages, including time and cost reduction due to shorter protocols and the omission of slow freezing equipment. Here, we provide a detailed protocol for ovarian tissue vitrification.

Keywords Cryopreservation, Fertility preservation, FertiPROTEKT, Ovarian tissue, Vitrification

1 Introduction

In recent years, significant advancements have been made in fertility preservation techniques within reproductive medicine, largely in response to the damaging effects of cancer therapies on reproductive health [1]. The rising incidence of cancer [2], coupled with improving survival rates, has shifted attention toward safeguarding patients' long-term quality of life—including the ability to have biological children [3, 4]. Among the available fertility preservation options for women undergoing gonadotoxic treatment, cryopreservation—including oocytes, embryos, and ovarian tissue—has become increasingly important.

Ovarian tissue cryopreservation (OTC) is especially beneficial when immediate cancer treatment is necessary and time does not allow for ovarian stimulation. Moreover, for prepubescent girls, OTC remains the only option for fertility preservation since hormonal stimulation is not possible. Furthermore, reimplanting preserved

Willem F. Wolkers and Harriëtte Oldenhof (eds.), *Cryopreservation and Freeze-Drying Protocols*, Methods in Molecular Biology, vol. 3060, https://doi.org/10.1007/978-1-0716-5416-3_23,

tissue may help restore endocrine function, potentially postponing the onset of menopause [5–8]. Accumulating evidence supports the effectiveness and safety of OTC, with numerous reports of successful pregnancies and live births following tissue transplantation. These outcomes confirm that OTC is a viable method not only for fertility restoration but also for reestablishing hormonal balance [9–14].

For OTC, vitrification as well as slow freezing is available. Slow freezing provides a less demanding protocol in terms of critical handling steps, whereas vitrification is a less time- and cost-consuming method [15, 16].

Slow freezing is currently more widely used for OTC. It aims to freeze the tissue in an equilibrated state using cryoprotectants in a controlled, slow freezing rate. This allows the water to leave the cell, to minimize but not avoid ice crystals [17]. However, they can still cause mechanical damage to the tissue. Moreover, different optimal cooling rates for variable cells, which are present in the tissue, come with compromises [17].

Vitrification is an emerging approach for OTC [16, 18, 19]. The fast cooling rate prevents ice crystal formation and reduces the risks of tissue damage. Meta-analyses show comparable results between OTC by slow freezing versus vitrification [20–22]. Compared to slow freezing, vitrification is less time- and cost-consuming. However, vitrification is more demanding in terms of manual handling and operator skills, since it is not performed in a device. Thus, a detailed protocol for ovarian tissue vitrification is provided.

2 Materials

All materials must be kept sterile and used in a contamination-controlled environment. Please note that the storage conditions must be observed according to the manufacturer. This also applies to aliquoted reagents.

2.1 Media for Processing, Transportation, Vitrification, and Rapid Warming of Ovarian Tissue

1. Custodiol solution (Dr. Franz Köhler Chemie, Germany) is a ready-to-use medium designed for tissue transportation and preparation.
2. Gx-MOPS™ PLUS (Vitrolife, Sweden) is a pH-stable handling medium for handling at atmospheric conditions supplemented with antioxidants.
3. Ethylene glycol (Merck, Germany).
4. Sucrose ACS reagent grade, sterile (Merck, Germany).
5. Polyvinylpyrrolidone (360 kDa) PVP (Merck, Germany).
6. Serum substitute supplement SSS (Irvine, Germany).

2.2 Materials Required for the Processing Procedure

1. Laminar flow hood (LAF) for sterile working conditions with ultraviolet radiation (UV).
2. Cool plate (2–4°C) integrated into the LAF (Fryka-Kältetechnik, Germany).
3. 2 mm biopsy punches (PFM Medical, Germany).
4. Sterile disposable scalpel (No. 22; Feather, Japan).
5. Petridish 92 × 16 mm (Sarstedt, Germany).
6. Surgical gloves (Kanam, India).

2.3 Materials Required for the Transportation Procedure

1. Individually customized transportation box (Consarctic, Germany).
2. Isolation inlet for customized transportation box (deltaT, Germany).
3. Cold packs for customized transportation box (deltaT, Germany).
4. Calibrated data logger for recording transportation temperature permanently (deltaT, Germany).
5. Sterile 30 mL Tube (GML Alfaplast, Germany).

2.4 Materials Required for the Vitrification Procedure

1. For aliquoting vitrification solution, use sterile 50 mL tubes (Vitrolife, Sweden).
2. Sterile, 25 mL serological pipettes (Sarstedt, Germany).
3. Sterile cup (Sarstedt, Germany).
4. Roller mixer (Allpax, Germany).
5. For rapid vertical vitrification, use IsoSafe (Eppendorf, Germany) with a manual fitted grid (Biozym, Germany) to hold 1.8 mL vials.
6. Sterile 1.8 mL vials (Thermo Scientific, Germany).
7. Use screens for CD-1 (Merck, Germany) to customize metal meshes to fit into the cap of 1.8 mL vials after autoclaving.
8. Sterilization container (Euronda, Germany) to facilitate the autoclaving of metal meshes.
9. Autoclave (Tuttnauer, Germany).
10. Sterile single-use forceps (medimex, Germany).
11. Laminar flow hood (LAF) for sterile working conditions with ultraviolet radiation (UV).
12. Nterilizer for the decontamination of liquid nitrogen (Nterilizer, Italy).
13. Sterile Acrodisc syringe filters with a pore size of 0.20 µm (PALL Life Sciences, USA).
14. Sterile 20 mL syringes (Braun, Germany).

15. Surgical gloves (Kanam, India).
16. Six-well plate (Sarstedt, Germany).

2.5 Materials Required for the Rapid Warming Procedure

1. For aliquoting vitrification solution, use sterile 50 mL tubes (Vitrolife, Sweden).
2. Sterile, 25 mL serological pipettes (Sarstedt, Germany).
3. Roller mixer (Allpax, Germany).
4. Sterile cup (Sarstedt, Germany).
5. Laminar flow hood (LAF) for sterile working conditions with ultraviolet radiation (UV).
6. Sterile Acrodisc syringe filters with a pore size of 0.20 μm (PALL Life Sciences, USA).
7. Sterile 20 mL syringes (Braun, Germany).
8. Surgical gloves (Kanam, India).
9. Six-well plate (Sarstedt, Germany).
10. Water bath (Memmert, Germany) or incubator (Labotect, Germany) to bring RWS to 37°C.
11. Sterile single-use forceps (Medimex, Germany).

2.6 Materials Required for the Assessment of the Vital Follicle Count

1. Collagenase, lyophilized powder (Merck, Germany).
2. Calcein acetoxymethyl ester (Merck, Germany).
3. DMSO (WAK-Chemie, Germany).
4. For aliquoting collagenase, calcein-AM, and DPBS, use 1.5 mL reaction tubes (Sarstedt, Germany).
5. AIM V medium (Gibco, Thermo Fisher Scientific, USA).
6. Four-well dish (Vitromed, Germany).
7. Incubator (Labotect, Germany).
8. Fluorescence microscope (Nikon, Japan).
9. DPBS (Thermo Fisher Scientific, USA).

2.7 Core Requirements, Operational and Technical Criteria for an Ovarian Tissue Bank

1. All steps—preparation, freezing, storage, and quality control—must follow current scientific and clinical guidelines.
2. To comply with legal regulations (i.a. Commission Directive 2006/86/EC, EU Guidelines to GMP—Annex 1 (2022), Directive 2004/23/EC), tissue handling must occur in a sterile class II laminar airflow cabinet with air quality grade A (equivalent to ISO Class 5) located within a cleanroom environment with a background of grade D (equivalent to ISO 8).
3. Ovarian tissue processing should take place in a sterile environment, specifically within a Class II laminar flow hood equipped with an integrated cooling platform (FRYKA-Kältetechnik GmbH, Germany), see Fig. 1.

4. Prior to tissue collection, donors must undergo screening and be confirmed negative for certain infectious diseases in line with food manufacturing practice (GMP) requirements (EMA, GMP Guideline on ATMPs). This includes testing for HIV types 1 and 2 (including p24 antigen), hepatitis B (surface antigen HBsAg and core antibody anti-HBc, with HBV-PCR if anti-HBc is reactive), hepatitis C (anti-HCV IgG), and syphilis (*Treponema pallidum*).
5. Fertility assessment: To evaluate reproductive potential and determine the appropriate number of tissue slices required for future transplantation, factors such as serum AMH levels, ovarian volume, antral follicle count (AFC), and donor age should be thoroughly assessed.
6. The use of validated methods to measure the density of follicles is essential to ensure that the tissue remains viable and uncompromised during the processes of collection, transportation, processing, cryopreservation, and thawing [23, 24].
7. Histological safety evaluation. Before any tissue can be used therapeutically, its quality and safety must be confirmed. A representative portion of the ovarian tissue should undergo histopathological examination to verify the absence of abnormalities or malignancy.
8. Tissue storage conditions. Tissue is kept in vapor-phase liquid nitrogen tanks at ≤−160°C with 24/7 temperature monitoring and backup systems. Inventory: A secure digital system tracks all samples using unique identifiers, ensuring traceability and compliance with privacy laws. Access control: Only authorized personnel may access storage areas. Regular audits ensure security and compliance [25, 26].
9. Trained staff. All procedures are performed by qualified, regularly trained personnel. SOPs: Clear protocols guide each

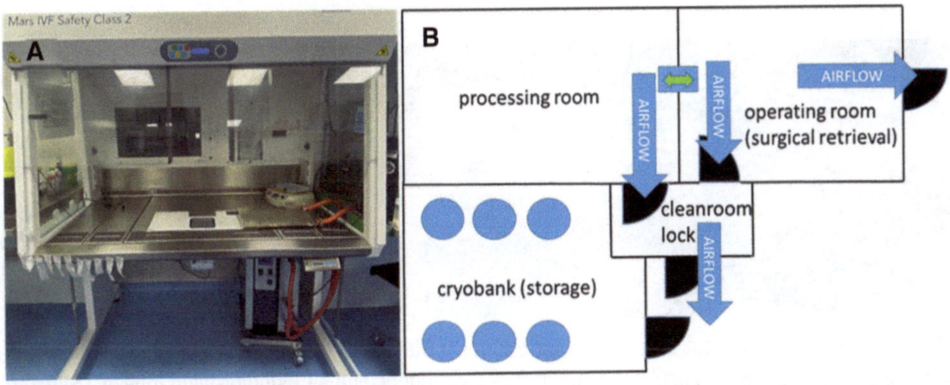

Fig. 1 Room setup for ovarian tissue processing. (**A**) Laminar flow hood with integrated cooling plate. (**B**) Operating room, processing room, and cryobank with a short distance for optimal workflow procedures

process step. Any deviations must be logged and reviewed. Ethical standards: Informed consent is mandatory. Use of tissue must align with donor wishes and legal requirements.

10. To prevent contamination and ensure the highest standards of quality and safety, sterile products must be used. Products from IVF suppliers are particularly suitable, as IVF consumables adhere to the most stringent standards, including CE marking (Conformité Européenne) and mouse embryo assay (MEA) testing. These certifications help ensure that materials are safe and nontoxic for use with human cells and tissues. In cases where certified products are not available, standardized MEA testing can be performed to verify the biocompatibility of materials such as plasticware and culture media.

3 Methods

3.1 Harvesting Ovarian Tissue and Transportation Prior to Cryostorage

Ovarian tissue is surgically collected prior to the initiation of gonadotoxic cancer therapy or in cases of medical conditions associated with early depletion of the ovarian reserve. The procedure is conducted laparoscopically as an outpatient intervention.

1. To preserve follicular viability, excise the tissue using sharp scissors without the use of electrocoagulation. Remove approximately 30%–50% of one ovary.
2. After excision, place the tissue in a tube containing custodiol solution maintained at 4°C as indicated in Fig. 2C.
3. Transport the tissue with delay to the processing laboratory at a controlled temperature of 4–8°C (Fig. 2A–C).
4. If tissue processing and storage cannot be performed on-site at the clinic where the surgery takes place, collaboration with

Fig. 2 Cooling setup for ovarian tissue transportation. (**A**) Stackable cold pack is shown. (**B**) Contents of an ovary transport box include cold packs, a temperature data logger, and a Custodiol-filled tube. (**C**) Stacked cold pack in the order as for transportation, which are inserted into the transport box shown in (**B**). (**D**) Ovarian tissue transportation to the central processing laboratory and cryobank must be conducted within 24 h

an external, specialized cryobank is a viable option for fertility preservation. In such cases, transport is conducted using certified shipping containers designed to maintain the required cold chain conditions, ensuring tissue integrity during transit.

5. For histopathological examination, fix a small fragment in formalin with particular attention to identifying malignant cells or metastatic involvement.

3.2 Media Preparation for Vitrification of Ovarian Tissue

1. Gather the following materials: one 10 mL serological pipette with a suiting pipette aid, three 50 mL sterile tubes labeled for VS1, VS2, and VS3. Make sure to note the preparation date.
2. Activate the ultraviolet (UV) light in the laminar flow hood for 30 min to ensure an additional level of sterilization before starting the procedure.
3. For preparation of VS1 (15 mL): add 12 mL of G-MOPS PLUS into the first 50 mL tube, 1.5 mL of SSS, and 1.5 mL of ethylene glycol.
4. For preparation of VS2 (15 mL): add 10.5 mL of G-MOPS PLUS, 3 mL of ethylene glycol, and 1.5 mL of SSS.
5. For preparation of VS3 (15 mL): add 8.5 mL of G-MOPS PLUS, 5.25 mL of ethylene glycol, 2.57 g of sucrose, and 0.75 g of PVP (*see* **Note 1** and **2**).

3.3 Media Preparation for Rapid Warming of Ovarian Tissue

1. Gather the following materials: one 10 mL serological pipette, one electronic pipette aid, three 50 mL sterile tubes labeled for rapid warming solution (RWS), equilibration solution (ES), and rinsing solutions (RS).
2. Activate the ultraviolet (UV) light for 30 min in the laminar flow hood to ensure an additional level of sterilization before starting the procedure.
3. For preparation of RWS (30 mL): add 27 mL of G-MOPS PLUS into the first 50 mL tube, 3 mL of SSS, and 8.16 g of sucrose.
4. For preparation of ES (15 mL): add 13.5 mL of G-MOPS PLUS, 1.5 mL of SSS, and 2.04 g of sucrose.
5. For preparation of RS1/2 (15 mL): add 13.5 mL of G-MOPS PLUS and 1.5 mL of SSS (*see* **Notes 1** and **2**).

3.4 Loading Device Setup Procedure

1. Gather surgical scissors and metal mesh sheets to begin the customization process. Using the scissors, cut the mesh into strips sized 25 mm by 8 mm, as shown in Fig. 3A.
2. Place the cut mesh strips into a sterilization container and autoclave them for 2 h (see Fig. 3B). After the autoclaving cycle is complete, transfer both the sterilization container and 1.8 mL vials into a laminar flow hood.

Fig. 3 Preparation of the vitrification device. (**A**) Metal cell mesh was cut into strips measuring 25 × 8 mm and (**B**) prepared for autoclaving. (**C**) Following sterilization, the meshes were connected to the caps of 1.8 mL vials under sterile conditions

3. Activate the ultraviolet (UV) light in the laminar flow hood and expose the materials to 254 nm UV radiation for 30 min to ensure additional sterilization.
4. Once UV treatment is complete, open the sterilization container and insert the sterilized metal strips into the caps of the 1.8 mL vials using sterile forceps, following the setup shown in Fig. 3C. Securely close the vials (*see* **Note 2**).

3.5 Preparation for Viable Follicle Count

1. Bring DMSO to room temperature overnight. Dissolve calcein–AM in DMSO at 10 mg/mL. Aliquot 3 μL of the dissolved calcein-AM per 1.5 mL reaction tube and store the vials at −20°C.
2. Place 7 mg of lyophilized collagenase per 1.5 mL tube and store it at −20°C.

3.6 Ovarian Tissue Handling: Vitrification

1. Carefully isolate the ovarian cortex by removing the medulla. The firmer texture of the cortex typically allows for easy differentiation and separation from the softer medulla.
2. Cut the cortical tissue into appropriately sized fragments (see Fig. 4), depending on the application: approximately 10 mm × 5 mm for transplantation purposes and tissue punches of 2 mm in diameter for assessing viable follicle count before and after vitrification.
3. Place a cell strainer into the first well of a six-well plate (as shown in Fig. 5) containing 5 mL of vitrification solution 1 (VS1). Add the tissue and incubate for 5 min to equilibrate on an orbital shaker.
4. Transfer the cell strainer with the tissue to the second well containing 5 mL of VS2 and continue equilibration for another 5 min on an orbital shaker.
5. Move the strainer to a third well filled with 5 mL of VS3 and equilibrate for 6 min, as shown in Fig. 5.

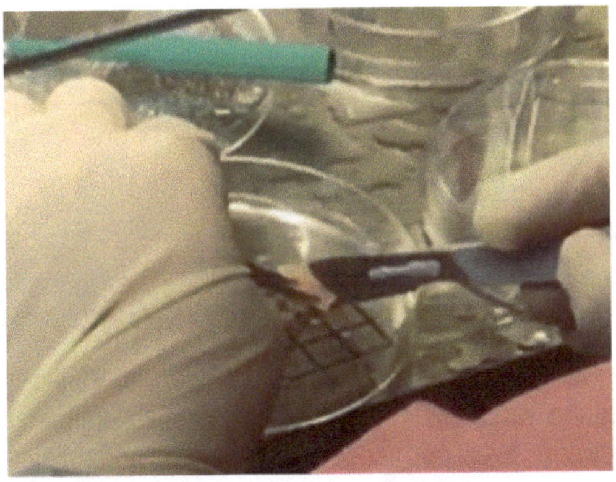

Fig. 4 Ovarian tissue processing. For orientation, a foil with a millimeter grid is placed under the Petri dish. This allows the ovarian tissue to be measured objectively and the fragments to be cut into evenly sized pieces

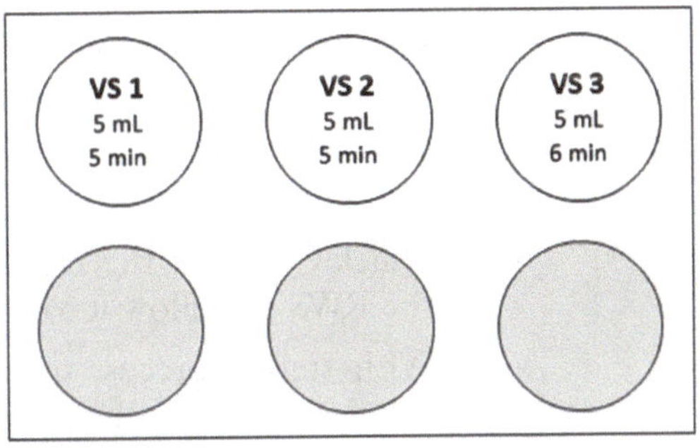

Fig. 5 Scheme of prepared vitrification solutions. The six-well plate is set up with the vitrification solutions. Wells contain information on volume and incubation time

6. For vitrification, use UV-decontaminated nitrogen as by the Nterilizer system or LAF bench UV light.
7. Open the cryovials and fill them with sterile liquid nitrogen. Place the vials in a cryo dewar filled with liquid nitrogen (see Fig. 6A).
8. Load a tissue piece onto individual metal grids of the vitrification device (see Fig. 6B, C) and immediately immerse the loaded metal grid in the cryovial containing liquid nitrogen. Repeat for every tissue piece within 1 min to avoid prolonged incubation in VS3 (*see* **Notes 2–4**).

3.7 Ovarian Tissue Handling: Rapid Warming

1. Set up the workplace in the laminar flow bench. Prepare a cup with 30 mL RWS (rapid warming solution) and a six-well plate with ES (equilibration solution) and RS (rinsing solution) as

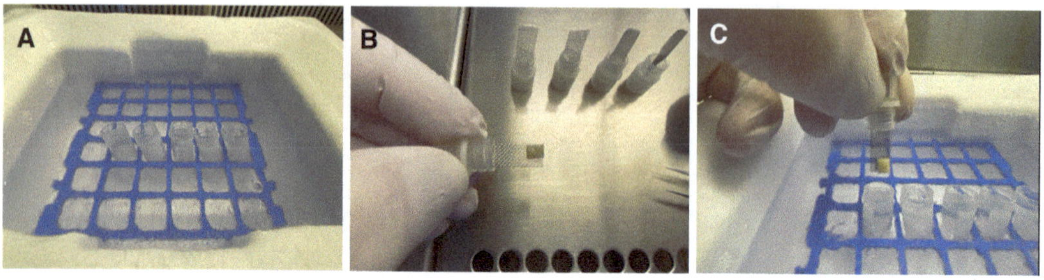

Fig. 6 Vitrification of ovarian tissue. (**A**) Cryovials are pre-filled with liquid nitrogen and placed in a customized, grid-patterned styrofoam platform, optimized to accommodate a high number of samples simultaneously. (**B**) Ovarian cortical tissue is positioned on a metal grid and (**C**) inserted vertically in liquid nitrogen designed for rapid vertical vitrification

shown in Fig. 7. Bring RWS to 37.2°C and the six-well plate containing ES and RS to room temperature (see Fig. 7).

2. Cryopreserved vials containing vitrified ovarian cortical tissue should be transported in liquid nitrogen to a laminar flow hood.
3. Open the vials while they remain partially submerged in liquid nitrogen.
4. Quickly transfer the mesh containing the vitrified tissue into the RWS and allow it to warm for 1 min at 37°C.
5. With sterile forceps, move the tissue into the equilibration solution (ES) and incubate it for 3 min with gentle agitation using an orbital shaker.
6. Wash the tissue at room temperature for 10 min each in RS1 and RS2, both on an orbital shaker.
7. Place the rewarmed tissue into a sterile beaker containing 5 mL of Custodiol. Keep this solution at 4°C until the tissue is transported to the transplantation site (*see* **Notes 2** and **4**).

3.8 Quality Control: Determination of Vital Follicular Count

1. Ovarian tissue punches are kept in the incubator at 37°C overnight in 500 μL AIM V medium in a four-well dish. For warmed samples, this allows the tissue to recover metabolically [27].
2. On the day of the assessment, add 997 μl of DPBS to the pre-aliquoted calcein and resuspend to thaw and dilute the calcein.
3. Transfer 1 mL of the calcein solution to 7 mg collagenase and vortex to obtain the working solution.
4. Use 500 μL of the working solution to digest the two ovarian cortex punches with a 2 mm in diameter at 37°C, shielded from light, in a four-well dishes. After 60 min and a further 70 min, resuspend the solution ten times. Let the cells and follicles sit for 5 min. Finally, determine the viable follicular count

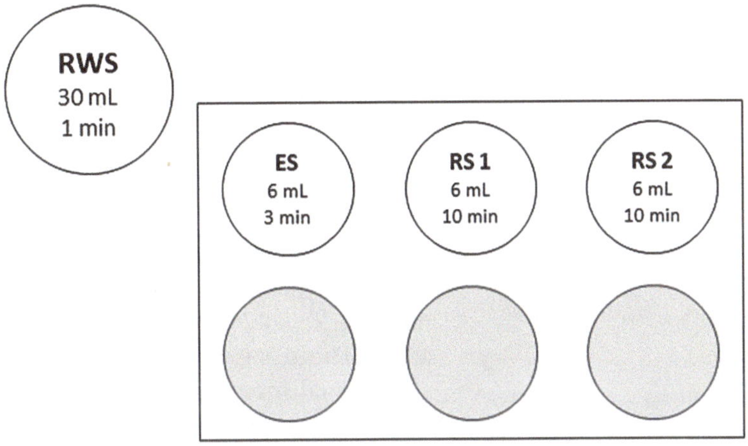

Fig. 7 Preparation of rapid warming solutions. Rapid warming solution (RWS), equilibration solution (ES), and rinsing solutions (RS) are prepared and transferred to a six-well plate. Note that RWS is maintained at 37.2°C on a heating plate. Vessels contain information on the volume of the respective solution and incubation time

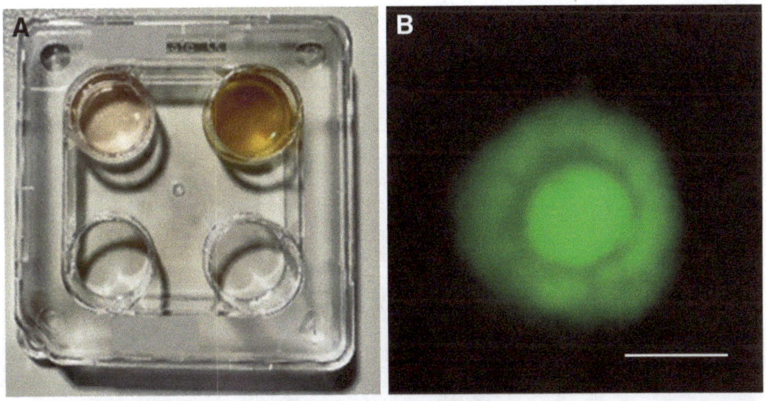

Fig. 8 Calcein–AM staining. (**A**) Four-well dish setup containing a well with AIM V on the left after overnight incubation and a well with the calcein–AM collagenase solution on the right with the tissue punches for digestion and staining. (**B**) Viable cells are stained green with calcein–AM. Viable follicles appear brighter and bigger than other cells in the suspension. Scale bar marks 50 μm

via fluorescence microscopy. Viable cells show a green fluorescence signal (see Fig. 8). Follicles appear brighter and larger than the surrounding cells in the suspension.

3.9 Cryobanking

1. Access control in a cryobank is essential to protect high-value biological specimens. Transponder-controlled keys, password-protected software access, and manual locks for storage containers maintain regulatory compliance and ensure

traceability, security, and safety. Whether storing stem cells, embryos, tissues, or DNA, controlling who can access, move, or modify samples is vital for both biosecurity and legal accountability.

2. Ovarian tissue should be stored in the vapor phase of a liquid nitrogen storage system to minimize the risk of contamination (see Fig. 9A).
3. The main storage tank for liquid nitrogen is located outside the main building to minimize the asphyxiation risk, as liquid nitrogen rapidly vaporizes into nitrogen gas at room temperature. This gas is colorless, odorless, and displaces oxygen. In an enclosed space, a leak can lead to oxygen deficiency, causing a serious asphyxiation hazard. Outdoors, the gas can disperse safely in the open air, reducing this risk (see Fig. 9B).
4. Monitoring liquid nitrogen (LN_2) levels in a cryobank is critical to ensure the safe storage of biological materials such as

Fig. 9 (**A**) 300-L cryotank for gas phase storage and (**B**) a main supply tank located outside. Safety precautions in working with cryotanks: (**C**) alert system for storage tank, (**D**) manual checks of nitrogen levels

cells, embryos, tissues, or DNA. A continuous monitoring and alarm system (see Fig. 9C) must be employed to detect deviations in temperature or liquid nitrogen levels, ensuring the integrity and safety of the stored tissue. Relying solely on automation is not considered sufficient in most critical environments, especially those dealing with irreplaceable biological samples, such as human tissues or embryos. Routine manual checks (see Fig. 9D) verify the accuracy of automated readings as technicians can spot abnormal signs (frost buildup, unusual tank noise, physical damage) that sensors cannot.

5. To enhance the efficiency and traceability of ovarian tissue cryobanking, implement a secure digital inventory management system. This system should include barcode or RFID tracking for each sample, enabling accurate identification, location tracking, and retrieval. It should also allow for the recording of metadata such as patient ID (coded for privacy), storage conditions, vitrification/warming protocols, and quality control results. Integration with temperature and liquid nitrogen level monitoring systems is recommended, enabling automated alerts and documentation to ensure compliance with biobanking standards and safety protocols.
6. A digital storage system (MedITEX, Germany) in cryobanking refers to a software-based platform that manages, tracks, and documents the storage, movement, and access of biological samples stored in cryogenic conditions. These systems are essential for ensuring sample integrity, regulatory compliance, and efficient operations to deal with rolling in/out, transfer, and disposal of samples, besides charging storage fees in real time [25].

4 Notes

1. Mix each solution thoroughly using a roller mixer.
2. All work must be conducted inside a Class II laminar flow hood. Prior to starting, disinfect the work surface using an appropriate disinfectant while wearing disposable, non-powdered laboratory gloves to maintain sterility.
3. Handle tissue delicately to prevent damage from excessive pressure. Sample characteristics may differ between individuals.
4. Please work conscientiously and promptly to ensure the rapid cooling and warming processes. Exceeding the time specifications or prolonged working steps can lead to severe tissue damage.

References

1. Preservation EGGoFF, Anderson RA, Amant F, Braat D, D'Angelo A, Chuva de Sousa Lopes SM, Demeestere I, Dwek S, Frith L, Lambertini M, Maslin C, Moura-Ramos M, Nogueira D, Rodriguez-Wallberg K, Vermeulen N (2020) ESHRE guideline: female fertility preservation. Hum Reprod Open 2020:hoaa052. https://doi.org/10.1093/hropen/hoaa052
2. Miller KD, Nogueira L, Mariotto AB, Rowland JH, Yabroff KR, Alfano CM, Jemal A, Kramer JL, Siegel RL (2019) Cancer treatment and survivorship statistics, 2019. CA Cancer J Clin 69:363–385. https://doi.org/10.3322/caac.21565
3. Michalczyk K, Cymbaluk-Ploska A (2021) Fertility preservation and long-term monitoring of gonadotoxicity in girls, adolescents and young adults undergoing cancer treatment. Cancers 13:202. https://doi.org/10.3390/cancers13020202
4. Anderson RA, Clatot F, Demeestere I, Lambertini M, Morgan A, Nelson SM, Peccatori F, Cameron D (2021) Cancer survivorship: reproductive health outcomes should be included in standard toxicity assessments. Eur J Cancer 144:310–316. https://doi.org/10.1016/j.ejca.2020.11.032
5. Donnez J, Dolmans MM (2017) Fertility preservation in women. N Engl J Med 377:1657–1665. https://doi.org/10.1056/NEJMra1614676
6. Jadoul P, Guilmain A, Squifflet J, Luyckx M, Votino R, Wyns C, Dolmans MM (2017) Efficacy of ovarian tissue cryopreservation for fertility preservation: lessons learned from 545 cases. Hum Reprod 32:1046–1054. https://doi.org/10.1093/humrep/dex040
7. Terenziani M, Meazza C, Massimino M, Vigano P, Gandola L, Mangili G, Raspagliesi F, Biasoni D, Podda M, Veneroni L, Filippi F, Riccipetitoni G, Somigliana E (2016) Female fertility preserving practices at a pediatric unit: a challenge of multiprofessional and multidisciplinary cooperation. Tumori 102:174–177. https://doi.org/10.5301/tj.5000452
8. Wallace WH, Kelsey TW, Anderson RA (2016) Fertility preservation in pre-pubertal girls with cancer: the role of ovarian tissue cryopreservation. Fertil Steril 105:6–12. https://doi.org/10.1016/j.fertnstert.2015.11.041
9. Kim S, Lee Y, Lee S, Kim T (2018) Ovarian tissue cryopreservation and transplantation in patients with cancer. Obstet Gynecol Sci 61:431–442. https://doi.org/10.5468/ogs.2018.61.4.431
10. Donnez J, Dolmans MM, Demylle D, Jadoul P, Pirard C, Squifflet J, Martinez-Madrid B, van Langendonckt A (2004) Livebirth after orthotopic transplantation of cryopreserved ovarian tissue. Lancet 364:1405–1410. https://doi.org/10.1016/S0140-6736(04)17222-X
11. Meirow D, Ra'anani H, Shapira M, Brenghausen M, Derech Chaim S, Aviel-Ronen S, Amariglio N, Schiff E, Orvieto R, Dor J (2016) Transplantations of frozen-thawed ovarian tissue demonstrate high reproductive performance and the need to revise restrictive criteria. Fertil Steril 106:467–474. https://doi.org/10.1016/j.fertnstert.2016.04.031
12. Sigismondi C, Papaleo E, Vigano P, Vailati S, Candiani M, Ottolina J, Di Mattei VE, Mangili G (2015) Fertility preservation in female cancer patients: a single center experience. Chin J Cancer 34:56–60. https://doi.org/10.5732/cjc.014.10252
13. Dolmans MM, von Wolff M, Poirot C, Diaz-Garcia C, Cacciottola L, Boissel N, Liebenthron J, Pellicer A, Donnez J, Andersen CY (2021) Transplantation of cryopreserved ovarian tissue in a series of 285 women: a review of five leading European centers. Fertil Steril 115:1102–1115. https://doi.org/10.1016/j.fertnstert.2021.03.008
14. Lotz L, Bender-Liebenthron J, Dittrich R, Haberle L, Beckmann MW, Germeyer A, Korell M, Sänger N, Kruessel JS, von Wolff M, FertiProtekt (2022) Determinants of transplantation success with cryopreserved ovarian tissue: data from 196 women of the FertiPROTEKT network. Hum Reprod 37:2787–2796. https://doi.org/10.1093/humrep/deac225
15. Schallmoser A, Einenkel R, Färber C, Huren V, Emrich N, John J, Sänger N (2023) Comparison of angiogenic potential in vitrified vs. slow frozen human ovarian tissue. Sci Rep 13: 12885. https://doi.org/10.1038/s41598-023-39920-x
16. Sänger N, John J, Einenkel R, Schallmoser A (2024) First report on successful delivery after retransplantation of vitrified, rapid warmed ovarian tissue in Europe. Reprod Biomed Online 49:103940. https://doi.org/10.1016/j.rbmo.2024.103940
17. De Santis L, Coticchio G (2011) Theoretical and experimental basis of slow freezing. Reprod Biomed Online 22:125–132. https://doi.org/10.1016/j.rbmo.2010.10.012
18. Silber SJ, DeRosa M, Goldsmith S, Fan Y, Castleman L, Melnick J (2018) Cryopreservation and transplantation of ovarian tissue: results from one center in the USA. J Assist Reprod Genet 35:2205–2213. https://doi.org/10.1007/s10815-018-1315-1

19. Suzuki N, Yoshioka N, Takae S, Sugishita Y, Tamura M, Hashimoto S, Morimoto Y, Kawamura K (2015) Successful fertility preservation following ovarian tissue vitrification in patients with primary ovarian insufficiency. Hum Reprod 30:608–615. https://doi.org/10.1093/humrep/deu353
20. Kong Q, Pei C, Rahimi G, Mallmann P, Isachenko V (2025) Comparison of the quality of ovarian tissue cryopreservation by conventional slow cryopreservation and vitrification-a systematic review and meta-analysis. J Ovarian Res 18: 62. https://doi.org/10.1186/s13048-024-01561-7
21. Behl S, Joshi VB, Larson NB, Young MC, Bilal M, Walker DL, Khan Z, Granberg CF, Chattha A, Zhao Y (2023) Vitrification versus slow freezing of human ovarian tissue: a systematic review and meta-analysis of histological outcomes. J Assist Reprod Genet 40:455–464. https://doi.org/10.1007/s10815-022-02692-w
22. Shi Q, Xie Y, Wang Y, Li S (2017) Vitrification versus slow freezing for human ovarian tissue cryopreservation: a systematic review and meta-anlaysis. Sci Rep 7: 8538. https://doi.org/10.1038/s41598-017-09005-7
23. Schallmoser A, John J, Einenkel R, Sänger N (2024) Vitrification of ovarian cortex tissue to achieve a glassy state of aggregation. J Vis Exp 2024:e66801. https://doi.org/10.3791/66801
24. Kristensen SG, Rasmussen A, Byskov AG, Andersen CY (2011) Isolation of pre-antral follicles from human ovarian medulla tissue. Hum Reprod 26:157–166. https://doi.org/10.1093/humrep/deq318
25. Schallmoser A, Einenkel R, Färber C, Huren V, Pougin A, Emrich N, John J, Sänger N (2023) Cryostorage of human ovarian tissue: evaluating the storage and disposal pattern over a 22-year period in 2475 patients. Reprod Biomed Online 47:103239. https://doi.org/10.1016/j.rbmo.2023.05.011
26. Emrich NLA, Einenkel R, Färber CM, Schallmoser A, Sänger N (2025) Ovarian tissue cryopreservation for fertility preservation: a two-decade single-center experience with 451 children and adolescents. Reprod Biol Endocrinol 23: 51. https://doi.org/10.1186/s12958-025-01388-x
27. Einenkel R, Schallmoser A, Sänger N (2022) Metabolic and secretory recovery of slow frozen-thawed human ovarian tissue in vitro. Mol Hum Reprod 28:gaac037. https://doi.org/10.1093/molehr/gaac037

Chapter 24

Cryopreservation of Platelets

Denese C. Marks and Lacey Johnson

Abstract

Platelets are transfused to stop bleeding from trauma or during surgery, and to prevent bleeding that may occur in various diseases, including cancer, bone marrow failure, and chronic blood disorders due to low or dysfunctional platelets. The platelet supply is often unpredictable, leading to shortages in hospitals, blood centers, and the military. Cryopreservation offers a solution by extending platelet shelf-life and enabling stockpiling for emergencies, rare donor phenotypes, and use in remote locations. In this chapter, cryopreservation of platelets below −65°C using 5%–6% DMSO with reconstitution of thawed platelets in plasma is described. The procedure requires standard blood banking equipment and a −80°C freezer. Platelets are stored in a −80°C freezer, as liquid nitrogen storage is impractical. Cryopreserved platelets remain viable for up to 12 years when stored below −65°C. Thawing is straightforward, though resuspension in plasma or saline is required, and the shelf life after thawing is limited to 4–8 h.

Keywords Cryopreservation, Platelets, Transfusion

1 Introduction

The platelet supply can be unpredictable, and many blood centers, hospitals, and the military have experienced shortages in recent years [1]. Platelet cryopreservation can address these shortages, ensuring an available platelet supply. Cryopreservation of platelets, as with other blood products, extends their shelf life well beyond that of fresh components, enabling creation of stockpiles for emergencies as well as inventories of frozen products from donors with rare phenotypes, and supply of blood components for use in austere environments where fresh components are not readily available. Published data suggest that cryopreserved platelets can be stored for up to 12 years when stored at −80°C [2].

It has been more than 50 years since Robert Valeri's method for plateletcryopreservation was developed, with the goal of providing platelets for far-forward military medical facilities [3–5]. The method is simple, requiring standard blood banking technology and

Willem F. Wolkers and Harriëtte Oldenhof (eds.), *Cryopreservation and Freeze-Drying Protocols*, Methods in Molecular Biology, vol. 3060, https://doi.org/10.1007/978-1-0716-5416-3_24,

equipment, and has changed little apart from a process modification to remove the cryoprotectant prior to freezing [4].

Both apheresis and pooled whole-blood-derived platelets can be used as the starting material [6–9]. However, most protocols use apheresis platelets. DMSO is the most widely used cryoprotectant, typically added to yield a final concentration of 5%–6%. However, the DMSO content must be reduced after addition to the platelets due to toxicity and associated side-effects [10, 11]. Other cryoprotectants have been examined, including trehalose and second messenger inhibitors [12], deep eutectic solvents, ice-recrystallization inhibitors, and calcium-chelators [13–15]. More recently, freezing with a reduced DMSO concentration, or in the absence of a cryoprotectant have also been examined [16, 17]. Some of these options appear to be quite effective, but have not yet been adopted as a standard protocol for manufacturing transfusable products.

Platelets are typically frozen by placing them directly in a −80°C mechanical freezer, in which they are subsequently stored. Due to the size and plasticized nature of blood bags, storage of cryopreserved platelets in liquid nitrogen is not practical or feasible. Controlled rate freezing is not typically required [18], although it has been found to be beneficial for retaining platelet viability in the absence of a cryoprotectant [17].

Thawing frozen platelets is a simple process. However, as platelets are frozen as a low-volume hyperconcentrate, they must be resuspended in a larger volume prior to transfusion. There is also some variation in the type and volume of the solutions used to resuspend the platelets after thawing. Many protocols use plasma [6, 8, 19], while others use saline for reconstitution [7, 20, 21]. Upon thawing, the shelf-life of the platelets is limited to between 4 and 8 h [19, 21, 22].

Despite the relative simplicity of the cryopreservation techniques, plateletcryopreservation was not widely adopted until it was taken up by the Netherlands Military Blood Bank, where cryopreserved platelets were prepared and shipped to battlefield hospitals and used in massive transfusion protocols to save lives in military trauma care [2, 19, 23]. This may have been due to cost, the need for dedicated freezer storage, and end-user training requirements, but also due to a lack of clinical evidence of efficacy and health economic benefit. Since then, however, plateletcryopreservation has undergone a resurgence, and the value of this product has become more apparent. Definitive clinical trials examining the safety and efficacy of cryopreserved platelets, primarily for bleeding, have only been initiated in the last decade [21, 22, 24–26], and regulatory approval remains limited to a few countries [27].

In this chapter, cryopreservation of platelets at −80°C using 5%–6% DMSO with reconstitution of thawed platelets in plasma will be described. Other variations of the method are summarized elsewhere [28]. This is based on the method used in our institute, which was adapted from the processes used by the Netherlands Military

Blood Bank, with minor modifications. Thawing would normally take place in a laboratory at the site of transfusion, due to the short post-thaw shelf life, rather than in a blood establishment. However, it is necessary to thaw a defined number of frozen components for quality control testing, and during stability testing to establish the frozen shelf life of the cryopreserved platelets.

2 Materials

2.1 Equipment

1. Biosafety cabinet (required if DMSO is not in packaging with tubing that can be sterile welded to blood bags).
2. Centrifuge for blood bags (e.g., Beckman Colter and Hettich).
3. Freezer −80°C.
4. Hemostat clamps.
5. Hematology analyzer.
6. Infrared thermometer.
7. Leucocart or I.V. infusion stand/pole.
8. Manual plasma extractor.
9. Platelet agitator or platform rocker for platelet mixing.
10. Retort stand.
11. Scissors.
12. Sterile connecting device (e.g., TerumoBCT).
13. Tube Sealer (e.g., TerumoBCT).
14. Vacuum sealer.
15. Water bath.

2.2 Reagents

1. Apheresis platelets.
2. Fresh frozen plasma (FFP) for reconstitution.
3. 27% dimethylsulfoxide (DMSO)/0.9% saline (*see* **Note 1**).
4. 70% ethanol solution (biosafety cabinet disinfection).
5. Viraclean (biosafety cabinet disinfection).

2.3 Disposables

1. 27-gauge, 0.4-in. sterile airway needle (for spiking if adding DMSO from a bottle).
2. 600 mL freezing bag (MacoPharma or other).
3. Alcohol swabs.
4. Copper wafers.
5. Corflute or cardboard storage box.

6. Fistula needle set, 17 gauge (transfusion set: if adding DMSO from a bottle by spiking with a needle).
7. Hemostat clamp.
8. Lint-free cloth.
9. Red cell transfer set (for thawing).
10. Quarantine and other labels.
11. Transfer bags for aseptic sampling (TerumoBCT).
12. Vacuum bag.

3 Methods

The starting material in this process is a double-dose platelet concentrate collected by plateletpheresis, with a minimum platelet content of 2×10^{11}/unit. A single dose may be used if components with a higher platelet concentration are available. Platelet concentrates suspended in 100% plasma or in a mixture of plasma and platelet additive solution are suitable (*see* **Note 2**). Platelets should be inspected for swirl and should not be frozen if any aggregates are observed. Blood group selection is dependent on inventory requirements. For emergency use inventories, Group O RhD-positive and -negative components are typically selected. Labeling requirements should be according to local guidelines.

3.1 Prepare the Platelets for Cryopreservation

If using a biosafety cabinet, prepare, disinfect, and use the cabinet according to local institutional protocols. Place the platform rocker inside the biosafety cabinet.

Typically, a defined volume of 27% DMSO/0.9% saline DMSO is added to the platelet unit, to give a final concentration of between 5% and 6% (target 5.5%) (*see* **Note 3**).

1. If using a double-dose apheresis platelet (two bags), sterilely connect the two platelet packs and transfer the contents of the two platelet bags into the primary pack.
2. Remove air from the primary pack, and if required, remove excess volume by transferring it into the attached secondary pack.
3. Weigh and record the weight of the unit. To obtain the net weight, subtract the weight of the empty bag (tare weight) from the gross weight.
4. Calculate the volume of the unit—divide the net weight by the specific gravity of the product (e.g., by 1.025 for platelets suspended in plasma; 1.015 for platelets suspended in plasma/platelet additive solution).

5. Perform a platelet count and record the pre-freeze platelet concentration.
6. Place the platelet unit on the platform rocker, set to 40 rpm, to mix the contents.

3.2 Addition of DMSO

The method here describes addition by spiking through the rubber seal of a DMSO bottle. However, if DMSO is available in a format that is compatible with sterile welding (i.e., a closed system), the process can be followed without the need for spiking or a biosafety cabinet.

1. Inside the biosafety cabinet, sterilely weld the tubing of the fistula set to the platelet unit to attach the needle.
2. To add DMSO, invert the bottle so that the rubber seal is facing downwards and hang the bottle via a retort stand.
3. Insert the needle of the fistula set into the rubber cap of the DMSO bottle.
4. Open the weld join.
5. Insert the airway needle into the same rubber cap of the DMSO bottle to facilitate air flow.
6. Allow DMSO to flow into the platelet unit whilst mixing on the platform rocker until the bottle is empty.
7. Remove the platelet unit from the biological safety cabinet, seal off the needle, and weigh the unit.
8. Record the weight and calculate the volume, using the specific gravity of the platelet, and for DMSO/saline.

3.3 Centrifugation to Separate Platelets from DMSO-Containing Supernatant

These steps are required to minimize the amount of DMSO received by the patient upon transfusion.

1. Sterilely weld the platelet unit to a freezing bag.
2. Open the weld join and allow the platelets to drain into the freezing bag.
3. Gently squeeze any air in the freezing bag back into the primary pack.
4. Make sure the line is clear of platelets and close the clip.
5. Keep the empty bag attached.
6. Pack the platelet unit and the empty (folded) platelet bag into a centrifuge cup.
7. Prepare appropriate balance for centrifuge.
8. Centrifuge at 1377×g for 10 min, 20–24°C. Acceleration and deceleration should be set low to minimize mixing and disruption of the platelet pellet.

3.4 Removal of DMSO-Containing Supernatant

1. Remove the unit from the centrifuge cup and place it into a manual plasma expressor, with the label facing away from you. Ensure the tube ports are above the plates so that they are not trapped in the press.
2. Allow almost all the supernatant to run into the original platelet bag until the platelet pellet begins to move up the side of the bag.
3. Place a hemostat clamp on the tubing and remove the bag from the expressor.
4. Place the platelet unit on the scale. Holding the original bag upside-down so that the supernatant can enter the line, open the hemostat clamp and carefully allow 20–30 mL supernatant to slowly re-enter the freezing bag (calculate weight using the bag tare weight).
5. Heat-seal the tubing to separate the original bag from the freezing bag.
6. Record the final weight.
7. Resuspend the platelets in the remaining supernatant by gently massaging the plateletfreezing bag in a circular motion with a lint-free cloth.
8. Ensure that the platelets are removed from the bottom of the bag and along the edges. Massage until no aggregates can be seen when held up to a light source.

3.5 Packaging and Storage

These steps are designed to protect the platelet unit during storage and any subsequent transport.

Direct as much of the platelet volume into the middle of the pack as possible, then fold the plateletfreezing bag so that:

1. The tubing is folded under the ports, without any bends or kinks.
2. Approximately 2 cm of the bottom of the bag is folded back and under.
3. The resulting package should be the size of the label on the front. This helps protect the ports and tubing from damage during freezing.
4. Place the platelet unit in a vacuum bag and vacuum seal.
5. Place the vacuum-packed platelet component into a corflute or cardboard box.
6. Place the box in a single layer in a −80°C freezer
7. Store below −65°C until required.

3.6 Thawing and Resuspension

Thawing would normally be performed in a laboratory at the site of transfusion, due to the short post-thaw shelf-life. However, it is also necessary for blood establishments to thaw a defined number of frozen components for quality control testing.

Plasma for reconstitution of platelets in our institution is whole-blood derived (*see* **Note 4**). Typically, ABO-compatible plasma can be used, but group AB plasma is also suitable to make a universally transfusible component [19].

Only a single cryopreserved platelet component should be thawed and reconstituted at a time to maintain line clearance. If plasma needs to be thawed, it should be thawed before the platelets.

1. Turn on the water bath and allow it to reach 37°C.
2. If thawing plasma, remove it from the freezer and thaw, checking for leaks as it thaws. A whole unit of plasma (250–300 mL) will take approximately 20 min to thaw.
3. While the plasma is thawing, remove the frozen platelet unit from the freezer and from the cardboard/corflute box.
4. Keep the platelet in the vacuum bag and place it in the water bath, ensuring the bag remains fully submerged.
5. Thaw for 4–5 min, remove from the water bath, and check for leaks by wiping with a lint-free disposable cloth.
6. Remove the platelet unit from the vacuum bag and measure the temperature of the thawed platelet component using an infrared thermometer. The platelet component should be within the range of 25–35°C; if not, return to the water bath for 30-s intervals.
7. Gently massage the platelets in a circular motion using a lint-free cloth for approximately 10 s so that the platelets are fully resuspended.
8. Place the platelet component on a platelet agitator for at least 10 min or until the plasma component is thawed (between 25 and 35°C).
9. Examine the plasma, checking for clots, turbidity, discoloration, lipaemia, or fibrin strands. Discard if present.
10. Remove the platelet unit from the agitator.
11. Connect the thawed platelet component to the thawed plasma component using a sterile welding device. If unavailable, connect the plasma and platelet components using a double-spiked tubing connector, e.g., red cell transfer set.
12. Once connected, hang the plasma bag on a hook, holding the thawed platelet pack in one hand, gently agitating the platelet bag, allowing the plasma to drain into the thawed platelet unit whilst constantly agitating the platelets in a circular motion.
13. Place a hemostat clamp on the tubing between the two bags and seal the tubing to the plasma bag.

14. Weigh the thawed platelet component and record the weight.
15. Calculate the volume using the net weight divided by the specific gravity of plasma (1.025) and record. This will be needed if determining post-thaw recovery as described below.
16. If required, collect a sample for quality control testing by sterilely connecting the platelet bag to a labeled transfer bag.
17. Inspect the unit for aggregates by holding it up to a light source.
18. If not required for transfusion immediately, place the reconstituted platelet unit on an agitator at 20–24°C until expiry.

3.7 Quality Control

The guide to the preparation, use, and quality assurance of blood components (EDQM) provides a component monograph for platelets cryopreserved with 6% DMSO [30]. These guidelines indicate that all manufactured platelets should undergo quality control testing. However, this is not practical when cryopreserved platelets are distributed to end-user sites in their frozen state for future thawing. Instead, a sampling plan can be devised, whereby a set number or proportion of units are thawed regularly for destructive testing, depending on the frequency at which they are manufactured.

While the minimum recommended recovery is >50%, the literature indicates that 70% recovery can be readily achieved, even after prolonged storage [2, 6, 17]. When calculating recovery, the platelet content/unit should be used, rather than the concentration. This is because the centrifugation and reconstitution steps change the volume of the product between freezing and thawing. Additionally, to calculate the recovery, a sample must be taken prior to freezing.

It is also useful to consider a minimum clinical dose per unit, such as that required by the EDQM guidelines for liquid stored platelets (2×10^{11}/unit) [29]. If the platelets are prepared using an open system, bacterial contamination screening of the final product should be considered as part of the quality control requirements.

The original platelet bag and the thawed platelet bag (after addition of FFP) are sampled by transferring a small volume to sterilely connected sample bags. The original platelet bag and the thawed platelet bag are weighed, and the net weight is recorded. The volume is calculated from the net weight divided by the specific gravity: 1.025 for plasma and platelets in plasma, and 1.015 for platelets in platelet additive solution (PAS).

1. Measure the platelet count using a hematology analyzer.
2. Calculate the platelet content as platelet concentration/volume.
3. Calculate the post-thaw recovery as post-thaw platelet content/pre-freeze platelet content × 100%.
4. If screening for bacterial contamination, sterilely connect a blood component sampling device to the thawed platelet component.

4 Notes

1. DMSO is also available in a bag that can be sterilely connected to the platelet bag by use of a sterile connection device, which simplifies the production of cryopreserved platelets, enabling maintenance of a closed system [2].
2. The platelets should not be frozen until all transfusion-transmitted disease screening has been performed and a sample has been taken for bacterial contamination screening. This is typically between 36 and 48 h post-collection, although laboratory studies have shown that platelets can be frozen at expiry [30].
3. It may be necessary to remove excess volume from the platelet bag, depending on the volume of DMSO that will be added.
4. Others have used plasma collected by apheresis [19].

References

1. Gehrie EA, Young PP, Basavaraju SV, Bracey AW, Cap AP, Culler L, Dunbar NM, Homer M, Isufi I, Macedo R, Petraszko T, Ramsey G, Tormey CA, Kaufman RM, Snyder EL (2024) Addressing platelet insecurity—a national call to action. Transfusion 64:2001–2013. https://doi.org/10.1111/trf.17987
2. Noorman F, Rijnhout TWH, de Kort B, Hoencamp R (2023) Frozen for combat: quality of deep-frozen thrombocytes, produced and used by The Netherlands Armed Forces 2001–2021. Transfusion 63:203–216. https://doi.org/10.1111/trf.17166
3. Valeri CR, Feingold H, Marchionni LD (1974) A simple method for freezing human platelets using 6 percent dimethylsulfoxide and storage at -80 degrees C. Blood 43:131–136. https://doi.org/10.1182/blood.V43.1.131.131
4. Valeri CR, Ragno G, Khuri S (2005) Freezing human platelets with 6 percent dimethyl sulfoxide with removal of the supernatant solution before freezing and storage at -80 degrees C without postthaw processing. Transfusion 45:1890–1898. https://doi.org/10.1111/j.1537-2995.2005.00647.x
5. Handin RI, Valeri CR (1972) Improved viability of previously frozen platelets. Blood 40:509–513. https://doi.org/10.1182/blood.V40.4.509.509
6. Johnson L, Reade MC, Hyland RA, Tan S, Marks DC (2015) In vitro comparison of cryopreserved and liquid platelets: potential clinical implications. Transfusion 55:838–847. https://doi.org/10.1111/trf.12915
7. Dumont LJ, Cancelas JA, Dumont DF, Siegel AH, Szczepiorkowski ZM, Rugg N, Pratt PG, Worsham DN, Hartman EL, Dunn SK, O'Leary M, Ransom JH, Michael RA, Macdonald VW (2013) A randomized controlled trial evaluating recovery and survival of 6% dimethyl sulfoxide-frozen autologous platelets in healthy volunteers. Transfusion 53:128–137. https://doi.org/10.1111/j.1537-2995.2012.03735.x
8. Crimmins D, Flanagan P, Charlewood R, Ruggiero K (2016) In vitro comparison between gamma-irradiated cryopreserved and Day 7 liquid-stored buffy coat-derived platelet components. Transfusion 56:2799–2807. https://doi.org/10.1111/trf.13763
9. Martinaud C, Sugier HHR, Javaudin O, Burin des Roziers N, Bégué S (2022) In vitro characteristics of cryopreserved platelet concentrates reconstituted by fresh frozen or lyophilized plasma. Transfus Clin Biol 29:118–123. https://doi.org/10.1016/j.tracli.2022.01.003
10. Awan M, Buriak I, Fleck R, Fuller B, Goltsev A, Kerby J, Lowdell M, Mericka P, Petrenko A, Petrenko Y, Rogulska O, Stolzing A, Stacey GN (2020) Dimethyl sulfoxide: a central player since the dawn of cryobiology, is efficacy balanced by toxicity. Regen Med 15:1463–1491. https://doi.org/10.2217/rme-2019-0145
11. Brayton CF (1986) Dimethyl sulfoxide (DMSO): a review. Cornell Vet 76:61–90
12. Waters L, Ben R, Acker JP, Padula MP, Marks DC, Johnson L (2020) Characterizing the ability of an ice recrystallization inhibitor to improve

platelet cryopreservation. Cryobiology 96:152–158. https://doi.org/10.1016/j.cryobiol.2020.07.003

13. Pedrazzoli P, Noris P, Perotti C, Schiavo R, Ponchio L, Belletti S, Da Prada GA, Balduini CL, Salvaneschi L, Robustelli Della Cuna G, Siena S (2000) Transfusion of platelet concentrates cryopreserved with ThromboSol plus low-dose dimethylsulphoxide in patients with severe thrombocytopenia: a pilot study. Br J Haematol 108:653–659. https://doi.org/10.1046/j.1365-2141.2000.01860.x
14. Johnson L, Bryant SJ, Lei P, Roan C, Marks DC, Bryant G (2024) A deep eutectic solvent is an effective cryoprotective agent for platelets. Cryobiology 116:104913. https://doi.org/10.1016/j.cryobiol.2024.104913
15. Waters L, Padula MP, Marks DC, Johnson L (2020) Calcium chelation: a novel approach to reduce cryopreservation-induced damage to frozen platelets. Transfusion 60:1552–1563. https://doi.org/10.1111/trf.15799
16. Johnson L, Lei P, Roan C, Marks DC (2025) Development of a simplified platelet cryopreservation method: an in vitro investigation of reducing the DMSO concentration to allow administration without its pre-transfusion removal. Vox Sang 120:284–292. https://doi.org/10.1111/vox.13789
17. Ehn K, Wikman A, Uhlin M, Sandgren P (2023) Cryopreserved platelets in a non-toxic DMSO-free solution maintain hemostatic function in vitro. Int J Mol Sci 2023:24. https://doi.org/10.3390/ijms241713097
18. Tynngård N, Bell A, Gryfelt G, Cvetkovic S, Wikman A, Uhlin M, Sandgren P (2021) Cryopreservation of buffy coat derived platelets: paired in vitro characterization using uncontrolled versus controlled freezing rate protocols. Transfusion 61:546–556. https://doi.org/10.1111/trf.16227
19. Lelkens CC, Koning JG, de Kort B, Floot IB, Noorman F (2006) Experiences with frozen blood products in the Netherlands military. Transfus Apher Sci 34:289–298. https://doi.org/10.1016/j.transci.2005.11.008
20. Slichter SJ, Dumont LJ, Cancelas JA, Jones M, Gernsheimer TB, Szczepiorkowski ZM, Dunbar NM, Prakash G, Medlin S, Rugg N, Kinne B, Macdonald VW, Housler G, Valiyaveettil M, Hmel P, Ransom JH (2018) Safety and efficacy of cryopreserved platelets in bleeding patients with thrombocytopenia. Transfusion 58:2129–2138. https://doi.org/10.1111/trf.14780
21. Whitman G, Kramer R, Tanaka K, Holcomb J, Fitzpatrick GM, Raphael J, Ness P (2023) Design of the randomized controlled trial comparing dimethyl sulfoxide cryopreserved platelets to liquid stored platelets in patients undergoing cardiopulmonary bypass surgery. JTCVS Open 13:232–241. https://doi.org/10.1016/j.xjon.2022.11.003
22. Reade MC, Marks DC, Bellomo R, Deans R, Faulke DJ, Fraser JF, Gattas DJ, Holley AD, Irving DO, Johnson L, Pearse BL, Royse AG, Wong J (2019) A randomized, controlled pilot clinical trial of cryopreserved platelets for perioperative surgical bleeding: the CLIP-I trial (Editorial, p. 2759). Transfusion 59:2794–2804. https://doi.org/10.1111/trf.15423
23. Noorman F, van Dongen TT, Plat MJ, Badloe JF, Hess JR, Hoencamp R (2016) Transfusion: −80°C frozen blood products are safe and effective in military casualty care. PLoS One 11:e0168401. https://doi.org/10.1371/journal.pone.0168401
24. Reade MC, Marks DC, Howe B, McGuinness S, Parke R, Navarra L, Charlewood R, Johnson L, McQuilten Z (2022) Cryopreserved platelets compared with liquid-stored platelets for the treatment of surgical bleeding: protocol for two multicentre randomised controlled blinded non-inferiority trials (the CLIP-II and CLIPNZ-II trials. BMJ Open 12:e068933. https://doi.org/10.1136/bmjopen-2022-068933
25. McGuinness S, Charlewood R, Gilder E, Parke R, Hayes K, Morley S, Al-Ibousi A, Deans R, Howe B, Johnson L, Marks DC, Reade MC (2022) A pilot randomized clinical trial of cryopreserved versus liquid-stored platelet transfusion for bleeding in cardiac surgery: the cryopreserved versus liquid platelet-New Zealand pilot trial. Vox Sang 117:337–345. https://doi.org/10.1111/vox.13203
26. Ang AL, Gan LSH, Tuy TT, Ang CH, Tan CW, Tan HH, Shu PH, Zhang Q, Cao Y, Moorakonda RB, Pokharkar Y, Lu J (2023) A randomized cross-over study of cryopreserved platelets in prophylactic transfusions of thrombocytopenic patients. Transfusion 63:1649–1660. https://doi.org/10.1111/trf.17503
27. Cohn CS, Dumont LJ, Lozano M, Marks DC, Johnson L, Ismay S, Bondar N, T'Sas F, Yokoyama APH, Kutner JM, Acker JP, Bohonek M, Sailliol A, Martinaud C, Pogłód R, Antoniewicz-Papis J, Lachert E, Pun PBL, Lu J, Cid J, Guijarro F, Puig L, Gerber B, Alberio L, Schanz U, Buser A, Noorman F, Zoodsma M, van der Meer PF, de Korte D, Wagner S, O'Neill M (2017) Vox Sanguinis International Forum on platelet cryopreservation. Vox Sang 112:e69–e85. https://doi.org/10.1111/vox.12533
28. Waters L, Marks DC, Johnson L (2025) Strategies to improve platelet cryopreservation:

a narrative review. Transfusion 65:740–749. https://doi.org/10.1111/trf.18204

29. Guide to the preparation, use and quality assurance of blood components. 22nd Edition. Council of Europe Press, Strasbourg, 2025

30. Johnson L, Waters L, Green S, Wood B, Marks DC (2020) Freezing expired platelets does not compromise in vitro quality: an opportunity to maximize inventory potential. Transfusion 60:454–459. https://doi.org/10.1111/trf.15616

Chapter 25

Cryopreserving Small Volumes of Red Blood Cells Using the High Glycerol Method

Tracey R. Turner, Carly Olafson, Olga Mykhailova and Jason P. Acker

Abstract

Cryopreservation of red blood cells (RBCs) plays a critical role in blood inventory management by enabling long-term frozen storage of rare blood types. This process requires the addition of glycerol, a cryoprotectant that mitigates cellular damage caused by ice crystal formation during freezing and thawing. A high glycerol method (HGM) is commonly used, which requires an approximate concentration of 40% glycerol and enables slow freezing rates (1–3°C/min) and extended storage in low-temperature freezers (≤−65°C). While automated cell processing systems have been developed for the glycerolization and deglycerolization of full-size RBC units for transfusion, there is no such technology for the small-scale cryopreservation of RBC aliquots. This chapter presents a high-recovery, small-scale model that mirrors current automated cell processing systems that employ the HGM for cryopreservation of RBCs. This technique is valuable for (1) research studies, particularly for investigations on cryopreservation of rare donor phenotypes or investigating novel cryoprotectants, (2) to provide control or reagent cells for antibody identification or compatibility testing, or (3) any other applications requiring small volumes of functional and intact RBCs.

Keywords Cryopreservation, Cryoprotectant, Deglycerolization, Glycerol, Glycerolization, Red blood cells, Small-scale

1 Introduction

1.1 Cryopreservation of Red Blood Cells Using Glycerol as a Cryoprotective Agent

Advancements in the preservation of red blood cells (RBCs) for transfusion have been ongoing since the early 1900s, beginning with the development of preservative solutions for in vitro storage [1–3]. Modern blood banking practices allow for refrigerated storage of RBC products for 35–42 days in most jurisdictions, dependent on processing methods and storage solutions used [4, 5]. However, the discovery of the cryoprotective effects of glycerol in the 1950s opened the door for frozen red cell storage [6]. Since then, significant advancements to RBCcryopreservation methods have expanded options for blood inventory management practices, allowing blood

Willem F. Wolkers and Harriëtte Oldenhof (eds.), *Cryopreservation and Freeze-Drying Protocols*, Methods in Molecular Biology, vol. 3060, https://doi.org/10.1007/978-1-0716-5416-3_25,

centers to establish banks of frozen red cell concentrates (RCCs) for transfusion; cryopreserved RCCs are particularly valuable in providing blood products for patients with rare phenotypes since they can be stored for up to 30 years [7–9].

Glycerol acts as a cryoprotective agent by mitigating damage caused by the formation of ice crystals during freezing [10, 11]. Two techniques for RBCcryopreservation using glycerol that are commonly used include the low glycerol method (LGM) and the high glycerol method (HGM). The LGM, with a 15%–20% glycerol concentration, requires rapid freezing rates (>100°C/min) and storage at ultralow temperatures (<−150°C) facilitated using liquid nitrogen. In contrast, the HGM, with an approximate 40% glycerol concentration, permits the use of slower freezing rates (1–3°C/min) and long-term storage in low-temperature freezers (≤−65°C) [9]. The HGM may also offer greater protection to RBCs when they are unintentionally exposed to temperatures above the recommended storage temperature [12]. For these reasons, the HGM is more commonly employed by blood centers to maintain frozen blood reserves for transfusion.

Successful RBCcryopreservation relies on the controlled addition and removal of glycerol to minimize lysis of the cells due to changes in the osmotic gradient. Historically, glycerolization has been performed using manual addition methods, while deglycerolization has either been performed manually or using open system cell processors, such as the COBE 2991 [7]. However, introduction of closed system automated cell processors, such as the Haemonetics ACP 215, improved the efficiency of both glycerolization and deglycerolization processes while maintaining sterility, allowing for an extended post-processing shelf life up to 14 days, compared to 24 h for open systems [13–16]. Advancements in RBC additive solutions (AS) have also provided flexibility in pre-cryopreservation processing. For example, the use of saline-adenine-glucose-mannitol (SAGM) AS with RBCs manufactured from citrate-phosphate-dextrose (CPD) whole blood collections extends pre-cryopreservation refrigerated storage time up to 21 days [16]. This offers logistical flexibility, particularly if samples require transport to an alternate site for processing. These incremental improvements in RBCcryopreservation methods have undoubtedly contributed to the widespread adoption of frozen blood banks.

Despite significant progress, gaps persist, particularly for donors with RBC disorders, where cells may respond differently to glycerolization and deglycerolization procedures. Such is the case with sickle cell or sickle cell trait RBCs, which exhibit clumping during deglycerolization, resulting in poor cell recovery and loss of valuable products [15]. Additionally, phenotypes associated with altered red cell membrane structure and function may be more susceptible to cryopreservation due to decreased membrane permeability and inability to endure osmotic stress. The McLeod and Gerbich phenotypes, for example, have demonstrated good post-cryopreservation and quality, but investigation of other rare phenotypes with RBC membrane irregularities remains limited [17].

Introducing small-scale models of glycerolization and deglycerolization techniques could significantly benefit investigations into rare donor phenotypes by reducing the volume of sample material required. Small-scale cryopreservation methods could also benefit blood centers and laboratories requiring reagent RBCs for antibody identification and compatibility testing. Moreover, there are several research applications, such as investigating novel additive solutions or cryoprotectants and evaluating the impact of transient warming effects on storage stability, which would benefit greatly from a small-scale model.

Currently, there are a few methods describing small-scale protocols and even fewer that achieve both high recovery and optimal RBC quality for these purposes. Lusianti et al. introduced a small-scale method aiming to mathematically optimize deglycerolization and reduce procedure time [18]. While successful in minimizing the time required to deglycerolize samples, their method suffered from low RBC recovery due to elevated hemolysis during the wash steps and limitations with mixing techniques. Moreover, their adapted small-scale method does not reflect current large-scale processes, making meaningful comparisons impractical. Rogers et al. also presented a small-scale method for evaluating RBC phenotype fidelity post-cryopreservation, focusing on improving method efficiency while maintaining high RBC recovery [19]. They highlighted the variability of the manual processing technique and the potential loss of cell populations vulnerable to osmotic and mechanical stress, which could hinder assessments of rare donor RBCs with pathologies rendering them susceptible to such stress. Finally, Eades emphasized the importance of both maximizing RBC recovery and selecting a method tailored to individual RBCcryopreservation requirements [20]. While they introduced two glycerolization and two deglycerolization techniques, no data were available to demonstrate the achievable RBC recoveries or method reproducibility, thereby impeding a comprehensive assessment of the practicality of the methods.

Our laboratory has refined small-scale glycerolization and deglycerolization processes, ensuring not only a high RBC recovery but also that they are representative of the large-scale processes utilized by blood banks for RBCcryopreservation. This chapter delineates our cryopreservation methodologies for RBCs obtained from EDTA whole blood (WB) collections and CPD SAGM RCCs and provides data to support their effectiveness and reproducibility.

1.2 Recovery of Functional RBCs After Cryopreservation Processing

When performed proficiently, the glycerolization and deglycerolization procedures detailed in this protocol should consistently yield high-quality RBCs. Once glycerolized, RBC samples from EDTA WB or CPD SAGM RCC collections should exhibit acceptable characteristics for hematocrit (Hct), hemoglobin (Hgb) content, and

Table 1

Characteristics of glycerolized RBCs by the glycerolization method. Sample characteristics of glycerolized RBCs

Characteristic	Method 1 (EDTA)	Method 2 (EDTA Volume Reduced)	Method 3 (RCC)	Method 4 (RCC Volume Reduced)
Aliquot volume (mL)	1	1	1	1
Pre-freeze				
RBC hemolysis (%)	N/A	0.64 ± 0.03	0.37 ± 0.06	0.43 ± 0.10
Post-thaw				
Total hemoglobin (mg/unit)	83.0 ± 2.8	148.0 ± 1.5	58.6 ± 2.9	137.0 ± 7.5
Hematocrit (%)	26 ± 1	51 ± 2	18 ± 1	47 ± 2
RBC hemolysis (%)	2.78 ± 0.30	N/A	1.17 ± 0.33	1.15 ± 0.22

[1]Data for Method 1 was collected from a single donor in triplicate ($n = 3$), for Method 2 was collected from a single donor in duplicate ($n = 2$), and for Method 3 and Method 4 were collected from six different donors ($n = 6$/method). Data are shown as mean ± standard deviation. *N/A* not available

for samples glycerolized RBC hemolysis (see Table 1). When glycerolized without pre-glycerolization volume reduction (removal of excess supernatant), glycerolized RBC samples will have lower Hct and Hgb content. However, RBC hemolysis values will be similar, with variations depending on the anticoagulant and/or pre-cryopreservation red cell additive solution used.

Post deglycerolization sample volume, Hct, and Hgb content for samples glycerolized without volume reduction will be lower than samples that underwent the volume reduction step prior to freezing; however, RBC hemolysis is elevated in these samples after 24 h of hypothermic storage (see Table 2). All samples had <3% hemolysis after deglycerolization and/or after 24 h of hypothermic storage. Samples glycerolized from RCCs maintained RBC hemolysis of ≤0.8% after 24 h of hypothermic storage, which is in line with regulatory standards for cryopreserved RCCs [21–23]. Hemoglobin recovery for samples glycerolized by all four methods was greater than 80% (see Table 2), indicating that these cryopreservation processes will not have significant loss of RBCs.

1.3 Small-Scale RBCCryopreservation Techniques for Process Development

The small-scale RBCcryopreservation techniques described here were originally developed to emulate the procedures used with the COBE 2991 and the ACP 215 cell processors. The development of this small-scale method is practical for reducing the volume of donor material required to develop novel cryopreservation reagents and post-thaw RBCstorage solutions. However, it can also be applied for biobanking, preservation of control materials, and many other research and development applications. Due to the wide range of

Table 2
Characteristics of deglycerolized RBCs by glycerolization method

Characteristic	Method 1 (EDTA)	Method 2 (EDTA volume reduced)	Method 3 (RCC)	Method 4 (RCC volume reduced)
Deglycerolized volume (mL)	0.66 ± 0.05	0.90 ± 0.06	0.63 ± 0.02	0.92 ± 0.03
Total hemoglobin (mg/unit)	69.4 ± 4.7	134.7 ± 0.0	53.6 ± 0.04	127.6 ± 8.9
Hematocrit (%)	33 ± 1	52 ± 4	31 ± 1	49 ± 4
RBC hemolysis immediate (%)	N/A	0.73 ± 0.24	0.52 ± 0.04	0.36 ± 0.09
RBC hemolysis 24 h of hypothermic storage (%)	0.91 ± 0.03	N/A	0.71 ± 0.13	0.51 ± 0.14
Hemoglobin recovery (%)[a]	88 ± 3	91 ± 11	92 ± 5	93 ± 7

[a]Percentage of hemoglobin that was retained after processing (glycerolization and deglycerolization)
Sample characteristics of deglycerolized RBCs after resuspension in an additive solution (AS-3). Data for Method 1 were collected from a single donor in triplicate (n =3), for Method 2 were collected from a single donor in duplicate (n = 2), and for Method 3 and Method 4 were collected from six different donors (n = 6/method). Data are shown as mean ± standard deviation. *N/A* not available.

practical uses, there are several considerations to be made when implementing this protocol. Firstly, the rate of addition of glycerol is critical to prevent osmotic shock and thereby RBC hemolysis. During glycerolization, a high-osmolality 57% glycerol solution is added to the RBCs, which must equilibrate to accommodate the osmotic shift. RBCs will shrink as water is released and then swell as glycerol enters; thus, to avoid osmotic shock and lysis, the rate of glycerol addition must be closely controlled. This is achieved by adding the solution in two aliquots with an incubation period for equilibration in between, and by using an automated rate of ≤15 mL/min. This rate of addition was determined experimentally and found to reduce RBC hemolysis (see Fig. 1). If automated methods are unavailable, visualization of fast-flowing individual drops is an acceptable alternative to gauge the addition rate. Additionally, continual mixing and agitation are critical to ensure even distribution of glycerol and reduce the severity of an osmotic gradient within the solution; this is further facilitated by using a 50 mL conical tube, which is an optimal size to allow for effective mixing (wider surface area when compared to a 15 mL conical tube, for example). Controlling the rate of addition and ensuring proper mixing is critical to reducing cell lysis and ensuring successful cryopreservation.

An additional factor influencing the efficacy of this cryopreservation protocol lies in the meticulous management of frozen samples. Preserving the samples at or below −65°C is critical as it serves to mitigate damage to cellular components caused by ice crystal growth. For example, if sample temperatures rise to −20°C, ice crystals within

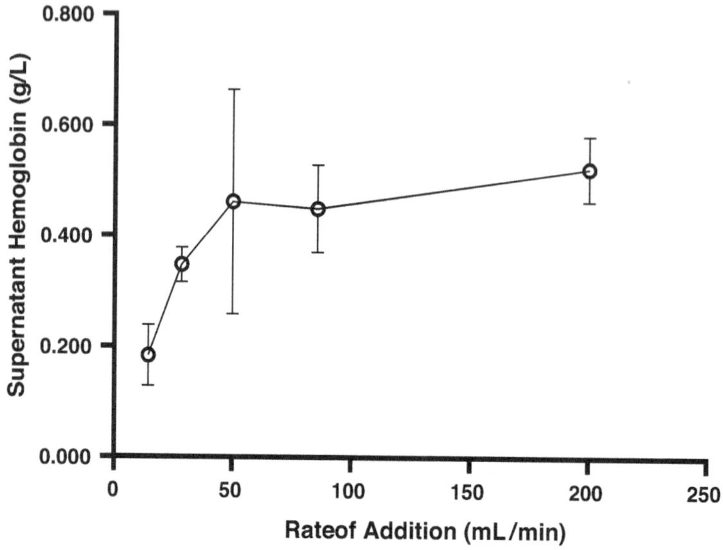

Fig. 1 Impact of the rate of addition of glycerol to RBCs during the glycerolization process (mean ± SD, n =3). Red blood cell samples from three different donor red cell concentrates were sampled, and 12 mL of cell material was aliquoted into 50 mL conical tubes. Glycerol was added to each donor sample in two parts; 7 and 21 mL with a 5-min incubation in between. Five different aliquots were tested per donor using different rates of addition for glycerol, controlled using a repeater pipette. Rates of addition included five settings on the repeater pipette, ranging from the lowest to the highest. Corresponding rates of addition with these settings (15, 29, 53, 86, and 200 mL/min, respectively) were measured gravimetrically with the glycerol solution in a separate study. Hemolysis was measured on all samples post-glycerolization addition and incubation. With manual mixing and automated glycerol addition, RBCs demonstrated increased sensitivity to glycerol addition at higher dispense rates. An addition rate for glycerol addition of ≤15 mL/min is recommended

the sample grow larger, potentially leading to the lysis of RBCs upon thawing. To minimize inadvertent warming exposures, a frozen sample handling procedure should be in place to safeguard the quality and integrity of the samples during storage. With the effective management of a frozen cell bank, small-volume glycerolized RBCs should exhibit similar storage stability as their large-scale cryopreserved RCC counterparts stored in conventional blood banks. Thus, it is expected that samples processed by glycerolization methods 3 (Sect. 3.1.2) or 4 (Sect. 3.1.3) for RCCs should be stable for at least 10 years, given that storage temperature is monitored, and unintentional warming events are avoided. This assertion is likely also applicable for samples prepared by glycerolization methods 1 and 2 for EDTA RBCs (Sect. 3.1.1), however, storage stability has not been investigated during development of these protocols.

Several pivotal measures must also be adhered to during the deglycerolization process. First, a round-bottom tube should be selected for the procedure, as it facilitates resuspension of the RBCs, which reduces RBC hemolysis, resulting in higher cell recovery when compared to conical bottom tubes. Next, as with glycerol addition, the removal of glycerol is also impacted by the rate of addition, in this

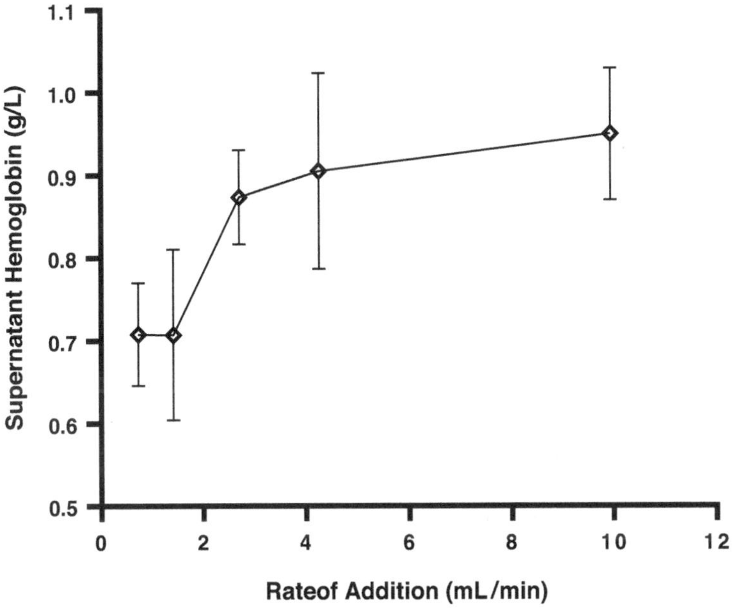

Fig. 2 Impact of rate of addition of 12% saline to RBCs during the deglycerolization process (mean ± SD, $n = 3$). Red blood cells from three donors, previously glycerolized, were thawed in a 37°C water bath and transferred to 14 mL round-bottom tubes. Using a repeater pipette, 12% saline was added to the glycerolized RBCs using one of five dispense settings ranging from the lowest setting to the highest. Corresponding rates of addition with these settings (0.73, 1.42, 2.71, 4.26, and 9.94 mL/min, respectively) were measured gravimetrically with the 12% saline solution in a separate study. Hemolysis was measured on all samples post-saline addition and incubation. With manual mixing and automated 12% saline addition, RBCs demonstrated increased sensitivity to the reagent when added at higher dispense rates. An addition rate for the 12% saline step of <1.5 mL/min is recommended

case for the deglycerolization solutions. High-concentration saline solution (12%) is added to the thawed RBCs, causing them to shrink and release glycerol into the extracellular space. Successive washes with solutions of decreasing osmolality are required to incrementally reduce glycerol to a final concentration of <1% and return RBCs to a physiological osmolality. The impact of rate of addition of the 12% saline solution in wash step 1 was evaluated and it was observed that, along with continuous mixing/agitation, a rate of <1.5 mL/min yielded the lowest level of hemolysis (see Fig. 2). When developing this procedure, hemoglobin loss was evaluated at each wash step and it was revealed that the introduction of 1.6% saline during wash step 2 was associated with a higher level of supernatant hemoglobin (see Figs. 3 and 4). Consequently, careful attention should be given to thorough mixing and wash solution addition in this step, thereby minimizing the risk of hemolysis. It should also be noted that the original protocol was developed as a four-step deglycerolization process to emulate the COBE 2991 procedure (Step 3.6 of "Methods"). However, to reduce the processing time, a three-step procedure was introduced, which also reduced operator burden and potential for deviations to occur (see Table 3). Both methods yield similar results, as shown in Table 4 (without supernatant reduction step) and Table 5 (with supernatant reduction step).

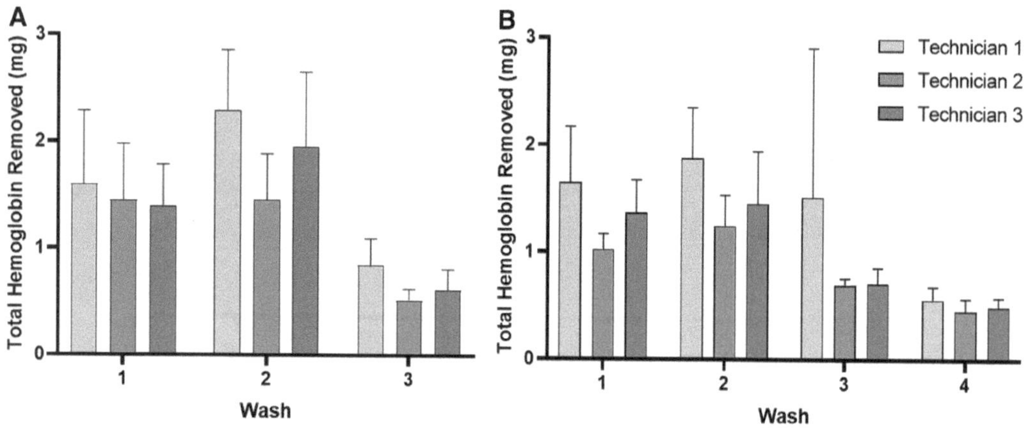

Fig. 3 Hemoglobin loss in supernatants by step during deglycerolization of RBCs glycerolized without supernatant removal (mean $\pm$ SD, $n = 6$). Thawed glycerolized RBCs from six different donors were deglycerolized using either the **A** three-step deglycerolization process or the **B** four-step deglycerolization process. Supernatants were retained and tested for hemoglobin content at each wash step. Total hemoglobin removed was calculated by multiplying the hemoglobin concentration by the volume of supernatant removed. This was performed by three different technicians and demonstrates that the addition of the second wash solution is more likely to increase RBC lysis during processing by either deglycerolization method

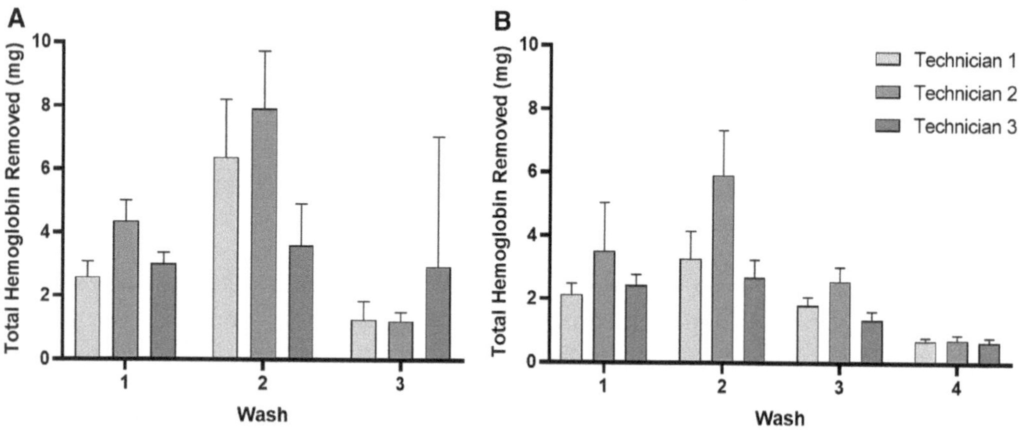

Fig. 4 Hemoglobin loss in supernatants by step during deglycerolization of RBCs glycerolized with supernatant removal (mean $\pm$ SD, $n = 6$). Thawed glycerolized RBCs from six different donors were deglycerolized using either the **A** three-step deglycerolization process or the **B** four-step deglycerolization process. Supernatants were retained and tested for hemoglobin content at each wash step. Total hemoglobin removed was calculated by multiplying the hemoglobin concentration by the volume of supernatant removed. This was performed by three different technicians and demonstrates that the addition of the second wash solution is more likely to increase RBC lysis during processing by either of the deglycerolization methods

Table 3
Comparison of estimated turnaround time for the manual three-step vs. four-step deglycerolization methods for single samples

Method	Thawing	Wash 1A and 1B addition, incubation, and centrifugation	Wash 2 addition, incubation, and centrifugation	Wash 3 addition, incubation, and centrifugation	Wash 4 addition, incubation, and centrifugation	Suspension in final hypothermic storage solution	Estimated turnaround time
Three step (min)	2–4	20–25	15–20	15–20	N/A	1	53–70
Four step (min)	2–4	20–25	15–20	15–20	15–20	1	68–90

Table 4

Comparison of post deglycerolization RBCs using the three-step or four-step deglycerolization methods for samples glycerolized without volume reduction steps

Glycerolization without volume reduction	Three-step deglycerolization method	Four-step deglycerolization method	Mean difference
<1 h Post deglycerolization (n=6)			
Volume (mL)	0.614 ± 0.038	0.613 ± 0.025	−0.001
Hematocrit (%)	31 ± 3	30 ± 1	−1
Hemoglobin (mg/unit)	55.8 ± 3.0	53.1 ± 2.4	−2.7
Supernatant hemoglobin (g/L)	0.701 ± 0.071	0.650 ± 0.079	-0.051
RBC hemolysis (%)	0.53 ± 0.05	0.53 ± 0.08	0.00
24h Post deglycerolization (n=6)			
Volume (mL)	0.633 ± 0.021	0.590 ± 0.017	−0.043
Hematocrit (%)	31 ± 1	32 ± 1	1
Hemoglobin (mg/unit)	57.1 ± 2.6	53.7 ± 2.4	−3.4
Supernatant hemoglobin (g/L)	0.775 ± 0.078	0.743 ± 0.136	−0.032
RBC hemolysis (%)	0.60 ± 0.07	0.56 ± 0.10	−0.04

Results displayed as mean ± standard deviation. Mean difference was calculated by subtracting mean results of the three-step deglycerolizationDeglycerolization method from the mean results of the four-step deglycerolizationDeglycerolization method. Samples from six different donors were glycerolized, aliquoted into cryovials (1 mL), and frozen. Two samples from each donor were thawed in a 37°C water bath and deglycerolized using either the three-step or the four-step deglycerolizationDeglycerolization process

Post-deglycerolization, RBCs are resuspended in an additive solution and stored refrigerated until required for their intended use. RBCstorage duration depends heavily on the AS selected, as well as aseptic handling of the product throughout the entire process (glycerolization and deglycerolization). An additive solution provides metabolism and membrane support to RBCs by counteracting or impeding the red cell storage lesion, while hypothermic storage is required to limit bacterial growth if contamination occurs during open system processing. This protocol recommends AS-3 be used for post-deglycerolizationRBCstorage. AS-3 is used in routine blood banking and has demonstrated its ability to support post-deglycerolized RBCs for up to 14 days in hypothermic storage conditions. Due to the manual processing of these small-scale samples, storage up to 14 days may not be achievable, as increases in hemolysis above the regulatory standards may occur [21–23]. Therefore, implementation of a biobank of this type should include characterization of RBC quality throughout the post-deglycerolizationstorage period of interest.

Table 5

Comparison of post deglycerolization RBCs using the three-step or four-step deglycerolization methods for samples glycerolized with volume reduction steps

Glycerolization with volume reduction	Three-step deglycerolization method	Four-step deglycerolization method	Mean difference
<1 h Post deglycerolization (n=6)			
Volume (mL)	0.905 ± 0.019	0.909 ± 0.020	0.004
Hematocrit (%)	50 ± 1	49 ± 2	−1
Hemoglobin (mg/unit)	131.3 ± 5.6	129.8 ± 5.3	−1.5
Supernatant hemoglobin (g/L)	0.820 ± 0.125	0.980 ± 0.117	0.160
RBC hemolysis (%)	0.28 ± 0.03	0.35 ± 0.05	0.07
24h Post deglycerolization (n=6)			
Volume (mL)	0.878 ± 0.016	0.869 ± 0.017	−0.009
Hematocrit (%)	50 ± 2	51 ± 2	1
Hemoglobin (mg/unit)	127.0 ± 6.8	127.8 ± 5.3	0.8
Supernatant hemoglobin (g/L)	1.193 ± 0.089	1.566 ± 0.163	0.373
RBC hemolysis (%)	0.42 ± 0.04	0.52 ± 0.04	0.10

Results displayed as mean ± standard deviation. Mean difference was calculated by subtracting mean results of the three-step deglycerolizationDeglycerolization method from the mean results of the four-step deglycerolizationDeglycerolization method. Samples from six different donors were glycerolized using the RCC glycerolizationGlycerolization method with post-glycerolizationGlycerolization volume reduction, aliquoted into cryovials (1 mL), and frozen. Two samples from each donor were thawed in a 37°C water bath and deglycerolized using either the three-step or the four-step deglycerolizationDeglycerolization process

In addition to storage stability, pre-implementation should include method variability studies. During process characterization and validation of the method described here, several investigations were conducted, including within-run (single lot) repeatability (see Fig. 5), across-run (lot-to-lot) repeatability (see Fig. 6), and technician-to-technician variability (see Fig. 7). Results from these investigations demonstrate that deglycerolized RBC product characteristics within run are highly reproducible, with similar volume, spun hematocrit, and RBC hemolysis values (see Table 6). The lot-to-lot repeatability was also highly reproducible (see Table 7). The main contributor to the observed within-lot and lot-to-lot variability was supernatant hemoglobin and RBC hemolysis. This variation was predominantly due to the addition method and rate of the deglycerolization solutions during manual processing, which could be reduced if these aspects of the method were automated. However, it is important to emphasize that variability will exist from one donor to another, as some RBCs may display increased susceptibility to the cryopreservation process. This susceptibility may be attributed

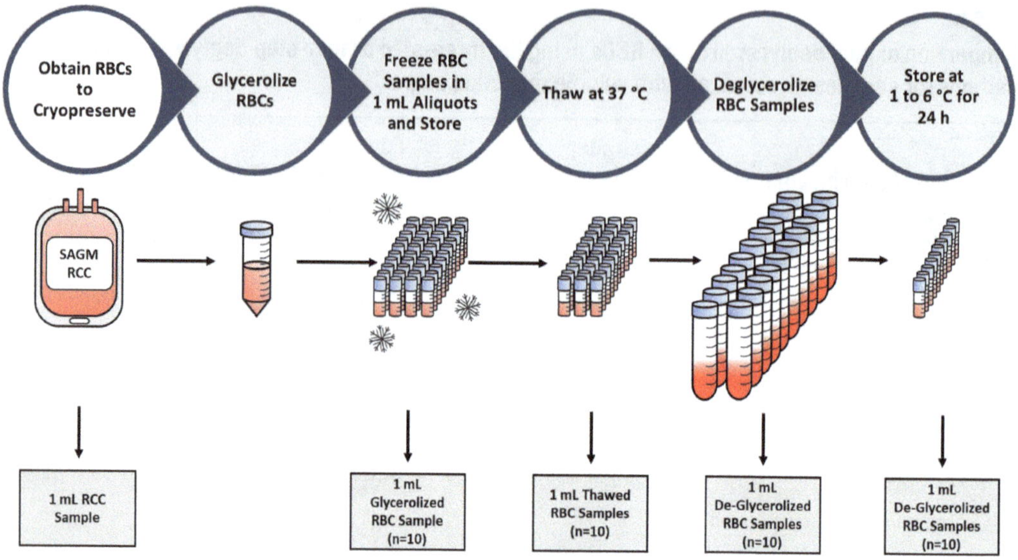

Fig. 5 Process validation study diagram: within run/lot repeatability. To evaluate the repeatability within a lot (RBC sample glycerolized together then aliquoted into small volumes), one RCC unit was sampled and glycerolized using the desired method (supernatant reduced or non-supernatant reduced). The glycerolized RBCs were then aliquoted into multiple vials, frozen, thawed, deglycerolized, and stored for 24 h. To evaluate the repeatability of the freeze/thaw process, samples were tested post-thaw immediately. To evaluate the repeatability of the deglycerolization process, samples were tested immediately post deglycerolization and after 24 h of hypothermic storage. All samples were tested using the RBC hemolysis method as previously described [24]. Metrics used for evaluation include hematocrit, total hemoglobin, supernatant hemoglobin, and hemolysis. Mean, standard deviation, and coefficient of variation were calculated for each metric

to specific RBC phenotypes or underlying disease states (i.e., sickle cell disease), but also arises due to donor characteristics such as age and sex. Lastly, technician-to-technician variability demonstrated that with sufficient training of operators, RBCs maintain similar quality characteristics regardless of the operator throughout the cryopreservation process (see Figs. 8–10).

The small-scale RBCcryopreservation methods presented here offer a reliable and reproducible alternative to previously published approaches. The efficient, high recovery methods reflect the large-scale processes employed at most blood centers and provide a valuable tool for biobanking and research initiatives that serve to further advance the fields of RBCcryopreservation and blood banking.

2 Materials

2.1 Reagents

1. 0.9%/0.2% Sodium chloride/dextrose solution (see Table 8).
2. 1.6% Sodium chloride solution (see Table 9).

Table 6
Within run repeatability results for the RCC glycerolization and RCC glycerolization with volume reduction methods. One sample from an RCC unit was glycerolized by both the glycerolization and the glycerolization with volume reduction methods. The glycerolized sample was then split into cryovials and frozen. Sets of ten cryovials were thawed in a 37°C water bath. Ten vials were used for post-thaw measurements, 20 vials were deglycerolized and split into two groups, one group of ten was used for <1 h post deglycerolization measurements, where the remaining ten cryovials were used for 24 h post-deglycerolization measurements after storage in a 1–6°C refrigerator. Hemolysis was performed as previously described [24]. All processing steps and sample testing were performed by one technician.

		Volume (mL)		Spun hematocrit (%)		Total Hgb (mg/unit)		Supernatant Hgb (g/L)		RBC hemolysis (%)	
Step	Sample description	Mean ± SD	CV (%)	Mean ± SD	CV (%)	Mean ± SD	CV (%)	Mean ± SD	CV (%)	Mean ± SD	CV (%)
Glycerolization method (without volume reduction) with the three-step deglycerolization method											
Pre-freeze	Glycerolized RBCs (n=1)	N/A	N/A	17	N/A	54.2	N/A	0.185	N/A	0.28	N/A
Post-thaw	Glycerolized RBCs (n=10)	N/A	N/A	17 ± 0	0	56.1 ± 1.1	1.9	0.753 ± 0.101	13.4	1.12 ± 0.16	13.9
<1 h Post-deglycerolization	Deglycerolized RBCs in AS-3 (n=10)	0.70 ± 0.04	5.4	26 ± 2	8	53.5 ± 1.6	3.0	0.459 ± 0.078	17.1	0.44 ± 0.04	9.5
24 h Post-deglycerolization	Deglycerolized RBCs in AS-3 (n=10)	0.59 ± 0.02	2.6	32 ± 2	6	53.8 ± 2.1	3.9	0.656 ± 0.107	16.3	0.49 ± 0.05	10.3
Glycerolization with post-glycerolization volume reduction with the three-step deglycerolization method											
Pre-freeze	Glycerolized RBCs (n=1)	N/A	N/A	47	N/A	130.1	N/A	1.065	N/A	0.43	N/A
Post-thaw	Glycerolized RBCs (n=10)	N/A	N/A	46 ± 0	0	137.3 ± 2.0	1.5	2.418 ± 0.081	3.3	0.95 ± 0.03	3.1

Continued

Table 6
Continued

Step	Sample description	Volume (mL)		Spun hematocrit (%)		Total Hgb (mg/unit)		Supernatant Hgb (g/L)		RBC hemolysis (%)	
		Mean ± SD	CV (%)	Mean ± SD	CV (%)	Mean ± SD	CV (%)	Mean ± SD	CV (%)	Mean ± SD	CV (%)
<1 h Post-deglycerolization	Deglycerolized RBCs in AS-3 (*n*=10)	0.91 ± 0.04	4.3	48 ± 2	4	124.6 ± 4.3	3.4	0.647 ± 0.096	14.8	0.25 ± 0.02	9.3
24 h Post-deglycerolization	Deglycerolized RBCs in AS-3 (*n*=10)	0.86 ± 0.03	3.1	49 ± 2	4	119.6 ± 2.3	1.9	0.890 ± 0.216	24.3	0.33 ± 0.07	22.4

Table 7

Across-run repeatability results for the RCC glycerolization and the RCC glycerolization with volume reduction methods. Twenty samples from an RCC unit were split into two study arms, where ten samples were processed using the RCC glycerolization method, and ten samples were processed using the glycerolization with post-glycerolization volume reduction. Samples from each preparation were aliquoted into cryovials (1 mL) and frozen. Ten samples from each process (one from each preparation) were tested before freezing, and an additional ten samples from each process (one from each preparation) were tested after thawing. Samples were thawed in a 37°C water bath. Twenty samples from each process (two from each preparation) were thawed in a 37°C water bath, deglycerolized using the three-step method, and tested either immediately (<1 h post deglycerolization) or after 24 h post deglycerolization. Hemolysis was performed as previously described [24]. All processing steps and sample testing were performed by one technician

		Volume (mL)		Spun Hematocrit (%)		Total Hgb (mg/unit)		Supernatant Hgb (g/L)		RBC Hemolysis (%)	
Step	**Sample Description**	**Mean ± SD**	**CV (%)**	**Mean ± SD**	**CV (%)**	**Mean ± SD**	**CV (%)**	**Mean ± SD**	**CV (%)**	**Mean ± SD**	**CV (%)**
Glycerolization method (without volume reduction) with the three-step deglycerolization method											
Pre-freeze	Glycerolized RBCs (n=1)	N/A	N/A	16 ± 0	2	55.7 ± 1.3	2.3	0.298 ± 0.034	11.5	0.45 ± 0.05	12.0
Post-thaw	Glycerolized RBCs (n=10)	N/A	N/A	17 ± 1	4	54.9 ± 2.3	4.2	0.968 ± 0.158	16.3	1.47 ± 0.23	15.9
<1 h Post-deglycerolization	Deglycerolized RBCs in AS-3 (n=10)	0.68 ± 0.05	6.9	26 ± 1	6	51.3 ± 2.3	4.5	0.456 ± 0.067	14.6	0.44 ± 0.04	9.3
24 h Post-deglycerolization	Deglycerolized RBCs in AS-3 (n=10)	0.65 ± 0.06	9.0	27 ± 2	9	51.2 ± 2.4	4.7	0.610 ± 0.109	17.9	0.56 ± 0.07	12.0
Glycerolization with post-glycerolization volume reduction with the three-step deglycerolization method											
Pre-freeze	Glycerolized RBCs (n=1)	N/A	N/A	43 ± 1	3	134.2 ± 1.9	1.4	1.226 ± 0.102	8.3	0.52 ± 0.04	7.5
Post-thaw	Glycerolized RBCs (n=10)	N/A	N/A	44 ± 1	3	132.6 ± 2.1	1.6	2.493 ± 0.287	11.5	1.05 ± 0.11	10.8

Continued

Table 7
Continued

Step	Sample Description	Volume (mL)		Spun Hematocrit (%)		Total Hgb (mg/unit)		Supernatant Hgb (g/L)		RBC Hemolysis (%)	
		Mean ± SD	CV (%)	Mean ± SD	CV (%)	Mean ± SD	CV (%)	Mean ± SD	CV (%)	Mean ± SD	CV (%)
<1 h Post-deglycerolization	Deglycerolized RBCs in AS-3 (n=10)	0.93 ± 0.06	6.8	44 ± 3	7	118.1 ± 4.2	3.5	0.745 ± 0.119	16.0	0.33 ± 0.03	8.2
24 h Post-deglycerolization	Deglycerolized RBCs in AS-3 (n=10)	0.97 ± 0.03	6.5	42 ± 3	6	121.4 ± 2.9	2.4	1.019 ± 0.100	9.8	0.47 ± 0.05	10.4

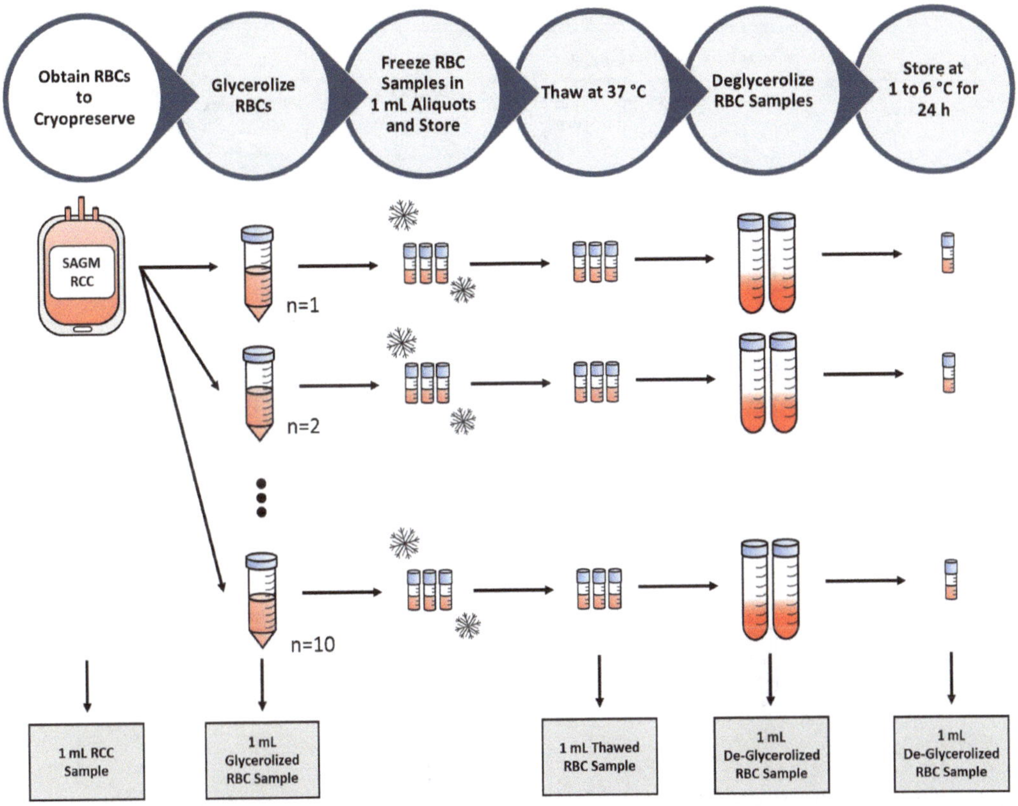

Fig. 6 Process validation study diagram: across run repeatability. To evaluate the repeatability of the whole process and impact of variability due to the manual glycerolization process, one RCC unit was sampled ten independent times on the same day and glycerolized using the desired method (supernatant reduced or non-supernatant reduced). To evaluate the repeatability of the glycerolization process a sample was tested after glycerol addition but prior to freezing. To evaluate the repeatability of the glycerolization and freeze/thaw process a sample was tested immediately post thaw. To evaluate the repeatability of the whole cryopreservation process, glycerolization to deglycerolization samples were tested immediately post deglycerolization and after 24 h of hypothermic storage. All samples were tested using the RBC hemolysis method as previously described [24]. Metrics used for evaluation include hematocrit, total hemoglobin, supernatant hemoglobin, and hemolysis. Mean, standard deviation, and coefficient of variation were calculated for each metric

3. 12% Sodium chloride solution (see Table 10).
4. AS-3 (Haemonetics, Massachusetts, USA) (see Table 11).
5. Glycerolyte 57 (Fresenius Kabi AG, Hessen, Germany) (see Table 12).

2.2 Consumables

1. 14 mL round-bottom tubes with caps (352059, Corning, New York, USA).
2. 50 mL conical tube with screw cap.
3. 1000 μL pipette tips.
4. 5000 μL pipette tips.

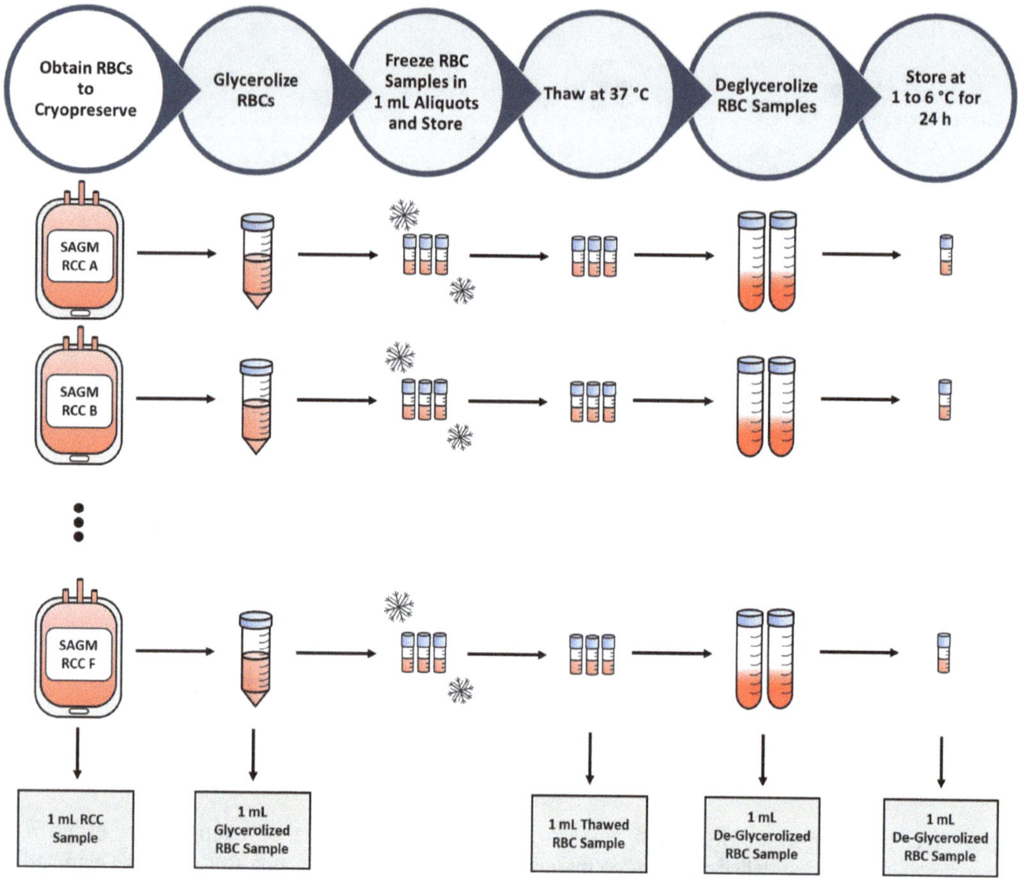

Fig. 7 Process validation study diagram: technician variability. Due to the manual nature of both the glycerolization and deglycerolization protocols, technician variability could lead to a shift in lot-to-lot characteristics. To evaluate technician-to-technician variability, samples from six different donor RCC units were cryopreserved by three highly skilled technicians who were trained in this protocol. Samples were tested after glycerolization, immediately after thawing, immediately post deglycerolization, and after 24 h of hypothermic storage. All samples were tested using the RBC hemolysis method as previously described [24]. Metrics used for evaluation include hematocrit, total hemoglobin, supernatant hemoglobin, and hemolysis

5. Cryovials, round bottom, 2 mL volume.
6. Storage boxes for cryovials.
7. Transfer pipettes.

2.3 Equipment

1. Centrifuge with EDTA tube insert and swing bucket rotor.
2. Centrifuge with 50 mL conical tube insert and swing bucket rotor.
3. Centrifuge with 14 mL round-bottom tube insert and swing bucket rotor.
4. 1000 μL pipette.

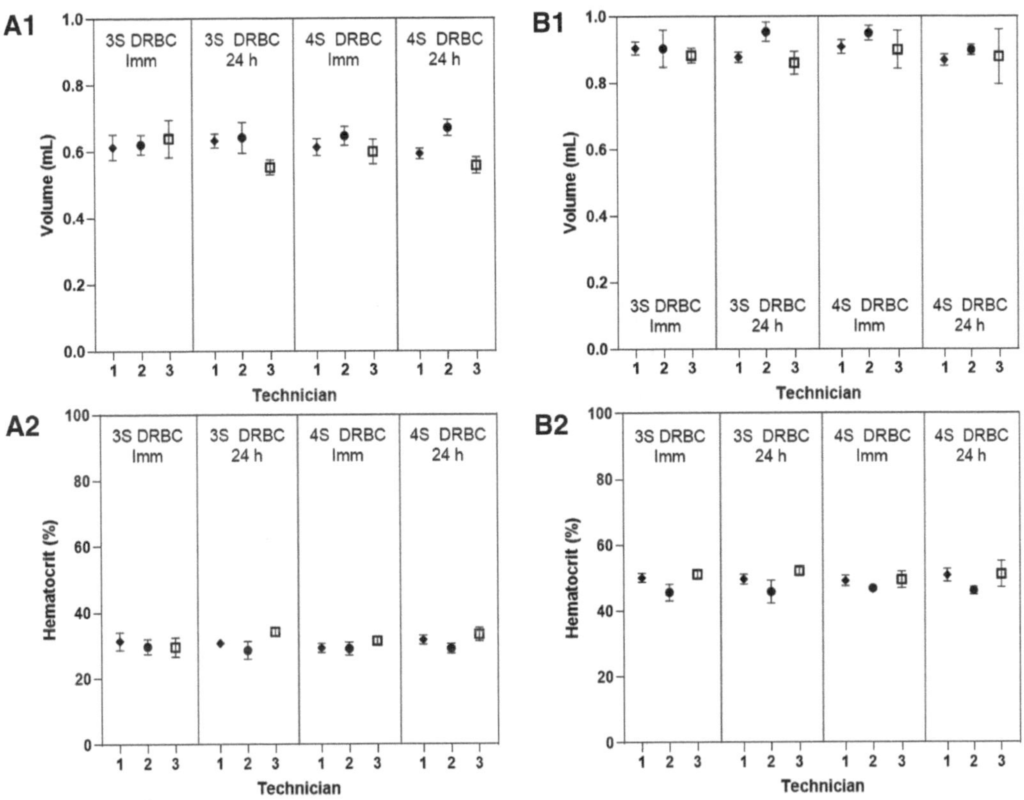

Fig. 8 Technician variability in volume and hematocrit for glycerolization Method 3 (**A**) and glycerolization Method 4 (**B**). Six CDP SAGM RBC donor units were cryopreserved and deglycerolized by three different technicians. The mean ± SD ($n = 6$) is displayed for each technician. Figure (A1) and (A2) display samples glycerolized using Method 3, and Figure (B1) and (B2) display samples glycerolized using Method 4. 3S DRBCs Imm = RBCs deglycerolized using the three-step method tested post deglycerolization immediately, 3S DRBC 24 h = RBCs deglycerolized using the three-step method tested 24 h after deglycerolization and hypothermic storage, 4 S DRBC Imm = RBCs deglycerolized using the four-step method tested post deglycerolization immediately, and 4S DRBC 24 h = RBCs deglycerolized using the four-step method tested 24 h after deglycerolization and hypothermic storage

5. 5000 μL pipette.
6. 50 mL conical tube rack.
7. 14 mL round-bottom tube rack.
8. 2 mL cryovial rack.
9. 1°C/min freezing device (options include: styrofoam cryovial rack that comes with some cryovials, CoolCell freezing container, or controlled rate freezer).
10. Biological safety cabinet.
11. Freezer, −80°C.
12. Water Bath, 37°C.

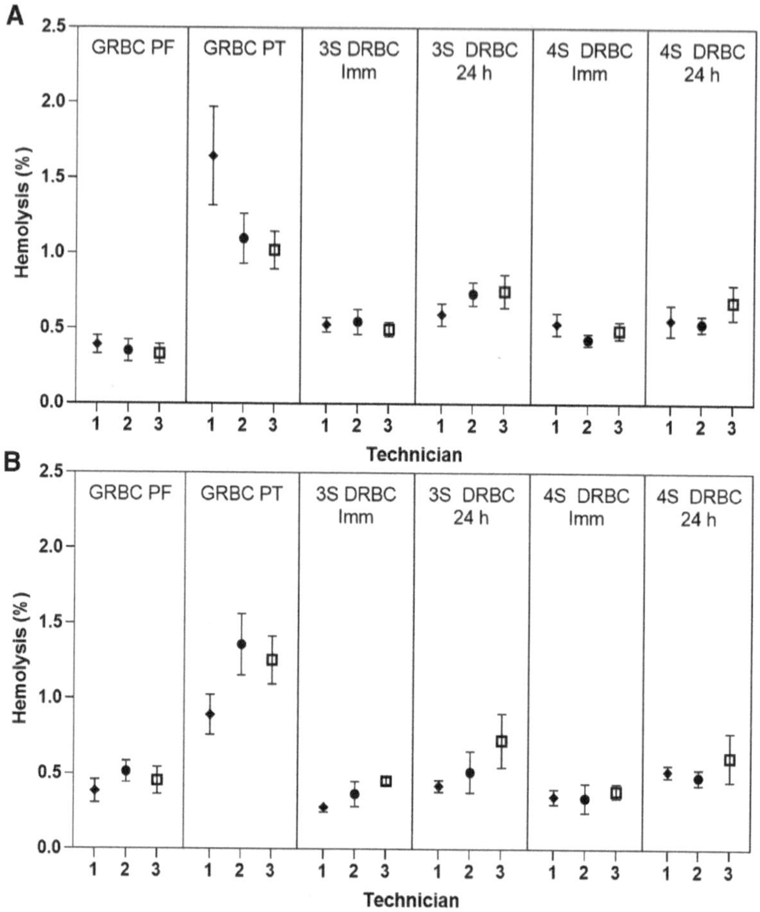

Fig. 9 Technician variability for RBC hemolysis throughout the cryopreservation process. Six CPD SAGM RBC donor units were cryopreserved and deglycerolized by three different technicians. The mean ± SD (n=6) is displayed for each technician. Figure **A** displays samples glycerolized using Method 3, and Figure **B** displays samples glycerolized using Method 4. GRBC PF = glycerolized RBCs pre-freeze, GRBC PT = glycerolized RBCs post-thaw, 3S DRBCs Imm = RBCs deglycerolized using the three-step method tested post deglycerolization immediately, 3 S DRBC 24 h = RBCs deglycerolized using the three-step method tested 24 h after deglycerolization and hypothermic storage, 4S DRBC Imm = RBCs deglycerolized using the four-step method tested immediately post deglycerolization, and 4S DRBC 24 h = RBCs deglycerolized using the four-step method tested 24 h after deglycerolization and hypothermic storage

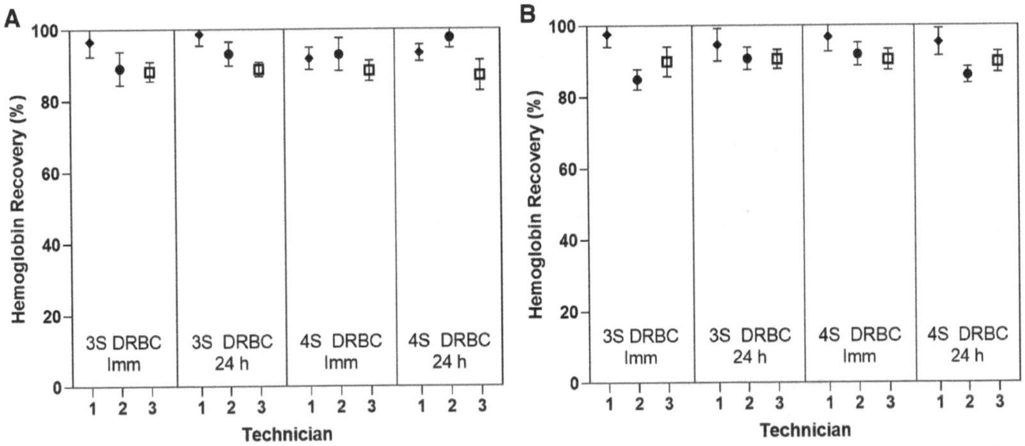

Fig. 10 Technician variability for hemoglobin recovery throughout the cryopreservation process. Six CPD SAGM RBC donor units were cryopreserved and deglycerolized by three different technicians. The mean ± SD (n=6) is displayed for each technician. Figure **A** displays samples glycerolized using Method 3, and Figure **B** displays samples glycerolized using Method 4. 3S DRBC Imm = RBCs deglycerolized using the three-step method immediately post deglycerolization, 3S DRBC 24 h = RBCs deglycerolized using the three-step method tested 24 h after deglycerolization, 4S DRBC Imm = RBCs deglycerolized using the four-step method testing immediately post deglycerolization, and 4S DRBC 24 h = RBCs deglycerolized using the four-step method testing 24 h after deglycerolization and hypothermic storage

Table 8
Components of 0.9% sodium chloride/0.2% dextrose solution

Component	Amount	Unit
Sodium chloride (NaCl)	0.9	g
Dextrose	0.2	g
Distilled/deionized water	100	mL

When formulating this reagent in-house, the resulting reagent should be sterilized before use

Table 9
Components of 1.6% sodium chloride solution

Component	Amount	Unit
Sodium chloride (NaCl)	1.6	g
Distilled/deionized water	100	mL

When formulating this reagent in-house, the resulting reagent should be sterilized before use

Table 10
Components of 12% sodium chloride processing solution

Component	Amount	Unit
Sodium chloride (NaCl)	12	g
Distilled/deionized water	100	mL

When formulating this reagent in-house, the resulting reagent should be sterilized before use

Table 11
Components of red blood cell additive solution (AS-3)

Component	Amount	Unit
Citric acid (monohydrate)	0.042	g
Monobasic sodium phosphate (monohydrate)	0.276	g
Sodium chloride	0.410	g
Adenine	0.030	g
Dextrose (anhydrous)	1.000	g
Sodium citrate (dihydrate)	0.588	g
Distilled/deionized water	100	mL

When formulating this reagent in-house, the resulting reagent should be sterilized before use

Table 12
Components of glycerolyte 57 solution, pH 6.8

Component	Amount	Unit
Glycerin	57	g
Potassium chloride	30	mg
Sodium lactate	1.6	g
Sodium phosphate dibasic	124.2	mg
Sodium phosphate monobasic	51.7	mg
Distilled/deionized water	100	mL

When formulating this reagent in-house, the resulting reagent should be sterilized before use

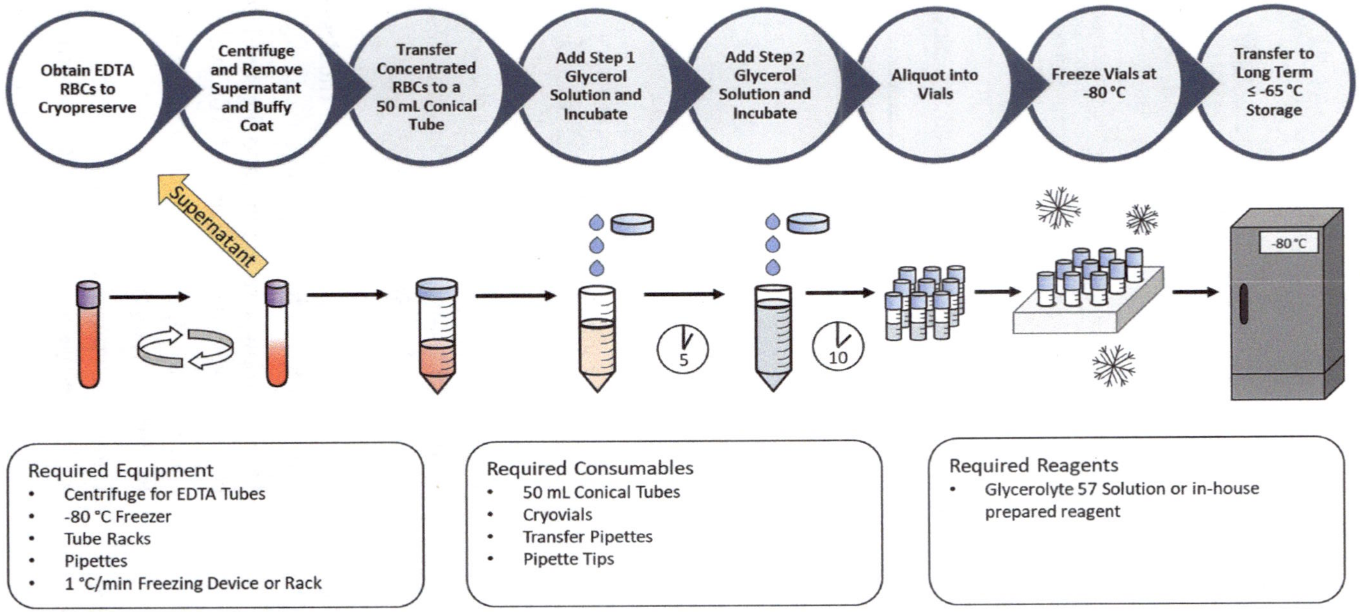

Fig. 11 Process diagram for glycerolization Method 1—EDTA RBCs without glycerol supernatant reduction

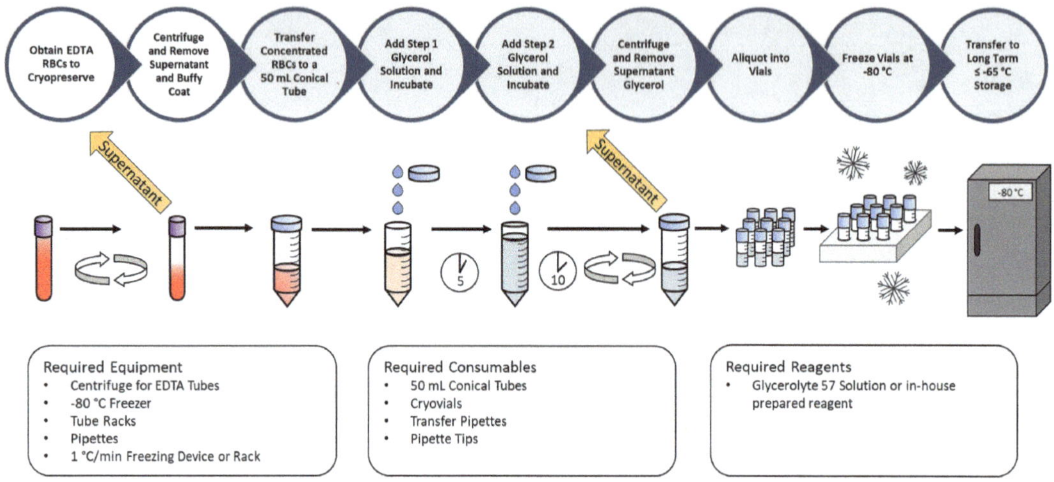

Fig. 12 Process diagram for glycerolization Method 2—EDTA RBCs with glycerol supernatant reduction

3 Methods

3.1 Preparation of RBCs for Cryopreservation (see Notes 1–4)

3.1.1 Obtain, Prepare, and Glycerolize RBCs from EDTA Whole Blood Collections (see Fig. 11, Glycerolization Method 1)

1. Centrifuge the EDTA whole blood sample at 1500×*g* for 10 min at room temperature (RT, 18–22°C) with fast acceleration and a slow break.
2. Using a sterile transfer pipette, remove the plasma and buffy coat layers. Swirl the tube to mix the concentrated RBCs.
3. Into a 50 mL conical tube, pipette the desired volume of concentrated RBCs to cryopreserve.
4. Using Table 13, identify the volume of 57% glycerol solution for step 1 and step 2 additions for the desired volume of concentrated RBCs to be cryopreserved.
5. Using a dropwise technique, add the step 1 volume of 57% glycerol solution to the concentrated RBCs with continuous mixing.
6. Cap the tube and swirl five to ten times to mix. Incubate at RT for 5 min undisturbed.
7. Using a dropwise technique, add the step 2 volume of the 57% glycerol solution to the RBCs with continuous mixing.
8. Cap the tube and swirl five to ten times to mix. Incubate at RT for 10 min undisturbed.
9. If minimal sample processing is preferred, RBCs may be frozen after step 8 (proceed to step 3.2 of Methods). If a higher concentration of RBCs per sample is required, continue to the volume reduction step (step 10, see Fig. 12 Glycerolization Method 2).

Table 13

Glycerol solution volume required for each addition step based on RBC input volume for glycerolization Method 1—EDTA RBCs without supernatant reduction, and Method 2—EDTA RBCs with supernatant reduction

RBC input volume (mL)[a]	57% Glycerol solution volume required (mL) [Eq. 1, Table 19]	Step 1: 57% Glycerol solution volume (mL) [Eq. 2, Table 19]	Step 2: 57% Glycerol solution volume (mL) [Eq. 3, Table 19]	Glycerolized RBC volume (mL)[b] [Eq. 4, Table 19]
0.50	1.25	0.30	0.95	1.75
1.00	2.25	0.55	1.70	3.25
1.50	3.50	0.90	2.60	5.00
2.00	4.75	1.20	3.55	6.75
2.50	5.75	1.45	4.30	8.25
3.00	7.00	1.75	5.25	10.00
3.50	8.25	2.05	6.20	11.75
4.00	9.25	2.30	6.95	13.25
4.50	10.50	2.65	7.85	15.00
5.00	11.75	2.95	8.80	16.75
5.50	12.75	3.20	9.55	18.25

[a]Use this table for RBC Input Volumes of 0.50 mL to 5.50 mL. For Volumes ≥ 6.0, refer to Table 15. [b]Final glycerol concentration range = 38% to 41% when using these RBC to glycerol volume ratios

10. Centrifuge the glycerolized RBCs at 2200×*g* for 10 min at RT with fast acceleration and slow break.
11. Using Table 14, identify the volume of supernatant glycerol to remove after centrifugation.
12. Using a pipette, remove the calculated volume of supernatant glycerol from the centrifuged glycerolized RBCs.
13. Cap and swirl the tube five to ten times to mix. Continue mixing as necessary to ensure all RBCs packed into the bottom of the tube are incorporated into the solution. Proceed to Step 3.2 of "Methods."

3.1.2 Obtain, Prepare, and Glycerolize RBCs from Red Cell Concentrate (RCC) Collections (see Fig. 13, Glycerolization Method 3)

1. Using a sterile sampling technique, extract a sample for cryopreservation from an RCC and allow RBCs to warm to RT on the benchtop (approximately 15 min) if previously refrigerated. Transfer a maximum of 15 mL to a 50 mL conical tube.
2. Using Table 15, identify the volume of 57% glycerol solution for steps 1–3 additions for the desired volume of concentrated RBCs to be cryopreserved.
3. Using a dropwise technique, add the step 1 volume of 57% glycerol solution to the RBCs with continuous mixing.

Table 14
Supernatant glycerol volume to remove from centrifuged glycerolized RBCs prior to freezing

Original RBC input volume (mL)	Glycerolized RBC volume (mL) [Eq. 4, Table 19]	Volume of supernatant glycerol to remove (mL) [Eq. 5, Table 19]	Supernatant reduced glycerolized RBC volume (mL) [Eq. 6, Table 19]
0.50	1.75	0.75	1.00
1.00	3.25	1.50	1.75
1.50	5.00	2.25	2.75
2.00	6.75	3.00	3.75
2.50	8.25	3.75	4.50
3.00	10.00	4.50	5.50
3.50	11.75	5.25	6.50
4.00	13.25	6.00	7.25
4.50	15.00	6.75	8.25
5.00	16.75	7.50	9.25
5.50	18.25	8.25	10.00

4. Cap the tube and swirl five to ten times to mix. Incubate at RT for 5 min undisturbed.
5. Using a dropwise technique, add the step 2 volume of 57% glycerol solution to the RBCs with continuous mixing. Cap the tube and invert five to ten times to mix.
6. Using a dropwise technique, add the step 3 volume of 57% glycerol solution to the RBCs with continuous mixing.
7. Cap the tube and invert five to ten times to mix. Incubate at RT for 10 min undisturbed. Proceed to Step 3.2 of "Methods."

3.1.3 Obtain, Prepare, and Glycerolize RBCs from RCC collections with supernatant reduction (see Fig. 14, Glycerolization Method 4)

1. Using a sterile sampling technique, extract a sample for cryopreservation from an RCC and allow RBCs to warm to RT on the benchtop (approximately 15 min) if previously refrigerated. Transfer a maximum of 23 mL to a 50 mL conical tube.
2. Centrifuge the sample at 2200×*g* for 10 min at RT with fast acceleration and slow break.
3. Using Table 16, identify the volume of supernatant to remove after centrifugation.
4. Using a pipette, remove the calculated supernatant volume. Ensure to remove the total volume indicated, even if some RBCs are lost (depending on the RBC concentration of the RCC product, some RBCs may be lost in this supernatant reduction step).
5. Cap the tube and swirl five to ten times to mix. Continue mixing if necessary to ensure RBCs are incorporated into the solution.

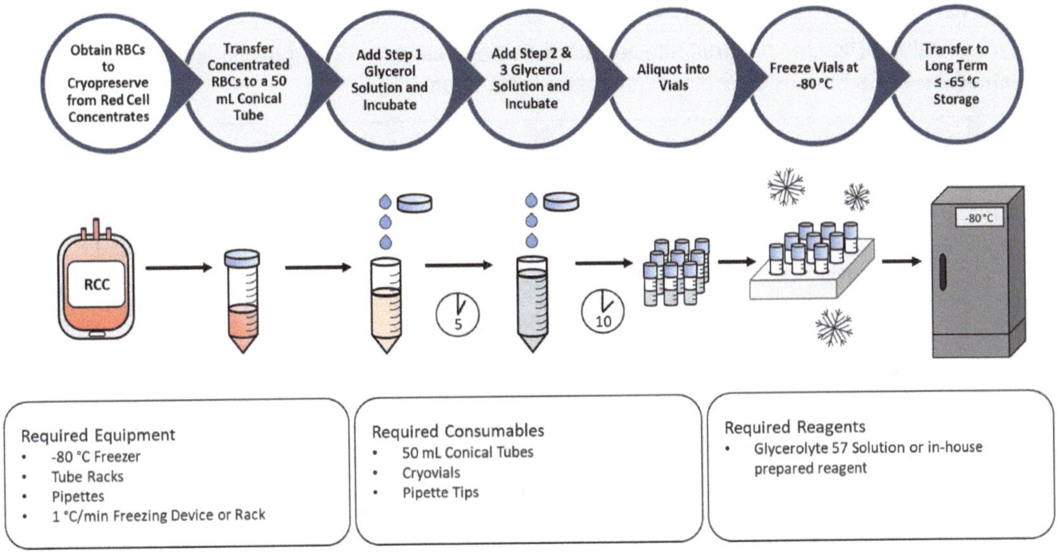

Fig. 13 Process diagram for glycerolization Method 3—RBCs from red cell concentrates (RCCs) without glycerol supernatant reduction

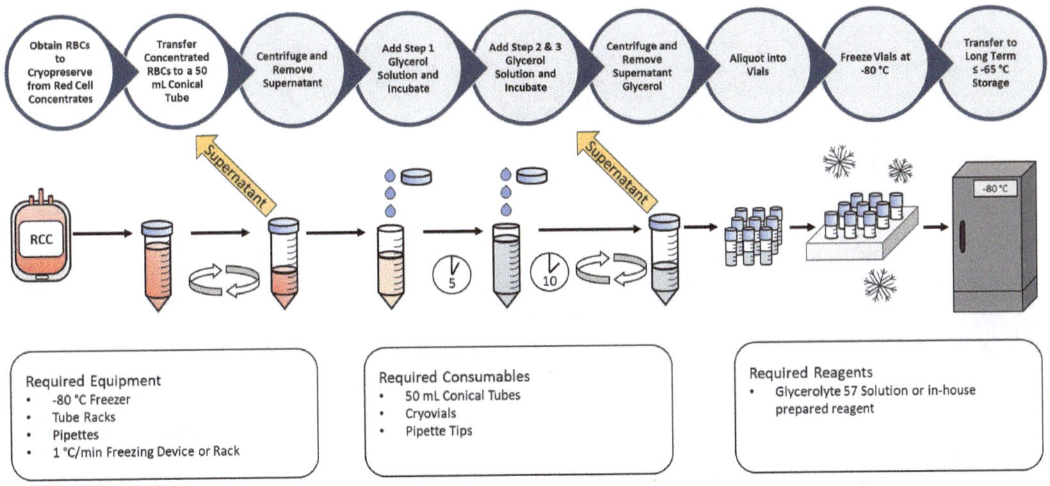

Fig. 14 Process diagram for glycerolization Method 4—RBCs from red cell concentrates (RCCs) with glycerol supernatant reduction

6. Using Table 17, identify the volume of 57% glycerol solution for steps 1–3 additions for the original input volume of RBCs to be cryopreserved.
7. Using a dropwise technique, add the step 1 volume of 57% glycerol solution to the RBCs with continuous mixing.
8. Cap the tube and invert five to ten times to mix. Incubate at RT for 5 min undisturbed.

Table 15

Glycerol solution volume required for each addition step based on RBC input volume for glycerolization Method 3—RBCs from RCCs without supernatant reduction

RBC input volume (mL)[a]	Total glycerol solution to be added to RBCs (mL) [Eq. 5, Table 19]	Step 1: 57% Glycerol volume (mL) [Eq. 6, Table 19]	Step 2: 57% Glycerol volume (mL) [Eq. 7, Table 19]	Step 3: 57% Glycerol volume (mL) [Eq. 8, Table 19]	Glycerolized RBC volume (mL)[b] [Eq. 9, Table 19]
6.0	14.0	3.5	10.0	0.5	20.0
6.5	15.0	4.0	10.0	1.0	21.5
7.0	16.5	4.0	10.0	2.5	23.5
7.5	17.5	4.5	10.0	3.0	25.0
8.0	18.5	4.5	10.0	4.0	26.5
8.5	20.0	5.0	10.0	5.0	28.5
9.0	21.0	5.5	10.0	5.5	30.0
9.5	22.0	5.5	10.0	6.5	31.5
10.0	23.5	6.0	10.0	7.5	33.5
10.5	24.5	6.0	10.0	8.5	35.0
11.0	25.5	6.5	10.0	9.0	36.5
11.5	27.0	7.0	10.0	10.0	38.5
12.0	28.0	7.0	10.0	11.0	40.0
12.5	29.0	7.5	10.0	11.5	41.5
13.0	30.5	7.5	10.0	13.0	43.5
13.5	31.5	8.0	10.0	13.5	45.0
14.0	32.5	8.0	10.0	14.5	46.5
14.5	34.0	8.5	10.0	15.5	48.5
15.0	35.0	9.0	10.0	16.0	50.0

[a]Use this table for RBC input volumes of 6.0 mL to 15.0 mL. For volumes ≤6.0, refer to Table 13. [b]Final glycerol concentration = 40% when using these RBC to glycerol volume ratios

9. Using a dropwise technique, add the step 2 volume of 57% glycerol solution to the RBCs with continuous mixing. Cap the tube and invert five to ten times to mix.
10. Using a dropwise technique, add the step 3 volume of 57% glycerol solution to the RBCs with continuous mixing.
11. Cap the tube and invert five to ten times to mix. Incubate at RT for 10 min undisturbed.
12. Centrifuge the glycerolized RBC samples at 2200×*g* for 10 min at RT with fast acceleration and slow break.

13. Allow samples to sit on the benchtop for 5–10 min to allow any RBCs suspended in the supernatant to settle (if necessary).
14. Using Table 18, identify the volume of supernatant glycerol to remove after centrifugation.
15. Using a pipette, remove the calculated volume of supernatant glycerol.
16. Cap the tube and invert it five to ten times to mix. Continue mixing if necessary to ensure RBCs are incorporated into the solution. Proceed to Step 3.2 of "Methods."

3.2 Aliquot, Freeze, and Store Glycerolized RBCs

1. Aliquot 1 mL of the glycerolized RBCs into labeled cryovials and cap. Keep tubes upright to prevent sample volume from accumulating in the cap of the tube.
   (a) Sample identification number, date frozen, RBC source, and glycerolization method are recommended as minimum label information for long-term storage.
2. Freeze the prepared aliquots using one of the following methods:
   (a) Place the cryovials into a prepared 1 °C/min freezing device and place them into a −80°C freezer for a minimum of 12 h undisturbed.
   (b) Place the cryovials into a controlled rate freezer programmed with a 1 °C/min freezing rate to −90°C and execute the protocol. Ensure the sample temperature has reached −80°C before transfer.
3. Transfer frozen vials to long-term storage boxes/containers. Use dry ice to transport samples from the freezing device to the storage device and avoid RT exposure to ensure samples remain at temperatures below −65°C.
4. If sample sorting after freezing is required, perform on dry ice to minimize sample exposure to temperatures above −65°C.

3.3 Thaw Glycerolized RBCs

1. Remove the cryovial of RBCs from frozen storage and place it immediately into a 37°C water bath for 2–3 min.
   (a) If the cryovial must be transported to a water bath, place the vial on dry ice to avoid exposure to room temperature/slow warming.
   (b) When thawing, place the cryovial so that the top of the frozen sample is submerged under the waterline, but the cap is above the water's surface.
2. Remove the cryovial from the water bath just after all ice has melted. Proceed to deglycerolization

Table 16
Supernatant removal volume (RBC additive solution) prior to glycerolization for RBCs cryopreserved with glycerolization Method 4.

RBC input volume (mL)	Volume of supernatant to remove (mL) [Eq. 12, Table 19]	Supernatant reduced RBC volume (mL) [Eq. 13, Table 19]
1.00	0.35	0.65
2.00	0.70	1.30
3.00	1.05	1.95
4.00	1.40	2.60
5.00	1.75	3.25
6.00	2.10	3.90
7.00	2.45	4.55
8.00	2.80	5.20
9.00	3.15	5.85
10.0	3.5	6.5
11.0	4.0	7.0
12.0	4.0	8.0
13.0	4.5	8.5
14.0	5.0	9.0
15.0	5.5	9.5
16.0	5.5	10.5
17.0	6.0	11.0
18.0	6.5	11.5
19.0	6.5	12.5
20.0	7.0	13.0
21.0	7.5	13.5
22.0	7.5	14.5
23.0	8.0	15.0

3.4 Deglycerolize the Thawed RBC Aliquot (see Fig. 15, Three-Step Deglycerolization Method)

1. Using a sterile transfer pipette, transfer the volume of thawed RBCs from the cryovial to a 14 mL round-bottom tube.
2. Perform the first deglycerolization step:
   (a) Using a dropwise technique, add 0.265 mL of 12% sodium chloride solution to the thawed RBCs with continuous mixing.
   (b) Cap the tube and swirl five to ten times to mix. Incubate at RT for 5 min undisturbed.

Table 17
Glycerol solution volume required for each addition step based on RBC input volume for glycerolization Method 4—RBCs from RCCs with supernatant reduction

Original RBC input volume (mL)	Supernatant reduced RBC volume (mL) [Eq. 13, See Table 19]	Total 57% Glycerol solution required (mL) [Eq. 14, Table 19]	Step 1: 57% Glycerol solution volume (mL) [Eq. 15, Table 19]	Step 2: 57% Glycerol solution volume (mL) [Eq. 16, Table 19]	Step 3: 57% Glycerol solution volume (mL) [Eq. 17, Table 19]	Glycerolized RBC volume (mL)[a] [Eq. 18, Table 19]
1.00	0.65	1.50	0.38	1.13	0	2.15
2.00	1.30	3.00	0.75	2.25	0	4.30
3.00	1.95	4.50	1.25	3.25	0	6.45
4.00	2.60	6.00	1.50	4.50	0	8.60
5.00	3.25	7.50	2.00	5.50	0	10.75
6.00	3.90	9.00	2.25	6.75	0	12.90
7.00	4.55	10.50	2.75	7.75	0	15.05
8.00	5.20	12.25	3.00	9.25	0	17.45
9.00	5.85	13.75	3.50	10.00	0.25	19.60
10.0	6.5	15.0	4.0	10.0	1.0	21.5
11.0	7.0	16.5	4.0	10.0	2.5	23.5
12.0	8.0	18.5	4.5	10.0	4.0	26.5
13.0	8.5	20.0	5.0	10.0	5.0	28.5
14.0	9.0	21.0	5.5	10.0	5.5	30.0
15.0	9.5	22.0	5.5	10.0	6.5	31.5
16.0	10.5	24.5	6.0	10.0	8.5	35.0
17.0	11.0	25.5	6.5	10.0	9.0	36.5
18.0	11.5	27.0	7.0	10.0	10.0	38.5
19.0	12.5	29.0	7.5	10.0	11.5	41.5
20.0	13.0	30.5	7.5	10.0	13.0	43.5
21.0	13.5	31.5	8.0	10.0	13.5	45.0
22.0	14.5	34.0	8.5	10.0	15.5	48.5
23.0	15.0	35.0	9.0	10.0	16.0	50.0

[a]Final glycerol concentration = 40% when using these RBC to glycerol volume ratios

(c) Using a dropwise technique, add 1.735 mL of 1.6% sodium chloride solution to RBCs with continuous mixing.

(d) Cap the tube and swirl five to ten times to mix. Incubate at RT for 5 min undisturbed.

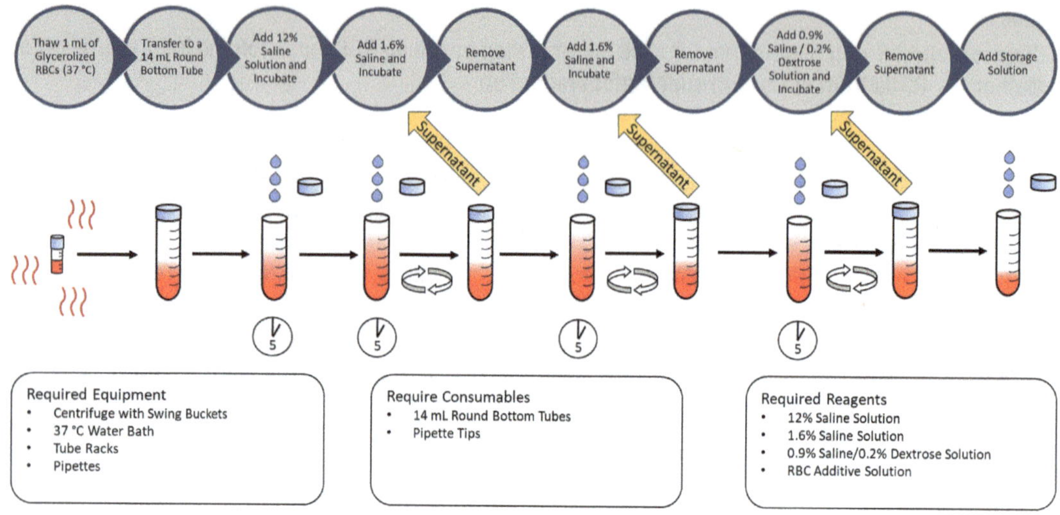

Fig. 15 Process flow diagram for the three-step deglycerolization method.

(e) Centrifuge the sample at 1000×*g* for 2 min at RT with fast acceleration and slow break.

(f) Using a clean pipette with sterile tips, remove 2400 mL of supernatant. Cap the tube and swirl five to ten times to mix. If RBCs are fully incorporated into the solution, proceed to step 3. If not, proceed to step 2g.

(g) Perform additional mixing by rocking the tube back and forth as required until no cell pellet (or minimal) is visible at the bottom of the tube. This is necessary to reduce RBC hemolysis during the addition of the second deglycerolization reagent.

3. Perform the second deglycerolization step:

(a) Using a dropwise technique, add 5 mL of 1.6% sodium chloride solution to the concentrated RBCs with continuous mixing.

(b) Cap the tube and invert with a rocking motion to mix the RBCs to ensure any residual cell pellet observed at the bottom of the tube is completely incorporated into the solution.

(c) Incubate at room temperature for 5 min undisturbed.

(d) Centrifuge the sample at 1000×*g* for 2 min at RT with fast acceleration and slow break.

(e) Using a clean pipette with sterile tips, remove 5 mL of supernatant. Cap the tube and swirl five to ten times to mix.

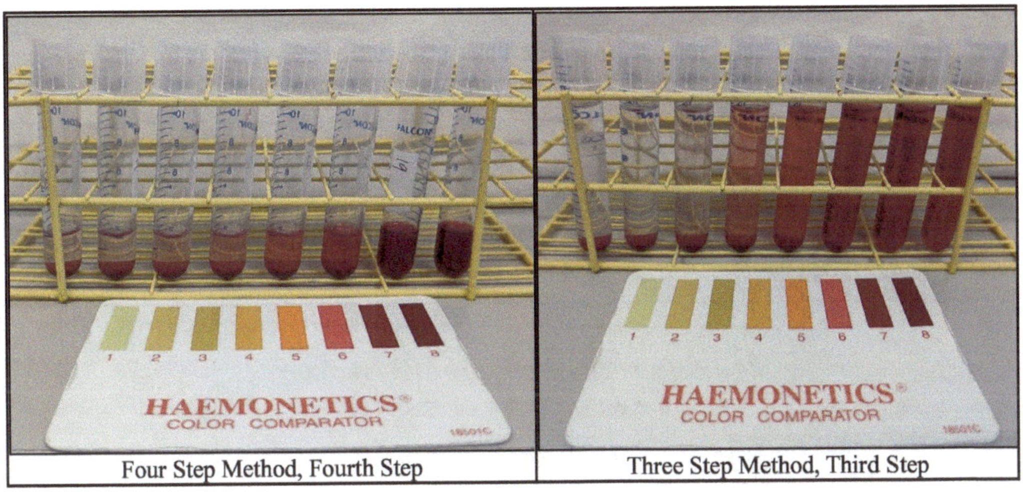

Fig. 16 Visual assessment of free hemoglobin content in deglycerolized RBCs after the 0.9%/0.2% saline dextrose wash process. If visual hemolysis in the wash solution is at level 4 or lower, no additional wash step is recommended

4. Perform the third deglycerolization step:
   (a) Using a dropwise technique, add 10 mL of 0.9% sodium chloride solution with 0.2% dextrose to the concentrated RBCs.
   (b) Cap the tube and invert with a rocking motion to mix the RBCs to ensure any remaining cell pellet observed at the bottom of the tube is completely incorporated into the solution.
   (c) Incubate at room temperature for 5 min undisturbed.
   (d) Centrifuge the sample at 1000×*g* for 2 min at RT with fast acceleration and slow break.
   (e) Using a clean pipette with sterile tips, remove all supernatant without removing any RBCs (approximately 10 ± 0.2 mL). Cap the tube and swirl five to ten times to mix.
   (f) If the supernatant in the previous step contains a large amount of free hemoglobin (visibly cherry red prior to removal, Fig. 16), another wash in 0.9% sodium chloride solution with 0.2% dextrose can be performed by repeating the third deglycerolization step (steps 4a to 4e).

3.5 Suspend the Concentrated RBCs in an RBCStorage Solution and Place Them into Hypothermic Storage

1. Select an appropriate RBCstorage solution for the intended use of the RBCs.
   (a) For short-term storage (up to 24 h), 0.9% sodium chloride with or without 0.2% dextrose may be used.
   (b) For long-term storage (>24 h), an RBC additive solution should be used, such as AS-3. This solution can also

Table 18

Supernatant glycerol volume to remove from centrifuged glycerolized RBCs prior to freezing for glycerolization Method 4—RBCs from RCCs with supernatant reduction

Original RBC input volume (mL)	Supernatant reduced RBC volume (mL) [Eq. 13, Table 19]	Glycerolized RBC volume (mL) [Eq. 18, Table 19]	Volume of supernatant glycerol to remove (mL) [Eq. 19, Table 19]	Glycerolized volume reduced RBC output Volume (mL) [Eq. 20, Table 19]
1.00	0.65	2.15	0.86	1.29
2.00	1.30	4.30	1.72	2.58
3.00	1.95	6.45	2.58	3.87
4.00	2.60	8.60	3.44	5.16
5.00	3.25	10.75	4.30	6.45
6.00	3.90	12.90	5.16	7.74
7.00	4.55	15.05	6.02	9.03
8.00	5.20	17.45	6.98	10.47
9.00	5.85	19.60	7.84	11.76
10.0	6.5	21.5	8.6	12.9
11.0	7.0	23.5	9.4	14.1
12.0	8.0	26.5	10.6	15.9
13.0	8.5	28.5	11.4	17.1
14.0	9.0	30.0	12.0	18.0
15.0	9.5	31.5	12.6	18.9
16.0	10.5	35.0	14.0	21.0
17.0	11.0	36.5	14.6	21.9
18.0	11.5	38.5	15.4	23.1
19.0	12.5	41.5	16.6	24.9
20.0	13.0	43.5	17.4	26.1
21.0	13.5	45.0	18.0	27.0
22.0	14.5	48.5	19.4	29.1
23.0	15.0	50.0	20.0	30.0

be used for short-term storage to improve RBC quality, depending on the intended use of the cells.

2. Using a dropwise technique, add 0.300 mL of the solution to the concentrated RBCs. Cap the tube and swirl five to ten times to mix.

3. Using a sterile transfer pipette, gently pipette up and down to ensure the pellet observed at the bottom of the tube is completely incorporated into the solution. Transfer the sample to an appropriate storage container.
4. Place samples into a refrigerator (typically 1–6°C) for storage. Samples can be stored for up to 24 h or up to the established storage stability/product expiry.

3.6 Supplemental Protocol: Four-Step Deglycerolization Method (see Fig. 17, Four-Step Deglycerolization Method)

1. Using a sterile transfer pipette, transfer the volume of thawed RBCs from the cryovial to a 14 mL round-bottom tube.
2. Perform the first deglycerolization step:
   (a) Using a dropwise technique, add 0.265 mL of 12% sodium chloride solution to the thawed RBCs with continuous mixing.
   (b) Cap the tube and swirl five to ten times to mix. Incubate at RT for 5 min undisturbed.
   (c) Using a dropwise technique, add 1.140 mL of 1.6% sodium chloride solution to RBCs with continuous mixing.
   (d) Cap the tube and swirl five to ten times to mix. Incubate at RT for 5 min undisturbed.
   (e) Centrifuge the sample at 1000×*g* for 2 min at RT with fast acceleration and a slow break.
   (f) Using a clean pipette with sterile tips, remove 1.800 mL of supernatant. Cap the tube and swirl five to ten times to mix. If RBCs are fully incorporated into the solution, proceed to step 3. If not, proceed to step 2g.
   (g) Perform additional mixing by rocking the tube back and forth as required until no cell pellet (or minimal) is visible at the bottom of the tube. This is necessary to reduce RBC hemolysis during the addition of the second deglycerolization reagent.
3. Perform the second deglycerolization step:
   (a) Using a dropwise technique, add 2 mL of 1.6% sodium chloride solution to the concentrated RBCs with continuous mixing.
   (b) Cap the tube and invert with a rocking motion to mix the RBCs to ensure any residual cell pellet observed at the bottom of the tube is completely incorporated into the solution.
   (c) Incubate at room temperature for 5 min undisturbed.
   (d) Centrifuge the sample at 1000×*g* for 2 min at RT with fast acceleration and a slow break.

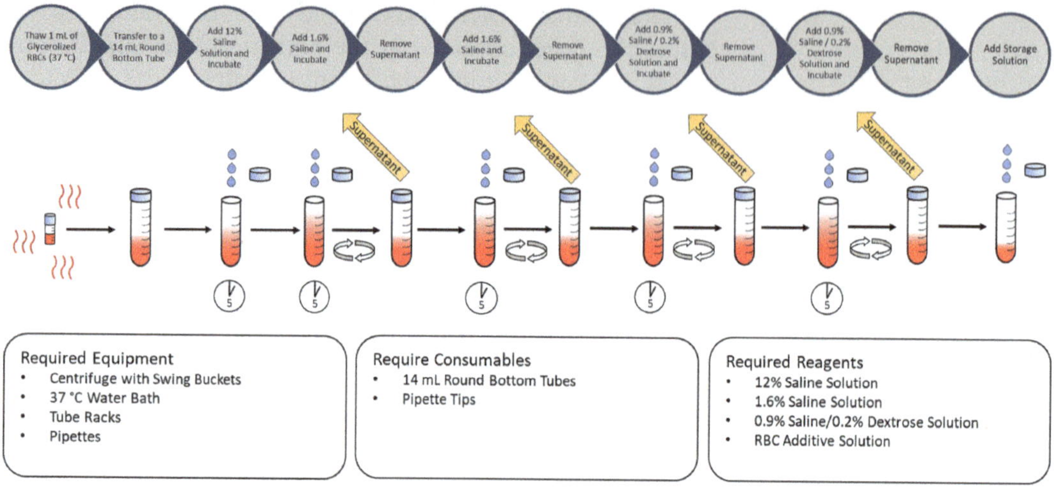

Fig. 17 Process flow diagram for the four-step deglycerolization method

(e) Using a clean pipette with sterile tips, remove 2 mL of supernatant. Cap the tube and swirl the tube five to ten times to mix.

4. Perform the third deglycerolization step:

   (a) Using a dropwise technique, add 2 mL of 0.9% sodium chloride solution with 0.2% dextrose to the concentrated RBCs.

   (b) Cap the tube and invert with a rocking motion to mix the RBCs to ensure any remaining cell pellet observed at the bottom of the tube is completely incorporated into the solution.

   (c) Incubate at RT for 5 min undisturbed.

   (d) Centrifuge the sample at 1000×*g* for 2 min at RT with fast acceleration and a slow break.

   (e) Using a clean pipette with sterile tips, remove 1.9 mL of supernatant. Cap the tube and swirl the tube five to ten times to mix.

5. Perform the fourth deglycerolization step:

   (a) Using a dropwise technique, add 2 mL of 0.9% sodium chloride solution with 0.2% dextrose to the concentrated RBCs.

   (b) Cap the tube and invert with a rocking motion to mix the RBCs to ensure any remaining cell pellet observed at the bottom of the tube is completely incorporated into the solution.

   (c) Incubate at RT for 5 min undisturbed.

Table 19
Equations

Number	Purpose	Equation	Notes
1	Total volume of glycerol solution to be added to the input RBCs.	57% Glycerol solution volume (mL) = RBC input volume (mL) × 2.333	Round result to nearest 0.250 mL increment.[a]
2	Volume of glycerol solution for step 1 in the addition process.	Step 1 Volume (mL) = 57% glycerol volume (mL) x 0.250	Round result to nearest 0.050 mL increment.[a]
3	Volume of glycerol solution for step 2 in the addition process.	Step 2 Volume (mL) = 57% glycerol volume (mL) – Step 1 volume (mL)	
4	Volume of glycerolized RBCs ready to be aliquoted, frozen, and stored.	Glycerolized RBCvolume (mL) = RBC input volume (mL) + Step 1 volume (mL) + Step 2 volume (mL)	
5	Supernatant removal volume to concentrate RBCs prior to freezing.	Supernatant removal volume (mL) = original RBC input volume (mL) × 1.5	
6	Volume of glycerolized RBCs ready to be aliquoted, frozen, and stored.	Supernatant reduced glycerolized RBCvolume (mL) = glycerolized RBC volume (mL) – supernatant removal volume (mL)	
7	Total volume of glycerol solution to be added to the input RBCs.	57% Glycerol solution volume (mL) = concentrated RBC volume (mL) × 2.333	Round result to nearest 0.5 mL increment.[a]
8	Volume of glycerol solution for step 1 in the addition process.	Step 1 Volume (mL) = 57% glycerol solution volume (mL) × 0.250	Round result to nearest 0.5 mL increment.[a]
9	Volume of glycerol for step 2 in the addition process.	Step 2 Volume (mL) = 10 mL	
10	Volume of glycerol solution for the step 3 addition process.	Step 3 Volume (mL) = 57% glycerol solution volume (mL) – Step 1 volume (mL) – Step 2 volume (mL)	
11	Volume of glycerolized RBCs ready to be aliquoted, frozen, and stored.	Glycerolized RBCvolume (mL) = RBC input volume (mL) + Step 1 volume (mL) + Step 2 volume (mL) + Step 3 volume (mL)	

Number	Purpose	Equation	Notes
12	Volume of supernatant to remove to concentrate RBCs prior to glycerolization.	Supernatant removal volume (mL) = RBC input volume (mL) × 0.35	For RBC input volumes of <10 mL or ≥10 mL, round to the nearest 0.25 mL or 0.5 mL increment respectively.[a]
13	Supernatant reduced the RBC volume to be glycerolized.	Supernatant reduced RBCVolume (mL) = RBC input volume (mL) – supernatant removal volume (mL)	
14	The total volume of glycerol solution to be added to the supernatant reduced the RBCs.	Total 57% glycerol solution volume (mL) = supernatant reduced RBC volume (mL) × 2.333	For supernatant reduced RBC volume of <10 mL or ≥10 mL, round to the nearest 0.25 mL or 0.5 mL increment respectively.[a]
15	Volume of glycerol solution for step 1 in the addition process.	Step 1 Volume (mL) = 57% glycerol solution volume (mL) × 0.25	For supernatant reduced RBC volume of <2 mL round to 2 decimal places; 2 mL to 10 mL round to the nearest 0.25 mL increment; ≥10 mL round to the nearest 0.5 mL increment.[a]
16	Volume of glycerol solution for step 2 in the addition process.	Step 2 Volume (mL) = 57% glycerol solution volume (mL) – Step 1 volume (mL)	If step 2 volume is >10 mL, step 2 = 10 mL, and calculation for step 3 volume is required.
17	Volume of glycerol solution for step 3 in the addition process (when required).	Step 3 Volume (mL) = total 57% glycerol solution volume (mL) – Step 1 volume (mL) – Step 2 volume (mL)	
18	Volume of glycerolized RBCs is read to be reduced.	Glycerolized RBC volume (ml) = supernatant reduced volume (mL) + Step 1 volume (mL) + Step 2 volume (mL) + Step 3 volume (mL)	
19	Supernatant removal volume to concentrate RBCs prior to freezing.	Supernatant glycerol volume (mL) = glycerolized RBC volume (mL) × 0.40	
20	Volume of glycerolized RBCs ready to be aliquoted, frozen, and stored.	Glycerolized RBCvolume (mL) = glycerolized RBC volume (mL) – supernatant glycerol volume (mL)	

[a]Round to simplify pipetting volumes

(d) Centrifuge the sample at 1000×*g* for 2 min at RT with fast acceleration and a slow break.

(e) Using a clean pipette with sterile tips, remove all supernatant without removing any RBCs (approximately 2 ± 0.2 mL). Cap the tube and swirl five to ten times to mix.

(f) If the supernatant in the previous step contains a large amount of free hemoglobin (visibly cherry red prior to removal, Fig. 16), another wash in 0.9% sodium chloride solution with 0.2% dextrose can be performed by repeating the fourth deglycerolization step (steps 5a–5e).

6. Proceed to step 3.5 of "Methods."

4 Notes

1. The preparation method will differ depending on the source of RBCs to cryopreserve. Section 3.1.1 is used for RBCs obtained from EDTA whole blood collections, where Sects. 3.1.2 and 3.1.3 are used for RBCs obtained from red cell concentrates.
2. All centrifugation steps should be performed using a swing bucket rotor to allow for easy visualization of the cell interface. This will facilitate supernatant removal with minimal impact on the cell pellet.
3. When a slow break is indicated for centrifugation steps, use the lowest setting (or a validated setting) above zero to avoid disturbing the cell interface during processing.
4. When a dropwise technique is indicated, a steady flow of individual drops should be seen during fluid addition.

Acknowledgments

The authors would like to acknowledge the contributions of Jayme Kurach (Canadian Blood Services), April Xu (Canadian Blood Services), Kiarra Durand (University of Alberta), and Celina Phan (University of Alberta) for assistance in formatting and with collecting and/or verifying characterization data on this method. Our research program received financial support from Canadian Blood Services, funded by the Federal (Health Canada), Provincial, and Territorial Ministries of Health. The views expressed herein do not represent the views of the federal government.

References

1. Rous P, Turner JR (1915) The preservation of living red blood cells in vitro: II the transfusion of kept cells. J Exp Med 23:239–248. https://doi.org/10.1084/jem.23.2.239
2. Rous P, Turner JR (1915) The preservation of living red blood cells in vitro: I. Methods of preservation. J Exp Med 23:219–237. https://doi.org/10.1084/jem.23.2.219
3. Mollison PL (2000) The introduction of citrate as an anticoagulant for transfusion and of glucose as a red cell preservative. Br J Haematol 108:13–18. https://doi.org/10.1046/j.1365-2141.2000.01827.x
4. Lagerberg JW, Korsten H, van der Meer PF, De Korte D (2017) Prevention of red cell storage lesion: a comparison of five different additive solutions. Blood Transfus 15:456–462. https://doi.org/10.2450/2017.0371-16
5. Chen D, Serrano K, Devine DV (2016) Introducing the red cell storage lesion. ISBT Sci Ser 11(S1):26–33. https://doi.org/10.1111/voxs.12222
6. Smith AU (1950) Prevention of haemolysis during freezing and thawing of red blood cells. Lancet 2:910–911. https://doi.org/10.1016/s0140-6736(50)91861-7
7. Turner TR, Acker JP (2020) A Canadian perspective on the use and preparation of cryopreserved red cell concentrates. Transfus Apher Sci 59:102853. https://doi.org/10.1016/j.transci.2020.102853
8. Goldman M, Croix LS (2016) Rare donor program: Canadian blood services. Immunohematology 32:15–16
9. Lagerberg JW (2021) Frozen blood reserves. In: Wolkers WF, Oldenhof H (eds) Cryopreservation and freeze-drying protocols. Methods in molecular biology 2180. Humana, New York, NY. https://doi.org/10.1007/978-1-0716-0783-1_26
10. Meryman HT (2007) Cryopreservation of living cells: principles and practice. Transfusion 47:935–945. https://doi.org/10.1111/j.1537-2995.2007.01212.x
11. Rowe AW (1994) Cryopreservation of red blood cells. Vox Sang 67(Suppl 3):201–206. https://doi.org/10.1111/j.1423-0410.1994.tb04576.x
12. Kurach J, Brandon-Coatham M, Olafson C, Turner TR, Phan C, Yazdanbakhsh M, Osmani R, Ehsani-Moghaddam B, Clarke G, Acker JP (2024) Exposure of cryopreserved red cell concentrates to real-world transient warming events has a negligible impact on quality. Transfusion 64:2353–2363. https://doi.org/10.1111/trf.18054
13. Meryman HT, Hornblower M (1977) A simplified procedure for deglycerolizing red blood cells frozen in a high glycerol concentration. Transfusion 17:438–442. https://doi.org/10.1046/j.1537-2995.1977.17578014580.x
14. Meryman HT, Hornblower M (1972) A method for freezing and washing red blood cells using a high glycerol concentration. Transfusion 12:145–156. https://doi.org/10.1111/j.1537-2995.1972.tb00001.x
15. Meryman HT, Hornblower M (1976) Freezing and deglycerolizing sickle-trait red blood cells. Transfusion 16:627–632. https://doi.org/10.1046/j.1537-2995.1976.16677060246.x
16. Howell A, Turner TR, Hansen A, Lautner LJ, Yi Q, Acker JP (2022) Closed system processing variables affect post-thaw quality characteristics of cryopreserved red cell concentrates. Transfusion 62:2577–2586. https://doi.org/10.1111/trf.17138
17. Turner TR, Clarke G, Denomme GA, Skeate R, Acker JP (2021) A. Effect of cryopreservation on a rare McLeod donor red blood cell concentrate. Immunohematology 37:78–83. https://doi.org/10.21307/immunohematology-2021-012
18. Lusianti RE, Benson JD, Acker JP, Higgins AZ (2013) Rapid removal of glycerol from frozen-thawed red blood cells. Biotechnol Prog 29:609–620. https://doi.org/10.1002/btpr.1710
19. Rogers SC, Dosier LB, McMahon TJ, Zhu H, Timm D, Zhang H, Herbert J, Atallah J, Palmer GM, Cook A, Ernst M, Prakash J, Terng M, Towfighi P, Doctor R, Said A, Joens MS, Fitzpatrick JAJ, Hanna G, Lin X, Reisz JA, Nemkov T, D'Alessandro A, Doctor A (2018) Red blood cell phenotype fidelity following glycerol cryopreservation optimized for research purposes. PLoS ONE 13:e0209201. https://doi.org/10.1371/journal.pone.0209201
20. Eades B (2020) Freezing and recovering rare red blood cells using glycerol. Immunohematology 36:85–88
21. Canadian Standards Association (CSA Group, 2020)
22. AABB Standards for blood banks and transfusion services 33rd edition. 30th edn (2022)
23. Council of Europe, Guide to the preparation, use and quality assurance of blood components, 21st edn. (Council of Europe Publishing, 2023)
24. Acker JP, Hansen AL, Kurach JDR, Turner TR, Croteau I, Jenkins C (2014) A quality monitoring program for red blood cell components: in vitro quality indicators before and after implementation of semiautomated processing. Transfusion 54:2534–2543. https://doi.org/10.1111/trf.12679

Chapter 26

Clinical-Grade Cryopreservation of Human-Induced Pluripotent Stem Cells

Binnur Eroglu, Stephanie Revoredo and Ali Eroglu

Abstract

Human-induced pluripotent stem cells (hiPSCs) can be derived from a variety of biopsy samples and have an unlimited capacity for self-renewal and differentiation into almost any cell type in the body. Therefore, hiPSCs offer unprecedented opportunities for patient-specific cell therapies, modeling of human diseases, biomarker discovery, and drug testing. To realize the full potential of clinical applications of hiPSCs, xeno-free and ideally chemically defined methods are required for their generation, expansion, and cryopreservation. In this chapter, we present a chemically defined and xeno-free slow freezing method for hiPSCs along with a chemically undefined protocol. Both approaches yield satisfactory post-thaw viability and cell growth.

Keywords Chemically defined, Clinical-grade, Cryopreservation, Human-induced pluripotent stem cells (iPSCs), Xeno-free

1 Introduction

The idea of having indefinitely expandable pluripotent cells that could be used for the treatment of various diseases, injuries, along with organ and tissue regeneration has propelled the progress of stem cell research over the decades [1, 2]. With the first discovery of embryonic stem cells (ESCs) in mice in 1981 [3, 4] and then the isolation of ESCs from human blastocysts in 1998 [1], there have been several advancements in technology that have allowed scientists to further understand the complexities and problems associated with the isolation and future use of stem cells. Although human ESCs hold great potential for many clinical and pharmaceutical applications, there are ethical and political issues surrounding the use of these cells due to their derivation from preimplantation embryos [5]. This obstacle led to a major breakthrough in the field of regenerative medicine with the discovery of reprogramming of mouse embryonic and adult fibroblasts into a pluripotent state by forced expression of

Willem F. Wolkers and Harriëtte Oldenhof (eds.), *Cryopreservation and Freeze-Drying Protocols*, Methods in Molecular Biology, vol. 3060, https://doi.org/10.1007/978-1-0716-5416-3_26,

four transcription factors (i.e., Oct4, Sox2, Klf4, and c-Myc) in 2006 [6]. These reprogrammed cells were called induced pluripotent stem (iPS) cells and were similar to mouse ESCs. Subsequently, different cell types, including human somatic cells, were reprogrammed to pluripotency by forced expression of the same or slightly different transcription factors [7, 8]. Unlike ESCs, iPSCs can be generated from small biopsies taken from patients, making them patient-specific. Therefore, properly generated and handled iPSCs are expected not to induce any immune response when transplanted for tissue repair and other treatments. Moreover, by reprogramming patients' own cells, it is possible to generate disease-specific iPSC models that can be used for drug testing and the development of personalized treatment plans for patients [9]. Considering their self-renewal and differentiation capacity into any cell type of all three germ layers, hiPSCs hold great potential for regenerative medicine, tissue engineering, and the treatment of degenerative diseases. While the benefits of using hiPSCs for clinical applications are evident, there is a lack of viable clinical-grade cryopreservation methods needed to preserve these cells, and thus, to realize their potential clinical applications. The main challenges associated with banking of human pluripotent stem cells (hPSCs) include their high sensitivity to cryopreservation-induced stresses and dissociation to single cells [10]. Although high serum concentrations and adding a selective Rho-associated kinase (ROCK) inhibitor such as Y-27632 to freeze–thaw and subsequent culture media somewhat improved the outcome of hPSC cryopreservation, the overall post-thaw recovery still remains unsatisfactory [11–14]. Fundamental studies on rational design of cryopreservation methods for hiPSCs are largely missing, except for a membrane permeability study involving dimethyl sulfoxide (DMSO), glycerol, and 1,2-propanediol [15]. Consequently, several aspects of hiPSC cryopreservation, such as addition and removal of cryoprotectants, remain non-optimized and highly variable. To add to this, clinical-grade cryopreservation protocols prohibit the use of serums and animal products due to the biosafety and reproducibility issues; therefore, development of a xeno-free and chemically defined method is essential for cryobanking stem cells [16]. Since serum contains numerous factors that are beneficial to cell attachment and growth, overcoming oxidative stresses, buffering pH, binding and neutralization of toxic molecules, and reduction of shear stresses, it might be critical to implement such beneficial properties of serum when developing chemically defined, xeno-free culture and cryopreservation methods. Indeed, mimicking some of the beneficial effects of serum by including antioxidants, amino acids, vitamins, and polymers in the freezing medium while combining low concentrations of two penetrating cryoprotectants to reduce their toxicity [17] resulted in the development of a successful xeno-free and chemically defined cryopreservation method for adipose-derived stem cells [16]. Using the same concept, we formulated a defined medium for hiPSCs with some modifications and tested it against the conventional undefined freezing medium (i.e., 10% DMSO, 20% KSR, and 10 μM ROCK inhibitor Y-27632 in DMEM/F-12) and unfrozen controls. Our experiments showed that the chemically defined, xeno-free freezing medium formulated for hiPSCs yields satisfactory results, although not completely comparable to the conventional undefined mediums and unfrozen controls (unpublished data).

In this chapter, we present a xeno-free and chemically defined protocol for clinical-grade cryopreservation of hiPSCs along with an undefined freezing protocol that can be used for research applications of hiPSCs.

2 Materials

2.1 Equipment

1. Class II biological safety cabinet.
2. −80°C mechanical freezer.
3. NALGENE® Mr. Frosty freezing container or similar 1°C/min freezing containers.
4. Pipette aid.
5. Water bath.
6. Centrifuge.
7. Tissue culture incubator.
8. Dissecting stereomicroscope.
9. Inverted phase-contrast microscope.
10. Hemocytometer.
11. Autoclave.
12. Liquid nitrogen tank.
13. Cell storage tank.
14. Liquid nitrogen dewar.
15. Eye protection.
16. Cryogloves (face mask or goggles).

2.2 Disposables

1. Conical centrifuge tubes, 15 mL, sterile.
2. Conical centrifuge tubes, 50 mL, sterile.
3. 6- and 12-well cell culture plates.
4. Cryovials.
5. 5-, 10-, 25-, and 50-mL plastic disposable pipettes, sterile.
6. 20-, 200-, and 1000-μL pipette tips, sterile.
7. 0.2-μm sterile filter units, different membranes.

8. 0.2-μm sterile syringe filters, different membranes.
9. Kim wipes.
10. Parafilm.
11. Liquid nitrogen.

2.3 Reagents, Solutions, and Media

1. Dimethyl sulfoxide (DMSO, Sigma).
2. Ethylene glycol (EG, Fluka).
3. Methylcellulose (Sigma).
4. Ethylene diamine tetraacetic acid tetrasodium salt (EDTA, Sigma).
5. Reduced glutathione (Sigma).
6. L-ascorbic acid 2-phosphate magnesium salt (Wako).
7. Primocin (InvivoGen).
8. ROCK Inhibitor (Y-27632, Res Gen).
9. Versene (Gibco).
10. TrypLE Select (Gibco).
11. Recombinant Human Vitronectin (BioLegend).
12. 100× Glutamax (Gibco).
13. 100× Nonessential Amino Acids (NEAA, Gibco).
14. 50× Essential Amino Acids (EAA, Gibco).
15. RPMI 1640 Vitamins Solution (100×, Sigma).
16. Dulbecco's calcium- and magnesium-free phosphate-buffered saline (DPBS−, Gibco).
17. Dulbecco's phosphate-buffered saline with calcium and magnesium (DPBS+, Gibco).
18. Hanks Balanced Salt Solution containing 1.26 mM calcium and 0.90 mM magnesium without phenol red (HBSS+, Gibco).
19. Hanks Balanced Salt Solution without calcium and magnesium (HBSS−, Gibco).
20. DMEM/F-12, HEPES (Gibco).
21. Essential 8 (E8) Flex Medium + Supplement (Gibco).
22. Knockout Serum Replacement (Gibco).
23. 10 mM ROCK inhibitor (Y-27632) stock solution: Dissolve 2 mg of Y-27632 in 0.62 mL of PBS to obtain a 10-mM stock solution, and then aliquot and store the stock solution at −20°C or −80°C. The stock solution is stable for at least 1 year (*see* **Note 1**).
24. 300 mM (100×) reduced glutathione stock solution: Dissolve 0.47 g reduced glutathione in 5 mL of PBS to prepare a

300 mM stock solution, and then aliquot and store the stock solution at −20 or −80°C.

25. 500 mM (100×) ascorbic acid 2-phosphate stock solution: Dissolve 0.76 g L ascorbic acid 2-phosphate (magnesium salt) in 5 mL of PBS to prepare a 500 mM stock solution, and then aliquot and store the stock solution at −20°C or −80°C.
26. 0.5 M EDTA stock solution (1000×): Dissolve 1.96 g EDTA tetrasodium salt in ultrapure water (final volume: 10 mL) by adjusting pH to 8 and vigorously stirring. Filter the resulting solution using a 0.2-μm membrane filter and store at 4°C for up to 6 months.
27. 0.5 mM EDTA dissociation buffer: Add 100 μL of 0.5 M EDTA (pH 8.0) and 0.18 g NaCl (to increase osmolarity to 340 mOsm) to 99.8 mL PBS (final volume: 100 mL), sterilize using a 0.2-μm membrane filter and store at 4°C for up to 6 months. Alternatively, a commercially available Versene solution that has a similar composition but slightly lower osmolality can be used instead of the 0.5 mM EDTA dissociation buffer.
28. Complete E8 flex medium: Slowly thaw the E8 flex supplement (50×, 10 mL) at room temperature or at 4°C overnight, mix the content, and then add to 489 mL E8 flex basal medium. Also, add 1 mL primocin solution to obtain a working concentration of 100 μL/mL.
29. Undefined cryopreservation medium at 2× concentration: Add sterile 20% (v/v) DMSO, 40% (v/v) Knockout Serum Replacement (KSR), 20 μM ROCK inhibitor Y-27632 to HEPES-buffered DMEM/F-12 and store at 4°C after mixing.
30. Xeno-free defined cryopreservation medium at 2× concentration: Add 10% DMSO, 10% EG, 1% methylcellulose, 6 mM reduced glutathione, 10 mM ascorbic acid 2-phosphate, 2% Glutamax, 2% Nonessential Amino Acids, 4% Essential Amino Acids, 2% RPMI 1640 Vitamins Solution, and 20 μM ROCK inhibitor Y-27632 to HBSS− (*see* **Note 2**).
31. Undefined thawing medium at 1× concentration: Add 0.1 M sucrose, 20% KSR, 10 μM ROCK inhibitor Y-27632 to HEPES-buffered DMEM/F-12 medium, filter under the hood using a 0.2-μm membrane filter, and store at 4°C (*see* **Note 3**).
32. Xeno-free chemically defined thawing medium at 1× concentration: Add 0.1 M sucrose, 0.5% methylcellulose, 3 mM reduced glutathione, 5 mM ascorbic acid 2-phosphate, 1% Glutamax, 1% Nonessential Amino Acids, 2% Essential Amino

Acids, 1% RPMI 1640 Vitamins Solution, and 10 μM ROCK inhibitor Y-27632 to HBSS− (*see* **Note 4**).

33. Xeno-free chemically defined thawing medium without sucrose and methylcellulose: Add 3 mM reduced glutathione, 5 mM ascorbic acid 2-phosphate, 1% Glutamax, 1% Nonessential Amino Acids, 2% Essential Amino Acids, 1% RPMI 1640 Vitamins Solution, and 10 μM ROCK inhibitor Y-27632 to HBSS−.

3 Methods

3.1 Coating Culture Plates With Recombinant Vitronectin

This step is important for promoting cell attachment and growing undifferentiated hiPSC colonies. Although complex extracellular matrix extracts such as Matrigel and Geltrex are widely used to successfully grow hiPSCs, they are chemically undefined and animal-derived coating materials. In contrast, human recombinant proteins, including vitronectin and laminin 521, provide a chemically defined culture matrix and support undifferentiated growth of hiPSCs, while the former is more affordable.

1. Dilute recombinant human vitronectin to a final concentration of 10 μg/mL by adding 1× PBS+ (*see* **Note 5**).
2. Add sufficient amounts from the diluted vitronectin solutions to each culture well. Typically, 1.5 and 0.5 mL are added to each well of 6-well and 12-well plates, respectively.
3. Gently shake the plates to spread the vitronectin solution across the entire surface of each well.
4. Incubate the plates at ambient temperature for at least 1 h or at 4°C overnight.
5. Aspirate the coating solution and plate iPSCs without allowing any well to dry out.

3.2 Culturing hiPSCs

Different chemically defined media have been developed for culturing hPSCs and are commercially available [18–20]. To date, we have used E8 medium (also called TeSR-E8), and, thus, are suggesting its use; however, other chemically defined media might be as good as E8 medium.

1. Before plating hiPSCs, aspirate the coating solution and quickly add E8 medium containing 10 μM Y-27632 to each well to prevent drying out of the coated surfaces.
2. Plate detached and partially dissociated hiPSC colonies on the coated wells upon suspension in E8 medium supplemented with 10 μM ROCK inhibitor Y-27632 (*see* **Note 6**).

3. Replace 2/3 of the medium volume with fresh E8 medium every day to maintain effective concentrations of the growth factors, and thus, to grow undifferentiated colonies (*see* **Note** 7).

3.3 Passaging of hiPSCs

hiPSC colonies should be passaged before the colonies start merging—usually when they are around 70%–80% confluent. Overgrown colonies start to spontaneously differentiate. hiPSCs can be passaged both enzymatically (e.g., using Accutase and TrypLE) and nonenzymatically (e.g., by manual cutting or EDTA treatment), although they are sensitive to dissociation into single cells. For better cell survival, partial dissociation to small cell clumps is a preferred choice, which can be achieved by a simple EDTA treatment. However, subtle handling differences in the cell detachment process can significantly affect the subsequent viability of hiPSCs by inducing apoptosis. It is critical to gently detach and harvest hiPSCs in the detachment solution using wide-bore pipette tips or 2-mL pipettes without adding any culture medium [21].

1. Before dissociation, replace E8 Flex medium with fresh medium containing 10 μM ROCK inhibitor (Y-27632) and incubate at 37°C for 2 h.
2. At the end of the incubation period, aspirate the medium and gently rinse the wells with 1 or 2 mL per well of prewarmed EDTA dissociation buffer to completely remove calcium and magnesium. The EDTA dissociation buffer should slowly be added to the wall of each well to avoid early dislodging hiPSC colonies.
3. Upon removal of the EDTA dissociation buffer, add fresh 1 or 2 mL per well of prewarmed EDTA dissociation buffer for the second time and hold the plate(s) within the biosafety hood for 10 min (longer incubation periods result in tiny clumps of a few cells).
4. At the end of the incubation period, gently dislodge hiPSC colonies by aspirating and releasing the EDTA dissociation buffer in each well using a 2-mL pipette; however, excess pipetting should be avoided so as not to end up with a single cell suspension.
5. Aspirate the detached hiPSC clumps, transfer to a centrifuge tube, pellet at a low speed (200–300×*g*), resuspend in E8 medium containing 10 μM ROCK inhibitor, and then distribute to the vitronectin-coated wells. When passaging, split ratios range from 1:3 to 1:6 depending on the density and growth characteristics of the cells. Usually, hiPSCs reach 70%–80% confluency within 3–5 days after passaging.

3.4 Freezing of hiPSCs

It is important that hiPSC colonies are healthy and in good shape at the time of cryopreservation. Furthermore, hiPSC colonies should be detached and harvested in the EDTA dissociation buffer to avoid considerable cell damage and apoptosis [21]. Independent of the dissociation approach, the cryopreservation-associated processes lead to disruption of cell-to-cell contacts, resulting in partially dissociated hiPSCs. Therefore, it is advisable to include 10 μM ROCK inhibitor in freezing and thawing media, as well as in culture medium before freezing (for ~1 h) and after thawing (for up to 48 h). The freezing and thawing steps are summarized in Fig. 1.

1. Prepare cryopreservation media in advance.
2. Determine the number of cryovials needed for freezing and label them with cell line name, passage number, type of cryopreservation medium, and date.
3. Place both defined and undefined cryopreservation media and related solutions on ice. Adding and removing cryoprotectants at a low temperature seem to be helpful in terms of reducing cryoprotectant toxicity to hiPSCs.
4. Harvest hiPSC colonies as described earlier (see Sect. 3.3) and determine cell density and viability using trypan blue and a hemocytometer or by commercially available kits.
5. After pelleting, resuspend the partially dissociated cell clumps in either cold HBSS– containing 10 μM ROCK inhibitor Y-27632 (for xeno-free chemically defined cryopreservation)

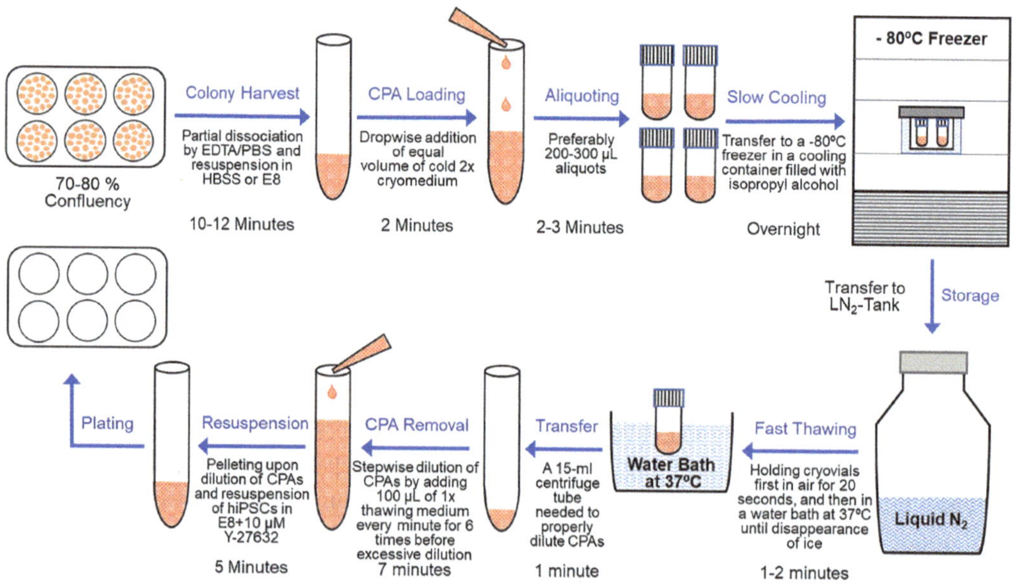

Fig. 1 Overview of freezing and thawing steps. Shown are critical steps with relevant processing times. See text for details

or in cold E8 medium supplemented with 10 μM ROCK inhibitor Y-27632 (for undefined cryopreservation).

6. Add dropwise an equal volume of either ice-cold 2× chemically defined cryomedium (for xeno-free chemically defined cryopreservation) or ice-cold 2× undefined cryomedium (for undefined cryopreservation) while mixing by tapping the tube.
7. Quickly aliquot the resulting cell suspension in either 1× chemically defined cryomedium or 1× undefined cryomedium into cryovials as 200- to 300-μL aliquots. Small aliquots can be thawed faster and more uniformly.
8. Transfer the cryovials to a precooled (4 °C) freezing container (e.g., Mr. Frosty) and immediately place the container in a −80°C freezer. hiPSCs seem to quickly equilibrate with cryoprotectants [15]. Thus, addition of cryopreservation media, aliquoting, and subsequent transferring to a −80°C freezer can be completed within 5 min to minimize cryoprotectant toxicity.
9. After slow cooling to −80°C overnight, the cryovials containing hiPSCs should be transferred to a liquid nitrogen tank for long-term storage.

3.5 Thawing of hiPSCs

1. Prepare vitronectin-coated plate(s) ready to seed thawed hiPSCs (see Sect. 3.1).
2. Remove a cryovial containing hiPSCs from the liquid nitrogen tank and hold in air for 20 s.
3. Next, partially immerse it in a 37°C water bath and gently swirl for fast thawing. Remove the cryovial from the water bath just before the ice has completely melted.
4. Before placing the cryovial under the biosafety hood, spray and wipe down the outside of it with 70% ethanol.
5. Under the biosafety hood, gently transfer the cryovial content to a 15-mL tube for dilution of cryoprotectants.
6. In the case of xeno-free chemically defined cryopreservation, add 100 μL of 1× chemically defined thawing medium every minute for a total of six times while gently mixing the tube content by tapping each time. For the final dilution step, add an excess amount (10 mL) of defined thawing medium without sucrose and methylcellulose.
7. In the case of undefined cryopreservation, perform the dilution steps described above using 1× undefined thawing medium. For the final dilution step, add 10 mL of HEPES-buffered DMEM/F-12 medium containing 20% KSR and 10 μM ROCK inhibitor.
8. Wait for 1 min, and then pellet hiPSCs at 300×*g* for 5 min.

9. Aspirate the supernatant without disturbing the cell pellet and resuspend the cells in complete E8 flex medium containing 10 μM ROCK inhibitor.
10. Take out 40 μL of cell suspension to determine cell viability, and then plate hiPSCs at a desired density (e.g., $5 \times 10^4 =$ cells/cm^2).
11. Leave the plate undisturbed in a humidified CO_2 incubator at 37°C for at least 1 day to ensure the attachment of viable cells. Subsequently, two-third of the culture medium can be changed daily.

4 Notes

1. Y-27632 (dihydrochloride) is less soluble in water (up to 30 mM) than in DMSO (up to 100 mM). The solvent can be added directly to the vial, which should then be vortexed vigorously for complete solubilization. When using DMSO as a solvent, the final DMSO concentration should not exceed 0.1% to avoid its toxicity. The stock solution should not undergo multiple freeze–thaw cycles and should be protected from light.
2. It is extremely difficult to sterilize the defined cryopreservation medium by filtering due to the presence of methylcellulose. Therefore, we recommend first dissolving methylcellulose in HBSS– with or without DMSO and EG, and then autoclaving the resulting solution using a liquid program with slow pressure release to prevent boil over. The aqueous solutions of DMSO and EG are heat-stable and can be autoclaved using a proper autoclave with an appropriate liquid program. Immediately after autoclaving, the methylcellulose solution may appear opaque but becomes clear and transparent upon cooling in a refrigerator. Once sterilized and cooled, the remaining sterile ingredients of the defined cryopreservation medium can be added under the biosafety hood without the need to filter.
3. KSR contains undefined lipids that can unintentionally be removed during filtration. Therefore, sucrose can first be dissolved in DMEM/F-12 and filtered before adding the remaining sterile components under the biosafety hood.
4. The sterilization of the defined thawing medium is also challenging. We recommend first dissolving methylcellulose and sucrose in HBSS–, and then autoclaving the resulting solution before aseptically adding the remaining sterile components under the biosafety hood.
5. Recombinant human vitronectin is available as a concentrated solution, usually at 0.5 mg/mL or lyophilized powder. When

received as lyophilized powder, an appropriate volume of PBS+ can be added directly to the vial to dissolve it. Next, the dissolved vitronectin can be transferred to a larger sterile tube to make a working solution at 10 μg/mL by adding more PBS+. Then, the working solution should be aliquoted and stored at −20 or −80°C to avoid repeated freeze/thaw cycles.

6. When dissociated into single cells, hiPSCs require the presence of a ROCK inhibitor to survive passaging. It is not necessary to supplement E8 medium with a ROCK inhibitor if hiPSC colonies are partially dissociated by EDTA treatment. Nevertheless, we prefer to supplement E8 medium with 10 μM ROCK inhibitor Y-27632 during the first 48 h of passaging, regardless of the extent to which hiPSC colonies are dissociated.

7. Particularly, native bFGF is highly unstable and quickly degraded under cell culture conditions, requiring changing the culture medium every day. Recently, a heat-stable recombinant human bFGF has been engineered and added to E8 medium (called E8 flex) to skip changing the culture medium on weekends. In our experience, skipping 1 day was okay, but hiPSCs seemed to be stressed when the E8 flex medium was not changed during the entire weekend. Therefore, we cannot recommend skipping 2 days.

References

1. Thomson JA, Itskovitz-Eldor J, Shapiro SS, Waknitz MA, Swiergiel JJ, Marshall VS, Jones JM (1998) Embryonic stem cell lines derived from human blastocysts. Science 282:1145–1147. https://doi.org/10.1126/science.282.5391.1145
2. Takahashi K, Yamanaka S (2016) A decade of transcription factor-mediated reprogramming to pluripotency. Nat Rev Mol Cell Biol 17:183–193. https://doi.org/10.1038/nrm.2016.8
3. Evans MJ, Kaufman MH (1981) Establishment in culture of pluripotential cells from mouse embryos. Nature 292:154–156. https://doi.org/10.1038/292154a0
4. Martin GR (1981) Isolation of a pluripotent cell line from early mouse embryos cultured in medium conditioned by teratocarcinoma stem cells. Proc Natl Acad Sci USA 78:7634–7638. https://doi.org/10.1073/pnas.78.12.7634
5. Lo B, Parham L (2009) Ethical issues in stem cell research. Endocr Rev 30:204–213. https://doi.org/10.1210/er.2008-0031
6. Takahashi K, Yamanaka S (2006) Induction of pluripotent stem cells from mouse embryonic and adult fibroblast cultures by defined factors. Cell 126:663–676. https://doi.org/10.1016/j.cell.2006.07.024
7. Takahashi K, Tanabe K, Ohnuki M, Narita M, Ichisaka T, Tomoda K, Yamanaka S (2007) Induction of pluripotent stem cells from adult human fibroblasts by defined factors. Cell 131:861–872. https://doi.org/10.1016/j.cell.2007.11.019
8. Yu J, Vodyanik MA, Smuga-Otto K, Antosiewicz-Bourget J, Frane JL, Tian S, Nie J, Jonsdottir GA, Ruotti V, Stewart R, Slukvin II, Thomson JA (2007) Induced pluripotent stem cell lines derived from human somatic cells. Science 318:1917–1920. https://doi.org/10.1126/science.1151526
9. Svendsen CN (2013) Back to the future: how human induced pluripotent stem cells will transform regenerative medicine. Hum Mol Genet 22:R32–R38. https://doi.org/10.1093/hmg/ddt37
10. Watanabe K, Ueno M, Kamiya D, Nishiyama A, Matsumura M, Wataya T, Takahashi JB, Nishikawa S, Nishikawa S, Muguruma K, Sasai Y (2007) A ROCK inhibitor permits survival of dissociated human embryonic stem cells.

Nat Biotechnol 25:681–686. https://doi.org/10.1038/nbt1310
11. Ha SY, Jee BC, Suh CS, Kim HS, Oh SK, Kim SH, Moon SY (2005) Cryopreservation of human embryonic stem cells without the use of a programmable freezer. Hum Reprod 20:1779–1785. https://doi.org/10.1093/humrep/deh854
12. Martin-Ibanez R, Unger C, Stromberg A, Baker D, Canals JM, Hovatta O (2008) Novel cryopreservation method for dissociated human embryonic stem cells in the presence of a ROCK inhibitor. Hum Reprod 23:2744–2754. https://doi.org/10.1093/humrep/den316
13. Li X, Meng G, Krawetz R, Liu S, Rancourt DE (2008) The ROCK inhibitor Y-27632 enhances the survival rate of human embryonic stem cells following cryopreservation. Stem Cells Dev 17:1079–1085. https://doi.org/10.1089/scd.2007.0247
14. Claassen DA, Desler MM, Rizzino A (2009) ROCK inhibition enhances the recovery and growth of cryopreserved human embryonic stem cells and human induced pluripotent stem cells. Mol Reprod Dev 76:722–732. https://doi.org/10.1002/mrd.21021
15. Xu Y, Zhang L, Xu J, Wei Y, Xu X (2014) Membrane permeability of the human pluripotent stem cells to Me(2)SO, glycerol and 1,2-propanediol. Arch Biochem Biophys 550-551:67–76. https://doi.org/10.1016/j.abb.2014.04.010
16. Lopez M, Bollag RJ, Yu JC, Isales CM, Eroglu A (2016) Chemically defined and xeno-free cryopreservation of human adipose-derived stem cells. PLoS ONE 11:e0152161. https://doi.org/10.1371/journal.pone.0152161
17. Szurek EA, Eroglu A (2011) Comparison and avoidance of toxicity of penetrating cryoprotectants. PLoS ONE 6:e27604. https://doi.org/10.1371/journal.pone.0027604
18. Chen G, Gulbranson DR, Hou Z, Bolin JM, Ruotti V, Probasco MD, Smuga-Otto K, Howden SE, Diol NR, Propson NE, Wagner R, Lee GO, Antosiewicz-Bourget J, Teng JM, Thomson JA (2011) Chemically defined conditions for human iPSC derivation and culture. Nat Methods 8:424–429. https://doi.org/10.1038/nmeth.1593
19. Rajala K, Lindroos B, Hussein SM, Lappalainen RS, Pekkanen-Mattila M, Inzunza J, Rozell B, Miettinen S, Narkilahti S, Kerkelä E, Aalto-Setälä K, Otonkoski T, Suuronen R, Hovatta O, Skottman H (2010) A defined and xeno-free culture method enabling the establishment of clinical-grade human embryonic, induced pluripotent and adipose stem cells. PLoS ONE 5:e10246. https://doi.org/10.1371/journal.pone.0010246
20. Yasuda SY, Ikeda T, Shahsavarani H, Yoshida N, Nayer B, Hino M, Vartak-Sharma N, Suemori H, Hasegawa K (2018) Chemically defined and growth-factor-free culture system for the expansion and derivation of human pluripotent stem cells. Nat Biomed Eng 2018 2:173–182. https://doi.org/10.1038/s41551-018-0200-7
21. Takahashi K, Okubo C, Nakamura M, Iwasaki M, Kawahara Y, Tabata T, Miyamoto Y, Woltjen K, Yamanaka S (2022) A stress-reduced passaging technique improves the viability of human pluripotent cells. Cell Rep Methods 2:100155. https://doi.org/10.1016/j.crmeth.2021.100155

Chapter 27

Microencapsulation for Cryopreservation of Mouse Preantral Follicles

Gang Zhao, Yue Cheng, Conghui Tian and Guangzhu Shang

Abstract

Recent advances in microencapsulation for cryopreservation have transformed cell preservation through innovative hydrogel-based strategies. By leveraging the ice-inhibiting properties of three-dimensional hydrogel networks, this approach significantly reduces reliance on toxic cryoprotectants while maintaining high cell viability. The technology has demonstrated remarkable success across diverse cell types, from human mesenchymal stem cells and erythrocytes to murine reproductive tissues. This chapter presents a comprehensive protocol using mouse follicles as a model system, detailing: (1) mechanical follicle isolation, (2) centrifugal microfluidic encapsulation in alginate hydrogels, (3) synthesis of magnetothermal nanoparticles, and (4) optimized vitrification/rewarming processes. The integrated methodology combines the protective benefits of physical encapsulation with advanced nanowarming techniques, establishing a robust platform for low-toxicity, high-efficiency cryopreservation.

Keywords 3D culture, Alginate hydrogel, Centrifugal microfluidics, Cryopreservation, Ice inhibition, Low-CPA vitrification, Magnetothermal rewarming, Microencapsulation, Mouse preantral follicles, Nanoparticle-enhanced cryopreservation

1 Introduction

Conventional cell cryopreservation methods, whether employing slow freezing or vitrification protocols, inherently subject cells to multiple stressors during cryoprotectant (CPA) loading, freezing-thawing cycles, and CPA removal. These processes induce three primary mechanisms of cellular injury: (i) ice crystal formation causing mechanical damage, (ii) osmotic shock during CPA exchange, and (iii) direct CPA toxicity - all of which compromise cell viability and function [1, 2]. Despite continuous optimization of CPA formulations and thermal protocols, fundamental limitations persist due to the direct exposure of cells to these stressors in their native, unprotected state.

Drawing inspiration from natural core-shell architectures, such as the zona pellucida in oocytes, the protective shells of walnuts,

Willem F. Wolkers and Harriëtte Oldenhof (eds.), *Cryopreservation and Freeze-Drying Protocols*, Methods in Molecular Biology, vol. 3060, https://doi.org/10.1007/978-1-0716-5416-3_27,

the resilient coats of ginkgo seeds, or the impermeable testa of lotus seeds, microencapsulation technology represents a paradigm shift in cryopreservation. This approach leverages three-dimensional hydrogel matrices (including alginate, chitosan, and supramolecular gels) to create a protective microenvironment that simultaneously addresses the tripartite challenge of cryopreservation: (i) physical confinement of ice crystals to subcritical sizes, with reported ice crystal diameters of approximately 150–170 μm [3, 4], (ii) dynamic buffering of osmotic gradients during CPA exchange, and (iii) dramatic reduction of CPA concentrations while maintaining equivalent cryoprotection [5–8]. Rigorous validation studies utilizing pancreatic islet β cells [9] (viability increase from 30% to 85%) and neural stem cells [10] (two- to threefold higher neurite outgrowth post-thaw) have conclusively demonstrated the technology's superiority over conventional approaches [11, 12].

The critical challenge in cell encapsulation-based cryopreservation lies in selecting optimal encapsulation methodologies, where microfluidic technology has emerged as a transformative platform. By enabling precise spatial control over cell-hydrogel prepolymer mixtures, microfluidics facilitates the generation of monodisperse, cell-laden microcapsules (10–200 μm diameter) with <5% size variation, which is a crucial feature for standardized cryopreservation protocols. Current state-of-the-art encapsulation techniques include: (i) PDMS-based microfluidic chips for high-throughput production [13], (ii) electrospray systems for scalable encapsulation [14], (iii) capillary microfluidic devices for superior droplet control [15, 16], and (iv) centrifugal microfluidics offering simplified operation [17]. Despite differing configurations, these systems share fundamental operational principles: (i) hydrodynamic focusing to generate uniform droplets, (ii) rapid crosslinking (ionic/photo-initiated) of hydrogel matrices (e.g., alginate, PEGDA), and (iii) formation of biomimetic ECM-like scaffolds that physically shield cells from both CPA toxicity and ice crystal penetration, thereby improving post-thaw viability by one–three fold compared to unprotected cells [18].

A pivotal advantage of encapsulation-based cryopreservation is its capacity to dramatically reduce dependence on high-concentration permeating cryoprotectants (CPAs). This is a critical advancement, particularly for vitrification protocols where conventional approaches typically require toxic CPA levels (≥6 M). This breakthrough enables clinically viable low-CPA vitrification (1–2 M) while maintaining exceptional post-warming survival rates, revolutionizing fertility preservation strategies. The technology has demonstrated remarkable success across reproductive specimens, including human oocytes, ovarian follicles (with a 33% increase in preantral follicle viability), and intact ovarian tissue (characterized by enhanced energy metabolism, reduced apoptosis, and increased cellular proliferation) [19–21].

This chapter details an optimized protocol combining: (i) centrifugal microfluidic encapsulation of mouse preantral follicles in alginate hydrogel, (ii) ultra-rapid cooling (cooling rate: ~20,000°C/min) using minimal CPA concentrations (0.75 M EG + 0.75 M PROH + 1 M trehalose), and (iii) precision electromagnetic rewarming (rewarming rate: ~5,500°C/min). The integrated approach achieves unprecedented follicle preservation outcomes, with 90% post-thaw viability and complete maintenance of follicular architecture, demonstrating the transformative potential of encapsulation-based low-CPA cryopreservation in reproductive medicine.

2 Materials

2.1 Equipment

1. Standard laboratory equipment and supplies.
2. Centrifuges, for encapsulating follicles.
3. Electromagnetic induction heating system.
4. Laser heating system.
5. Fiber optic temperature sensor.
6. Thermal camera.
7. Vibrating sample magnetometer (VSM).
8. X-ray diffraction (XRD) spectrometer.
9. Transmission electron microscopy (TEM) system,
10. Fourier-transform infrared spectrometer,
11. Raman spectrometer.

2.2 Reagents

1. Reagents needed for assessment of encapsulated follicles are listed in Table 1.
2. For Fe_3O_4 nanoparticles; $FeCl_3.6H_2O$, $FeCl_2.4H_2O$, NH_3.
3. For Graphene oxide (GO) nanoparticles; H_2SO_4, $NaNO_3$ (1.5 g), graphite powder, $KMnO_4$, H_2O_2.

2.3 Laboratory Animal

1. SPF-grade female Kunming mice (KM mice, 3-week-old) were used for tissue sampling, and ensured all operations and research strictly complied with the regulations of the Animal Ethics Committee.

2.4 Preparation of Reagents

1. PAF Isolation Medium: L-15 medium, 10% (v/v) fetal bovine serum, 1% (v/v) penicillin-streptomycin.
2. PAF Culture Medium: α-MEM-GlutaMAX medium, 20% (v/v) fetal bovine serum, 1×ITS+100 mIU/mL FSH, 1% (v/v) penicillin-streptomycin.
3. PAF Incubation Medium: PAF culture medium, 0.3% (w/v) Fe_3O_4 nanoparticles, 0.03% (w/v) GO nanoparticles.

Table 1
Reagents

Name of reagent	Catalog No.	Supplier
Phosphate buffer	C10010500BT	Gibco, USA
$CaCl_2$ Powder	1000581922	Sinopharm, China
Sodium alginate	S100127	Aladdin, China
Paraformaldehyde	P6148-500G	Sigma-Aldrich, USA
Paraffin	411663	Sigma-Aldrich, USA
Hematoxylin - Eosin	MHS32	Sigma-Aldrich, USA
TRIzol	B511311	Sangon-biotech, China
DNase	EN0523	Thermo Scientific, USA
iScript cDNA Synthesis Kit	AE301-02	TransGen, China
SYBR Green master mix	Q111-02	Vazyme, China

4. CPA carrier solution: α-MEM-GlutaMAX medium, 20% (v/v) fetal bovine serum, 1% (v/v) penicillin-streptomycin.
5. CPA #1: 0.50 M EG, 0.50 M PROH, 1 M trehalose.
6. CPA #2: 0.75 M EG, 0.75 M PROH, 1 M trehalose (*see* **Note 1**).
7. CPA #3: 1.00 M EG, 1.00 M PROH, 1 M trehalose.
8. Preparation of Rewarming Solutions. Using PAF Culture Medium as the carrier solution, prepare gradient rewarming solutions at trehalose concentrations of 0.75 M, 0.5 M, and 0.25 M, respectively (*see* **Note 2**).

3 Methods

3.1 Isolation of Mouse PAF

Use a mechanical isolation method to isolate PAF with a diameter of 100–190 μm from the ovaries of KM mice. The specific steps are briefly described as follows:

1. Euthanize a 3-week-old female KM mouse by cervical dislocation.
2. Incise the abdomen with surgical scissors to remove the ovaries, which then need to be washed twice in PBS preheated at 37°C for 30 min in an incubator.
3. Place the ovaries in PAF isolation solution, and isolate mouse PAF using two 1 mL syringes under a stereomicroscope.

4. Transfer the isolated PAF to PAF culture medium preheated at 37°C for 30 min and incubate in a 37°C cell incubator for 1 h.
5. Select PAF with intact morphology for toxicity testing (*see* **Note 3**).

3.2 Generation of Mouse PAF-Laden Microcapsules via Centrifugal Microfluidics

Use the centrifugal microfluidic method to generate hydrogel microcapsules embedding mouse PAF. The schematic diagram of the encapsulation process is shown in Fig. 1. The device consists of a 1.5 mL EP tube and a nozzle (inner diameter: 340 μm) fixed on the tube cap. The specific experimental procedures are briefly described as follows:

1. Add 500 μL of $CaCl_2$ solution (0.15 M) and 250 μL of sodium alginate solution (1% w/v) containing mouse PAF into the centrifuge tube and nozzle (*see* **Note 4**).
2. Place the device in a centrifuge and centrifuge at 100×*g* for 2 min. The sodium alginate solution with mouse PAF is ejected from the needle tip at the lower end of the nozzle, falls into the $CaCl_2$ solution, and crosslinks with Ca^{2+} to form hydrogel microcapsules (*see* **Note 5**).
3. Collect the generated hydrogel microcapsules encapsulating mouse PAF into an EP tube for subsequent experiments.

3.3 Morphological Characterization of Alginate Hydrogel Microcapsules

1. The morphology of the generated alginate hydrogel microcapsules can be observed under a bright-field inverted microscope, and representative images can be captured to evaluate their uniformity, size, and structural integrity.

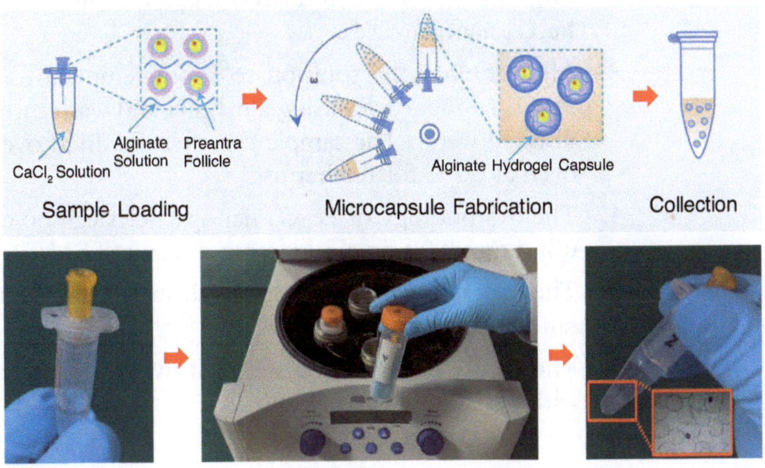

Fig. 1 Schematic illustration of centrifugal microfluidic fabrication of mouse preantral follicle (PAF)-laden hydrogel microcapsules.(Source: Adapted from [19], under a Creative Commons Attribution 4.0 International License (http://creativecommons.org/licenses/by/4.0/))

2. Acquired images of the hydrogel microcapsules can be processed using Image-Pro Plus 6.0 software to measure the microcapsule area and calculate their diameters.
3. The hydrogel microcapsules can be collected into an EP tube and incubated in 1.5 M combination CPA solution for 15 min. Subsequently, they need to be quickly loaded into 0.25 mL plastic straws and plunged into liquid nitrogen for vitrification. The equilibration time in liquid nitrogen is at least 10 min to ensure the straw temperature reaches −196°C (*see* **Note 6**).
4. After vitrification, the microcapsules need to be rapidly rewarmed by immersing the straws in a 37°C water bath (*see* **Note 7**). The rewarmed hydrogel microcapsules can be transferred into an EP tube, washed with deionized (DI) water to remove residual CPA, and then stored at −20°C for 24 h. On the following day, the EP tube can be rapidly transferred to a vacuum freeze dryer for lyophilization over 48 h, after which the microcapsules can be subjected to scanning electron microscopy (SEM) imaging.

3.4 Synthesis and Characterization of Nanoparticles

3.4.1 Fe_3O_4 Nanoparticles

Fe_3O_4 nanoparticles can be synthesized using a co-precipitation method. The specific steps are as follows:

1. Add $FeCl_3{\cdot}6H_2O$ and $FeCl_2{\cdot}4H_2O$ to deionized water at a molar ratio of 1:2 under vigorous stirring and continuous nitrogen gas flow at 70°C.
2. Add $NH_3{\cdot}H_2O$ in a dropwise fashion under continuous stirring until the solution turns black, indicating the formation of Fe_3O_4 nanoparticles.
3. After cooling the solution to room temperature, collect the Fe_3O_4 nanoparticles using a magnet and wash three times with distilled water. The sample can then be dried overnight in an oven at 60°C for further use.
4. The morphology of Fe_3O_4 nanoparticles can be characterized using transmission electron microscopy (TEM).
5. The magnetic properties can be characterized using a vibrating sample magnetometer (VSM).
6. The composition can be analyzed by X-ray diffraction (XRD) within a 2θ range of 10°–90°.

3.4.2 Graphene Oxide (GO) Nanoparticles

Graphene oxide (GO) nanoparticles can be synthesized using a modified Hummers' method with natural graphite powder as the raw material. The detailed steps are as follows:

1. Mix H_2SO_4 (70 mL), $NaNO_3$ (1.5 g), and graphite powder (3.0 g) in an ice bath with continuous magnetic stirring.
2. Slowly add $KMnO_4$ (9.0 g) under the condition that the temperature remains below 20°C. Maintain the mixture in a 34°C water bath for 24 h.
3. Afterward, slowly add 30% (w/v) H_2O_2 (15 mL). Subsequently, wash the mixture with deionized water several times until the pH is neutral, and then purify by dialysis for one week.
4. The morphology of GO nanoparticles can be observed using TEM.
5. The chemical composition of GO can be analyzed by Fourier-transform infrared spectroscopy (FTIR) in the range of 400–4000 cm^{-1}.
6. Raman spectra of GO nanoparticles can be obtained using a confocal microprobe Raman system with an excitation wavelength of 514.54 nm.

3.5 Cytotoxicity Assessment of Nanoparticles and Hydrogel on Mouse PAF

The viability of mouse PAF can be assessed using fluorescence staining. The detailed procedure is as follows:

1. Divide mouse PAF into two groups: a fresh group (control, without any treatment) and a 1% alginate encapsulation group (containing nanoparticles).
2. Culture the PAF in either PAF culture medium or PAF incubation medium at 37°C in a cell incubator for 0, 6, 12, 24, 36, 48, and 60 h.
3. Stain the PAF with acridine orange/ethidium bromide (AO/EB, 1:1, v/v) staining solution in the dark for 2 min, and capture images using an inverted fluorescence microscope (*see* **Note 8**). Representative fluorescence images illustrating the viability of PAFs after incubation for different durations are shown in Fig. 2.

3.6 Residual Detection of Nanoparticles

3.6.1 Scanning Electron Microscopy (SEM), Energy-Dispersive X-ray Spectroscopy (EDS)

The presence of residual nanoparticles on the surface and inside the hydrogel microcapsules after incubation with nanoparticles can be examined by scanning electron microscopy (SEM) combined with energy-dispersive X-ray spectroscopy (EDS). The detailed procedure is as follows:

1. For the control group, place hydrogel microcapsules in 1.5 mL EP tubes filled with double-distilled water (ddH_2O) and incubate at room temperature for 24 h. For the experimental group, hydrogel microcapsules were incubated in ddH_2O containing

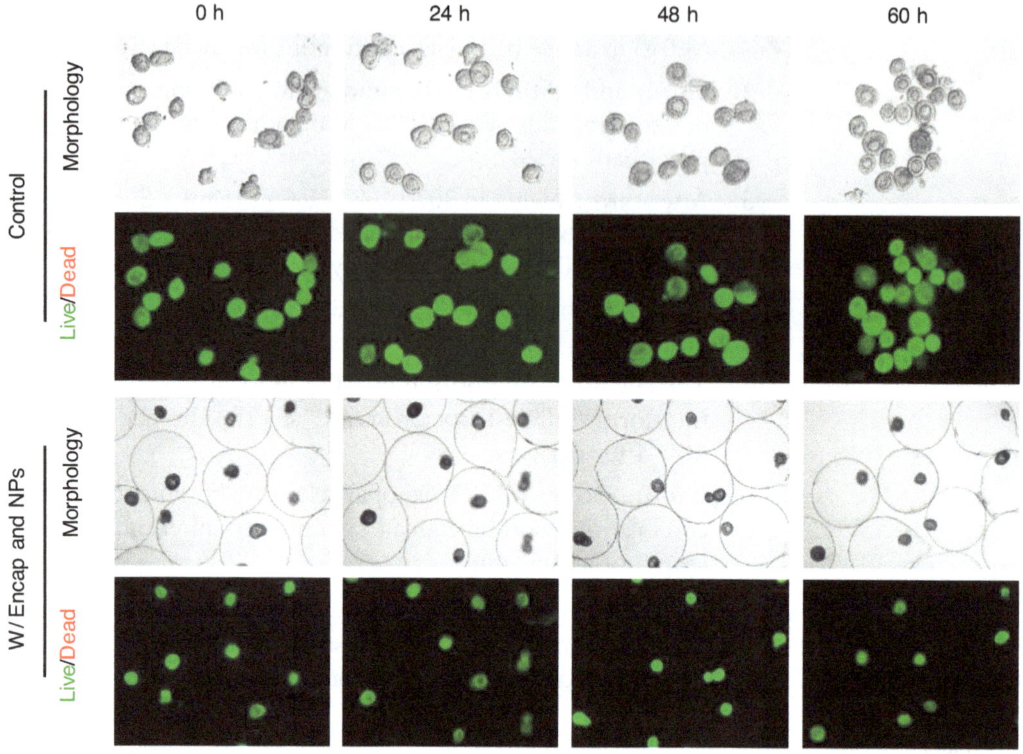

Fig. 2 Viability assessment of encapsulated mouse preantral follicles (PAFs) under nanoparticle exposure. Fluorescence microscopy images of PAFs encapsulated in 1% (w/v) alginate hydrogel and incubated with Fe_3O_4 (0.3% w/v) and graphene oxide (GO, 0.03% w/v) nanoparticles at 37°C for 0, 24, 48, and 60 h. Viability was assessed using [specific fluorescent dye, e.g., Calcein-AM (green, live) and propidium iodide (red, dead)] Source: Adapted from [19], under a Creative Commons Attribution 4.0 International License (http://creativecommons.org/licenses/by/4.0/)

Fe_3O_4 (3%, w/v) and GO (0.03%, w/v) under the same conditions. Divide the experimental group equally into two EP tubes.

2. Use a magnet to remove Fe_3O_4 nanoparticles from one of the experimental EP tubes, and remove residual GO nanoparticles by multiple washes.
3. Transfer all three EP tubes (control, with nanoparticles, and after magnetic removal) to a –80°C freezer for 24 h.
4. Quickly transfer the tubes to a vacuum freeze dryer and lyophilize at –80°C for 24 h.
5. The lyophilized microcapsules can be sputter-coated with gold and subjected to SEM imaging and EDS analysis.

3.6.2 Transmission Electron Microscopy (TEM)

To further verify whether nanoparticles are completely removed, use transmission electron microscopy (TEM) to examine whether nanoparticles have entered and remained inside the PAF. The procedure is as follows:

1. Incubate hydrogel microcapsules encapsulating mouse PAF at 37°C for 24 h in PAF culture medium containing Fe_3O_4 (3%, w/v) and GO (0.03%, w/v).
2. Release the encapsulated PAF by incubating in 75 mM sodium citrate solution for 5–10 min, then collect and fix in 3% glutaraldehyde at 4°C overnight.
3. The next day, incubate the samples in a solution of 0.1 M potassium ferrocyanide and 1% (w/v) osmium tetroxide for 40 min, followed by three washes with 0.1 M cacodylate buffer.
4. After buffer washing, post-fix the samples with 1% (w/v) osmium tetroxide for another 40 min.
5. Stain the samples with 2% (w/v) uranyl acetate for 1 h, then wash three times with cacodylate buffer and ddH_2O.
6. Dehydrate samples using a graded ethanol series (30%, 50%, 70%, 80%, 95%, 100%, 100%), each for 15 min.
7. Infiltrate the samples sequentially with 25%, 50%, 100%, and 100% Epon resin for 3, 5, 12, and 24 h, respectively.
8. Polymerize the resin-embedded samples in an oven at 45°C for 48 hours; section using an ultramicrotome, and image with TEM (*see* **Note 9**).

3.7 Vitrification of Mouse PAF Encapsulated in Hydrogel Microcapsules

Use low-toxicity vitrification cryoprotectants to preserve mouse PAF encapsulated in hydrogel microcapsules. The detailed procedure is as follows (Fig. 3):

1. Equilibrate the PAF in CPA solution containing permeating cryoprotectants (EG + PROH) for 3 min.
2. Transfer to the CPA solution containing 0.5 M trehalose and equilibrate for 3 min.

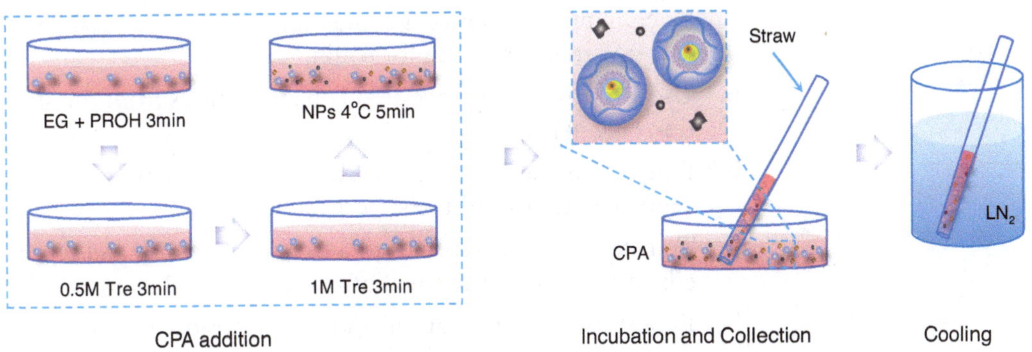

Fig. 3 Schematic of vitrification protocol for mouse preantral follicles (PAFs) encapsulated in hydrogel microcapsules.(Source: Adapted from [19], under a Creative Commons Attribution 4.0 International License (http://creativecommons.org/licenses/by/4.0/))

3. Transfer to the CPA solution containing 1 M trehalose and equilibrate for another 3 min.
4. Finally, transfer to CPA containing different concentrations of nanoparticles (Fe_3O_4: 0.2%, 0.3%, 0.4%, w/v; GO: 0.02%, 0.03%, 0.04%, w/v) and equilibrate at 4°C for 5 min (*see* **Notes 10** and **11**).
5. Quickly load the PAF with the final CPA mixture into a 250 μL straw.
6. Seal the straw and immediately plunge it into liquid nitrogen for vitrification for 30 min (see **Note 12**).

3.8 Nanowarming Process for Mouse PAF Encapsulated in Hydrogel Microcapsules

1. Water bath group (control): Retrieve the straw from liquid nitrogen and immediately immerse it in a 37°C water bath with gentle shaking for 2–3 min.
2. Electromagnetic induction heating (MIH) group: Retrieve the straw and immerse it in a 50 mL centrifuge tube filled with 37°C water (the tube is placed within the copper coil). Heating is controlled at 375 kHz by adjusting the current (5 A, 15 A, 25 A) or Fe_3O_4 concentration in CPA (0.2%, 0.3%, 0.4%, w/v).
3. Laser-induced heating (LIH) group: Place the straw at the laser irradiation point in a 37°C water bath. Heating is controlled by adjusting laser power (3 W/cm^2, 4 W/cm^2, 5 W/cm^2) or GO concentration in CPA (0.02%, 0.03%, 0.04%, w/v).
4. Combined MIH + LIH group: Place the straw into a 50 mL centrifuge tube filled with 37°C water (the tube is placed within the copper coil), while simultaneously turning on the laser directed at the straw (Fig. 4). Set the current to 15 A and laser power to 3 W/cm^2. The CPA contains 0.3% (w/v) Fe_3O_4 and 0.03% (w/v) GO.
5. After warming, remove CPA through four sequential steps by equilibrating in media containing 0.75 M, 0.5 M, 0.25 M, and 0 M trehalose, each for 5 min.
6. Use a magnet to remove Fe_3O_4 nanoparticles from the solution (Fig. 4).
7. Pipette out the hydrogel microcapsules containing PAF and centrifuge to discard the supernatant.
8. Transfer the microcapsules to a 6 cm culture dish and repeatedly wash with culture medium to remove GO nanoparticles.
9. Repeat the centrifugation and resuspension process three times, then resuspend in fresh PAF culture medium for further experiments.

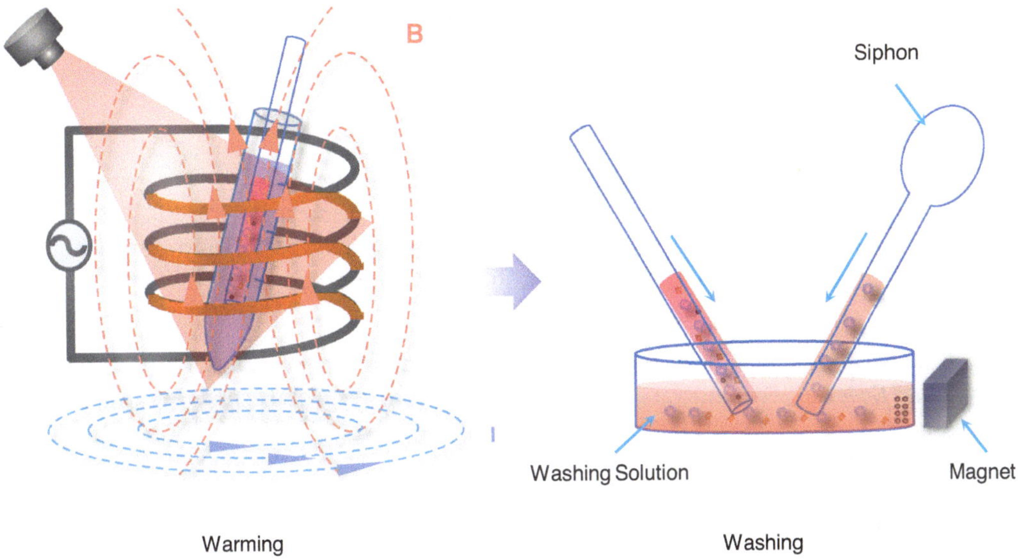

Fig. 4 Schematic of nanowarming system for cryopreserved samples.(Source: Adapted from [19], under a Creative Commons Attribution 4.0 International License (http://creativecommons.org/licenses/by/4.0/))

10. Use a fiber optic temperature sensor to monitor the temperature change of CPA (containing nanoparticles) from liquid nitrogen to 37°C.
11. Acquire thermal images during the nanowarming process using an infrared thermal camera.
12. CPA solutions containing different concentrations of nanoparticles (Fe_3O_4: 0.2%, 0.3%, 0.4%, w/v; GO: 0.02%, 0.03%, 0.04%, w/v) were loaded into 1.5 mL centrifuge tubes. Starting at 36.6°C, the tubes were heated for 45 s either within the coil or under the laser spot, and thermal images and temperatures were recorded.
13. Assess the photothermal effect of GO by observing laser propagation and penetration depth in DI water, carbon black solution, and nanoparticle solutions (*see* **Note 13**).

3.9 Viability Assessment of Mouse PAF After Nanowarming

Assess the viability of mouse PAF after nanowarming using acridine orange/ethidium bromide (AO/EB) fluorescence staining. The detailed procedure is as follows:

1. Add 3 μL AO/EB staining solution (v/v = 1:1) to 60 μL culture medium containing hydrogel microcapsules with PAF.
2. Incubate at room temperature in the dark for 3 min, then capture fluorescence images of PAF using a fluorescence microscope.

3. Incubate with 75 mM sodium citrate solution for 10 min to release the PAF, then capture fluorescence images again (*see* **Notes 14** and **15**). The viability of PAF after warming under different conditions—including with (W/ Encap) or without (W/O Encap) encapsulation, magnetic induction heating (MIH), laser-induced heating (LIH), and combined MIH + LIH—was evaluated. Representative DIC and fluorescence images clearly showing live (green) and dead (red) PAF under each condition are presented in Fig. 5.

3.10 Histological Analysis and Gene Expression of Mouse PAF After Nanowarming

The morphology, viability, and gene expression of mouse PAF after nanowarming can be assessed through DIC imaging, fluorescence staining, histological staining, and qRT-PCR. The detailed procedure is as follows:

1. Take nanowarmed hydrogel microcapsules containing PAF. Add 3 μL AO/EB staining solution (v/v = 1:1) to 60 μL culture medium. Incubate at room temperature in the dark for 3 min. Capture DIC and fluorescence images of PAF within

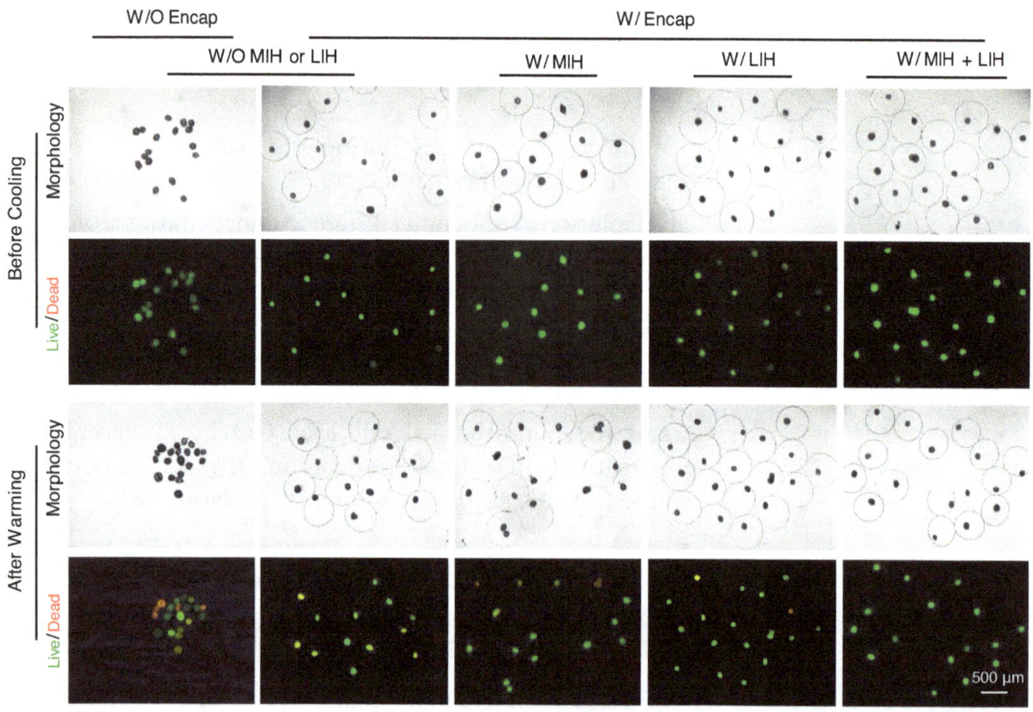

Fig. 5 Viability assessment of mouse preantral follicles (PAFs) under different warming conditions. DIC and fluorescence images of mouse PAF after warming under different conditions. W/ Encap: With alginate encapsulation. W/O Encap: Without alginate encapsulation. W/ LIH: With laser-induced heating. W/ MIH: With magnetic induction heating. W/ MIH + LIH: With combined magnetic and laser heating. Live cells are shown in green (AO staining), while dead cells are shown in red (EB staining).(Source: Adapted from [19], under a Creative Commons Attribution 4.0 International License (http://creativecommons.org/licenses/by/4.0/))

hydrogel microcapsules using a fluorescence microscope. This step reflects the viability and morphology of PAF in the encapsulated state (see Fig. 6a).

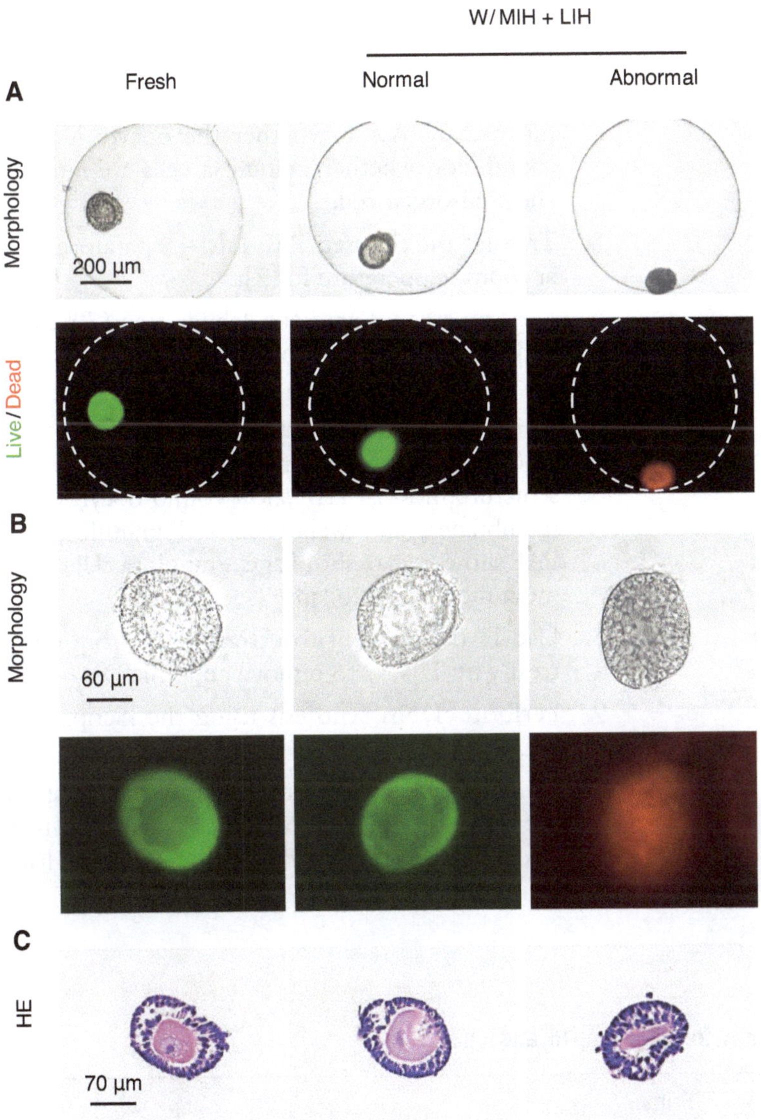

Fig. 6 Morphological, Viability, and Histological Analysis of Mouse PAF After Nanowarming. (**a**) Representative DIC and fluorescence images of PAFs within alginate hydrogel microcapsules after nanowarming. Green fluorescence indicates live PAF, and red indicates dead PAF. (**b**) Representative DIC and fluorescence images of PAFs after hydrogel dissolution. Normal follicles exhibit a round oocyte and intact granulosa cells; abnormal follicles show oocyte shrinkage and granulosa cell damage. (**c**) Typical H&E-stained histological images of PAF after nanowarming. Normal follicles maintain complete follicular architecture, while abnormal follicles display oocyte condensation, granulosa cell disruption, and follicular collapse.(Source: Adapted from [19], under a Creative Commons Attribution 4.0 International License (http://creativecommons.org/licenses/by/4.0/))

2. Incubate the stained hydrogel microcapsules in 75 mM sodium citrate solution for 10 min to dissolve the hydrogel and release the encapsulated PAF.
3. Immediately after release, capture DIC and fluorescence images of free PAF to assess morphology and viability in the non-encapsulated state. This step provides a comparison between encapsulated and free PAF post-warming (*see* Fig. 6b).
4. Observe both encapsulated and released PAF under DIC microscopy. Assess whether the oocyte is centrally located, round, and whether granulosa cells are intact, multilayered, and well-organized.
5. Transfer the released PAF into 4% paraformaldehyde and fix at room temperature for 2 h.
6. Perform ethanol gradient dehydration (70%, 80%, 90%, 95%, and 100%), embed in paraffin, and section the paraffin blocks into 5 μm slices.
7. Stain the paraffin sections with hematoxylin and eosin (H&E). Observe under light microscopy to assess follicular structure: normal PAF show round oocytes, intact basement membranes, and well-organized granulosa cells; abnormal PAF show oocyte shrinkage, granulosa cell rupture, and basement membrane collapse (see Fig. 6c).
8. Use TRIzol reagent to extract RNA from released PAF and treat with DNase to remove genomic DNA (*see* **Note 16**).
9. Perform cDNA synthesis using the iScript cDNA Synthesis Kit.
10. Conduct qRT-PCR using SYBR Green master mix. Analyze the expression levels of follicle- and oocyte-related genes (Gdf9, Fgf8, Bmp15). Primer sequences are listed in Table 2.

Table 2
Primer Sequences for Gdf9, Bmp15, and Fgf8

Gene	Primer
Gdf9	Forward primer: CTTCAGCCCCTAGTGACCTC
	Reverse primer: GAGGCATGCTAAACACTCCG
Bmp15	Forward primer: GAAGCTCTGGAATCGCAAGG
	Reverse primer: CCCTGAACTCTGTCATCGGT
Fgf8	Forward primer: GGAAGCAGAGTCCGAGTTC
	Reverse primer: GCCGTTGCTCTTGGCAATTA

3.11 3D Culture, in vitro Maturation, Fertilization, and Development of Mouse PAF After Nanowarming

1. After nanowarming, transfer the hydrogel microcapsules encapsulating PAF into a 96-well culture plate, with one microcapsule per well in 200 μL of culture medium.
2. The culture medium consists of α-MEM-GlutaMAX supplemented with 20% (v/v) fetal bovine serum (FBS), 1×ITS, 100 mIU/mL follicle-stimulating hormone (FSH), and 1% (v/v) penicillin-streptomycin.
3. Place the 96-well plate in a cell incubator at 37°C with 5% CO_2 for 13 days. Replace half of the medium with fresh medium every other day to maintain optimal growth conditions.
4. During the culture period, monitor the growth of follicles from the preantral stage to the antral stage by DIC microscopy. Typical developmental images of PAFs after vitrification and nanowarming are shown in Fig. 7a.
5. On day 13 of culture, the hydrogel microcapsules can be dissolved in 75 mM sodium citrate solution to release the antral follicles.

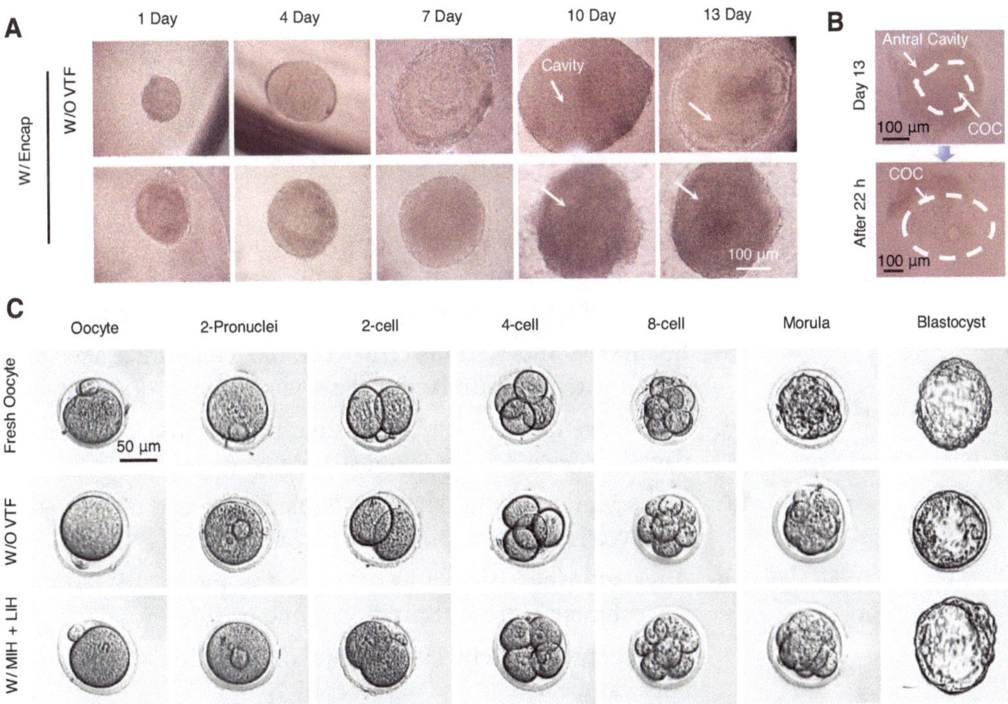

Fig. 7 Developmental competence of vitrified-nanowarmed mouse preantral follicles (PAFs) during 3D in vitro culture. (**a**) Representative images showing the growth of PAFs from preantral follicles to antral follicles during 3D in vitro culture. (**b**) Images of cumulus-oocyte complexes (COCs) collected after follicle maturation. (**c**) Representative images of embryo development in vitro, from MII oocytes through the blastocyst stage. (Source: Adapted from [19], under a Creative Commons Attribution 4.0 International License (http://creativecommons.org/licenses/by/4.0/))

6. Transfer the released antral follicles into 200 μL maturation medium composed of α-MEM-GlutaMAX supplemented with 5 μg/mL epidermal growth factor (EGF), 2.5 IU/mL human chorionic gonadotropin (HCG), 20% (v/v) FBS, 1 × ITS, 100 mIU/mL FSH, and 1% (v/v) penicillin-streptomycin.
7. Incubate the follicles in a 5% CO_2 incubator at 37°C for 48 h to induce oocyte maturation.
8. Collect mature cumulus-oocyte complexes (COCs) by mechanically puncturing antral follicles using a syringe needle, as shown in Fig. 7b.
9. Treat the COCs with 80 IU/mL hyaluronidase for 5 min to remove cumulus cells and obtain metaphase II (MII) oocytes.
10. Sperm were collected from the epididymis of 8-week-old male KM mice. Wash the sperm with equilibrated washing solution, then incubate at 37°C with 5% CO_2 for 30 min to allow motile sperm to swim out.
11. Incubate motile sperm in fresh washing solution for 2 h to enhance viability.
12. In vitro fertilization (IVF) can be performed by incubating MII oocytes with sperm (final concentration: 5×10^4 sperm/mL) in fertilization medium (K-SIFM-50, Cook Medical) for 6 h.
13. After fertilization, transfer embryos at the 2-pronuclei (2PN) stage to cleavage medium (K-SICM-50, Cook Medical) and culture them at 37°C with 5% CO_2.
14. Monitor embryo development under DIC microscopy, recording stages from 2-cell to blastocyst. Representative images of embryo development in vitro are shown in Fig. 7c.
15. For in vivo embryo transfer, 8-week-old female mice need to be mated with vasectomized males to induce pseudopregnancy.
16. After anesthesia, the reproductive tract of the pseudopregnant females can be exposed surgically.
17. Make a micro-incision in the fallopian tube, and transfer 9–16 embryos into the ampulla using a fine pipette.
18. After suturing the incision, return the mouse to a suitable environment to recover and continue pregnancy.
19. After approximately 19–21 days of gestation, the birth of pups can be monitored and recorded.
20. Wean offspring at 3 weeks, and separate males and females into different cages.
21. At 9 weeks of age, a female and a male mouse from the offspring can be paired to assess fertility in the next generation.

22. The birth of second-generation pups can be recorded approximately 3 weeks later, confirming the reproductive competence of the offspring derived from vitrified-warmed PAFs.

4 Notes

1. The CPA solutions with nanoparticles are all prepared using CPA #2.
2. Performing the warming process on a 37°C heating stage yields better results.
3. Select pre-antral follicles with an intact basement membrane, at least two layers of surrounding granulosa cells, and at a developmental stage close to the antral follicle stage for the experiment.
4. To avoid contamination, the microfluidic device should be sequentially washed with 75% (v/v) ethanol and PBS, irradiated under UV light for 30 min, and both $CaCl_2$ solution and sodium alginate solution should be filtered through a 0.22-μm sterile filter before use.
5. When the nozzle diameter is fixed, the size of microcapsules can be controlled by adjusting the alginate concentration and rotation speed.
6. Load 5–8 hydrogel microcapsules into each straw. Avoid overloading, as excessive volume negatively affects vitrification efficiency.
7. During warming in a 37°C water bath, rapidly shake the straw to enhance warming uniformity.
8. Live cells are indicated by green fluorescence, while dead cells are indicated by red fluorescence. PAF viability = (Number of green fluorescent PAF) / (Total number of PAF).
9. To observe the surface and internal structure of hydrogel microcapsules after nanoparticle incubation, gently crush a portion of the lyophilized microcapsules using weighing paper. Avoid cutting with a blade, as this can damage the internal structure of the microcapsules.
10. For improved CPA equilibration, the CPA containing follicles can be placed on a shaker during incubation.
11. To prevent nanoparticle aggregation, the CPA solutions containing Fe_3O_4 and GO should be treated with an ultrasonic cleaner for a few minutes before use to ensure uniform dispersion of nanoparticles.

12. To enhance vitrification efficiency, pre-cool the straw for 5 s in liquid nitrogen vapor before quickly plunging it into liquid nitrogen.
13. To achieve more accurate warming, turn on the laser emitter 1 min in advance. This allows easier alignment of the sample-loaded section of the straw with the laser heating spot, improving the warming outcome.
14. Green fluorescence indicates live cells, and red fluorescence indicates dead cells. PAF viability = (Number of green fluorescent PAF) / (Total number of PAF).
15. To ensure more stable follicle viability, it is recommended to perform viability staining 2 h after warming.
16. Use RNase-free pipette tips and EP tubes throughout the procedure, and maintain a clean and sterile environment.

References

1. Edgar DH, Gook DA (2012) A critical appraisal of cryopreservation (slow cooling versus vitrification) of human oocytes and embryos. Hum Reprod Update 18:536–554. https://doi.org/10.1093/humupd/dms016
2. Rienzi L, Gracia C, Maggiulli R, LaBarbera AR, Kaser DJ, Ubaldi FM, Vanderpoel S, Racowsky (2017) Oocyte, embryo and blastocyst cryopreservation in ART: systematic review and meta-analysis comparing slow-freezing versus vitrification to produce evidence for the development of global guidance. Hum Reprod Update 23:139–155. https://doi.org/10.1093/humupd/dmw038
3. Grenier J, Duval H, Lv P, Barou F, Le Guilcher C, Aid R, David B, Letourneur D (2022) Interplay between crosslinking and ice nucleation controls the porous structure of freeze-dried hydrogel scaffolds. Biomater Adv 139:212973. https://doi.org/10.1016/j.bioadv.2022.212973
4. Grenier J, Duval H, Barou F, Lv P, David B (2019) Mechanisms of pore formation in hydrogel scaffolds textured by freeze-drying. Acta Biomater 94:195–203. https://doi.org/10.1016/j.actbio.2019.05.070
5. Cao K, Shen L, Guo X, Wang K, Hu X, Ouyang X, Zhao G (2020) Hydrogel microfiber encapsulation enhances cryopreservation of human red blood cells with low concentrations of glycerol. Biopres Biobank 18:228–234. https://doi.org/10.1089/bio.2020.0003
6. Huang H, Choi JK, Rao W, Zhao S, Agarwal P, Zhao G, He X (2015) Alginate hydrogel microencapsulation inhibits devitrification and enables large-volume low-CPA cell vitrification. Adv Funct Mater 25:6839–6850. https://doi.org/10.1002/adfm.201503047
7. Qin X, Chen Z, Shen L, Liu H, Ouyang X, Zhao G (2024) Core-shell microfiber encapsulation enables glycerol-free cryopreservation of RBCs with high hematocrit. Nano Micro Lett 16:3. https://doi.org/10.1007/s40820-023-01213-3
8. Zheng YY, Zhao G (2016) Integrating micro- or nanoscale encapsulation technology with vitreous cryopreservation: a new strategy to improve biopreservation. Cryo Letters 37:421–426
9. Cheng Y, Yu Y, Zhang Y, Zhao G, Zhao Y (2019) Cold-responsive nanocapsules enable the sole-cryoprotectant-trehalose cryopreservation of β cell-laden hydrogels for diabetes treatment. Small 15:e1904290. https://doi.org/10.1002/smll.201904290
10. Zhao G, Liu X, Zhu K, He X (2017) Hydrogel encapsulation facilitates rapid-cooling cryopreservation of stem cell-laden core-shell microcapsules as cell-biomaterial constructs. Adv Healthc Mater 6:1700988. https://doi.org/10.1002/adhm.201700988
11. Chen W, Shu Z, Gao D, Shen AQ (2016) Sensing and sensibility: single-islet-based quality control assay of cryopreserved pancreatic islets with functionalized hydrogel microcapsules. Adv Healthc Mater 5:223–231. https://doi.org/10.1002/adhm.201500515
12. Pravdyuk AI, Petrenko YA, Fuller BJ, Petrenko AY (2013) Cryopreservation of alginate encapsulated mesenchymal stromal cells. Cryobiology 66:215–222. https://doi.org/10.1016/j.cryobiol.2013.02.002

13. Kashaninejad N, Shiddiky MJA, Nguyen NT (2018) Advances in microfluidics-based assisted reproductive technology: from sperm sorter to reproductive system-on-a-chip. Adv Biosyst 2:1700197. https://doi.org/10.1002/adbi.201700197
14. Li Y, Memon K, Memon K, Zheng Y, Cheng Y, Mbogba MK, Wang P, Ouyang X, Zhao G (2019) Microencapsulation facilitates low-cryoprotectant vitrification of human umbilical vein endothelial cells. ACS Biomater Sci Eng 5:5273–5283. https://doi.org/10.1021/acsbiomaterials.9b00726
15. Tian C, Zhang X, Zhao G (2019) Vitrification of stem cell-laden core-shell microfibers with unusually low concentrations of cryoprotective agents. Biomater Sci 7:889–900. https://doi.org/10.1039/c8bm01231h
16. Zhu K, Yu Y, Cheng Y, Tian C, Zhao G, Zhao Y (2019) All-aqueous-phase microfluidics for cell encapsulation. ACS Appl Mater Interfaces 11:4826–4832. https://doi.org/10.1021/acsami.8b19234
17. Cheng Y, Zhang X, Cao Y, Tian C, Li Y, Wang M, Zhao Y, Zhao G (2018) Centrifugal microfluidics for ultra-rapid fabrication of versatile hydrogel microcarriers. Appl Mater Today 13:116–125. https://doi.org/10.1016/j.apmt.2018.08.012
18. Zhang CC, Zhou Y, Zhang L, Wu L, Chen Y, Xie D, Chen W (2018) Hydrogel cryopreservation system: an effective method for cell storage. Int J Mol Sci 19:3330. https://doi.org/10.3390/ijms19113330
19. Tian C, Shen L, Gong C, Cao Y, Shi Q, Zhao G (2022) Microencapsulation and nanowarming enables vitrification cryopreservation of mouse preantral follicles. Nature Commun 13:7515. https://doi.org/10.1038/s41467-022-34549-2
20. Feng Q, Wang J, Yi D, Li S, Cheng Y, Chen Z, Zhao G (2025) Epigallocatechin gallate cooperated with hydrogel encapsulation enables high-performance cryopreservation of mouse ovaries. Mater Today Bio 32:101883. https://doi.org/10.1016/j.mtbio.2025.101883
21. Li X, Zhang S, Zhang Y, Zhou X (2024) Visualization of ice crystal behavior in mouse oocytes during high-speed quench cooling and ice inhibition by antifreezing hydrogels. Biopres Biobank 22:404–412. https://doi.org/10.1089/bio.2023.0108

Chapter 28

Protocol for Cryopreservation of Endothelial Monolayers: An Update

Leah A. Marquez-Curtis, Ming Han Yu, Nasim Eskandari and Janet A. W. Elliott

Abstract

Understanding the cryobiological responses of endothelial cells is important to ensure the successful cryopreservation of vascularized tissues and organs. The endothelium, a monolayer of endothelial cells comprising the inner lining of blood vessels and other tissues such as the cornea, is sensitive to cryoinjury. In this chapter, we update our published cryopreservation protocol, which we have since validated for several more cell types. Our optimized protocol involves culturing cells on Rinzl plastic coverslips, using a combination of a permeating cryoprotectant (5% dimethyl sulfoxide) and a non-permeating cryoprotectant (6% hydroxyethyl starch), addition of 2% chondroitin sulfate, controlled-rate slow cooling, storage in liquid nitrogen, and removal of cryoprotectants immediately after thaw. Our post-thaw assessments include membrane integrity and a functional assay immediately after thaw and after extended incubation in culture conditions.

Keywords Cryopreservation, Dimethyl sulfoxide, Hydroxyethyl starch, Chondroitin sulfate, Graded freezing, Membrane integrity, AlamarBlue reduction, Fluorescent microscopy, Rinzl plastic, Delayed-onset cryoinjury

1 Introduction

A monolayer of endothelial cells (also called the endothelium) constitutes the inner lining of blood vessels, lymphatic vessels, and other tissues, including the cornea. The endothelium was initially perceived as a passive structure separating the blood from surrounding tissues, but is now recognized as a key player in regulating blood fluidity, platelet aggregation, leukocyte trafficking, inflammation, angiogenesis, and tissue homeostasis and regeneration [1–5]. For example, in the cornea, the endothelial cell monolayer acts as a barrier and pump to regulate stromal hydration, because excess water in the stroma causes swelling of the cornea, resulting in blurred vision and eventual blindness [6]. Age-dependent deterioration of the corneal endothelium or its impairment by disease or injury constitutes the

Willem F. Wolkers and Harriëtte Oldenhof (eds.), *Cryopreservation and Freeze-Drying Protocols*, Methods in Molecular Biology, vol. 3060, https://doi.org/10.1007/978-1-0716-5416-3_28,

most commonly cited indication for corneal transplants [7]. Transplant of allogeneic corneal tissue from donor sources is the standard of treatment; however, current tissue storage methods can only guarantee one month of optimal preservation of donor corneas, leading to global scarcity [8, 9]. The shortage of donor corneas has prompted the alternative use of tissue-engineered grafts or cell-based therapies [10]. Endothelial cells are also a core component of the blood–brain barrier (BBB), a semipermeable barrier with high selectivity in the transport of substances between the blood and the central nervous system. Dysfunctions of the BBB have been associated with Alzheimer's disease, Parkinson's disease, and multiple sclerosis [11, 12]. To investigate how drug delivery to the brain can be enhanced for treating these diseases, in vitro models of the BBB are increasingly being explored. Endothelial cells also play a crucial role in the heart, where they influence coagulation, inflammation, and tissue repair [13–16]. Dysfunctions of blood vessel endothelium in the heart are implicated in cardiovascular disease, the primary cause of death worldwide [17–19]. Thus, in vitro models of heart endothelium are increasingly valuable. Tissue-engineered corneas, cell-based corneal grafts, and BBB and cardiac in vitro models can all benefit from the availability of on-demand cryopreserved endothelial monolayers.

Endothelial cells, even in suspension, are susceptible to cryoinjury. For example, the routine cryopreservation procedure for cells in suspension of cooling at 1 °C/min in the presence of 10% dimethyl sulfoxide (DMSO) leaves only about 66% of human umbilical vein endothelial cells (HUVECs) and 76% of porcine corneal endothelial cells (PCECs) viable [20, 21]. We exploited the observation that there exist critical sub-zero temperature ranges where interrupting the cooling process improves post-thaw cell recovery [22]. Our group has designed interrupted freezing procedures that differentiate the cell damage incurred during slow cooling to this critical temperature range from the damage incurred during further rapid cooling to the liquid nitrogen storage temperature [23]. Our interrupted slow-cooling procedure (also known as graded freezing) not only identifies the occurrence and extent of cryoinjury, but also provides an efficient method for optimizing the degree of protection conferred by permeating and non-permeating cryoprotective agents (CPAs) [24, 25]. In this experimental technique, the response of cells is studied as a function of temperature throughout the cooling protocol. At various temperatures along a slow-cooling profile (e.g., 1 °C/min), cells are either directly thawed and assessed for membrane integrity to gain information about slow-cooling injury or plunged into liquid nitrogen and then thawed and assessed to identify rapid-cooling injury. Slow-cooling injury is mitigated by adding permeating CPAs; rapid-cooling injury is mitigated by adding non-permeating CPAs. By iteratively using graded freezing, we greatly improved the cryopreservation of endothelial cells in suspension using 5% DMSO and 6%

hydroxyethyl starch (HES) as CPAs. This resulted in 94% post-thaw membrane integrity relative to fresh control for HUVECs, and 91% and 93% relative to fresh control, respectively, for porcine and human corneal endothelial cells; cells were also fully functional post-thaw [21, 26]. Similarly, cryopreservation of human cerebral microvascular endothelial cells/D3 clone (hCMEC/D3) in suspension using the combination of 5% DMSO and 6% HES resulted in functional cells with post-thaw viability of 94% relative to fresh control [27, 28]. As a permeating CPA, DMSO colligatively reduces the amount of ice formed at a given temperature and mitigates solute-effects injury [24]. HES, a non-permeating CPA, osmotically draws water out of the cells and mitigates damage due to intracellular ice formation, and also stabilizes the cell membrane [29].

Compared to cells in suspension, cells in monolayers are extremely susceptible to intracellular ice formation due to the presence of cell–cell and cell–matrix interactions [30]. Because the extracellular environment and the surface upon which the cells are cultured contract and expand with temperature during the freezing and thawing processes, we hypothesized that a culture surface with a coefficient of linear thermal expansion similar to that of ice would reduce mismatch strain during the cryopreservation process and prevent detachment of cells from the surface [31]. We subsequently found that such a culture surface also has an effect on cell viability of the cells that remain attached [32]. In addition to DMSO and HES, our CPA solution also includes chondroitin sulfate (CS), which acts as a cell-contact stabilizer and antioxidant [33]. For HUVEC and PCEC monolayers, we reported >90% post-thaw membrane integrity with metabolic function after overnight incubation [32, 34]. We also demonstrated cryopreservation of monolayers of hCMECs/D3 (87% post-thaw viability, metabolic activity similar to control, and expression of the tight junction proteins claudin-5 and ZO-1), monolayers of primary human astrocytes (87% post-thaw viability), and monolayers of co-cultured hCMECs/D3 and primary human astrocytes (viability >90% of fresh control) [27]. Membrane integrity and metabolic activity of cryopreserved monolayers of hCMEC/D3 and astrocyte single cultures or co-cultures were maintained following overnight incubation post-thaw [27], thereby ruling out any appreciable delayed-onset cryoinjury (e.g., apoptosis and secondary necrosis) [34]. Moreover, this protocol worked as well for the cryopreservation of monolayers of a mouse cardiac endothelial cell line resulting in functional cells with post-thaw viability of 94% [35].

Briefly, our optimized protocol for cryopreservation of endothelial monolayers involves culturing cells to confluency on fibronectin-coated Rinzl coverslips, loading of CPA (5% DMSO + 6% HES + 2% CS wt/wt in cell medium) for 15 min on ice, ice nucleation at −5°C, cooling to −40°C (at a controlled cooling rate of 1 °C/min), then plunging into and storage in liquid nitrogen. The frozen monolayers are rapidly thawed in a 37°C water bath, and the CPAs are removed by serial dilution. To assess the success of the cryopreservation procedure and the occurrence of delayed-onset cryoinjury, membrane integrity, as a surrogate of cell viability, is used as a rapid assay that can be carried out immediately and after extended incubation of the post-thaw samples. A broadly applicable functional assay for metabolic activity described here is by AlamarBlue reduction [27, 36, 37]. We refer the reader to other specific fit-for-purpose functional assays that may also be performed, including immunocytochemical staining of tight junction proteins such as Zonula Occludens (ZO)-1 [26–28], tube formation on Matrigel to examine angiogenic potential [26–28, 38], and intracellular calcium signaling [35].

2 Materials

2.1 Equipment

1. Analytical balance.
2. Biological safety cabinet.
3. CO_2 incubator.
4. Centrifuge (Eppendorf 5810 R tabletop centrifuge, Hamburg, Germany).
5. Coulter counter (Beckman Coulter Z2 particle count and size analyzer) or hemocytometer.
6. Fluorescent microscope (Leitz, Dialux 22) with Infinity3 camera and Infinity Capture software (if membrane integrity and ZO-1 expression are to be assessed).
7. Liquid nitrogen (LN_2) Dewar.
8. Phase-contrast microscope (Labovert, Leitz) with Diractor camera and imaging software (Pixera, Santa Clara, CA, USA).
9. Pipette-aid and pipettors.
10. Plate reader (if metabolic activity is to be assessed).
11. Programmable methanol cooling bath (FTS Systems, Stone Ridge, NY, USA).
12. Timer.
13. T-type thermocouple and OMB-DAQ-55 data acquisition module (OMEGA Engineering Inc., Stamford, CT, USA).
14. Water bath (set at 37°C for rapid thawing).

2.2 Supplies and Reagents

1. Rinzl coverslips (18 × 18 mm, Electron Microscopy Sciences, Hatfield, PA, USA).
2. Petri dishes (100 mm).
3. Fine-point forceps or tweezers.

4. Standard dissecting scissors.
5. 70% ethanol or isopropanol.
6. 1× phosphate-buffered saline (PBS).
7. Fetal bovine serum.
8. Tissue culture well plates (24-wells; 96-wells).
9. Fibronectin (from bovine plasma, Sigma-Aldrich, Oakville, ON, Canada).
10. Microcentrifuge tubes (1.5 mL).
11. Pipettes and tips (20 μL, 200 μL, 1000 μL).
12. Pasteur pipettes.
13. Glass shell vials with lid (15×45 mm, Kimble Chase, Rochester, NY, USA).
14. Styrofoam floaters.
15. Dimethyl sulfoxide (DMSO).
16. Hydroxyethyl starch (HES, Bristol-Myers Squibb), or, alternatively, Pentastarch (20% w/v HES solution, PST0002, Preservation Solutions, Inc., Elkhorn, WI, USA).
17. Chondroitin sulfate A sodium salt (from bovine trachea, C9819, Sigma-Aldrich).
18. Methanol.
19. Liquid nitrogen.
20. SYTO 13 (5 mM solution in DMSO, S7575, Molecular Probes, Eugene, OR, USA)/GelRed (41003, Biotium, Scarborough, ON, Canada) live/dead dual stain (if membrane integrity is to be assessed).
21. AlamarBlue reagent (DAL1025, Invitrogen) (if metabolic activity is to be assessed).
22. Long metal forceps.
23. Styrofoam box with at least a one-inch wall thickness.

3 Methods

See **Note 1** for safety recommendations.

3.1 Cell Cultures on Rinzl Plastic Coverslips

Cells may be isolated from tissues or purchased. If purchased, the cells are usually received cryopreserved in suspension shipped on dry ice and should be immediately placed in liquid nitrogen storage until use. The cells should be thawed and cultured according to the manufacturer's instructions using the recommended media and reagents for passaging. Cells isolated from tissue are also typically cultured prior to cryopreservation. Follow culture confluency and passage recommendations according to cell type (*see* **Note 2**). Ensure that cells have reached appropriate confluency by examining the cultures under a phase-contrast microscope before trypsinization. Over-confluency is indicated by floating cells, which do not have any more surface to attach to. Cryopreserve cells when they are at an appropriate early passage and have the typical morphology of healthy growing cells. Overgrown and late-passage cells tend to differentiate or approach senescence and are not recommended for cryopreservation (see Fig. 1 for examples of different endothelial cell monolayer cultures).

3.1.1 Preparing Coverslips

1. Rinzl coverslips come in 18 × 18 mm configuration and need to be cut into 8 × 8 mm squares using standard dissecting scissors (*see* **Note 3**).
2. In the biological safety cabinet (BSC), transfer the coverslips to a sterile 100 mm Petri dish using fine-point sterile metal forceps. Pour 70% ethanol or isopropanol, making sure all coverslips are fully covered. Put the lid in place, seal with tape, and leave in the BSC for at least 30 min before use. This step sterilizes the plastic coverslips, because they cannot be autoclaved.
3. Put enough sterile 1 × PBS in a sterile 100 mm Petri dish, then using fine point forceps (sterilized by dipping in a beaker containing 70% ethanol or isopropanol and allowed to dry), transfer the coverslips from the dish containing 70% ethanol or isopropanol into the dish with 1 × PBS. Ensure that the coverslips are fully covered. Leave in PBS for 15 min in the BSC. This step washes off residual alcohol from the sterilization step.
4. Use sterile forceps to pick one coverslip at a time and remove excess buffer by touching the edge of the coverslip to the side of the Petri dish. Transfer to a sterile 24-well plate, one coverslip in each well. Remove excess buffer by aspiration using sterile Pasteur pipettes.
5. Prepare fibronectin solution in the BSC by adding 982 μL of 1 × PBS to a sterile microtube and adding 18 μL of 1 mg/mL stock solution of fibronectin. These volumes may be adjusted accordingly depending on how many coverslips need to be coated. Mix thoroughly by inverting the tube several times. Add 100 μL of fibronectin solution to each coverslip in the well. The fibronectin solution forms a bead on top of the Rinzl and covers most of its surface. It is best not to spread the drop using the pipette tip because spilling the fibronectin solution over the edge of the cover slip can result in uneven coating. If spilling does occur, add additional fibronectin solution to the well until the coverslip is completely covered. Remove the fibronectin solution after incubation for at least 30 min at room temperature. Allow the water to evaporate and the fibronectin

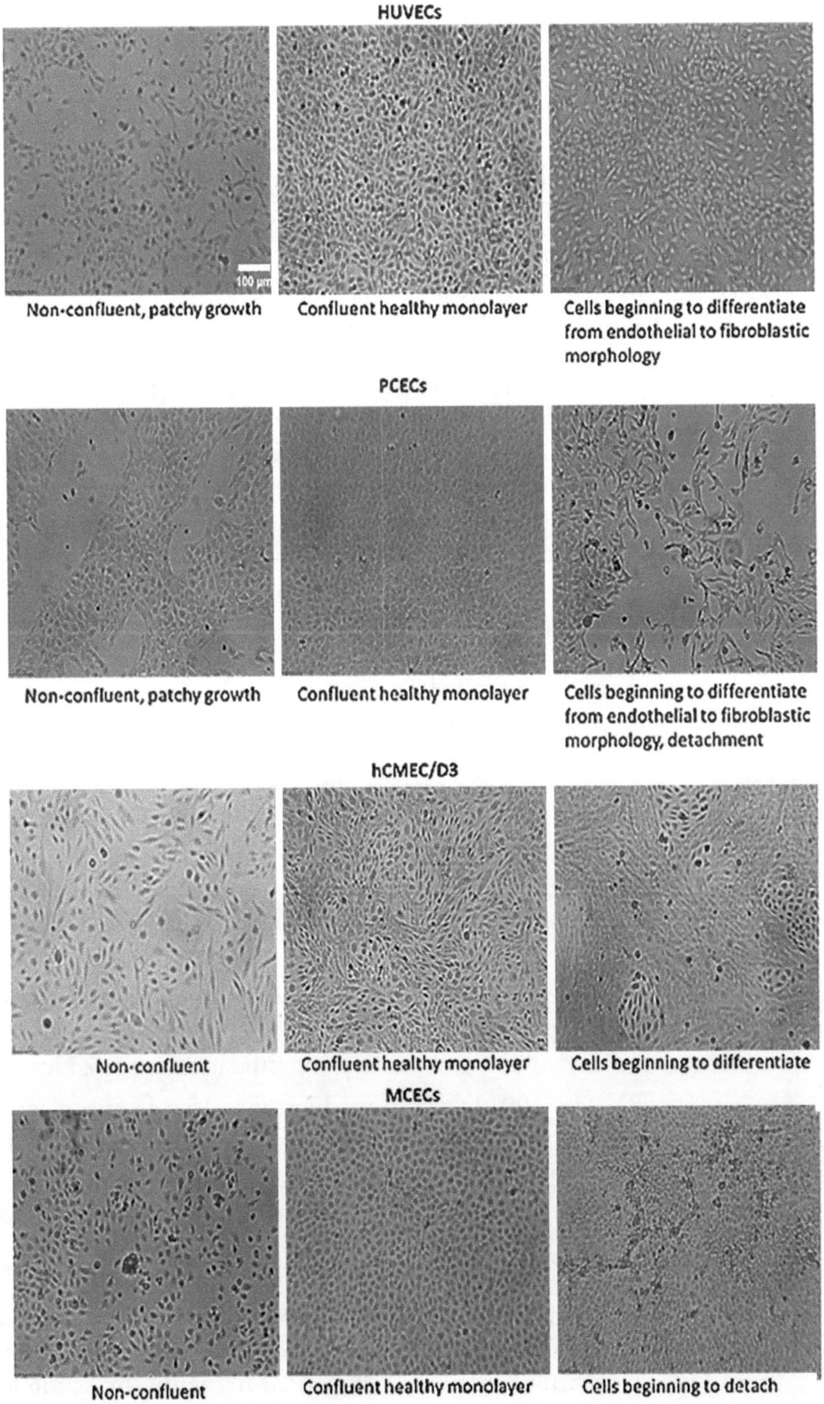

Fig. 1 Example phase-contrast images of monolayers of human umbilical vein endothelial cells (HUVECs), porcine corneal endothelial cells (PCECs), human cerebral microvascular endothelial cells/D3 cell line (hCMECs/D3), and mouse cardiac endothelial cell line (MCECs) at different levels of confluency and showing cell differentiation or detachment. Healthy confluent monolayers are shown in the middle panels. The scale bar at the top left-most image applies to all images

to adhere to the Rinzl surface by leaving the plate uncovered in the BSC.

3.1.2 Seeding Endothelial Cells on Coverslips—This Step Should be Done in the BSC

1. Count the cells in suspension using a cell counter or hemocytometer. Seed the cells in the recommended cell-type-specific medium at the recommended cell density (usually 5000–10,000 cells per cm^2 on the fibronectin-coated Rinzl coverslips at a final volume of 500 μL per well of a 24-well plate) (*see* **Note 4**).
2. Dispense and distribute the cells evenly by sliding the plate on a flat surface from side-to-side and back-and-forth several times after seeding. Avoid swirling the plate in a circular motion because this can cause clustering of the cells at the center of the well. Make sure the coverslips are not floating in the medium by gently pressing them down with a sterile pipette tip. Check that there is no clustering of cells by viewing under a phase-contrast microscope. Put the plate in a humidified incubator (37°C and 5% CO_2). The day of seeding is considered as day 0. Examine cell growth and morphology every day using a phase-contrast microscope. Replace the cell medium every other day. Proceed with the graded freezing of the cell monolayers when they have reached confluency (see example images in Fig. 1).

3.2 Freezing Cell Monolayers (see Schematic Depiction in Fig. 2)

1. Turn on the methanol bath and set the temperature to −5°C. Ensure that the methanol level is about three-quarters full and that the magnetic bar is spinning (stirrer speed at setting 70). Monitor both the temperature of the bath and the temperature inside a surrogate glass vial containing 200 μL medium using calibrated T-type thermocouples and an OMB-DAQ-55 data acquisition module.
2. Prepare 10% (wt/wt) DMSO, 12% (wt/wt) HES, and 4% (wt/wt) CS in cell-type-specific medium (*see* **Note 5**).
3. Put 100 μL of cell medium in a 15 × 45 mm glass vial. Take the plate of confluent monolayers from the incubator and using fine-point forceps transfer a coverslip from each well and place it cell-side up in each vial containing medium (*see* **Note 6**). Take extra care to avoid scratching the coverslips during transfer by picking them up by the edge.
4. Add 100 μL of the CPA cocktail to each vial containing the coverslip in 100 μL of medium (final concentrations: 5% DMSO, 6% HES, 2% CS). Mix gently by swirling the vial, then cover with the lid in such a way that closure will prevent leaking but would also allow its easy removal during the rapid-thaw step.
5. Keep the vials in an ice-slush bath (at 0°C) for 15 min to allow DMSO to permeate the cell monolayer.

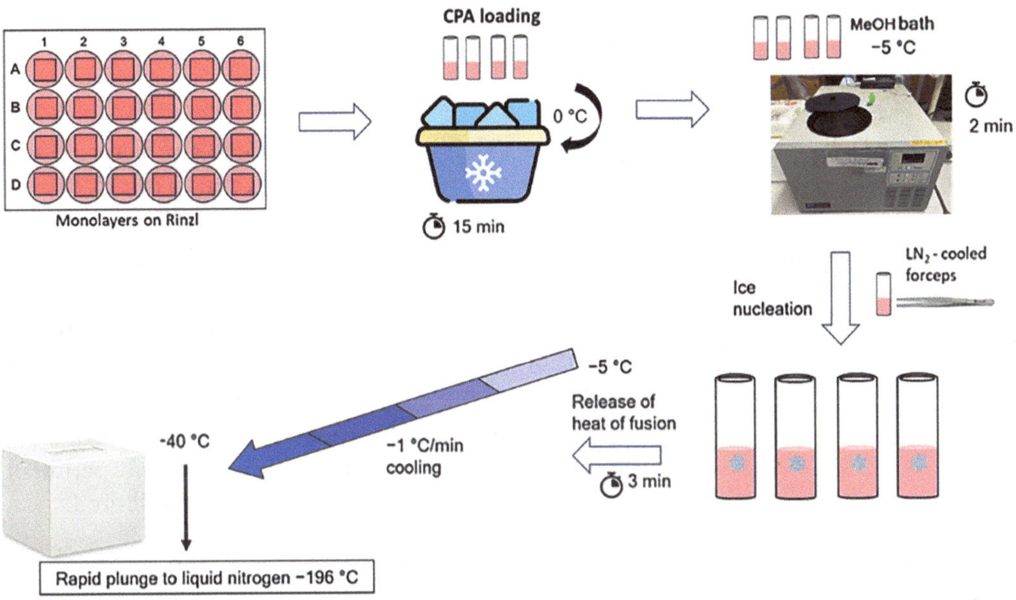

Fig. 2 Schematic depiction of the protocol for freezing cell monolayers cultured on fibronectin-coated Rinzl coverslips. CPAs are loaded at 0°C for 15 min, then the monolayers are equilibrated at the ice nucleation temperature (–5°C) in a methanol bath. Extracellular ice formation is induced using pre-cooled forceps, then held at –5°C to release the latent heat of fusion. The monolayers are cooled at 1°C/min to –40°C, then plunged into liquid nitrogen for storage

6. Put each vial in a Styrofoam floater and place it in the methanol bath, making sure that the bottom of the vial containing medium is immersed. Set the timer for 2 min to allow the cell monolayer to attain equilibrium with the methanol-bath temperature (–5°C).
7. Induce ice nucleation in the solution by quickly touching the vial with cold metal forceps (previously immersed in liquid nitrogen). The contact area should be near the top of the medium meniscus, away from the bottom of the vial where the coverslip is located. The nucleation process is fast (less than 1 s) and is clearly indicated by a sudden increase in opacity in the solution due to ice formation at the contact area (*see* **Note 7**).
8. Return the glass vial to the methanol bath and let it stand for 3 min to release the latent heat of fusion.
9. Set the cooling rate in the methanol bath to 1°C/min and the final temperature to –40°C.
10. When the temperature reaches –40°C, remove the vial from the methanol bath and transfer it to liquid nitrogen. This step should be done quickly to prevent transient warming.

11. Store in liquid nitrogen for at least 30 min, which is enough time for the samples to reach liquid nitrogen temperature (−196°C).

3.3 Thawing Cell Monolayers

1. After removing a vial from liquid nitrogen and before placing it in the water bath, loosen the lid of the vial to release pressure (*see* **Note 8**).
2. Hold the vial in a 37°C water bath, making sure that the vial lid is kept above the water level to minimize the risk of contamination, and gently shake the vial in the bath to enable rapid thawing. Complete thawing is indicated when all of the solution within the vial has regained clear transparency.
3. Remove the CPA from the monolayer as described below. The monolayers are now ready for downstream experiments, including viability and functional assessments if desired.

3.4 CPA Removal by Serial Dilution

1. Without removing the CPA solution, add 200 μL of 20% fetal bovine serum (FBS) in 1×PBS to the glass vial and incubate for 2 min at room temperature.
2. Add 200 μL of 10% FBS in 1×PBS and let stand for 2 min at room temperature. Repeat this step two more times (the monolayer would now be in 1000 μL solution).
3. Gently remove the CPAs and wash solution completely using a 1-mL pipette tip. Replace with 190 μL 1×PBS for immediate membrane integrity assessment, or with 500 μL of specific cell-type medium for extended incubation under cell culture conditions, as described below.

3.5 If Desired, Membrane Integrity Assessment by Fluorescent Microscopy Immediately Post-Thaw and After Extended Incubation

1. To 262.5 μL of 1×PBS in a microtube, add 10 μL GelRed stock solution (10,000× in water) and 10 μL SYTO 13 stock solution (5 mM in DMSO). These volumes may be adjusted accordingly, as long as the final concentrations are: 350× GelRed and 0.18 mM SYTO 13. Keep the stain solution cold at 0°C and in the dark to prevent photobleaching by placing it inside a covered black bucket containing ice-water slush. Other live/dead stains may be used but precautions must be taken as to their applicability to cell monolayers (*see* **Note 9** and sample fluorescent images in Fig. 3).
2. Turn on the fluorescent microscope, being mindful that the mercury lamp takes at least 15 min to warm up.
3. Add 10 μL of SYTO 13/GelRed solution to each coverslip in 190 μL of 1×PBS, then incubate at room temperature for 5 min in the dark.
4. Transfer the coverslip (monolayer-side down) using fine-point tweezers onto a glass slide and touch the edge with a Kimwipe to drain off any excess dye.

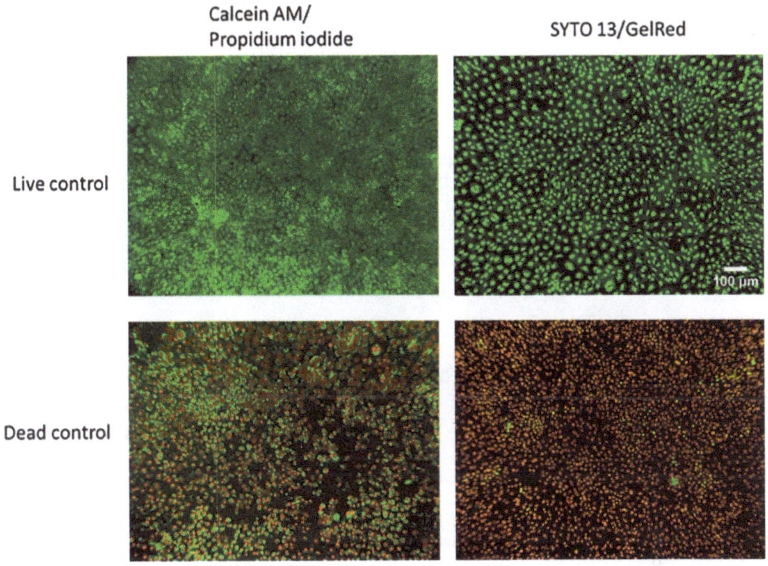

Fig. 3 Representative fluorescent images of mouse cardiac endothelial cell monolayers labeled with different fluorescent stains. Live controls consist of monolayers loaded with CPAs but did not go through ice nucleation and slow cooling. Dead controls consist of monolayers loaded with CPAs, then immediately plunged into liquid nitrogen. Left panels show monolayers labeled with Calcein AM + propidium iodide (PI); right panels show monolayers labeled with SYTO 13 + GelRed. Calcein AM (Corning, 354216) is a non-fluorescent dye that, upon entry to live cells, is converted by acetoxymethyl ester hydrolysis to green fluorescent Calcein in the cytoplasm [39]. SYTO 13 binds to nucleic acids and exhibits green fluorescence after permeating live or dead cells. PI and GelRed are both unable to permeate the membrane of live cells but can bind very strongly to nucleic acids upon entry into membrane-compromised (dead) cells, causing the nuclei to fluoresce red. The live control staining by Calcein AM shows poor resolution caused by the cytoplasmic fluorescence, whereas the nuclear staining by SYTO 13 has very good resolution that enables more accurate counting of membrane-intact cells. Moreover, the dead control staining by Calcein AM/PI results in cytoplasmic green fluorescence that overlaps and confounds the nuclear red fluorescence of dead cells. On the other hand, GelRed shows unambiguous staining of dead cells. The scale bar of the image at the top right corner applies to all images

5. Capture at least 9 images per coverslip using a fluorescent microscope (Dialux 22, Leitz, Germany) at 100× magnification with an Infinity 3 camera and Infinity Capture software (Lumenera Corporation, Ottawa, ON, Canada). To prevent bias, capturing images should follow a systematic scanning pattern over the entire coverslip to avoid capturing from only a few selective regions. Automated cell counting can be carried out using the Viability3 program (custom software by Locksley McGann, Version 3.2, The Great Canadian Computer Company, Spruce

Grove, AB, Canada), which outputs data on the total number of cells and the number of green-fluorescing and red-fluorescing cells. The percentage (%) relative membrane integrity or relative viability is calculated as:

$$\text{Relative viability}(\%) = \frac{\text{number of green–fluorescing cells on a coverslip}}{\text{total number of cells on a coverslip}} \times 100\%$$

The % relative viability does not account for cell detachment and can give an overestimation of membrane integrity when the average total cell count on the coverslip is decreased due to detachment of cells; therefore, the % absolute viability is also calculated as follows:

$$\text{Absolute viability}(\%) = \frac{\text{number of green–fluorescing cells on an experimental covers}}{\text{total cell count on a pre–experiment(positive control)coversl}} \times 100\%$$

6. For assessment of delayed-onset cryoinjury, after CPA removal by serial dilution, post-thaw monolayers are re-incubated in 500 μL of culture medium under cell culture conditions of 37°C and 5% CO_2. Cell membrane integrity for these samples is then assessed the following day after culturing for 18–24 h, depending on the cell doubling time (*see* **Note 2**).

3.6 If Desired, Metabolic Activity Assessment by AlamarBlue Reduction (see Note 10)

1. Allow the AlamarBlue reagent to come to room temperature before use.
2. To the monolayers of cells on Rinzl coverslips in empty wells of a 24-well plate, add 250 μL of cell-specific medium to each well.
3. Use 4–8 empty wells of the 24-well plate for negative control (medium only, no cell monolayer) by adding 250 μL of cell-specific medium to each well.
4. Aseptically add 25 μL of AlamarBlue reagent to each well.
5. Place the plate in a humidified incubator at 37°C and 5% CO_2 and leave for 4 h.
6. After the incubation time ends, remove the plate and measure the absorbance at 570 nm and 600 nm wavelengths using a plate reader.
7. Calculate the % reduction of AlamarBlue using the following formula [36]:

$$\%\text{AlamarBlue reduction} = \frac{(\varepsilon_{ox})_{\lambda_2} A_{\lambda_1} - (\varepsilon_{ox})_{\lambda_1} A_{\lambda_2}}{(\varepsilon_{red})_{\lambda_1} A'_{\lambda_2} - (\varepsilon_{red})_{\lambda_2} A'_{\lambda_1}} \times 100\%$$

where $\lambda_1 = 570$ nm and $\lambda_2 = 600$ nm; ε_{ox} is the molar extinction coefficient of AlamarBlue in oxidized form (blue) $\left((\varepsilon_{ox})_{\lambda_1} = 80,586 \text{ and } (\varepsilon_{ox})_{\lambda_2} = 117,216\right)$; ε_{red} is the molar extinction coefficient of AlamarBlue in reduced form (red) $\left((\varepsilon_{red})_{\lambda_1} = 155,677 \text{ and } (\varepsilon_{red})_{\lambda_2} = 14,652\right)$; A is the absorbance of test wells (containing cells); and A' is the absorbance of negative control wells (i.e., no-cell control).

8. Compare the metabolic activity (i.e., % reduction of AlamarBlue) between the fresh, unfrozen monolayers and the cryopreserved, thawed monolayers prepared and cultured under the same conditions.

4 Notes

1. Safety recommendations:
   (a) Liquid nitrogen: Exposure to extreme cold liquid nitrogen temperature (−196°C) can rapidly freeze skin tissue and eye fluid, resulting in cold burns, frostbite, or permanent eye damage. Liquid nitrogen expands 695 times in volume when it vaporizes and has no warning properties, such as odor or color. Hence, if sufficient liquid nitrogen is vaporized so as to reduce the oxygen level to <19.5%, then there is a risk of oxygen deficiency, which may cause dizziness, nausea, and unconsciousness. Liquid nitrogen can leak into vials during storage, which may result in the vial exploding/spraying its contents when the vial is warmed, and the liquid nitrogen expands from the liquid to the gas phase. Use only approved containers/cryovials that can withstand extreme low temperature and that do not become brittle with time. Always wear basic personal protective equipment (gloves, safety glasses, and a lab gown) along with closed impervious shoes and floor-length pants with no cuffs, worn outside shoes. In addition, wear a full-face shield when dispensing liquid nitrogen to minimize injuries associated with splash or explosion and carry out transfers in a well-ventilated area. In case of a spill or leak, isolate the area and stop the source of the leak, if possible; ventilate the area and move

exposed personnel to fresh air. Also, wear loose-fitting insulated gloves so that they may be quickly removed if liquid nitrogen is spilled on arms and hands. Insulated gloves only provide short-term protection from accidental contact with the liquid and should not be dipped into liquid nitrogen; use long tongs to retrieve samples, if needed. In case of accidental skin contact, immerse the affected area in warm water (37°C) but do not use dry heat.

(b) Handling of cells carries the potential of exposure to biohazardous materials; for example, pipetting and vortexing cell suspensions can produce aerosols, which pose biosafety hazards. Always wear basic personal protective equipment and carry out culture and passaging procedures in a biological safety cabinet. Follow proper biohazard handling and disposal procedures.

(c) Methanol is flammable and should be kept away from open flames or other ignition sources. Keep stock in a designated flammable-solvent storage cabinet. Keep the methanol bath covered when not in use. Methanol is toxic if swallowed, absorbed through the skin, or inhaled. Basic personal protective equipment (gloves, safety glasses, lab gown, full-length pants, and closed-toed shoes) is required. In case of skin contact, wash with copious amounts of water. In case of inhalation, move the affected person to a ventilated area.

(d) Dimethyl sulfoxide is a combustible liquid that may cause skin, eye, and respiratory tract irritation. It readily penetrates the skin and may carry other dissolved materials into the body. Wear basic personal protective equipment; in case of skin contact, wash with copious amounts of water.

(e) GelRed® was used in our membrane integrity assessments because it was designed specifically to be cell-membrane impermeable, and is non-toxic and non-mutagenic, unlike ethidium bromide. SYTO 13 is non-hazardous but contains dimethyl sulfoxide as a solvent.

(f) Glassware can cause cuts and lacerations when it breaks. Inspect glassware for cracks before use. Always wear gloves and safety glasses. Loosen the lid to relieve pressure build-up when transferring the glass vial from liquid nitrogen to the 37°C water bath, and carry out the procedure at arm's length, away from the face.

(g) Water baths can get contaminated with bacteria and/or fungi. Add anti-bacterial and anti-fungal drops during routine lab maintenance and always wear gloves. Aquaguard-2 Solution (01-916-1E, FroggaBio) is a safe and effective reagent for preventing contamination by microorganisms in water baths. The active ingredient in Aquaguard-2 is biodegradable and safe to humans and does not cause any irritating effects to the skin when used in the recommended concentration.

2. It is recommended that a cell-growth kinetics experiment be performed to determine the population doubling time in order to know the number of days the cells need to be cultured to confluency. The population doubling time is also useful in designing experiments for evaluating delayed-onset cryoinjury, because different cells may take different time periods to complete a cell cycle.

$$\text{Doubling time} = \text{Duration} \times \ln(2) / \ln\left(\frac{\text{final concentration}}{\text{initial concentration}}\right)$$

3. It is best to trim two sides of the Rinzl coverslips by 2 mm into 16×16 mm, which can then be cut into four 8x8 mm squares. This size fits the bottom of the glass vials that we used for the graded freezing experiments.
4. The surface area of each well of the 24-well plate is 2 cm^2. Thus, if all 24 wells are to be used, prepare about 12.5 mL (0.5 mL extra) of cell suspension with a concentration of 10,000–20,000 cells/mL (depending on the recommended seeding density) and load 500 μL per well. It is judicious to prepare extra coverslips in the event of mishaps in the experimental procedures.
5. The constituents of the CPA solution, prepared in the cell-type-specific medium, are weighed rather than measured in terms of volume, because volume is temperature-dependent. The concentrations of each component are twice that of their final desired concentration. The HES powder as supplied by Bristol-Myers Squibb and used in our earlier studies [21, 26, 32] has been discontinued. As an alternative, we have validated Pentastarch (20%, w/v, Preservation Solutions, Inc.) and used it in our new protocols [40, 41].
6. It is best to pick the coverslip up by one corner, tilt the glass vial at an angle, and slide the coverslip in, ensuring that the cell side is facing up. This technique helps prevent accidentally dropping the coverslips and minimizes scratching the monolayer.
7. If there are several samples to be ice-nucleated, two metal forceps may be immersed in liquid nitrogen and used in an alternating manner. To avoid frostbite from handling the cold forceps, an extra layer of gloves should be used.

8. Loosening of the lid to release trapped liquid nitrogen in the vial should be carried out inside the BSC. Be aware that the warming during rapid thaw may cause any liquid nitrogen trapped inside the vial to expand and increase pressure inside the vial, which may cause an explosion or the vial to break. Hold the vial in the water bath away from the face.
9. SYTO 13 permeates all cells (both membrane-intact and membrane-damaged) and complexes with nucleic acids to emit a green fluorescence; GelRed permeates only cells with damaged or compromised membranes but is a stronger nucleic-acid binding dye and emits an intense red fluorescence, which masks the SYTO 13 green fluorescence. Membrane-intact (live) cells fluoresce green, while membrane-damaged (dead) cells fluoresce red (see Fig. 3 for representative images). We found that using Calcein AM/propidium iodide was not appropriate for live/dead staining of cell monolayers as shown (see Fig. 3).
10. AlamarBlue contains the cell-permeable and non-toxic indicator dye called resazurin. Resazurin (oxidized form, blue) detects cell metabolic activity by converting to resorufin (reduced form, red) in response to chemical reduction of the growth medium resulting from cell growth. The colorimetric signal detected at 570 nm is proportional to the number of cells with innate metabolic activity in the sample. Because there is considerable overlap in the absorption spectra of the oxidized and reduced forms of AlamarBlue, absorbance is measured at two wavelengths, namely 570 nm and 600 nm. The AlamarBlue dye present in the supplied reagent and its reduced form converted during the assay are light sensitive. Prolonged exposure of the reagent to light will result in increased background fluorescence in the assay and decreased sensitivity.

Acknowledgments

Development of this protocol and preparation of this book chapter were funded by the Canadian Institutes of Health Research (CIHR MOP 133684 and PS 166125). N Eskandari received a scholarship from the University of Alberta. JAW Elliott held a Canada Research Chair in Thermodynamics from 2001 to 2025.

Conflict of Interest Eskandari, N., Elliott, J. A. W., McGann, L. E., Nychka, J. A., & Marquez-Curtis, L. A. U.S. Patent 12,532,880 'Cryopreservation of cell monolayers.' (2026).

References

1. Rafii S, Gomez-Salinero JM (2018) Endothelial cell adaptation in regeneration. Science 362:1116–1117. https://doi.org/10.1126/science.aar4800
2. Kvietys PR, Granger DN (1997) Endothelial cell monolayers as a tool for studying microvascular pathophysiology. Am J Physiol - Gastrointest Liver Physiol 273:G1189–G1199. https://doi.org/10.1152/ajpgi.1997.273.6.g1189
3. Larionov A, Hammer CM, Fiedler K, Filgueira L (2024) Dynamics of endothelial cell diversity and plasticity in health and disease. Cells 13:1276. https://doi.org/10.3390/cells13151276
4. Gomez-Salinero JM, Redmond D, Rafii S (2025) Microenvironmental determinants of endothelial cell heterogeneity. Nat Rev Mol Cell Biol 26:476–495. https://doi.org/10.1038/s41580-024-00825-w
5. Neubauer K, Zieger B (2022) Endothelial cells and coagulation. Cell Tissue Res 387:391–398. https://doi.org/10.1007/s00441-021-03471-2
6. Ng XY, Peh G, Morales-Wong F, Gabriel R, Soong PL, Lin KH, Mehta JS (2024) Towards clinical application: calcium waves for in vitro qualitative assessment of propagated primary human corneal endothelial cells. Cells 13:2012. https://doi.org/10.3390/cells13232012
7. Gain P, Jullienne R, He Z, Aldossary M, Acquart S, Cognasse F, Thuret G (2016) Global survey of corneal transplantation and eye banking. JAMA Ophthalmol 134:167–173. https://doi.org/10.1001/jamaophthalmol.2015.4776
8. Rodríguez-Fernández S, Álvarez-Portela M, Rendal-Vázquez E, Piñeiro-Ramil M, Sanjurjo-Rodríguez C, Castro-Viñuelas R, Sánchez-Ibáñez J, Fuentes-Boquete I, Díaz-Prado S (2021) Analysis of cryopreservation protocols and their harmful effects on the endothelial integrity of human corneas. Int J Mol Sci 22:12564. https://doi.org/10.3390/ijms222212564
9. Tidke SC, Tidake P (2022) A review of corneal blindness: causes and management. Cureus 14:e30097. https://doi.org/10.7759/cureus.30097
10. Nishant P, Gurnani B, Singh P, Sinha S, Kaur K, Kumar A, Sinha RK (2025) Current concepts and recent trends in endothelial keratoplasty. World J Transplant 15:102507. https://doi.org/10.5500/wjt.v15.i2.102507
11. Liu J, Wang T, Dong J, Lu Y (2025) The blood–brain barriers: novel nanocarriers for central nervous system diseases. J Nanobiotechnology 23: 146. https://doi.org/10.1186/s12951-025-03247-8
12. Obermeier B, Daneman R, Ransohoff RM (2013) Development, maintenance and disruption of the blood-brain barrier. Nat Med 19:1584–1596. https://doi.org/10.1038/nm.3407
13. Hemanthakumar KA, Kivelä R (2020) Angiogenesis and angiocrines regulating heart growth. Vasc Biol 2:R93–R104. https://doi.org/10.1530/vb-20-0006
14. Moccia F, Negri S, Shekha M, Faris P, Guerra G (2019) Endothelial Ca2+ signaling, angiogenesis and vasculogenesis: just what it takes to make a blood vessel. Int J Mol Sci 20:3692. https://doi.org/10.3390/ijms20163962
15. Tombor LS, Dimmeler S (2022) Why is endothelial resilience key to maintain cardiac health. Basic Res Cardiol 117:35. https://doi.org/10.1007/s00395-022-00941-8
16. Dalal PJ, Muller WA, Sullivan DP (2020) Endothelial cell calcium signaling during barrier function and inflammation. Am J Pathol 190:535–542. https://doi.org/10.1016/j.ajpath.2019.11.004
17. Li C, Tan Z, Li H, Yao X, Peng C, Qi Y, Wu B, Zhao T, Li C, Shen J, Wang H (2024) Elevated microRNA-187 causes cardiac endothelial dysplasia to promote congenital heart disease through inhibition of NIPBL. J Clin Invest 135:e178355. https://doi.org/10.1172/JCI178355
18. Cardiovascular Diseases (CVDs) (2025). World Health Organization. https://www.who.int/news-room/fact-sheets/detail/cardiovascular-diseases-(cvds)
19. Martin SS, Aday AW, Almarzooq ZI et al (2024) 2024 Heart disease and stroke statistics: a report of US and global data from the American Heart Association. Circulation 2024;149(19):e1164. https://doi.org/10.1161/CIR.0000000000001247, 10.1161/CIR.0000000000001344
20. Polchow B, Kebbel K, Schmiedeknecht G, Reichardt A, Henrich W, Hetzer R, Lueders C (2012) Cryopreservation of human vascular umbilical cord cells under good manufacturing practice conditions for future cell banks. J Transl Med 10:98–114. https://doi.org/10.1186/1479-5876-10-98
21. Sultani AB, Marquez-Curtis LA, Elliott JAW, McGann LE (2016) Improved cryopreservation of human umbilical vein endothelial cells: a systematic approach. Sci Rep 6: 34393. https://doi.org/10.1038/srep34393
22. McGann LE (1979) Optimal temperature ranges for control of cooling rate. Cryobiology 16:211–216. https://doi.org/10.1016/0011-2240(79)90033-6

23. Ross-Rodriguez LU, Elliott JAW, McGann LE (2010) Investigating cryoinjury using simulations and experiments: 2. TF-1 cells during graded freezing (interrupted slow cooling without hold time). Cryobiology 61:46–51. https://doi.org/10.1016/j.cryobiol.2010.04.005
24. McGann LE (1978) Differing actions of penetrating and non-penetrating cryoprotective agents. Cryobiology 15:382–390. https://doi.org/10.1016/0011-2240(78)90056-1
25. Meryman HT (1971) Cryoprotective agents. Cryobiology 8:173–183. https://doi.org/10.1016/0011-2240(71)90024-1
26. Marquez-Curtis LA, McGann LE, Elliott JAW (2017) Expansion and cryopreservation of porcine and human corneal endothelial cells. Cryobiology 77:1–13. https://doi.org/10.1016/j.cryobiol.2017.04.012
27. Marquez-Curtis LA, Bokenfohr R, McGann LE, Elliott JAW (2021) Cryopreservation of human cerebral microvascular endothelial cells and astrocytes in suspension and monolayers. PLoS One 16:e0249814. https://doi.org/10.1371/journal.pone.0249814
28. Mohammed L, Marquez-Curtis LA, Elliott JAW (2023) Cryopreservation of human cerebral microvascular endothelial cells with glycerol. Cryobiology 113:104551. https://doi.org/10.1016/j.cryobiol.2023.104551
29. Stolzing A, Naaldijk Y, Fedorova V, Sethe S (2012) Hydroxyethylstarch in cryopreservation - mechanisms, benefits and problems. Transfus Apher Sci 46:137–147. https://doi.org/10.1016/j.transci.2012.01.007
30. Acker JP, Larese A, Yang H, Petrenko A, McGann LE (1999) Intracellular ice formation is affected by cell interactions. Cryobiology 38:363–371. https://doi.org/10.1006/cryo.1999.2179
31. Rutt T, Eskandari N, Zhurova M, Elliott JAW, McGann LE, Acker JP, Nychka JA (2019) Thermal expansion of substrate may affect adhesion of Chinese hamster fibroblasts to surfaces during freezing. Cryobiology 86:134–139. https://doi.org/10.1016/j.cryobiol.2018.10.006
32. Eskandari N, Marquez-Curtis LA, McGann LE, Elliott JAW (2018) Cryopreservation of human umbilical vein and porcine corneal endothelial cell monolayers. Cryobiology 85:63–72. https://doi.org/10.1016/j.cryobiol.2018.10.001
33. Crisol M, Wu K, Laouar L, Elliott JAW, Jomha NM (2020) Antioxidant additives reduce reactive oxygen species production in articular cartilage during exposure to cryoprotective agents. Cryobiology 96:114–121. https://doi.org/10.1016/j.cryobiol.2020.07.008
34. Yu MH, Marquez-Curtis LA, Elliott JAW (2024) Cryopreservation-induced delayed injury and cell-type-specific responses during the cryopreservation of endothelial cell monolayers. Cryobiology 115:104857. https://doi.org/10.1016/j.cryobiol.2024.104857
35. Marquez-Curtis LA, Elliott JAW (2025) Interrupted slow cooling of mouse cardiac endothelial cell monolayers. Cryobiology 121:105319. https://doi.org/10.1016/j.cryobiol.2025.105319
36. AlamarBlue Cell Viability Assay Reagent (2016) G-Biosciences. https://cdn.gbiosciences.com/pdfs/protocol/786-921_protocol.pdf. Accessed 15 Mar 2025
37. AlamarBlue Cell Viability Reagent (2008). Invitrogen MP01025. https://assets.thermofisher.com/TFS-Assets/LSG/manuals/AlamarBluePIS.pdf. Accessed 24 Mar 2025
38. Marquez-Curtis LA, Sultani AB, McGann LE, Elliott JAW (2016) Beyond membrane integrity: assessing the functionality of human umbilical vein endothelial cells after cryopreservation. Cryobiology 72:183–190. https://doi.org/10.1016/j.cryobiol.2016.05.005
39. Calcein AM Fluorescent Dye (2013). Corning life sciences. https://ecatalog.corning.com/life-sciences/b2b/CA/en/Cell-Culture/Supplements-and-Reagents/Fluorescent-Dyes/Corning®-Fluorescent-Dye/p/354216?clear=true. Accessed 27 Apr 2022
40. Marquez-Curtis LA, Eskandari N, McGann LE, Elliott JAW (2020) Protocol for cryopreservation of endothelial monolayers. Methods Mol Biol 2180:581–591. https://doi.org/10.1007/978-1-0716-0783-1_30
41. Marquez-Curtis, LA, Sultani, AB, McGann, LE, Elliott, JAW, 2018. Protocol for cryopreservation of endothelial cells in suspension, In: Preservation of cells: a practical manual. Wiley-Blackwell, pp. 119–124

Check for updates

Chapter 29

Cryopreservation of Reproductive Organoids

Riley E. Thompson-Brandhagen

Abstract

Three-dimensional organoid cell cultures are an in vitro model that is more similar to the in vivo environment than other cell culture methods. Because of this, organoids are a powerful alternative to live animal research studies. Cryopreservation of organoids is essential when working with a tissue source that is not readily available at all times, and a cryo-repository of organoids allows these cell cultures to be performed as needed. The cryopreservation and thawing protocols must be optimized for each species and tissue type to ensure the optimal cellular viability and proliferation after freeze-thawing. The protocol provided here has been used to cryopreserve various types of reproductive cell cultures (oviduct and endometrium) in several domestic and wildlife species.

Keywords 3D, Cell culture, Endometrium, Fallopian tube, In vitro, Organoid, Oviduct, Uterus

1 Introduction

Organoids are a type of three-dimensional cell culture that can proliferate long-term while maintaining the normal function of the original organ, and organoids can be grown using stem cells or tissue-specific progenitor cells [1]. While several methods for growing organoids have been reported, the classical method involves suspending cells in an extracellular matrix (ECM), like Matrigel, that is fluid when cold and more gel-like when warmed. The droplets of ECM containing cells are placed in a well plate, warmed, and then overlaid with culture medium that diffuses through the ECM to support organoid proliferation. One organoid is comprised of many individual cells, and one droplet of ECM contains hundreds of organoids. The expected organoid morphology is tissue- and species-dependent; however, most reproductive organoids are round structures that are considered 'cystic', indicating that they have a central lumen [2–7].

The protocol and technique for cryopreservation and thawing of organoids are critical for cellular survivability. As with all cryopreservation, the type of cryoprotectant used affects cryopreservation, and most protocols use dimethyl sulfoxide (DMSO) for reproductive

Willem F. Wolkers and Harriëtte Oldenhof (eds.), *Cryopreservation and Freeze-Drying Protocols*, Methods in Molecular Biology, vol. 3060, https://doi.org/10.1007/978-1-0716-5416-3_29,

organoid cryopreservation [2, 3, 5]. The freezing medium that contains the cryoprotectant must be added to the organoids slowly to allow time for cellular equilibration. However, after adding the freezing medium, organoids should begin the slow freezing process immediately to prevent excessive dehydration of cells caused by the cryoprotectant [8]. The thawing protocol is equally important to prevent cells from swelling and bursting in response to the osmotic changes associated with adding the thawing medium.

Cryopreservation and thawing protocols must be optimized for each new species and tissue type. For example, the three freezing media reported in the following protocol have been tested for the organoid cryopreservation of equine endometrium, feline oviduct, and bovine oviduct. While no differences were detected in growth and proliferation of either equine endometrial or feline oviductal organoids in response to the freezing media (FM1, FM2, Recovery Cell Culture Freezing Medium), bovine oviductal organoids demonstrated poor proliferation after thawing in response to cryopreservation with Recovery Cell Culture Freezing Medium compared to either FM1 or FM2 (see Fig. 1). Thus, testing several freezing protocols is essential to ensure the greatest organoid viability and proliferation after thawing and subsequent culture. This protocol provides detailed technical notes on how to prepare the media required for organoid freezing and thawing, prepare and cryopreserve organoids, and thaw the organoids for optimal post-thaw viability.

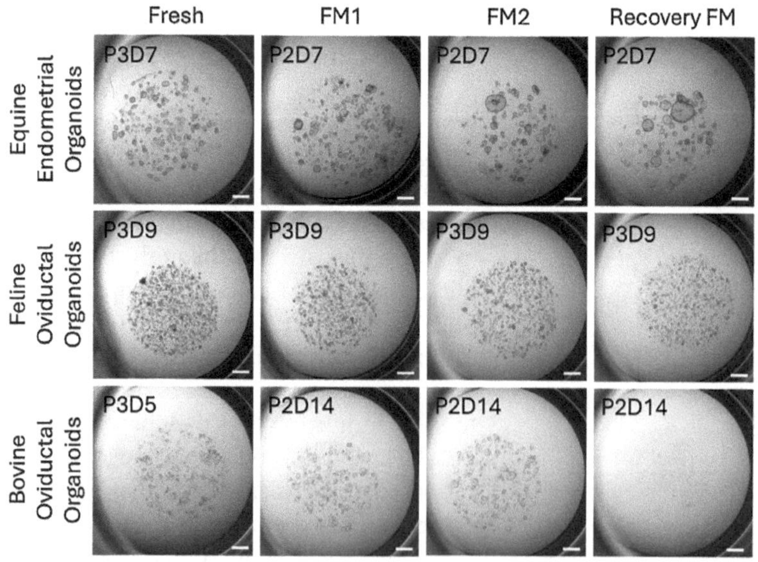

Fig. 1 Fresh and frozen-thawed organoids derived from equine endometrial, feline oviductal, and bovine oviductal tissue demonstrating differences in proliferation after cryopreservation with different freezing media. FM = freezing medium; P = passage; D = day. Scale bar = 1 mm

2 Materials

1. Water bath.
2. −80°C freezer.
3. Microcentrifuge.
4. Cool Cell (Corning) OR Mr. Frosty (Thermo Scientific).
5. Liquid nitrogen and a liquid nitrogen tank.
6. Cryo-forceps.
7. Styrofoam cooler for liquid nitrogen transport.
8. 200 and 1000 μL pipettors and tips.
9. Ice.
10. 2 mL cryovials.
11. 20 mL syringes.
12. 0.22 μm syringe filters.
13. 15 mL conical tubes.
14. 1.7 mL microcentrifuge tubes.
15. Organoid Harvesting Solution (R&D Systems) OR Cell Recovery Solution (Corning).
16. DMEM/F12.
17. MEM Eagle with Earle's Salts and sodium bicarbonate.
18. HEPES.
19. Penicillin/streptomycin.
20. Sodium pyruvate.
21. Glutamax.
22. Fetal bovine serum (FBS).
23. Dimethyl sulfoxide (DMSO).
24. Recovery Cell Culture Freezing Medium (Gibco).

3 Methods

3.1 Preparation of Media

1. Prepare base medium (see Table 1). Filter medium using 0.22 μm syringe filter.
2. Prepare freezing medium of choice (see Tables 2 and 3 or Recovery Cell Culture Freezing Medium) and post-thaw medium (see Table 4; *see* **Note 1**). Filter media using 0.22 μm syringe filter.
3. Warm freezing medium and post-thaw medium to 37°C.

Table 1
Preparation of base medium

Component	Amount
MEM Eagle with Earle's Salts and sodium bicarbonate	18.9 mL
HEPES (1 M)	0.5 mL
Penicillin/Streptomycin	200 μL
Sodium Pyruvate (100 mM)	200 μL
Glutamax (100×)	200 μL

Table 2
Preparation of freezing medium 1

Component	Amount
Base Medium	7 mL
Fetal Bovine Serum (FBS)	2 mL
Dimethyl Sulfoxide (DMSO)	1 mL

Table 3
Preparation of freezing medium 2

Component	Amount
Fetal Bovine Serum (FBS)	9 mL
Dimethyl Sulfoxide (DMSO)	1 mL

Table 4
Preparation of post-thaw medium

Component	Amount
Base Medium	16 mL
Fetal Bovine Serum (FBS)	4 mL

3.2 Preparation of Organoids for Cryopreservation

1. Remove the 48-well plate containing organoids from the incubator (*see* **Note 2**).
2. Remove culture medium from each well leaving the organoids in the extracellular matrix (ECM).
3. Add 250 μL Organoid Harvesting Solution (*see* **Notes 3 and 4**).
4. Transfer the contents of the wells to microcentrifuge tubes.
5. Place on ice for 1 h.

6. Centrifuge at 1000×*g* for 10 min. Remove supernatant.
7. Resuspend the pellet in 1 mL DMEM/F12.
8. Centrifuge at 1000×*g* for 10 min. Remove supernatant.

3.3 Cryopreservation of Organoids

1. Add freezing medium to the organoid pellet dropwise while gently flicking the tube to mix (*see* **Notes 5 and 6**; see Fig. 2).
2. Transfer 250 μL organoids in freezing medium to 2 mL cryovial.
3. Transfer cryovials to a slow freezing container, such as Cool Cell (Corning) or Mr. Frosty (Thermo Scientific), that reduces temperature 1°C per minute (*see* **Note 7**).
4. Place the slow freezing container in −80°C overnight.
5. Transfer cryovials to the tank containing liquid nitrogen for storage.

3.4 Thawing Organoids

3.4.1 Thawing Protocol 1 (see Note 8; see Fig. 3)

1. Remove cryovial from liquid nitrogen tank, and transport cryovial in a container with liquid nitrogen.
2. Remove cryovial from liquid nitrogen with pre-cooled cryo-forceps, and thaw in the air for 1 min (*see* **Notes 9 and 10**).
3. Place cryovial in 25°C water bath for 1 min (see **Note 11**).
4. Add 1 mL pre-warmed post-thaw medium (37°C) dropwise over 5 min to cryovial while gently flicking tube to mix.
5. Transfer the solution to a 15 mL conical tube and add an additional 5 mL post-thaw medium.
6. Centrifuge at 200×*g* for 10 min. Remove supernatant.
7. Add 1 mL DMEM/F12 to the pellet.
8. Centrifuge at 1000×*g* for 10 min. Remove supernatant.
9. Plate organoids using standard procedures [2–5, 7].

3.4.2 Thawing Protocol 2 (see Note 12)

1. Remove cryovial from liquid nitrogen tank, and transport cryovial in a container with liquid nitrogen.
2. Remove cryovial from liquid nitrogen with pre-cooled cryo-forceps (*see* **Note 8**) and transfer to 37°C water bath for 1 min (*see* **Notes 13 and 14**).
3. Add 1 mL pre-warmed post-thaw medium (37°C) dropwise over 5 min to cryovial while gently flicking tube to mix.
4. Transfer the solution to a 15 mL conical tube and add an additional 5 mL post-thaw medium.
5. Centrifuge at 200×*g* for 10 min. Remove supernatant.
6. Add 1 mL DMEM/F12 to the pellet.
7. Centrifuge at 1000×*g* for 10 min. Remove supernatant.
8. Plate organoids using standard procedures [2–5, 7].

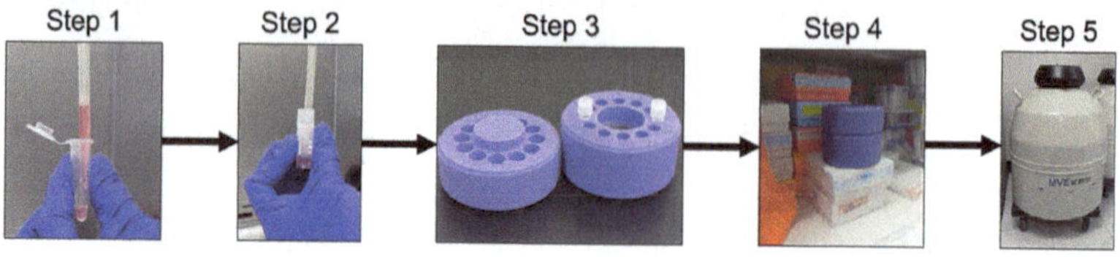

Fig. 2 Flow diagram depicting the steps for organoid cryopreservation corresponding to the steps outlined in Sect. 3.3

4 Notes

1. Can prepare base medium, freezing medium, and/or post-thaw medium in advance and store in −20°C until use.
2. Protocol describes organoids using standard culture in 48-well plate using 25 μL droplets of the extracellular matrix (ECM) Matrigel or Cultrex UltiMatrix [2–5] but can be modified for other organoid culture systems.
3. Alternative to Organoid Harvesting Solution is Cell Recovery Solution (Corning).
4. If using a different amount of ECM per well, use 10× the Organoid Harvesting Solution to the amount of ECM (i.e., if 50 μL ECM droplets, add 500 μL Organoid Harvesting Solution).
5. Typically, cryopreserving four wells of organoids per 250 μ L freezing medium is an appropriate ratio. However, if not many organoids are present in each well, increase the number of wells that are combined per 250 μL of freezing medium.
6. Different freezing medium compositions may be appropriate for different species and tissue types. If working with a new species and/or tissue type, consider trialing all three freezing media mentioned and perform a growth trail to determine which freezing medium is best. To perform a growth trial for organoids, dissociate organoids to single cells using TrypLE Express Enzyme, count the number of live cells using a hemocytometer and trypan blue, and plate a finite number of cells per well in ECM (such as 5000 or 10,000 cells per well), and compare which freezing medium results in the greatest number and size of organoids following culture.
7. Be efficient following the addition of freezing medium and placing the slow freezing container with the cryovials in −80°C. Extended times in the freezing medium prior to freezing can reduce cellular viability.
8. This is the preferred thawing protocol if using freezing medium 1.

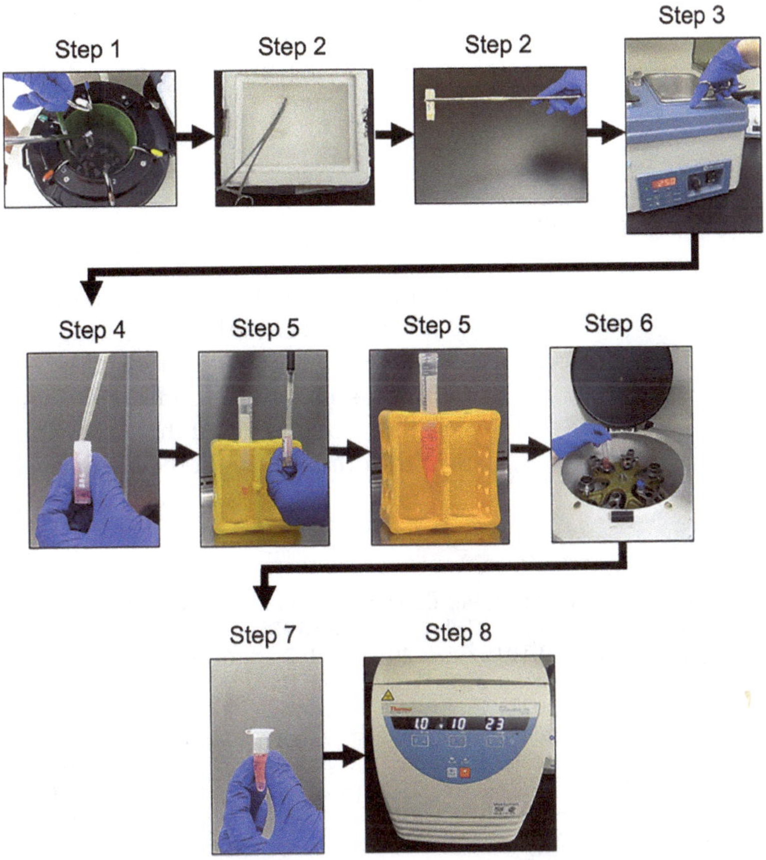

Fig. 3 Flow diagram depicting the steps for thawing organoids corresponding to the steps outlined in Sect. 3.4.1

9. Cool tips of cryo-forceps by placing them in liquid nitrogen until the liquid nitrogen no longer bubbles around them.
10. Carefully wave the cryovial in the air; can loosen and tighten the cryovial cap while avoiding touching hot fingers over the frozen organoids if liquid nitrogen may have entered the cryovial to release pressure.
11. Constantly move/agitate the cryovial in the water bath to avoid creating a gradient of colder water surrounding the thawing tube. Ensure the cap is tightly closed to avoid water entering the tube.
12. This is the preferred thawing protocol if using freezing medium 2 or Recovery Cell Culture Freezing Medium (Gibco).
13. Transfer the cryovial to the water bath efficiently – the organoids are thawing in the air if moving too slowly.

14. Ideally, the cryovial will be mostly thawed with a few pieces of ice remaining when removed from the water bath. Can increase the amount of time in the water bath if not thawed to this stage.

Acknowledgments

The author thanks Dr. Budhan Pukazhenthi, Dr. Fiona Hollinshead, and Mindy Meyers for their work in developing these protocols. The author is supported by the Office of the Director, National Institutes of Health, under award number K01OD032455. The content is solely the responsibility of the author and does not necessarily represent the official views of the National Institutes of Health.

References

1. Clevers H (2016) Modeling development and disease with organoids. Cell 165:1586–1597. https://doi.org/10.1016/j.cell.2016.05.082
2. Thompson RE, Meyers MA, Pukazhenthi BS, Hollinshead FK (2021) Evaluation of growth, viability, and structural integrity of equine endometrial organoids following cryopreservation. Cryobiology 104:56–62. https://doi.org/10.1016/j.cryobiol.2021.11.003
3. Thompson RE, Meyers MA, Veeramachaneni DNR et al. (2022) Equine oviductal organoid generation and cryopreservation. Methods Protoc 5:51. https://doi.org/10.3390/mps5030051
4. Thompson RE, Johnson AK, Dini P, Turco MY, Prado TM, Premanandan C, Burton GJ, Ball BA, Whitlock BK, Pukazhenthi BS (2020) Hormone-responsive organoids from domestic mare and endangered Przewalski's horse endometrium. Reproduction 160:819–831. https://doi.org/10.1530/REP-20-0266
5. Thompson RE, Meyers MA, Premanandan C, Hollinshead FK (2023) Generation and cryopreservation of feline oviductal organoids. Theriogenology 196:167–173. https://doi.org/10.1016/j.theriogenology.2022.11.020
6. Menjivar NG, Gad A, Thompson RE, Meyers MA, Hollinshead FK, Tesfaye D (2023) Bovine oviductal organoids: a multi-omics approach to capture the cellular and extracellular molecular response of the oviduct to heat stress. BMC Genomics 24: 646. https://doi.org/10.1186/s12864-023-09746-y
7. Turco MY, Gardner L, Hughes J, Cindrova-Davies T, Gomez MJ, Farrell L, Hollinshead M, Marsh SGE, Brosens JJ, Critchley HO, Simons BD, Hemberger M, Koo BK, Moffett A, Burton GJ (2017) Long-term, hormone-responsive organoid cultures of human endometrium in a chemically defined medium. Nat Cell Biol 19:568–577. https://doi.org/10.1038/ncb3516
8. Mazur P (1963) Kinetics of water loss from cells at subzero temperatures and the likelihood of intracellular freezing. J Gen Physiol 47:347–369. https://doi.org/10.1085/jgp.47.2.347

Check for updates

Chapter 30

Protocol for Cryopreservation of Particulated Articular Cartilage by Vitrification

Mohammadhamed Shahsavari, Kezhou Wu, Nadia Shardt, Nadr M. Jomha and Janet A.W. Elliott

Abstract

Transplantation of particulated articular cartilage (AC) is an established surgical procedure that produces promising clinical results while potentially preventing joint deterioration into osteoarthritis. This chapter documents a patented vitrification protocol that is able to cold preserve particulated AC tissue, thereby maintaining cell viability and matrix integrity. Mathematical modeling was used to develop this two-step, dual-temperature, multi-cryoprotectant loading protocol. The cryoprotectant loading consists of exposing the particulated AC to two sequential cryoprotectant cocktails at different temperatures and for different durations. The fresh particulated AC is incubated in the first cryoprotectant solution that contains 3 M dimethyl sulfoxide and 3 M ethylene glycol at 0°C for 10 min. Subsequently, the particulated AC is transferred to the second cryoprotectant solution that consists of 3 M dimethyl sulfoxide, 3 M ethylene glycol, and 2 M propylene glycol at −10°C for an additional 20 min. After cryoprotectant loading, the particulated AC is loaded in a cryovial and plunged directly into liquid nitrogen for vitrification and storage. Rewarming is performed using a 37°C water bath for approximately 30 s and the cryoprotectants are washed from the rewarmed sample using 30 mL of DMEM/F-12-HEPES for 45 min at 4°C on a shaker.

Keywords Cartilage repair, Cryoprotectant cocktail, Particulated articular cartilage, Vitrification

1 Introduction

Articular cartilage (AC) defects are a frequently encountered disorder that can lead to joint pain, functional disability, and reduced quality of life [1, 2]. Cartilage repair surgery is frequently necessary to treat focal AC defects especially in young, active patients [3]. One recently developed treatment option is transplanting fresh particulated AC (< 1-mm^3 cubes) allograft using viable juvenile AC fragments to fill AC defects (e.g., DeNovo NT Natural Tissue Graft [Zimmer Biomet]) [4–6]. Between 2003 and 2013, more than 8700 patients were successfully treated using particulated AC grafts with no reports of graft rejection [6–8]. Thus, fresh particulated AC allograft is a

Willem F. Wolkers and Harriëtte Oldenhof (eds.), *Cryopreservation and Freeze-Drying Protocols*, Methods in Molecular Biology, vol. 3060, https://doi.org/10.1007/978-1-0716-5416-3_30,

good option to treat AC defects. However, widespread use of fresh particulated AC grafts to repair AC defects has been hindered primarily by cost and the limited lifetime of stored graft material. A comprehensive review of AC-related tissue banking in the United States [7] documented that although approximately 4700 osteochondral allografts were used in 2018, the percentage of allografts that expired before use ranged from 20%–29% across four tissue banks. This wastage is observed even after the average time from procurement to implantation was extended to 38.5 days [7]. Extended storage time at 4°C is extremely detrimental to tissue quality and cells being implanted, with significant biochemical changes noted within 7 days and cell death within 14 days of storage [9, 10]. With longer holding periods, these key tissue parameters only deteriorate further, resulting in the implantation of compromised tissue. Current commercially available human fresh particulated AC allograft (DeNovo) is recommended for use within 45 days, although the criteria for determining storage time are not clear [11]. In addition, commercially available products are very expensive (US$4440 per packet in 2013 [8]; CAD$3820 per packet in 2025). Considering the significant limitations in currently available products and their associated high costs, developing long-term particulated AC storage techniques is urgently required to improve cell viability and matrix integrity, while decreasing tissue wastage and costs.

Long-term preservation of cells and tissues beyond the times achievable at hypothermic temperatures (i.e., 4°C) is possible with vitrification. Vitrification is an "ice-free" cell and tissue preservation method that requires high concentrations of cryoprotectants, also known as cryoprotective agents (CPAs), combined with an appropriate cooling approach, that prevents ice formation that could kill cells and damage tissue structure [12–14]. With temperatures lower than −130°C, storage can be indefinite because biochemical reactions cease. Our group has developed a patented vitrification procedure that is able to vitrify AC tissue and maintain cell viability and matrix integrity [15]. Under the patented procedure, we have designed a specific vitrification protocol for particulated AC that is a two-step dual-temperature method that uses multiple CPAs [16]. This protocol was designed using mathematical modeling to determine optimal CPA exposure times and temperatures (via calculations of spatially and temporally resolved CPA concentration, solution freezing point, and vitrifiability, while qualitatively minimizing cytotoxicity with an appropriate selection of CPA type and concentration in the loading solution and not unnecessarily exceeding required loading times) [16]. For both porcine and human particulated AC, we showed that the vitrification protocol resulted in high post-warming metabolic activity and in situ chondrocyte membrane integrity and the ability of post-warming isolated chondrocytes to migrate and produce glycosaminoglycans [16]. In a pig transplant model, we showed that 6 months post-transplantation, the repair effect of vitrified particulated AC was equivalent to the repair effect of fresh particulated AC that had been stored at 4°C for 21 days in terms of morphology, chondrocyte membrane integrity, and histological assessment for production of collagen type I and II and glycosaminoglycans [17]. We also showed that chondrocyte membrane integrity is maintained for 7 days post-warming of vitrified porcine particulated AC [18].

Our methods for vitrification of particulated AC are given in detail below [16–18].

2 Materials

2.1 Tissue Preparation

This will depend on the type of tissue being investigated. In this chapter, porcine articular cartilage (AC) is the tissue of interest. The experimental use of animal tissues and cells for research was approved by the Research Ethics Office at the University of Alberta (RES0036331).

1. Cordless electric saw.
2. Surgical scalpel (e.g., #20 feather stainless steel surgical blade).
3. Tissue forceps.
4. Chemical spatula.
5. Digital caliper.
6. Conical tube (e.g., 50 mL).
7. Plastic cooler (e.g., 28-quart cooler).
8. Dissection board (e.g., 18 × 24 in. cutting board).
9. Absorbent underpads (e.g., 36 × 20 in.).
10. 70% ethanol (prepared from 100% ethanol, anhydrous ethyl alcohol).
11. Paper towels.
12. 4°C refrigerator.
13. Lab beaker (e.g., 250 mL).
14. Balance (e.g., analytical balance).

2.2 Cryoprotective Agent Loading/Unloading and Rewarming

1. Mini shaker (e.g., VWR Mini shaker).
2. Bottle-top filter.
3. Sterile chemical spoon.
4. 50-mL sterile conical tube.
5. 50-mL sterile plastic cup.
6. Ice-water bath (4-liter plastic bucket).

7. Methanol/ethanol cooling bath.
8. Mesh strainer.
9. 1.8-mL cryovial.
10. Cryovial rack.
11. Liquid nitrogen tank.
12. Water bath (37°C).
13. Thermocouple thermometer (e.g., Cole-Parmer, thermocouple thermometer dual type JKTE).
14. Digital timer.
15. Magnetic stirrer (e.g., VWR Mini stirrer).
16. Lab beaker (e.g., 100 mL).
17. Sterile pipettes (e.g., 5 and 25 mL sterile serological pipettes).
18. Pipette controller.
19. Adjustable pipette set (e.g., 30–300 μL Finnpipette™ manual pipettes).
20. Sterile pipette tips.
21. Balance (e.g., analytical balance).
22. Magnetic stirrer bar.

2.3 Chemicals

1. CPAs:
   (a) Dimethyl sulfoxide (>99.5%).
   (b) Ethylene glycol (1, 2-ethanediol, >99.5%).
   (c) Propylene glycol (1, 2-propanediol, >99.5%).
2. 1× Dulbecco's phosphate-buffered saline (PBS).
3. Dulbecco's Modified Eagle Medium F12 (DMEM-F12, Gibco).
4. Newborn calf serum (Sigma-Aldrich).
5. Antimicrobial agent solution (e.g., Antibiotic–Antimycotic, 100×, including liquid 10,000 U/mL penicillin, 10,000 μg/mL streptomycin, 25 μg/mL amphotericin B, 100 mL, Gibco).
6. Methanol/ethanol (99.8%).

3 Methods

3.1 Preparation of Articular Cartilage

Safe laboratory practices should be used at all times (*see* **Note 1**). The objective of the specimen preparation is to have pure tissue with well-defined dimensions (*see* **Note 2**).

1. Prepare articular cartilage (AC) in a fume hood (or in an area designated for animal work).
2. Label several large lab beakers (250 mL) with the condyle number.
3. Place absorbent underpads beneath the dissection board, remove the porcine hind leg from the plastic cooler, and place it on the board.
4. Use a sterile surgical scalpel to dissect the porcine joint and expose the femoral condyles.
5. Use the electric saw to remove the femoral condyles from the femur bone.
6. Immerse the femoral condyles in a beaker (250 mL) containing 150 mL of 1 × PBS solution to keep the AC moist, avoiding prolonged exposure of the AC to air.
7. Shave off AC from the femoral condyles using a sterile surgical scalpel. The harvested chips should just contain AC and no subchondral bone.
8. Cut AC slices into approximately 1-mm^3 cubes using a sterile surgical scalpel.
9. Weigh 0.2 g (1 dose) of particulated AC using a sterile chemical spatula and an analytical balance.
10. Clean each dose of particulated AC by immersing in a beaker containing 150 mL of sterile 1 × PBS solution supplemented with antibiotic–antimycotic (100 U/mL penicillin, 100 µg/mL streptomycin, and 0.25 µg/mL amphotericin B) for a 15 min disinfection step in a biological safety cabinet under sterile conditions.
11. Transfer each dose of particulated AC to complete medium (DMEM complete: Dulbecco's Modified Eagle Medium F12 supplemented with 10% heat-inactivated newborn calf serum, 100 U/mL penicillin, 100 µg/mL streptomycin, 0.25 µg/mL amphotericin B, and 1 mM sodium pyruvate), and store at 4°C in a refrigerator prior to CPA exposure. The antimicrobial agent solution is used to prevent bacterial or fungal growth with longer holding periods, while the DMEM complete is important to provide nutrients and help keep the cells viable.
12. Use the AC within 24 h after dissection.
13. Clean the fume hood and biological safety cabinet with 70% ethanol and paper towels.

3.2 Setup of Ice Bath

1. Use a 4-liter plastic bucket and fill it with ice cubes and distilled water.
2. Set the shaker speed to 180 rpm.

3. Put the filled bucket with ice and water on the shaker for 30 min.
4. Measure the temperature of the ice bath using a thermocouple thermometer to record 0°C.

3.3 Setup of Methanol/Ethanol Bath

1. Fill the bath with methanol/ethanol.
2. Set the temperature to −10°C.
3. Start the cooling process and wait until bath temperature reaches −10°C (60 min).

3.4 Cryoprotective Agent Loading

CPA loading consists of exposing the particulated AC to two sequential CPA cocktail solutions at different temperatures and for different durations.

3.4.1 Making CPA Solutions

1. Solution 1:
   (a) Prepare the 50-mL final volume CPA Solution 1 [3 M dimethyl sulfoxide (DMSO; BPD1281-Thermo Fisher Scientific, Canada) and 3 M ethylene glycol (EG; E1784-Thermo Fisher Scientific) in Dulbecco's modified Eagle medium (DMEM/F-12) containing 15 mM of HEPES (Gibco, USA)] by adding 11.720 g of DMSO and 9.311 g of EG and topping up to 50-mL volume with DMEM/F-12 with HEPES (or adding 30.955 g of DMEM/F-12 with HEPES at 22°C). Use a 100 mL lab beaker, a magnetic stirrer bar, and a magnetic stirrer to mix CPAs completely for 10 min.
   (b) After 10 min of mixing CPA Solution 1 using a magnetic stirrer, filter the solution with a 0.2-mm bottle-top filter into a 50-mL sterile conical tube.
   (c) Incubate the conical tube in a 0°C ice-water bath for 30 min while the ice-water bath is shaking on the shaker (*see* **Note 3**).
   (d) Measure the temperature of CPA Solution 1 using a thermocouple thermometer to record 0°C (*see* **Notes 4 and 5**).
2. Solution 2:
   (a) Prepare the 50-mL final volume CPA Solution 2 [3 M DMSO, 3 M EG, and 2 M propylene glycol (PG; P3554-Thermo Fisher Scientific, Canada) in DMEM/F-12 containing 15 mM of HEPES] by adding 11.720 g of DMSO, 9.311 g of EG, and 7.610 g of PG, and topping up to 50 mL volume with DMEM/F-12 with HEPES (or adding 23.624 g DMEM/F-12 with HEPES at 22°C). Use a

100-mL lab beaker, a magnetic stirrer bar, and a magnetic stirrer to mix CPAs completely.

(b) After 10 min of mixing CPA Solution 2 using a magnetic stirrer, filter with a 0.2-μm bottle-top filter into a 50-mL sterile conical tube.

(c) Incubate the conical tube in −10°C methanol/ethanol cooling bath for 45 min (*see* **Note 5**).

(d) Measure the temperature of CPA Solution 2 using a thermocouple thermometer to record −10°C (*see* **Notes 4–6**).

3.4.2 Exposing Particulated AC to CPA Solutions

1. Use a sterile chemical spatula to transfer one dose of particulated AC to a 50-mL conical tube containing 50 mL of pre-cooled CPA Solution 1 and incubate it in the ice-water bath at 0°C for 10 min while the ice-water bath is shaking on the shaker (180 rpm).
2. Use a mesh strainer and sterile chemical spatula to transfer particulated AC to the tube containing 50 mL of pre-cooled solution 2 and incubated at −10°C (in a methanol/ethanol cooling bath) for an additional 20 min (*see* **Note 7**).

3.5 Vitrification and Storage

1. After CPA loading, remove the AC from the tubes using a sterile chemical spatula and a mesh strainer and transfer it to a 1.8-mL cryovial.
2. Close the cryovial screw cap tightly.
3. Load the cryovial in the cryovial rack.
4. Plunge the cryovial rack directly into liquid nitrogen for vitrification and storage.

3.6 Rewarming and CPA Unloading

1. Remove the cryovial tube containing vitrified AC quickly.
2. Vent the cryovial by slightly loosening the lid (*see* **Note 8**).
3. Warm the cryovial in a 37°C water bath for approximately 30 s until the surrounding CPA glass is fully liquefied.
4. Use a sterile chemical spatula to remove the AC cubes from the cryovial and transfer them to a 50 mL sterile plastic cup containing 30 mL of DMEM/F-12-HEPES for 45 min at 4°C on a shaker (180 rpm).
5. Use a sterile chemical spatula to transfer the AC cubes from the first washing solution to the next sterile plastic cup containing 30 mL of DMEM/F-12-HEPES and antimicrobial agent solution (100 units/mL penicillin, 100 mg/mL streptomycin, and 0.25 mg/mL amphotericin B) and store at 4°C until tissue is needed.

4 Notes

1. Guidelines for safe laboratory practices:
   (a) For safety reasons, make sure to wear appropriate personal protective equipment (lab gown, safety goggles, gloves, full-length pants, and closed-toe shoes) before starting any procedure.
   (b) Laboratory chemicals can be harmful to the human body. The researcher should always consult the Safety Data Sheet provided by the manufacturer. Methanol/ethanol is a flammable liquid; it should be stored away from open flames. Methanol is toxic if swallowed, inhaled, absorbed through the skin, or in contact with the eyes. When cooling with a methanol/ethanol-based low-temperature bath, always keep the bath covered. Cryoprotectants (CPAs) such as dimethyl sulfoxide (DMSO), ethylene glycol (EG), and propylene glycol (PG) are irritative and toxic if swallowed, inhaled, absorbed through the skin, or in contact with the eyes. Wash with plenty of water in case of any eye or skin contact; refer to medical service if needed.
   (c) Be cautious when using electric equipment, including the shaker, magnetic stirrer, methanol/ethanol cooling bath, and electric saw, which may cause electric shock to the user if not handled properly.
   (d) When dissecting animal samples, use the cordless saw with proper care, and operate in a fume hood or a designated area for animal work. Carry out the cutting procedure at arm's length. Dissection with a surgical blade should be performed under supervision or after training. Handling the electric saw can be dangerous because of both the electricity and the sharpness of the blade.
   (e) Discard biohazardous laboratory waste following university/company environmental health and safety and environmental regulations. Sharp blades and glass can cause cuts and should be kept in the appropriate sharps container and labeled biohazardous after use.
2. In our previous research [16], hind legs with joints from sexually mature pigs (aged over 54 weeks) were obtained from meat processing plants, and femoral condyles with intact articular cartilage (AC) were dissected for research. Porcine hind legs were collected and transported to the laboratory in a plastic cooler in 30 min after processing. This research was provided an Exemption from the Animal Use and Care Committee at the University of Alberta Research Ethics Office.

3. The sample may only reach a temperature of 0°C in the ice-water bath when the conical tube is completely submerged (up to the cap) into the ice water and with constant shaking on a shaker. The ice-water bath should always be replenished with new ice cubes to keep the temperature of the ice water constant at 0°C.
4. To make this cryopreservation protocol reproducible and practical for future transplant applications, all steps need to be sterile. To achieve this, the probe of the thermocouple thermometer must be sterile when it is used for measuring the temperature of CPA solutions. Also, it is recommended to use a different probe for each CPA solution because they consist of different types and concentrations of CPAs.
5. The temperature of the 50-mL conical tube that contains 50 ml of CPA solution could be affected by room temperature when it is out of the ice-water bath or methanol/ethanol cooling bath for a few seconds. Therefore, to record a correct temperature using the thermocouple thermometer, the process should be very quick and preferably take place while the conical tube is still submerged in the ice water or methanol/ethanol.
6. The sample may only reach a temperature of −10°C in the methanol/ethanol cooling bath when the conical tube is completely submerged (up to the cap) into the methanol/ethanol, which is stirred constantly.
7. All processes of CPA loading, vitrification, storage, and rewarming are very sensitive to any unplanned temperature variations. To minimize and avoid this, it is recommended to transfer the sample between different solutions and environments as quickly as possible. A practical strategy to achieve this is preparing all required materials and tools before starting the process and reading through the protocol in advance to make sure all required elements are met. Also, the locations of equipment and tools could affect the required time for tissue handling. It is recommended to have the refrigerator, ice-water bath, methanol/ethanol cooling bath, and liquid nitrogen tank in one room with an ideal 1–2 meter distance between them.
8. Loosening a cryovial cap slightly before warming can prevent pressure buildup from expanding liquid nitrogen that may potentially cause the vial to explode. This is especially important if liquid nitrogen may have penetrated into the vial during storage.

Acknowledgments

The AC research projects referred to in this manuscript were funded by the Edmonton Orthopaedic Research Committee. Janet A. W. Elliott held a Canada Research Chair in Thermodynamics from 2001 to 2025. Kezhou Wu was funded by the Li Ka Shing Sino-Canadian Exchange Program between the University of Alberta and Shantou University, and the First Affiliated Hospital of Shantou University Medical College (YCTJ-2022-07). Nadia Shardt acknowledges funding from the Natural Sciences and Engineering Research Council of Canada (NSERC), Alberta Innovates and Alberta Advanced Education, the Government of Alberta, and the University of Alberta.

References

1. Heir S, Nerhus TK, Røtterud JH, Løken S, Ekeland A, Engebretsen L, Arøen A (2010) Focal cartilage defects in the knee impair quality of life as much as severe osteoarthritis: a comparison of knee injury and osteoarthritis outcome score in 4 patient categories scheduled for knee surgery. Am J Sports Med 382:231–237. https://doi.org/10.1177/0363546509352157
2. Rorick CB, Mitchell JA, Bledsoe RH, Floren ML, Wilkins RM (2020) Cryopreserved, thin, laser-etched osteochondral allograft maintains the functional components of articular cartilage after 2 years of storage. J Orthop Surg Res 15:521. https://doi.org/10.1186/s13018-020-02049-y
3. Mameri ES, Kerzner B, Obioha OA, McCormick JR, Dasari SP, Khan ZA, Fortier LM, Jackson GR, Chahla J (2023) Revision lateral femoral condyle osteochondral allograft transplantation with the snowman technique after failed previous oblong osteochondral allograft. Arthrosc Tech 13:e363–e370. https://doi.org/10.1016/j.eats.2022.11.016
4. Buckwalter JA, Bowman GN, Albright JP, Wolf BR, Bollier M (2014) Clinical outcomes of patellar chondral lesions treated with juvenile particulated cartilage allografts. Iowa Orthop J 34:44–49. https://pmc.ncbi.nlm.nih.gov/articles/PMC4127721/
5. Farr J, Yao JQ (2011) Chondral defect repair with particulated juvenile cartilage allograft. Cartilage 2:346–353. https://doi.org/10.1177/1947603511405838
6. Ryan PM, Turner RC, Anderson CD, Groth AT (2018) Comparative outcomes for the treatment of articular cartilage lesions in the ankle with a DeNovo NT natural tissue graft: open versus arthroscopic treatment. Orthop J Sports Med 6:2325967118812710. https://doi.org/10.1177/2325967118812710
7. Goodfriend B, Essilfie AA, Jones IA, Vangsness TC (2019) Fresh osteochondral grafting in the US: the current status of tissue banking processing. Cell Tissue Bank 20:331–337. https://doi.org/10.1007/s10561-019-09768-5
8. Tompkins M, Adkisson HD, Bonner KF (2013) DeNovo NT allograft. Oper Tech Sports Med 21:82–89. https://doi.org/10.1053/j.otsm.2013.03.005
9. Ball ST, Amiel D, Williams SK, Tontz W, Chen AC, Sah RL, Bugbee WD (2004) The effects of storage on fresh human osteochondral allografts. Clin Orthop Relat Res 418:246–252. https://doi.org/10.1097/00003086-200401000-00043
10. Williams SK, Amiel D, Ball ST, Allen RT, Wong VW, Chen AC, Sah RL, Bugbee WD (2003) Prolonged storage effects on the articular cartilage of fresh human osteochondral allografts. J Bone Joint Surg Am 85:2111–2120. https://doi.org/10.2106/00004623-200311000-00008
11. Yanke AB, Tilton AK, Wetters NG, Merkow DB, Cole BJ (2015) DeNovo NT particulated juvenile cartilage implant. Sports Med Arthrosc 23:125–129. https://doi.org/10.1097/JSA.0000000000000077
12. Jomha NM, Law GK, McGann LE (2004) Hypothermic storage of porcine articular cartilage. Cryobiology 49:294–295. https://doi.org/10.1016/j.cryobiol.2004.10.001
13. Brockbank KGM, Chen ZZ, Song YC (2010) Vitrification of porcine articular cartilage. Cryobiology 60:217–221. https://doi.org/10.1016/j.cryobiol.2009.12.003
14. Abazari A, Jomha NM, Elliott JAW, McGann LE (2013) Cryopreservation of articular cartilage. Cryobiology 66:201–209. https://doi.org/10.1016/j.cryobiol.2013.03.001
15. Jomha NM, Elliott JAW, Law GK, Maghdoori B, Forbes JF, Abazari A, Adesida AB, Laouar

L, Zhou X, McGann LE (2012) Vitrification of intact human articular cartilage. Biomaterials 33:6061–6068. https://doi.org/10.1016/j.biomaterials.2012.05.007
16. Wu K, Shardt N, Laouar L, Elliott JAW, Jomha NM (2021) Vitrification of particulated articular cartilage via calculated protocols. NPJ Regen Med 6:1–13. https://doi.org/10.1038/s41536-021-00123-5
17. Wu K, Yong KW, Ead M, Sommerfeldt M, Skene-Arnold TD, Westover L, Duke K, Laouar L, Elliott JAW, Jomha NM (2022) Vitrified particulated articular cartilage for joint resurfacing: a swine model. Am J Sports Med 50:3671–3680. https://doi.org/10.1177/03635465221123045
18. Crisol M, Wu K, Congdon Jr B, Skene-Arnold TD, Laouar L, Elliott JAW, Jomha NM (2024) Chondrocyte viability of particulated porcine articular cartilage is maintained in tissue storage after cryoprotectant exposure, vitrification, and tissue warming. Cartilage 15:139–146. https://doi.org/10.1177/19476035221118656

Check for updates

Chapter 31

Cryopreservation and Freeze-Drying of Heart Valves for Replacement Surgery

Kelvin G. M. Brockbank, Willem F. Wolkers and Shangping Wang

Abstract

Malfunctioning heart valves can cause severe health problems, which, if left untreated, can lead to death. Heart valves can be replaced with biological heart valve constructs obtained from human or animal origin that closely resemble the properties of native heart valves. Tissue banking protocols are needed to ensure off-the-shelf availability in clinical settings. Heart valve tissue constructs may contain cells, or they can be decellularized to remove the immunogenic factors associated with cellular components. In order to preserve cells in the tissues, vitrification or freezing cryopreservation strategies can be used, whereas freeze-drying is particularly suitable to preserve decellularized heart valves. Here, we provide freezing, vitrification, and freeze-drying protocols to preserve heart valve conduits for replacement surgery. The protocols are described for porcine heart valves, but procedures can easily be adapted for material obtained from other species.

Keywords Cryopreservation, Freeze-drying, Heart valves, Matrix implants, Vitrification

1 Introduction

Malfunctioning heart valves can cause various health problems, which eventually may result in death. Heart valve dysfunction is typically treated by replacing the diseased valve with mechanical or biological prostheses. Side effects associated with mechanical valves include hemorrhage and thromboembolism, whereas bioprosthetic valves, which are made by treating bovine or porcine tissue with glutaraldehyde, have been implicated in fibrosis, calcification, degeneration, and immunogenic complications [1, 2]. Heart valves can also be replaced with biological heart valve constructs obtained from human or animal origin that more closely resemble the properties of native heart valves. These constructs may contain cells, or cells can be removed by decellularization to eliminate immunogenic factors. If heart valves are stored in liquid solutions, they are prone to degradation reactions and bacterial contamination. Therefore, appropriate preservation methods are needed to ensure off-the-shelf availability.

Willem F. Wolkers and Harriëtte Oldenhof (eds.), *Cryopreservation and Freeze-Drying Protocols*, Methods in Molecular Biology, vol. 3060, https://doi.org/10.1007/978-1-0716-5416-3_31,

Standard freezing cryopreservation techniques, using dimethyl sulfoxide (DMSO) and controlled rate freezing, have been traditionally used to preserve human heart valves for replacement procedures [2–5]. In Protocol 1, we provide a standard cryopreservation protocol. Frozen valves are currently less utilized because of failures in young patients [6], particularly infants [7, 8]. In some allografts, rapid deterioration has been observed, attributed to disruptive interstitial ice damage during cryopreservation, which subsequently led to accelerated valve degeneration upon implantation.

In our initial heart valve vitrification studies, we utilized pieces of porcine aortic valves and intact rat aortic valves in small volumes of a vitrification solution (VS) called to reflect its 55% (w/v) of cryoprotectant solutes [9]. VS55 consists of an 8.4 M mixture of 1,2-propanediol, formamide, and dimethyl sulfoxide in EuroCollins (EC) solution. This protocol is suitable for small-volume tissue samples where relatively rapid cooling and warming rates can be achieved to avoid ice formation in the tissue. Figure 1 shows a hypothesized mechanism indicating how cryopreservation methods affect the immune compatibility of human cardiovascular matrices.

Use of Protocols 2 and 3 with VS83 permits the use of small and large sample volumes and heart valve tissue storage at −80°C at the expense of cell viability. The potential advantages of VS83 heart valve storage at −80°C employing Protocol 4 includes reduced infrastructural needs for preservation, storage, and shipping in comparison with traditional freezing methods while maintaining extracellular matrix integrity and material properties. The loss of cell viability is probably a benefit contributing to the reduction of immunogenicity observed in vivo [11–14]. Development of vitrification methods for larger samples with retention of cell viability resulted in a new protocol, Protocol 5 (see Figs. 2 and 3).

The use of decellularized native heart valves (see Fig. 4) to replace a malfunctioning heart valve reduces the risk of immunogenic rejection, and anti-coagulation therapy is generally not required [16, 17]. We have established methods to freeze-dry decellularized heart valves using sucrose as lyoprotectant [18–20]. Sucrose treatment preserves the extracellular matrix (ECM) histoarchitecture and biomechanical characteristics during freeze-drying and rehydration (see Fig. 5). Pre-treatment with very high sucrose concentrations is needed (≥40% w/v) to avoid pore formation in the tissue and alterations in tissue stiffness [20]. We have demonstrated that sucrose-protected freeze-dried decellularized pulmonary heart valves show excellent clinical performance and cell repopulation potential in juvenile sheep (see Fig. 4). In Protocol 6, we provide a freeze-drying protocol for decellularized heart valves. In a comparative study on the effects of different preservation techniques on characteristic properties of decellularized pericardium patches, we have shown that both

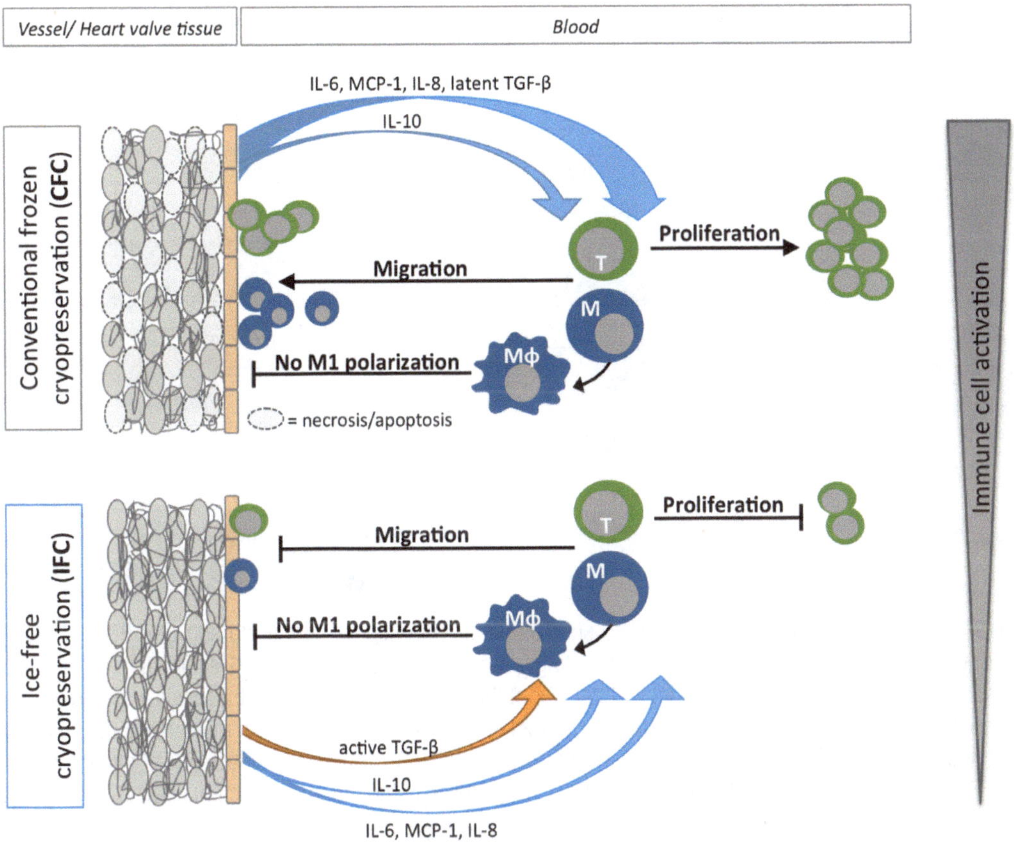

Fig. 1 Hypothesized mechanism indicating how the cryopreservation methods affect the immune compatibility of human cardiovascular matrices. Potential effects of conventional freezing cryopreservation (CFC) and ice-free vitrified cryopreservation (IFC) treatment employing VS83 on the human immune response to cardiovascular tissue (vessel or heart valves) are illustrated. After freezing, some tissue cells are apoptotic or necrotic, but the tissue still releases high amounts of cytokines such as IL-6, MCP-1, IL-8, latent TGF-β, and smaller amounts of IL-10. These cytokines lead to a strong activation of blood immune cells. Monocytes and T cells are attracted to the tissue graft, where they can infiltrate or receive signals to proliferate. In contrast, after VS83 vitrification (IFC) very few viable cells remain due to cryoprotectant cytotoxicity, the tissues demonstrate low metabolic activity and minimal, if any, necrosis or apoptosis. The cytokines IL-6, MCP-1, IL-8, IL-10, and TGF-β are released only in small amounts and TGF-β is secreted in its biologically active form. The decreased levels of cytokine release and the presence of active TGF-β lead to an attenuated activation of immune cells. Particularly, migration and infiltration of T cells and monocytes are reduced, and T cell proliferation is blocked, which results in a diminished human immune response to the vitrified tissue compared to conventional frozen tissue with 10% DMSO. (Adapted from [10])

freeze-dried and vitrified patches closely resemble control patches, whereas cryopreservation by freezing alters tissue structure and biomechanical characteristics [21].

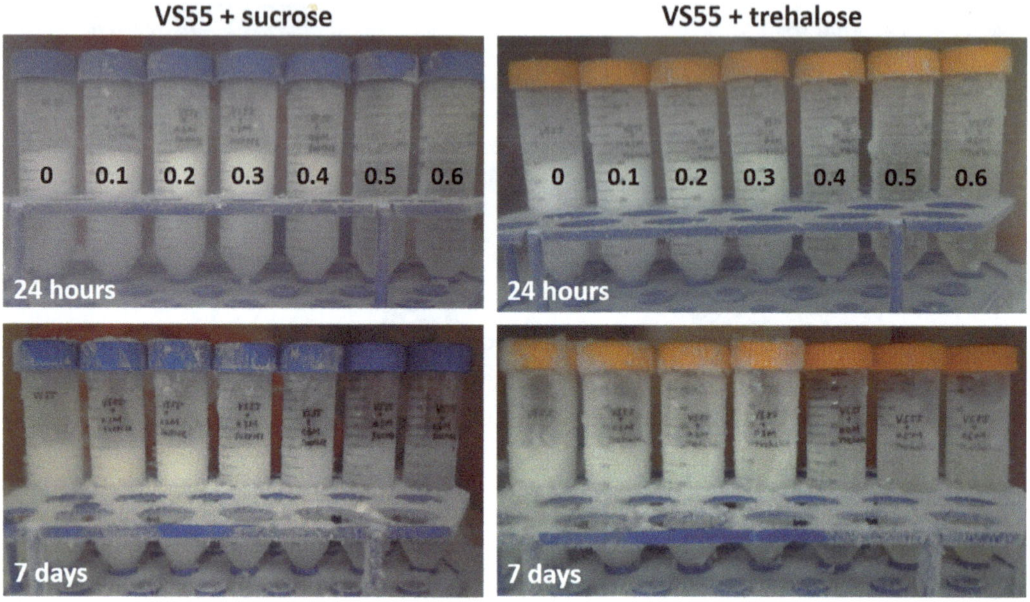

Fig. 2 Freedom from ice formation in VS55 supplemented with sugars during storage at −80°C. Top left: VS55 with 0.0–0.6 M sucrose at 24 h demonstrating freedom from ice at 0.5 and 0.6 M concentrations. Lower sucrose concentrations demonstrate ice (white). Top right: VS55 supplemented with trehalose demonstrated freedom from ice at 0.4–0.6 M concentrations after 24 h. Bottom left: VS55 with sucrose after 7 days. Bottom right: VS55 with trehalose after 7 days. After a week at −80°C, the highest concentrations of sugars are still free of ice although it is hard to tell due to frosting on the outside of the tubes that accumulated during storage

2 Materials

Prepare all solutions using deionized water (prepared with an activated carbon filter, mixed bed working deionizer, and a mixed bed polishing deionizer to attain a sensitivity of 10 MΩ cm at room temperature) in a laminar flow hood. The solutions are made at room temperature and stored at 4°C for up to 4 weeks. The solution formulations described are for 1 L batches.

2.1 Solutions and Equipment for Slow Freezing Cryopreservation

1. Solutions: (1) Dulbecco's Modified Eagle Medium (DMEM), (2) 5% (v/v) DMSO, and (3) 10% (v/v) DMSO in DMEM.
2. Automatic freezer, e.g., an IceCube 14 M programmable controlled rate freezer (German Cryo, Jüchen, Deutschland) or equivalent equipment.
3. Polystyrene bags suitable for freezing (50 mL or greater).

2.2 Preparation of EuroCollins Solution

1. For preparing EuroCollins solution (EC), put 0.5 L of water in a container (graduated cylinder or glass beaker).
2. Add a stir bar followed by the addition of each component in the following order: 174.76 g Dextrose, 10.2 g KH_2PO_4, 36.5 g K_2HPO_4, 5.6 g KCl, 4.2 g $NaHCO_3$. Make up to 1 L with water to make a 5× stock solution.

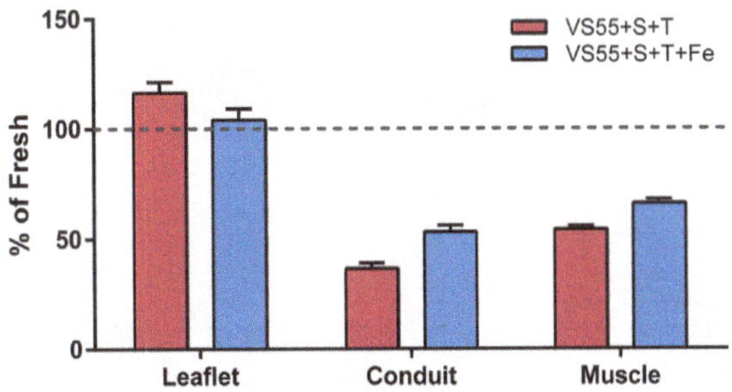

Fig. 3 Ice-free vitrification of pulmonary heart valves with VS55 supplemented with 0.3 M sucrose and 0.3 M trehalose in 30 mL total volume. The tissue components (leaflets, conduit, and fibrous muscle band) were rewarmed by convection warming (as described for Protocol 1) or by nanowarming [15], and viability was evaluated using alamarBlue. There was no significant difference between nanowarmed and convection-warmed leaflets; both the conduit and fibrous muscle band were significantly higher ($p < 0.05$, t-test). Data from ≥125 replicates from N = 7–8 experiments

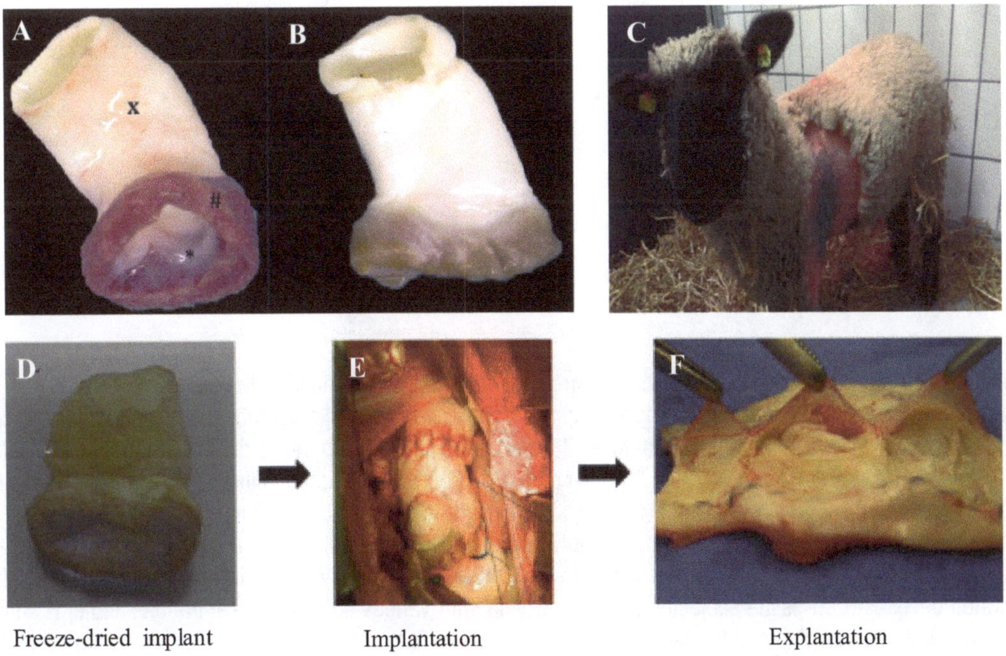

Fig. 4 Porcine pulmonary valve conduits in native (**A**) and decellularized (**B**) state, composed of pulmonary artery parts (x), myocardial rim (#), and properly pulmonary heart valve comprising three semilunar-shaped leaflets. Morphological appearance of a freeze-dried valve before rehydration (**D**), after rehydration and implantation in the pulmonary valve position in a sheep (**C** and **E**), and 6 months post-operation after explantation (**F**)

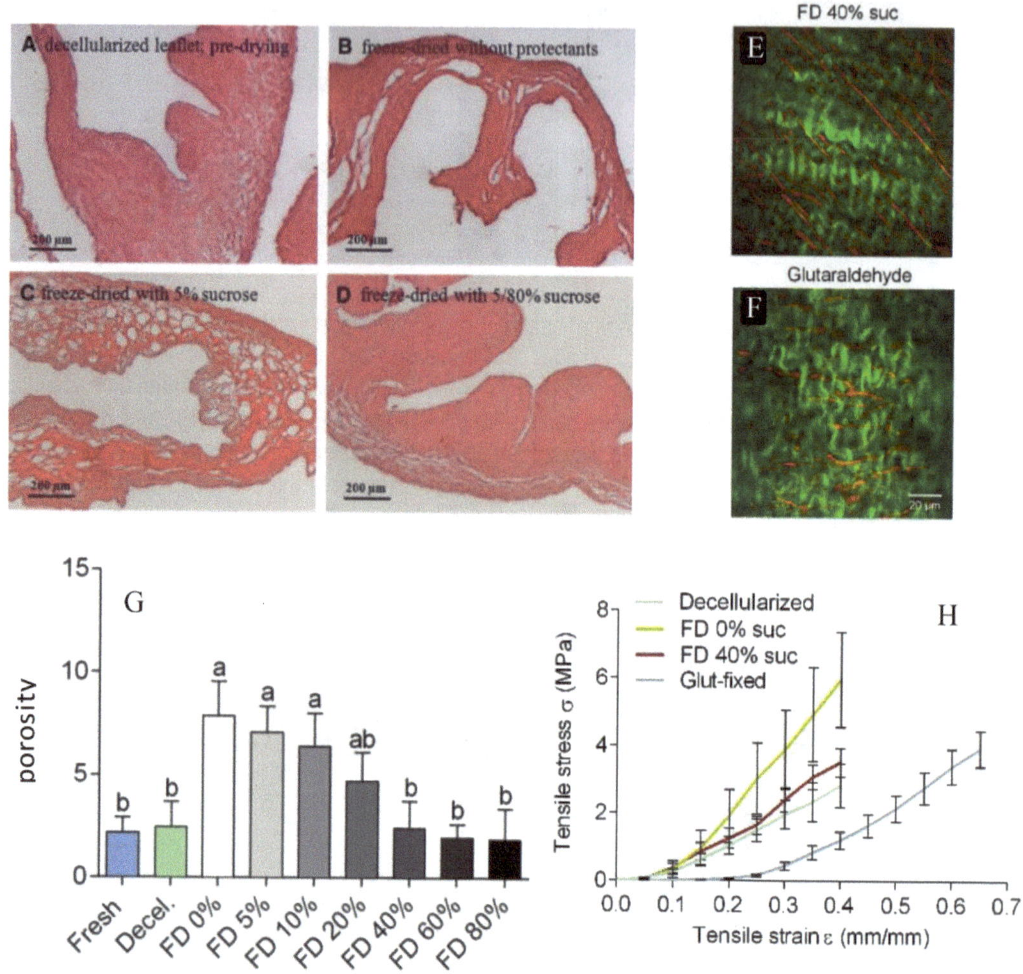

Fig. 5 Micrographs of histological sections prepared from pulmonary heart valve tissues, before as well as after decellularization and freeze-drying in sections of the valve cusps. Micrographs are presented for decellularized tissue (**A**), as well as rehydrated decellularized tissue subjected to freeze drying without protectant (**B**), freeze-dried with 5% sucrose (**C**), and 80% sucrose (**D**). TPEF/SHG multiphoton images show the appearance of collagen bundles (green) and elastic fibers (red) for rehydrated decellularized cusps subjected to freeze-drying with 40% sucrose (**E**) and for a glutaraldehyde-fixed cusp (**F**). Panel G shows the porosity and pore size distribution of pulmonary heart valve cusp tissues, before and after decellularization as well as for rehydrated decellularized tissue subjected to freeze-drying with various sucrose concentrations. Panel H shows stress-strain curves obtained for decellularized cusps (green line), rehydrated decellularized cusps subjected to freeze-drying without protectants (yellow line) and with 40% sucrose (red line), as well as for glutaraldehyde-fixed cusps (light blue). (Data adapted from [19, 20])

Table 1
VS protocols, formulations, and comments

Protocol (#)	Formulation	Comments
1	VS55	• Effective for small volumes
		• Requires rapid cooling and warming to prevent ice nucleation
		• Retains about 80% cell viability and ECM integrity
2/3	VS83	• Effective for large samples
		• Minimal cell viability
		• ECM retention
		• Recipient immune reaction minimized
4	VS55 + sugars	• Effective for large samples
		• Requires up to 0.6 M of sucrose and/or trehalose combined
		• Retains both cell viability and ECM integrity
		• Research in progress to determine recipient immune reaction

Table 2
VS formulation, quantities

	VS55	VS83
5× EuroCollins (mL)	200	200
HEPES buffer (g)	2.39	2.39
Propylene glycol[a] (g)	168.38	252.57
Formamide[a] (g)	139.56	209.34
Me_2SO[a] (g)	242.14	363.21
Water (mL)	to 1 L	to 1 L
pH (units)	7.9–8.1	7.9–8.1
Osmolality[b] (mOsm/kg)	460 ± 5	670 ± 5

[a]Liquids measured by weight [b]20× dilution = 50 μL in 950 μL of distilled water

3. Stir continuously during formulation, when the solution is clear of particles it should be filtered using 0.2 μm filters. We do not test the osmolality or pH of 5× EC.
4. Dilute 1:5 with water to make 1× EC and check the pH (~7.4) and osmolality (365 ± 5 mOsm/kg). Store at 4°C.

2.3 Preparation of Vitrification Solutions

1. Vitrification solutions VS55 and VS83 (Table 1) are formulated by placing a stir bar and then each component, in the order that they appear in the list below (see Table 2), in a graduated container.
2. Stir continuously, when finished, and if it is clear of particles, it should be filtered. We use bottle top 0.2 μm filters. It takes a long time (hours) to filter due to the high viscosity of the solutions.
3. Check the final pH and make a 1:20 dilution for assessment of osmolality. Store at 4°C.

2.4 Preparation of Addition and Removal Solutions

1. Sequential vitrification solution addition and removal solutions are used in certain protocols (*see* **Note 1**). The VS chemical component concentration can be increased from ⅛, ¼, ½, ¾ to 100% of the final full-strength VS formulation being employed.
2. Prepare diluted VS solutions by the addition of appropriate volumes of 1× EC.
3. The solutions for removal are similar to those for addition, except that 200–400 mM mannitol is added to the formulation as an osmotic buffer. For example, using seven steps, the VS CPAs plus mannitol are decreased from full strength to 0 M, plain EC, and finally Dulbecco's Modified Eagle Medium (DMEM).

2.5 Solutions and Equipment for Decellularization

1. Braunol disinfection solution (7.5 % povidone-iodine).
2. Decellularization solutions: (1) 0.5% (v/v) Triton X-100, and (2) 0.5% (w/v) sodiumdodecylsulfate (SDS) in filter-sterilized water.
3. Phosphate-buffered saline solution (PBS): 137 mM NaCl, 27 mM KCl, 10 mM Na_2HPO_4, 1.8 mM KH_2PO_4, pH 7.4.
4. PBS+: PBS supplemented with 1% (w/v) penicillin/streptomycin, 1% (w/v) partricin, and 1% (w/v) gentamycin.
5. Shaking incubator for continuously shaking during decellularization and washing.

2.6 Solutions and Equipment for Freeze-Drying

1. Lyoprotectant incubation and freeze-drying medium.
   (a) PBS/sucrose: 5% (w/v) sucrose in PBS. Prepare 100 mL (for processing of one heart valve), filter-sterilize using 0.22-μm filters, and store at 4°C until use.
   (b) PBS/sucrose: 80% (w/v) sucrose in PBS. Prepare 300 mL (for processing of one heart valve), filter-sterilize using 0.22-μm filters, and store at 4°C until use (*see* **Note 10**)
2. Water bath or incubator for loading the scaffolds with protectants at elevated temperatures (37°C).

3. Freeze dryer (e.g., Virtis Advantage Plus Benchtop freeze dryer from SP scientific, Warminster, PA, USA) (*see* **Note 11**).
4. Vacuum sealer and vacuum sealing bags to store freeze-dried samples.
5. Refrigerator.

2.7 Histological Evaluation of Tissue Structure

1. Formalin (3.5 % v/v) solution for fixation of tissue specimens.
2. Ethanol (70, 75, 80, 85, 90, 95, and 100% v/v-solutions in distilled water), isopropanol (100%), and Roticlear (Carl Roth GmbH, Karlsruhe, Germany) for dehydration of tissue specimens, and liquid paraffin for embedding.
3. Leica TP 1020 automatic tissue processor (Leica, Wetzlar, Germany) for dehydration of tissue samples with increasing alcohol concentrations and final embedding in liquid paraffin.
4. Microtome (e.g., manual rotary Reichert-Jung 2040 autocut microtome, Leica, Wetzlar, Germany) for preparing tissue sections.
5. Corbit-Balsam mounting medium (Hecht, Hamburg, Germany).
6. Hematoxylin-eosin staining solution. Filter solution prior to use.

3 Methods

3.1 Protocol 1: Frozen Cryopreservation

1. Transfer one heart valve into a polystyrene bag and expose to 50 mL 5% (v/v) DMSO for 10 mi. Then change to 50 mL 10% DMSO for 10 min at 22°C. Work in a laminar flow hood.
2. Freeze the bag(s) to −80°C in a controlled rate freezer at 1°C/min followed by slow cooling to ≤−135°C.
3. Store the bag at −135°C in a mechanical storage freezer or at ≤−135°C near the top in a vapor phase nitrogen freezer.
4. Rewarm in a two-stage process: slow warming to −100°C followed by rapid warming to ~4°C.
5. Remove the cryopreservation solution in one to four sequential, 10 min washing steps by exposing the tissue to DMEM solution (100 mL).

3.2 Protocol 2: Vitrified Cryopreservation

Suitable for small heart valve tissue samples (≤3 mL total volume, sample plus solution) with maximum retention of cell viability and matrix integrity [9].

1. Gradually infiltrate tissues with VS55 at 4°C (*see* **Note 1**). Precooled diluted VS solutions are employed in six 15 min steps at 4°C of increasing CPA concentration (*see* **Note 2**).
2. Cover the top of the vitrification solution with 2-methylbutane (isopentane; *see* **Note 3**).
3. Cool the samples rapidly to −100°C followed by slow cooling to ≤−135°C (*see* **Note 4**).
4. Store the samples at −135°C in a mechanical storage freezer or at ≤−135 °C near the top in a vapor phase nitrogen freezer.
5. Rewarming of vitrified tissues is done in a two-stage process, including slow warming to −100°C and then rapid warming to ~4°C (*see* **Note 5**). Wipe the container with 70% ethanol to minimize the risk of microbial contamination and use sterile pickups to transfer the tissue to the first VS removal solution.
6. Remove the vitrification solution in a stepwise manner at 4°C in six or seven sequential, 15 min steps of increasingly dilute VS solutions containing mannitol (*see* **Note 1**). Finally, the tissues are placed in DMEM culture medium.

3.3 Protocol 3: Vitrified Cryopreservation

Suitable for large heart valve tissue samples (up to 100 mL total volume) with preservation of matrix integrity and minimal residual cell viability.

1. Gradually infiltrate tissues with VS83 in six 15 min steps of increasing concentration at 4°C.
2. After the final step, place the tissues individually in sterile polyester bags containing 70–80 mL of the vitrification solution. Evacuate each bag of air by squeezing (*see* **Note 3**) and heat-sealing.
3. Cool the bags containing tissues rapidly to −100°C in a precooled 2-methylbutane bath and then transfer for storage (*see* **Note 4**).
4. Store at −80 °C in a mechanical storage freezer (*see* **Note 6**).
5. Rewarming should be performed by placing each bag in a 37°C water bath until the solution moves freely.
6. Transfer the bag onto ice and clean the bag externally with 70% ethanol to reduce the risk of microbial contamination. Cut the bag open with sterile scissors and use sterile pickups to transfer the tissue to the first VS removal solution.
7. A multistep VS removal protocol was employed in our first unpublished sheep study, similar to that used in Protocol 1, above. Simpler washing methods have been employed in a subsequent sheep study [20, 22] and a study with arterial patches [12] (*see* **Note 7**).

3.4 Protocol 4: Vitrified Cryopreservation

Simplified version of Protocol 3 in development for heart valves.

1. Place the tissues in sterile polyethylene bags containing 80 mL of VS83 at room temperature (*see* **Note 1**).
2. Evacuate the air and heat seal the bags (*see* **Note 3**).
3. Incubate the tissues with continuous shaking for at least 1 h at room temperature prior to cooling.
4. Cool the samples by placing the bags for 10 min in a pre-cooled bath of 2-methylbutane (<−100°C; *see* **Note 4**).
5. The tissues can then be placed at −80 °C for storage.
6. Rewarming of the tissue can be performed by submersion of the tissue in its bag in a 37°C water bath.
7. Dry the bag and externally clean it with 70% ethanol to reduce the risk of microbial contamination. Cut the bag open with sterile scissors and use sterile pickups to transfer the tissue to the first wash step.
8. Washing to remove CPAs has been performed using five wash steps of 5 min duration each with ice Euro-Collins solution or with several washes with Lactated Ringer's solution containing 5% dextrose (*see* **Note 8**).

3.5 Protocol 5: Vitrified Cryopreservation

Modified version of Protocol 2 in development for heart valves for retention of tissue cell viability in relatively large samples (>30 mL) [9].

1. Gradually infiltrate tissues with VS55 at 4°C (*see* **Note 1**). Precooled diluted VS solutions are employed in six 15 min steps at 4°C of increasing CPA concentration (*see* **Note 2**). Sugars, sucrose, or trehalose or a mixture, are added in the final addition step at 0.1–0.6 M concentrations (*see* **Note 9**).
2. Evacuate the air and heat seal the bags (*see* **Note 3**).
3. The cooling process can then be achieved by placing the bags for 10 min in a pre-cooled bath of 2-methylbutane (<−100°C) (*see* **Note 4**).
4. The tissues can then be placed at −135°C for storage.
5. Rewarming of the tissue is done by submersion of the tissue in its bag as described in Protocol 2, Step 5.
6. Dry the bag and externally clean it with 70% ethanol to reduce the risk of microbial contamination. Cut the bag open with sterile scissors and use sterile pickups to transfer the tissue to the first wash step.
7. The vitrification solution is then removed in a stepwise manner at 4°C in six or seven sequential, 15 min steps of increasingly

dilute VS solutions containing mannitol (*see* **Note 1**). Finally, the tissues are placed in DMEM culture medium.

3.6 Protocol 6: Freeze-Drying of Decellularized Tissues

3.6.1 Decellularization

1. Obtain porcine hearts from a local slaughterhouse and dissect the heart valves from the hearts.
2. Disinfect by incubation in Braunol solution for 5 min, and then remove this solution by incubation in PBS+ for 20 min.
3. Place tissues into Triton-X100 decellurization solution for 24 h at room temperature, while continuously shaking the solution and replacing the Triton-X100 solution with fresh solution after 12 h.
4. Place tissues into SDS decellularization solution for 24 h at room temperature, while continuously shaking the solution and replacing the SDS solution with fresh solution after 12 h.
5. Remove the detergents by washing two times 12 h in distilled water.
6. Conduct another 12 washing cycles in PBS+, while continuously shaking and changing the solution every 12 h (*see* **Note 12**).
7. Store decellularized heart valves in PBS+ at 4°C until further processing.

3.6.2 Sucrose Loading and Freeze-Drying

1. Incubate decellularized heart valves in 100 mL 5% sucrose solution, at 37°C for 4 h (*see* **Note 13**).
2. Transfer the heart valves in 100 mL 80% sucrose solution, at 37°C for 30 min, changing the sucrose solution with fresh sucrose solution every 10 min (*see* **Notes 14 and 15**).
3. Dip the valves in tissue paper to remove excess solution. Transfer the heart valves to an open or semi-open container (i.e., a Petri-dish with knuckles allowing vapor transport) and put on the freeze-dryer shelves with the shelve temperature set at 20°C.
4. Start the freeze-drying protocol.
   (a) Freeze the samples by cooling the shelves at a cooling rate of 1°C/min until the shelve temperature reaches −40°C, and keep at −40°C for 3 h.
   (b) Increase shelf temperature to −30°C at a rate of 1°C/min.
   (c) For primary drying; decrease the chamber pressure to 60 mTorr while maintaining the shelf temperature at −30°C, and dry for 48 h.

(d) For secondary drying; increase the shelf temperature from −30 to 40°C at a rate of 0.1°C/min, while maintaining the pressure at 60 mTorr. Keep the shelf temperature at 40°C for 30 min, and then reduce to 5°C at a rate of 1°C/min and keep there until the next step (*see* **Note 16**).

5. After freeze-drying, transfer the dried valves into a dry environment or seal with a vacuum sealer, and store in a standard refrigerator or at room temperature (*see* **Note 17**).
6. Rehydrate dried valves in a large amount (>100 mL per valve) of PBS solution for minimally 1 h changing the PBS solution with fresh solution several times (*see* **Note 18**).

3.6.3 Histological Analysis

1. Rehydrated valves should be kept in water for at least 1 h at room temperature prior to histological inspection.
2. Fix hydrated valve tissue by incubating in formalin solution for 24 h at room temperature.
3. Transfer the tissue samples into distilled water (twice), 25% ethanol (twice), and 50% ethanol (twice), respectively, keeping the samples for 30 min in each solution.
4. Transfer tissue samples into the automatic tissue processor (*see* **Note 19**) to dehydrate by exposing the sample to ethanol solutions with increasing ethanol concentration. After dehydration in ethanol, infiltrate in liquid paraffin. Immerse the samples in the following solutions:
   (a) 75% ethanol, 30 min.
   (b) 80% ethanol, 30 min.
   (c) 85% ethanol, 30 min.
   (d) 90% ethanol, 30 min.
   (e) 95% ethanol, 45 min.
   (f) 100% ethanol, 45 min.
   (g) isopropanol, 45 min.
   (h) isopropanol, 45 min.
   (i) Roticlear, 45 min.
   (j) Roticlear, 45 min.
   (k) paraffin, 6 h.
   (l) paraffin, 24 h.
5. Transfer tissue samples infiltrated with paraffin into embedding templates and pour more hot paraffin wax (60°C) over the samples. Cool the paraffin blocks on a cold plate (at −10°C).
6. Make serial paraffin sections of 10 μm thickness using a manual rotary microtome. Transfer sections onto glass slides.

7. Dry and flatten specimens on a heated plate (40°C) for several hours.
8. In order to remove paraffin from the sections, first incubate the slides at 60°C for 30 min. To rehydrate the specimens, subject the slides to the following incubations at room temperature:
   (a) xylene, 10 min.
   (b) xylene, 10 min.
   (c) 100% ethanol, 2 min.
   (d) 90% ethanol, 2 min.
   (e) 80% ethanol, 2 min.
   (f) 70% ethanol, 2 min.
   (g) distilled water, 0.5 min.
9. Stain the slides with hematoxylin-eosin staining solution to visualize nuclei and the general tissue architecture (*see* **Note 20**).
   (a) Incubate slides in hematoxylin solution for 8 min, followed by 10 min washing under running tap water.
   (b) Incubate slides for 5 min in eosin solution.
   (c) Incubate the slides in 80% and 90% ethanol, 2 min each.
   (d) Incubate the slides two times in xylene, 5 min each.
   (e) Mount the specimens in mounting medium.
10. Inspect specimens using light microscopy.

4 Notes

1. Step-wise VS addition and removal procedures are employed in Protocol 1 in order to minimize damage to cells as they respond to osmotic changes in their environment as a consequence of being placed in VS solutions with high concentrations of CPAs. This is not an issue in Protocols 2 and 3 because we are not trying to maximize cell viability. Loss of cell viability is considered advantageous because it appears to reduce immunogenicity in vivo [11, 12].
2. In the last step, the tissues are placed in vials, we usually use glass scintillation vials (25 × 60 mm), containing 1–2 mL of precooled full-strength VS solution. In circumstances where unexpected cytotoxicity is observed, the last one or two steps can be performed at subzero temperatures in a cold bath to minimize cytotoxicity.

3. We use 0.7–1.0 mL of 2-methylbutane for the glass vials described in *see* **Note 1**. The 2-methylbutane has a freezing point of −160°C and a density of 0.62. The 2-methylbutane layer prevents direct air contact minimizing the risk of ice nucleation. Alternatively, the vial can be purged with dry gas, such as nitrogen, or a container without excess air space can be used.
4. Rapid cooling to <−100°C is done by placing the vials in a precooled (−135°C) 2-methylbutane bath, slow cooling is performed by placing the vials in air in the top of storage freezer, either a −135°C mechanical freezer or near the top of a vapor-phase nitrogen cooled freezer. More rapid cooling can be obtained, if needed, using liquid ethane at ~−183°C. However, this strategy may result in the development of thermal stresses within the sample because the edges will cool rapidly and the inside more slowly. Such stress may lead to cracking during storage or upon rewarming.
5. A dummy sample prepared under the same conditions as the preserved samples with a thermocouple is used to monitor cooling and warming. Make sure that the thermocouple has been calibrated using a NIST standard. The dummy sample is placed in close proximity to the samples to be rewarmed in the storage freezer. Plug the dummy sample thermocouple into a reader. Place the samples and dummy sample at the top of the storage freezer and hold just inside the lid. When the temperature reaches −100°C (~1 min), initiate the rapid warming step. Place both vials in a room temperature 30% DMSO bath and swirl the vials manually until the glass begins to melt (the VS starts to move). Open the vial and remove the VS with a transfer pipette and add the first dilution solution. Start timer for 15 min.
6. The samples can be stored in vapor-phase nitrogen but VS solution fractures may be observed. Although the visual impact of the fracturing is disturbing, we have rarely seen tissue fractures and on the rare occasions when tissue cracking has occurred, it has been during use of nitrogen dry shippers for transport. If samples are stored in vapor phase nitrogen, the two-step rewarming should be performed as described in Protocol 1.
7. Ice-free artery patches were rewarmed by placing the bag containing the tissue in a 37°C water bath for 1 min and then washed by transferring the contents of the bag to an AlloFlow™ continuous gradient washout chamber with 1 L of room temperature isotonic saline over 20 min. Ice-free cryopreserved valves were rinsed briefly in ice-cold EC-Solution containing 200 mM mannitol. Then 3 sequential rinses were

performed for 15 min each with continuous careful agitation in ice-cold EC-Solution with 200 mM mannitol, followed by EC-solution alone and finally DMEM. The patches were stored briefly on ice in DMEM culture medium until implantation or evaluation.

8. We have evaluated the use of Lactated Ringer's solution containing 5% dextrose (LRD5) for washing because it is commonly available in surgery suites where ice-free cryopreserved tissues may be used. There were no statistically significant viability differences comparing endothelial cell-seeded decellularized vessels and ice-free preserved tissue-engineered blood vessels that were not decellularized after 2, 4, or 8 washes in cell culture. The endothelial cells formed confluent monolayers on all surfaces of the washed tissues, regardless of the number of post-rewarming washes, demonstrating that any residual cryoprotectants present were not cytotoxic.
9. Addition of sugars to VS55 has a stabilizing effect during slow cooling and warming of vitrified tissue samples. Comparison of VS55 solutions supplemented with 0.1–0.6 M trehalose or sucrose has demonstrated concentration-dependent inhibition of nucleation (patent pending) (see Fig. 2). Results for porcine pulmonary heart valves cryopreserved using equimolar concentrations of sucrose and trehalose, 0.3 M of each, are shown in Fig. 3.
10. Clinical trials with freeze-dried heart valves were conducted using 80% (w/v) sucrose in PBS. Later studies indicated that 40% (w/v) sucrose is sufficient to avoid pore formation and changes in biomechanical properties caused by the freeze-drying process [20].
11. A freeze-dryer system with temperature-controlled shelves and control of chamber pressure is preferred. Freeze-drying, however, can also be done with a simpler benchtop manifold freeze-drying system lacking temperature-controlled shelves or control of the chamber pressure. In this case, freezing can be done by placing the samples in a −80°C freezer. The frozen sample can be transported on dry ice to the freeze-dryer to avoid thawing of the sample.
12. This step is done to remove residual detergents in the tissue. Residual detergents may inhibit the regrowth of cells in vivo.
13. It takes less than 4 h to load porcine heart valve tissue with sucrose [19].
14. Incubation in 80% sucrose solution is particularly done to partially dehydrate the tissue to avoid ice crystal formation during subsequent cooling. The solution should be replaced with a fresh solution two times because the sucrose solution will be diluted with water coming from the tissue.

15. Trehalose can likely be used as an alternative to sucrose, but the highly concentrated solutions that are needed for processing would make the procedure more expensive.
16. Secondary drying is done to remove unfrozen water from the sample. The water content of the sample depends on the end temperature of the shelves that is selected in the secondary drying cycle.
17. Most freeze-drying systems have a stoppering device allowing sealing the sample under vacuum. There are no suitable commercial containers (vial + stopper) available for freeze-drying of heart valves. Such containers, however, can be constructed, which would allow to seal the samples under vacuum. Alternatively, a vacuum sealer can be used to seal the freeze-dried samples. Samples can be stored at room temperature, in a standard refrigerator, or in a −20°C freezer. In any case, samples should be stored in a dry environment to avoid moisture uptake during storage, which can result in a decrease of the glass transition temperature causing degradation reactions in the sample.
18. Buffer can be replaced several times with a fresh buffer to remove the protectants. Sucrose, however, is a relatively harmless protectant that does not need to be completely removed prior to transplantation.
19. Alternatively, samples can be manually transferred in the different solutions.
20. Hematoxylin has a deep blue-purple color and stains proteins nonspecifically. In a typical tissue, nuclei are stained blue, whereas cytoplasm and extracellular matrix have varying degrees of pink staining.

Acknowledgments

This work was supported by funding from the German Research Foundation (DFG, Deutsche Forschungsgemeinschaft) for the Cluster of Excellence 'From Regenerative Biology to Reconstructive Therapy' (REBIRTH).

References

1. Hammermeister K, Sethi GK, Henderson WG, Grover FL, Oprian C, Rahimtoola SH (2000) Outcomes 15 years after valve replacement with a mechanical versus a bioprosthetic valve: final report of the veterans affairs randomized trial. J Am Coll Cardiol 36:1152–1158. https://doi.org/10.1016/S0735-1097(00)00834-2
2. Jayakrishnan A, Jameela SR (1996) Glutaraldehyde as a fixative in bioprostheses and drug delivery matrices. Biomaterials

17:471–484. https://doi.org/10.1016/0142-9612(96)82721-9
3. Angell WW, DeLanerolle P, Shumway NE (1973) Valve replacement: present status of homograft valves. Prog Cardiovasc Dis 15: 589–622. https://doi.org/10.1016/S0033-0620(73)80025-8
4. Stelzer P, Elkins RC (1989) Homograft valves and conduits: applications in cardiac surgery. Curr Probl Surg 26:381–452. https://doi.org/10.1016/0011-3840(89)90017-8
5. Angell WW, Oury JH, Lamberti JJ, Koziol J (1989) Durability of the viable aortic allograft. J Thorac Cardiovasc Surg 98:48–56. https://doi.org/10.1016/S0022-5223(19)34460-5
6. O'Brien MF, Stafford EG, Gardner MAH, Pohlner PF, Tesar PJ, Cochrane AD, Mau TK, Gall KL, Smith SE (1995) Allograft aortic valve replacement: long-term follow-up. Ann Thorac Surg 60:565–570. https://doi.org/10.1016/0003-4975(95)00223-8
7. Clarke DR, Campbell DN, Hayward AR, Bishop DA (1993) Degeneration of aortic valve allografts in young recipients. J Thorac Cardiovasc Surg 105:934–942. https://doi.org/10.1016/S0022-5223(19)34168-6
8. Yankah AC, Alexi-Meskhishvili V, Weng Y, Schorn K, Lange RE, Hetzer R (1995) Accelerated degeneration of allografts in the first two years of life. Ann Thorac Surg 60:71–77. https://doi.org/10.1016/0003-4975(95)00289-W
9. Brockbank KGM, Chen Z, Greene ED, Campbell LH (2021) Vitrification of heart valve tissues. In: Wolkers WF, Oldenhof H (eds) Methods in cryopreservation and freeze-drying, Methods in Molecular Biology, Springer, pp 593–605. https://doi.org/10.1007/978-1-4939-2193-5_20
10. Schneider M, Stamm C, Brockbank KGM, Stock UA, Seifert M (2017) The choice of cryopreservation method affects immune compatibility of human cardiovascular matrices. Sci Rep 7:17027. https://doi.org/10.1038/s41598-017-17288-z
11. Lisy M, Pennecke J, Brockbank KGM, Fritze O, Schleicher M, Schenke-Layland K, Kaulitz R, Riemann I, Weber CN, Braun J, Mueller KE, Fend F, Scheunert T, Gruber AD, Albes JM, Ziemer G, Stock UA (2010) The performance of ice-free cryopreserved heart valve allografts in an orthotopic pulmonary sheep model. Biomaterials 31:5306–5311. https://doi.org/10.1007/978-1-0716-0783-1_31
12. Brockbank KGM, Schenke-Layland K, Greene ED, Chen Z, Fritze O, Schleicher M, Kaulitz R, Riemann I, Fend F, Albes JM, Stock UA, Lisy M (2012) Ice-free cryopreservation of heart valve allografts: better extracellular matrix preservation in vivo and preclinical results. Cell Tissue Bank 13:663–671. https://doi.org/10.1007/s10561-011-9288-7
13. Biermann AC, Marzi J, Brauchle E, Schneider M, Kornberger A, Abdelaziz S, Wichmann JL, Arendt CT, Nagel E, Brockbank KGM, Martina Seifert M, Schenke-Layland K, Stock UA (2018) Impact of T-cell mediated inflammation on xenograft heart valve transplantation: short-term success and mid-term failure. Eur J Cardiothoracic Surg 53:784–792. https://doi.org/10.1093/ejcts/ezx396
14. Biermann AC, Marzi J, Brauchle E, Wichmann JL, Arendt CT, Puntmann V, Nagel E, Abdelaziz S, Winter AG, Brockbank KGM, Layland S, Schenke-Layland K, Stock UA (2019) Improved long-term durability of allogeneic heart valves in the orthotopic sheep model. Eur J Cardiothorac Surg 55:484–493. https://doi.org/10.1093/ejcts/ezy292
15. Manuchehrabadi N, Gao Z, Zhang J, Ring HL, Shao Q, Liu F, McDermott M, Fok A, Rabin Y, Brockbank KGM, Garwood M, Haynes CL, Bischof J (2017) Improved tissue cryopreservation using nanowarming: inductive heating of magnetic nanoparticles. Sci Transl Med 9:eaah4586. https://doi.org/10.1126/scitranslmed.aah4586
16. Cebotari S, Lichtenberg A, Tudorache I, Hilfiker A, Mertsching H, Leyh R, Breymann T, Kallenbach K, Maniuc L, Batrinac A, Repin O, Maliga O, Ciubotaru A, Haverich A (2006) Clinical application of tissue engineered human heart valves using autologous progenitor cells. Circulation 114(1 Suppl):I132–I137. https://doi.org/10.1161/CIRCULATIONAHA.105.001065
17. Cebotari S, Tudorache I, Ciubotaru A, Boethig D, Sarikouch S, Goerler A, Lichtenberg A, Cheptanaru E, Barnaciu S, Cazacu A, Maliga O, Repin O, Maniuc L, Breymann T, Haverich A (2011) Use of fresh decellularized allografts for pulmonary valve replacement may reduce the reoperation rate in children and young adults: early report. Circulation 124(11 Suppl):S115–S123. https://doi.org/10.1161/CIRCULATIONAHA.110.012161
18. Wang S, Goecke T, Meixner C, Haverich A, Hilfiker A, Wolkers WF (2012) Freeze-dried heart valve scaffolds. Tissue Eng Part C Methods 18:517–525. https://doi.org/10.1089/ten.tec.2011.0398
19. Wang S, Oldenhof H, Goecke T, Ramm R, Harder M, Haverich A, Hilfiker A, Wolkers WF (2015) Sucrose diffusion in decellularized heart valves for freeze-drying. Tissue Eng Part C Methods 21:922–931. https://doi.org/10.1089/ten.TEC.2014.0681
20. Vásquez-Rivera A, Oldenhof H, Dipresa D, Goecke T, Kouvaka A, Will F, Haverich A,

Korossis S, Hilfiker A, Wolkers WF (2018) Use of sucrose to diminish pore formation in freeze-dried heart valves. Sci Rep 8:12982. https://doi.org/10.1038/s41598-018-31388-4
21. Zouhair S, AguiariP, Iop L, Vásquez-Rivera A, Filippi A, Romanato F, Korossis S, Wolkers WF, Gerosa G (2019) Preservation strategies for decellularized pericardial scaffolds for off-the-shelf availability. Acta Biomater 84:208–221. https://doi.org/10.1016/j.actbio.2018.10.026
22. Goecke T, Theodoridis K, Tudorache I, Ciubotaru A, Cebotaria S, Ramm R, Höffler K, Sarikouch S, Vásquez-Rivera A, Haverich A, Wolkers WF, Hilfiker A (2018) In vivo performance of freeze-dried decellularized pulmonary heart valve allo- and xenografts orthotopically implanted into juvenile sheep. Acta Biomater 68:41–52. https://doi.org/10.1016/j.actbio.2017.11.041

Check for updates

Chapter 32

Freeze-Drying and Room Temperature Preservation of Mammalian Sperm

Sayaka Wakayama, Daiyu Ito and Teruhiko Wakayama

Abstract

Freeze-drying technology enables the safe preservation of genetic resources at room temperature, even in situations where liquid nitrogen cannot be replenished following a major disaster. In addition, genetic resource storage becomes easier, and international transport becomes possible at low cost. In addition, freeze-drying has been found to confer resistance to high temperatures and radiation, and research into the safe and permanent preservation of genetic resources outside the Earth is now underway. It has also become possible to generate cloned individuals from freeze-dried somatic cells, enabling the recovery and preservation of genetic resources even from infertile animals. However, the birth rate using freeze-dried sperm is significantly lower than that achieved with fresh sperm, and long-term storage has not yet been demonstrated. Therefore, this technology has not yet gained reliability and has not been put into practical use, although these issues are being addressed by the latest research. This chapter first provides an overview of current sperm freeze-drying methods, followed by a detailed freeze-drying protocol.

Keywords Clone, Freeze-drying, Genetic resources, ICSI, Preservation, Room temperature, Spermatozoa

1 Introduction

The routine storage of animal genetic resources in liquid nitrogen (LN_2) for extended periods requires a constant supply of LN_2, which incurs significant expense. In addition, it also presents challenges such as the continuous need for LN_2 production and the difficulty of handling genetic resources at extremely low temperatures. Furthermore, if liquid nitrogen cannot be replenished due to a major disaster, such as a large earthquake or massive hurricane, all stored genetic resources will thaw and be lost [1, 2]. Despite these limitations, at present this method remains the most common method of sperm storage.

However, it would be ideal if genetic resources could be stored indefinitely at ambient temperature. Consequently, numerous attempts have been made to design such a protocol. If successful,

Willem F. Wolkers and Harriëtte Oldenhof (eds.), *Cryopreservation and Freeze-Drying Protocols*, Methods in Molecular Biology, vol. 3060, https://doi.org/10.1007/978-1-0716-5416-3_32,

this approach would enable safe and low-cost preservation of genetic resources even during major disasters, making it feasible in developing countries. In addition, international transport would become possible at very low cost, facilitating collaborative research using the same genetically modified mice.

Polge et al. first attempted to freeze-dry (FD) spermatozoa by combining fowl sperm with glycerol, followed by freeze-drying while removing 90% of the water content [3]. Up to 50% of spermatozoa regained motility after rehydration, but their fertility was not determined. However, many critical technical details were omitted from the paper; in addition, glycerol cannot dry, remaining liquid at room temperature and, more importantly, is highly toxic. Subsequent attempts to freeze-dry viable human and bull spermatozoa largely failed, with a few exceptions [4, 5]. In a later paper, Yushchenko reported the birth of 12 normal rabbit offspring using freeze-dried spermatozoa [6], and Meryman et al. [7, 8] reported pregnancies in cows following artificial insemination with freeze-dried spermatozoa. However, to date these results have not been replicated by other groups [9].

In previous studies, we attempted to completely dehydrate mouse sperm in glass ampoules before storing them at room temperature or in a refrigerator (without glycerol) for periods as long as several months. Due to the absence of a cryoprotectant, the cell membranes of the FD sperm ruptured, and all spermatozoa died during freezing or shortly after the addition of water [10]. However, since the development of mouse intracytoplasmic sperm injection (ICSI) techniques in 1995 [11], it has become possible to produce offspring using ICSI even from dead spermatozoa [12, 13]. Surprisingly, although all sperm die after freeze-drying, their DNA remains intact. Thus, healthy offspring can be produced when these sperm are injected into oocytes. The development of ICSI marked the first successful generation of offspring from mammalian (mouse) spermatozoa preserved at room temperature for over one month. This technique has since been used to produce offspring from the FD sperm of rabbits [14], rats [15], horses [16], and hamsters [17] and basic research on cattle [18], pigs [19], sheep [20], dogs [21], cats [22], and rhesus monkeys [23]. Further study on human sperm [24] is also underway. In a previous paper, our lab developed a technique for preserving freeze-dried sperm using commonly used, low-cost PCR tubes [25] and thin, unbreakable plastic sheets [26] instead of glass ampoules. Although this sheet method has not yet succeeded in long-term storage at room temperature, it may eventually enable storage of thousands of mouse strains in a single book or card holder [27]. Furthermore, another method of storing (sheep) sperm in stainless steel minicapsules has also been developed (see Table 1) [28].

However, these preservation methods have not yet become widely adopted since FD sperm have a significantly lower birth rate

Table 1
List of important papers on research into the freeze-drying preservation of sperm and cells

Date of publication	Purpose	Results	Reference
1998	Availability of FD for mammalian sperm	First success for production of offspring (mouse)	Wakayama and Yanagimachi [10]
2003	Storage period	Without cryoprotectant	Ward et al. [29]
2004	Radiation damage	Chromosome integrity	Kusakabe et al. [30]
2004	Other animal	First success of rabbit	Liu et al. [14]
2005	Storage period	Accelerated degradation kinetics	Kawase et al. [31]
2005	Other animal	First success of rat	Hirabayashi et al. [15]
2008	Other cell type	Somatic cell	Loi et al. [32]
2011	Other animal	First success of hamster	Muneto et al. [17]
2011	Other animal	First success of horse	Choi et al. [16]
2014	Storage period	2 years at RT	Liu et al. [33]
2014	Other animal	Preliminary study of wild animals	Kaneko et al. [34]
2017	Effect of space radiation	Preserved on international space station for 1 year	Wakayama et al. [35]
2019	Tolerance to temperature	From −196°C to 150°C	Wakayama et al. [36]
2021	Effect of space radiation	Possible to store FD sperm on ISS over 200 years	Wakayama et al. [37]
2021	Shipping of sperm	Sperm can be sent by postcard	Ito et al. [26]
2022	Other cell type	Somatic cell	Wakayama et al. [38]
2022	Other cell type	Round spermatid	Wakayama et al. [39]
2023	Container	Plastic microtube	Yang et al. [25]
2023	Container	Stainless steel minicapsules	Palazzese et al. [28]
2025	Storage period	6 years at RT	Kamada et al. [40]
2025	DNA damage	Removed abnormal micronuclei	Shibazaki et al. [41]
2025	Effect of monosodium glutamate	Improved birth rate	Ushigome et al. [42]

than fresh sperm after fertilization, and their reliability for long-term storage has not been demonstrated. One cause of reduced birth rates after freeze-drying is that FD sperm commonly contain chromosomal abnormalities, which can lead to miscarriage [29]. Although the cause of these abnormalities remains unclear, a technique has been developed to extract individual chromosomes with separation defects [41], and the underlying mechanism will likely be clarified in the future. Although new efforts, such as the identification of novel cryoprotectants and improvements in freeze-drying methods, have been made to increase the birth rate, to date no method to remediate this damage has been validated [2, 43]. Recently, we discovered that adding water to FD sperm at temperatures above 50°C can increase the birth rate by approximately 1.5 times [44]. Although this still does not match the birth rate achieved with fresh sperm, it may represent a step toward practical application. At the same time, the maximum storage period of FD sperm at room temperature has been extended using cryoprotective agents, deoxygenators, and desiccants. For example, when high-vacuum ampoules containing FD sperm were selected using non-destructive testing for long-term storage, FD sperm were successfully preserved in a desk drawer for over six years without a decrease in success rate (see Table 2) [40, 45]. Importantly, we successfully produced offspring from sperm stored for six years in all individuals, indicating high reproducibility and reliability.

In general, sperm and oocytes are necessary for the preservation of genetic resources. However, it is generally not possible to collect sperm or oocytes from elderly, immature, or infertile individuals. However, using somatic cell cloning technology, it is possible to produce offspring from somatic cells, even from infertile animals [46, 47]. Consequently, somatic cells are now regarded as a genetic resource. Therefore, research began on freeze-drying somatic cells in the same way as sperm for room temperature storage [32, 48]. However, nuclear transfer using FD somatic cells proved difficult, and it was assumed that sperm may be the only mammalian cell that can be freeze-dried before producing viable offspring [49]. To test this hypothesis, we first conducted an experiment using round spermatids, which are progenitor and haploid cells like sperm but do not yet have a condensed nucleus formed by protamine, like somatic cells, to demonstrate that FD is possible in cells other than sperm [39]. Based on these results, after much trial and error, we successfully cloned mice from freeze-dried somatic cell nuclei [38, 50]. Although we have not yet succeeded in producing cloned mice from FD somatic cells stored at room temperature, this study demonstrates the potential to extract genetic resources from any animal and store them safely at room temperature.

The development of freeze-drying protocols to conserve genetic resources was originally intended to enable preservation without the use of liquid nitrogen. However, it has since been discovered that freeze-dried sperm possess distinct advantages over fresh spermatozoa. For example, FD sperm can survive brief exposure to temperatures as hot as 150°C (see Table 2) [36]. Similarly, the radiation resistance of freeze-dried spermatozoa has been reported to exceed that of fresh spermatozoa, and healthy offspring were produced

Table 2
Extreme environment tolerance and long-term storage of freeze-dried sperm

Experiment	Tolerance condition or preservation periods	No. of oocytes surviving after ICSI	No. (%) of fertilized embryo	No. of two-cell embryo developed (%) or transferred	No. (%) of offspring	Reference
Tolerance against high temperature	95°C for 6 h	65	53 (81)	52 (80)[a]	4 (7)	[14]
	150°C for 3 min	87	82 (94)	82 (94)[a]	51 (61)	
Tolerance against X-ray	5 Gy	98	78 (79)	78 (79)[a]	22 (28)	[37]
	30 Gy	330	274 (83)	270 (82)[a]	3 (1)	
Preservation on Space station	1 years at −95°C	1076	1017 (95)	784 (77)[a]	73 (9)	[35]
	3 years at −95°C	2411	2321 (96)	1981 (85)[a]	243 (12)	[37]
	6 years at −95°C	1624	1446 (89)	1301 (90)[a]	168 (13)	
Preservation at room temperature	6 years for ICR strain	937	724 (77)	251[b]	34 (14)	[40]
	5 years for BDF1 strain	432	368 (85)	212[b]	69 (33)	
	5 years for B6 strain	563	457 (81)	223[b]	40 (18)	
	4 y and 8 m for C3H strain	421	322 (77)	117[b]	21 (18)	

[a]All two-cell-stage embryos were transferred to recipient mice [b]Since two-cell-stage embryos were randomly collected for embryo transfer from each experiment, the developmental rate to the two-cell stage cannot be shown in here

even after 30 Gy of irradiation [37]. Based on these findings, in a previous paper, we proposed storing genetic resources underground on the moon [35]. In Japan, the massive 2011 earthquake caused both flooding and a nuclear accident, affecting even regions once considered safe from any disaster [51]. This suggests that nowhere on Earth is entirely secure. In contrast, genetic resources stored underground on the moon would be protected safely and permanently, regardless of conditions on Earth. Until recently, transporting genetic material into space was not feasible because rockets could not carry liquid nitrogen. In addition, resources transported to the lunar surface would be exposed to intense deep-space radiation and abrupt temperature fluctuations. Nevertheless, prior studies have shown that FD sperm are not only compatible with rocket transport but can also tolerate radiation and heat [36, 37]. We have already delivered FD sperm to the ISS and initiated studies examining the effects of cosmic radiation on sperm integrity [35, 37]. In the future, as humans expand into space, it will be essential to deliver large quantities of genetic material to other planets to prevent inbreeding within species [52, 53]. Moreover, since FD genetic resources can be stored in compact containers, they are both lightweight and space-efficient, allowing even a single spacecraft to carry substantial volumes. Thus, the freeze-drying of mammalian germ cells represents a critical technology for the long-term survival of humans. In this chapter, we introduce a method for freeze-drying and preserving mammalian spermatozoa.

2 Materials

2.1 Stock Solutions

1. Hyaluronidase: 10% (100× stock solution). Hyaluronidase (0.1 mg) was first dissolved in 1 mL of HEPES-buffered Chatot, Ziomek, and Bavister (H-CZB) medium [12, 54]. This solution was then aliquoted into small tubes (5–10 μL) and stored at −30°C. The final concentration was 0.1%.
2. PMSG and hCG dissolved individually in normal saline and stored in aliquots at −30°C. PMSG: Pregnant mare serum gonadotropin; hCG: human chorionic gonadotropin; both: 50 IU/mL.
3. Medetomidine (1.875 mL; 1 mg/mL), midazolam (2 mL; 5 mg/mL), and butorphanol (2.5 mL; 5 mg/mL) dissolved in 18.625 mL of normal saline in a final volume of 25 mL. We used 0.1 milliliters of stock solution per 10 g body weight. Atipamezole (0.75 mg/kg; 5 mg/mL) in normal saline was used to wake up mice. Anesthetic stock solution: 0.75/4/5 mg/kg medetomidine/midazolam/butorphanol.

2.2 Culture Media

1. Oocyte and embryo culture media: Chatot, Ziomek, and Bavister (CZB) medium [54] was used for all oocyte/embryo cultures. H-CZB medium [12] was used for gamete handling and ICSI, which was performed in air. H-CZB is not suitable for use in a CO_2 incubator.
2. Sperm diffusion medium for ICSI: 12% polyvinylpyrrolidone (PVP: 360 kDa) in H CZB.
3. Sperm freeze-drying medium: any medium can be used, including saline solution, CZB, human tubal fluid (HTF) [55], or Tris-EGTA [56]. Note that the presence of reducing sugars like glucose may negatively affect long-term storage stability of chromatin, whereas non-reducing disaccharides (e.g., sucrose or trehalose) enhance chromatin stability during storage by

increasing the glass transition temperature of the dried sample [57].

2.3 Equipment and Materials for Freeze-Drying

1. Freeze dryer, e.g., model FDU-2200 from EYELA, Tokyo, Japan) for use with glass ampoules or a FreeZone 2.5 L Benchtop Freeze Dryer (Labconco, MO, USA) for freeze-drying of sperm on thin sheets.
2. Tesla coil leak detector. (Sanko Electronic Laboratory Co., Ltd., Tokyo, Japan).
3. Glass ampoules. Capable of withstanding high vacuum and can be melted and sealed with a gas burner.

3 Methods

3.1 Spermatozoa Collection

1. Euthanize mature male mice. Ensure that mice are at least nine weeks old and capable of spermatogenesis.
2. Collect spermatozoa from both epididymides and cut ducts using a pair of sharp scissors (Fig. 1-1) (*see* **Note 1**).
3. Place a few drops of dense spermatozoa into a 1.5-mL centrifuge tube containing 1 mL of sperm freeze-drying medium (*see* **Note 2**).
4. Incubate the sample in a CO_2 incubator at 37°C for 30 min to promote capacitation (*see* **Note 3**).

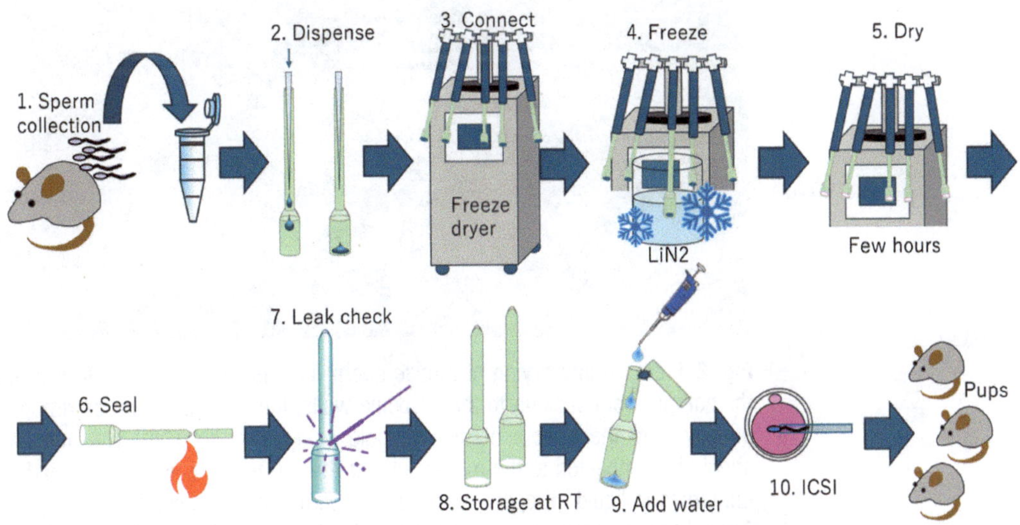

Fig. 1 Schematic of a freeze-drying method of mouse sperm using a glass ampoule

5. Optional: count the number of sperm and adjust the concentration of the sperm suspension (>1 × 10^6 sperm/mL) by adding HTF medium (*see* **Note 4**).

3.2 Preparation of Freeze-Dried Spermatozoa Using Glass Ampoules (for Long-Term Preservation at Room Temperature)

1. Dispense 50 μL aliquots of the spermatozoa into lyophilization-resistant glass ampoules (Figs. 1-2 and 2a) (*see* **Note 5**).
2. Cover the ampoule with a simple aluminum cap before use.
3. Next, the freeze dryer should be turned on. The freeze-dryer should operate under standard conditions with a trap temperature of −80°C and a vacuum pressure of 3.0 Pa (*see* **Note 6**).
4. Connect glass ampoules to the tube of the freeze-dryer (Fig. 1-3) (*see* **Note 7**).
5. Freeze the glass ampoules containing sperm by immersion in liquid nitrogen after attachment to the freeze-dryer (Figs. 1-4 and 2b) (*see* **Note 8**).
6. After ensuring that all samples are sufficiently frozen, open the flow valve of the freeze-dryer while keeping the samples immersed in liquid nitrogen (*see* **Note 9**).

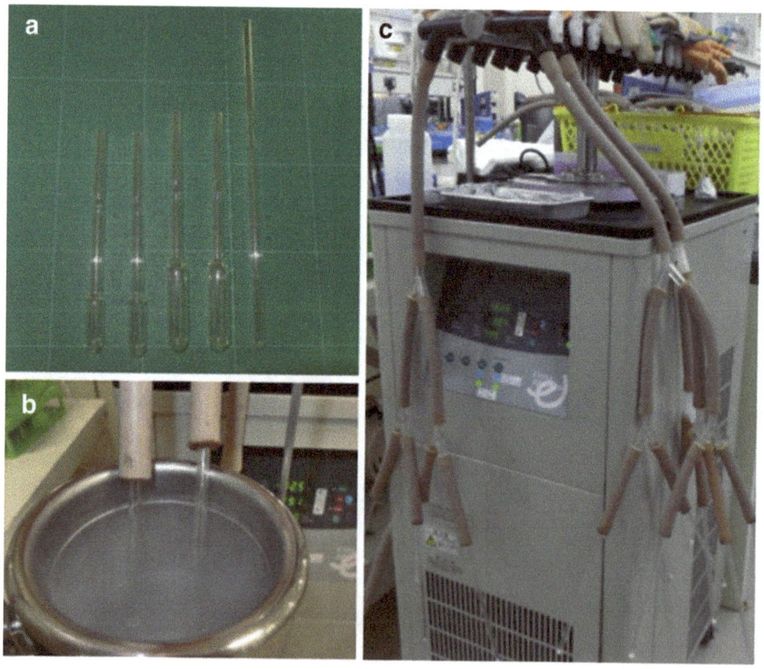

Fig. 2 Freezing and drying of mouse sperm in glass ampoules. (**a**) To prevent the sample from sticking to the ampoule walls, the sample should be dropped directly onto the bottom of the ampoule using a long tip. (**b**) The ampoule should be connected to the rubber tube of the freeze-dryer and then should be immersed in liquid nitrogen to freeze the sperm. Then, the suction valve of the freeze-dryer is opened. (**c**) When drying begins, ampoules should be removed from liquid nitrogen and dried at room temperature for 3–6 h

7. After confirming that the air pressure is sufficiently decreased, remove the samples from the liquid nitrogen in which they are immersed (Figs. 1-5 and 2C). All samples remain frozen due to the heat of vaporization (*see* **Note 10**).
8. Seal the glass ampoules. The sealing operation takes about 3–6 h, and is completed once after confirming the sample to be completely dry (*see* **Note 11**).
9. Use a gas burner to heat, melt, and seal frozen ampoules (Figs. 1-6 and 3a, b). Ensure that all glass ampoules are sealed by a screw top (*see* **Notes 12–14**).

3.3 Preservation of Freeze-Dried Spermatozoa for Extended Periods at Room Temperature

We previously reported that if air remains in the ampoules, sperm can deteriorate over time due to oxidation and humidity. This results in a significant decrease in birth rate [45]. Therefore, we strongly recommend checking the vacuum level before storing or opening glass ampoules containing FD sperm.

1. Check the vacuum level of the glass ampoules using a Tesla coil leak detector (Figs. 1-7 and 3c), which is usually used to measure the vacuum of a fluorescent lamp, was used to check the vacuum level of glass ampoules [45].
2. Fix ampoules on the desk in a standing position: We found it easier to make measurements if the ampoules to be tested were lined up in a standing position.
3. Turn on the Tesla leak detector and bring the electric spark at the tip of the detector close to the ampoule (Fig. 3d) (*see* **Note 15**).
4. Preservation of FD Sperm at Room Temperature. Next, wrap each ampoule in a protective sheet and store in a storage container in a warehouse or desk drawer at room temperature (15–25°C) in the laboratory (Fig. 3e, f) until further use (*see* **Note 16**).

3.4 Preparation of Freeze-Dried Spermatozoa Using Plastic Sheets (for Thin Sheet Preservation)

Storing sperm in thin sheets [27] is not yet suitable for long-term storage at room temperature. However, it can be an effective method for transporting sperm by air mail due to the durability and thinness of the sheets [26]. The procedure described below is illustrated in Fig. 4h.

1. Turn on the freeze-dryer (Fig. 4a) (*see* **Note 17**).
2. Prepare the weighing paper and the glued plastic sheet. The size of the weighing paper should be approximately 10 mm square, and the plastic sheet should be cut larger than the weighing paper (Fig. 4b).
3. Prepare an aluminum boat and float it in liquid nitrogen (LN2) contained within a Styrofoam box.

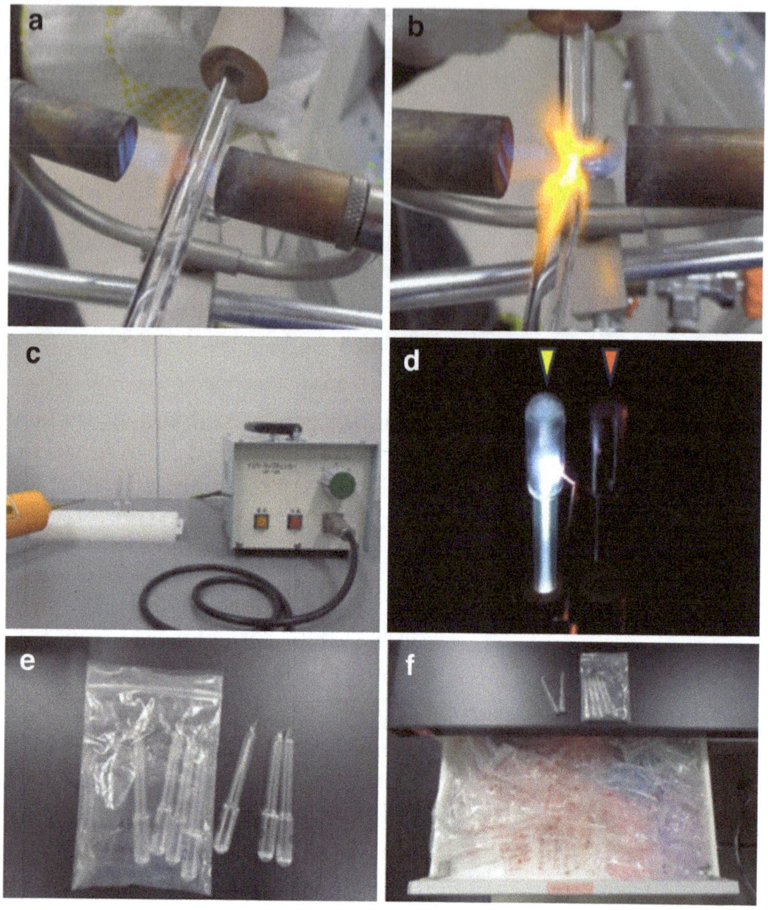

Fig. 3 Sealing of vacuum ampoules and storage of freeze-dried mouse sperm at room temperature. (**a**) After the complete freeze-drying procedure, we recommend heating ampoules with a double-sided burner while they are still connected to the freeze-dryer. Be careful not to burn yourself. (**b**) When the glass has softened, quickly twist the ampoule to seal it. Note that if an inexperienced person attempts this, the glass may often break, thereby ruining the ampoule. It is therefore necessary to practice this technique well in advance. (**c**) The Tesla coil leak detector measures the strength of the vacuum in the samples. (**d**) When the Tesla coil tip at minimum power was brought near vacuum ampoules, the tip sparked around the glass and ionized the very low-pressure gas inside the ampoule. Therefore, ampoules that turned bright (yellow arrow heads) with the Tesla detector were determined to be effectively vacuumed. (**e**) Glass ampoules containing FD mouse sperm. These are approximately the same length as the yellow tip. (**f**) It is possible to store mouse sperm for several years at room temperature, even in a drawer in an office desk

4. Place the weighing paper and plastic sheet on the aluminum boat. Ensure that the weighing paper is sufficiently cold inside the Styrofoam box before adding the sperm suspension.

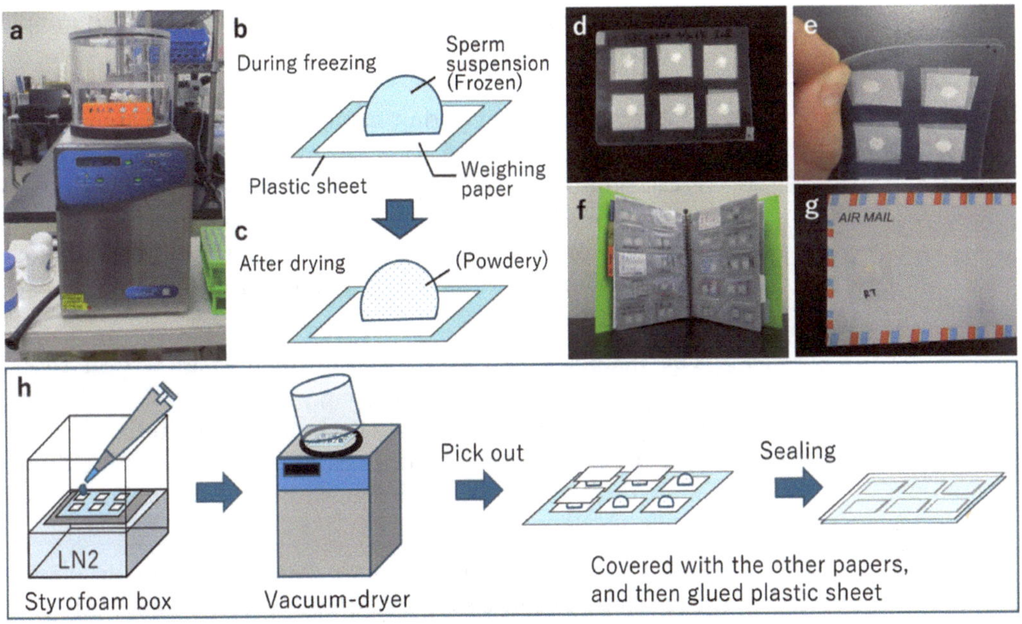

Fig. 4 Preparation of freeze-dried spermatozoa using plastic sheets. (**a**) The freeze-dryer equipment, FreeZone 2.5 L Benchtop Freeze Dryer. (**b** and **c**) Enlarged view of the sperm suspension, weighing paper, and plastic sheet. Sperm suspensions are dropped on weighing paper floating on LN2. After drying treatment, the drops become powdery. (**d** and **e**) FD sperm were stored between plastic sheets. These sheets are soft and do not break even when bent. (**f**) A sperm book using a card holder. This setup can compactly store many freeze-dried sperm sheets. (**g**) A plastic sheet of FD sperm was placed in the envelope and dropped directly into the mailbox without further protection. (**h**) Schematic of the freeze-drying method of mouse sperm using a thin plastic sheet

5. Next, dispense 50 μL aliquots of sperm suspension (see Sect. 3.1) onto the weighing paper in the Styrofoam box (Fig. 4b, c) (*see* **Note 18**).
6. Close the Styrofoam box with the lid during freezing, i.e., at least 10 min.
7. Transfer frozen sperm along with the aluminum boat to the freeze-drying chamber as quickly as possible, and start the vacuum drying process (*see* **Note 19**).
8. Finally, turn off the freeze dryer and remove the FD sperm pellets from the chamber. Cover with additional pieces of weighing paper and seal with a sticky plastic sheet (Fig. 4d, e) (*see* **Notes 20 and 21**).

3.5 Preservation and Shipping of Freeze-Dried Spermatozoa in Thin Sheets

1. Store the thin sheets of FD sperm in a file book (e.g., an album; see Fig. 4f) at −30°C until ICSI (*see* **Note 22**).
2. Attach FD sheets, which were stored in the freezer, to envelopes or postcards on the day of mailing (Fig. 4g) (*see* **Note 23**).

3. After shipping by domestic airmail, the store shipped samples in a freezer until further use for ICSI (*see* **Note 24**).

3.6 Preparation of Media Dishes

1. Oocyte culture dish during manipulation: Place a minimum of 20 droplets (e.g., 4×5) of CZB medium (approximately 10μ L each) into 6-cm cell culture dishes. Cover dishes with sterile mineral oil. Now they can be used for oocyte collection and recovery for ICSI. Accordingly, they must be prepared at the beginning of the protocol and warmed in a 37°C incubator with 5% CO_2.
2. The manipulation chamber: The manipulation dish should include two types of media, e.g., H CZB and 12% polyvinylpyrrolidone (PVP). Cover each droplet of 10–15μL in volume with mineral oil. Draw lines on the outside of the dish to allow the user to easily distinguish between different types of media.
3. Culture dish: Place a minimum of 20 CZB medium droplets on a cell culture dish. Use several droplets (each droplet is 10–15μ L in volume) on the top for embryo washing, and use the last droplet for long-term culture of the embryos until transfer to recipient mice. Cover the dish with mineral oil.

3.7 Micromanipulator Setup

1. Pull a glass pipette using a pipette puller.
2. Cut the tip to an appropriate diameter (*see* **Note 25**).
3. Bent all pipettes close to the tip (approximately 300μm away) at a 10–25° angle using a microforge. Here, the desired angle depends on the setup of the micromanipulator.
4. Pipettes can be stored in a 10cm dish at room temperature for several months.
5. Insert a small volume of PMM Operation Liquid (Prime Tech Ltd, Japan) into the back of a pipette using a 1-mL syringe and a 30-G needle (approximately 1–2 mm-long column). This is done to enhance the power of the piezo unit (*see* **Note 26**).
6. Attach the holding pipette to one side of the micromanipulator.
7. On the opposite side, attach the sperm injection pipette to the pipette holder. This pipette holder is also attached to the piezo unit. Tightly screw the top of the pipette holder (*see* **Note 27**).
8. Expel air and oil from the injection pipette in PVP medium.
9. Wash the inside and outside of the pipette with PVP medium using a few drops of PMM Operation Liquid until all oil is removed (*see* **Note 28**).

3.8 Oocyte Preparation

1. PMSG injection: BDF1 or ICR female mice (8–10 weeks) need to be stimulated by an injection of 5 IU of PMSG into

the abdominal cavity three days before the experiment. This is typically performed between 5 and 6 p.m.

2. hCG injection: inject the mice with hCG (5 IU) 48 h later (i.e., one day before the experiment). This is typically performed between 5 and 6 p.m.
3. Oocyte collection: retrieve oocyte–cumulus cell complexes from the oviductal ampullae 14–15 h after hCG injection.
4. Hyaluronidase treatment: To remove cumulus cells from oocytes, first transfer oocyte–cumulus complexes into a 50-μL droplet of H-CZB containing 0.1% hyaluronidase (100× stock solution) for 5 min.
5. Oocyte selection: Oocytes from the hyaluronidase medium were washed twice using H-CZB then placed onto an oocyte culture dish prepared as described above (*see* **Note 29**). The number of oocytes contained within a drop depends on the skill level of the experimenter and the type of experiment.

3.9 Rehydration of Freeze-Dried Spermatozoa

Although samples do not currently become motile (swim) upon rehydration, it is possible to produce offspring from FD samples via ICSI. Here, we briefly describe an ICSI protocol that uses freeze-dried samples.

1. Glass ampoule method:
   (a) Break the neck of the ampoule.
   (b) Add 50 μL of sterile distilled water.
   (c) Mix the contents with a pipette (*see* **Note 30**).
2. Thin sheet method:
   (a) Cut and peel off the sheet to collect FD sperm pellets with tweezers (*see* **Note 31**).
   (b) Insert FD sperm pellets into PCR tubes.
   (c) Immediately add 50 μL of sterile distilled water, and mix contents with a pipette.
3. The rehydrated samples can now be used for ICSI (*see* **Note 32**).
4. The ICSI procedure must be completed within 30 min to 1 h, since hydrated sperm samples deteriorate over time (*see* **Note 33**).

3.10 Intracytoplasmic Sperm Injection (ICSI)

1. First set up a sperm injection pipette using 6–7 μm pipettes.
2. Mix sperm suspension (2–3 μL) thoroughly with 20 μL of H-CZB medium containing 12% (w/v) PVP (360 kDa) in a micromanipulation chamber (*see* **Note 34**).
3. Transfer a group (10–20) to H-CZB medium. The number of oocytes per droplet depends on the skill level of the individual

experimenter. Each group of injections must be completed within 15 min (*see* **Note 35**).

4. Next, the sperm head should be collected using a pipette. After freeze-drying treatment, many spermatozoa have separated into separate heads and tails.
5. Identify the metaphase II spindle on the surface of the oocyte. This can be easily recognized by the bump. The oocyte is then rotated to place the spindle between 11 o'clock and 1 o'clock or between 5 o'clock and 7 o'clock. This is then firmly attached to the holding pipette.
6. Cut the zona pellucida using several strong piezo pulses (*see* **Note 36**). The negative pressure inside the pipette enhances the piezo power.
7. Reduce the power of the piezo unit (*see* **Note 37**). The oolemma is weaker than the zona pellucida, and the survival rate of oocytes after injection increases if the power is reduced at this point.
8. Push one sperm head forward until it is near the tip of the pipette, and advance the pipette until it almost reaches the opposite side of the cortex of the oocyte (*see* **Note 38**).
9. Apply a weak piezo pulse to puncture the oolemma at the pipette tip. Rapid relaxation of the oocyte membrane can be observed. The sperm head needs to be immediately expelled from the oocyte cytoplasm using a minimal amount of PVP medium.
10. Gently withdraw the pipette from the oocyte (*see* **Note 39**).
11. Maintain sperm-injected oocytes for at least 10 min to recover the membrane. At this point, the oocytes can be transferred into the CZB culture medium. During this period, any injection damage to the oocyte will be completely fixed (*see* **Note 40**).
12. When all sperm heads have been injected into oocytes, the injection pipette can be washed in a PVP dish by expelling some PMM Operation Liquid while applying power from the piezo unit. This washing step is important to prevent the pipette from becoming sticky (*see* **Note 28**).
13. Check for pronuclear formation 6 h after the start of the ICSI protocol. Embryos can then be transferred to a new CZB medium dish for long-term culturing until reaching the two-cell or blastocyst stage.

3.11 Embryo Transfer

1. Pseudopregnant (recipient) mothers can be prepared by mating estrous ICR female mice with vasectomized males on the same day as the ICSI experiment [58, 59].

2. One day later, six to ten embryos can be transferred into each oviduct of a pseudopregnant female.
3. On day 18 or 19 of gestation, offspring are either delivered naturally or by Cesarian section. Offspring are then allowed to mature naturally.

4 Notes

1. It is better to squeeze the condensed sperm mass from the epididymis using fingers.
2. We used Tris-EGTA as the freeze-drying medium. In addition, the dense sperm mass was cultured in HTF until all sperm had dispersed. The supernatant was collected and centrifuged at 5000 revolutions per minute (rpm) for 10 min. After centrifugation, the supernatant was removed, and the sperm pellet was suspended in Tris-EGTA.
3. Before freeze-drying sperm, it is advisable to check their quality and motility using an inverted microscope.
4. If the concentration of a sperm sample is too low to be used for further experiments (i.e., $<1 \times 10^6$ sperm/mL), the sperm suspension was centrifuged, and the volume of the containing medium was reduced to increase sperm concentration.
5. Fluid volume can vary depending on experimental conditions. However, experimenters should ensure that an adequate amount is dispensed to allow thorough drying within a few hours. To do so, we recommend using a pipette tip with a long reach to avoid transferring sperm to the wall of the ampoule, since any liquid on the wall will reduce the drying yield.
6. The freeze-dryer platform should be started 1–2 h before attaching samples to ensure that samples reach a sufficiently low temperature. We then closed all open and closed valves on the ampoule attachment of the freeze-dryer.
7. Ensure that the valve is firmly closed before attaching the ampoule, and insert the ampoule securely into the corresponding rubber tube.
8. After attachment to the tube of the freeze-dryer, glass ampoules should be directly immersed in containers filled with liquid nitrogen for more than 1 min.
9. Opening the valve causes a rapid rise in air pressure within the freeze-dryer.

10. If the sample volume is too large, the frozen sample may melt before drying. This can damage the DNA of the sperm.
11. To maintain the sample at room temperature for prolonged periods, it is crucial to ensure a strong vacuum pressure within the glass ampoules. To do so, sealing should be performed while removing air from the ampoules and the freeze-dryer.
12. Application of a gas burner to ampoules while removing air with the freeze-dryer can cause the glass ampoules to break; this is a common beginner mistake.
13. If the ampoule breaks, air can enter all other glass ampoules attached to the freeze-dryer. If breakage occurs, immediately close the valve connected to the broken ampoule. A vacuum pump should then be used to remove the air. After confirming that the air pressure has dropped sufficiently (i.e., 10–20 min later), sample sealing should be performed again.
14. If the vacuum of the ampoule is measured immediately after creation, some of the positive ampoules may be negative a few days later.
15. Light intensity can vary depending on the degree of vacuum. In a perfect vacuum, light shines brightly inside ampoules, similar to a fluorescent lamp. Ampoules that do not glow (and therefore show an incomplete vacuum) should be eliminated at this point.
16. The sealed part of the ampoule is made of thin glass and is therefore easily broken. Accordingly, if the ampoules contact each other, there is a high possibility of breakage. It is safer to store each ampoule wrapped in a protective sheet.
17. We recommend using a shelf-type freeze-dryer, not a hanging device used for glass ampoules.
18. Sperm suspensions should be placed quickly onto weighing paper, since they freeze quickly inside the yellow tip.
19. For 50 μL sperm samples, we recommend allowing them to dry in the chamber for at least 6 h.
20. FD sperm pellets should be immediately placed between sheets. Here, the experimenter should take care to avoid static electricity since this absorbs moisture and shrivels up.
21. We recommend removing as much air as possible between plastic sheets.
22. Files containing pockets are useful for keeping sheets; this can make satisfying the management requirements of several experiments and mouse strains simpler.
23. Ensure that sheets are securely attached to the envelope or postcard using tape to prevent damage and/or loss. In

addition, sheets should be protected from light and scratches by wrapping them in aluminum foil.

24. Domestic air mail is typically completed within a few days. In general, transport should be completed within three days, as FD sheets can be preserved at 15–25°C for only a few days.
25. For the holding pipette, we recommend an outside diameter (OD) that is smaller than the oocyte (e.g., OD, 70–80 μ m; inner diameter [ID], 10 μm). Modified pipettes can be ordered from several companies to save time.
26. The PMM operation liquid should be injected into the pipette immediately before use since it will evaporate within a day.
27. If the screw closure is not secured tightly, the cutting of the zona pellucida of the oocyte by the piezo pulses will not be smooth.
28. The condition of the pipette is very important; it affects not only the oocyte survival rate but also embryo development after ICSI. However, the pipette can become sticky during micromanipulation. To avoid this problem, we washed pipettes in PVP medium. PVP can cover both the inside and outside of the pipette to keep the surface slick. Without this step, pipette soils rapidly develop and pipettes will need to be changed frequently.
29. Learning to select good oocytes requires practice. In general, good oocytes have a larger cytoplasm than bad ones.
30. Add the same amount of water as the liquid used to prepare the sample to avoid changing the concentration of the final culture medium.
31. If the FD sperm are attached to the weighing paper, attempt to scrape them off using tweezers or switch to an alternate sheet.
32. The head and tail of freeze-dried sperm are harder than those of fresh sperm. Therefore, care must be taken when performing ICSI. Some samples may require activation after insemination.
33. If it takes more than an hour for ICSI, store ampoules of FD sperm in a refrigerator and replace sperm samples every 30 min.
34. When spermatozoa are transferred to PVP droplets, they are mixed with PVP medium using sharp tweezers for at least 30 s. If the spermatozoa are not mixed thoroughly with the PVP medium, they will aggregate. If this happens, it will be very difficult to isolate a single sperm. We also recommend replacement of the spermatozoa suspension every 30 min during the ICSI procedure.

35. If you are a beginner, place approximately 5 oocytes per drop. More experienced experimenters can place as many as 20 oocytes per drop.

36. To avoid damaging the oocyte, ensure a large space between the zona pellucida and oolemma.

37. Typically, the power level required to cut the zona pellucida is 2–3, and the speed is 5–8. If the piezo unit is properly set up, the zona pellucida can be easily cut even at power level 2. Moreover, when using the piezo unit for the first time, it is necessary to test whether the zona pellucida can be cut smoothly. If this is not possible, check the connection between the pipette and the pipette holder. The top of the pipette holder should be screwed tightly. We expelled all oil from the pipette because its presence may reduce the power transmission of the piezo unit. Moreover, a slight negative pressure within the pipette can enhance the power of the piezo unit. In contrast, if the power is too strong, even at the lowest setting, a few drops of oil should be added inside the pipette.

38. The piezo pulses should not be applied until the pipette is near the opposite side. If the piezo pulse is applied when the tip of the pipette is in the middle of the oocyte, most oocytes die following the injection procedure.

39. It may be difficult to release the sperm head from the pipette if the pipette is too dirty. Hence, pipettes should be washed frequently using PVP medium. This can be done by extracting some PMM Operation Liquid from the injection pipette and applying power from the piezo unit to the PVP medium. Otherwise, the pipette should be changed.

40. There are many factors capable of inducing oocyte lysis after injection. These include a pipette with a large internal diameter, high room temperature, or shallow pipette insertion. A large pipette greatly increases the power of the piezo unit and thereby increases the rate of oocyte lysis. The ICSI process should be performed at room temperature (25–26°C). Do not use a warm plate for the microinjection process. For best results, insert the injection pipette deeply into the oocyte before applying the piezo pulse. The rate of oocyte lysis will decrease with training. Typically, after one month of training, we observe a survival rate of approximately 50%. After 1 year of practice, we observe a rate of approximately 80%. Sperm-injected oocytes should be placed in H-CZB medium for at least 10 min. If they are transferred into CZB medium immediately after injection, nearly 15% of the cells may undergo lysis due to damage caused by the injection.

References

1. Loi P, Palazzese L, Moncada M, Sterzo ML, Iuso D, Czernik M, Wakayama T, Kikawada T (2025) Advances in induced anhydrobiosis for cell and gamete storage. Trends Biotechnol 43:1599–1608. https://doi.org/10.1016/j.tibtech.2025.02.010
2. Loi P, Anzalone DA, Palazzese L, Dinnyés A, Saragusty J, Czernik M (2021) Dry storage of mammalian spermatozoa and cells: state-of-the-art and possible future directions. Reprod Fertil Dev 33:82–90. https://doi.org/10.1071/RD20264
3. Polge C, Smith AU, Parkes AS (1949) Revival of spermatozoa after vitrification and dehydration at low temperatures. Nature 164:666. https://doi.org/10.1038/164666a0
4. Sherman JK (1954) Freezing and freeze-drying of human spermatozoa. Fertil Steril 5:357–371. https://doi.org/10.1016/s0015-0282(16)31685-5
5. Bialy G, Smith VR (1957) Freeze-drying of bovine spermatozoa. J Dairy Sci 40:739–745
6. Yushchenk NP (1957) Proof of the possibility of spermatozoa in dried state. Proc Lenin Acad Agr Sci 22:37–40
7. Meyman H, Kafig E (1959) Survival of spermatozoa following drying. Nature 184:470–471. https://doi.org/10.1038/184470a0
8. Meryman HT (1960) Drying of living mammalian cells. Ann NY Acad Sci 85:729–734. https://doi.org/10.1111/j.1749-6632.1960.tb49993.x
9. Saacke RG, Almquist JO (1961) Freeze-drying of bovine spermatozoa. Nature 192:995–996. https://doi.org/10.1038/192995a0
10. Wakayama T, Yanagimachi R (1998) Development of normal mice from oocytes injected with freeze-dried spermatozoa. Nat Biotechnol 16:639–641. https://doi.org/10.1038/nbt0798-639
11. Wakayama T, Ogura A (2024) In memory of Dr. Ryuzo Yanagimachi (Yana) (1928-2023). J Reprod Dev 70:i–iv. https://doi.org/10.1262/jrd.2024-E01
12. Kimura Y, Yanagimachi R (1995) Intracytoplasmic sperm injection in the mouse. Biol Reprod 52:709–720. https://doi.org/10.1095/biolreprod52.4.709
13. Wakayama T, Whittingham DG, Yanagimachi R (1998) Production of normal offspring from mouse oocytes injected with spermatozoa cryopreserved with or without cryoprotection. J Reprod Fertil 112:11–17. https://doi.org/10.1530/jrf.0.1120011
14. Liu JL, Kusakabe H, Chang CC, Suzuki H, Schmidt DW, Julian M, Pfeffer R, Bormann CL, Tian XC, Yanagimachi R, Yang X (2004) Freeze-dried sperm fertilization leads to full-term development in rabbits. Biol Reprod 70:1776–1781. https://doi.org/10.1095/biolreprod.103.025957
15. Hirabayashi M, Kato M, Ito J, Hochi S (2005) Viable rat offspring derived from oocytes intracytoplasmically injected with freeze-dried sperm heads. Zygote 13:79–85. https://doi.org/10.1017/s096719940500300x
16. Choi YH, Varner DD, Love CC, Hartman DL, Hinrichs K (2011) Production of live foals via intracytoplasmic injection of lyophilized sperm and sperm extract in the horse. Reproduction 142:529–538. https://doi.org/10.1530/REP-11-0145
17. Muneto N, Horiuchi T (2011) Full-term development of hamster embryos produced by injecting freeze dried spermatozoa into oocytes. J Mamm Ova Res 28:32–39. https://doi.org/10.1274/jmor.28.32
18. Keskintepe L, Pacholczyk G, Machnicka A, Norris K, Curuk MA, Khan I, Brackett BG (2002) Bovine blastocyst development from oocytes injected with freeze-dried spermatozoa. Biol Reprod 67:409–415. https://doi.org/10.1095/biolreprod67.2.409
19. Men NT, Kikuchi K, Nakai M, Fukuda A, Tanihara F, Noguchi J, Kaneko H, Linh NV, Nguyen BX, Nagai T, Tajima A (2013) Effect of trehalose on DNA integrity of freeze-dried boar sperm, fertilization, and embryo development after intracytoplasmic sperm injection. Theriogenology 80:1033–1044. https://doi.org/10.1016/j.theriogenology.2013.08.001
20. Olaciregui M, Gil L (2017) Freeze-dried spermatozoa: a future tool?. Reprod Domest Anim 52(Suppl 2):248–254. https://doi.org/10.1111/rda.12838
21. Olaciregui M, Luño V, Gonzalez N, De Blas I, Gil L (2015) Freeze-dried dog sperm: dynamics of DNA integrity. Cryobiology 71:286–290. https://doi.org/10.1016/j.cryobiol.2015.08.001
22. Tsujimoto Y, Kaneko T, Yoshida T, Kimura K, Inaba T, Sugiura K, Hatoya S (2020) Development of feline embryos produced using freeze-dried sperm. Theriogenology 147:71–76. https://doi.org/10.1016/j.theriogenology.2020.02.021
23. Sanchez-Partida LG, Simerly CR, Ramalho-Santos J (2008) Freeze-dried primate sperm retains early reproductive potential after intracytoplasmic sperm injection. Fertil Steril 89:742–745. https://doi.org/10.1016/j.fertnstert.2007.02.066
24. Gianaroli L, Magli MC, Stanghellini I, Crippa A, Crivello AM, Pescatori ES, Ferraretti

AP (2012) DNA integrity is maintained after freeze-drying of human spermatozoa. Fertil Steril 97:1067–1073. https://doi.org/10.1016/j.fertnstert.2012.02.014 e1

25. Yang LL, Ito D, Ushigome N, Wakayama S, Ooga M, Wakayama T (2023) A novel, simplified method to prepare and preserve freeze-dried mouse sperm in plastic microtubes. J Reprod Dev 69:198–205. https://doi.org/10.1262/jrd.2023-034
26. Ito D, Wakayama S, Emura R, Ooga M, Wakayama T (2021) Mailing viable mouse freeze-dried spermatozoa on postcards. iScience 24:102815. https://doi.org/10.1016/j.isci.2021.102815
27. Ito D, Wakayama T (2021) Protocol to preserve mouse freeze-dried spermatozoa in the thin plastic sheets. STAR Protoc 2:100933. https://doi.org/10.1016/j.xpro.2021.100933
28. Palazzese L, Turri F, Anzalone DA, Saragusty J, Bonnet J, Colotte M, Tuffet S, Pizzi F, Luciani A, Matsukawa K, Czernik M, Loi P (2023) Reviving vacuum-dried encapsulated ram spermatozoa via ICSI after 2 years of storage. Front Vet Sci 10:1270266. https://doi.org/10.3389/fvets.2023.1270266
29. Ward MA, Kaneko T, Kusakabe H, Biggers JD, Whittingham DG, Yanagimachi R (2003) Long-term preservation of mouse spermatozoa after freeze-drying and freezing without cryoprotection. Biol Reprod 69:2100–2108. https://doi.org/10.1095/biolreprod.103.020529
30. Kusakabe H, Kamiguchi Y (2004) Chromosomal integrity of freeze-dried mouse spermatozoa after 137Cs gamma-ray irradiation. Mutat Res 556:163–168. https://doi.org/10.1016/j.mrfmmm.2004.08.001
31. Kawase Y, Araya H, Kamada H, Jishage K, Suzuki H (2005) Possibility of long-term preservation of freeze-dried mouse spermatozoa. Biol Reprod 72:568–573. https://doi.org/10.1095/biolreprod.104.035279
32. Loi P, Matsukawa K, Ptak G, Clinton M, Fulka JrJ, Nathan Y, Arav A (2008) Freeze-dried somatic cells direct embryonic development after nuclear transfer. PLOS One 3:e2978. https://doi.org/10.1371/journal.pone.0002978
33. Liu J, Lee GY, Lawitts JA, Toner M, Biggers JD (2014) Live pups from evaporatively dried mouse sperm stored at ambient temperature for up to 2 years. PLOS One 9:e99809. https://doi.org/10.1371/journal.pone.0099809
34. Kaneko T, Ito H, Sakamoto H, Onuma M, Inoue-Murayama M (2014) Sperm preservation by freeze-drying for the conservation of wild animals. PLOS One 9:e113381. https://doi.org/10.1371/journal.pone.0113381
35. Wakayama S, Kamada Y, Yamanaka K, Kohda T, Suzuki H, Shimazu T, Tada MN, Osada I, Nagamatsu A, Kamimura S, Nagatomo H, Mizutani E, Ishino F, Yano S, Wakayama T (2017) Healthy offspring from freeze-dried mouse spermatozoa held on the International Space Station for 9 months. Proc Natl Acad Sci USA 114:5988–5993. https://doi.org/10.1073/pnas.1701425114
36. Wakayama S, Ito D, Kamada Y, Yonemura S, Ooga M, Kishigami S, Wakayama T (2019) Tolerance of the freeze-dried mouse sperm nucleus to temperatures ranging from −196°C to 150°C. Sci Rep 9:5719. https://doi.org/10.1038/s41598-019-42062-8
37. Wakayama S, Ito D, Kamada Y, Shimazu T, Suzuki T, Nagamatsu A, Araki R, Ishikawa T, Kamimura S, Hirose N, Kazama K, Yang L, Inoue R, Kikuchi Y, Hayashi E, Emura R, Watanabe R, Nagatomo H, Suzuki H, Yamamori T, Tada MN, Osada I, Umehara M, Sano H, Kasahara H, Higashibata A, Yano S, Abe M, Kishigami S, Kohda T, Ooga M, Wakayama T (2021) Evaluating the long-term effect of space radiation on the reproductive normality of mammalian sperm preserved on the International Space Station. Sci Adv 7:eabg5554. https://doi.org/10.1126/sciadv.abg5554
38. Wakayama S, Ito D, Hayashi E, Ishiuchi T, Wakayama T (2022) Healthy cloned offspring derived from freeze-dried somatic cells. Nat Commun 13:3666. https://doi.org/10.1038/s41467-022-31216-4
39. Wakayama S, Ito D, Ooga M, Wakayama T (2022) Production of mouse offspring from zygotes fertilized with freeze-dried spermatids. Sci Rep 12:18430. https://doi.org/10.1038/s41598-022-22850-5
40. Kamada Y, Yamaji K, Ushigome N, Ito D, Wakayama S, Hiraoka K, Hayashi M, Kawai K, Wakayama T (2025) Method for long-term room temperature storage of mouse freeze-dried sperm. Sci Rep 15:303. https://doi.org/10.1038/s41598-024-83350-2
41. Shibasaki I, Sugiyama H, Kamada Y, Nagatomo H, Ito D, Wakayama S, Ooga M, Kasai T, Kohda T, Wakayama T (2025) Extracting and analyzing micronuclei from mouse two-cell embryos fertilized with freeze-dried spermatozoa. Commun Biol 8:6. https://doi.org/10.1038/s42003-024-07358-0
42. Ushigome N, Miyashita M, Ito D, Wakayama S, Wakayama T (2025) Monosodium glutamate enhances freeze-dried mouse sperm quality during room-temperature storage. Sci Rep 15:30994. https://doi.org/10.1038/s41598-025-16635-9

43. Saragusty J, Loi P (2019) Exploring dry storage as an alternative biobanking strategy inspired by nature. Theriogenology 126:17–27. https://doi.org/10.1016/j.theriogenology.2018.11.027
44. Yamaji K, Wakayama S, Ushigome N, Ito D, Wakayama T (2026) Improved birth rates via rehydration of mouse freeze-dried spermatozoa using high-temperature ultrapure water. PLoS One 21(1):e0333682. https://doi.org/10.1371/journal.pone.0333682
45. Kamada Y, Wakayama S, Shibasaki I, Ito D, Kamimura S, Ooga M, Wakayama T (2018) Assessing the tolerance to room temperature and viability of freeze-dried mice spermatozoa over long-term storage at room temperature under vacuum. Sci Rep 8:10602. https://doi.org/10.1038/s41598-018-28896-8
46. Wakayama T, Perry AC, Zuccotti M, Johnson KR, Yanagimachi R (1998) Full-term development of mice from enucleated oocytes injected with cumulus cell nuclei. Nature 394:369–374. https://doi.org/10.1038/28615
47. Wakayama S, Kishigami S, Van Thuan N, Ohta H, Hikichi T, Mizutani E, Yanagimachi R, Wakayama T (2005) Propagation of an infertile hermaphrodite mouse lacking germ cells by using nuclear transfer and embryonic stem cell technology. Proc Natl Acad Sci USA 102:29–33. https://doi.org/10.1073/pnas.0408548102
48. Ono T, Mizutani E, Li C, Wakayama T (2008) Nuclear transfer preserves the nuclear genome of freeze-dried mouse cells. J Reprod Dev 54:486–491. https://doi.org/10.1262/jrd.20112
49. Iuso D, Czernik M, Toschi P, Fidanza A, Zacchini F, Feil R, Curtet S, Buchou T, Shiota H, Khochbin S, Ptak GE, Loi P (2015) Exogenous expression of human protamine 1 (hPrm1) remodels fibroblast nuclei into spermatid-like structures. Cell Rep 13:1765–1771. https://doi.org/10.1016/j.celrep.2015.10.066
50. Tanaka K, Namiki A, Ushigome N, Yamaji K, Ito D, Wakayama S, Wakayama T (2026) Monosodium glutamate protects DNA of freeze–dried somatic cells and improves development of somatic cell nuclear transfer embryos. J Reprod Dev 72(4):660-670 https://doi.org/10.1262/jrd.2026-036
51. Nollet KE, Komazawa T, Ohto H (2016) Transfusion under triple threat: lessons from Japan's 2011 earthquake, tsunami, and nuclear crisis. Transfus Apher Sci 55:177–183. https://doi.org/10.1016/j.transci.2016.09.009
52. Wakayama S, Wakayama T (2025) Can humanity thrive beyond the galaxy?. J Reprod Dev 71:10–16. https://doi.org/10.1262/jrd.2024-099
53. Wakayama S, Wakayama T (2026) The challenges and prospects of mammalian reproduction in extraterrestrial environments. J Reprod Dev 72(3):105-114. https://doi.org/10.1262/jrd.2025-080
54. Chatot CL, Ziomek CA, Bavister BD, Lewis JL, Torres I (1989) An improved culture medium supports development of random-bred 1-cell mouse embryos in vitro. J Reprod Fert 86:679–688. https://doi.org/10.1530/jrf.0.0860679
55. Quinn P, Moinipanah R, Steinberg JM, Weathersbee PS (1995) Successful human in vitro fertilization using a modified human tubal fluid medium lacking glucose and phosphate ions. Fertil Steril 63:922–924. https://doi.org/10.1016/s0015-0282(16)57504-9
56. Kaneko T, Nakagata N (2006) Improvement in the long-term stability of freeze-dried mouse spermatozoa by adding of a chelating agent. Cryobiology 53:279–282. https://doi.org/10.1016/j.cryobiol.2006.06.004
57. Oldenhof H, Zhang M, Narten K, Bigalk J, Sydykov B, Wolkers WF, Sieme H (2017) Freezing-induced uptake of disaccharides for preservation of chromatin in freeze-dried stallion sperm during accelerated aging. Biol Reprod 97:892–901. https://doi.org/10.1093/biolre/iox142
58. Ito D, Wakayama S, Kamada Y, Shibasaki I, Kamimura S, Ooga M, Wakayama T (2019) Effect of trehalose on the preservation of freeze-dried mice spermatozoa at room temperature. J Reprod Dev 65:353–359. https://doi.org/10.1262/jrd.2019-058
59. Inoue R, Harada K, Wakayama S, Ooga M, Wakayama T (2020) Improvement of a twice collection method of mouse oocytes by surgical operation. J Reprod Dev 66:427–433. https://doi.org/10.1262/jrd.2020-059

Check for updates

Chapter 33

Functional and Structural Alterations Associated with Dehydration of Lactic Acid Bacteria

Nelson Romano, Monica Marro, Maria Marsal, Pablo Loza-Álvarez and Andrea Gomez-Zavaglia

Abstract

The dehydration of lactic acid bacteria for technological purposes induces multilevel damage to bacterial cells. This chapter provides protocols for analyzing functional and structural alterations following the dehydration of *Lactobacillus delbrueckii* subsp. *bulgaricus* CIDCA 333 using vacuum drying technology. Protocols involving conventional microbiological assays (plate counting, metabolic activity), fluorescence (membrane integrity), and Raman spectroscopies (DNA conformational analysis) are presented, along with experimental tips and suggestions.

Keywords Bacterial dehydration, Fluorescence spectroscopy, Functionality, Raman spectroscopy, Structural changes

1 Introduction

The important role of lactic acid bacteria in the food and biotechnological industries highlights the need for adequate dehydration methods to minimize viability and functionality losses during processing and storage [1]. Different dehydration methods, including freeze-drying, spray-drying, and vacuum-drying, are commonly used at an industrial level to facilitate the supply chain of lactic acid bacteria, reduce storage costs, and extend product shelf-life [2]. Despite their effectiveness, these processes can lead to functional alterations, mainly arising from structural damages on lipid membrane, protein denaturation, conformational changes of nucleic acids [3].

Lipid membranes are particularly vulnerable to damage during bacterial dehydration [4, 5] because their fluidity strongly depends on environmental factors such as temperature, pH, and osmolarity [6]. A first approach to assessing membrane integrity can be performed using propidium iodide (PI), a fluorescent probe that binds to DNA but does not cross intact (undamaged) lipid membranes. Therefore,

Willem F. Wolkers and Harriëtte Oldenhof (eds.), *Cryopreservation and Freeze-Drying Protocols*, Methods in Molecular Biology, vol. 3060, https://doi.org/10.1007/978-1-0716-5416-3_33,

PI-positive cells are those with damaged membranes [7]. Certain fluorescent probes, such as Laurdan (1-[6-(dimethylamino)-2-naphthalenyl]-1-dodecanone), are sensitive to the hydration states of lipid membranes. In response to changes in the hydration environment (*e.g.*, the presence of hydrogen bonding), the probe undergoes shifts in its emission spectrum. In the gel phase, phospholipids are tightly packed, creating a more structured membrane state with fewer embedded water molecules, known as the liquid-ordered phase (L_o). When lipids are in the L_o phase, the probe exhibits a blue shift, with a maximum emission at 440 nm. Conversely, in the liquid-crystalline or liquid-disordered phase (L_d), increased water penetration into the interfacial region (glycerol region) allows for greater probe interaction. As a result, Laurdan fluorescence undergoes a significant red shift, with emission increasing at 490 nm [8]. Using confocal fluorescence microscopy allows for the direct and high-resolution observation of these changes. In lactic acid bacteria, Romano et al. recently implemented this approach to evaluate membrane fluidity after vacuum drying processes [3].

Damaged membranes act such as open doors, allowing structural changes (*e.g.*, proteins, DNA) that ultimately impact bacterial functionality [3]. To investigate these alterations, Raman spectroscopy is a valuable tool, mainly because its molecular sensitivity and ease of implementation. It provides complementary information to other spectroscopic methods and can be used to gather multiple molecular signatures from bacterial cells in a label-free and non-destructive manner [9]. When combined with data science tools, multivariate analysis enables in-depth analysis of bacterial structures.

This chapter presents the protocols for assessing functional and structural changes of *L. delbrueckii* subsp. *bulgaricus* CIDCA 333 following vacuum drying. Detailed protocols for plate counting, *lag* time determination, and acidification capacity are provided as tools for analyzing bacterial functionality. Protocols for PI uptake and steady-state fluorescence using Laurdan probes are included to assess membrane alterations. Finally, Raman spectroscopy coupled with Multivariate Curve Resolution (MCR) analysis is proposed as a methodology to evaluate the structural changes associated to dehydration processes. All protocols can be applied to microorganisms dehydrated in 0.85% w/v NaCl and in protective solutions, such as sucrose and fructo-oligosaccharides (FOS).

2 Materials

2.1 Reagents

1. Non-fat milk solids (Difco, Detroit, MI, USA).
2. MRS broth (Difco, Detroit, MI, USA).
3. NaCl.

4. FOS Orafti Beneo p95 (Mannheim, Germany).
5. Sucrose.
6. MRS agar (Difco, Detroit, MI, USA).
7. Propidium iodide (Merck, Darmstadt, Germany).
8. Laurdan (1-[6-(dimethylamino)-2-naphthalenyl]-1-dodecanone) (Cayman Chemical, MI, USA).
9. Isopropanol (Merck, Darmstad, Germany).
10. N,N-dimethylformamide (DMF) (Merck, Darmstadt, Germany).
11. KH_2PO_4.
12. Na_2HPO_4.
13. Cover glass (Cellvis, CA, USA).
14. Poly-L-lysine (Merck, Darmstadt, Germany).

2.2 Equipment

1. Sorvall centrifuge (Thermo Fisher, Waltham, MA, USA).
2. Vacuum centrifuge (DNA Concentrator DNA-23050, Genevac Ltd, Ipswich, UK).
3. Aqualab water activity instrument (Aqualab, Model Series 3TE, USA).
4. Orbital shaker (Orbi-Shaker, Merck, Darmstadt, Germany).
5. Microplate reader UV-vis-fluorescence (Synergy 2 HTX Microplate Readers; BioTek, Winooski, VT, USA).
6. Spectrophotometer UV-vis (Spectronic 200 UV-Vis ThermoFiher, Waltham, MA, USA).
7. Confocal microscope TCS SP8 equipped with a HeNe 405 nm laser, and a HC PL APO CS2 100×/1.40 OIL objective and high-efficient single molecule detectors (SMD-HyD) (Leica, Wetzlar, Germany).
8. Raman microscope (In Via, Renishaw, Apply Innovation, Gloucestershire, UK).
9. ImageJ 1.53c software (Wayne Rasband NIH, USA).
10. Renishaw WiRE software (Wotton-under-Edge, UK).
11. MatLab 9.4 software (MathWorks, MA, USA).

3 Methods

3.1 Bacterial Growth and Dehydration

1. Reactivate frozen *Lactobacillus delbrueckii* subsp. *bulgaricus* CIDCA 333 in 120 g/L non-fat milk solids at 37°C (*see* **Note 1**).

2. Inoculate MRS broth with the reactivated microorganisms (2% inoculum) and incubate at 37°C in aerobic conditions for 16 h (stationary phase).
3. Harvest microorganisms in the stationary phase (*ca.* 10^9 CFU/mL) by centrifugation at 14,000×*g* for 3 min at 25°C and wash twice with NaCl 0.85% w/v by centrifugation in the same conditions (*see* **Note 2**).
4. Suspend the resulting pellets in NaCl 0.85% w/v (control), or 20% w/v aqueous solutions of sucrose and commercial FOS (*see* **Note 3**).
5. Incubate the suspensions for 15 min at 25°C and centrifuge as described above.
6. Dehydrate the pellets at 45°C for 80 min in a vacuum centrifuge at 250×*g* and 10.0 mbar (7.5 Torr) (see Fig. 1).
7. Determine the a_w of the dried pellets in a water activity equipment calibrated using standard salt solutions.

3.2 Assessment of Culturability

1. Rehydrate the vacuum dried microorganisms in NaCl 0.85% w/v.
2. Make serial dilutions in NaCl 0.85% w/v.
3. Plate the suspension on MRS agar.
4. Incubate at 37°C in aerobic conditions for 48 h.
5. Express results as log N/N_0, where N_0 and N represent the plate counts (CFU/mL) before and after vacuum-drying, respectively.

3.3 Determination of the Lag Time and Acidification Kinetics

1. Rehydrate the vacuum dried microorganisms to the original volume by adding 1 mL NaCl 0.85% w/v (*see* **Note 4**).
2. Inoculate the rehydrated microorganisms into MRS broth (2% v/v inoculum) in a 48 wells cell culture microplate and incubate at 37°C for 48 h (*see* **Note 5**). Use fresh cultures as a control.
3. Read the absorbance of each well every 30 min at 600 nm (*see* **Note 6**).
4. Adjust experimental data using Eq. 1.

$$Abs_{600nm}(t) = \frac{Abs_{600_i} - Abs_{600_f}}{1 + \frac{t}{c}^{\,p}} + Abs_{600_f}, \qquad (1)$$

where t is the time in hours, Abs_{600i} is the absorbance at 600 nm of the culture medium at time equal to 0, Abs_{600f} is the absorbance at 600 nm once attained the stationary phase, c is the time at the inflection point and p is an exponential fitting factor.

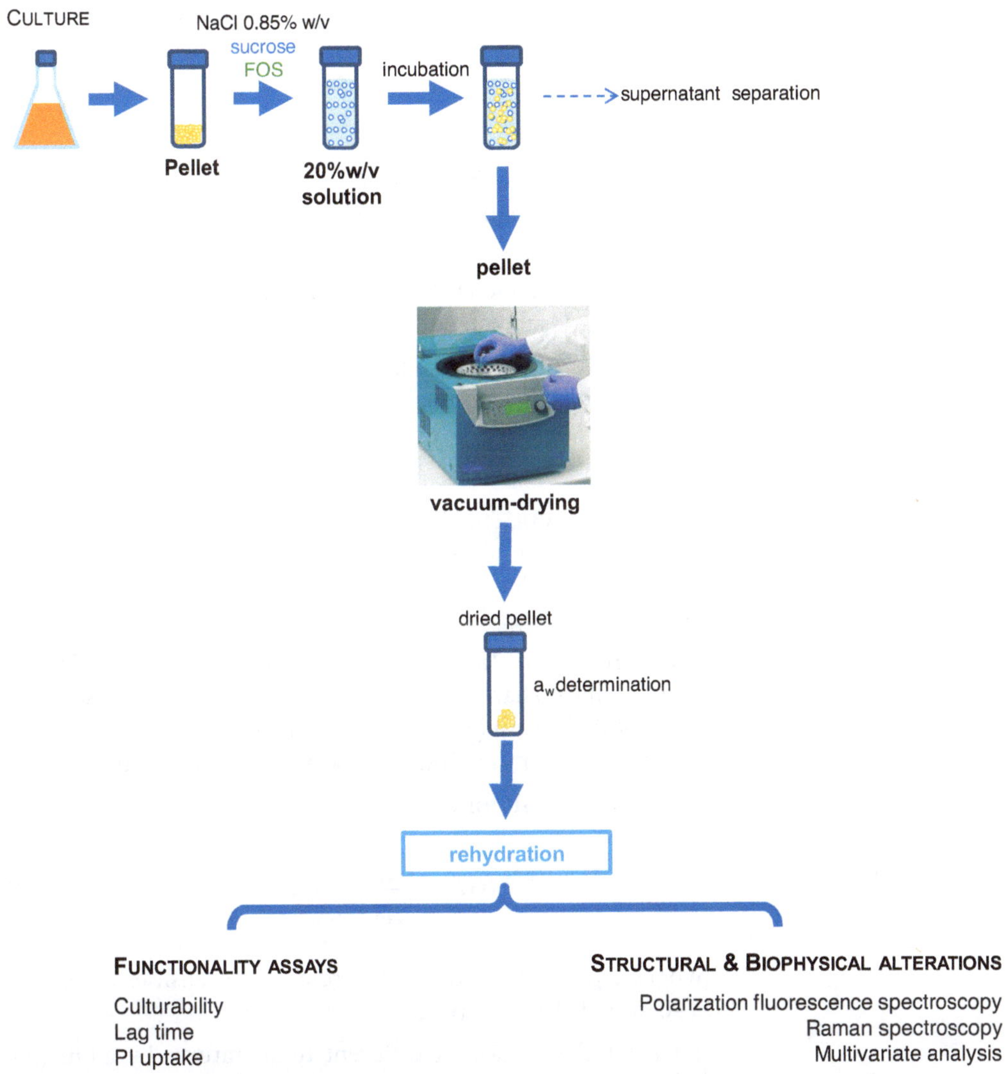

Fig. 1 Scheme of the experimental protocol

5. Calculate the *lag* time as the intersection between the tangent line at t=c and $Abs_{600\,i}$. The growth rate (m) during the exponential phase can be calculated as the module of the slope of the tangent line.

3.4 Determination of Membrane Damage: Propidium Iodide Staining

1. Rehydrate the vacuum dried microorganisms with NaCl 0.85% w/v to achieve an Abs_{600} of 1.0 absorbance units (a.u).
2. Add 50 μL of a propidium iodide (PI) stock solution (15 μM) to 100 μL of bacterial suspensions (PI final concentration: 5 μM) (*see* **Note 7**).
3. Incubate the samples at 25°C for 60 min in the dark.

4. Read fluorescence (λ_{ex}=488/420 nm; λ_{em}=645/440 nm) in a microplate reader. Use thermally dead cells (80°C for 30 min and isopropanol) and fresh cultures stained with PI as controls (*see* **Note 8**).

3.5 Determination of Membrane Fluidity: Laurdan Sample Labeling and Generalized Polarization (GP)

1. Solubilize a stock solution of Laurdan 200 μM in N, N-dimethylformamide (*see* **Note 9**).
2. Rehydrate the vacuum dried bacteria in 20 mM phosphate-buffer saline (PBS) (*see* **Note 10**) and adjust the Abs_{600} to 0.4 a.u. Use fresh cultures as controls.
3. Add 2.5 μL of the Laurdan stock solution to 1 mL aliquots of all samples (Laurdan final concentration 0.5 μM).
4. Incubate at 37°C for 90 min in the dark, with orbital shaking (200 rpm).
5. Transfer aliquots of 200 μL of unlabeled and Laurdan labeled samples (see above, Steps 1–4) to a 96-well fluorescence microplate reader.
6. Register the emission spectra in a microplate reader as follows: preheat at (25°C, 30°C, 37°C, 40°C, 43°C, or 45°C) for 10 min, linear shaking for 30 s (λ_{ex}=350 nm), and λ_{em} screening from 380–620 nm (Step: 10 nm). Remove the background fluorescence by subtracting a blank of PBS and Laurdan.
7. Calculate the GP for each condition according to Eq. 2:

$$GP = \frac{I_{440} - I_{490}}{I_{440} + I_{490}}, \qquad (2)$$

where I_{440} and I_{490} indicate fluorescence intensities emission at 440 and 490 nm, respectively.

8. Fit the GP variation at different temperatures by adjusting a sigmoidal regression function, according to Eq. 3:

$$GP_{(T)} = GP_1 + \frac{GP_2 - GP_1}{1 + 10^{p \times (T_m - T)}}, \qquad (3)$$

where $GP_{(T)}$ correspond to the measured GP at a given temperature (T), Tm is the apparent thermal transition temperature and p, the Hill's slope factor. GP_1 and GP_2 are the bottom and top asymptote values of GP. Figure 2 presents a typical plot of this adjustment, where organized membranes with lower fluidity (fresh cultures) exhibit higher GP values, while destabilized membranes with increased fluidity correspond to lower GP values (as observed in bacteria dehydrated without protective compounds). An intermediate scenario is observed when protective solutions (*e.g.*, sucrose) are used.

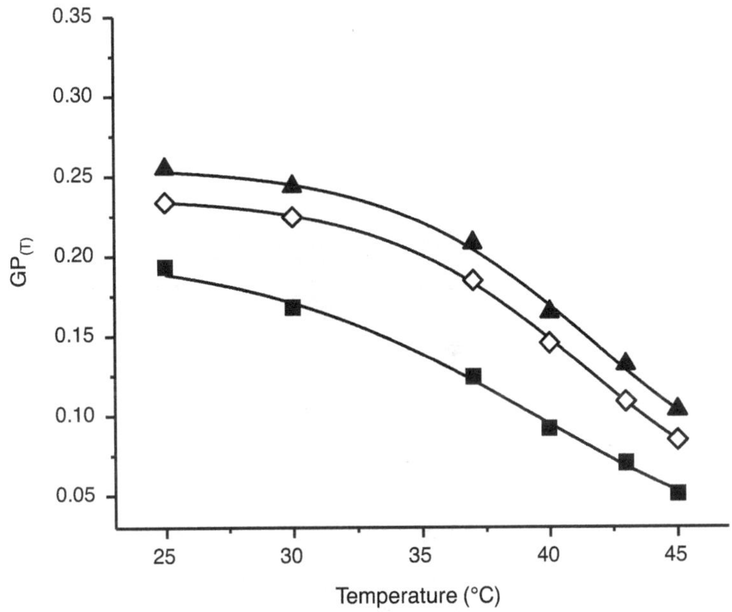

Fig. 2 Laurdan generalized polarization (GP) *vs* Temperature (T) for fresh cultures (black filled up-triangles), vacuum-dried samples with sucrose (open diamonds), and vacuum-dried samples without a protectant (black filled squares). The data were fitted to a sigmoidal regression model (solid lines)

3.6 Confocal Imaging

1. Load a 29 mm glass bottom dish (100 mm microwell 1#1.5 cover glass) coated with poly-L-lysine with 100 μL of unlabeled and Laurdan labeled samples (see Sect. 3.5).
2. Incubate in the dark at 25°C for 30 min.
3. Place the samples in a chamber of a confocal microscope at 25°C (λ_{ex}=405 nm, λ_{em}=430–480 nm).
4. Scan the selected areas at 400 Hz with a final pixel size of 70 nm.
5. Use the Image J 1.53 software to generate the color-coded GP images with a custom written macro [10] (*see* **Note 11**). Figure 3 exhibits a representative GP image of *Lactobacillus delbrueckii* subsp. bulgaricus CIDCA 333, showing high values for bacteria with lipid membranes in the liquid-ordered (L_o) (see Fig. 3A), and low values for those in the liquid-disordered (L_d) state (see Fig. 3B).

3.7 Raman Spectroscopy Measurements and Data Analysis

1. Rehydrate the vacuum dried microorganisms in NaCl 0.85% w/v and use fresh cells (non-vacuum dried) as controls.
2. Remove excess of water in the coverslip to ensure microorganisms are hydrated but not moving or floating in the liquid.

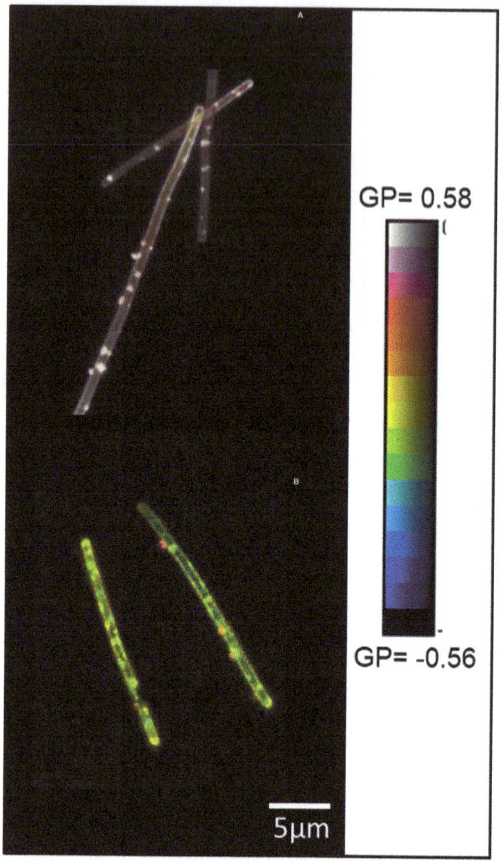

Fig. 3 Laurdan synthetic images representing the GP distribution over the bacterial cells, obtained from the fluorescence emission confocal imaging of *Lactobacillus delbrueckii* subsp. *bulgaricus* CIDCA 333 stained with Laurdan. (**A**): bacteria in the LO state (high GP values), (**B**): bacteria in the L_d state (low GP values)

3. Acquire the Raman data using a 532 nm laser at 25 mW power, focusing the beam using a 100× objective and recording the spectra with a deep-depletion charge-coupled device (CCD) detector, digitized and displayed on a personal computer using Renishaw WiRE software (*see* **Note 12**).
4. Preprocess all data by subtracting the cosmic rays and autofluorescence background before the application of multivariate analysis [11] using MATLAB software.
5. Construct a matrix with data to be introduced in the partial least squares (PLS) toolbox from Eigenvector Research.
6. Use PLS toolbox to apply a multiplicative signal/scatter correction (MSC)-median correction [12].

7. Make a first exploration of the spectral data set using principal component analysis (PCA) and remove outliers or non-informative data (*see* **Note 13**).
8. Perform a multivariate curve resolution (MCR) algorithm and plot the loadings and scores. The number of components chosen to decompose the spectra by MCR should be around the number of PCs that explain more than the 95% of variance.
9. Loadings are associated with the spectra of the molecular components that exhibit the most significant changes in the spectral dataset. Scores correspond to the abundance of their respective molecular component in each sample measurement. The spectral bands in the loadings will be compared with the Raman spectra of molecular components present in lactic acid bacteria, as published in Raman databases or reference articles [3]. The association of a loading with a molecular component will be established when most of the bands coincide with the Raman spectra of the molecule or group of molecules. Figure 4 presents a representative example of a loading plot profile, where spectral variations are attributed to DNA conformational changes (A and B forms) and result from the differing efficiency of carbohydrates (sucrose and FOS) in protecting microorganisms during dehydration.

4 Notes

1. Culture stock shall be frozen at −80°C in 120 g/L non-fat milk solids as cryoprotectant.

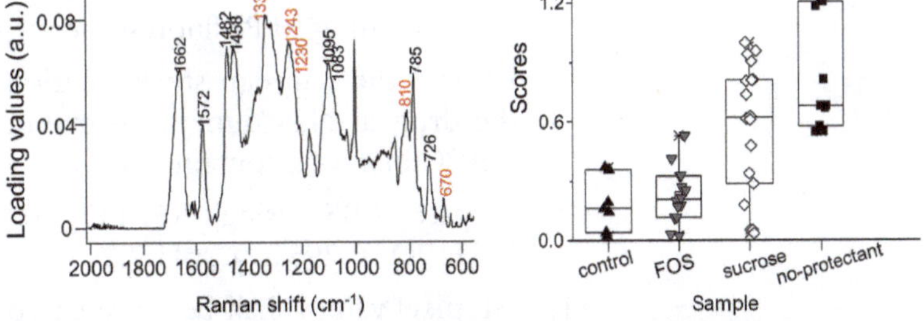

Fig. 4 Raman spectroscopy results. (**A**). Loading plot showing the spectral profile of a molecular component that contains bands corresponding to DNA: DNA A-like conformation (red) and B-conformation (black). (**B**). Scores plot showing the abundance of DNA A-like component for fresh cultures (control), bacteria dehydrated without protectants, and bacteria dehydrated in the presence of sucrose or FOS (protective compounds)

2. After harvest, microorganisms should be plate counted by serially diluting them in NaCl 0.85% w/v, plating the suspensions on MRS agar, incubating at 37°C in aerobic conditions and count the colonies for each dilution. Consider the dilutions to determine the CFU/mL.
3. Sugar solutions should be sterilized by filtration using 0.2 μm sterile filters.
4. The rehydration process can be facilitated by incubating suspensions at 25°C for 30 min and shaking (1000 rpm).
5. To avoid evaporation and/or contamination, cover the microplate with an acetate sealing film.
6. As an additional functional parameter, the evolution of pH along the process can be assessed (acidification kinetics). In this case, either in parallel or as an alternative to the registration of Abs_{600}, register the pH every 30 min for each condition tested and adjust the experimental data to the following equation (Eq. 4), derived from Eq. 1:

$$pH\,(t) = \frac{pH_0 - pH_f}{1 + \frac{t}{c}^{\,p}} + pH_f \tag{4}$$

where t is the time in hours, pH_0 is the pH of the culture medium at time equal to 0, pH_f is the pH at the stationary phase, c is the time corresponding to the inflection point and p is an exponential fitting factor. Calculate the *lag* time as the intersection between the tangent line at t=c and pH_0. Obtain the acidification rate during the exponential phase by determining the module of the slope of the tangent line.
7. Prepare a PI stock solution by diluting the drug in PBS buffer.
8. A blank containing 50 μL of PI stock solution and 100 μL 0.85% w/v NaCl shall be used to subtract the background corresponding to PI fluorescence.
9. Prepare the Laurdan stock solution (200 μM) by dissolving the drug in N,N-dimethylformamide. Store the solution at −20°C in a dark container until use.
10. Prepare PBS using KH_2PO_4 0.144 g/L, NaCl 9 g/L, Na_2HPO_4 0.795 g/L, pH 7.2.
11. GP pixel values shall be represented in synthetic images as a linear color-coded GP map, whereby blue represents low GP values (high fluidity, L_d) and red represents high GP values (low fluidity, L_o).
12. For acquisition of the Raman spectra, the system should consist in a coherent 532 nm laser that supplies an excitation

beam of about 10 mW power, which is focused onto the sample via a microscope with 100× objective.

13. The number of components selected will be the number in which the cumulative variance explained (the higher the better).

Acknowledgments

This project has received funding from the European Union's Horizon 2020 research and innovation programme under grant agreement N° 777657, from the Spanish Ministry of Economy and Competitiveness through the "Severo Ochoa" program for Centres of Excellence in R&D (CEX2019-000910-S), from Fundació Privada Cellex, Fundación Mig-Puig, from Generalitat de Catalunya through the CERCA program and "Laser lab Europe" (No. 654148), from the Argentinean Agency for the Scientific and Technological Promotion (ANPCyT) [Projects PICT(2020)/0482; PICT/2021/0239] and the Argentinean Research Council (CONICET) (project PIBAA 0754). N.R. and A.G.-Z. are members of the research career CONICET.

References

1. Romano N, Santos M, Mobili P, Vega R, Gómez-Zavaglia A (2016) Effect of sucrose concentration on the composition of enzymatically synthesized short-chain fructo-oligosaccharides as determined by FTIR and multivariate analysis. Food Chem 202:467–475. https://doi.org/10.1016/j.foodchem.2016.02.002
2. Terpou A, Papadaki A, Lappa IK, Kachrimanidou V, Bosnea LA, Kopsahelis N (2019) Probiotics in food systems: significance and emerging strategies towards improved viability and delivery of enhanced beneficial value. Nutrients 11:1591. https://doi.org/10.3390/nu11071591
3. Romano N, Marro M, Marsal M, Loza-Álvarez P, Gomez-Zavaglia A (2021) Fructose derived oligosaccharides prevent lipid membrane destabilization and DNA conformational alterations during vacuum-drying of *Lactobacillus delbrueckii* subsp. *bulgaricus*. Food Res Int 143:110235. https://doi.org/10.1016/j.foodres.2021.110235
4. Capozzi V, Weidmann S, Fiocco D, Rieu A, Hols P, Guzzo J, Spano G (2011) Inactivation of a small heat shock protein affects cell morphology and membrane fluidity in *Lactobacillus plantarum* WCFS1. Res Microbiol 162:419–425. https://doi.org/10.1016/j.resmic.2011.02.010
5. Tymczyszyn EE, Díaz R, Pataro A, Sandonato N, Gómez-Zavaglia A, Disalvo EA (2008) Critical water activity for the preservation of *Lactobacillus bulgaricus* by vacuum drying. Int J Food Microbiol 128:342–347. https://doi.org/10.1016/j.ijfoodmicro.2008.09.009
6. Bessa LJ, Ferreira M, Gameiro P (2018) Evaluation of membrane fluidity of multidrug-resistant isolates of *Escherichia coli* and *Staphylococcus aureus* in presence and absence of antibiotics. J Photochem Photobiol B: Biology 181:150–156. https://doi.org/10.1016/j.jphotobiol.2018.03.002
7. Santos MI, Araujo-Andrade C, Esparza-Ibarra E, Tymczyszyn E, Gómez-Zavaglia A (2014) Galacto-oligosaccharides and lactulose as protectants against desiccation of *Lactobacillus delbrueckii* subsp. *bulcaricus*. Biotechnol Prog 30:1231–1238. https://doi.org/10.1002/btpr.1969
8. Jay AG, Hamilton JA (2017) Disorder amidst membrane order: standardizing Laurdan generalized polarization and membrane fluidity terms. J Fluoresc 27:243–249. https://doi.org/10.1007/s10895-016-1951-8
9. Santos MI, Gerbino E, Tymcszyszyn EE, Gomez-Zavaglia A (2015) Applications of infrared and Raman spectroscopies to probiotic

investigation. Foods 4:283–305. https://doi.org/10.3390/foods4030283
10. Owen DM, Rentero C, Magenau A, Abu-Siniyeh A, Gaus K (2012) Quantitative imaging of membrane lipid order in cells and organisms. Nat Protoc 7:24–35. https://doi.org/10.1038/nprot.2011.419
11. Mikhailyuk IK, Razzhivin AP (2003) Background subtraction in experimental data arrays illustrated by the example of Raman spectra and fluorescent gel electrophoresis patterns. Instrum Exp Tech 46:765–769. https://doi.org/10.1023/B:INET.0000008930.87741.b9
12. Parnell JR, Yager P (1995) Application of multiplicative signal correction (MSC) to Raman spectra for use in an anesthetic sensor. Proc of SPIE—The Int Soc for Opt Eng 2388:481–492. https://doi.org/10.1117/12.208525

Check for updates

Chapter 34

Thermostabilization of Vaccines by Cryopreservation

Eric R. James

Abstract

For thermostabilization of vaccines that incorporate live eukaryotic organisms (e.g., parasites) or cells, cryopreservation is the preferred and usually only method for long-term storage. Parasites include the unicellular protozoa and multicellular helminths (nematodes, trematodes, and cestodes). The PfSPZ live sporozoite vaccines against malaria are cryopreserved. Additionally, many parasites are transmitted by vectors (e.g., mosquitoes, ticks), and these can be targets for vaccination or used in the production of vaccines. Veterinary live vaccines against bovine babesiosis, East Coast Fever, anaplasmosis, several poultry parasites and some viruses, are cryopreserved. Cellular vaccines comprising activated, cytokine-secreting or antigen presenting cells are stabilized by cryopreservation. Cryopreserved vaccines may be stored short-term in ultralow freezers, but typically are distributed via liquid nitrogen (LN_2) or LN_2 vapor-phase (LNVP) cold chains at below −150 °C.

Keywords Cellular vaccines, Cryopreservation, Helminth parasites, Protozoan parasites

1 Introduction

Vaccines comprised of live viruses or bacteria can be stabilized and stored as liquid formulations, or frozen or lyophilized. With almost no exceptions, these methods do not work with parasites, which are eukaryotic organisms requiring stabilization by cryopreservation. The two major groupings of parasites are the unicellular protozoa and the multicellular helminths. They are causative agents of many important human and veterinary diseases, including malaria, toxoplasmosis, Chagas Disease, and leishmaniasis of humans, for all of which there have been at least attempts to produce live vaccines, and the veterinary diseases babesiosis, anaplasmosis, coccidiosis, and East Coast Fever (ECF) for which live vaccines are available in certain markets. Helminth parasites include the nematodes, (such as hookworms, cattle lungworms, filariae, and *Onchocerca*, agent of river blindness); the trematodes (responsible for schistosomiasis/Bilharzia, liver and lung fluke infections); and the cestodes (adult tapeworm infections, and the cysticercus and Hydatid larval stages).

Willem F. Wolkers and Harriëtte Oldenhof (eds.), *Cryopreservation and Freeze-Drying Protocols*, Methods in Molecular Biology, vol. 3060, https://doi.org/10.1007/978-1-0716-5416-3_34,

Several of these helminths have been the subject of experimental vaccines, however, only those targeting cattle and sheep lungworm are in current production and these are the exception to the rule and are distributed at 2–8°C.

Many parasitic (and viral and bacterial) diseases are transmitted by vectors—these include mosquitoes and ticks. There are two licensed vaccines against tick infestation of cattle, TickGARD and Gavac, however, both are based on recombinant Bm86, a protein expressed in the tick gut wall [1] and formulated as an oil emulsion stored and distributed at 2–8°C. Ticks, however, are the vectors of, and used in the manufacture of, live vaccines against babesiosis (*Babesia bovis* and *B. bigemina*), and East Coast Fever (*Theileria parva*) of cattle. These livestock vaccines, together with live vaccines against anaplasmosis, coccidiosis, ovine toxoplasmosis, Newcastle Disease and Marek's Disease are cryopreserved and distributed using liquid nitrogen (LN_2)-based or LN_2 vapour phase (LNVP) cold chains. Meeusen et al. [2] provide an overview of all classes of veterinary vaccines.

The first vaccine against a helminth disease was a live attenuated vaccine against cattle lungworm developed by the Glasgow University School of Veterinary Medicine in the 1950s. The immunogen, the third stage larva (L3) of *Dictyocaulus viviparus* harvested from the feces of a herd of infected cattle, was purified, irradiated with an attenuating dose of UV light, vialed as a liquid formulation and shipped at ambient temperature for oral administration. The original vaccine was Dictol®, now manufactured by MSD Animal Health as Bovilis® Huskvac. The Dictol concept of a radiation-attenuated live vaccine has been applied to other diseases of humans and animals, most notably against dog hookworm [3], malaria [4], and human and animal schistosomiasis. The gamma radiation-attenuated cryopreserved anti-schistosome vaccine against *Schistosoma bovis* [5], developed by Martin Taylor and colleagues at the London School of Hygiene and Tropical Medicine, was stored and distributed using a LNVP cold chain and field-tested in cattle in Sudan. While a parallel vaccine against *S. japonicum* developed by Hsu and Hsu [6] in China was tested in water buffalo, reservoir hosts for this human pathogen. Clinical trial testing of the live attenuated cryopreserved malaria vaccines [7, 8] against *Plasmodium falciparum* (Pf) also stored and distributed through a LNVP cold chain, are ongoing and have now been tested in over 40 trials at sites on four continents. Examples of the preparation and cryopreservation of live anti-parasite vaccines are described below for the protozoan agent of ECF, *Theileria parva*, and for the helminth parasite, *Schistosoma* spp.

The field of anti-cancer vaccines and cellular therapies is expanding rapidly. To date, three FDA-approved vaccines have been developed to combat late stage prostate cancer (Sipuleucel-T, liquid formulation), early stage bladder cancer (BCG, lyophilized), and advanced melanoma (Talimogene laherparepvec, frozen liquid stored and transported at −90°C to −70°C) [9–11]. Vaccines against papillomavirus (Gardasil and Cervarix), and hepatitis B virus-induced liver cancer, are in liquid formulations with alum adjuvant stored and distributed at 2–8°C in prefilled syringes. Most autologous and allogeneic vaccines comprising dendritic cells (DC) and cytokine and antigen presenting cells, and the chimeric antigen receptor T-cell (CAR-T) products such as Kymriah [12] and Yescarta [13] are cryopreserved and distributed using a LNVP cold chain. Preparation and cryopreservation of the experimental DC-001 vaccine [14, 15] for acute myeloid leukemia is outlined below.

2 Materials

2.1 Reagents

2.1.1 Reagents for the Protozoan Parasite Theileria parva

1. Cattle, *Bos indicus* (or rabbits), ticks, frozen stabilate of *T. parva*.
2. Cloth ear bags.
3. Culture media (RPMI 1640, Eagle's), PBS, glycerol, bovine serum albumin (BSA), L-glutamine, antibiotics (pen/strep), ice.
4. Pipette, 1.5 mL Eppendorf tubes, gridded counting slides.
5. Straws 0.5 mL (e.g., high security from IMV Technologies/CBS).

2.1.2 Reagents for the Helminth Parasite Schistosoma *spp.*

1. Cercariae produced from *S. bovis*-infected *Bulinus* spp. snails.
2. Conditioned Cl-free water for snail maintenance; oven dried lettuce as food.
3. Culture media M199 with Earle's salts and pH indicator.
4. Tubes: 50 mL centrifuge.
5. Syringes: 20 or 50 mL Luer, 21 G needles.
6. Ethanediol CPA.
7. Glass coverslips cut to slivers (5 mm × 15 mm) using glass cutter and rule.
8. Cold aluminium block at 0°C.
9. Cryovials, standard 2.0 mL.

2.1.3 Reagents for Dendritic Cell Vaccines

1. Culture media for peripheral blood mononuclear cells (PBMC).
2. Cytokines including interleukin-4 (IL-4), granulocyte macrophage-colony stimulating factor (GM-CSF), IL-1ß, IL-6, tumor necrosis factor alpha (TNF-α).
3. Trypan blue or Alamar blue etc for viability assessment.

4. Cryopreservation medium, e.g., Cryostor CS10 (Biolife Solutions) with autologous plasma.
5. Cryoprotectant additive (CPA) dimethylsulfoxide (DMSO)10%.
6. Standard cryovials, 2.0 mL.

2.2 Equipment

2.2.1 Equipment for the Protozoan Parasite Theileria parva

1. Dissecting microscope, forceps.
2. Homogenizer or pestle and mortar, centrifuge, incubator.
3. Straw printing, loading (vacuum manifold) and sealing machines (e.g., SYMS I, Diva).
4. Freezer −80°C, controlled-rate freezer (e.g., Planer, IMV Technologies).
5. Dewars, LN_2, LNVP freezer, Visotubes, daisy goblets, goblet racks, or canisters.

2.2.2 Equipment for the Helminth Parasite Schistosoma *spp.*

1. Temperature-controlled room for snail rearing.
2. Aquarium tanks approx. 100 L, aeration pumps.
3. Filtration system (e.g., 47 mm Cole-Parmer), lamp, vacuum pump.
4. Insulated transport container.
5. Access to gamma irradiator.
6. Centrifuge.

2.2.3 Equipment for Dendritic Cell Vaccines

1. Leucopheresis equipment.
2. Centrifuge.
3. Culture flasks, incubator, rotator.
4. Microscope, counting chambers/slides.
5. Pipettors for samples and aliquots.
6. Controlled rate freezer (e.g., Planer Kryo 560-16, Cytiva VIA freeze).
7. LNVP freezer for storage.
8. Water bath or cryovial thawing device (e.g., ThawSTAR, Biolife Solutions).

3 Methods

3.1 Cryopreserved Vaccine for the Protozoan Parasite Theileria parva

Cryopreservation is routinely used to create and maintain stocks as stabilates of many species and strains of protozoa for research and for stabilization of live organisms that can be used in controlled infections to induce a protective immune response (*see* **Note 1**). The ECF vaccine is prepared as follows:

1. Infect cattle by SC injection with stabilates from three strains of *T. parva*: Muguga, Kiambu 5, and Serengeti.
2. At 7–10 days postinfection apply ticks in ear bags to the infected cattle, leave for 3–5 days.
3. Incubate ticks for 30–40 days at 26°C and 93% RH.
4. Collect ticks, surface sterilize with ethanol.
5. Chill the ticks on ice and transfer into cold media or PBS, 100–150 ticks per batch.
6. Dissect in cold media to collect salivary glands in Eppendorf tubes (*see* **Note 2**).
7. Pool the glands and disrupt the salivary glands using homogenizer or pestle and mortar to release the sporozoites.
8. Centrifuge for 15 min, 4°C at 200×*g* to pellet cell debris.
9. By microscopy, count sporozoites in supernatant to determine concentration.
10. Prepare straws—direct print information onto straws.
11. Prepare 2× CPA: Eagle's MEM with 7.0% v/v BSA and 15% v/v glycerol.
12. To make the "Muguga Cocktail" blend supernatants containing sporozoites from the three *T. parva* strains according to defined ratios.
13. Add 2× CPA to the sporozoite suspension 50:50 to create final product.
14. Load final product into 0.5 mL straws and thermally seal both ends. At scale, each straw may hold 30–40 doses.
15. Place straws on rack in CRF and cool at 1°C min^{-1} to −150°C or below or collect straws into tubes and place into −80°C freezer overnight with transfer to LNVP or LN_2 storage (heat-sealed straws are compatible with submersion in liquid phase LN_2).
16. Quality control: test at intervals during manufacture for potency, sterility, pH, and osmolarity, confirm viability via test inoculations in cattle, identity via PCR, and during storage for stability.
17. ECF vaccine straws are transported in Dewars in LN_2, or in LNVP dry shippers. Straws are thawed for 2–3 min on a steel tray in a 37°C water bath and the contents extracted and diluted.

3.2 Cryopreserved Vaccine for the Helminth Parasite Schistosoma *spp.*

Following the same principal as used for producing the successful Dictol vaccine against cattle lungworm (which can be cryopreserved [16]), Thomas Miller [3] developed a vaccine against dog hookworm using the live radiation-attenuated third-stage larvae L3 of

Ancylostoma caninum (*see* **Note 3**). The steps to produce the *S. bovis* live radiation-attenuated schistosomula vaccine [17, 18] including the vitrification method for cryopreservation [19] are described below:

1. Harvest cercariae from snails in water exposed to light.
2. Concentrate on filtration device under light stimulation, with vacuum.
3. Replace water with M199 medium in filtration device, concentrate.
4. Decant cercariae in media into 50 mL tube(s).
5. Using syringe and needle aspirate and expel cercaria in media ~7 times to separate cercaria tails from heads (heads = schistosomula).
6. Incubate at 37°C in insulated container during transport to irradiator.
7. Irradiate with 20 kRad gamma irradiation.
8. Allow schistosomula to sediment and remove supernatant containing tails.
9. Continue 37°C incubation to total of 90 min (to complete transformation of schistosomula by shedding of water- and CPA-impermeable surface glycocalyx coat).
10. Add ethanediol to 10% v/v and incubate at 37°C for 10 min (CPA permeation).
11. Cool sample on ice to 0°C and equilibrate for 5 min.
12. Increase ethanediol concentration to 35% v/v (CPA induces dehydration of schistosomula, increasing internal CPA concentration to 35% with reduced toxicity at lower temperature) = vaccine.
13. Arrange glass slivers on 0°C cold block, add 20 μL drops of schistosomula final product, and spread to ~4 x 10 mm.
14. Plunge glass slivers into LN_2 (cooling rate ~5,000°C/min), or into slush LN_2.
15. Store in cryotubes in LNVP cryotanks, transport to veterinary clinic in LNVP dry shippers.
16. To thaw, drop one sliver into 1 mL 42°C M199 and shake, remove glass sliver.
17. Aspirate vaccine into 1 mL syringe with 21G needle for intramuscular injection.

3.3 Cryopreserved Dendritic Cell Anti-Cancer Vaccine

1. Obtain PBMCs by leukopheresis and centrifugation.
2. Culture in DC medium and stimulate with cytokines (interleukin-4, granulocyte macrophage-colony stimulating factor) to

generate the DCs, pulse with immunostimulant then in maturation cytokine cocktails.

3. Assess mature DCs for purity, sterility and viability, and immunophenotype by flow cytometry.
4. Pellet by centrifugation and resuspend in chilled CPA, then aliquot into cryovials.
5. Cool using the CRF at 1°C min^{-1} to −150°C.
6. Store in LNVP freezer below −150°C.
7. Thaw cryovial in water bath at 37°C or thawing device (e.g., ThawSTAR, Biolife Solutions) to last ice.
8. Assess for viability.
9. Wash in serum-supplemented media to remove DMSO, formulate in syringe and inject.

4 Notes

1. This approach is used in leishmanization where infection with a non-virulent strain of *Leishmania major* protects against pathogenic strains [20]. Similarly, infections of cryopreserved and thawed virulent parasites controlled by simultaneous drug treatment can induce high levels of protection as with PfSPZ-CVac, a vaccine against falciparum malaria of humans [21] and with the cattle disease, East Coast Fever (ECF). The tick-transmitted *Theileria parva* parasite is endemic in 13 countries in eastern Africa with a high level of morbidity and mortality. The ECF vaccine [22–33] is produced from the non-attenuated and fully infectious sporozoite stage given with a co-administered long-acting oxytetracycline that controls infection and disease symptoms and allows protective immunity to develop—termed the infection-treatment method (ITM) of vaccination. To produce the *T. parva* sporozoites, nymphs of the cattle tick, *Rippicephalus appendiculatus*, are applied to infected calves when the parasites are at the schizont and piroplasm stages in erythrocytes. Blood-fed ticks harbouring *T. parva* are removed and incubated until they have moulted to adults and the *T. parva* sporozoites are present in the salivary glands.
2. Note: for large scale preparation of ECF vaccine the ticks are surface sterilized, homogenized and the homogenate filtered, then clarified by centrifugation, and the sporozoites counted.
3. The vaccine had limited commercial success and was eventually discontinued, in part due to its requirement for storage at 8–10°C and short shelf live even though the L3 can be cryopreserved relatively easily [34]. The schistosomula larval

stage of schistosomes can be cryopreserved [19, 35], and experimental vaccines against the human parasites, *Schistosoma mansoni* [36], *S. haematobium*, and *S. japonicum* and cattle and sheep parasites, *S. bovis* [18, 37] and *S. mattheei* were developed in the 1970s–1980s. These were based on the live radiation-attenuated cryopreserved schistosomula concept. However, although significant protection was achieved and a strong economic benefit, in terms of increased milk and meat yields was demonstrated for the *S. bovis*vaccine [38], interest in this approach waned and has been replaced by a recombinant protein immunogen approach [39], which would be simpler to produce and would not require a LNVP cold chain.

References

1. Bishop LJ, Stutzer C, Maritz Olivier C (2023) More than three decades of Bm86: what we know and where to go. Pathogens 12:1071. https://doi.org/10.3390/pathogens12091071
2. Meeusen ENT, Walker J, Peters A, Pastoret PP, Jungersen G (2007) Current status of veterinary vaccines. Clin Microbiol Rev 20:489–510. https://doi.org/10.1128/CMR.00005-07
3. Miller TA (1971) Vaccination against the canine hookworm diseases. Adv Parasitol 9:153–183. https://doi.org/10.1016/s0065-308x(08)60161-x
4. Nussenzweig RS, Vanderberg J, Most H, Orton C (1967) Protective immunity produced by the injection of x-irradiated sporozoites of *Plasmodium berghei*. Nature 216:160–162. https://doi.org/10.1038/216160a0
5. Taylor MG, James ER, Bickle Q, Andrews BJ, Dobinson AR, Nelson GS, Hussein MF (1979) Immunization of sheep against *Schistosoma bovis* using an irradiated schistosomula vaccine. J Helminthol 53:1–5. https://doi.org/10.1017/S0022149X00005666
6. Hsu SY, Hsu HF, Xu ST, Shi FH, He YX, Shen W, Osborne JW, Clarke WR (1983) Vaccination against bovine schistosomiasis japonica with highly X-irradiated schistosomula. Am J Trop Med Hyg 32:367–370. https://doi.org/10.4269/ajtmh.1983.32.367
7. Hoffman SL, Billingsley PF, James ER, Richman A, Loyevsky M, Li T, Chakravarty S, Gunasekera A, Chattopadhyay R, Li M, Stafford R, Ahumada A, Epstein JE, Sedegah M, Reyes S, Richie TL, Lyke KE, Edelman R, Laurens MB, Plowe CV, Sim BKL (2010) Development of a metabolically active, non-replicating sporozoite vaccine to prevent *Plasmodium falciparum* malaria. Hum Vaccines 6:97–106. https://doi.org/10.4161/hv.6.1.10396
8. Richie TL, Billingsley PF, Sim BKL, James ER, Chakravarty S, Epstein JE, Lyke KE, Mordmüller B, Alonso P, Duffy PE, Doumbo OK, Sauerwein RW, Tanner M, Abdulla S, Kremsner PG, Seder RA, Hoffman SL (2015) Progress with *Plasmodium falciparum* sporozoite (PfSPZ)-based malaria vaccines. Vaccine 33:7452–7461. https://doi.org/10.1016/j.vaccine.2015.09.096
9. https://www.fda.gov/media/78511/download?utm_source=chatgpt.com
10. FDA Package insert BCG. https://www.fda.gov/files/vaccines%2C%20blood%20%26%20biologics/published/Package-Insert-BCG-Vaccine.pdf
11. FDA package insert IMLYGIC: https://www.fda.gov/media/94129/download
12. FDA package insert KYMRIA: http://www.fda.gov/media/107296/download
13. FDA package insert YESCARTA: https://www.fda.gov/media/108377/download
14. van de Loosdrecht AA, van Wetering S, Santegoets SJAM, Singh SK, Eeltink CM, den Hartog Y, Koppes M, Kaspers J, Ossenkoppele GJ, Kruisbeek AM, de Gruijl TD (2018) A novel allogeneic off-the-shelf dendritic cell vaccine for post-remission treatment of elderly patients with acute myeloid leukemia. Cancer Immunol Immunother 67:1505–1518. https://doi.org/10.1007/s00262-018-2198-9
15. Granato AM, Pancisi E, Piccinini C, Stefanelli M, Pignatta S, Soldati V, Carloni S, Fanini F, Arienti C, Bulgarelli J, Tazzari M, Scarpi E, Passardi A, Tauceri F, La Barba G, Maimone G, Baravelli S, de Rosa F, Ridolfi L, Petrini M (2024) Dendritic cell vaccines as cancer treatment: focus on 13 years of manufacturing and quality control experience in advanced therapy medicinal products. Cytotherapy

26:1547–1555. https://doi.org/10.1016/j.jcyt.2024.07.005

16. James ER, Peacock R (1986) Preliminary studies on the cryopreservation of *Dictyocaulus viviparus (Nematoda)* third stage larvae. J Helminthol 60:65–73. https://doi.org/10.1017/s0022149x00008269
17. James ER, Taylor MG (1976) Transformation of cercariae to schistosomula: a quantitative comparison of transformation techniques and of infectivity by different injection routes of the organisms produced. J Helminthol 50:223–233. https://doi.org/10.1017/s0022149x0002664x
18. James ER (1980) Production and cryopreservation of schistosomula. Int J Nucl Med Biol 7:125–132. https://doi.org/10.1016/0047-0740(80)90030-3
19. James ER (1981) Schistosoma mansoni: cryopreservation of schistosomula by two-step addition of ethanediol and rapid cooling. Exp Parasitol 52:105–116. https://doi.org/10.1016/0014-4894(81)90066-7
20. Mohebali M, Nadim A, Khamesipour A (2019) An overview of leishmanization experience: a successful control measure and a tool to evaluate candidate vaccines. Acta Tropica 200:105173. https://doi.org/10.1016/j.actatropica.2019.105173
21. Mordmüller B, Surat G, Lagler H, Chakravarty S, Ishizuka AS, Lalremruata A, Gmeiner M, Campo JJ, Esen M, Ruben AJ, Held J, Calle CL, Mengue JB, Gebru T, Ibáñez J, Sulyok M, James ER, Billingsley PF, KC N, Manoj A, Murshedkar T, Gunasekera A, Eappen AG, Li T, Stafford RE, Li M, Felgner PL, Seder RA, Richie TL, Sim BKL, Hoffman SL, Kremsner PG (2017) Sterile protection against human malaria by chemoattenuated PfSPZ vaccine. Nature 542:445–449. https://doi.org/10.1038/nature21060
22. Cunningham MP, Brown CG, Burridge MJ, Purnell RE (1973) Cryopreservation of infective particles of *Theileria parva*. Int J Parasitol 3:583–587. https://doi.org/10.1016/0020-7519(73)90082-9
23. Di Giulio G, Lynen G, Morzaria S, Oura C, Bishop RP (2009) Live immunization against East Coast fever—an update. Trends Parasitol 25:85–92. https://doi.org/10.1016/j.pt.2008.11
24. Manangi M (1999) ECF immunization in Zambia. In: Morzaria S, Williamson S (eds) Live vaccines for *Theileria parva*. Nairobi, Kenya, ILRI, pp 30–33
25. Marsh TL, Yoder J, Deoch T, McElwain TF, Palmer GH (2016) Livestock vaccinations translate into increased human capital and school attendance by girls. Sci Adv 2:e1601410. https://doi.org/10.1126/sciadv.1601410
26. Bishop RP, Odongo D, Ahmed J, Mwamuye M, Fry LM, Knowles DP, Nanteza A, Lubega G, Gwakisa P, Clausen PH, Obara I (2020) A review of recent research on *Theileria parva*: implications for the infection and treatment vaccination method for control of East Coast fever. Transbound Emerg Dis 67(Suppl 1):56–67. https://doi.org/10.1111/tbed.13325
27. Lynen G, Yrjö-Koskinen AE, Bakuname C, Di Giulio G, Mlinga N, Khama I, Hanks J, Taylor NM, James AD, McKeever D, Peters AR, Rushton J (2012) East Coast fever immunization field trial in crossbred dairy cattle in Hanang and Handeni districts in northern Tanzania. Trop Anim Health Prod 44:567–572. https://doi.org/10.1007/s11250-011-9936-8
28. Patel E, Mwaura S, Kiara H, Morzaria S, Peters A, Toye P (2016) Production and dose determination of the infection and treatment method (ITM) Muguga cocktail vaccine used to control East Coast fever in cattle. Ticks Tick Borne Dis 7:306–314. https://doi.org/10.1016/j.ttbdis.2015.11.006
29. Radley DE, Brown CGD, Cunningham MP, Kimber CD, Musisi FL, Payne RC, Purnell RE, Stagg SM, Young AS (1975) East Coast fever: 3. chemoprophylactic immunization of cattle using oxytetracycline and a combination of theilerial strains. Vet Parasitol 1:51–60. https://doi.org/10.1016/0304-4017(75)90007-2
30. Nene V, Kiara H, Lacasta A, Pelle R, Svitek N, Steinas L (2016) The biology of *Theileria parva* and control of East Coast fever—current status and future trends. Ticks Tick Borne Dis 7:549–564. https://doi.org/10.1016/j.ttbdis.2016.02.001
31. Patel E, Mwaura S, diGiulio G, Cook EAJ, Lynen G, Toye P (2019) Infection and treatment method (ITM) vaccine against East Coast fever: reducing the number of doses per straw for use in smallholder dairy herds by thawing, diluting and refreezing already packaged vaccine. BMC Vet Res 15:46. https://doi.org/10.1186/s12917-019-1787-y
32. Peters AR, Toye P, Spooner P, diGuilio G, Lynen G (2020) Registration of the East Coast fever infection and treatment method vaccine (Muguga cocktail) in East Africa. Gates Open Research. https://doi.org/10.21955/gatesopenres.1116653.1
33. https://www.youtube.com/watch?v=7C1G0NXWJM8
34. Nolan TJ, Bhopale VM, Megyeri Z, Schad GA (1994) Cryopreservation of human hookworms. J Parasitol 80:648–650
35. James ER, Farrant J (1977) Recovery of infective *Schistosoma mansoni* schistosomula from liquid nitrogen: a step towards a live schistosomiasis vaccine. Trans R Soc Trop Med Hyg 71:498–500.

https://doi.org/10.1016/0035-9203(77)90143-2

36. Bickle QD, James ER (1978) Resistance against *Schistosoma mansoni* induced by immunization of mice with cryopreserved schistosomula. Trans R Soc Trop Med Hyg 72:677–678. https://doi.org/10.1016/0035-9203(78)90043-3
37. James ER, Dobinson AR, Lucas SB, Andrews BJ, Bickle QD, Taylor MG, Ham PJ (1985) Protection of sheep against *Schistosoma bovis* using cryopreserved radiation-attenuated schistomomula. J Helminthol 59:51–55. https://doi.org/10.1017/s0022149x00034489
38. Bushara HO, Hussein MF, Saad AM, Taylor MG, Dargie JD, Marshall TFdeC, Nelson GS (1978) Immunization of calves against *Schistosoma bovis* using irraadiated cercariae or schistosomula of *S. bovis*. Parasitology 77:303–311. https://doi.org/10.1017/s0031182000050265
39. Molehin AJ (2020) Schistosomiasis vaccine development: update on human clinical trials. J Biomed Sci 27:28. https://doi.org/10.1186/s12929-020-0621-y

Chapter 35

Thermostabilization of Vaccines by Lyophilization, Spray Drying, and Foam Drying

Eric R. James

Abstract

Vaccines reduce morbidity and mortality, by interrupting transmission of infectious agents and/or reducing the severity of clinical disease. Vaccines can comprise killed organisms, toxins, proteins, RNA, or live immunogens and are mostly stabilized either in liquid culture, by drying, by freezing, or by cryopreservation. The advantage of dry preservation is that this permits storage and transport at higher temperatures compared to liquid, frozen, or cryopreserved vaccines. Dry stabilization methods vary with the immunogen, use of adjuvants and excipients, the anticipated shelf life and the logistics of distribution. Drying techniques include freeze-drying, spray drying, and foam drying. In this chapter, methods to preserve live vaccines developed against viruses and bacteria in the dried state using lyophilization, spray drying, and foam drying are described.

Keywords Antiviral vaccines, Antibacterial vaccines, Foam drying, Freeze-drying, Spray drying

1 Introduction

The traditional, and critical, role played by vaccines has been in public health, reducing morbidity and mortality, and interrupting transmission of infectious agents. Human vaccines have saved many millions of lives and are joined by new vaccines in the struggle to contain emerging diseases such as Ebola and SARS-CoV-2. As of this writing, 80 vaccines against 41 human viral and bacterial disease agents are currently licensed by the US Food and Drug Administration (FDA) [1] and recommended by the US Centers for Disease Control and Prevention [2], and 59 vaccines to 38 disease agents have marketing authorization by the European Medicines Agency (EMA). The US Department of Agriculture (USDA) lists ~2,200 vaccines to ~250 agents [3] causing diseases in >35 species of livestock, fish, and companion animals. Vaccines can comprise killed organisms, toxins, proteins, RNA, or live immunogens and are mostly stabilized in liquid culture, by lyophilization, by freezing, or by cryopreservation, and

Willem F. Wolkers and Harriëtte Oldenhof (eds.), *Cryopreservation and Freeze-Drying Protocols*, Methods in Molecular Biology, vol. 3060, https://doi.org/10.1007/978-1-0716-5416-3_35,

the newer experimental methods of spray drying and foam drying. There is also a growing list of vaccines targeting cancers. All stabilization methods aim to minimize potency loss during manufacture and subsequent losses during storage, distribution, and in-clinic use. Methods of stabilization vary according to the immunogen, use of adjuvants and excipients, the anticipated shelf life and the logistics of distribution. Liquid or lyophilized formulations are usually stored and distributed at 2–8°C utilizing refrigerators. Other vaccines are stored at −15°C or −50°C (e.g., measles), or −60 to −90°C (e.g., Ebola, COVID-19 mRNA). Eukaryotic whole cell vaccines including several of the veterinary vaccines (as well as CAR-T and cellular therapies) require cryopreservation with storage and distribution below −150°C. Little to no loss in potency occurs with cryopreserved vaccines stored below Tg, compared to ~1 log loss over 12 months typical of lyophilized vaccines stored at 2–8°C. Cryogenic storage may be favored for some RNA-based and DNA-based vaccines and for extending shelf-life by stockpiling as vaccine antigen concentrate (VAC) (e.g., for foot-and-mouth disease vaccine) in anticipation of future need [4].

Lyophilization methods for different vaccines are similar but do vary both in terms of ingredients of the lyoprotective additive (LPA) cocktails and the process steps of freezing, primary drying, and secondary drying [5–8]. The end products are generally similar: a uniform "cake" with low water content in the 1.5%–3% range that requires a diluent for reconstitution before injection. An exception is Vivotif, the live attenuated *Salmonella*vaccine [6] that is finished by milling the lyophilized cake into a powder and packaging into enteric-coated capsules for oral administration. Generally, storage and distribution require a 2–8°C cold chain; however, the VAR (chickenpox) and MMRV (Measles, Mumps, Rubella, and Vaccinia) lyophilized vaccines can also be stored frozen at −50°C or −20°C [5]. Lyophilized products suffer a reduction in potency due to the manufacturing process (process loss) and a subsequent decline during storage. The WHO stability requirement for a freeze-dried measles vaccine is less than 1 log loss in viral activity after 1 week of accelerated storage at 37°C [9, 10].

While lyophilization is the most used method for stabilizing solid product vaccines targeting viral and bacterial diseases, the process loss, reliance on a 2–8°C cold chain and often relatively short shelf lives of lyophilized products have stimulated investigation of alternative thermostabilization methods and particularly those that have potential to provide acceptable levels of potency at ambient storage and distribution temperatures. Though no licensed vaccines are yet stabilized by spray drying or foam drying methods, these methods have both been the subject of considerable experimentation.

Spray drying produces a solid powder product in a single step by injecting a suspension of liquid vaccine particles with excipients through a nozzle into a heated (usually nitrogen gas) environment. The liquid is aerosolized as it exits the nozzle and the fine droplets are dried by the rapid evaporation of water of solution resulting in a powder that is separated from the gas in a cyclone collection vessel. Potential damage from heat and shear stresses are a consideration. Spray drying is best suited for stabilizing smaller pharmaceuticals and viruses and generates products that may be administered by inhalation or by the oral route.

Foam drying (or vacuum foam drying) converts an aqueous solution into a stable dry foam product through application of low pressure at controlled ambient temperatures. The foam drying process avoids freezing and the potential for ice crystal damage associated with lyophilization and also the high temperatures and shear stresses of spray drying. Foam drying is a viable option for stabilizing more complex biologics and pharmaceuticals, including live viral and bacterial vaccines.

Examples of the methods employed for lyophilization, spray drying, and foam drying are given below for viral (measles) and bacterial (*Salmonella Typhi*) vaccine.

2 Materials

2.1 Reagents

2.1.1 Reagents for Lyophilization of Viral Vaccines

1. Measles virus, Edmonston-Enders strain, live attenuated through serial passage in chick embryo cells.
2. Lyoprotectant additives (LPA): trehalose or sucrose, 3.8 mg/mL. sorbitol, 29 mg/mL. gelatin as stabilizer, partially hydrolyzed, 29 mg/mL. human serum albumin (HSA), 0.6 mg/mL. sodium phosphate buffer, pH 7.0.
3. Neomycin, 50 μg/mL.
4. Water for injection (WFI), USP, sterile.
5. Glass vials (2–10 mL), pyrogen-free, sterile.
6. Rubber (2-leg) lyophilization stoppers, matched to vials, sterile.
7. Aluminium caps/seals, crimp type.

2.1.2 Reagents for Lyophilization of Bacterial Vaccines

1. Master and working seed banks of *Salmonella Typhi* Ty21a stored at −80°C.
2. Growth media: glucose (5–10 g/L), NaCl (1–2 g/L), $(NH_4)_2SO_4$ (1–2 g/L), KH_2PO_4 (1–3 g/L), K_2HPO_4 (1–3 g/L), $MgSO_4 \bullet 7H_20$ (0.2-0.5 g/L), yeast extract (2–5 g/L), biotin (0.1–1.0 mg/L), thiamine (0.1–1.0 mg/L), trace elements, water.
3. LPA ingredients: sucrose (5%–10%), gelatin (hydrolysed; 0.3%–1.0%), ascorbic acid (0.01%–0.05%), monosodium

glutamate (0.5%–2.0%). Concentrations based on published information, but actual concentrations are proprietary.

4. Buffers: phosphate-buffered saline (PBS), pH 7.2 (Na_2HPO_4: 8–10 mM, KH_2PO_4: 1.5–2.5 mM, NaCl: 130–150 mM, KCl: 2–5 mM).
5. Gelatin capsules enteric coated for acid-resistance and blister materials (for forming, lidding).

2.1.3 Reagents for Spray Drying of Viral Vaccines

1. Attenuated measles virus (Edmonston-Zagreb), 5, 4 log10 (TCID50).
2. Vero cells, 1×10^5 cells/mL.
3. Excipients (spray drying protective additive cocktail: SDPA): sucrose-trehalose, 21.0% w/v, gelatin (Gelita), 6.25% w/v, L-arginine, 4% w/v, glycerol 1.25% w/v, human serum albumin (HSA) 4%, calcium chloride, 2 mM; zinc chloride, 2 mM.
4. Argon gas, filling environment.

2.1.4 Reagents for Foam Drying of Bacterial Vaccines

1. *Salmonella Typhi* Ty21a expressing recombineered protective antigens, seed bank −80°C.
2. Stirred-tank bioreactor.
3. Typtic soy agar (TSA) plates.
4. Water for injection (WFI).
5. CY medium for Ty21a cultures, pH 7.2: yeast extract, 12 mg/mL, Hy-Case, 20 mg/mL, pepticase, 12 mg/mL, NaH_2PO_4, 1.25 mg/mL, NaCl, 3.3 mg/mL, glucose, 2 mg/mL, galactose, 0.05 mg/mL.
6. Foam drying protective additive (FDPA), made at 2×, concentrations in final product: trehalose 25% w/w, gelatin (Gelita), 5% w/w, methionine, 0.5% w/w, potassium phosphate pH 8.0, 0.34%.

2.2 Equipment

2.2.1 Equipment for Lyophilization of Viral Vaccines

1. Cleanroom suite, Class A/B.
2. Biological Safety Cabinet (BSC), Laminar Flow Hood (LFH), or Isolator.
3. Filling system, piston or peristaltic, fluid reservoirs, automated.
4. Lyophilizer with controller for programming temperature and vacuum steps.
5. Crimping machine.
6. Karl Fischer titrator, pH meter.
7. Tissue culture infectious dose-50 ($TCID_{50}$) assessment capability.

2.2.2 Equipment for Lyophilization of Bacterial Vaccines

1. Cleanroom suite, Class A/B.
2. BSC, LFH, or Isolator.
3. Lyotray with filling system, piston or peristaltic, fluid reservoirs, automated.
4. Lyophilizer with controller for programming temperature and vacuum steps.
5. Low energy impact Jet, Cone or Pin mill.
6. Karl Fischer titrator, pH meter.
7. Semi- or fully-automated capsule filling equipment.
8. Blister packaging equipment (e.g., Uhlmann, Bosch) and ancillary equipment for forming, feeding, sealing, label printing, and packaging.

2.2.3 Equipment for Spray Drying of Viral Vaccines

1. Cleanroom suite, Class A/B.
2. Biological safety cabinet (BSC), Class II, LFH, or Isolator.
3. Stirred-tank bioreactor.
4. Refrigerated centrifuge.
5. Filtration unit.
6. pH meter, conductivity meter.
7. Balance, analytical.
8. Reservoir stirrer, magnetic.
9. Pump, syringe or peristaltic, 0.5 mL/min feed.
10. Spray dryer with inert gas (nitrogen) system (e.g., Büchi B-290 with inert loop) including:
    (a) Heated drying chamber.
    (b) Atomizer, ultrasonic: 15 psi atomization, outlet temperature 40°C, particle size <7 μm.
    (c) Collecting chamber, cyclone.
11. HEPA filtration and dehumidifier (RH <10%, <30°C).

2.2.4 Equipment for Foam Drying of Bacterial Vaccines

1. Biological safety cabinet (BSC), Class II, LFH, or Isolator.
2. Culture flask, 500 mL or stirred-tank bioreactor.
3. Refrigerated centrifuge.
4. Filtration unit.
5. pH meter, conductivity meter.
6. Balance, analytical.
7. Lyophilizer (e.g., Vertis Advantage XL70) with vacuum pump and stoppering.
8. Karl-Fischer apparatus with Mettler-Toledo volumetric titrator and HYDRANAL standard.

3 Methods

3.1 Lyophilization of Viral Vaccines

Lyophilized viral vaccines include those for BCG, Cholera, Dengue, Haemophilus b, Meningiococcal groups, Measles, Mumps, Rubella, Varicella, Rabies, Yellow fever, Rotavirus, Smallpox, and COVID-19 (Adenovirus platform) (*see***Notes 1–7**). A protocol is given below for lyophilization of measles vaccine.

1. Maintain sterile/aseptic procedures.
2. Harvest virus from cell culture supernatant (cell-associated virus may be released by gentle lysis).
3. Remove cell debris by clarifying the culture.
4. Treat with endonuclease to remove extraneous cell nucleic acids.
5. Purify and concentrate virions using gentle filtration (tangential flow filtration) or centrifugation.
6. Standardize virions to desired potency and concentration.
7. Prepare LPA cocktail in culture medium (e.g., M199, Eagle's Minimum Essential) that will yield concentrations of ingredients in the final formulation of:
   (a) Trehalose or sucrose: 5%–7% w/v.
   (b) Sorbitol: 2%–3% w/v.
   (c) Gelatin: 0.5% w/v.
   (d) Sodium phosphate buffer, pH 7.0, 10 mM.
8. Mix bulk virus with LPA under aseptic conditions.
9. Check pH (target: 7.0 ± 0.2).
10. Aseptically fill 0.5–1.0 mL of the formulated vaccine into sterile vials.
11. Partially insert elastomer stoppers.
12. Load vials onto lyophilizer trays, pre-cooled.
13. Load lyophilizer trays into the lyophilizer and activate the freeze-drying program, for example:
    (a) −45°C to −50°C for at least 2 h to freeze the product.
    (b) Bring shelf temperature up to −20°C and reduce pressure to 100–200 mTorr for primary drying over 24 h.
    (c) Elevate shelf temperature to +20°C to +25°C and run for 10–20 h: time predetermined from test runs assaying for moisture content.
    (d) Stopper the vials under vacuum or following an inert gas (N or Ar) backfill.

14. Crimp seal the vial stoppers with the aluminium caps.
15. Inspect all vials for integrity and for cake color, consistency, and uniformity.
16. Label and store vials, protected from light, between −50°C and +8°C [1].
17. Postprocessing: assess for residual moisture using the Karl-Fischer method; target moisture content is 1.5%–3.0%.
18. Test for sterility (USP <71>).
19. Test for potency by plaque assay or TCID50.
20. Test for pH: 6.8–7.2.
21. Test for reconstitution with diluent (supplied in prefilled syringe): <60 s and clear.

3.2 Lyophilization of Bacterial Vaccines

FDA-licensed anti-bacterial vaccines include those targeting *Salmonella*, *Bacillus anthracis*, *Bordetella pertussis*, *Haemophilus influenzae*, and *Clostridium tetani* (*see* **Note 8**). The steps in production and fill-finish of the Vivotif vaccine are described below (*see* **Note 9**).

1. Maintain sterile/aseptic good manufacturing practice (GMP) procedures.
2. Thaw vial from working cell bank, add contents to medium (70 mL in 250 mL flask), grow at 37°C with shaking at 250 rpm.
3. Inoculate culture into a stainless steel stirred-tank bioreactor (>100 L scale) in media with anti-foam agent (e.g., Dow-Corning, Momentive, Wacker: 10–100 ppm), aeration, and agitation.
4. Incorporate real-time sensors for dissolved oxygen (DO), pH, temperature, agitation speed, and pressure.
5. Maintain pH 7.0 with automated NaOH/HCl addition, temperature 37°C, DO 20%–30%.
6. Monitor growth by OD600 to early stationary phase (typically 12–18 h).
7. Harvest bacteria, centrifuge to concentrate, wash in media, resuspend, add LPA and mix.
8. Load product into pre-cooled open lyophilizer trays.
9. Load lyophilizer trays into the lyophilizer and activate the freeze-drying program:
   (a) −45°C to −50°C for at least 2 h to freeze the product.
   (b) Bring shelf temperature up to −20°C and reduce pressure to 100–200 mTorr for primary drying over 24–36 h.

(c) Elevate shelf temperature to +20°C to +25°C and run for 10–20 h: time predetermined from test moisture content assessments.

10. Remove cake from lyophilizer tray, transfer to the closed-system mill.
11. Mill (at <25°C and <30% RH) to uniform and optimum particle size.
12. Gelatin capsules (enteric coated, size 0 or 1) opened mechanically, milled powder (2–10 mg) added, and capsules closed/sealed under low humidity or nitrogen.
13. Capsules screened by weight and samples tested for CFU (target 1×10^{10} to maintain $>2\times10^{9}$ by end of 3-year shelf life stored at 2–8°C).
14. Quality control assays: Cell count ($>1\times10^{10}$ CFU/capsule), Disintegration test (USP<2040>, none at pH 1.2, release at pH 6.8), Residual moisture (Karl Fischer, <3%), Capsule weight (±5% of target), Purity and identity (PCR, culture), Stability (2–8°C and accelerated).

3.3 Spray Drying of Viral Vaccines

Viruses that have been stabilized experimentally by spray drying include measles, influenza, vesicular stomatitis virus (VSV), some adenovirus-vectored vaccines, *Herpes simplex* and hepatitis B [11–13]. An outline of the optimal materials and methods used in spray-drying of measles vaccine based on the process development of Ohtaki et al (2010) [13] is given below. Special precautions and aseptic technique must be followed for handling of live virus.

1. Harvest virus by disruption of infected Vero cells grown in a bioreactor at 10^{6}–10^{7} TCID50/mL, and clarify by centrifugation to remove cell debris.
2. Mix virus with SDPA to create final product (feed solution), pass through a 22 μm filter and load into reservoir for spray drying.
3. Pump feed solution through atomizer into hot nitrogen gas (inlet 60–80°C; outlet 40°C), collect dried powder in a cyclone separator collection chamber and transfer to glass or cyclic olefine vials under low humidity and argon gas, stopper and seal.
4. Conduct post-process testing for moisture (target 3%), process loss (TCID50/mL; target 0.0), particle size (<7 μm), sterility and endotoxin, store at 2–8°C and 37°C for accelerated stability.

3.4 Foam Drying of Bacterial Vaccines

Foam drying was developed initially in the food industry for the dry preservation of milk [14, 15] and other products [16, 17], and has been adapted, through sequential advances in equipment and methodology, for use in stabilization of other products, including pharmaceuticals. [18, 19] Foam drying has been used successfully experimentally for stabilization of viral and bacterial live vaccines [20–27]. Vaccines that have experimentally been successfully foam dried include measles, influenza, rotavirus, adenovirus and LaSota virus [20–25], and *Salmonella* [26–28]. The advantages of foam drying over the standard lyophilization are principally that the material is not frozen, and the water is "boiled off" rather than sublimated, so that damage due to ice crystal formation is circumvented and drying times can be shorter. The foaming process provides a greater specific surface area (SSA) producing a more efficient evaporative water loss than occurs with lyophilization. The SSA is higher with spray drying, but that comes at the cost of the potentially damaging effects of heating and the shear stresses imposed by aerosolization. Below is outlined the protocol for foam-dried *Salmonella Typhi* Ty21a constructs stably expressing the protective antigens of anthrax [26] or SARS-CoV-2 [27].

1. Thaw a seed bank vial of *Salmonella Typhi* Ty21a and streak onto a TSA plate.
2. Grow overnight at 37°C and pick 6 colonies.
3. Inoculate each colony into 70 mL CY medium in a 500 mL flask.
4. Incubate 5.5 h at 37°C with shaking, 250 rpm, remove 70 mL and initiate two overnight cultures with 600 mL CY medium in 3 L flasks. Or transfer to a 10 L stirred-cell bioreactor.
5. Incubate at 37°C with 250 rpm shaking for 16 h, or to OD 5–6.
6. Centrifuge at 6000 × *g* for 15 min, discard supernatant, save and weigh pellet.
7. Combine bacterial pellet 1:1 with 2× FDPA, using pipette to mix and aliquot 3 mL into 250 mL glass vials. Weigh each vial (with stopper) before and after filling.
8. Transfer glass vials onto the shelf, precooled to 15°C, of the Vertis lyophilizer.
9. Execute the following program in the lyophilizer:
   (a) Atmospheric pressure for 15 min at 15°C.
   (b) 5 Torr for 20 min at 15°C to degass.
   (c) 1 Torr for 2 h at 22°C for foam formation.
   (d) 100 mTorr for >48 h at 15°C for secondary drying.
10. Combine contents of vials, mill to fine powder and revial.
11. Stopper vials after argon purge.
12. Store at 4°C and at 22°C, and 37°C for accelerated stability.

13. Assay for water content of 100 mg samples by Karl Fischer method post foam drying.
14. Assay for potency/viability by reconstituting with WFI to original prefoam drying weight, and plating serially diluted samples on tryptic soy agar. Following incubation at 37°C for 20 h, count colonies and express survival as CFU.

4 Notes

1. Other antiviral vaccines may use alternate materials and methods [6]. These will have been determined and selected as a result of extensive empirical testing.
2. In addition to serial passage, live virus can be attenuated using heat, formaldehyde, β-propiolactone, or other chemical treatments.
3. Virus may be grown in different cell types, such as Vero, MRC-5, fibroblasts, or in chicken eggs (used for influenza and yellow fever vaccines).
4. Alternative components of the LPA cocktail may include other sugars such as mannitol or lactose; amino acids such as glycine, alanine, or methionine; human serum albumin (HSA), and/or polymers such as polyvinylpyrrolidone (PVP).
5. Alternative buffers include phosphate-buffered saline (PBS), Tris, or citrate buffer.
6. Live viral vaccines do not typically incorporate adjuvants such as aluminium hydroxide (used for inactivated rabies vaccine) since these are generally incompatible with lyophilization. Instead, adjuvants may be included in the diluent used for reconstitution.
7. All ingredients and steps in the lyophilization program need to be optimized for specific vaccines.
8. With the exception of Vivotif, the vaccine to protect against typhoid fever, the final product is a lyophilized cake in a stoppered and crimped glass vial that is reconstituted with diluent and drawn into a syringe for injection (as above for the lyophilized ani-viral vaccines). Vivotif is processed further to generate a powder, which is then packaged in enteric coated gelatin capsules for oral dosing [29–34].
9. The immunogen in the Vivotif vaccine manufactured by Crucell/Janssen is the live attenuated Ty21a strain of *Salmonella Typhi*. Lyophilization materials and methods for other bacteria, toxoids (tetanus, diphtheria) and polysaccharides or conjugates (Hib, meningococcal, pneumococcal vaccines), follow

the same basic outline [30]; however, those final products are freeze-dried in glass vials which are then capped and sealed.

References

1. https://www.fda.gov/vaccines-blood-biologics/vaccines/vaccines-licensed-use-united-states
2. https://www.cdc.gov/vaccines/by-disease/index.html
3. https://www.aphis.usda.gov/veterinary-biologics/product-summaries
4. Lombard M, Füssel AE (2007) Antigen and vaccine banks: technical requirements and the role of the European Antigen Bank in emergency foot and mouth disease vaccination. Rev Sci Tech 26:117–34
5. FDA package insert MMR II. https://www.fda.gov/media/147563/download
6. FDA package insert Vivotif. https://www.fda.gov/media/75988/download
7. Maa YF, Hsu CC (1996) Protein denaturation by freeze-drying and methods of minimizing it. Pharm Dev Technol 2:253–262
8. Franks F (1998) Freeze-drying of vaccines. Process Biochem 45:221–229
9. WHO TRS 1011 (Annex 3): Guidelines on stability evaluation of vaccines. https://cdn.who.int/media/docs/default-source/biologicals/vaccine-quality/guidelines-on-stability-evaluation-of-vaccinese-c802a21-dc04-455b-b09c-b7ff76c4d160.pdf?sfvrsn=9a3a8dcb_1
10. Watson JC, Hadler SC, Dykewicz CA, Reef S, Phillips L (1998) Measles, mumps and rubella - vaccine use and strategies for elimination of measles, mumps and rubella, and congenital rubella syndromes and control of mumps: recommendations of the advisory committee on immunization practices (ACIP). MMWR Recommended Rep 47(RR–8):1-57
11. Kanojia G, ten Have R, Soema PC, Frijlink H, Amorij J-P, Kersten G (2017) Developments in the formulation and delivery of spray dried vaccines. Hum Vaccin Immunother 13:2364–2378. https://doi.org/10.1080/21645515.2017.1356952
12. Françon A, Brass O, Chouvene P, Leleu A (2012) US Patent 8,142,7955 B2, "Stabilizer and vaccine composition comprising oneor more live attenuatd flaviviruses"
13. Ohtake S, Martin RA, Yee L, Chen D, Kristensen DD, Lechuga-Ballesteros D, Truong-Le V (2010) Heat-stable measles vaccine produced by spray drying. Vaccine 28:1275–1284. https://doi.org/10.1016/j.vaccine.2009.11.024
14. Campbell CH (1917) US patent 1.250,427, "Drying milk"
15. Schoppet EF, Panzer CC, Talley FB, Sinnamon HI (1974) Continuous vacuum foam-drying of whole milk. VI. Iron enrichment. Dairy Sci 57:1256–1257. https://doi.org/10.3168/jds.s0022-0302(74)85047-2
16. Sramek M, Schweiggert RM, van Kampen A, Carle R, Kohlus RJ (2015) Preparation of high-grade powders from tomato paste using a vacuum foam drying method. Food Sci 80:E1755–1762. https://doi.org/10.1111/1750-3841.12965
17. Ong SP, Law CL (2010) Hydrothermal properties of various foods, vegetables and fruits. In: Jangam SV, Law CL (eds) Drying of foods, vegetables and fruits, vol 1. pp 31-58
18. Langford A, Bhatnagar B, Walters R, Tchessalov S, Ohtake S (2018) Drying of biopharmaceuticals: recent developments, new technologies and future direction. Jpn J Food Eng 19:15–24. https://doi.org/10.11301/jsfe.18514
19. Osanlóo DT, Mahlin D, Bjerregaard S, Bergenståhl B, Millqvist-Fureby A (2024) Exploring vacuum foam drying as an alternative to freeze-drying and spray drying for a human lipase. Int J Pharm 667:124883. https://doi.org/10.1016/j.ijpharm.2024.124883
20. Lyu F, Zhao YH, Zuo XX, Nyide B, Deng BH, Zhou MX, Hou J, Jiao JJ, Zeng MQ, Jie HY, Olaniran A, Lu Y, Khoza T (2024) Thermostable vacuum foam dried Newcastle disease vaccine: process optimization and pilot-scale study. Appl Microbiol Biotechnol 108:359. https://doi.org/10.1007/s00253-024-13174-7
21. Lovalenti PM, Anderl J, Yee L, Nguyen V, Ghavami B, Ohtake S, Saxena A, Voss T, Truong-Le V (2016) Stabilization of live attenuated influenza vaccines by freeze drying, spray drying, and foam drying. Pharm Res 33:1144–1160. https://doi.org/10.1007/s11095-016-1860-1
22. Pisal S, Wawde G, Salvankar S, Lade S, Kadam S (2006) Vacuum foam drying for preservation of LaSota virus: effect of additives. AAPS Pharm Sci Tech 7:60. https://doi.org/10.1208/pt070360
23. Chen D, Stinchcomb DT, Silengo SJ, Arguello J, IR, Tary-Lehmann M, Powell TD, Livengood JA, Kinney RM, Huang CY-H, Stinchcomb DT (2010) Thermostable formulations of a live attenuated tetravalent

dengue vaccine using vacuum drying. Vaccine 29:7456–7462. https://doi.org/10.1016/j.vaccine.2011.07.054

24. Dormitzer PR, Mandl CW, Rappuoli R (2011) Replicating vaccines: a new generation. Springer Basel, Basel
25. Lv F, Lu Y, Hao ZL, Zhao YH, Zhang LH, Feng L, Chen J, Wang LL, Rui R, Hou JB (2016) Preparation and heat resistance study of porcine reproductive and respiratory syndrome virus sugar glass vaccine. Vaccine 34:3746–3750. https://doi.org/10.1016/j.vaccine.2016.06.003
26. Sim BKL, Li M, Osorio M, Yun W, Wai T, Peterson JW, James ER, Chakravarty S, Gao L, Xu R, KC N, Stafford RE, Lawrence WS, Yeager LA, Peel JE, Sivasubramani SK, Chopra AK, Filippova S, Hoffman SL (2017) Protection against inhalation Anthrax by immunization with ^Ty21a stably producing protective antigen of *Bacillus anthracis.* Vaccines 2:17. https://doi.org/10.1038/s41541-017-0018-4
27. Koutzoumis D, Wai T, Inbar E, Sim BKL, James ER, Hoffman SL (2025) Foam drying of an oral vaccine for SARS-CoV-2 expressed in *Salmonella enterica Typhi* Ty21a. Drying Technology 43:2083-2089 https://doi.org/10.1080/07373937.2025.2565676
28. Ohtake S, Martin R, Saxena A, Pham B, Chiueh G, Osorio M, Kopecko D, Xu D, Lechuga-Ballesteros D, Truong-Le V (2011) Room temperature stabilization of oral, live attenuated *Salmonella enterica serovar Typhi*-Vectored vaccines. Vaccine 29:2761–2771. https://doi.org/10.1016/j.vaccine.2011.01.093
29. Formal B, Baron LS, Kopecko DJ (1986) US patent 4,632,830, "Oral vaccine for immunization against enteric disease"
30. FDA package insert vivotif. https://www.fda.gov/media/75988/download
31. WHO TRS No. 658: guidelines on production and quality control of live oral typhoid vaccine
32. WHO TRS 926 Annex 1 (Live bacterial vaccine manufacture)
33. Levine MM, Black RE, Ferreccio C, Germanier R (1987) Large-scale field trial of Ty21a live oral typhoid vaccine in enteric-coated capsule formulation. Lancet 329:1049–1052. https://doi.org/10.1016/S0140-6736(87)90480-6 Chilean Typhoid Committee
34. Simanjuntak CH, Totosudirjo H, Haryanto P, Suprijanto E, Paleologo FP, Punjabi NH, Witham ND, Darmowigoto R, Soeprawoto, Hoffman SL (1991) Oral immunisation against typhoid fever in Indonesia with Ty21a vaccine. Lancet 338:1055–1059. https://doi.org/10.1016/0140-6736(91)91910-m

Chapter 36

Development of Freeze-Dried Formulations for Biopharmaceutical Drug Products Using a Design of Experiments Approach: A Case Study on Antibody-Drug Conjugates

Maite Harguindeguy, Andrea Hawe and Tim Menzen

Abstract

The successful development of a lyophilized biopharmaceutical formulation relies on a systematic, science- and risk-based approach. By using a mock antibody-drug conjugate (ADC) as an example, this chapter outlines a structured pathway beginning with the definition of the quality target product profile (QTPP) and a thorough understanding of the active pharmaceutical ingredient's (API) physicochemical properties. Formulation components are selected based on their stabilization roles, guided by prior knowledge and literature, as well as a Design of Experiments (DoE methodology, which supports efficient screening of excipient combinations). An initial conservative freeze-drying cycle ensures stability of the API and drying efficiency. Analytical characterization, using techniques to detect critical quality attributes, which indicate stability and degradation of the API, enables the comparison of formulation candidates and the selection of a lead formulation. This lead formulation, paired with a well-designed initial freeze-drying process, serves as a solid foundation for further optimization within a quality by design (QbD) framework.

Keywords Analytical characterization, Antibody-drug conjugate (ADC), Biopharmaceuticals, Design of experiments (DoE), Excipients, Formulation, Freeze-drying, Quality by design (QbD), Quality target product profile (QTPP), Stability study

1 Introduction

1.1 Lyophilization of Biopharmaceutical Drug Products

Biopharmaceuticals have revolutionized modern medicine by offering targeted and effective treatments for a wide range of diseases. Among them, antibodies and antibody-related drug products play a key role, enabling therapies for various conditions and having become one of the most prominent classes of biopharmaceuticals on the market.

Biopharmaceutical drug products are typically administered parenterally, e.g., in the form of intravenous infusion or subcutaneous injection. Therefore, biopharmaceutical drug products are

Willem F. Wolkers and Harriëtte Oldenhof (eds.), *Cryopreservation and Freeze-Drying Protocols*, Methods in Molecular Biology, vol. 3060, https://doi.org/10.1007/978-1-0716-5416-3_36,

preferably manufactured as liquids in vials or prefilled syringes. Nevertheless, therapeutic proteins are complex molecules with various routes of physicochemical instability and degradation. Particularly antibody-drug conjugates (ADCs), which combine an antibody for target-specificity with a small molecule such as a chemotherapeutic drug, are very sensitive. In case of ADCs, degradation may occur at the antibody, the linker, or the conjugated small molecule (payload). Thus, product quality is endangered over time during manufacturing, storage, and administration.

Freeze-drying, or lyophilization, is one frequently used process to remove the water and thereby increase protein stability. Consequently, freeze-drying of biopharmaceuticals, particularly in the case of ADCs, is one option to preserve product quality, allowing storage stability sometimes even under room temperature conditions. Currently, 14 out of 16 commercially available ADC drug products are lyophilized [1].

Nonetheless, freeze-drying comes with its own challenges in terms of freezing- and drying-induced stresses. To address them, (i) formulations are designed to protect and stabilize the active pharmaceutical ingredient (API) during freeze-drying, and (ii) the freeze-drying process must be carefully designed to ensure product quality and economic efficiency. To help elucidate the formulation development process, this chapter presents a mock case study focused on the formulation development of a freeze-dried ADC drug product. However, the formulation concept described here can be applied broadly to other biopharmaceutical modalities—such as peptides, enzymes, protein hormones, fusion proteins, and other bioconjugates—by tailoring protocol parameters to the specific molecule and desired quality attributes, as discussed in the following sections.

This chapter will provide a general overview of a freeze-drying process. For more details regarding lyophilization cycles parameters, the reader is referred to Chap. 37 of this book.

1.2 Quality by Design in Formulation Development

Ensuring product quality is of utmost importance in the (bio)pharmaceutical industry. Traditional quality control methods, such as quality by testing (QbT), rely on randomly sampling finished drug products, which increases the risk of distributing medicines that do not meet the desired specifications. In contrast, quality by design (QbD) approach implies building quality into the product from early discovery through market authorization. This approach supports the development of safer and more effective therapeutics by emphasizing that both product and process characteristics should be scientifically engineered already in the early development stages to meet predefined objectives, rather than relying on trial-and-error batch testing. Recognizing these benefits, the US FDA has promoted QbD as a modern, risk-based approach to pharmaceutical manufacturing, integrating it into updated guidelines to ensure more consistent, high-quality drug products [2–4].

QbD principles involve the systematic development of both the formulation and the manufacturing process to ensure the final product meets the predefined quality target product profile (QTPP). The QTPP outlines critical product characteristics such as dosage form, strength, purity, stability, and delivery method. The QTPP is defined and refined through ongoing risk assessment during drug development and clinical trials. Some characteristics, such as the concentration of API and choice of primary container, are often predefined for Clinical Phase I/II development to simplify early formulation and testing. In contrast, at the commercial stage, these characteristics are optimized for patient-friendly handling and administration [5].

1.3 Lyophilized Formulation Components and Their Stabilizing and Protecting Effects

Excipients play essential roles in protecting lyophilized biopharmaceuticals from physical and chemical stresses during freezing, drying, storage, and reconstitution.

Manufacturing a drug product, whether liquid or lyophilized, involves steps such as compounding, filtration, and aseptic filling. These processes can already represent significant stress to the API. Further details regarding the freeze-drying process for biopharmaceuticals are explained in Chap. 37. In brief, typically the vials which are filled with the formulation are placed in trays that are loaded onto the freeze-dryer shelves for processing. These shelves are temperature controlled and used for heat-exchange with the vials and consequently the formulation in them. The freeze-dryer chamber is designed to be under vacuum and to carry out the freeze-drying process.

Once the vials are loaded into the freeze-dryer, the following steps are carried out in sequence: (i) Freezing: formulation is frozen, typically below the glass transition temperature of the maximally freeze-concentrated solution (Tg'). (ii) Primary-drying: heat is provided to the formulation (typically still at subzero temperatures) via the shelves of the freeze-dryer, while pressure is decreased well below the triple point for water, to promote sublimation of the unbound frozen water in the formulation. (iii) Secondary drying: additional heat is provided (usually around room temperature) to remove the remaining bound water. (iv) Stoppering: inert gas is injected into the chamber and the vials are closed with their stoppers to protect the formulation from external air and moisture upon unloading. (v) Unloading: product is crimped, labeled, and stored until further transportation and shipping.

All these steps may subject the API to different types of stress. Different formulation components are used to protect the API from these stress factors and thus improve its stability. The following sections discuss the main excipients used to address different stress factors that arise from each processing step during lyophilization.

1.3.1 Buffering Agents

The first stress event is the freezing process. As freezing progresses, water crystallizes into ice, leaving the remaining formulation increasingly concentrated (cryo-concentration). Buffers are used in the formulation to maintain pH stability in the liquid state, but also during freezing and drying, preventing acid/base-induced protein denaturation and aggregation. However, buffer crystallization during freezing can cause large pH shifts with negative effects on API stability [6]. Some buffers such as sodium phosphate selectively crystallize, causing pH drops of 2–4 units while potassium phosphate system in contrast, shows a modest pH increase of less than one unit during freezing. The addition of excipients like sucrose can minimize this effect by inhibiting buffer salt crystallization, as they promote the formation of an amorphous phase that effectively traps the salts and preserves buffering capacity [7].

Buffers such as acetate and ammonium-based buffers are volatile and may be partially or completely removed during drying, resulting in pH shifts during reconstitution. In contrast, citrate and histidine buffers resist crystallization and better maintain pH stability during freezing. Another relevant consideration is that pH shifts tend to be minimal when low buffer concentrations are used [6, 8]. Buffers with chelating properties, such as citrate, may affect the stability of metalloproteins or ADCs with linker metal coordination formulations.

When selecting a buffer, it is important to consider the target pH range, its potential impact on the lyophilization process, and the requirements of the final application. The drug product pH should ideally be close to physiological, which is most tolerable for patients, but the pH should ensure proper folding of the API structure [9]. One important factor defining at which pH range the API is more stable is its isoelectric point (pI). The isoelectric point is the pH at which the molecule has a zero net charge, making it more prone to aggregation. Thus, pH ranges away from this pI are preferred.

Each buffer system has its pKa, which is the pH at which the concentrations of the weak acid and its conjugate base are equal. Meaning that the buffer system can better resist pH changes around this pH. It is advisable to stay within ± 1 pH unit from the selected buffer's pKa. Thus, the choice of buffer system strongly depends on the desired pH range for the drug product [10].

In lyophilized formulations, histidine, phosphate, and citrate buffers are most commonly used, generally covering pH ranges from approximately 5.0–8.0. Succinate is also employed, particularly at lower pH values around 5.0–6.0 [11, 12]. High buffer concentrations can lower the glass transition temperature of the freeze concentrate, which requires lower drying temperatures, increasing the total process time (more details see Chap. 37). Thus, lower buffer concentrations are usually preferred for lyophilized formulations [13]. Regarding the final administration, citrate buffers are usually associated with higher pain perception, whereas formulations containing histidine or acetate are generally perceived as less painful upon administration [14]. Although tris buffers are also commonly used, their use is not advisable in peptide and protein formulations intended for storage at elevated temperatures. Tris can degrade over time, releasing formaldehyde, which may compromise the stability and safety of the formulation [9].

1.3.2 Surfactants

Surfactants are commonly used in proteinaceous drug products. Surfactants can serve multiple roles due to their surface activity. In solution, they accumulate at interfaces, preventing loss of API due to adsorption to containers, filters, and other surfaces. This function is particularly important for drug products with low API concentrations. Moreover, surfactants can protect proteins from surface-induced unfolding and aggregation [15]. The formation of the ice-water interfaces during freezing can induce protein adsorption and subsequent surface-induced unfolding on these surfaces. Such structural perturbations expose critical bonds (e.g., hydrophobic interactions, hydrogen bonds) that maintain the protein's native conformation, rendering them vulnerable to stress-induced damage [13]. Surfactants can be used to minimize protein adsorption at ice–liquid and air–liquid interfaces, thereby reducing aggregation by shielding hydrophobic regions of the protein during freezing and reconstitution. However, they are prone to both hydrolytic and oxidative damage, which can result in release of free fatty acids (FFAs) to the formulation that can subsequently form subvisible and visible particles when the solubility limit is exceeded [16]. Additionally, high surfactant concentrations can introduce impurities, such as peroxides, which can in turn compromise protein stability [13, 17]. Thus, the surfactant concentration needs to be optimized during drug product development.

1.3.3 Cryoprotectant

Conformational stability is a function of temperature and proteins typically unfold at high and low temperatures. Low temperatures weaken hydrophobic interactions, destabilizing protein's native structures, which are then susceptible to further degradation. This cold denaturation is particularly relevant for oligomeric proteins, which may dissociate below their thermal stability range, i.e., the temperature range in which a protein maintains its native, functional, structure. This loss of the native structure can lead to irreversible aggregation or loss of activity upon rehydration, even if individual subunits remain intact [8, 9]. Cryoprotectant are formulation components that can inhibit ice-induced denaturation and prevent protein aggregation during freezing. One of the proposed mechanisms for protein stabilization of sugars as cryoprotectants is preferential exclusion. Sucrose and trehalose are excluded from the protein surface, stabilizing native conformation by increasing the free energy of unfolding [18]. Another stabilizing mechanism is vitrification, where high concentrations of the excipient (5%–10% w/v) form viscous matrices that reduce molecular mobility, increasing protein stability during freezing [8]. It should be noted that sorbitol and glycerol, despite the fact that they are very good cryoprotectors, are not recommended for lyophilized formulations, because they substantially lower the Tg', increasing the risk of collapsed cakes.

1.3.4 Tonicity Agents

Tonicity agents are formulation components that help achieving a physiologically compatible osmolality when administering the drug product, such as sucrose, sorbitol, and sodium chloride (NaCl) [11]. Additionally, proteins may be sensitive to high or low ionic strength, which can become an important factor to consider during formulation development. However, this needs to be carefully evaluated when designing a formulation for a freeze-dried product. Dissolved excipients and ions are cryoconcentrated during freezing, increasing the formulation's ionic strength and osmolality. A major increase in ionic strength weakens hydrophobic interactions, destabilizing charge-sensitive proteins [19], which can then lose their native state. In addition, salts such as NaCl lower the Tg' of formulations, increasing the risk of collapse and requiring longer and less economical lyophilization cycles.

1.3.5 Lyoprotectants

During freezing, water molecules form hydrogen bonds with polar groups on the protein surface. Drying disrupts these hydrogen bonds between water and proteins, which can then affect the protein structure. Disaccharides such as sucrose and trehalose can replace water molecules and form hydrogen bonds with the protein surface, mimicking hydration shells [13, 20]. A 360:1 molar ratio of lyoprotectant (sucrose or trehalose) to protein prevents aggregation, deamidation, and oxidation while lower ratios can lead to instability [21]. The higher ratio maintains the native-like structure during drying and storage. Vitrification also contributes to protein stabilization by immobilizing them within a rigid glassy matrix, stabilizing them and slowing degradation reactions such as deamidation later during storage [8].

1.3.6 Collapse Temperature Modifiers

The temperature at which the maximally freeze-concentrated amorphous phase of a formulation transitions from a brittle glass to a viscous fluid (Tg') is a critical parameter in lyophilization. Below Tg', molecular mobility is restricted, stabilizing proteins and preventing collapse during primary drying. However, exceeding Tg' during primary drying causes structural collapse, trapping moisture and reducing stability. The temperature at which this collapse takes place is the collapse temperature (Tc) [18]. Formulation components can be selected or added to increase the overall collapse temperature of the formulation. Common examples include sugars like sucrose or trehalose, sugar alcohols like mannitol and polymers such as dextran, gelatine, hydroxyethyl starch, β-cyclodextrin, and polyvinylpyrrolidone (PVP). These excipients help preventing collapse while enabling higher primary drying temperatures, shortening cycle times and enhancing process robustness [17, 22, 23]. Other excipients, such as sorbitol, glycerol, and NaCl (see above) have the opposite effect; they substantially lower the Tg' and are therefore not recommended for lyophilized formulations, unless unavoidable due to specific stability requirements of the particular API.

1.3.7 Bulking Agents

Once the product is freeze-dried, the dried cake structure is not only a cosmetic attribute but also an indicator of product quality. Collapsed cakes may exhibit higher moisture content, which can compromise product stability. Additionally, the product is reconstituted prior to use, thus, a cake structure that enables rapid dissolution is desired. Bulking agents provide mechanical support to the lyophilized cake, ensuring elegant appearance, reduced loss by "blow out" during drying, rapid reconstitution, and stability. Crystalline agents such as mannitol, lactose, and glycine form rigid scaffolds but require precise freezing protocols to avoid hydrate formation. Amorphous agents such as sucrose enhance stability but risk collapse if used alone. A 4:1 mannitol:sucrose ratio balances structural integrity and protein protection while also improving reconstitution [24, 25].

1.3.8 Other Stabilizers

Additional excipients may be used to improve drug administration and formulation stability after reconstitution. Antioxidants such as methionine may be used (typically below 0.5 mg/mL) due to their ability to scavenger peroxides from polysorbate degradation [23]. Chelating agents, e.g., EDTA (typically below 0.1 mg/mL) can also be used to mitigate metal ion-induced oxidation [13]. Preservatives such as phenol inhibit microbial growth in multidose vials and are typically employed at concentrations below 0.5 mg/mL [23]. Arginine, glycine, proline, and lysine may be used as a viscosity reducer in high-concentration antibody formulations (typically in the range of 10–100 mg/mL) [11, 26].

1.3.9 Conclusion on Formulation Components

Excipients may be selected to address different stability issues during freezing, drying, storage, and reconstitution. Many excipients serve multiple roles, such as sucrose (cryoprotectant, lyoprotectant, tonicity agent, bulking agent, collapse temperature modifier), which is useful when designing lyophilized formulations and explains why it is largely employed [11]. From a regulatory perspective and economic reasons, an ideal formulation employs a minimum number of formulation components while ensuring API stability. Table 1 summarizes the main components used in lyophilized formulations and lists some typical concentration ranges from the literature.

1.4 Freeze-Drying Process

In the early stages of lyophilized drug product development, the primary objective is to achieve a stable product for testing in first-in-human clinical trials. To maximize the likelihood of successful drying

Table 1
Summary of main formulation components used in lyophilized biopharmaceuticals

Category	Function	Examples	Typical concentration range	Desired trait
Buffers	Prevent pH shifts	Histidine, citrate, phosphate and succinate	10–50 mM [6, 23]	Resist crystallization
Surfactants	Reduce interfacial stress	Polysorbate 80, polysorbate 20, poloxamer 188	0.005%–0.05% w/v [23]	Shield hydrophobic surfaces
Cryoprotectant	Inhibit ice-induced denaturation	Sucrose, Trehalose	2%–10% w/v [20, 27]	Preferential exclusion, vitrification
Tonicity agents	Adjust osmolality	Glycine, mannitol	1%–10% w/v	Adjust osmolality and ionic strength in a desired way
Lyoprotectants	Replace water, form glassy matrix	Sucrose, trehalose,	2%–10% w/v [17]	Hydrogen bonding, vitrification
Collapse temperature modifiers	Increase Tc	Mannitol, sucrose, trehalose, dextran	2%–10% w/v [22, 23]	Form rigid crystalline matrix/increase high-Tg' constituents in the formulation
Bulking agents	Structural support	Mannitol, sucrose, trehalose, glycine	2%–10% w/v [17, 24]	Crystalline/amorphous balance

across diverse compositions of multiple formulations in a screening, a conservative freeze-drying cycle is typically employed, characterized by extended durations and low temperatures. This approach ensures that all tested formulations achieve sufficiently low residual moisture content by the end of secondary drying.

To define the appropriate thermal limits for primary drying, critical temperature parameters such as the glass transition temperature of the maximally freeze-concentrated solution (Tg′) or the collapse temperature (Tc) are determined using techniques such as differential scanning calorimetry (DSC) or freeze-drying microscopy (FDM) as discussed in Chap. 37. These values indicate the maximum allowable product temperature during primary drying to prevent structural collapse due to viscous flow. Additional considerations and detailed guidance on freeze-drying process parameters are provided in Chap. 37.

1.5 Formulation Evaluation by Analytical Characterization

To evaluate product stability both the lyophilized cake and the reconstituted product are analyzed. Typically, several analytical methods are required to investigate the stability of the API in solution, targeting different degradation pathways. In case of proteins, e.g., size-exclusion chromatography (SEC) can be used to detect protein aggregation (high-molecular-weight species, HMW) and fragmentation (low-molecular-weight species, LMW). Ion exchange chromatography (IEX), either cation exchange chromatography (CEX) or anion exchange chromatography (AEX), can be used to assess protein's charge variants. Reversed phase HPLC (RP-HPLC) can be used to identify and quantify hydrophobic degradation products, typically after enzymatic digestion of the protein (peptide mapping). Capillary electrophoresis (CE) provides insight into charge heterogeneity and degradation products of the protein. UV-Vis spectroscopy based methods are used to quantify protein concentration. Fourier-transform infrared spectroscopy (FTIR) can assess secondary structure, both in solution and in solid state. Differential scanning calorimetry (DSC) can be used to measure thermal transitions (e.g., unfolding temperature, Tm) during a heating ramp. In the case of an ADC, additional methods are required that are related to the conjugated small molecule drug. Important parameters are the drug-antibody-ratio (DAR), which can be determined by UV spectroscopy, hydrophobic interaction chromatography (HIC), reverse phase high-performance liquid chromatography (RP–HPLC) or liquid chromatography–mass spectrometry (LC–MS), and the drug load distribution, as well as the drug load distribution, where LC–MS is often the method of choice. Moreover, the residual contents of free, unconjugated small molecule drug and antibody need to be determined by, e.g., RP–HPLC and LC–MS, respectively. Potency assays for ADCs address target binding and effector function of the antibody (depending on the mode of action) and cytotoxicity.

Last but not least, particle analytics, e.g., light obscuration (LO) and flow imaging microscopy (FIM) quantifies and characterizes subvisible particles (>1 μm), whereas dynamic light scattering (DLS) can be used to assess particle size and polydispersity of submicron particles (<1 μm). Those particles may be aggregated API or other particulate impurities originating from the manufacturing process, the packaging material or other extrinsic sources.

Solid-state properties of the lyophilized cake are monitored using residual moisture determination (e.g., Karl Fischer titration), differential scanning calorimetry (DSC) for thermal transitions, X-ray powder diffraction (XRPD) to detect crystallinity in excipients, and scanning electron microscopy (SEM) for cake morphology and pore structure. Visual inspection remains a critical qualitative assessment for lyophilized cake appearance.

The choice of analytical method is typically dictated by the API's degradation pathways and regulatory expectations (stage of drug product development), and orthogonal methods are frequently used in combination to improve data interpretation.

2 Materials

2.1 Freeze-Drying System

1. Freeze-dryer with precise shelf temperature and pressure control, with a nitrogen (or inert gas) backfill line, a low-temperature condenser and process monitoring tools such as temperature and pressure sensors.

2.2 Formulation Components

1. Active pharmaceutical ingredient (API).
2. Highly purified water.
3. Pharmaceutical-grade excipients, typically including:
   (a) Buffering agents (e.g., histidine, citrate, phosphate).
   (b) Cryo- and lyoprotectants (e.g., sucrose, trehalose).
   (c) Surfactants (e.g., polysorbates).
   (d) Other excipients depending on the specific stability needs of the API (see Sect. 1.3 for details).

2.3 Laboratory Instruments

1. Calibrated precision balances for accurate weighing.
2. Calibrated pH meters for pH adjustment and control.
3. Calibrated osmometers for osmolality measurements.
4. UV-Vis spectrophotometer for protein concentration measurements.
5. Lab setup and materials for sterile filtration and aseptic filling (e.g., LAFs, isolators).

6. Analytical instruments and related consumables to evaluate critical quality attributes of interest of the API (see Sect. 1.5 for details).

2.4 Other Items

1. Controlled temperature and relative humidity chambers or refrigerators for stability studies.
2. Primary packaging materials, including vials, stoppers, seals, or prefilled syringes.
3. Equipment for washing and sterilizing primary packaging material, if necessary.

2.5 Specific Software

1. Design of experiments (DoE) software for planning of formulation studies, analysis of variable effects, and process optimization.

3 Methods

The following steps are exemplified for a fictitious ADC molecule called *MyADC1* but are applicable across multiple modalities.

3.1 Preparations and Formulation Design

1. Extract specifications and requirements from the quality target product profile (QTTP) such as target concentration, dosage, and route of administration (*see* **Note 1**).
2. Check scientific literature, formulations of similar marketed drug products, as well as previous knowledge with the API to plan formulation selection (*see* **Note 2**).
3. Define analytical methods for characterization of relevant quality attributes (*see* **Note 3**).
4. Define stability study conditions (*see* **Note 4**).
5. Define Design of Experiment (DoE) parameters (*see* **Note 5**).
6. Define DoE model (*see* **Note 6**).
7. Define DoE formulation list (*see* **Note 7**).
8. Define DoE response variables (*see* **Note 8**).

3.2 Formulation Preparation

1. Calculate the volume of formulation needed. The amount of formulation to be prepared needs to be calculated based on the desired resulting formulation for all planned analytical evaluations and time points and including excess volumes to compensate losses during preparation.
2. Prepare formulation buffer stock solutions.
3. Buffer exchange *MyADC1* from its current formulation buffer into the intended set of formulation buffers to be tested by, e.g., dialysis or tangential flow filtration.

4. Sterile filter each formulation and fill into washed and sterilized glass vials under aseptic conditions.

3.3 Lyophilization

1. Measure the glass transition temperature of the maximally freeze-concentrated solutions (Tg′) of each liquid formulation by DSC to define lyophilization process conditions (*see* **Note 9**). For details, see Chap. 37.
2. Record weight of selected, individual vials of each formulation before lyophilization for subsequent determination of reconstitution volume (*see* **Note 10**).
3. Program the freeze-dryer by using a conservative, i.e., long and cold lyophilization process based on the Tg' values measure for all formulations to ensure complete drying of all formulations (*see* **Note 11**).
4. Load vials on precooled shelves and place thermocouples for process monitoring into selected vials for each formulation.
5. Start lyophilization process and monitor process by using process analytical tools (PAT) such as product temperature and chamber pressure (*see* **Note 11**).
6. Unload the vials at the end of secondary drying under nitrogen atmosphere.
7. Record weight of vials after freeze-drying (*see* **Note 10**).

3.4 Stability Testing and Data Evaluation

1. Store samples under intended storage conditions as outlined in the study design (*see* **Note 4**).
2. Perform analysis of the samples taken before and after lyophilization as benchmark for stability assessment.
3. Perform analysis for each subsequent time point (*see* **Note 4**).
4. Analyze data and feed results into DoE software as response to generate first model (*see* **Note 12**).
5. Optimize DoE model (*see* **Note 13**).
6. Conclude optimal formulation from the DoE model (*see* **Note 13**).

3.5 Further Steps

1. Surfactant screening to test alternative surfactants and/or further optimize surfactant concentrations by performing mechanical stress testing such as shaking.
2. Formulation optimization to test additional excipients or narrow concentration ranges to further optimize protein stability.
3. Lyophilization cycle development to fine-tune the lyophilization process based on the requirements of the one, final lead formulation in order to for instance shorten primary or secondary drying times. For this purpose, mechanistic models of the freeze-drying process provide powerful tools to evaluate

the design space and optimize shelf temperature and chamber pressure [28] (*see* **Note 14**).

4. After the therapeutic success has been demonstrated in clinical phase I and II trials, the drug products need further formulation development to meet the QTPP requirements for clinical phase III and subsequent commercialization.

4 Notes

1. The QTTP describes product attributes and requirements to be achieved during formulation and manufacturing process development. A shortened, condensed exemplary QTPP for the mock antibody drug conjugate *MyADC1* is shown in Table 2.
2. Figure 1 shows that histidine buffer with sucrose and polysorbate at pH 5–8 were the most frequently used formulation

Table 2
Shortened, exemplary quality target product profile (QTPP) for *MyADC1* for clinical phase I/II studies

Property	Phase I/II target
Indication	Chemotherapy
Patient populations	Adults
Route of administration	Intravenous
Way of administration	Clinical personnel
Dosage form	Lyophilized, single use
Dose	1 mg/kg
Content per container	100 mg as extractable dose
Concentration	20 mg/mL
Average drug-antibody-ratio	4
Primary container (including the device constituent part if a device)	20 R type 1 borosilicate glass vials, laminated stopper
Target shelf life and storage conditions	≥2 years at 2°C–8°C
Pharmacopeial compliance	Meets pharmacopeial requirements, colorless to slightly yellow, practically free of visible particles and meets USP<787> criteria for subvisible particles
Target osmolality	250–380 mOsm/kg
Target pH	Appropriate for the formulation to support product stability and compatible with the route of administration

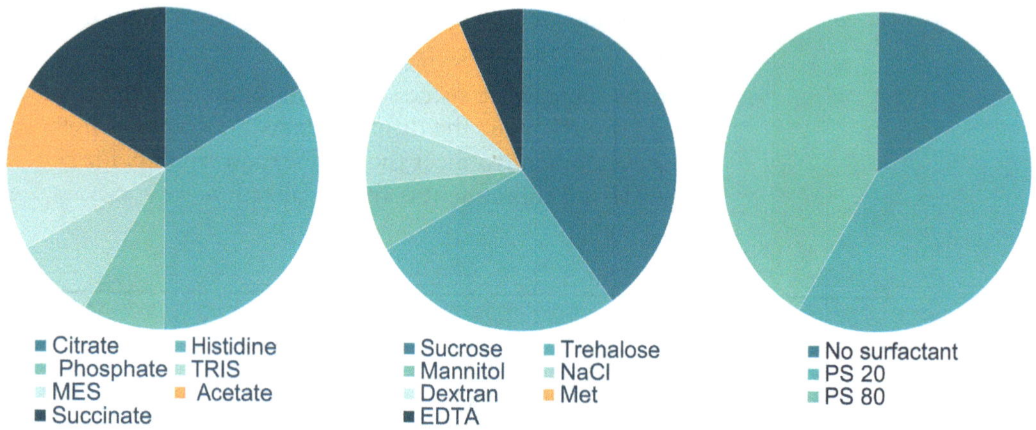

Fig. 1 Formulation overview of commercially available lyophilized ADCs showing buffers used (left), excipients (middle), and surfactants (right) used at 5.0–8.0 pH range. The concentration ranges from 0.25–50 mg/mL, with the majority of the formulations contain 5–20 mg/mL ADC. Three pie charts summarizing the composition of commercial lyophilized ADC pharmaceutical formulations. The first chart shows the distribution of buffering agents including citrate, phosphate, MES, histidine, tris, and acetate. The second chart displays the distribution of stabilizers and bulking agents such as sucrose, trehalose, mannitol, NaCl, dextran, EDTA, and methionine. The third chart presents the use of surfactants, divided among no surfactant, polysorbate 20 (PS 20), and polysorbate 80 (PS 80)

Table 3
Excipients and ranges selected for the lyophilized formulation screening for *MyADC1*

Formulation components evaluated	Tested concentration range	Function	Comment
MyADC1	20 mg/mL	Active pharmaceutical ingredient	According to QTPP
Histidine/ citrate at 5.5–7 pH range	20 mM	Buffer	Compatible with freeze-drying Frequently used for lyophilized ADCs
Sucrose	2%–10% w/v	Cryoprotectant, tonicity agent, lyoprotectant, collapse temperature modifier, bulking agent	Widely used excipient freeze-dried proteins and ADCs
Polysorbate 80 (PS80)	0.02% w/v	Surfactant	Frequently used in biopharmaceutical drug products

condition of the currently marketed, lyophilized ADC drug products. Citrate and succinate are also used in lyophilized ADC formulations, while citrate is also widely used for other modalities. The conditions summarized in Table 3 were selected for the initial lyophilized formulation screening of *MyADC1* (*MyADC1* has a pI of 8.5).

3. Analytical characterization methods are defined based on the modality of the therapeutic molecule. Some specific methods are required for ADCs and may target the whole ADC complex, the payload or the antibody [29]. An overview of exemplary methods and investigated quality attributes for *MyADC1* are provided in Table 4.

4. An exemplary design of a stability study for an initial lyophilized formulation screening for *MyADC1* is presented in Table 5. To assess the baseline values of *MyADC1*'s quality attributes, stability is analyzed before and after lyophilization. In addition, samples are stored for 1 month at 40°C /75% r.h. (International Council for Harmonisation, ICH conditions) to accelerate the degradation. A control sample is stored at intended storage conditions at 2°C–8°C as benchmark. The properties of the lyophilized cake as well as the stability of the API are investigated. Depending on the sample amount, the development stage and timelines, the duration of the stability study could be extended to include, e.g., 3 and 6 months storage and more conditions, e.g., 25°C /60% r.h.

5. During the first lyophilized formulation screening of *MyADC1* by DoE, three factors are tested at two levels: buffer type (histidine or citrate), pH (5.5–7) and sucrose concentration (2%–10% w/v). The setup of the first DoE includes formulations with osmolality values lower than the typical target range

Table 4
Exemplary list of quality attributes and analytical methods for characterization of *MyADC1*

Quality attribute	Analytical methods	Analyte
Drug-antibody-ratio (DAR)	HIC, UV spectroscopy, RP–HPLC, LC–MS	ADC
Drug load distribution	HIC, icIEF, RP–HPLC, LC–MS	ADC
Conjugation site	LC–MS, peptide mapping	ADC
Residual unconjugated antibody (DAR0)	LC–MS	Antibody
Residual free drug	RP–HPLC	Payload
Size variants	SEC	ADC and antibody
Charge variants	IEX	ADC and antibody
Potency	Binding, cytotoxicity, effector functions (depending on mechanism of action)	ADC (Binding: Antibody part is relevant)
Particulate impurities	FIM, LO, visual inspection	n.a

Table 5
Exemplary stability study design for first lyophilization screening for *MyADC1*

Time point	Explanation	Analytical methods
T_before_lyo	After formulation preparation and before lyophilization	*MyADC1* stability (Table 4)
T_after_lyo	After lyophilization	Reconstituted *MyADC1* stability (Table 4) Lyo cake properties (DSC, Karl-Fischer)
T_4w_40C	After lyophilization and storage for 1 month at 40°C /75% r.h. (accelerated degradation)	Reconstituted *MyADC1* stability (Table 4) Lyo cake properties (DSC, Karl-Fischer)
T_4w_5C	After lyophilization and storage for 1 month at 2°C–8°C (intended storage condition)	Reconstituted *MyADC1* stability (Table 4)Lyo cake properties (DSC, Karl-Fischer)

of approximately 280–320 mOsmol/kg. Osmolality can be adjusted and optimized in subsequent formulation steps. Among these, buffer type is a qualitative factor, meaning each level (histidine and citrate) must be tested independently, as no intermediate level exists between them. The remaining variables, pH and sucrose, are quantitative, with defined test ranges that may allow interpolation or further optimization. *MyADC1* concentration at 20 mg/mL, buffer concentration at 20 mM and polysorbate 80 concentration at 0.02% w/v remain fixed.

6. A full factorial DoE was designed by using MODDE® software (version 13.0, Umetrics) to investigate the influence of formulation variables on product stability. Three factors were evaluated at two levels each, including both qualitative and quantitative parameters. The experimental design encompassed all possible combinations of factor levels to enable a comprehensive assessment of main effects and interactions. This resulted in ten formulations: two buffer types, times two pH levels, times two sucrose levels, plus two center points, one for each buffer type. The software also generates a random run order for evaluating the formulations, to minimize the impact of uncontrolled variables and biases that can distort the results.
7. The formulation list is created based on the DoE, including a low and high level of each of the tested factors. The resulting DoE formulations are listed in Table 6. These formulations allow exploring the tested ranges and evaluating potential interaction effects between them.

Table 6
Resulting tested formulations in the DoE study using MODDE® software

Formulation	Run order	Buffer type	pH	Sucrose concentration (% w/v)
F1	1	Histidine	5.5	2.0
F2	7	Citrate	5.5	2.0
F3	8	Histidine	7.0	2.0
F4	4	Citrate	7.0	2.0
F5	2	Histidine	5.5	10.0
F6	10	Citrate	5.5	10.0
F7	6	Histidine	7.0	10.0
F8	3	Citrate	7.0	10.0
F9	9	Histidine	6.25	6.0
F10	5	Citrate	6.25	6.0

8. The analyzed response of the DoE may depend on the observed degradation profile. For simplicity, *MyADC1*'s SEC relative main peak area and CEX relative main peak area evaluated in this exemplary case study.
9. Proteins such as *MyADC1* typically increase the glass transition temperature of the maximally freeze-concentrated solution (Tg′). However, at the relatively low API concentration of 20 mg/mL, the Tg′ values of the sucrose-based formulations are predominantly influenced by sucrose and were approximately −32°C.
10. The weight loss for each formulation, before and after freeze-drying, is recorded to note the correct reconstitution volume using water to achieve the desired target concentration.
11. The freeze-drying cycle is designed based on the Tg' values measured for each formulation (*see* **Note 9**). A freezing step at −45°C is followed by primary drying at a shelf temperature of −35°C under a chamber pressure of 10 Pa. The process is shown in Fig. 2.
12. Exemplary results from *MyADC1* analysis at T_after_lyo and T4w_40C by SEC and CEX are provided in Table 7, and a graphical representation of these results in given in Figs. 3 and 4.
13. After entering the experimental results as response into the MODDE® DoE software, partial least squares (PLS) regression modeling is performed. This multivariate approach

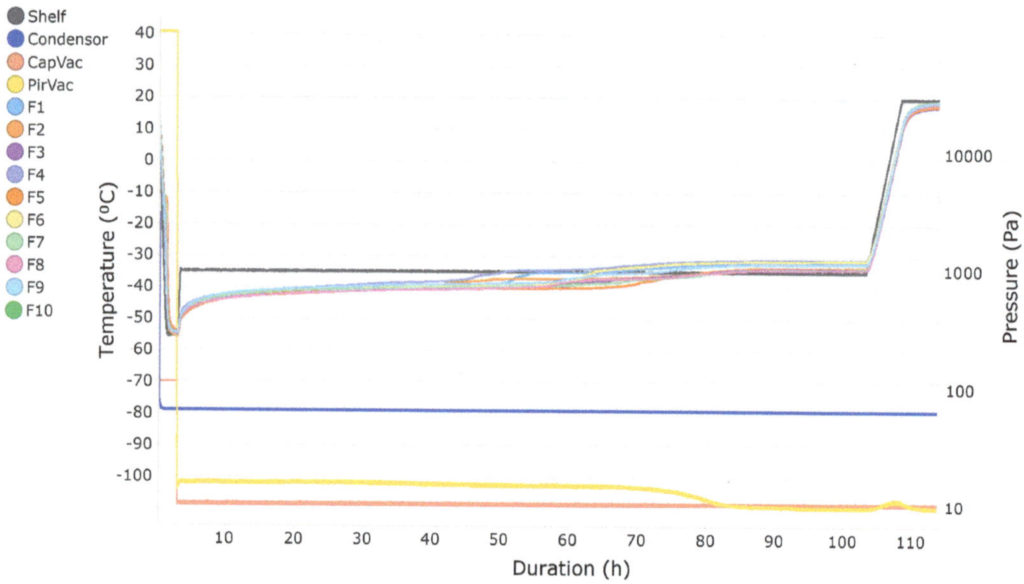

Fig. 2 Freeze-drying cycle process parameters (shelf and condenser temperature, Pirani and capacitance pressure gauges) and product temperatures for all ten *MyADC1* formulations. A cold (shelf temperature below lowest Tg') and long, i.e., "conservative" freeze-drying process is designed to ensure that all tested formulations reach acceptable, low residual moisture at the end of secondary drying. CapVac and PirVac refer to capacitance manometer and Pirani sensor used to monitor the lyo process. Graph showing the temperature and pressure profile of a lyophilization cycle over 100 h. Temperature data are plotted for the shelf, condenser, and ten product thermocouples (F1–F10), with pressure readings from CapVac and Pirani sensors. The plot illustrates key lyophilization phases: initial freezing, primary drying with sustained low shelf temperature and pressure, and secondary drying marked by a temperature rise

allowed for the identification of statistically significant factors and interactions affecting the selected critical quality attributes, as well as estimation of optimal formulation conditions. By using MODDE®, the model can be adjusted to predict the response the evaluated factors at the tested levels. Figure 5 shows the summary of the resulting model parameters for the DoE PLS for the SEC and CEX response variables, still not optimized. In the nonoptimized models, many model coefficients (Fig. 5) have low significance and thus are not good predictors of the data. These parameters can be removed, so the model is simplified, and the model fit is improved. Figure 6 shows the PLS model coefficients after optimization by removing model coefficients with low significance. Surface plots from DoE analysis are presented in Fig. 7 for SEC and Fig. 8 for CEX. The surface plot analysis indicates that formulations buffered with histidine outperformed those with citrate for *MyADC1*. Interestingly, lower pH values and higher sucrose concentrations improved SEC results while higher pH values improved CEX results. A formulation with 20 mg/

Table 7
Tabulated exemplary results for the 4-week at 40°C /75% r.h. stability study of *MyADC1*

				SEC main peak area, %		CEX main peak area, %	
Formulation	Buffer type	pH	Sucrose concentration, % w/v	T_after_lyo	T_4w_40C (40°C /75% r.h.)	T_after_lyo	T_4w_40C (40°C /75% r.h.)
F1	Histidine	5.5	2	98.2	94.2	69.4	66.8
F2	Citrate	5.5	2	96.6	92.3	69.2	65.9
F3	Histidine	7	2	97.1	92.6	70.7	67.3
F4	Citrate	7	2	95.8	91.2	69.8	66.5
F5	Histidine	5.5	10	98.7	93.7	70.0	66.8
F6	Citrate	5.5	10	97.3	92.6	69.3	66.0
F7	Histidine	7	10	97.7	93.0	70.9	67.5
F8	Citrate	7	10	96.3	91.6	69.8	66.4
F9	Histidine	6.25	6	99.0	93.4	70.9	67.2
F10	Citrate	6.25	6	97.8	92.3	70.1	66.2

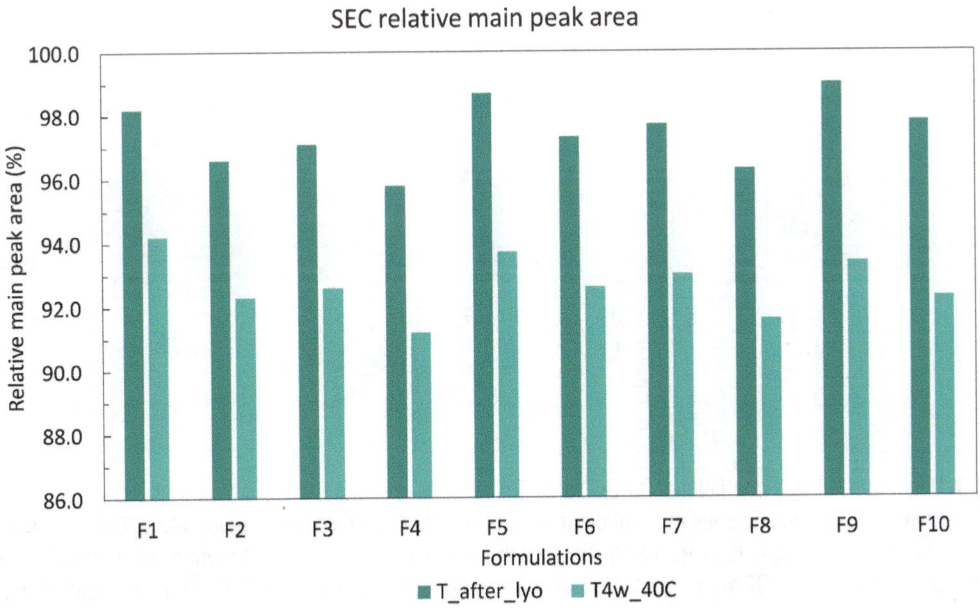

Fig. 3 Graphical SEC results for the 4-week at 40°C /75% r.h. accelerated stability study using DoE. Bar chart comparing the relative main peak area percentages of SEC analyses for ten lyophilized formulations (F1–F10) at initial time (T0) and after 4 weeks (T4w). The chart illustrates stability differences between SEC main peaks across formulations during storage, relevant for formulation development of lyophilized products. SEC values remain high (around 95%–100%) for all formulations and show minimal change over time

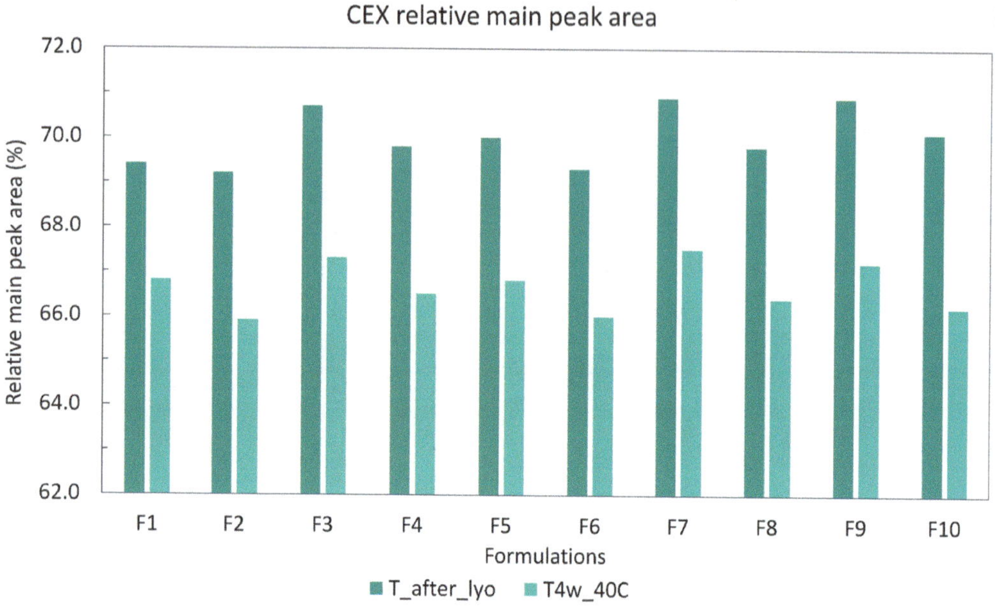

Fig. 4 Graphical CEX results for the 4-week at 40°C /75% r.h. accelerated stability study using DoE. Bar chart comparing the relative main peak area percentages of CEX analyses for ten lyophilized formulations (F1–F10) at initial time (T0) and after 4 weeks (T4w). The chart illustrates stability differences between CEX main peaks across formulations during storage, relevant for formulation development of lyophilized products. CEX values remain around 65%–70% and show minimal change over time

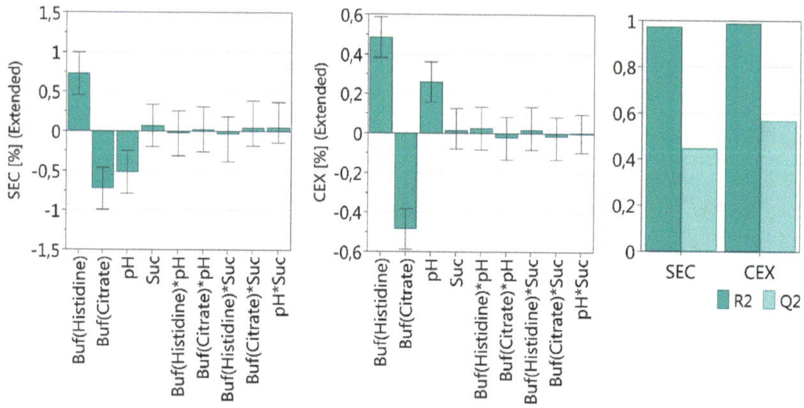

Fig. 5 Nonoptimized PLS coefficients for SEC and CEX response results (left) and their respective coefficients of determination (R2) and Predictive Coefficient of Determination (Q2), obtained using MODDE® software as part of the DoE analysis. A high R2 value may result from overfitting, especially when many variables are included, whereas a low Q2 value indicates poor predictive ability of the model. Three-panel chart displaying non-optimized PLS (partial least squares) regression coefficients from a Design of Experiments study for lyophilized formulation development. The first panel shows the effects of buffer type (Histidine and Citrate), pH, and sucrose (Suc) on SEC relative main peak area with histidine having a strong positive effect and citrate a strong negative effect. The second panel shows the same formulation factors on CEX relative main peak area, with histidine again positive and citrate negative. The third panel compares R2 and Q2 values for SEC and CEX models, both indicating high model fit but low predictive ability. Error bars represent coefficient uncertainty

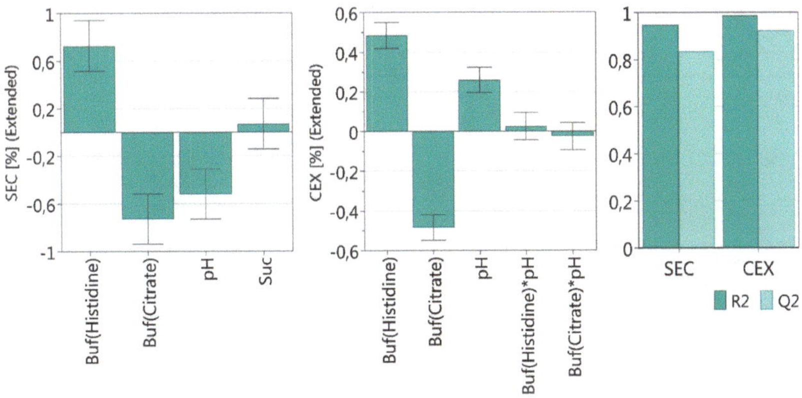

Fig. 6 Optimized PLS coefficients for SEC and CEX response results (left) and their respective coefficients of determination (R2) and Predictive Coefficient of Determination (Q2), obtained using MODDE® software as part of the DoE analysis. Three-panel chart displaying optimized PLS (Partial Least Squares) regression coefficients from a Design of Experiments study for lyophilized formulation development. The first panel shows the effects of buffer type (Histidine and Citrate), pH, and sucrose (Suc) on SEC relative main peak area with histidine having a strong positive effect and citrate a strong negative effect. The second panel shows the same formulation factors on CEX relative main peak area, with histidine again positive and citrate negative. The third panel compares R2 and Q2 values for SEC and CEX models, both indicating high model fit and predictive ability. Error bars represent coefficient uncertainty

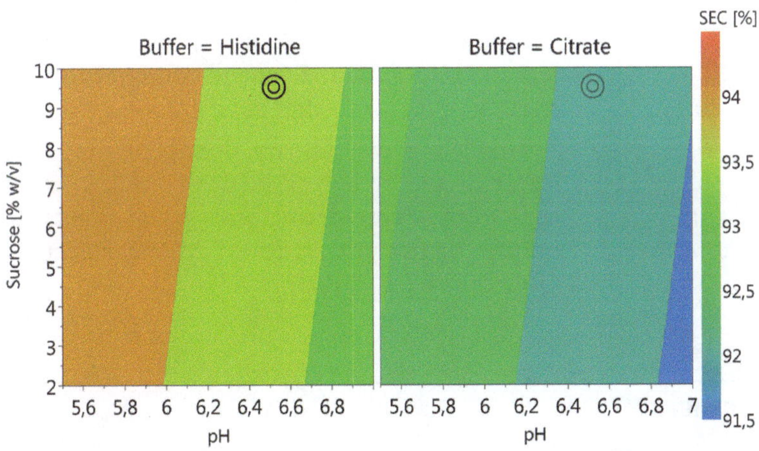

Fig. 7 DoE surface plots for relative main peak area (%) from SEC for histidine (left) and for citrate (right) buffers. The dark circles indicate the calculated optimum point for the tested factors and response variables. Response contour plots showing predicted SEC percentage results from a Design of Experiments (DoE) analyzed using a PLS model. The x-axis represents pH (5.6–7), and the y-axis represents sucrose concentration (2%–10% w/w). The left plot corresponds to formulations with histidine buffer, and the right plot to citrate buffer. Color gradients indicate SEC relative main peak area, with higher SEC values shown in warm colors (towards lower pH and higher sucrose concentration) and lower values in cold colors. Optimal regions with higher predicted SEC relative main peak area are marked with black target circles near pH 6.5 and 9.5% sucrose

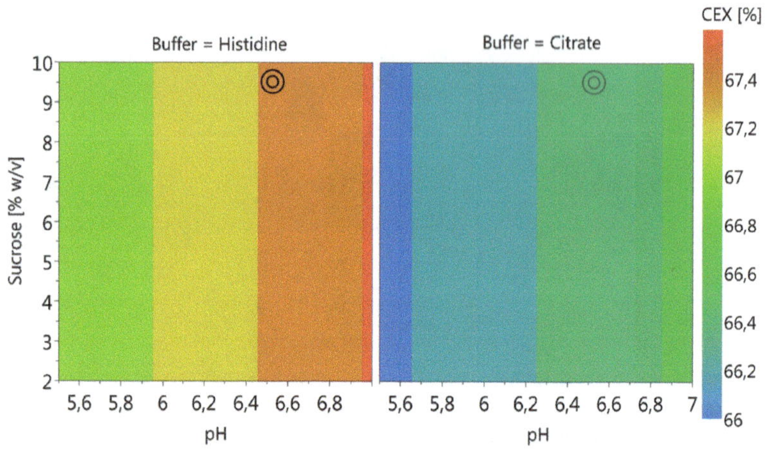

Fig. 8 DoE surface plots for relative main peak area (%) from CEX for histidine (left) and for citrate (right) buffers. The dark circles indicate the calculated optimum point for the tested factors and response variables. Response contour plots showing predicted SEC percentage results from a Design of Experiments (DoE) analyzed using a PLS model. The x-axis represents pH (5.6–7), and the y-axis represents sucrose concentration (2%–10% w/w). The left plot corresponds to formulations with histidine buffer, and the right plot to citrate buffer. Color gradients indicate CEX relative main peak area, with higher CEX values shown in warm colors (towards higher pH) and lower values in cold colors. Optimal regions with higher predicted CEX relative main peak area are marked with black target circles near pH 6.5 and 9.5% sucrose

mL *MyADC1* containing 20 mM histidine buffer at pH 6.5, 9.5% w/v sucrose, 0.02% polysorbate 80 was identified as the optimum point (Figs. 7 and 8) based on the DoE parameters and response variables and consequently could serve as a starting point for further optimization.

14. By modeling the primary drying process, fast and robust freeze-drying cycles can be achieved [28]. Moreover, modeling of the freeze-drying process also facilitates scale-up and transfer of the process into GMP manufacturing following QbD principles [30].

Acknowledgments

The authors would like to thank Brecht Vanbillemont from Coriolis Pharma for his valuable suggestions during preparation of this chapter.

References

1. Wen L, Zhang Y, Sun C, Wang SS, Gong Y, Jia C, Luo J (2025) Fundamental properties and principal areas of focus in antibody–drug conjugates formulation development. Antib Ther 8(2):99–110. https://doi.org/10.1093/abt/tbaf005

2. Jameel F, Hershenson S, Khan MA Martin-Moe S (eds) (2015) Quality by design for biopharmaceutical drug product development. In: AAPS advances in the pharmaceutical sciences series. Springer New York, New York, NY
3. Beg S, Hasnain MS, Rahman M et al (2019) Introduction to quality by design (QbD): fundamentals, principles, and applications. In: Pharmaceutical quality by design: principles and applications, 1st edn. CCXLII. Academic Press, pp 1–17
4. Awotwe-Otoo D, Agarabi C, Wu GK, Casey E, Read E, Lute S, Brorson KA, Khan MA, Shah RB (2012) Quality by design: impact of formulation variables and their interactions on quality attributes of a lyophilized monoclonal antibody. Int J Pharm 438:167–175. https://doi.org/10.1016/j.ijpharm.2012.08.033
5. Ibnidris A, Liaskos N, Eldem E, Gunn A, Streffer J, Gold M, Rea M, Teipel S, Gardiol A, Boccardi M (2024) Facilitating the use of the target product profile in academic research: a systematic review. J Transl Med 22: 693. https://doi.org/10.1186/s12967-024-05476-1
6. Mehmood Y, Farooq U (2015) Excipients use in parenteral and lyophilized formulation development.. Open Sci J Pharm Pharmacol 3:19–27
7. Szkudlarek BA (1997) Selective crystallization of phosphate buffer components and pH changes during freezing: implications to protein stability. Ph.D., University of Michigan
8. Wang W (2000) Lyophilization and development of solid protein pharmaceuticals. Int J Pharm 203:1–60. https://doi.org/10.1016/s0378-5173(00)00423-3
9. Song JG, Lee SH, Han H-K (2017) The stabilization of biopharmaceuticals: current understanding and future perspectives. J Pharm Investig 47:475–496. https://doi.org/10.1007/s40005-017-0341-9
10. Samuelsen L, Holm R, Lathuile A, Schönbeck C (2019) Buffer solutions in drug formulation and processing: how pKa values depend on temperature, pressure and ionic strength. Int J Pharm 560:357–364. https://doi.org/10.1016/j.ijpharm.2019.02.019
11. Strickley RG, Katz JS, Schoneker DR (2021) Open scientific discussion: excipients in sterile products. AAPS Excipients and Sterile Products Communities
12. Cui Y, Cui P, Chen B, Li S, Guan H (2017) Monoclonal antibodies: formulations of marketed products and recent advances in novel delivery system. Drug Dev Ind Pharm 43:519–530. https://doi.org/10.1080/03639045.2017.1278768
13. Carpenter JF, Pikal MJ, Chang BS, Randolph TW (1997) Rational design of stable lyophilized protein formulations: some practical advice. Pharm Res 14:969–975. https://doi.org/10.1023/a:1012180707283
14. Mathias N, Huille S, Picci M, Mahoney RP, Pettis RJ, Case B, Helk B, Kang D, Shah R, Ma J, Bhattacharya D, Krishnamachari Y, Doucet D, Maksimovikj N, Babaee S, Garidel P, Esfandiary R, Gandhi R (2024) Towards more tolerable subcutaneous administration: review of contributing factors for improving combination product design. Adv Drug Deliv Rev 209:115301. https://doi.org/10.1016/j.addr.2024.115301
15. Arsiccio A, Pisano R (2018) Surfactants as stabilizers for biopharmaceuticals: an insight into the molecular mechanisms for inhibition of protein aggregation. Eur J Pharm Biopharm 128:98–106. https://doi.org/10.1016/j.ejpb.2018.04.005
16. Manning MC, Holcomb RE, Payne RW, Stillahn JM, Connolly BD, Katayama DS, Liu H, Matsuura JE, Murphy BM, Henry CS, Crommelin DJA (2024) Stability of protein pharmaceuticals: recent advances. Pharm Res 41:1301–1367. https://doi.org/10.1007/s11095-024-03726-x
17. Haeuser C, Goldbach P, Huwyler J, Friess W, Allmendinger A (2020) Excipients for room temperature stable freeze-dried monoclonal antibody formulations. J Pharm Sci 109:807–817. https://doi.org/10.1016/j.xphs.2019.10.016
18. Pikal MJ, Rigsbee DR (1997) The stability of insulin in crystalline and amorphous solids: observation of greater stability for the amorphous form. Pharm Res 14:1379–1387
19. Franks F (1998) Freeze-drying of bioproducts: putting principles into practice. Eur J Pharm Biopharm 45:221–229. https://doi.org/10.1016/S0939-6411(98)00004-6
20. Crowe JH, Oliver AE, Hoekstra FA, Crowe LM (1997) Stabilization of dry membranes by mixtures of hydroxyethyl starch and glucose: the role of vitrification. Cryobiology 35:20–30. https://doi.org/10.1006/cryo.1997.2020
21. Cleland JF, Lam X, Kendrick B, Yang J, Yang TH, Overcashier D, Brooks D, Hsu C, Carpenter JF (2001) A specific molar ratio of stabilizer to protein is required for storage stability of a lyophilized monoclonal antibody. J Pharm Sci 90:310–321
22. Schwegman JJ, Hardwick L, Akers M (2005) Practical formulation and process development of freeze-dried products. Pharm Dev Tech 10:151–173. https://doi.org/10.1081/PDT-200056308
23. Baheti A, Bansal A (2010) Excipients used in lyophilization of small molecules. J Excip Food Chem 1:41–54

24. Johnson RE, Kirchhoff CF, Gaud HT (2002) Mannitol-sucrose mixtures—versatile formulations for protein lyophilization. J Pharm Sci 91:914–922. https://doi.org/10.1002/jps.10094
25. Liao X, Krishnamurthy R, Suryanarayanan R (2007) Influence processing conditions on the physical state of mannitol—implications in freeze drying. Pharm Res 24:370–376
26. Ren S (2023) Effects of arginine in therapeutic protein formulations: a decade review and perspectives. Antib Ther 6:265–276. https://doi.org/10.1093/abt/tbad022
27. Allison SD, Randolph TW, Manning MC, Middleton K, Davis A, Carpenter JF (1998) Effects of drying methods and additives on structure and function of actin: mechanisms of dehydration-induced damage and its inhibition. Arch Biochem Byophys 358:171–181
28. Vanbillemont B, Greiner AL, Ehrl V, Menzen T, Friess W, Hawe A (2023) A model-based optimization strategy to achieve fast and robust freeze-drying cycles. Int J Pharm X 5:100180. https://doi.org/10.1016/j.ijpx.2023.100180
29. Duerr C, Friess W (2019) Antibody-drug conjugates-stability and formulation. Eur J Pharm Biopharm 139:168–176. https://doi.org/10.1016/j.ejpb.2019.03.021
30. Vanbillemont B, Greiner AL, Ehrl V, Menzen T, Friess W, Hawe A (2025) Optimization and transfer of robust primary drying protocols for biopharmaceuticals through lyophilization process modeling. AAPS Open 11: 4573. https://doi.org/10.1186/s41120-025-00115-5

Check for updates

Chapter 37

Freeze-Drying of Proteins

Paul Kropp, Theresa Goettling and Wolfgang Friess

Abstract

Freeze-drying is a well-established method for preserving unstable proteins. This guide details how to freeze dry proteins, comprising equipment selection, primary packaging, and sample preparation, including proper filling, vial placement and closure. The chapter outlines reliable freezing, primary drying and secondary drying protocols, advising on temperature and pressure parameters. Finally, a range of physicochemical characterization methods, such as visual inspection, freeze-drying microscopy, DSC, Karl-Fischer titration, NIR, XRD, and BET are presented to ensure product quality and stability.

Keywords Equipment, Freeze-drying, Freezing, Lyophilization cycle, Physicochemical characterization, Primary drying, Primary packaging, Process monitoring, Proteins, Secondary drying

1 Introduction

Freeze-drying is a common process to preserve proteins that show insufficient stability in aqueous solutions. It is widely used by the biopharmaceutical industry to turn instable therapeutic proteins into stable drugs. A dogma in the freeze-drying community claims: the process is the product. The process starts with selecting the right equipment, the adequate packaging material, the best configuration of container placement, the process parameters considering various input factors, and it ends with characterizing the final product. Here, we will give practical advice on how to successfully setup a freeze-drying cycle for proteins.

2 Materials

This section lists general preparatory steps required for successful freeze-drying of proteins.

2.1 Freeze Drier

1. The freeze drier should consist of a set of shelves for tray placement, a condenser with sufficient capacity, and a

Willem F. Wolkers and Harriëtte Oldenhof (eds.), *Cryopreservation and Freeze-Drying Protocols*, Methods in Molecular Biology, vol. 3060, https://doi.org/10.1007/978-1-0716-5416-3_37,

high-performance vacuum pump. Lab flange systems and freeze-drying systems comprising freezing in a separate device should be avoided. We must be able to precisely set shelf temperature T_s and pressure in the drying chamber, which ultimately rule product temperature Tp and sublimation rate. The shelf cooling system must cover a temperature range from −50°C or below for the freezing step up to approx. 50°C for the secondary drying step. Predefined cooling and heating rates must be possible. To keep the process running upon primary and secondary drying, the condenser must reach temperatures of below −60°C. Furthermore, the system must be able to maintain a vacuum of 0.07 mbar or below. Keep in mind that nominal equipment limits for condenser and Ts, as well as vacuum pump pressure limits might not be suitable as controllable process setpoints. Calibrated temperature sensors [e.g., T-type thermocouples, wireless thermocouples, resistance temperature detectors (RTD)] are necessary for process monitoring. Direct Tp measurement through small and freely placeable temperature sensors is highly recommended. Likewise precise pressure monitoring in the freeze-drying chamber must be assured. As standard a capacitance manometer is included. For other purposes, a Pirani or piezo manometer may be built in. The freeze drier should allow stoppering of vials through lifting the shelves either manually or hydraulically. Having the option to fill the vacuum chamber with filtered ambient air or nitrogen gas through a backfill valve up to a target chamber pressure is useful.

2.2 Selecting the Suitable Primary Packaging Material

1. Primary packaging for freeze dried proteins consists of three components: a glass vial, a stopper, and a cap. Picking the right combination of those three has important implications on protein stability, appearance of the lyophilized cake and container closure integrity.
2. For protein solutions, preferably use glass type I vials made from tubing glass. The vial should be sufficiently large to accommodate your sample solution up to fill height of 12–15 mm. Larger fill heights are disfavored due to significantly longer drying times since the sublimation front moves from top to bottom of the vial. Very low fill heights may results in cakes, which break loose and a less reliable product temperature measurement. Before usage, bulk vials, if not supplied as ready for use, must be washed—ideally using a lab dishwasher—rinsed with highly purified water and dried, potentially including a heat sterilization step.
3. Commonly used stoppers are made of bromobutyl or chlorobutyl rubber, for 2 R vials of 13 mm, for 6 R and 10 R vials of 20 mm diameter, respectively. Stoppers differ in their shape.

Common shapes are igloo and 2-leg. Igloo stoppers usually sit better in the vial, whereas 2-leg stoppers have bigger cavities that allow for a better water vapor flow during drying. Stoppers may have a fluoropolymer coating, which facilitates vial closing, reduces the occurrence of particles, as well as leachables and extractables from the rubber material and limits protein-stopper material interactions. Stoppers can be bought in prewashed and presterilized ready to use (RTU) configurations. Alternatively, stoppers are rinsed with highly purified water, if needed autoclaved, and finally dried to remove residual moisture, e.g., for several hours at 110°C. Note that the drying may be critical as already little residual moisture can have a substantial impact on product stability when released upon storage. For stoppering, place the stopper shank halfway into the vial neck of all vials as displayed in Fig. 1. Often when placing stoppers manually, a first pressure point indicates the correct stopper position.

4. After freeze-drying, stoppered vials are sealed with aluminium-plastic flip-off caps (Fig. 2) using a manual or pneumatic crimping device. The cap is necessary to fixate the stopper sealing the vial. Reconstitution is possible after removal of the plastic lid. Complete cap removal, known as decrimping, necessitates a specific decrimping device (Fig. 3).

Fig. 1 Left: Semi stoppered vial; right: fully stoppered vial. Two transparent glass vials are placed next to each other. The left vial has a gray igloo stopper placed halfway into the vial shank. The right vial has a gray igloo stopper placed completely into the vial shank

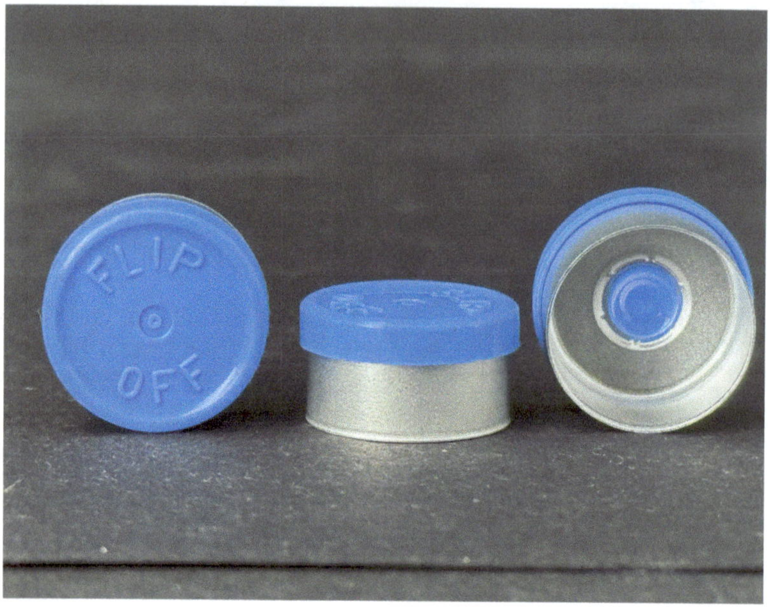

Fig. 2 Flip-off crimping caps. Three of the same flip off caps placed in different orientations. Flip off cap have an aluminium cylindrical middle part and a blue, removable plastic lid on top

Fig. 3 Left: 13 mm manual crimping device, right: 13 mm manual decrimping device. A manual device for crimping (on the left) and decrimping (on the right)

2.3 Selecting a Lyoprotective Formulation

A lyoprotective formulation for proteins consists of excipients to stabilize the protein by forming hydrogen bonds with the protein, and creating a stable, glassy matrix that limits molecular mobility. Finding the optimal formulation involves balancing different excipients and often requires testing multiple candidates (see Chap. x). Examples of compounds in a formulation are:

1. Sugars (e.g., sucrose, trehalose), polyols (e.g., mannitol, sorbitol) forming H-bonds, an amorphous matrix, and potentially a crystalline scaffold structure, providing osmolality.
2. Amino acids (e.g., glycine, arginine, histidine) acting as bulking agents, modulating protein–protein interactions, and offering buffering capacity.
3. Surfactants protecting against interfacial stress.
4. Buffering compounds maintaining an optimal pH but consider pH shift upon cooling and freezing.
5. Salts controlling protein self-interaction.

3 Methods

3.1 Setting Up the Lyophilizer

1. The optimized sample solution (see Chap. 36) should be filtered prior to filling using a 0.45 μm low protein binding filter to avoid the presence of particles. If your sample vials do not cover the entire tray area, use placebo vials to fill it up. Prepare a placebo solution of similar Tg' (glass transition temperature of maximally freeze concentrated solution) and total solid content. You may use model proteins such as bovine serum albumin.
2. Filling. Fill the same volume of sample solution into every sample vial. Proceed for the placebo solutions accordingly. Avoid splashing and wetting the inner vial walls. The filling process should take place in a controlled environment containing low concentrations of airborne particles.
3. Tray placement. Start by placing empty vials on the entire tray in a hexagonal array to keep them from falling over during the filling process. Heat transfer during freeze-drying depends on vial position and vial environment. Vials located at the tray edge or corner exhibit higher product temperatures due to radiation from the chamber wall and missing neighboring shielding vials [1]. To ensure similar mass transfer within the sample vials, they should be ideally shielded by one to two rows of sample or placebo vials as shown exemplarily in Fig. 4. Empty space between tray wall should be filled with empty vials for fixation.
4. Setting up thermocouples. Usage of thermocouples is highly recommended to monitor the freeze-drying process. Place the tip of the temperature probe center bottom in the vial as shown

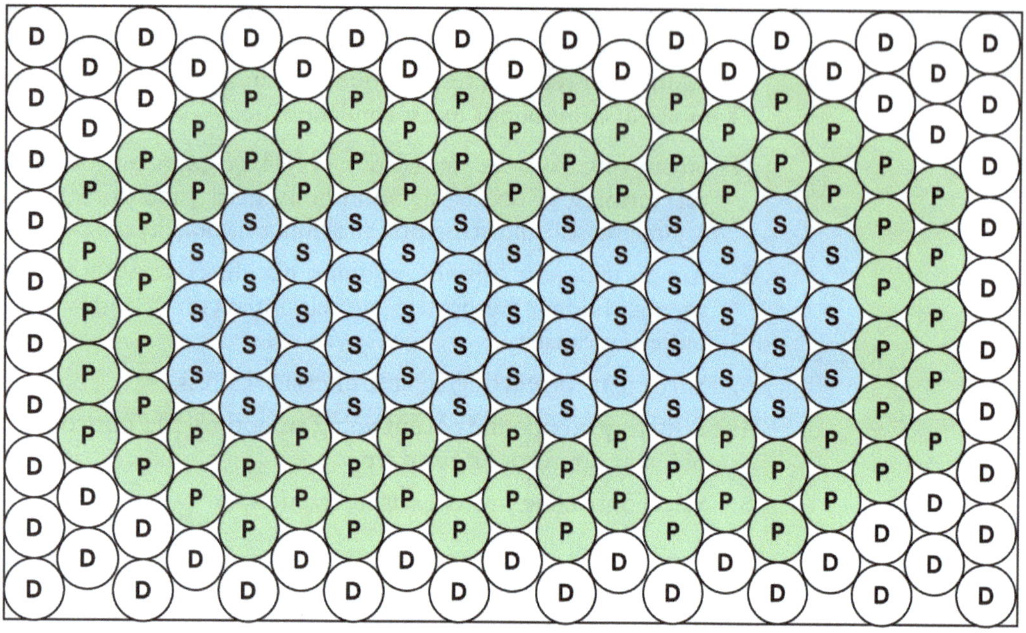

Fig. 4 Exemplary schematic vial placement on a tray (S: sample vial; P: placebo vial; D: dummy vial). Schematic view from the top of a lyo tray. Center of tray contains sample vials, around them are placebo vials and around them up to the tray borders are dummy vials

in Fig. 5. Wireless thermocouples usually allow easier handling but are less common and more expensive than T-type thermocouples. If your sample is scarce, place the temperature probes in representative placebo vials. Use multiple thermocouples and place them at different positions on the tray as heat transfer does not take place homogenously and progression of the sublimation front and probe positioning vary. Vials equipped with thermocouples are not properly closed after freeze-drying, the probes have to be removed, and the product quality is impaired. Thermocouples can lead to early nucleation and insert additional heat into the vial, which makes them not fully representative for vials without a probe.

5. Stoppering and loading. Place stoppers into all vials in the semi stoppered position and double check the stopper placement (Fig. 1). The cavity must be large enough to ensure free flow of water vapor from the vial into the drying chamber. Place the lyophilization tray onto the middle shelf of the freeze drier. This shelf is best shielded by other shelves, has the most representative vapor flow and is less impacted by the closer vicinity of the condenser chamber. Load your samples at room temperature or, if liquid stability is a concern, at 5°C. Lower loading temperatures should be avoided as atmospheric vapor could condensate or sublimate on the shelves. Avoid loading

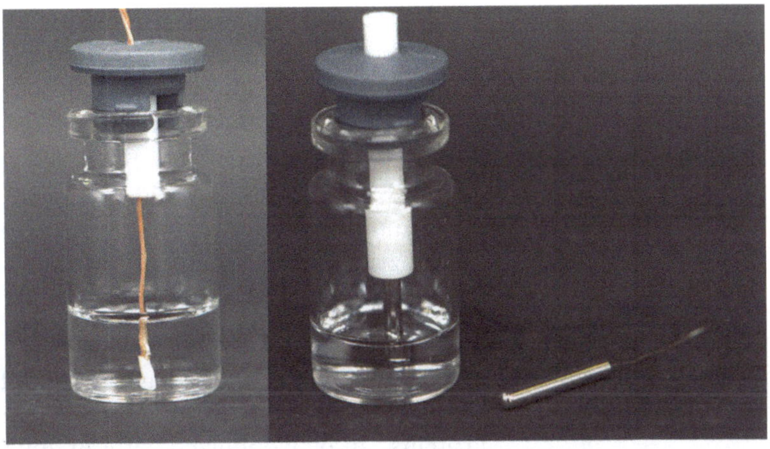

Fig. 5 Left: T-type thermocouple; right: wireless thermocouple. Left T-type thermocouple, right wireless thermocouple

frozen samples. Make sure to connect the temperature probes to the freeze drier and check their position again after loading. If possible, remove the bottom of the tray. Close the freeze drier door and check if the drain outlet and other valves are closed. The filling, stoppering, and loading process should be performed in a timely manner to avoid sample evaporation, particle contamination, and fogging (film of product on the inner vial wall) due to Marangoni flow.

3.2 Freeze-Drying Protocol

A freeze-drying protocol consists of freezing, primary drying, and secondary drying.

3.2.1 Freezing

In the freezing step, all freezable water molecules should be transformed into ice and kinetically arrested.

1. Consider including one or two soaking steps to homogenize the T-distribution within and in between the vials to obtain more homogenous freezing. For example, vials are cooled from loading temperature to 5°C–10°C and held for 30–60 min for equilibration. Another soak for 30–60 min at, e.g., −5°C may be implemented; the product is in a supercooled state and the intervial and intravial freezing homogeneity is increased when subsequently further lowering the shelf temperature [2, 3].
2. Freeze well below your target product temperature defined by Tg' or Tc (collapse temperature. *See* **Notes 1 and 2** for Tg' and Tc determination. In most cases, a final freezing Tp of −40°C to −50°C is recommended [4]. Due to lag time and offset between Ts and Tp and edge vial effects set Ts to −45°C to −55°C.

3. Typical freezing rates range from 0.1°C/min to 1°C/min. Faster freezing rates lead to a more directional freezing, heterogenous nucleation temperatures and subsequently a higher vial-to-vial heterogeneity [5].
4. Freezing hold time should be set depending on liquid fill height, e.g. 60 min or 120 min for fill heights below or above 10 mm, respectively [4]. If your formulation contains crystallizing bulking agents like mannitol or glycine, consider an additional annealing step (*see* **Note 3**).

3.2.2 Primary Drying

During primary drying, frozen water and the majority of the unfrozen water is removed by sublimation and desorption, respectively. Proteins can be lyophilized without any stabilizing excipients, as it is often the case for research or diagnostic purposes, if the solid content is sufficiently high. This results in a high Tg', allowing for straightforward drying at e.g. Tp of −20°C or −25°C. For lower protein contents, when physical stability issues arise during freezing or upon long-term storage, amorphous matrix formers, most commonly non-reducing disaccharides such as sucrose, are typically added to the formulation. These excipients lower the Tg', making careful consideration of the Tp during drying necessary. Above glass transition temperature Tg, amorphous systems exhibit viscous flow. That is why Tp should generally remain below Tg during lyophilization to prevent structural collapse. Yet, it has been shown that collapse itself does not necessarily compromise product quality but hampers water removal [6]. Formulations with high protein content above 10 g/L often retain sufficient mechanical strength above Tg' to avoid collapse for the time of the drying process. In those formulations, Tc, which typically lies 1–2°C above Tg', can be significantly higher, with a difference of up to 10°C between Tg' and Tc [7]. The addition of crystalline bulking agents helps to prevent cake collapse by providing a rigid framework. This structure supports the amorphous phase, which may soften, but without compromising the overall integrity of the lyophilized cake [7]. In principle, Tp should always stay a few degrees Celsius below the products Tc during primary drying. Tp is indirectly adjusted via setting Pc and Ts as follows. In addition, it is influenced by multiple factors such as formulation composition and container type. For transition from freezing to primary drying (Fig. 6), proceed with the following steps:

1. Reduce Pc. To promote a high sublimation rate, Pc must be kept below the vapor pressure of ice corresponding to the desired Tp (see Table 1) plus a safety margin. For sucrose-based formulations with low to medium protein concentration (up to approximately 10 g/L) and total solid content (up to 10%), a Pc between 0.07 and 0.25 mbar is typically used. 0.1 mbar is

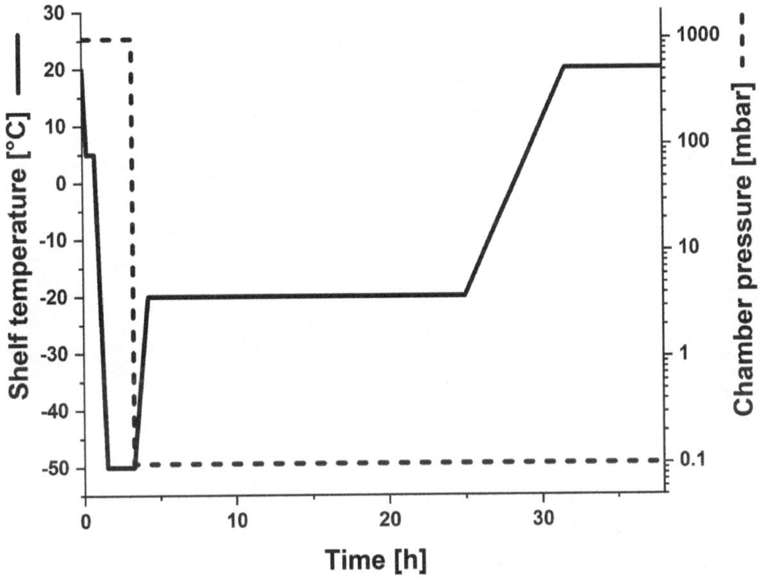

Fig. 6 Exemplary freeze-drying cycle. Line plot of a freeze-drying cycle. X-axis is time. Primary y-axis depicts shelf temperature and secondary y-axis chamber pressure. Typical temperature and pressure settings for freezing, primary drying and secondary drying are shown

generally suitable providing a compromise between mass and heat transfer, which becomes limited at lower pressures.

2. Increase Ts. To ensure efficient sublimation, enough heat must be supplied to the product from the shelves. Apply a ramp rate of 0.5°C–1°C/min after freezing to reach the setpoint for Ts in primary drying. Tp at the sublimation front is well below Ts, which can typically be set to approximately 20°C above Tc. If time permits, use a conservative, lower Ts. If faster drying is required, try drying at a higher Ts but make sure to monitor Tp.
3. Condenser temperature must be at −60°C or less in order to assure consistent vapor flow [5].
4. Determine the endpoint of primary drying by evaluating Tp data obtained by temperature probes. With sublimation coming to an end, a steep increase of Tp up to Ts indicates the end of primary drying. A safety margin of at least 10% of the primary drying time should be added, as the temperature measured in a single vial is not representing the whole batch. For once, thermocouples can influence ice nucleation and subsequently affect pore structure and drying behavior. Furthermore, vial position within the freeze-dryer has an impact: vials located near chamber walls or the door may dry faster due to increased radiation, whereas more shielded vials in the center typically

Table 1

The vapor pressure of ice at subzero temperatures, i.e., the pressures/temperatures at which ice will sublime directly into a gas

Temperature [°C]	Vapor pressure of ice [mbar]	Temperature [°C]	Vapor pressure of ice [mbar]
−10	2.599	−38	0.161
−12	2.173	−40	0.128
−14	1.812	−42	0.102
−16	1.506	−44	0.081
−18	1.249	−46	0.064
−20	1.032	−48	0.050
−22	0.851	−50	0.039
−24	0.699	−52	0.031
−26	0.572	−54	0.024
−28	0.468	−56	0.018
−30	0.380	−58	0.014
−32	0.308	−60	0.011
−34	0.249	−62	0.0082
−36	0.200	−64	0.0062

run at lower temperatures during primary drying and require longer drying times. Too early transition to secondary drying comes with the risk of incomplete sublimation, leading to product defects due to meltback of remaining ice. Additionally, there is always the possibility that probes may be mispositioned or may have moved in the course of loading and freezing [8]. *See* **Note 4** for additional methods to determine the primary drying endpoint.

3.2.3 Secondary Drying

The goal of secondary drying is to reduce the residual moisture (RM) through desorption by applying higher temperatures. Be aware that already during primary drying not only ice is removed but also most of the water bound in the freeze concentrated solutes matrix is removed. Typically, RM levels of the final product below 1% are targeted to ensure a long-term stable product [4].

1. Keep Pc of the primary drying step. In contrast to primary drying, Pc during secondary drying has less influence on the drying rate or final RM [9].
2. Increase Ts. The product still contains high RM after primary drying, depending on the Ts and the Pc settings, primary

drying duration, as well as formulation composition and total solid content. Fast ramping into secondary drying may cause collapse; rates between 0.1 and 0.5°C/min are commonly applied. For formulations containing amorphous excipients, a slow temperature ramp of approximately 0.1°C/min is recommended. Formulations with crystalline components generally tolerate faster ramps, such as 0.3°C–0.4°C/min [5]. Ts is typically set between 25°C and 40°C for protein formulations. Amorphous products and those with higher solid content (i.e., >10% w/v) typically require longer drying times. Durations of 3–6 h at the final temperature are common [5]. If time is not a limiting factor, secondary drying can be extended by a few more hours. However, low RM is primarily achieved through high Ts rather than extended drying at lower temperatures. Typically, higher Ts and thus Tp such as 40°C, are not detrimental to protein stability in the solid state over a few hours [10].

3. Monitor the drying progress through Tp data. The endpoint of the secondary drying is indicated by Tp approaching Ts. Yet, the sensitivity is not sufficient to trace slow evaporation of little amounts of water from the product. Therefore, after Tp reached Ts and stagnates at this temperature, add a longer drying period. If the freeze drier is equipped with a Pirani sensor, *see* **Note 5**.

3.3 Post-Processing

1. After completion of secondary drying, an optional holding step at storage temperature (e.g., 5°C) may be added only if the product is not immediately removed from the freeze drier to prevent additional uncontrolled drying and stress on the product. Otherwise, either reduce the temperature to room temperature or maintain the secondary drying temperature and terminate the process.
2. Just before removing the samples, backfill the chamber with an inert gas (e.g., nitrogen) or filtered air. Consider a partial backfill to 750 mbar and stoppering under those conditions which keep the stopper in place and assure container closure integrity.
3. If the freeze drier allows, stopper the vials within the drying chamber by raising the shelves to press the rubber stoppers into place.
4. If only partial vacuum was applied, increase Pc to atmospheric level.
5. After atmospheric pressure is reached, open the freeze dryer and remove vials from the drying chamber.
6. Remove temperature probes carefully from the vials and if not done already, stopper the vials.

7. Seal vials with crimp caps. Figure 7 shows, from left to right: a correctly stoppered vial, a vial with a loose crimp cap positioned on top and a properly sealed vial after crimping.
8. Finally, open all valves, remove the product and optically inspect the product and the chamber for possible vial breakage, contamination, or moisture. Ensure proper defrosting of the condenser.

3.4 Physicochemical Characterization of the Product

To assess the quality of the freeze-dried product, a variety of analytical methods can be applied:

(a) Visual inspection to identify macroscopic cake defects which include shrinkage (reduction in cake volume), macroscopic collapse (loss of porous structure), splashing (product on vial wall due to filling process), fogging (film of product on the inner vial wall), or cake lifting (detachment from the vial bottom) [11]. In addition, visual inspection is also a critical part of verifying container closure integrity (CCI). This includes checking for properly seated stoppers and intact crimp caps.

(b) Stereo microscopy or scanning electron microscopy (SEM) are used for a more detailed structural analysis including the interior of the cake. After careful removal of the cake from the vial, if necessary, by cutting or fracturing the vial, as well as breaking or cutting the cake, the pore structure can be examined,

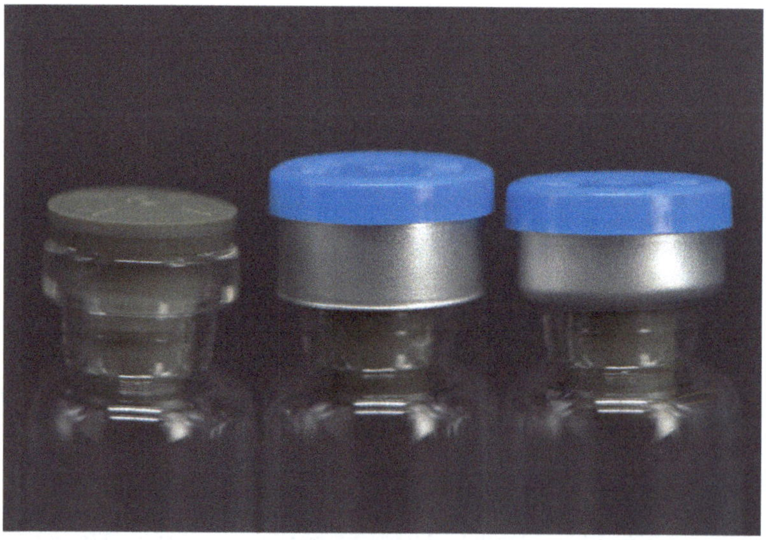

Fig. 7 From left to right: correctly stoppered vial, crimp cap placed on vial, crimped vial. Three glass vials. First glass vial has a gray stopper correctly and fully placed in the shank. Second glass vial is fully stoppered and has a loose blue aluminium flip off cap on top. Third glass vial is fully stoppered and fully capped with a blue aluminium flip off cap

and collapse in the inner and potential microcollapse can be identified.

(c) Differential scanning calorimetry (DSC) is used to characterize melting, glass transition, recrystallization or phase transitions events upon heating to understand the phase characteristics of the product. Furthermore, DSC can be used to indirectly estimate RM, as the Tg of amorphous systems decreases with increasing RM levels and a shallow water evaporation endotherm may be seen. For analysis, approximately 20 mg of the freeze-dried product are weighed into an aluminium pan under controlled low-humidity conditions, for example, inside a glove box. The pan is sealed hermetically. Typical heating rates are 5°C–10°C/min.

(d) Karl-Fischer titration is the method of choice to quantify RM in lyophilizates. Samples are either titrated directly after dispersion or dissolution in the titration cell medium, or the moisture is extracted and transferred into the solvent by heating out the sample in oven systems.

(e) Near-infrared spectroscopy (NIR) can serve as a nondestructive alternative for RM estimation. It is advantageous for nondestructive routine quality control but requires robust calibration [12].

(f) X-ray diffraction (XRD) is used to evaluate crystalline phases in the lyophilized matrix. This is particularly relevant if crystallizing bulking agents like mannitol or glycine are employed, which may not crystallize depending on process and formulation, or only partially, and can form different polymorphs or hydrates. Furthermore, crystallization of other formulation components such as sodium chloride can be detected depending on their content [13]. Amorphous materials typically render broad halos, whereas crystalline phases generate distinct Bragg peaks.

(g) Less frequently, the specific surface area (SSA) is measured by gas adsorption based on the Brunauer–Emmett–Teller (BET) method. The SSA typically reflects the initial ice structure formed upon freezing. Thus, it allows to assess the freezing pattern and homogeneity. A low SSA may indicate cake collapse.

(h) For analysis of protein integrity see Chap. 36.

4 Notes

1. Tg' is determined by differential scanning calorimetry (DSC) and appears as a step change in the heat flow curve. The sample is rapidly cooled to approximately −60°C and after a few

minutes hold heated at 5°C or 10°C/min. During heating Tg' is analyzed.

2. Tc is determined using freeze-drying microscopy (FDM), which allows to monitor the structural changes of a frozen sample during controlled warming under vacuum. Tc is identified by the onset and progression of structural loss as sublimation proceeds.
3. An annealing step can be beneficial for formulations containing crystalline bulking agents such as mannitol or glycine to allow for efficient and controlled crystallization thereof. Uncontrolled excipient crystallization during primary drying may lead to vial breakage [14]. Annealing also results in ice crystal growth, which may enable faster sublimation due to reduced dry product resistance ultimately resulting in a lower Tp. For annealing, after reaching the final Ts during freezing, increase Ts and thus Tp 10°C–20°C above Tg' and hold for several hours before returning to the initial final freezing temperature [4].
4. If a Pirani sensor is available, in addition to the standard capacitance manometer, the endpoint of primary drying can be determined by differential pressure measurement. The Pirani sensor shows elevated readings during active sublimation because its readout depends on the gas composition and is higher for water vapor compared to nitrogen or oxygen gas. Once sublimation ceases and the water vapor content of the chamber gas phase becomes drastically reduced, the Pirani reading approaches that of the capacitance manometer. Primary drying can be considered complete when the pressure difference between the two sensors became less than 10%–15%. Alternative methods for endpoint detection include the pressure rise test (PRT), where isolating the product chamber from the condenser causes the chamber pressure to rise if sublimation of remaining frozen water continues or remains stable if drying is complete. Manometric temperature measurement (MTM) is a variant of the PRT with a higher resolution by including a mathematical model. In principle, other parameters, such as shelf inlet and outlet temperature, vacuum pump activity, or mass spectrometry, can also be utilized [8], but their detailed discussion is beyond the scope of this chapter.
5. If a Pirani sensor is included, a small peak may be observed at the beginning of secondary drying. Yet, the sensitivity to the gas phase composition reaches its limits and an extended drying period is advised after the Pirani readout became constant [15].

References

1. Pikal MJ, Bogner R, Mudhivarthi V, Sharma P, Sane P (2016) Freeze-drying process development and scale-up: scale-up of edge vial versus center vial heat transfer coefficients, Kv. J Pharm Sci 105:3333–3343. https://doi.org/10.1016/j.xphs.2016.07.027
2. Kasper JC, Winter G, Friess W (2013) Recent advances and further challenges in lyophilization. Eur J Pharm Biopharm 85:162–169. https://doi.org/10.1016/j.ejpb.2013.05.019
3. Liu J, Viverette T, Virgin M, Anderson M, Paresh D (2005) A study of the impact of freezing on the lyophilization of a concentrated formulation with a high fill depth. Pharm Dev Technol 10:261–272. https://doi.org/10.1081/PDT-54452
4. Tang X(Charlie), Pikal MJ (2004) Design of freeze-drying processes for pharmaceuticals: practical advice. Pharm Res 21:191–200. https://doi.org/10.1023/B:PHAM.0000016234.73023.75
5. Tchessalov S, Maglio V, Kazarin P et al (2023) Practical advice on scientific design of freeze-drying process: 2023 update. Pharm Res 40:2433–2455. https://doi.org/10.1007/s11095-023-03607-9
6. Schersch K, Betz O, Garidel P et al (2010) Systematic investigation of the effect of lyophilizate collapse on pharmaceutically relevant proteins I: stability after freeze-drying. J Pharm Sci 99:2256–2278. https://doi.org/10.1002/jps.22000
7. Colandene JD, Maldonado LM, Creagh AT, Vrettos JS, Goad KG, Spitznagel TM (2007) Lyophilization cycle development for a high-concentration monoclonal antibody formulation lacking a crystalline bulking agent. J Pharm Sci 96:1598–1608. https://doi.org/10.1002/jps.20812
8. Patel SM, Doen T, Pikal MJ (2010) Determination of end point of primary drying in freeze-drying process control. AAPS Pharm Sci Tech 11:73–84. https://doi.org/10.1208/s12249-009-9362-7
9. Yoon K, Narsimhan V (2023) Comparison of vial heat transfer coefficients during the primary and secondary drying stages of freeze-drying. Int J Pharm 635:122746. https://doi.org/10.1016/j.ijpharm.2023.122746
10. Schersch K, Betz O, Garidel P et al (2012) Systematic investigation of the effect of Lyophilizate collapse on pharmaceutically relevant proteins, part 2: stability during storage at elevated temperatures. J Pharm Sci 101:2288–2306. https://doi.org/10.1002/jps.23121
11. Patel SM, Nail SL, Pikal MJ, Geidobler R, Winter G, Hawe A, Davagnino J, Rambhatla Gupta S (2017) Lyophilized drug product cake appearance: what is acceptable. J Pharm Sci 106:1706–1721. https://doi.org/10.1016/j.xphs.2017.03.014
12. Clavaud M, Lema-Martinez C, Roggo Y, Bigalke M, Guillemain A, Hubert P, Ziemons E, Allmendinger A (2020) Near-infrared spectroscopy to determine residual moisture in freeze-dried products: model generation by statistical design of experiments. J Pharm Sci 109:719–729. https://doi.org/10.1016/j.xphs.2019.08.028
13. Hawe A, Fries W (2006) Impact of freezing procedure and annealing on the physico-chemical properties and the formation of mannitol hydrate in mannitol–sucrose–NaCl formulations. Eur J Pharm Biopharm 64:316–325. https://doi.org/10.1016/j.ejpb.2006.06.002
14. Shalaev E, Franks F (2002) Solid-Liquid state diagrams in pharmaceutical lyophilisation: crystallization of solutes. In: Levine H (ed) Amorphous food and pharmaceutical systems. The Royal Society of Chemistry, pp 200–215
15. Nail S, Tchessalov S, Shalaev E, Ganguly A, Renzi E, Dimarco F, Wegiel L, Ferris S, Kessler W, Pikal M, Sacha G, Alexeenko A, Thompson TN, Reiter C, Searles J, Coiteux P (2017) Recommended best practices for process monitoring instrumentation in pharmaceutical freeze drying. AAPS Pharm Sci Tech 18:2379–2393. https://doi.org/10.1208/s12249-017-0733-1

Index

Willem F. Wolkers and Harriëtte Oldenhof (eds.), *Cryopreservation and Freeze-Drying Protocols*, Methods in Molecular Biology, vol. 3060, https://doi.org/10.1007/978-1-0716-5416-3,

T

U

V

W

Y

Kolibri 360 GmbH
Ferdinand-Jühlke-Straße 7
99095 Erfurt, Deutschland
produktsicherheit@kolibri360.de